Acta Numerica 2005

Acta Numerica

Volume 14   2005

CAMBRIDGE
UNIVERSITY PRESS

CAMBRIDGE UNIVERSITY PRESS
Cambridge, New York, Melbourne, Madrid, Cape Town,
Singapore, São Paulo, Delhi, Tokyo, Mexico City

Cambridge University Press
The Edinburgh Building, Cambridge CB2 8RU, UK

Published in the United States of America by Cambridge University Press, New York

www.cambridge.org
Information on this title: www.cambridge.org/9780521174282

First published 2005
First paperback edition 2011

A catalogue record for this publication is available from the British Library

ISBN 978-0-521-85807-6 Hardback
ISBN 978-0-521-17428-2 Paperback

Contents

Contents

Acta Numerica (2005), pp. 1–137
DOI: 10.1017/S0962492904000212

Numerical solution of
saddle point problems

Michele Benzi*

Department of Mathematics and Computer Science,
Emory University, Atlanta, Georgia 30322, USA
E-mail: benzi@mathcs.emory.edu

Gene H. Golub[†]

Scientific Computing and Computational Mathematics Program,
Stanford University, Stanford,
California 94305-9025, USA
E-mail: golub@sccm.stanford.edu

Jörg Liesen[‡]

Institut für Mathematik, Technische Universität Berlin,
D-10623 Berlin, Germany
E-mail: liesen@math.tu-berlin.de

We dedicate this paper to Gil Strang on the occasion of his 70th birthday

Large linear systems of saddle point type arise in a wide variety of applications throughout computational science and engineering. Due to their indefiniteness and often poor spectral properties, such linear systems represent a significant challenge for solver developers. In recent years there has been a surge of interest in saddle point problems, and numerous solution techniques have been proposed for this type of system. The aim of this paper is to present and discuss a large selection of solution methods for linear systems in saddle point form, with an emphasis on iterative methods for large and sparse problems.

* Supported in part by the National Science Foundation grant DMS-0207599.
† Supported in part by the Department of Energy of the United States Government.
‡ Supported in part by the Emmy Noether Programm of the Deutsche Forschungsgemeinschaft.

CONTENTS

1. Introduction

In recent years, a large amount of work has been devoted to the problem of solving large linear systems in saddle point form. The reason for this interest is the fact that such problems arise in a wide variety of technical and scientific applications. For example, the ever-increasing popularity of mixed finite element methods in engineering fields such as fluid and solid mechanics has been a major source of saddle point systems (Brezzi and Fortin 1991, Elman, Silvester and Wathen 2005c). Another reason for this surge in interest is the extraordinary success of interior point algorithms in both linear and nonlinear optimization, which require at their heart the solution of a sequence of systems in saddle point form (Nocedal and Wright 1999, Wright 1992, Wright 1997).

Because of the ubiquitous nature of saddle point systems, methods and results on their numerical solution have appeared in a wide variety of books, journals and conference proceedings, justifying a comprehensive survey of the subject. The purpose of this article is to review many of the most promising solution methods, with an emphasis on iterative methods for large and sparse problems. Although many of these solvers have been developed with specific applications in mind (for example, Stokes-type problems in fluid dynamics), it is possible to discuss them in a fairly general setting using standard numerical linear algebra concepts, the most prominent being perhaps the Schur complement. Nevertheless, when choosing a preconditioner (or developing a new one), knowledge of the origin of the particular problem at hand is essential. Although no single 'best' method exists, very effective solvers have been developed for some important classes of problems. We therefore devote some space to a discussion of saddle point problems arising in a few selected applications.

It is hoped that the present survey will prove useful to practitioners who are looking for guidance in the choice of a solution method for their own application, to researchers in numerical linear algebra and scientific computing, and especially to graduate students as an introduction to this very rich and important subject.

Notation. We have used boldface to denote vectors in sections describing fluid dynamics applications, where this is traditional, but we have otherwise followed the standard practice of numerical linear algebra and employed no special notation.

1.1. Problem statement and classification

The subject of this paper is the solution of block 2×2 linear systems of the form

$$\begin{bmatrix} A & B_1^T \\ B_2 & -C \end{bmatrix} \begin{bmatrix} x \\ y \end{bmatrix} = \begin{bmatrix} f \\ g \end{bmatrix}, \quad \text{or} \quad \mathcal{A}u = b, \tag{1.1}$$

$$A \in \mathbb{R}^{n \times n}, \quad B_1, B_2 \in \mathbb{R}^{m \times n}, \quad C \in \mathbb{R}^{m \times m} \quad \text{with} \quad n \geq m. \tag{1.2}$$

It is obvious that, under suitable partitioning, *any* linear system can be cast in the form (1.1)–(1.2). We explicitly exclude the case where A or one or both of B_1, B_2 are zero. When the linear system describes a (generalized) saddle point problem, the constituent blocks A, B_1, B_2 and C satisfy one or more of the following conditions:

C1 A is symmetric: $A = A^T$
C2   the *symmetric part* of A, $H \equiv \frac{1}{2}(A + A^T)$, is positive semidefinite
C3 $B_1 = B_2 = B$
C4 C is symmetric ($C = C^T$) and positive semidefinite
C5 $C = O$ (the zero matrix)

Note that C5 implies C4. The most basic case is obtained when *all* the above conditions are satisfied. In this case A is symmetric positive semidefinite and we have a *symmetric* linear system of the form

$$\begin{bmatrix} A & B^T \\ B & O \end{bmatrix} \begin{bmatrix} x \\ y \end{bmatrix} = \begin{bmatrix} f \\ g \end{bmatrix}. \tag{1.3}$$

This system arises as the first-order optimality conditions for the following equality-constrained quadratic programming problem:

$$\min \; J(x) = \frac{1}{2} x^T A x - f^T x \tag{1.4}$$

$$\text{subject to} \;\; Bx = g. \tag{1.5}$$

In this case the variable y represents the vector of Lagrange multipliers. Any solution (x_*, y_*) of (1.3) is a saddle point for the Lagrangian

$$\mathcal{L}(x, y) = \frac{1}{2} x^T A x - f^T x + (Bx - g)^T y,$$

hence the name 'saddle point problem' given to (1.3). Recall that a saddle point is a point $(x_*, y_*) \in \mathbb{R}^{n+m}$ that satisfies

$$\mathcal{L}(x_*, y) \le \mathcal{L}(x_*, y_*) \le \mathcal{L}(x, y_*) \quad \text{for any } x \in \mathbb{R}^n \text{ and } y \in \mathbb{R}^m,$$

or, equivalently,

$$\min_x \max_y \mathcal{L}(x, y) = \mathcal{L}(x_*, y_*) = \max_y \min_x \mathcal{L}(x, y).$$

Systems of the form (1.3) also arise in nonlinearly constrained optimization (sequential quadratic programming and interior point methods), in fluid dynamics (Stokes' problem), incompressible elasticity, circuit analysis, structural analysis, and so forth; see the next section for a discussion of applications leading to saddle point problems.

Another important special case is when conditions C1–C4 are satisfied, but not C5. In this case we have a block linear system of the form

$$\begin{bmatrix} A & B^T \\ B & -C \end{bmatrix} \begin{bmatrix} x \\ y \end{bmatrix} = \begin{bmatrix} f \\ g \end{bmatrix}. \tag{1.6}$$

Problems of this kind frequently arise in the context of *stabilized* mixed finite element methods. Stabilization is used whenever the discrete variables x and y belong to finite element spaces that do not satisfy the Ladyzhenskaya–Babuška–Brezzi (LBB, or inf-sup) condition (Brezzi and Fortin 1991). Another situation leading to a nonzero C is the discretization of the equations describing slightly compressible fluids or solids (Braess 2001, Chapter 6.3). Systems of the form (1.6) also arise from regularized, weighted least-squares problems (Benzi and Ng 2004) and from certain interior point methods in optimization (Wright 1992, Wright 1997). Often the matrix C has small norm compared to the other blocks.

In the literature, the phrase *generalized saddle point problem* has been used primarily to allow for the possibility of a nonsymmetric matrix $\mathcal{A}$ in (1.1). In such problems either $A \ne A^T$ (with condition C2 usually satisfied), or $B_1 \ne B_2$, or both. The most important example is perhaps that of the linearized Navier–Stokes equations, where linearization has been obtained by Picard iteration or by some variant of Newton's method. See Ciarlet Jr., Huang and Zou (2003), Nicolaides (1982) and Szyld (1981) for additional examples. We note that our definition of generalized saddle point problem as a linear system of the form (1.1)–(1.2), where the blocks A, B_1, B_2 and C satisfy one or more of the conditions C1–C5, is the most general possible, and it contains previous definitions as special cases.

In the vast majority of cases, linear systems of saddle point type have real coefficients, and in this paper we restrict ourselves to the real case. Complex coefficient matrices, however, do arise in some cases; see, *e.g.*, Bobrovnikova and Vavasis (2000), Mahawar and Sarin (2003) and Strang (1986, page 117). Most of the results and algorithms reviewed in this paper admit straightforward extensions to the complex case.

1.2. Sparsity, structure and size

Although saddle point systems come in all sizes and with widely different structural and sparsity properties, in this paper we are mainly interested in problems that are both large and sparse. This justifies our emphasis on iterative solvers. Direct solvers, however, are still the preferred method in optimization and other areas. Furthermore, direct methods are often used in the solution of subproblems, for example as part of a preconditioner solve. Some of the algorithms considered in this paper are also applicable if one or more of the blocks in $\mathcal{A}$ happen to be dense, as long as matrix-vector products with $\mathcal{A}$ can be performed efficiently, typically in $O(n+m)$ time. This means that if a dense block is present, it must have a special structure (*e.g.*, Toeplitz, as in Benzi and Ng (2004) and Jin (1996)) or it must be possible to approximate its action on a vector with (nearly) linear complexity, as in the fast multipole method (Mahawar and Sarin 2003).

Frequently, the matrices that arise in practice have quite a bit of structure. For instance, the A block is often block diagonal, with each diagonal block endowed with additional structure. Many of the algorithms discussed in this paper are able to exploit the structure of the problem to gain efficiency and save on storage. Sometimes the structure of the problem suggests solution algorithms that have a high degree of parallelism. This last aspect, however, is not emphasized in this paper. Finally we mention that in most applications n is larger than m, often much larger.

2. Applications leading to saddle point problems

As already mentioned, large-scale saddle point problems occur in many areas of computational science and engineering. The following is a list of some fields where saddle point problems naturally arise, together with some references:

- computational fluid dynamics (Glowinski 1984, Quarteroni and Valli 1994, Temam 1984, Turek 1999, Wesseling 2001)
- constrained and weighted least squares estimation (Björck 1996, Golub and Van Loan 1996)
- constrained optimization (Gill, Murray and Wright 1981, Wright 1992, Wright 1997)

- economics (Arrow, Hurwicz and Uzawa 1958, Duchin and Szyld 1979, Leontief, Duchin and Szyld 1985, Szyld 1981)
- electrical circuits and networks (Bergen 1986, Chua, Desoer and Kuh 1987, Strang 1986, Tropper 1962)
- electromagnetism (Bossavit 1998, Perugia 1997, Perugia, Simoncini and Arioli 1999)
- finance (Markowitz 1959, Markowitz and Perold 1981)
- image reconstruction (Hall 1979)
- image registration (Haber and Modersitzki 2004, Modersitzki 2003)
- interpolation of scattered data (Lyche, Nilssen and Winther 2002, Sibson and Stone 1991)
- linear elasticity (Braess 2001, Ciarlet 1988)
- mesh generation for computer graphics (Liesen, de Sturler, Sheffer, Aydin and Siefert 2001)
- mixed finite element approximations of elliptic PDEs (Brezzi 1974, Brezzi and Fortin 1991, Quarteroni and Valli 1994)
- model order reduction for dynamical systems (Freund 2003, Heres and Schilders 2005, Stykel 2005)
- optimal control (Battermann and Heinkenschloss 1998, Battermann and Sachs 2001, Betts 2001, Biros and Ghattas 2000, Nguyen 2004)
- parameter identification problems (Burger and Mühlhuber 2002, Haber and Ascher 2001, Haber, Ascher and Oldenburg 2000).

Quite often, saddle point systems arise when a certain quantity (such as the energy of a physical system) has to be minimized, subject to a set of linear constraints. In this case the Lagrange multiplier y usually has a physical interpretation and its computation is also of interest. For example, in incompressible flow problems x is a vector of velocities and y a vector of pressures. In the complementary energy formulation of structural mechanics x is the vector of internal forces, y represents the nodal displacements of the structure. For resistive electrical networks y represents the nodal potentials, x being the vector of currents.

In some cases, such as fluid dynamics or linear elasticity, saddle point problems result from the discretization of systems of partial differential equations with constraints. Typically the constraints represent some basic conservation law, such as mass conservation in fluid dynamics. In other cases, such as resistive electrical networks or structural analysis, the equations are discrete to begin with. Now the constraints may correspond to the topology (connectivity) of the system being studied. Because saddle point equations can be derived as equilibrium conditions for a physical system, they are sometimes called *equilibrium equations*. See Strang (1986, 1988) for a very nice discussion of equilibrium equations throughout applied mathematics. Another popular name for saddle point systems, especially in the optimization literature, is 'KKT system', from the Karush–Kuhn–Tucker first-order optimality conditions; see Nocedal and Wright (1999, page 328) for precise definitions, and Golub and Greif (2003) and Kjeldsen (2000) for historical notes.

Systems of the form (1.1)–(1.2) also arise from non-overlapping domain decomposition when interface unknowns are numbered last, as well as from FETI-type schemes when Lagrange multipliers are used to ensure continuity at the interfaces; see for instance Chan and Mathew (1994), Farhat and Roux (1991), Hu, Shi and Yu (2004), Quarteroni and Valli (1999) and Toselli and Widlund (2004).

It is of course not possible for us to cover here all these different applications. We choose instead to give some details about three classes of problems leading to saddle point systems. The first comes from the field of computational fluid dynamics, the second from least squares estimation, and the third one from interior point methods in constrained optimization.

2.1. Incompressible flow problems

We begin with the (steady-state) Navier–Stokes equations governing the flow of a Newtonian, incompressible viscous fluid. Let $\Omega \subset \mathbb{R}^d$ $(d = 2, 3)$ be a bounded, connected domain with a piecewise smooth boundary Γ. Given a force field $\mathbf{f} : \Omega \to \mathbb{R}^d$ and boundary data $\mathbf{g} : \Gamma \to \mathbb{R}^d$, the problem is to find a velocity field $\mathbf{u} : \Omega \to \mathbb{R}^d$ and a pressure field $p : \Omega \to \mathbb{R}$ such that

$$-\nu\Delta\mathbf{u} + (\mathbf{u} \cdot \nabla)\,\mathbf{u} + \nabla p = \mathbf{f} \quad \text{in} \ \ \Omega, \tag{2.1}$$

$$\nabla \cdot \mathbf{u} = 0 \quad \text{in} \ \ \Omega, \tag{2.2}$$

$$\mathcal{B}\mathbf{u} = \mathbf{g} \quad \text{on} \ \ \Gamma, \tag{2.3}$$

where $\nu > 0$ is the kinematic viscosity coefficient (inversely proportional to the Reynolds number Re), Δ is the Laplace operator in $\mathbb{R}^d$, ∇ denotes the gradient, $\nabla\cdot$ is the divergence, and $\mathcal{B}$ is some type of boundary operator (e.g., a trace operator for Dirichlet boundary conditions). To determine p uniquely we may impose some additional condition, such as

$$\int_\Omega p \, d\mathbf{x} = 0.$$

Equation (2.1) represents conservation of momentum, while equation (2.2) represents the incompressibility condition, or mass conservation. Owing to the presence of the convective term $(\mathbf{u} \cdot \nabla)\,\mathbf{u}$ in the momentum equations, the Navier–Stokes system is nonlinear. It can be linearized in various ways. An especially popular linearization process is the one based on Picard's iteration; see, e.g., Elman et al. (2005c, Section 7.2.2). Starting with an initial guess $\mathbf{u}^{(0)}$ (with $\nabla \cdot \mathbf{u}^{(0)} = 0$) for the velocity field, Picard's iteration constructs a sequence of approximate solutions $(\mathbf{u}^{(k)}, p^{(k)})$ by solving the linear *Oseen problem*

$$-\nu\Delta\mathbf{u}^{(k)} + (\mathbf{u}^{(k-1)} \cdot \nabla)\,\mathbf{u}^{(k)} + \nabla p^{(k)} = \mathbf{f} \quad \text{in } \Omega, \tag{2.4}$$

$$\nabla \cdot \mathbf{u}^{(k)} = 0 \quad \text{in } \Omega, \tag{2.5}$$

$$\mathcal{B}\mathbf{u}^{(k)} = \mathbf{g} \quad \text{on } \Gamma \tag{2.6}$$

$(k = 1, 2, \ldots)$. Note that no initial pressure needs to be specified. Under certain conditions on ν (which should not be too small) and $\mathbf{f}$ (which should not be too large in an appropriate norm), the steady Navier–Stokes equations (2.1)–(2.3) have a unique solution $(\mathbf{u}_*, p_*)$ and the iterates $(\mathbf{u}^{(k)}, p^{(k)})$ converge to it as $k \to \infty$ for any choice of the initial velocity $\mathbf{u}^{(0)}$. We refer to Girault and Raviart (1986) for existence and uniqueness results and to Karakashian (1982) for a proof of the global convergence of Picard's iteration.

Hence, at each Picard iteration one needs to solve an Oseen problem of the form

$$-\nu\Delta\mathbf{u} + (\mathbf{v} \cdot \nabla)\,\mathbf{u} + \nabla p = \mathbf{f} \quad \text{in } \Omega, \tag{2.7}$$

$$\nabla \cdot \mathbf{u} = 0 \quad \text{in } \Omega, \tag{2.8}$$

$$\mathcal{B}\mathbf{u} = \mathbf{g} \quad \text{on } \Gamma \tag{2.9}$$

with a known, divergence-free coefficient $\mathbf{v}$. Discretization of (2.7)–(2.9) using, *e.g.*, finite differences (Peyret and Taylor 1983) or finite elements (Elman *et al.* 2005c, Quarteroni and Valli 1994) results in a generalized saddle point system of the form (1.6), in which x represents the discrete velocities and y the discrete pressure. Here $A = \mathrm{diag}(A_1, \ldots, A_d)$ is a block diagonal matrix, where each block corresponds to a discrete convection-diffusion operator with the appropriate boundary conditions. Note that A is nonsymmetric, but satisfies condition C2 when an appropriate (conservative) discretization is used. The rectangular matrix B^T represents the discrete gradient operator while B represents its adjoint, the (negative) divergence operator. A nonzero C may be present if stabilization is used.

The important special case $\mathbf{v} = \mathbf{0}$ corresponds to the (steady-state) Stokes equations:

$$-\Delta\mathbf{u} + \nabla p = \mathbf{f} \quad \text{in } \Omega, \tag{2.10}$$

$$\nabla \cdot \mathbf{u} = 0 \quad \text{in } \Omega, \tag{2.11}$$

$$\mathcal{B}\mathbf{u} = \mathbf{g} \quad \text{on } \Gamma. \tag{2.12}$$

Note that without loss of generality we have set $\nu = 1$, since we can always divide the momentum equation by ν and rescale the pressure p and the forcing term $\mathbf{f}$ by ν. The Stokes equations can be interpreted as the Euler–Lagrange partial differential equations for the constrained variational problem

$$\min \ J(\mathbf{u}) = \frac{1}{2} \int_\Omega \|\nabla \mathbf{u}\|_2^2 \, d\mathbf{x} - \int_\Omega \mathbf{f} \cdot \mathbf{u} \, d\mathbf{x} \tag{2.13}$$

$$\text{subject to } \ \nabla \cdot \mathbf{u} = 0 \tag{2.14}$$

(see, *e.g.*, Gresho and Sani (1998, page 636)). Throughout this paper, $\|\mathbf{u}\|_2 = \sqrt{\mathbf{u} \cdot \mathbf{u}}$ denotes the Euclidean norm of the vector $\mathbf{u}$. Here the pressure p plays the role of the Lagrange multiplier. The Stokes equations describe the flow of a slow-moving, highly viscous fluid. They also arise as subproblems in the numerical solution of the Navier–Stokes equations by operator splitting methods (Glowinski 2003, Quarteroni and Valli 1994) and as the first step of Picard's iteration when the initial guess used is $\mathbf{u}^{(0)} = \mathbf{0}$.

Appropriate discretization of the Stokes system leads to a symmetric saddle point problem of the form (1.3) where A is now a block diagonal matrix, and each of its d diagonal blocks is a discretization of the Laplace operator $-\Delta$ with the appropriate boundary conditions. Thus, A is now symmetric and positive (semi-)definite. Again, a nonzero C may be present if stabilization is used. Typical sparsity patterns for $\mathcal{A}$ are displayed in Figure 2.1.

An alternative linearization of the Navier–Stokes equations can be derived on the basis of the identity

$$(\mathbf{u} \cdot \nabla)\mathbf{u} = \frac{1}{2}\nabla(\|\mathbf{u}\|_2^2) - \mathbf{u} \times (\nabla \times \mathbf{u}).$$

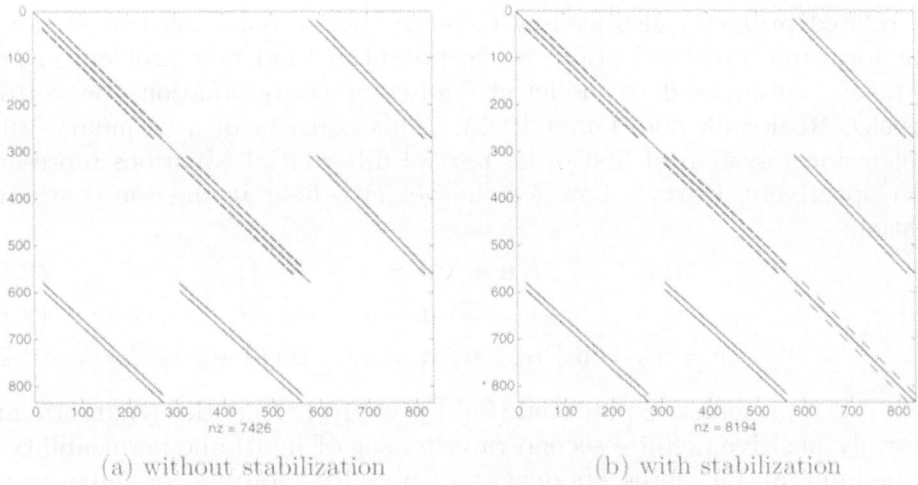

(a) without stabilization   (b) with stabilization

Figure 2.1. Sparsity patterns for two-dimensional Stokes problem (leaky lid-driven cavity) using Q1-P0 discretization.

See, for instance, Landau and Lifschitz (1959, page 5) or Chorin and Marsden (1990, page 47) as well as the discussion in Gresho (1991) and Gresho and Sani (1998). The corresponding linearized equations take the form

$$-\nu\Delta\mathbf{u} + \mathbf{w}\times\mathbf{u} + \nabla P = \mathbf{f} \quad \text{in } \Omega, \tag{2.15}$$

$$\nabla\cdot\mathbf{u} = 0 \quad \text{in } \Omega, \tag{2.16}$$

$$\mathcal{B}\mathbf{u} = \mathbf{g} \quad \text{on } \Gamma, \tag{2.17}$$

where, for the two-dimensional case,

- $(\mathbf{w}\times) = \begin{pmatrix} 0 & w \\ -w & 0 \end{pmatrix}$
- $w = \nabla\times\mathbf{v} = -\frac{\partial v_1}{\partial x_2} + \frac{\partial v_2}{\partial x_1}$
- $P = p + \frac{1}{2}\|\mathbf{v}\|_2^2$ (the so-called *Bernoulli pressure*)

Here the divergence-free vector field $\mathbf{v}$ again denotes the approximate velocity from the previous Picard iteration. See Olshanskii (1999) for the three-dimensional case. Note that when the 'wind' function $\mathbf{v}$ is irrotational ($\nabla\times\mathbf{v} = 0$), equations (2.15)–(2.17) reduce to the Stokes problem. It is worth stressing that the linearizations (2.7)–(2.9) and (2.15)–(2.17), although both conservative (Olshanskii 1999, page 357), are not mathematically equivalent. The so-called *rotation form* (2.15) of the momentum equations, although popular in fluid mechanics, has not been widely known among numerical analysts until the recent work by Olshanskii and co-workers (Lube and Olshanskii 2002, Olshanskii 1999, Olshanskii and Reusken 2002), which showed its advantages over the standard (convective) form. We return to this in Section 10.3.

A related problem, also leading to large sparse linear systems in saddle point form upon discretization, is the potential fluid flow problem in porous media, often used to model groundwater contamination (Bear 1972, Maryška, Rozložník and Tůma 1995). This consists of a boundary value problem for a system of first-order partial differential equations representing, respectively, Darcy's Law for the velocity field $\mathbf{u}$ and the continuity equation:

$$K\mathbf{u} + \nabla p = 0 \quad \text{in } \Omega, \tag{2.18}$$

$$\nabla\cdot\mathbf{u} = q \quad \text{in } \Omega, \tag{2.19}$$

$$p = p_D \quad \text{on } \Gamma_D, \quad \mathbf{u}\cdot\mathbf{n} = u_N \quad \text{on } \Gamma_N, \tag{2.20}$$

where p is a piezometric potential (fluid pressure), K is the symmetric and uniformly positive definite second-rank tensor of hydraulic permeability of the medium, and q represents density of potential sources (or sinks) in the medium. Here Γ_D and Γ_N are subsets of the boundary Γ of the bounded connected flow domain Ω, with $\Gamma = \bar{\Gamma}_D\cup\bar{\Gamma}_N$, $\Gamma_D \neq \emptyset$, and $\Gamma_D\cap\Gamma_N = \emptyset$; $\mathbf{n}$ is the outward normal vector defined (a.e.) on Γ. When discretized by mixed finite elements (Raviart–Thomas elements being a very popular choice for this problem), a linear system of the type (1.3) is obtained. The symmetric positive definite matrix A is now a discretization of the linear operator 'multiplication by K', a zeroth-order differential operator. The conditioning properties of A are independent of the discretization parameter h (for most discretizations), and depend only on properties of the hydraulic permeability tensor K. The matrix $-B$ represents, again, a discrete divergence operator and B^T a discrete gradient. We note that modelling the interaction between surface and subsurface flows leads to coupled (Navier–) Stokes and Darcy systems (Discacciati, Miglio and Quarteroni 2002, Discacciati and Quarteroni 2004). Problem (2.18)–(2.20) is just one example of a first-order system formulation of a second-order linear elliptic PDE (Brezzi and Fortin 1991). Saddle point systems also arise from mixed formulation of fourth-order (biharmonic) elliptic problems (Glowinski and Pironneau 1979).

In the course of this brief discussion we have restricted ourselves to stationary (steady-state) problems. The unsteady (time-dependent) case leads to sequences of saddle point systems when fully implicit time-stepping schemes are used, for example when the time derivative $\mathbf{u}_t$ is discretized using backward Euler or Crank–Nicolson schemes; see, *e.g.*, Turek (1999, Chapter 2). In the case of Stokes and Oseen, the resulting semi-discrete systems are often referred to as *generalized* Stokes and Oseen problems. The literature on numerical methods for incompressible flow problems is vast; see, *e.g.*, Elman *et al.* (2005c), Fortin (1993), Glowinski (2003), Gresho and Sani (1998), Gunzburger (1989), Quarteroni and Valli (1994), Temam (1984), Turek (1999) and Wesseling (2001).

2.2. Constrained and weighted least squares

Linear systems of saddle point type commonly arise when solving least squares problems. Consider the following least squares problem with linear equality constraints:

$$\min_{x} \; \|c - Gy\|_2 \tag{2.21}$$

$$\text{subject to} \;\; Ey = d, \tag{2.22}$$

where $c \in \mathbb{R}^p$, $G \in \mathbb{R}^{p \times m}$, $y \in \mathbb{R}^m$, $E \in \mathbb{R}^{q \times m}$, $d \in \mathbb{R}^q$ and $q < m$. Problems of this kind arise, for instance, in curve or surface fitting when the curve is required to interpolate certain data points; see Björck (1996, Chapter 5). The optimality conditions for problem (2.21)–(2.22) are

$$\begin{bmatrix} I_p & O & G \\ O & O & E \\ G^T & E^T & O \end{bmatrix} \begin{bmatrix} r \\ \lambda \\ y \end{bmatrix} = \begin{bmatrix} c \\ d \\ 0 \end{bmatrix}, \tag{2.23}$$

where I_p is the $p \times p$ identity matrix and $\lambda \in \mathbb{R}^q$ is a vector of Lagrange multipliers. Clearly, (2.23) is a special case of the symmetric saddle point problem (1.3).

Next we consider the generalized linear least squares problem

$$\min_x (f - Gx)^T W^{-1}(f - Gx), \tag{2.24}$$

where $f \in \mathbb{R}^n$, $G \in \mathbb{R}^{n \times m}$ with $m < n$, and the (scaled) covariance matrix $W \in \mathbb{R}^{n \times n}$ is symmetric positive definite; see Björck (1996, Chapter 4). It is readily seen that (2.24) is equivalent to a standard saddle point system of the form (1.3) with $A = W$, $B = G^T$ and $g = 0$. This saddle point formulation of the generalized least squares problem is often referred to as the *augmented system* formulation. The matrix W is diagonal when the errors in the observations are uncorrelated. When $W = I$ (the $n \times n$ identity matrix) we have the usual linear least squares problem. An advantage of the augmented system formulation over the standard one is that the former allows for the case in which W is singular, which is important in some applications. In other words, W could be positive semidefinite rather than definite. Some applications even lead to an indefinite weight matrix W; see Bojanczyk, Higham and Patel (2003), Chandrasekaran, Gu and Sayed (1998), Hassibi, Sayed and Kailath (1996).

Finally, if the original problem is ill-posed and Tikhonov regularization is applied, one obtains a saddle point system of the form (1.6) with $C = \gamma L^T L$ where $\gamma > 0$ is the regularization parameter and L is either the $m \times m$ identity or some type of smoothing operator, such as a first-order finite difference operator. See Benzi and Ng (2004) and Gonzales and Woods (1992) for applications in image processing.

2.3. Saddle point systems from interior point methods

Here we show how saddle point systems arise when interior point methods are used to solve constrained optimization problems. Our presentation is based on the nice synopsis given in Bergamaschi, Gondzio and Zilli (2004). Consider a convex nonlinear programming problem,

$$\min \ f(x) \tag{2.25}$$

$$\text{subject to} \ \ c(x) \le 0, \tag{2.26}$$

where $f : \mathbb{R}^n \to \mathbb{R}$ and $c : \mathbb{R}^n \to \mathbb{R}^m$ are convex and twice differentiable. Introducing a nonnegative slack variable $z \in \mathbb{R}^m$, we can write the inequality constraint as the system of equalities $c(x) + z = 0$, and we can introduce the associated barrier problem:

$$\min \ f(x) - \mu \sum_{i=1}^{m} \ln z_i \tag{2.27}$$

$$\text{subject to} \ \ c(x) + z = 0. \tag{2.28}$$

The corresponding Lagrangian is

$$\mathcal{L}(x, y, z; \mu) = f(x) + y^T (c(x) + z) - \mu \sum_{i=1}^{m} \ln z_i.$$

To find a stationary point of the Lagrangian we set

$$\nabla_x \mathcal{L}(x, y, z; \mu) = \nabla f(x) + \nabla c(x)^T y = 0, \tag{2.29}$$
$$\nabla_y \mathcal{L}(x, y, z; \mu) = c(x) + z = 0, \tag{2.30}$$
$$\nabla_z \mathcal{L}(x, y, z; \mu) = y - \mu Z^{-1} e = 0, \tag{2.31}$$

where $Z = \text{diag}(z_1, z_2, \ldots, z_m)$ and $e = [1 \, 1 \, \ldots \, 1]^T$. Introducing the diagonal matrix $Y = \text{diag}(y_1, y_2, \ldots, y_m)$, the first-order optimality conditions for the barrier problem become

$$\nabla f(x) + \nabla c(x)^T y = 0, \tag{2.32}$$
$$c(x) + z = 0, \tag{2.33}$$
$$Y Z e = \mu e, \tag{2.34}$$
$$y, z \geq 0. \tag{2.35}$$

This is a nonlinear system of equations with nonnegativity constraints and it can be solved by Newton's method. The barrier parameter μ is gradually reduced so as to ensure convergence of the iterates to the optimal solution of problem (2.25)–(2.26). At each Newton iteration, it is necessary to solve a linear system of the form

$$\begin{bmatrix} H(x, y) & B(x)^T & O \\ B(x) & O & I \\ O & Z & Y \end{bmatrix} \begin{bmatrix} \delta x \\ \delta y \\ \delta z \end{bmatrix} = \begin{bmatrix} -\nabla f(x) - B(x)^T y \\ -c(x) - z \\ \mu e - Y Z e \end{bmatrix}, \tag{2.36}$$

where

$$H(x, y) = \nabla^2 f(x) + \sum_{i=1}^{m} y_i \nabla^2 c_i(x) \in \mathbb{R}^{n \times n} \quad \text{and} \quad B(x) = \nabla c(x) \in \mathbb{R}^{m \times n}.$$

Here $\nabla^2 f(x)$ denotes the Hessian of f evaluated at x. The linear system (2.36) can be reduced to one of smaller dimensions by using the third equation to eliminate $\delta z = \mu Y^{-1} e - Z e - Z Y^{-1} \delta y$ from the second equation. The resulting system is

$$\begin{bmatrix} -H(x, y) & B(x)^T \\ B(x) & Z Y^{-1} \end{bmatrix} \begin{bmatrix} \delta x \\ -\delta y \end{bmatrix} = \begin{bmatrix} \nabla f(x) + B(x)^T y \\ -c(x) - \mu Y^{-1} e \end{bmatrix}. \tag{2.37}$$

Apart from the sign, (2.37) is a saddle point system of the form (1.6). If the objective function $f(x)$ and the constraints $c_i(x)$ are convex, the symmetric matrix $H(x, y)$ is positive semidefinite, and it is positive definite if $f(x)$ is strictly convex. The diagonal matrix $Z Y^{-1}$ is obviously positive semidefinite. The coefficient matrix in (2.37) depends on the current approximation (x, y), and it changes at each Newton step.

Similar linear systems arise when interior point methods are used to solve linear and quadratic programming problems. Now the systems to be solved at each Newton iteration are of the form

$$\begin{bmatrix} -H - D & B^T \\ B & O \end{bmatrix} \begin{bmatrix} \delta x \\ \delta y \end{bmatrix} = \begin{bmatrix} \xi \\ \eta \end{bmatrix},$$

where the $n \times n$ matrix H is symmetric positive semidefinite if the problem is convex and D is a (positive) diagonal matrix. Now H and B remain constant ($H \equiv O$ in linear programming), while D changes at each Newton iteration.

There are many interesting linear algebra problems arising from the use of interior point methods; see in particular Bergamaschi *et al.* (2004), Czyzyk, Fourer and Mehrotra (1998), Forsgren, Gill and Shinnerl (1996), Fourer and Mehrotra (1993), Frangioni and Gentile (2004), Freund and Jarre (1996), Gill, Murray, Poncele\u00f3n and Saunders (1992), Nocedal and Wright (1999), Oliveira and Sorensen (2005), Wright (1992) and Wright (1997).

3. Properties of saddle point matrices

This section is devoted to establishing basic algebraic properties of the saddle point matrix $\mathcal{A}$ such as existence of various factorizations, invertibility, spectral properties, and conditioning. Knowledge of these properties is important in the development of solution algorithms.

3.1. Block factorizations and the Schur complement

If A is nonsingular, the saddle point matrix $\mathcal{A}$ admits the following block triangular factorization:

$$\mathcal{A} = \begin{bmatrix} A & B_1^T \\ B_2 & -C \end{bmatrix} = \begin{bmatrix} I & O \\ B_2 A^{-1} & I \end{bmatrix} \begin{bmatrix} A & O \\ O & S \end{bmatrix} \begin{bmatrix} I & A^{-1} B_1^T \\ O & I \end{bmatrix}, \quad (3.1)$$

where $S = -(C + B_2 A^{-1} B_1^T)$ is the *Schur complement* of A in $\mathcal{A}$. A number of important properties of the saddle point matrix $\mathcal{A}$ can be derived on the basis of (3.1): we do this in the next three subsections.

Also useful are the equivalent factorizations

$$\mathcal{A} = \begin{bmatrix} A & O \\ B_2 & S \end{bmatrix} \begin{bmatrix} I & A^{-1} B_1^T \\ O & I \end{bmatrix} \quad (3.2)$$

and

$$\mathcal{A} = \begin{bmatrix} I & O \\ B_2 A^{-1} & I \end{bmatrix} \begin{bmatrix} A & B_1^T \\ O & S \end{bmatrix}. \quad (3.3)$$

The assumption that A is nonsingular may appear to be rather restrictive, since A is singular in many applications; see, *e.g.*, Haber and Ascher (2001). However, one can use augmented Lagrangian techniques (Fortin and Glowinski 1983, Glowinski and Le Tallec 1989, Golub and Greif 2003, Greif, Golub and Varah 2005) to replace the original saddle point system with an equivalent one having the same solution but in which the $(1,1)$ block A is now nonsingular. Hence, no great loss of generality is incurred. We shall return to augmented Lagrangian techniques in Section 3.5.

Besides being useful for deriving theoretical properties of saddle point matrices, the decompositions (3.1)–(3.3) are also the basis for many of the most popular solution algorithms for saddle point systems, as we shall see.

3.2. Solvability conditions

Assuming A is nonsingular, it readily follows from any of the block decompositions (3.1)–(3.3) that $\mathcal{A}$ is nonsingular if and only if S is. Unfortunately, very little can be said in general about the invertibility of the Schur complement $S = -(C + B_2 A^{-1} B_1^T)$. It is necessary to place some restrictions on the matrices A, B_1, B_2 and C.

Symmetric case

We begin with the standard saddle point system (1.3), where A is symmetric positive definite, $B_1 = B_2$ and $C = O$. In this case the Schur complement reduces to $S = -BA^{-1}B^T$, a symmetric negative semidefinite matrix. It is obvious that S, and thus $\mathcal{A}$, is invertible if and only if B^T has full column rank (hence, if and only if rank$(B) = m$), since in this case S is symmetric negative definite. Then both problems (1.3) and (1.4)–(1.5) have a unique solution: if (x_*, y_*) is the solution of (1.3), x_* is the unique solution of (1.4)–(1.5). It can be shown that x_* is the A-orthogonal projection of the solution $\hat{x} = A^{-1}f$ of the unconstrained problem (1.4) onto the constraint set $\mathcal{C} = \{x \in \mathbb{R}^n \,|\, Bx = g\}$. Here A-orthogonal means orthogonal with respect to the inner product $\langle v, w \rangle_A \equiv w^T A v$. We will discuss this in more detail in Section 3.3.

Next we consider the case where A is symmetric positive definite, $B_1 = B_2 = B$, and $C \neq O$ is symmetric positive semidefinite. Then again $S = -(C + BA^{-1}B^T)$ is symmetric negative semidefinite, and it is negative definite (hence, invertible) if and only if $\ker(C) \cap \ker(B^T) = \{0\}$. Obvious sufficient conditions for invertibility are that C be positive definite or that B have full row rank. We can summarize our discussion so far in the following theorem.

Theorem 3.1. Assume A is symmetric positive definite, $B_1 = B_2 = B$, and C is symmetric positive semidefinite. If $\ker(C) \cap \ker(B^T) = \{0\}$, then the saddle point matrix $\mathcal{A}$ is nonsingular. In particular, $\mathcal{A}$ is invertible if B has full rank.

Now we relax the condition that A be positive definite. If A is indefinite, the following simple example shows that $\mathcal{A}$ may be singular, even if B has full rank:

$$\mathcal{A} = \left[\begin{array}{cc|c} 1 & 0 & -1 \\ 0 & -1 & 1 \\ \hline -1 & 1 & 0 \end{array}\right] = \begin{bmatrix} A & B^T \\ B & O \end{bmatrix}.$$

However, $\mathcal{A}$ will be invertible if A is positive definite on $\ker(B)$. When A is symmetric positive semidefinite, we have the following result (see, $e.g.$, the discussion of quadratic programming in Hadley (1964) or Luenberger (1984, page 424)). Although this is a well-known result, we include a proof to make our treatment more self-contained.

Theorem 3.2. Assume that A is symmetric positive semidefinite, $B_1 = B_2 = B$ has full rank, and $C = O$. Then a necessary and sufficient condition for the saddle point matrix $\mathcal{A}$ to be nonsingular is $\ker(A) \cap \ker(B) = \{0\}$.

Proof. Let $u = \begin{bmatrix} x \\ y \end{bmatrix}$ be such that $\mathcal{A}u = 0$. Hence, $Ax + B^Ty = 0$ and $Bx = 0$. It follows that $x^TAx = -x^TB^Ty = -(Bx)^Ty = 0$. Since A is symmetric positive semidefinite, $x^TAx = 0$ implies $Ax = 0$ (see Horn and Johnson (1985, page 400)), and therefore $x \in \ker(A) \cap \ker(B)$, thus $x = 0$. Also, $y = 0$ since $B^Ty = 0$ and B^T has full column rank. Therefore $u = 0$, and $\mathcal{A}$ is nonsingular. This proves the sufficiency of the condition.

Assume now that $\ker(A) \cap \ker(B) \neq \{0\}$. Taking $x \in \ker(A) \cap \ker(B)$, $x \neq 0$ and letting $u = \begin{bmatrix} x \\ 0 \end{bmatrix}$ we have $\mathcal{A}u = 0$, implying that $\mathcal{A}$ is singular. Hence, the condition is also necessary. $\qquad\qquad\qquad\qquad\square$

Remark. It is clear from the proof of this theorem that the requirement that A be positive semidefinite can be somewhat relaxed: it suffices that A be definite on $\ker(B)$. In fact, all we need is that $x^TAx \neq 0$ for $x \in \ker(B)$, $x \neq 0$. This implies that A is either positive definite or negative definite on $\ker(B)$. In any case, the rank of A must be at least $n - m$ for $\mathcal{A}$ to be nonsingular.

How restrictive is the assumption that B has full rank? A rank deficient B signifies that some of the constraints are redundant. It is generally easy to eliminate this redundancy. For instance, in the Stokes and Oseen case, where B^T represents a discrete gradient, a one-dimensional subspace (containing all the constant vectors) is often present. Hence $\mathcal{A}$ has one zero eigenvalue, corresponding to the so-called *hydrostatic pressure mode*, due to the fact that the pressure is defined up to a constant. A similar situation occurs with electric networks, where y, the vector of nodal potentials, is also defined up to an additive constant. The rank deficiency in B can easily be removed by 'grounding' one of the nodes, that is, by specifying the value of the potential (or of the pressure) at one point. One problem with this approach is that the resulting linear system may be rather ill-conditioned; see Bochev and Lehoucq (2005). Fortunately, since the system $\mathcal{A}u = b$ is consistent by construction, it may not be necessary to remove the singularity of $\mathcal{A}$. Iterative methods like GMRES (Saad and Schultz 1986) are largely unaffected by the presence of a single eigenvalue exactly equal to zero, at least when using a zero initial guess, $u_0 = 0$. The reader is referred to Elman *et al.* (2005*c*, Section 8.3.4) for a detailed discussion of this issue in the context of fluid flow problems; see further the remarks in Olshanskii and Reusken (2004, Section 4) and Zhang and Wei (2004, Section 4).

General case
When $C = O$, a *necessary* condition for invertibility is provided by the following theorem, a slight generalization of a similar result for the case $B_1 = B_2$: see Gansterer, Schneid and Ueberhuber (2003).

Theorem 3.3. If the matrix

$$\mathcal{A} = \begin{bmatrix} A & B_1^T \\ B_2 & O \end{bmatrix}$$

is nonsingular, then $\mathrm{rank}(B_1) = m$ and $\mathrm{rank}\left(\begin{smallmatrix} A \\ B_2 \end{smallmatrix}\right) = n$.

Proof. If $\mathrm{rank}(B_1) < m$ then there exists a nonzero vector $y \in \mathbb{R}^m$ with $B_1^T y = 0$. Therefore, letting $u = \begin{bmatrix} 0 \\ y \end{bmatrix}$, we get $\mathcal{A}u = 0$, a contradiction.

If $\mathrm{rank}\left(\begin{smallmatrix} A \\ B_2 \end{smallmatrix}\right) < n$ then there exists a nonzero vector $x \in \mathbb{R}^n$ such that $\left(\begin{smallmatrix} A \\ B_2 \end{smallmatrix}\right)x = 0$. Letting $u = \begin{bmatrix} x \\ 0 \end{bmatrix}$, we get $\mathcal{A}u = 0$, a contradiction. $\square$

It is easy to show that these conditions are not sufficient to ensure the invertibility of $\mathcal{A}$. Some additional conditions are needed. Recall that for any matrix $A \in \mathbb{R}^{n \times n}$ we can write $A = H + K$ where $H = \frac{1}{2}(A + A^T)$ and $K = \frac{1}{2}(A - A^T)$ are the symmetric and skew-symmetric part of A, respectively. The following result provides a necessary and a sufficient condition for $\mathcal{A}$ to be invertible when $B_1 = B_2$.

Theorem 3.4. Assume that H, the symmetric part of A, is positive semidefinite, $B_1 = B_2 = B$ have full rank, and C is symmetric positive semidefinite (possibly zero). Then

(i) $\ker(H) \cap \ker(B) = \{0\} \Rightarrow \mathcal{A}$ invertible,
(ii) $\mathcal{A}$ invertible $\Rightarrow \ker(A) \cap \ker(B) = \{0\}$.

The converses of (i)–(ii) do not hold in general.

Proof. The proof of part (i) is similar to the sufficiency proof in Theorem 3.1, and can be found in Benzi and Golub (2004, Lemma 1.1). The proof of part (ii) is exactly as the necessity proof in Theorem 3.1.

Now we show that the converses of (i) and (ii) are false in general. To see that the converse of (i) is not true in general, consider the matrix

$$\mathcal{A} = \left[\begin{array}{ccc|c} 1 & -1 & 0 & 0 \\ 1 & 0 & 0 & 0 \\ 0 & 0 & 0 & 1 \\ \hline 0 & 0 & 1 & 0 \end{array}\right] = \begin{bmatrix} A & B^T \\ B & O \end{bmatrix}.$$

Here we have $\ker(H) \cap \ker(B) = \operatorname{span}([0\,1\,0]^T) \neq \{0\}$ and yet $\mathcal{A}$ is invertible.

To see that the converse of (ii) is not generally true, consider the matrix

$$\mathcal{A} = \left[\begin{array}{cc|c} 0 & -1 & 0 \\ 1 & 1 & 1 \\ \hline 0 & 1 & 0 \end{array}\right] = \begin{bmatrix} A & B^T \\ B & O \end{bmatrix}.$$

This matrix is manifestly singular, yet A is nonsingular and thus $\ker(A) \cap \ker(B) = \{0\}$.   □

We note that when H is positive semidefinite, the inclusion $\ker(A) \subset \ker(H)$ holds. For if $x \in \ker(A)$, then $Ax = Hx + Kx = 0$ and thus $x^T A x = x^T H x = 0$. But since H is symmetric positive semidefinite the last equality implies $Hx = 0$, and therefore $x \in \ker(H)$. That this is a proper inclusion can be seen from the simple example

$$A = \begin{bmatrix} 0 & 1 \\ -1 & 0 \end{bmatrix}.$$

3.3. *The inverse of a saddle point matrix*

If A is nonsingular, then we know that $\mathcal{A}$ is invertible if and only if $S = -(C + B_2 A^{-1} B_1^T)$ is nonsingular, and we have the following explicit expression for the inverse:

$$\mathcal{A}^{-1} = \begin{bmatrix} A & B_1^T \\ B_2 & -C \end{bmatrix}^{-1} = \begin{bmatrix} A^{-1} + A^{-1} B_1^T S^{-1} B_2 A^{-1} & -A^{-1} B_1^T S^{-1} \\ -S^{-1} B_2 A^{-1} & S^{-1} \end{bmatrix}. \quad (3.4)$$

If A is singular but C is nonsingular, an analogous expression can be given if we assume that the matrix $A + B_1^T C^{-1} B_2$, the Schur complement of C in $\mathcal{A}$, is nonsingular. However, such an expression is of limited interest in the numerical solution of saddle point problems. See Lu and Shiou (2002) for additional expressions for the inverse of a block 2×2 matrix.

An interesting special case arises when A is symmetric positive definite, $B_1 = B_2 = B$, $C = O$, $S = -BA^{-1}B^T$ is nonsingular, and $g = 0$. Then the explicit expression for $\mathcal{A}^{-1}$ shows that the solution (x_*, y_*) of (1.3) is given by

$$\begin{bmatrix} x_* \\ y_* \end{bmatrix} = \begin{bmatrix} \left(I + A^{-1}B^T S^{-1} B\right) A^{-1} f \\ S^{-1} B A^{-1} f \end{bmatrix}. \tag{3.5}$$

It is easy to see that the matrix

$$\Pi \equiv -A^{-1}B^T S^{-1} B = A^{-1}B^T (BA^{-1}B^T)^{-1}B$$

satisfies $\Pi^2 = \Pi$, *i.e.*, Π is a projector. Moreover, the relations

$$\Pi v \in \mathrm{range}(A^{-1}B^T) \quad \text{and} \quad v - \Pi v \perp \mathrm{range}(B^T), \quad \text{for all} \quad v \in \mathbb{R}^n,$$

show that Π represents an (oblique) projector onto $\mathrm{range}(A^{-1}B^T)$ and orthogonal to $\mathrm{range}(B^T)$. The first component in (3.5) can then be written as

$$x_* = (I - \Pi)\,\hat{x},$$

where $\hat{x} \equiv A^{-1}f$ is the solution of the unconstrained problem (1.4). Hence x_* is orthogonal to $\mathrm{range}(B^T)$. Furthermore, $\hat{x} = \Pi\hat{x} + x_*$, which means that the solution of the unconstrained problem is decomposed into a part that is in $\mathrm{range}(A^{-1}B^T)$ and a part that is orthogonal to $\mathrm{range}(B^T)$. By the nature of Π, this decomposition generally is oblique, and is orthogonal only when $\mathrm{range}(A^{-1}B^T) = \mathrm{range}(B^T)$ (the latter being true particularly for $A = I$). Next note that since $f - B^T y_* = Ax_*$, and $Bx_* = 0$,

$$0 = Bx_* = (BA^{-1})(Ax_*) = (A^{-1}B^T)^T\,(f - B^T y_*). \tag{3.6}$$

By assumption, A^{-1} is symmetric positive definite, so the function $\langle v, w \rangle_{A^{-1}} \equiv w^T A^{-1} v$ is an inner product. Then (3.6) shows that the vector $f - B^T y_* \in f + \mathrm{range}(B^T)$ is orthogonal with respect to the A^{-1}-inner product (A^{-1}-orthogonal) to the space $\mathrm{range}(B^T)$. But this means that y_* is the solution of the (generalized) least squares problem $B^T u \approx f$ with respect to the A^{-1}-norm, $\|v\|_{A^{-1}} \equiv (\langle v, v \rangle_{A^{-1}})^{1/2}$, *i.e.*,

$$\|f - B^T y_*\|_{A^{-1}} = \min_u \|f - B^T u\|_{A^{-1}}. \tag{3.7}$$

The relation (3.7) is also derived in Benbow (1999, Section 2), where it is used to compute the solution of the standard saddle point problem (1.3) with $g = 0$ by first solving the generalized least squares problem with the matrix B^T and the right-hand side f (giving y_*), and then computing $x_* = A^{-1}(f - B^T y_*)$. To solve the generalized least squares problem (3.7) numerically, a generalized version of the LSQR method of Paige and Saunders (1982) is developed in Benbow (1999). The numerical experiments in Benbow (1999) show that this approach to computing y_* is often superior to applying the MINRES method (Paige and Saunders 1975) (see Section 9 below for details about this method) to the (symmetric) Schur complement system $S\,y = BA^{-1}f$, which is the second component in (3.5).

Another important special case arises when A and C are both symmetric positive definite and $B_1 = B_2 = B$. Then the corresponding (symmetric) saddle point matrix $\mathcal{A}$ is said to be *quasidefinite*, regardless of what the rank of B may be (Vanderbei 1995). These matrices result from interior point methods in constrained optimization; see the discussion in Section 2.3. Properties of symmetric quasidefinite matrices have been studied, *e.g.*, in George, Ikramov and Kucherov (2000), Gill, Saunders and Shinnerl (1996) and Vanderbei (1995). One of the basic properties is that if $\mathcal{A}$ is quasidefinite, so is $\mathcal{A}^{-1}$ (under the same 2×2 block partitioning), as one can immediately see from (3.4). In George and Ikramov (2002), this property has been extended to the class of invertible *weakly quasidefinite* matrices, in which A and C are only assumed to be symmetric positive semidefinite.

An alternative expression of the inverse that does not require A to be invertible is the following. Assume that $B_1 = B_2 = B$ has full rank and $C = O$. Denote by $Z \in \mathbb{R}^{n \times (n-m)}$ any matrix whose columns form a basis for $\ker(B)$. If H, the symmetric part of A, is positive semidefinite, then it is easy to see that condition (i) in Theorem 3.4 implies that the $(n-m) \times (n-m)$ matrix $Z^T A Z$ is invertible; indeed, its symmetric part $Z^T H Z$ is positive definite. Letting $W = Z(Z^T A Z)^{-1} Z^T$, we have the following expression for the inverse of $\mathcal{A}$:

$$
\mathcal{A}^{-1} = \begin{bmatrix} A & B^T \\ B & O \end{bmatrix}^{-1} \tag{3.8}
$$
$$
= \begin{bmatrix} W & (I - WA)B^T(BB^T)^{-1} \\ (BB^T)^{-1}B(I - AW) & -(BB^T)^{-1}B(A - AWA)B^T(BB^T)^{-1} \end{bmatrix},
$$

which can be easily proved keeping in mind that $B^T(BB^T)^{-1}B = I - ZZ^T$; see Gansterer *et al.* (2003).

These explicit expressions for $\mathcal{A}^{-1}$ are of limited practical use, and their interest is primarily theoretical. See, however, Powell (2004) for a situation where the inverse of $\mathcal{A}$ is explicitly needed, and for a careful discussion of the problem of updating the inverse of $\mathcal{A}$ when a few of its rows and corresponding columns are modified.

It must be mentioned that in the finite element context the mere nonsingularity of $\mathcal{A}$ is not sufficient to ensure meaningful computed solutions. In order for the discrete problem to be well-posed it is essential that the saddle point matrix remain *uniformly invertible* as h, the mesh size parameter, goes to zero. This means that an appropriate (generalized) condition number of $\mathcal{A}$ remains bounded as $h \to 0$. Sufficient conditions for this to happen include the already-mentioned discrete LBB (or inf-sup) conditions; see Brezzi (2002), Brezzi and Fortin (1991), Ciarlet Jr. *et al.* (2003) and Nicolaides (1982). Some discussion of conditioning issues from a linear algebra standpoint can be found in Section 3.5.

Finally, we point out that *singular* saddle point systems have been studied in Gansterer *et al.* (2003) and Wu, Silva and Yuan (2004); other relevant references are Wei (1992a) and Zhang and Wei (2004). Explicit forms for various generalized inverses of saddle point matrices arising in constrained optimization can be found in Campbell and Meyer Jr. (1979, Section 3.5); see also George and Ikramov (2002).

3.4. Spectral properties of saddle point matrices

In this section we collect a few facts on the spectral properties of saddle point matrices which are relevant when solving the equations by iterative methods. We also introduce an alternative formulation of the saddle point equations leading to a (nonsymmetric) positive definite coefficient matrix.

Eigenvalues: The symmetric case

Assume that A is symmetric positive definite, $B_1 = B_2 = B$ has full rank, and C is symmetric positive semidefinite (possibly zero). Then from (3.1) we obtain

$$\begin{bmatrix} I & O \\ -BA^{-1} & I \end{bmatrix} \begin{bmatrix} A & B^T \\ B & -C \end{bmatrix} \begin{bmatrix} I & -A^{-1}B^T \\ O & I \end{bmatrix} = \begin{bmatrix} A & O \\ O & S \end{bmatrix} \tag{3.9}$$

where $S = -(C + BA^{-1}B^T)$ is symmetric negative definite. Hence $\mathcal{A}$ is congruent to the block diagonal matrix $\begin{bmatrix} A & O \\ O & S \end{bmatrix}$. It follows from Sylvester's Law of Inertia (see Horn and Johnson (1985, page 224)) that $\mathcal{A}$ is indefinite, with n positive and m negative eigenvalues. The same is of course true if B is rank deficient, as long as S remains negative definite. Clearly, in case S is rank deficient, say rank$(S) = m - r$, $\mathcal{A}$ has n positive, $m - r$ negative and r zero eigenvalues. A simple limiting argument shows that this result remains true if A is only assumed to be positive semidefinite, provided that the usual condition $\ker(A) \cap \ker(B) = \{0\}$ is satisfied. We refer the reader to Forsgren (2002), Forsgren and Murray (1993), Gill, Murray, Saunders and Wright (1991) and Gould (1985) for additional results on the inertia of symmetric saddle point matrices under various assumptions on A, B and C. Generally speaking, unless m is very small (which is seldom the case in practice), the matrix $\mathcal{A}$ is *highly indefinite*, in the sense that it has many eigenvalues of both signs.

The following result from Rusten and Winther (1992) establishes eigenvalue bounds for an important class of saddle point matrices.

Theorem 3.5. Assume A is symmetric positive definite, $B_1 = B_2 = B$ has full rank, and $C = O$. Let μ_1 and μ_n denote the largest and smallest eigenvalues of A, and let σ_1 and σ_m denote the largest and smallest singular values of B. Let $\sigma(\mathcal{A})$ denote the spectrum of $\mathcal{A}$. Then

$$\sigma(\mathcal{A}) \subset I^- \cup I^+,$$

where

$$I^- = \left[\frac{1}{2}\left(\mu_n - \sqrt{\mu_n^2 + 4\sigma_1^2} \right), \frac{1}{2}\left(\mu_1 - \sqrt{\mu_1^2 + 4\sigma_m^2} \right) \right]$$

and

$$I^+ = \left[\mu_n, \frac{1}{2}\left(\mu_1 + \sqrt{\mu_1^2 + 4\sigma_1^2} \right) \right].$$

An extension of this result to the case where $C \neq O$ (with C symmetric and positive semidefinite) can be found in Silvester and Wathen (1994), while the case where A is positive semidefinite with $\ker(A) \cap \ker(B) = \{0\}$ has been treated in Perugia and Simoncini (2000). These bounds can be used to obtain estimates for the condition number of $\mathcal{A}$ in specific cases. In turn, these estimates can be used to predict the rate of convergence of iterative methods like MINRES (Paige and Saunders 1975); see Fischer, Ramage, Silvester and Wathen (1998), Maryška, Rozložník and Tůma (1996), Wathen, Fischer and Silvester (1995) and Section 9 below. Eigenvalue bounds are also important when assessing the (inf-sup) stability of mixed finite element discretizations; see, *e.g.*, Bitar and Vincent (2000) and Malkus (1981).

Eigenvalues: The general case
In the general case, not much can be said about the eigenvalues of $\mathcal{A}$. However, in most cases of interest the convex hull of the eigenvalues of $\mathcal{A}$ contains the origin. If we consider for example the case where $A \neq A^T$, $B_1 = B_2 = B$, and $C = C^T$ (as in the Oseen problem, for example) then we have that the symmetric part of $\mathcal{A}$ is

$$\frac{1}{2}(\mathcal{A} + \mathcal{A}^T) = \begin{bmatrix} H & B^T \\ B & -C \end{bmatrix},$$

where H is the symmetric part of A. If H is positive definite and C positive semidefinite (as in the Oseen problem), then the symmetric part of $\mathcal{A}$ is indefinite and therefore $\mathcal{A}$ has eigenvalues on both sides of the imaginary axis. Figure 3.1(a) displays the eigenvalues of the discrete Oseen operator obtained from a Q1-P0 finite element approximation of problem (2.7)–(2.9) with $\nu = 0.01$ and $\Omega = [0,1] \times [0,1]$. The matrix was generated using the IFISS software package (Elman, Ramage, Silvester and Wathen 2005*b*).

Algorithms for solving both standard and generalized eigenvalue problems for saddle point matrices have been studied in the literature, particularly for investigating the stability of incompressible flows and in electromagnetism. See Arbenz and Geus (2005), Arbenz, Geus and Adam (2001), Cliffe, Garratt and Spence (1994), Graham, Spence and Vainikko (2003), Lehoucq and Salinger (2001) and Meerbergen and Spence (1997).

An alternative formulation

Eigenvalue distributions such as that shown in Figure 3.1(a) are generally considered unfavourable for solution by Krylov subspace methods and indeed, without preconditioning, Krylov subspace methods tend to converge poorly when applied to the corresponding linear system. It has been observed by several authors (*e.g.*, Benzi and Golub (2004), Fischer *et al.* (1998), Polyak (1970) and Sidi (2003), in addition to Glowinski (1984, page 20) and Quarteroni and Valli (1994, page 304)) that a simple transformation can be used to obtain an equivalent linear system with a coefficient matrix whose spectrum is entirely contained in the half-plane $\mathrm{Re}(z) > 0$. (Here we use $\mathrm{Re}(z)$ and $\mathrm{Im}(z)$ to denote the real and imaginary part of $z \in \mathbb{C}$.) Indeed, assuming that $B_1 = B_2 = B$, we can rewrite the saddle point system in the equivalent form

$$\begin{bmatrix} A & B^T \\ -B & C \end{bmatrix} \begin{bmatrix} x \\ y \end{bmatrix} = \begin{bmatrix} f \\ -g \end{bmatrix}, \quad \text{or} \quad \hat{\mathcal{A}} u = \hat{b}. \tag{3.10}$$

Note that $\hat{\mathcal{A}} = \mathcal{J}\mathcal{A}$ where

$$\mathcal{J} = \begin{bmatrix} I_n & O \\ O & -I_m \end{bmatrix} \tag{3.11}$$

and therefore $\hat{\mathcal{A}}$ is nonsingular if $\mathcal{A}$ is. Moreover, we have the following result.

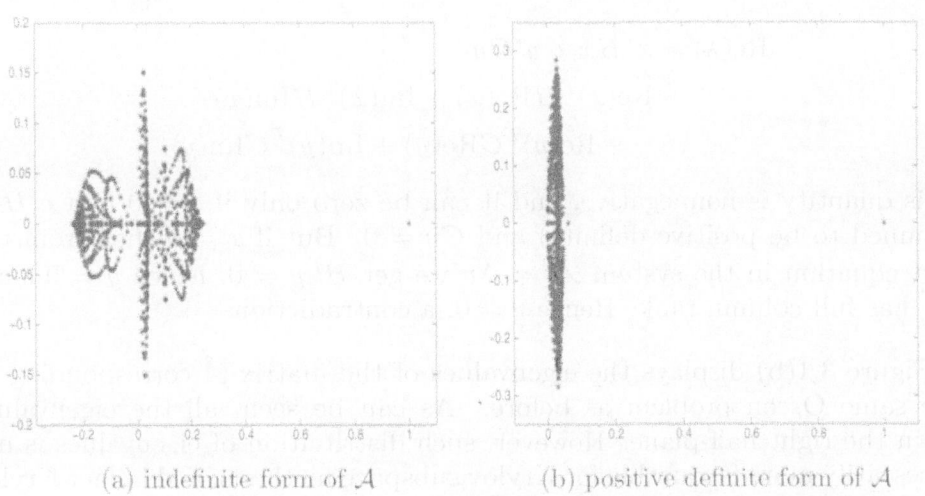

(a) indefinite form of $\mathcal{A}$ (b) positive definite form of $\mathcal{A}$

Figure 3.1. Eigenvalues for two-dimensional Oseen problem (leaky lid-driven cavity) using Q1-P0 discretization.

Theorem 3.6. Let $\hat{A}$ be the coefficient matrix in (3.10). Assume $H = \frac{1}{2}(A + A^T)$ is positive semidefinite, $B_1 = B_2 = B$ has full rank, $C = C^T$ is positive semidefinite, and $\ker(H) \cap \ker(B) = \{0\}$. Let $\sigma(\hat{A})$ denote the spectrum of $\hat{A}$. Then

(i) $\hat{A}$ is positive semidefinite, in the sense that $v^T \hat{A} v \geq 0$ for all $v \in \mathbb{R}^{n+m}$,

(ii) $\hat{A}$ is *positive semistable*; that is, the eigenvalues of $\hat{A}$ have nonnegative real part: $\mathrm{Re}(\lambda) \geq 0$ for all $\lambda \in \sigma(\hat{A})$,

(iii) if in addition $H = \frac{1}{2}(A + A^T)$ is positive definite, then $\hat{A}$ is *positive stable*: $\mathrm{Re}(\lambda) > 0$ for all $\lambda \in \sigma(\hat{A})$.

Proof. To prove (i) we observe that for any $v \in \mathbb{R}^{n+m}$ we have $v^T \hat{A} v = v^T \mathcal{H} v$, where

$$\mathcal{H} \equiv \tfrac{1}{2}(\hat{A} + \hat{A}^T) = \begin{bmatrix} H & O \\ O & C \end{bmatrix}$$

is the symmetric part of $\hat{A}$. Clearly $\mathcal{H}$ is positive semidefinite, so $v^T \hat{A} v \geq 0$.

To prove (ii), let (λ, v) be an eigenpair of $\mathcal{A}$, with $\|v\|_2 = 1$. Then $v^* \hat{A} v = \lambda$ and $(v^* \hat{A} v)^* = v^* \hat{A}^T v = \bar{\lambda}$. Therefore $\frac{1}{2} v^* (\hat{A} + \hat{A}^T) v = \frac{\lambda + \bar{\lambda}}{2} = \mathrm{Re}(\lambda)$. To conclude the proof, observe that

$$v^* (\hat{A} + \hat{A}^T) v = \mathrm{Re}(v)^T (\hat{A} + \hat{A}^T) \mathrm{Re}(v) + \mathrm{Im}(v)^T (\hat{A} + \hat{A}^T) \mathrm{Im}(v),$$

a real nonnegative quantity.

To prove (iii), assume (λ, v) is an eigenpair of $\hat{A}$ with $v = \begin{bmatrix} x \\ y \end{bmatrix}$. Then

$$\begin{aligned}
\mathrm{Re}(\lambda) &= x^* H x + y^* C y \\
&= \mathrm{Re}(x)^T H \mathrm{Re}(x) + \mathrm{Im}(x)^T H \mathrm{Im}(x) \\
&\quad + \mathrm{Re}(y)^T C \mathrm{Re}(y) + \mathrm{Im}(y)^T C \mathrm{Im}(y).
\end{aligned}$$

This quantity is nonnegative, and it can be zero only if $x = 0$ (since H is assumed to be positive definite) and $Cy = 0$. But if $x = 0$ then from the first equation in the system $\hat{A} v = \lambda v$ we get $B^T y = 0$, hence $y = 0$ since B^T has full column rank. Hence $v = 0$, a contradiction. $\square$

Figure 3.1(b) displays the eigenvalues of the matrix $\hat{A}$ corresponding to the same Oseen problem as before. As can be seen, all the eigenvalues lie in the right half-plane. However, such distribution of eigenvalues is not necessarily more favourable for Krylov subspace methods, and in fact Krylov subspace methods without preconditioning perform just as poorly as they do on the original problem. We revisit this topic repeatedly in the course of this survey.

When $A = A^T$ and $C = C^T$ the matrix $\mathcal{A}$ is symmetric indefinite whereas $\hat{\mathcal{A}}$ is nonsymmetric positive (semi-)definite. Moreover, $\hat{\mathcal{A}}$ satisfies

$$\mathcal{J}\hat{\mathcal{A}} = \hat{\mathcal{A}}^T \mathcal{J}, \tag{3.12}$$

with $\mathcal{J}$ defined as in (3.11); that is, $\hat{\mathcal{A}}$ is $\mathcal{J}$-symmetric (or *pseudosymmetric*: see Mackey, Mackey and Tisseur (2003)). In other words, $\hat{\mathcal{A}}$ is symmetric with respect to the indefinite inner product defined on $\mathbb{R}^{n+m}$ by $[v, w] \equiv w^T \mathcal{J} v$. Conversely, any $\mathcal{J}$-symmetric matrix is of the form $\begin{bmatrix} A & B^T \\ -B & C \end{bmatrix}$ for some $A \in \mathbb{R}^{n \times n}$, $B \in \mathbb{R}^{m \times n}$, and $C \in \mathbb{R}^{m \times m}$ with $A = A^T$ and $C = C^T$. Note that the set of all $\mathcal{J}$-symmetric matrices

$$\mathbb{J} = \left\{ \begin{bmatrix} A & B^T \\ -B & C \end{bmatrix} \middle| A = A^T \in \mathbb{R}^{n \times n}, B \in \mathbb{R}^{m \times n}, C = C^T \in \mathbb{R}^{m \times m} \right\}$$

is closed under matrix addition and under the so-called *Jordan product*, defined as

$$\mathcal{F} \cdot \mathcal{G} \equiv \frac{1}{2}(\mathcal{F}\mathcal{G} + \mathcal{G}\mathcal{F}).$$

The triple $(\mathbb{J}, +, \cdot)$ is a non-associative, commutative algebra over the reals. It is known as the *Jordan algebra* associated with the real Lie group $O(n, m, \mathbb{R})$ of $\mathcal{J}$-orthogonal (or *pseudo-orthogonal*) matrices, *i.e.*, the group of all matrices $\mathcal{Q} \in \mathbb{R}^{n+m}$ that satisfy the condition $\mathcal{Q}^T \mathcal{J} \mathcal{Q} = \mathcal{J}$; see Ammar, Mehl and Mehrmann (1999) and Mackey *et al.* (2003). The spectral theory of these matrices has been investigated by several authors. A Schur-like decomposition for matrices in $\mathbb{J}$ has been given in Ammar *et al.* (1999, Theorem 8), and properties of invariant subspaces of $\mathcal{J}$-symmetric matrices have been studied in Gohberg, Lancaster and Rodman (1983).

Besides being mathematically appealing, these algebraic properties have implications from the point of view of iterative methods: see for example Freund, Golub and Nachtigal (1992, page 80), where it is shown how $\mathcal{J}$-symmetry can be exploited to develop transpose-free variants of basic Krylov methods using short recurrences. This is of course not enough to justify using the nonsymmetric form $\hat{\mathcal{A}}$ when $\mathcal{A}$ is symmetric, since in this case one may as well use a symmetric Krylov solver on the original (symmetric) formulation; see Fischer and Peherstorfer (2001), Fischer *et al.* (1998) and Section 9 below. Nevertheless, there are some advantages in using the transformed linear system $\hat{\mathcal{A}}v = \hat{b}$ instead of the original one, especially when certain preconditioners are used; see Benzi and Golub (2004), Sidi (2003) and Section 10.3 below. It can be shown that when A and C are symmetric, at most $2m$ of the $n + m$ eigenvalues of $\hat{\mathcal{A}}$ can have a nonzero imaginary part (Simoncini 2004*b*); furthermore, in some important special cases it turns out that the eigenvalues of $\hat{\mathcal{A}}$ are all real and positive. This implies the existence of a nonstandard inner product on $\mathbb{R}^{n+m}$ with respect to which $\hat{A}$ is symmetric positive definite, a desirable property from the point of view of iterative methods. The following result gives a sufficient condition (easily checked in many cases) for the eigenvalues of $\hat{A}$ to be real.

Theorem 3.7. Assume that A is symmetric positive definite, $B_1 = B_2 = B$ has full rank, and $C = O$. Let $S = BA^{-1}B^T$, and let μ_n denote the smallest eigenvalue of A. If $\mu_n \geq 4\|S\|_2$, then all the eigenvalues of the matrix $\hat{A}$ in (3.10) are real and positive.

Proof. See Simoncini and Benzi (2005). ☐

We note that the conditions expressed in Theorem 3.7 are satisfied, for instance, for the stationary Stokes problem under a variety of finite differences and finite element discretization schemes.

A more detailed analysis is available in the special situation that $A = \eta I$, where $\eta > 0$ is a positive scaling parameter, $B_1 = B_2 = B$, and $C = O$. Denote the resulting usual saddle point matrix by $\mathcal{A}_\eta^+$, and the alternative formulation with negative (2,1) block by $\mathcal{A}_\eta^-$, i.e.,

$$\mathcal{A}_\eta^\pm = \begin{bmatrix} \eta I & B^T \\ \pm B & O \end{bmatrix}. \tag{3.13}$$

The following theorem characterizes the influence of the choice of $+$ or $-$ as well as η on the eigenvalues of the matrices $\mathcal{A}_\eta^\pm$.

Theorem 3.8. Suppose that the matrix B has rank $m - r$, and denote the nonzero singular values of B by $\sigma_1 \geq \cdots \geq \sigma_{m-r}$.

1 The $n + m$ eigenvalues of $\mathcal{A}_\eta^+$ in (3.13) are given by

    (i) zero with multiplicity r,

    (ii) η with multiplicity $n - m + r$,

    (iii) $\frac{1}{2}\left(\eta \pm \sqrt{4\sigma_k^2 + \eta^2}\right)$ for $k = 1, \ldots, m - r$.

2 Furthermore, if $\sigma_1 \geq \cdots \geq \sigma_t > \frac{\eta}{2} \geq \sigma_{t+1} \geq \cdots \geq \sigma_{m-r}$, then the $n + m$ eigenvalues of $\mathcal{A}_\eta^-$ in (3.13) are given by

    (i) zero with multiplicity r,

    (ii) η with multiplicity $n - m + r$,

    (iii) $\frac{1}{2}\left(\eta \pm \sqrt{\eta^2 - 4\sigma_k^2}\right)$ for $k = t+1, \ldots, m - r$,

    (iv) $\frac{1}{2}\left(\eta \pm i\sqrt{4\sigma_k^2 - \eta^2}\right)$ for $k = 1, \ldots, t$.

Proof. See Fischer *et al.* (1998, Section 2). ☐

This result shows that the eigenvalues of the symmetric indefinite matrix $\mathcal{A}_\eta^+$ (except for the multiple eigenvalues zero and η) always lie in two intervals symmetric about the point $\eta/2$. Changing η only leads to a scaling of these two intervals. On the other hand, the choice of η has a significant effect on the eigenvalues of the nonsymmetric matrix $\mathcal{A}_\eta^-$. For example, if $\eta > 2\sigma_1$, then all eigenvalues of $\mathcal{A}_\eta^-$ are real, while for $\eta < 2\sigma_{m-r}$ all eigenvalues (except zero and η) are purely imaginary. For intermediate values of η, the eigenvalues (except zero and η) form a cross in the complex plane with midpoint $\eta/2$. One is immediately tempted to determine what eigenvalue distribution is the most favourable for the solution by Krylov subspace methods; see Sidi (2003). We discuss this topic in Section 10.1, where the matrices $\mathcal{A}_\eta^\pm$ arise naturally as a result of block diagonal pre-conditioning of a symmetric saddle point matrix $\mathcal{A}$ with A positive definite and $C = O$.

3.5. Conditioning issues

Saddle point systems that arise in practice can be very poorly conditioned, and care must be taken when developing and applying solution algorithms. It turns out that in some cases the special structure of the saddle point matrix $\mathcal{A}$ can be exploited to avoid or mitigate the effect of ill-conditioning. Moreover, the structure of the right-hand side b in (1.1) also plays a role. Indeed, it is frequently the case that either f or g in (1.1) is zero. For instance, $f = 0$ in structural analysis (in the absence of dilation) and in mixed formulations of Poisson's equation, while $g = 0$ in incompressible flow problems and weighted least-squares. So if g (say) is zero, the $(1,2)$ and $(2,2)$ blocks in $\mathcal{A}^{-1}$ (see (3.4) and (3.8)) have no influence on the solution $u = \mathcal{A}^{-1}b$. In particular, any ill-conditioning that may be present in these blocks will not affect the solution, an important fact that should be taken into account in the development of robust solution algorithms; see Duff (1994), Gansterer et al. (2003) and Vavasis (1994).

Let us consider, for the sake of simplicity, a standard saddle point problem where $A = A^T$ is positive definite, $B_1 = B_2 = B$ has full rank, and $C = O$. In this case $\mathcal{A}$ is symmetric and its spectral condition number is given by

$$\kappa(\mathcal{A}) = \frac{\max |\lambda(\mathcal{A})|}{\min |\lambda(\mathcal{A})|}.$$

From Theorem 3.5 one can see that the condition number of $\mathcal{A}$ grows unboundedly as either $\mu_n = \lambda_{\min}(A)$ or $\sigma_m = \sigma_{\min}(B)$ goes to zero (assuming that $\lambda_{\max}(A)$ and $\sigma_{\max}(B)$ are kept constant). For mixed finite element formulations of elliptic PDEs, both μ_n and σ_m go to zero as h, the mesh size parameter, goes to zero, and the condition number of $\mathcal{A}$ grows like $O(h^{-p})$ for some positive value of p; see, e.g., Maryška et al. (1996), Wathen et al. (1995). This growth of the condition number of $\mathcal{A}$ means that the rate of convergence of most iterative solvers (like Krylov subspace methods) deteriorates as the problem size increases. As discussed in Section 10, preconditioning may be used to reduce or even eliminate this dependency on h in many cases. Similar considerations apply to nonsymmetric saddle point problems.

A different type of ill-conditioning is encountered in saddle point systems from interior point methods. Consider, for instance, the case of linear programming, where the $(1,1)$ block A is diagonal. As the iterates generated by the interior point algorithm approach the solution, many of the entries of A tend to zero or infinity, and thus A becomes very ill-conditioned (the constraint matrix B remains constant throughout the iteration process). In particular, the norm of the inverse Schur complement $S^{-1} = -(BA^{-1}B^T)^{-1}$ goes to infinity. However, Stewart (1989) and Todd (1990) (see also Forsgren (1996)) have shown that the norm of the matrices $X = S^{-1}BA^{-1}$ (a *weighted pseudo-inverse* of B^T) and $B^T X$ (the associated oblique projector onto the column space of B^T) are bounded by numbers that are independent of A. This important observation has been exploited in a series of papers by Vavasis and collaborators (Bobrovnikova and Vavasis 2001, Hough and Vavasis 1997, Vavasis 1994, Vavasis 1996) to develop stable algorithms for certain saddle point problems with a severely ill-conditioned $(1,1)$ block A.

When using direct methods based on triangular factorization, Björck (1996, Sections 2.5.3 and 4.4.2) has noted the importance of scaling the $(1,1)$ block A by a positive scalar quantity. Suitable tuning of this scaling factor can be interpreted as a form of preconditioning and has a dramatic impact on the accuracy attainable by sparse direct solvers (Arioli, Duff and De Rijk 1989, Duff 1994). On the other hand, such scaling seems to have little or no effect on the convergence behaviour of Krylov subspace methods (Fischer *et al.* 1998).

Another possible approach for dealing with an ill-conditioned or even singular $(1,1)$ block A is the augmented Lagrangian method; see Fortin and Glowinski (1983), Glowinski and Le Tallec (1989), Hestenes (1969), Powell (1969) and the more general treatment in Golub and Greif (2003) and Greif *et al.* (2005). Here we assume that $A = A^T$ (possibly singular), $B_1 = B_2 = B$ has full rank, and $C = O$. The idea is to replace the saddle point system (1.3) with the equivalent one

$$
\begin{bmatrix} A + B^T W B & B^T \\ B & O \end{bmatrix} \begin{bmatrix} x \\ y \end{bmatrix} = \begin{bmatrix} f + B^T W g \\ g \end{bmatrix}.
\tag{3.14}
$$

The $m \times m$ matrix W, to be suitably determined, is symmetric positive semidefinite. The simplest choice is to take $W = \gamma I$ ($\gamma > 0$). In this case the $(1,1)$ block in (3.14) is nonsingular, and indeed positive definite, provided that A is positive definite on $\ker(B)$. The goal is to choose W so that system (3.14) is easier to solve than the original one, particularly when using iterative methods. When $W = \gamma I$ is used, the choice $\gamma = \|A\|_2/\|B\|_2^2$ has been found to perform well in practice, in the sense that the condition number of both the $(1, 1)$ block and of the whole coefficient matrix in (3.14) are approximately minimized. This choice also results in rapid convergence of classical iterative schemes like the method of multipliers; see Golub and Greif (2003) and Section 8.2 below.

The conditioning of equality-constrained and weighted least squares problems has been studied in depth by several authors; see Gulliksson, Jin and Wei (2002), Wei (1992b) and the references therein.

Conditioning properties of quasidefinite and saddle-point matrices arising from interior-point methods in linear programming have also been investigated in George and Ikramov (2000) and Korzak (1999). Finally, we mention that a numerical validation method for verifying the accuracy of approximate solutions of symmetric saddle point problems has been presented in Chen and Hashimoto (2003).

4. Overview of solution algorithms

Besides the usual (and somewhat simplistic) distinction between direct and iterative methods, solution algorithms for generalized saddle point problems can be subdivided into two broad categories, which we will call *segregated* and *coupled* (or 'all at once') methods. Segregated methods compute the two unknown vectors, x and y, separately; in some cases it is x to be computed first, in others it is y. This approach involves the solution of two linear systems of size smaller than $n + m$ (called *reduced systems*), one for each of x and y; in some cases a reduced system for an intermediate quantity is solved. Segregated methods can be either direct or iterative, or involve a combination of the two; for example, one of the reduced systems could be solved by a direct method and the other iteratively. The main representatives of the segregated approach are the Schur complement reduction method, which is based on a block LU factorization of $\mathcal{A}$, and the null space method, which relies on a basis for the null space for the constraints.

Coupled methods, on the other hand, deal with the system (1.1) as a whole, computing x and y (or approximations to them) simultaneously and without making explicit use of reduced systems. These methods include both direct solvers based on triangular factorizations of the global matrix $\mathcal{A}$, and iterative algorithms like Krylov subspace methods applied to the entire system (1.1), typically with some form of preconditioning. As we shall see, preconditioning tends to blur the distinction between direct and iterative solvers, and also that between segregated and coupled schemes. This is because direct solvers may be used to construct preconditioners, and also because preconditioners for coupled iterative schemes are frequently based on segregated methods.

In the next sections we review a number of solution methods, starting with direct solvers and continuing with stationary iterative methods, Krylov subspace solvers, and preconditioners. We also include a brief discussion of multilevel methods, including multigrid and Schwarz-type algorithms. Within each group, we discuss segregated as well as coupled schemes and the interplay between them. It is simply not possible to cover every method that has been described in the literature; instead, we have striven to include, besides all of the 'classical' algorithms, those among the more recent methods that appear to be the most widely applicable and effective.

5. Schur complement reduction

Consider the saddle point system (1.1), or

$$Ax + B_1^T y = f, \quad B_2 x - Cy = g.$$

We assume that both A and $\mathcal{A}$ are nonsingular; by (3.1) this implies that $S = -(C + B_2 A^{-1} B_1^T)$ is also nonsingular. Pre-multiplying both sides of the first equation by $B_2 A^{-1}$, we obtain

$$B_2 x + B_2 A^{-1} B_1^T y = B_2 A^{-1} f.$$

Using $B_2 x = g + Cy$ and rearranging, we find

$$(B_2 A^{-1} B_1^T + C)\, y = B_2 A^{-1} f - g, \tag{5.1}$$

a reduced system of order m for y involving the (negative) Schur complement $-S = B_2 A^{-1} B_1^T + C$. Note that unless $f = 0$, forming the right-hand side of (5.1) requires solving a linear system of the form $Av = f$.

Once y_* has been computed from (5.1), x_* can be obtained by solving

$$Ax = f - B_1^T y_*, \tag{5.2}$$

a reduced system of order n for x involving the (1,1) block, A. Note that this is just block Gaussian elimination applied to (1.1). Indeed, using the block LU factorization (3.3) we get the transformed system

$$\begin{bmatrix} I & O \\ -B_2 A^{-1} & I \end{bmatrix} \begin{bmatrix} A & B_1^T \\ B_2 & -C \end{bmatrix} \begin{bmatrix} x \\ y \end{bmatrix} = \begin{bmatrix} I & O \\ -B_2 A^{-1} & I \end{bmatrix} \begin{bmatrix} f \\ g \end{bmatrix},$$

that is,

$$\begin{bmatrix} A & B_1^T \\ O & S \end{bmatrix} \begin{bmatrix} x \\ y \end{bmatrix} = \begin{bmatrix} f \\ g - B_2 A^{-1} f \end{bmatrix}.$$

Solving this block upper triangular system by block backsubstitution leads to the two reduced systems (5.1) and (5.2) for y and x. These systems can be solved either directly or iteratively. In the important special case where A and $-S$ are symmetric positive definite, highly reliable methods such as Cholesky factorization or the conjugate gradient (CG) method can be applied.

The solution strategy outlined above is commonly used in the complementary energy formulation of structural mechanics, where it is known as the *displacement* method, since the vector of nodal displacements y is computed first; the reduction to the Schur complement system (5.1) is known as *static condensation*, and the Schur complement itself is called the *assembled stiffness matrix* (McGuire and Gallagher 1979). In electrical engineering it is known as the *nodal analysis* method, and in optimization as the *range-space* method (Vavasis 1994). In all these applications, A is symmetric positive (semi)definite, $B_1 = B_2$, and $C = O$.

This approach is attractive if the order m of the reduced system (5.1) is small and if linear systems with coefficient matrix A can be solved efficiently. The main disadvantages are the need for A to be nonsingular, and the fact that the Schur complement $S = -(BA^{-1}B^T + C)$ may be completely full and too expensive to compute or to factor. Numerical instabilities may also be a concern when forming S, especially when A is ill-conditioned (Vavasis 1994). Dense Schur complements occur in the case of Stokes and Oseen problems, where A corresponds to a (vector) differential operator. Other examples include problems from optimization when B contains one or more dense columns. Note, however, that when B contains no dense columns and A^{-1} is sparse (*e.g.*, A is diagonal or block diagonal with small blocks), then S is usually quite sparse. In this case efficient (graph-based) algorithms can be used to form S, and it is sometimes possible to apply the Schur complement reduction recursively and in a way that preserves sparsity through several levels, in the sense that the number of nonzeros to be stored remains nearly constant throughout the successive reduction steps; see Maryška, Rozložník and Tůma (2000) for an example arising from the solution of groundwater flow problems.

In cases where A is positive semidefinite and singular, Schur complement reduction methods may still be applied by making use of augmented Lagrangian techniques (3.14), which replace the original saddle point system with an equivalent one with a nonsingular (1,1) block. If S is too expensive to form or factor, Schur complement reduction can still be applied by solving (5.1) by iterative methods that do not need access to individual entries of S, but only need S in the form of matrix-vector products

$$p = -Sy = (B_2 A^{-1} B_1^T + C)y.$$

The action of S on y can be computed by means of matrix-vector products with B_1^T, B_2 and C and by solving a linear system with matrix A. If the latter can be performed efficiently and the iteration converges sufficiently fast, this is a viable option. The Schur complement system (5.1), however, may be rather ill-conditioned, in which case preconditioning will be required. Preconditioning the system (5.1) is nontrivial when S is not explicitly available. Some options are discussed in Section 10.1 below, in the context of block preconditioners.

6. Null space methods

In this section we assume that $B_1 = B_2 = B$ has full rank and $C = O$. Furthermore, we assume that $\ker(H) \cap \ker(B) = \{0\}$, where H is the symmetric part of A. The saddle point system is then

$$Ax + B^T y = f, \quad Bx = g.$$

The null space method assumes that the following are available:

(1) a particular solution $\hat{x}$ of $Bx = g$;
(2) a matrix $Z \in \mathbb{R}^{n \times (n-m)}$ such that $BZ = O$, that is, range$(Z) = \ker(B)$ (the columns of Z span the null space of B).

Then the solution set of $Bx = g$ is described by $x = Zv + \hat{x}$ as v ranges in $\mathbb{R}^{n-m}$. Substituting $x = Zv + \hat{x}$ in $Ax + B^T y = f$, we obtain $A(Zv + \hat{x}) = f - B^T y$. Pre-multiplying by the full-rank matrix Z^T and using $Z^T B^T = O$, we get

$$Z^T A Z v = Z^T (f - A\hat{x}), \tag{6.1}$$

a reduced system of order $n - m$ for the auxiliary unknown v. This system is nonsingular under our assumptions. Once the solution v_* has been obtained, we set $x_* = Zv_* + \hat{x}$; finally, y_* can be obtained by solving

$$BB^T y = B(f - Ax_*), \tag{6.2}$$

a reduced system of order m with a symmetric positive definite coefficient matrix BB^T. Of course, (6.2) is just the normal equations for the overdetermined system $B^T y = f - Ax_*$, or

$$\min_y \|(f - Ax_*) - B^T y\|_2,$$

which could be solved, *e.g.*, by LSQR (Paige and Saunders 1982) or a sparse QR factorization (Matstoms 1994). Just as the Schur complement reduction method can be related to expression (3.4) for $\mathcal{A}^{-1}$, the null space method is related to the alternative expression (3.8). It is interesting to observe that when A is invertible, the null space method is just the Schur complement reduction method applied to the *dual* saddle point problem

$$\begin{bmatrix} A^{-1} & Z \\ Z^T & O \end{bmatrix} \begin{bmatrix} w \\ v \end{bmatrix} = \begin{bmatrix} -\hat{x} \\ -Z^T f \end{bmatrix}.$$

This strategy subsumes a whole family of *null space methods*, which differ primarily in the way the matrix Z (often called a *null basis*) is computed;

see the discussion below. Null space methods are quite popular in optimization, where they are usually referred to as *reduced Hessian* methods; see Coleman (1984), Fletcher (1987), Gill *et al.* (1981), Nocedal and Wright (1999) and Wolfe (1962). In this setting the matrix A is the $(n \times n)$ Hessian of the function to be minimized subject to the constraint $Bx = g$, and $Z^T A Z$ is the reduced $((n-m) \times (n-m))$ Hessian, obtained by elimination of the constraints. When Z has orthonormal columns, the reduced system (6.1) can also be seen as a projection of the problem onto the constraint set. The null space approach has been extensively used in structural mechanics where it is known under the name of *force* method, because x, the vector of internal forces, is computed first; see, *e.g.*, Berry and Plemmons (1987), Heath, Plemmons and Ward (1984), Kaneko, Lawo and Thierauf (1982), Kaneko and Plemmons (1984), Plemmons and White (1990) and Robinson (1973). Other application areas where the null space approach is used include fluid mechanics (under the somewhat misleading name of *dual variable* method, see Amit, Hall and Porsching (1981), Arioli and Manzini (2002, 2003), Arioli, Maryška, Rozložník and Tůma (2001), Gustafson and Hartmann (1983), Hall (1985), Sarin and Sameh (1998)) and electrical engineering (under the name of *loop analysis*; see Chua *et al.* (1987), Strang (1986), Tropper (1962), Vavasis (1994)).

The null space method has the advantage of not requiring A^{-1}. In fact, the method is applicable even when A is singular, as long as the condition $\ker(H) \cap \ker(B) = \{0\}$ is satisfied. The null space method is often used in applications that require the solution of a sequence of saddle point systems of the type

$$\begin{bmatrix} A_k & B^T \\ B & O \end{bmatrix} \begin{bmatrix} x \\ y \end{bmatrix} = \begin{bmatrix} f_k \\ g_k \end{bmatrix}, \quad k = 1, 2, \ldots,$$

where the A_k submatrix changes with k while B remains fixed. This situations arises, for instance, in the solution of unsteady fluid flow problems, and in the *reanalysis* of structures in computational mechanics; see, *e.g.*, Batt and Gellin (1985), Hall (1985) and Plemmons and White (1990). Another example is the analysis of resistive networks with a fixed connectivity and different values of the resistances. In all these cases the null basis matrix Z needs to be computed only once.

Null space methods are especially attractive when $n - m$ is small. If A is symmetric and positive semidefinite, then $Z^T A Z$ is symmetric positive definite and efficient solvers can be used to solve the reduced system (6.1). If Z is sparse then it may be possible to form and factor $Z^T A Z$ explicitly, otherwise iterative methods must be used, such as conjugate gradients or others.

The method is less attractive if $n - m$ is large, and cannot be applied if $C \neq O$. The main difficulty, however, is represented by the need for a null basis Z for B. We note that computing a particular solution for $Bx = g$ is usually not a difficult problem, and it can be obtained as a byproduct of the computations necessary to obtain Z. In the case where $g = 0$ (arising for instance from the divergence-free condition in incompressible flow problems), the trivial solution $\hat{x} = 0$ will do. Hence, the main issue is the computation of a null basis Z. There are a number of methods that one can use, at least in principle, to this end. In the large and sparse case, graph-based methods invariably play a major role.

Let P denote a permutation matrix chosen so that $BP = \begin{bmatrix} B_b & B_n \end{bmatrix}$, where B_b is $m \times m$ and nonsingular (this is always possible, since B is of rank m). Then it is straightforward to verify that the matrix

$$Z = P \begin{bmatrix} -B_b^{-1} B_n \\ I \end{bmatrix}, \tag{6.3}$$

where I denotes the identity matrix of order $n - m$, is a null basis for B. This approach goes back to Wolfe (1962); a basis of the form (6.3) is called a *fundamental basis*. Quite often, the matrix $B_b^{-1} B_n$ is not formed explicitly; rather, an LU factorization of B_b is computed and used to perform operations involving B_b^{-1}. For instance, if an iterative method like CG is used to solve (6.1), then matrix-vector products with $Z^T A Z$ can be performed by means of forward and backsubstitutions with the triangular factors of B_b, in addition to matrix-vector products with B_n, B_n^T, and A.

Since there are in principle many candidate submatrices B_b, (*i.e.*, many permutation matrices P) it is natural to ask whether one can find a matrix B_b with certain desirable properties. Ideally, one would like B_b to be easy to factor, well-conditioned, and to satisfy certain sparsity requirements (either in B_b, or in its factors, or in $B_b^{-1} B_n$). Another desirable property could be some kind of diagonal dominance. In the literature, this is known as the *nice basis* problem. This is a very difficult problem in general. Consider first the sparsity requirement. Unfortunately, not all sparse matrices admit a sparse null basis. To see this, consider the matrix $B = \begin{bmatrix} I & e \end{bmatrix}$, where e is the column vector all of whose components are equal to 1; clearly, there is no explicit sparse representation for its one-dimensional null space (Gilbert and Heath 1987). Moreover, even if a sparse null basis exists, the problem of computing a null basis Z (fundamental or not) with minimal number of nonzero entries has been shown to be NP-hard (Coleman and Pothen 1986, Pothen 1984). In spite of this, there are important situations where a sparse null basis exists and can be explicitly obtained. As we shall see, there may be no need to explicitly factor or invert any submatrix of B.

An example of this is Kirchhoff's classical method for finding the currents in a resistive electrical network (Kirchhoff 1847). Our discussion closely follows Strang (1986). For this problem, B is just the node–edge incidence matrix of the network, or directed graph, describing the connectivity of the network. More precisely, if the network consists of $m+1$ nodes and n edges, let B_0 be the $(m+1) \times n$ matrix with entries b_{ij} given by -1 if edge j starts at node i and by $+1$ if edge j ends at node i; of course, $b_{ij} = 0$ if edge j does not meet node i. Hence, B_0 behaves as a discrete divergence operator on the network: each column contains precisely two nonzero entries, one equal to $+1$ and the other equal to -1. Matrix B_0 can be shown to be of rank m; note that $B_0^T e = 0$. A full-rank matrix B can be obtained by dropping the last row of B_0; that is, by 'grounding' the last node in the network: see Strang (1986, page 112).

A null space for B can be found using Kirchhoff's Voltage Law, which implies that the sum of the voltage drops around each closed loop in the network must be zero. In other words, for current flowing around a loop there is no buildup of charge. In matrix terms, each loop current is a solution to $By = 0$. Since B has full rank, there are exactly $n - m$ independent loop currents, denoted by $z_1, z_2, \ldots z_{n-m}$. The loop currents can be determined by a procedure due to Kirchhoff, which consists of the following steps.

(1) Find a *spanning tree* for the network (graph); this is a connected subgraph consisting of the $m + 1$ nodes and just m edges, so that between any two nodes there is precisely one path, and there are no loops. As shown by Kirchhoff, there are exactly $t = \det BB^T$ spanning trees in the network (which is assumed to be connected).

(2) Once a spanning tree has been picked, the remaining $n - m$ edges can be used to construct the $n - m$ loop currents by noticing that adding any of these edges to the spanning tree will create a loop. For each of these *fundamental loops* we construct the corresponding column z_i of Z by setting the jth entry equal to ± 1 if edge j belongs to the loop, and equal to 0 otherwise; the choice of sign specifies the orientation of the edge.

The resulting $Z = [z_1, z_2, \ldots, z_{n-m}]$ is then called an edge–loop matrix, and is a basis for the null space of B. As a simple example, consider the directed graph of Figure 6.1, with the spanning tree on the right. In this example, $m = 4$ and $n = 7$. The node–edge incidence matrix B_0 for this graph is

$$B_0 = \begin{bmatrix} -1 & -1 & 0 & -1 & 0 & 0 & 0 \\ 0 & 1 & 1 & 0 & 0 & 0 & 0 \\ 1 & 0 & 0 & 0 & -1 & 0 & -1 \\ 0 & 0 & -1 & 1 & 0 & 1 & 1 \\ 0 & 0 & 0 & 0 & 1 & -1 & 0 \end{bmatrix}.$$

Note that $\text{rank}(B_0) = 4$; the matrix B obtained from B_0 by grounding node 5 (*i.e.*, by dropping the last row of B_0) has full row rank, equal to 4.

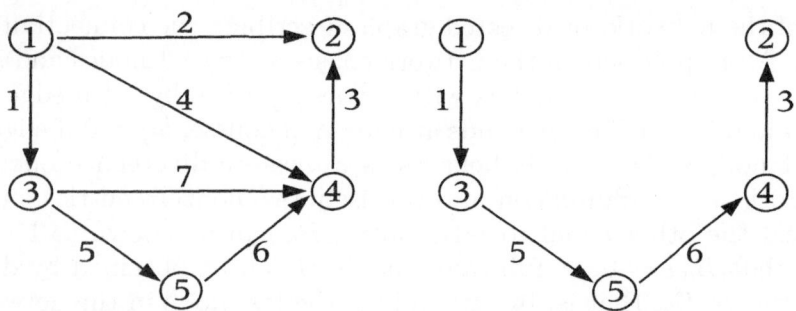

Figure 6.1. A directed graph with one of its spanning trees.

Consider now the spanning tree on the right of Figure 6.1. By adding the remaining edges (numbered 2, 4 and 7) to the tree we obtain, respectively, the following loops:

(1) $1 \rightarrow 3 \rightarrow 5 \rightarrow 4 \rightarrow 2 \rightarrow 1$, for the edge sequence $(1, 5, 6, 3, -2)$;

(2) $1 \rightarrow 3 \rightarrow 5 \rightarrow 4 \rightarrow 1$, for the edge sequence $(1, 5, 6, -4)$;

(3) $3 \rightarrow 5 \rightarrow 4 \rightarrow 3$, for the edge sequence $(5, 6, -7)$.

Note that an edge was given the negative sign whenever its orientation required it. It follows that the edge–loop matrix for the network under consideration is

$$Z = \begin{bmatrix} 1 & 1 & 0 \\ -1 & 0 & 0 \\ 1 & 0 & 0 \\ 0 & -1 & 0 \\ 1 & 1 & 1 \\ 1 & 1 & 1 \\ 0 & 0 & -1 \end{bmatrix}.$$

and it is straightforward to check that $BZ = O$.

It turns out that this elegant method is fairly general and can be applied to other problems besides the analysis of resistive networks. An important example is fluid dynamics, in particular the Darcy, Stokes and Oseen problems, where B represents a discrete divergence operator. In this case the null basis Z is called a *solenoidal basis*, since the columns of Z span the subspace of all discrete solenoidal (*i.e.*, divergence-free) functions. In other words, C can be regarded as a discrete curl operator (Chang, Giraldo and Perot 2002). In the case of finite differences on a regular grid, B is just the incidence matrix of a directed graph associated with the grid, and the cycles (loops) in this graph can be used to construct a sparse Z; see Amit *et al.* (1981), Burkardt, Hall and Porsching (1986), Chang *et al.* (2002), Hall

(1985), Sameh and Sarin (2002) and Sarin and Sameh (1998). Note that in this context, solving the system (6.2) for y amounts to solving a Poisson equation. This methodology is not restricted to simple finite difference discretizations or to structured grids; see Alotto and Perugia (1999), Arioli and Manzini (2002, 2003), Arioli *et al.* (2001), Hall, Cavendish and Frey (1991), Sarin (1997) and Sarin and Sameh (2003) for applications to a variety of discretization methods on possibly unstructured grids.

We note that no floating-point arithmetic is needed to form Z. Furthermore, the sparsity pattern of the matrix $Z^T A Z$ can be easily determined and the matrix $Z^T A Z$ assembled rather cheaply, as long as it is sufficiently sparse. The sparsity will depend on the particular spanning tree used to form Z. Finding a tree that minimizes the number of nonzeros in Z is equivalent to finding the tree for which the sum of all the lengths of the fundamental loops is minimal, which is an NP-hard problem. Nevertheless, many efficient heuristics have been developed; see Tarjan (1983) for the fundamental concepts and algorithms. The relative size of n, m and $n - m$ depends on the discretization scheme used, and on whether the underlying fluid flow problem is posed in 2D or 3D. For lowest-order discretizations in 2D, $n - m$ and m are comparable, whereas $n - m$ is much larger than m in 3D or for certain mixed finite element discretizations. If sparse direct solvers are used to solve for the dual variable in (6.1), this makes the null space approach not viable in 3D. In this case iterative solvers must be used, and the spectral properties of $Z^T A Z$ determine the convergence rate. When the matrix $Z^T A Z$ is not formed explicitly, finding appropriate preconditioners for it requires some cleverness. Some work in this direction can be found in Coleman and Verma (2001) and in Nash and Sofer (1996) for constrained optimization problems. See also Saint-Georges, Notay and Warzée (1998) for closely related work in the context of constrained finite element analyses, and Barlow, Nichols and Plemmons (1988), James (1992) and James and Plemmons (1990) for earlier work on the use of preconditioned CG methods in the context of *implicit null space algorithms* – *i.e.*, null space algorithms in which the matrix Z is not formed explicitly.

For many mixed finite element formulations of second-order elliptic problems, A is symmetric positive definite and has condition number bounded independently of the discretization parameter h. In this case, fast CG convergence can be obtained by using incomplete factorization preconditioners based on $Z^T Z$: see Alotto and Perugia (1999). Point and block Jacobi preconditioners constructed without explicitly forming $Z^T A Z$ have been tested in the finite element solution of the potential fluid flow problem (2.18)–(2.20) in Arioli and Manzini (2003).

Null space methods have been used for a long time in structural optimization. Some relevant references in this area include Cassell, Henderson and Kaveh (1974), Henderson and Maunder (1969), Kaveh (1979, 1992, 2004) and Pothen (1989). In this case B is an equilibrium matrix associated with Newton's Third Law (*i.e.*, action and reaction are equal and opposite). Now B is no longer an incidence matrix, but many of the above-described concepts can be extended to this more general situation. Algorithms to find a null basis for B developed by Coleman and Pothen (1987) consist of two phases, a combinatorial one and a numerical one. In the first phase, a maximum matching in the bipartite graph of B is used to locate the nonzero entries in the null basis. In the second phase (not needed when B is an incidence matrix), the numerical values of the nonzero entries in the basis are computed by solving certain systems of equations.

When additional structure, such as bandedness, is present in B, it is usually possible to exploit it so as to develop more efficient algorithms. Banded equilibrium matrices often arise in structural engineering. The so-called *turnback algorithm* (Topcu 1979) can be used to compute a banded Z; see also Kaneko *et al.* (1982), where an interpretation of the turnback algorithm in terms of matrix factorizations is given, and Gilbert and Heath (1987) for additional methods motivated by the turnback algorithm. We also mention Plemmons and White (1990) for approaches based on different graph-theoretic concepts with a focus on parallel implementation aspects, and Chow, Manteuffel, Tong and Wallin (2003) for another example of how structure in the constraints can be exploited to find a null basis resulting in a sparse reduced matrix $Z^T A Z$.

One problem that may occur in the null space method is that a computed null basis matrix Z may be very ill-conditioned, even numerically rank deficient. One way to avoid this problem is to compute a Z with orthonormal columns, which would be optimally conditioned. An orthonormal null basis for B can be computed by means of the QR factorization as follows. Let

$$B^T = Q \begin{bmatrix} R \\ O \end{bmatrix},$$

where Q is $n \times n$ orthogonal and R is $m \times m$, upper triangular and nonsingular. Then the first m columns of Q form an orthonormal basis for range(B^T) and the remaining $n - m$ columns form an orthonormal basis for range(B^T)$^\perp$ = ker(B). Therefore, if q_i denotes the ith column of Q, then

$$Z = \begin{bmatrix} q_{m+1} & q_{m+2} & \cdots & q_n \end{bmatrix}$$

is the desired matrix. Of course, in the sparse case special ordering techniques must be utilized in order to maintain sparsity in Z (Amestoy, Duff and Puglisi 1996, Matstoms 1994). The fact that the columns of Z are orthonormal is advantageous not only from the point of view of conditioning, but also for other reasons. For example, in the computation of thin-plate splines it is often required to solve saddle point systems of the form

$$\begin{bmatrix} A + \rho I & B^T \\ B & O \end{bmatrix} \begin{bmatrix} x \\ y \end{bmatrix} = \begin{bmatrix} f \\ 0 \end{bmatrix}, \tag{6.4}$$

where A is symmetric, $\rho > 0$ is a smoothing parameter, and B has full row rank; see Sibson and Stone (1991). Usually, problem (6.4) needs to be solved for several values of ρ. If Z is a null basis matrix with orthonormal columns, the coefficient matrix in the reduced system (6.1) is $Z^T(A+\rho I)Z = Z^T A Z + \rho I$. If $n - m$ is so small that a spectral decomposition $Z^T A Z = U \Lambda U^T$ of $Z^T A Z$ can be computed, then for any ρ we have $Z^T(A + \rho I)Z = U(\Lambda + \rho I)U^T$ and the reduced linear systems can be solved efficiently.

One further advantage of having an orthonormal null basis is that the reduced system (6.1) is guaranteed to be well-conditioned if A is. For example, if A is symmetric positive definite and has condition number bounded independently of mesh size, the same is true of $Z^T A Z$ and therefore the CG method applied to (6.1) converges in a number of iterations independent of mesh size, even without preconditioning; see Arioli and Manzini (2002) for an example from groundwater flow computations. This property may fail to hold if Z does not have orthonormal columns, generally speaking; see Arioli and Manzini (2003) and Perugia and Simoncini (2000).

Sparse orthogonal schemes have been developed by Berry, Heath, Kaneko, Lawo, Plemmons and Ward (1985), Gilbert and Heath (1987), Heath *et al.* (1984) and Kaneko and Plemmons (1984) in the context of structural optimization, and by Arioli (2000), Arioli and Manzini (2002) and Arioli *et al.* (2001) in the context of mixed-hybrid finite element formulations of the potential fluid flow problem (2.18)–(2.20). A parallel orthogonal null space scheme has been presented by Psiaki and Park (1995) for trajectory optimization problems in quadratic dynamic programming. One limitation of the QR factorization approach is that the null bases obtained by this method are often rather dense compared to those obtained by other sparse schemes; indeed, the sparsest orthogonal null basis may be considerably less sparse than an arbitrary null basis: see Coleman and Pothen (1987) and Gilbert and Heath (1987). Hence, there is a trade-off between good conditioning properties and sparsity.

Error analyses of various null space methods have been carried out by Cox and Higham (1999*a*) for dense problems, and by Arioli and Baldini (2001) for the sparse case. See further Barlow (1988), Barlow and Handy (1988), Björck and Paige (1994), Cox and Higham (1999*b*), Fletcher and Johnson (1997), Gulliksson (1994), Gulliksson and Wedin (1992), Hough and Vavasis (1997), Sun (1999) and Vavasis (1994) for stable implementations and other numerical stability aspects of algorithms for saddle point problems, in particular for equality-constrained and weighted least squares problems. Finally, appropriate stopping criteria for the CG method applied to the reduced system (6.1) in a finite element context have been given by Arioli and Manzini (2002).

7. Coupled direct solvers

In this section we give a brief overview of direct methods based on triangular factorizations of $\mathcal{A}$. Our discussion is limited to the symmetric case ($A = A^T$, $B_1 = B_2$ and $C = C^T$, possibly zero). As far as we know, no specialized direct solver exists for nonsymmetric saddle point problems. Although such problems are often *structurally* symmetric, in the sense that the nonzero pattern of $\mathcal{A}$ is symmetric, some form of numerical pivoting is almost certainly going to be needed for stability reasons; such pivoting would in turn destroy symmetry. See Duff, Erisman and Reid (1986) for a treatment of direct methods for general sparse matrices.

There are several ways to perform Gaussian elimination on a symmetric, possibly indefinite matrix in a way that exploits (and preserves) symmetry. A factorization of the form

$$\mathcal{A} = \mathcal{Q}^T \mathcal{L} \mathcal{D} \mathcal{L}^T \mathcal{Q}, \tag{7.1}$$

where $\mathcal{Q}$ is a permutation matrix, $\mathcal{L}$ is unit lower triangular, and $\mathcal{D}$ a block diagonal matrix with blocks of dimension 1 and 2 is usually referred to as an LDL^T factorization. The need for pivot blocks of size 2 is made clear by the following simple example:

$$\mathcal{A} = \left[\begin{array}{cc|c} 0 & 1 & 1 \\ 1 & 0 & 1 \\ \hline 1 & 1 & 0 \end{array} \right],$$

for which selecting pivots from the main diagonal is impossible. Diagonal pivoting may also fail on matrices with a zero-free diagonal due to instabilities. The use of 2×2 pivot blocks dates back to Lagrange (1759). In 1965, W. Kahan (in correspondence with R. de Meersman and L. Schotsmans) suggested that Lagrange's method could be used to devise stable factorizations for symmetric indefinite matrices. The idea was developed by Bunch and Parlett (1971), resulting in a stable algorithm for factoring symmetric indefinite matrices at a cost comparable to that of a Cholesky factorization for positive definite ones. The Bunch–Parlett pivoting strategy is akin to complete pivoting; in subsequent papers (Bunch 1974, Bunch and Kaufman 1977), alternative pivoting strategies requiring only $O(n^2)$ comparisons for a dense $n \times n$ matrix have been developed; see also Fletcher (1976b). The Bunch–Kaufman pivoting strategy (Bunch and Kaufman 1977) is widely accepted as the algorithm of choice for factoring dense symmetric indefinite matrices. In the sparse case, the pivoting strategy is usually relaxed in order to maintain sparsity in $\mathcal{L}$. Several sparse implementations are available; see Duff, Gould, Reid, Scott and Turner (1991), Duff and Reid (1983, 1995) and Liu (1987). In this case the permutation matrix $\mathcal{Q}$ is the result of symmetric row and column interchanges aimed at preserving sparsity in the factors as well as numerical stability. While the Bunch–Kaufman algorithm is normwise backward stable, the resulting factors can have unusual scaling, which may result in a degradation of the accuracy of computed solutions. As reported in Ashcraft, Grimes and Lewis (1998), such difficulties have been observed in the solution of saddle point systems arising in sparse non-linear optimization codes. We refer the reader to Ashcraft *et al.* (1998) for a thorough discussion of such accuracy issues and ways to address these problems; see also Vavasis (1994).

We note that when A is positive definite and B has full rank, the saddle point matrix $\mathcal{A}$ admits an LDL^T factorization with $\mathcal{D}$ diagonal and $\mathcal{Q} = I$ (*i.e.*, no pivoting is needed). Indeed, since A is positive definite it can be decomposed as $A = L_A D_A L_A^T$ with L_A unit lower triangular and D_A diagonal (and positive definite); furthermore the Schur complement $S = -(C + BA^{-1}B^T)$ is negative definite and therefore it can be decomposed as $S = -L_S D_S L_S^T$. Hence, we can write

$$\mathcal{A} = \begin{bmatrix} A & B^T \\ B & -C \end{bmatrix} = \begin{bmatrix} L_A & O \\ L_B & L_S \end{bmatrix} \begin{bmatrix} D_A & O \\ O & -D_S \end{bmatrix} \begin{bmatrix} L_A^T & L_B^T \\ O & L_S^T \end{bmatrix} = \mathcal{L}\mathcal{D}\mathcal{L}^T, \quad (7.2)$$

where $L_B = BL_A^{-T}D_A^{-1}$; note that $BA^{-1}B^T = L_B D_A L_B^T$. In practice, however, the factors will be rather dense with the original ordering, and symmetric permutations have to be used in order to preserve sparsity. Note that L_S and L_B will be completely full if the Schur complement is. However, not all sparsity-preserving permutations are acceptable. It can be shown that there exist permutation matrices $\mathcal{Q}$ such that $\mathcal{Q}\mathcal{A}\mathcal{Q}^T$ does not have an LDL^T factorization with $\mathcal{D}$ diagonal. Furthermore, some permutations may lead to numerical instability problems.

For many symmetric indefinite codes the factorization consists of two phases, a symbolic and a numeric one. In the symbolic phase, an initial fill-reducing ordering is computed based on the structure of $\mathcal{A}$ only. This is often some variant of minimum degree or nested dissection (Duff *et al.* 1986). In the numeric phase, the actual factorization is computed. Frequently in the course of this phase, the pivot order from the symbolic phase may have to be altered for numerical stability reasons. There are, however, a few exceptions to this rule. An important one is the quasidefinite case discussed in Section 3.3, *i.e.*, when C (as well as A) is symmetric positive definite. In this case $\mathcal{Q}\mathcal{A}\mathcal{Q}^T$ always has an LDL^T factorization with $\mathcal{D}$ diagonal, regardless of the choice of $\mathcal{Q}$; see Vanderbei (1995). This is an important result: it suggests that the fill-reducing ordering computed in the symbolic phase of the factorization will not need to be altered in the course of the numeric phase because of stability concerns. Since no pivoting is used in the numeric phase, it is possible to exploit all the features of modern supernodal sparse Cholesky factorization codes (Ng and Peyton 1993). The resulting algorithm is more efficient than performing a Bunch–Parlett or Bunch–Kaufman factorization. Numerical stability considerations in Vanderbei (1995) suggest that the resulting factorization is usually sufficiently accurate. A stability analysis was given in Gill *et al.* (1996), where the close relationship between $\mathcal{A}$ and its nonsymmetric positive definite form (3.10) was used – together with results in Golub and Van Loan (1979) – to derive stability conditions.

A further exception (with $C = O$) has been identified by Tůma (2002). For a large class of saddle point matrices arising from mixed and hybrid finite element discretizations it is possible to prove the existence of static, fill-reducing pre-orderings $\mathcal{Q}$ such that the permuted matrix $\mathcal{Q}\mathcal{A}\mathcal{Q}^T$ has the LDL^T factorization with $\mathcal{D}$ diagonal. Such pre-orderings are characterized in terms of conditions on the resulting elimination tree. The factorization can be carried out in three phases: a first, symbolic phase in which an initial fill-reducing ordering is computed and the corresponding elimination tree is built; a second phase, also symbolic, where the initial ordering is modified so that the permuted matrix satisfies the conditions that guarantee the existence of the factorization; and a final, numeric phase where the LDL^T factorization itself is computed. The numerical experiments in Tůma (2002) show that this is an effective approach. As in the quasidefinite case, no numerical stability problems have appeared in practice; however, a formal error analysis has not yet been carried out. We mention that examples of saddle point systems that cause difficulties for symmetric indefinite factorization algorithms have been pointed out in Vavasis (1994).

For the general case, sophisticated strategies for computing sparse LDL^T factorizations with 1×1 and 2×2 pivot blocks have been developed over many years by Duff and Reid together with several collaborators; see Duff (1994), Duff *et al.* (1986), Duff *et al.* (1991), Duff and Pralet (2004), Duff and Reid (1983, 1995, 1996). This work has led to a series of widely used codes that are part of the HSL library; see Section 12 for information on how to access these codes. The first is MA27, developed in the early 1980s; the second is MA47, a code geared towards symmetric indefinite systems in saddle point form (with $C = O$); later came MA57 and, recently, the MA67 code. All these codes, except for MA67, are multifrontal codes. The need for codes specifically designed for saddle point systems (with $C = O$) is clear when one considers the presence of the zero block in position $(2,2)$. Clearly, any form of symmetric pivoting must be restricted so that pivots are not chosen from the zero block. Failure to do so during the symbolic phase leads to a very large number of pivot order alterations during the numeric factorization phase, dramatically slowing down the computation. Furthermore, the structure of the matrix during the subsequent factorization steps must also be taken into account in order to avoid excessive fill-in. The code MA47 has been designed with both of these goals in mind; see Duff (1994) and Duff and Reid (1996) for details. The code MA67, also geared towards saddle point systems, is based on design concepts that are quite different from those of the previous codes; as already mentioned it is not a multifrontal code, and furthermore it does not have separate symbolic and numeric phases. Instead, the numerical values of the entries are taken into account during the selection of the pivots. A Markowitz-type strategy is used to balance sparsity and numerical stability needs. Unfortunately, the extensive comparison of HSL codes performed in Gould and Scott (2004) indicates that MA67 is generally inferior to its predecessors.

Other sparse direct solvers for symmetric indefinite systems, based on different design principles, exist; see for instance the recent reports (Meshar and Toledo 2005, Schenk and Gärtner 2004), and Section 12 below. While these codes have not been developed specifically for saddle point matrices, they may work quite well on such problems. For instance, Schenk and Gärtner (2004) report that their code factors a saddle point matrix from optimization of order approximately 2 million (with 6 million nonzeros) in less than a minute on a 2.4 GHz Intel 32-bit processor, producing a factor with about 1.4×10^8 nonzeros.

Although fairly reliable in practice, sparse LDL^T factorization methods are not entirely foolproof. Besides the examples given in Vavasis (1994), a few failures to compute acceptably accurate solutions have been reported in Gould and Scott (2004), even with the use of iterative refinement; see also Schenk and Gärtner (2004). Nevertheless, sparse LDL^T methods are the solvers of choice in various sparse optimization codes, where they are often preferred to methods based on Schur complement reduction ('normal equations methods') for both stability and sparsity reasons. Sparse direct solvers have been less popular in the numerical solution of PDE problems because of their intrinsic storage and computational limitations, although these solvers can be quite competitive for 2D problems; see, *e.g.*, Perugia *et al.* (1999). For saddle point systems arising from PDE problems on 3D meshes, it is necessary to turn to iterative methods.

8. Stationary iterations

We begin our discussion of iterative algorithms with stationary schemes. These methods have been popular for years as 'standalone' solvers, but nowadays they are most often used as preconditioners for Krylov subspace methods (equivalently, the convergence of these stationary iterations can be accelerated by Krylov subspace methods.) Another common use for stationary iterations is as smoothers for multigrid methods; we return to this in Section 11.

8.1. The Arrow–Hurwicz and Uzawa methods

The first iterative schemes for the solution of saddle point problems of a rather general type were the ones developed by the mathematical economists Arrow, Hurwicz and Uzawa (see Arrow *et al.* (1958)). The original papers addressed the case of inequality constraints; see Polyak (1970) for an early study of these methods in the context of the equality-constrained problem (1.4)–(1.5).

The Arrow–Hurwicz and Uzawa methods are stationary schemes consisting of simultaneous iterations for both x and y, and can be expressed in terms of splittings of the matrix $\mathcal{A}$. By elimination of one of the unknown vectors, they can also be interpreted as iterations for the reduced (Schur complement) system. Hence, these algorithms may be regarded both as coupled and as segregated solvers.

We start with Uzawa's method (Uzawa 1958), which enjoys considerable popularity in fluid dynamics, especially for solving the (steady) Stokes problem (Fortin and Glowinski 1983, Glowinski 1984, Glowinski 2003, Temam 1984, Turek 1999). For simplicity, we assume A is invertible and we describe the algorithm in the case $B_1 = B_2 = B$ and $C = O$. Generalization to problems with $B_1 \neq B_2$ or $C \neq O$ is straightforward. Starting with initial guesses x_0 and y_0, Uzawa's method consists of the following coupled iteration:

$$\begin{cases} Ax_{k+1} = f - B^T y_k, \\ y_{k+1} = y_k + \omega(Bx_{k+1} - g), \end{cases} \tag{8.1}$$

where $\omega > 0$ is a relaxation parameter. As noted in Golub and Overton (1988, page 591) (see also Saad (2003, page 258)), this iteration can be written in terms of a matrix splitting $\mathcal{A} = \mathcal{P} - \mathcal{Q}$, *i.e.*, as the fixed-point iteration

$$\mathcal{P}u_{k+1} = \mathcal{Q}u_k + b,$$

where

$$\mathcal{P} = \begin{bmatrix} A & O \\ B & -\frac{1}{\omega}I \end{bmatrix}, \quad \mathcal{Q} = \begin{bmatrix} O & -B^T \\ O & -\frac{1}{\omega}I \end{bmatrix}, \quad \text{and} \quad u_k = \begin{bmatrix} x_k \\ y_k \end{bmatrix}. \tag{8.2}$$

Note that the iteration matrix is

$$\mathcal{T} = \mathcal{P}^{-1}\mathcal{Q} = \begin{bmatrix} O & -A^{-1}B^T \\ O & I - \omega BA^{-1}B^T \end{bmatrix},$$

and therefore the eigenvalues of $\mathcal{T}$ are all real (and at least n of them are exactly zero).

On the other hand, if we use the first equation in (8.1) to eliminate x_{k+1} from the second one we obtain

$$y_{k+1} = y_k + \omega \left(BA^{-1}f - g - BA^{-1}B^T y_k \right), \tag{8.3}$$

showing that Uzawa's method is equivalent to a stationary Richardson iteration applied to the Schur complement system

$$BA^{-1}B^T y = BA^{-1}f - g. \tag{8.4}$$

If A is symmetric and positive definite, so is $BA^{-1}B^T$. Denoting the smallest and largest eigenvalues of $BA^{-1}B^T$ by $\lambda_{\min}$ and $\lambda_{\max}$, respectively, it is well known that Richardson's iteration (8.3) converges for all ω such that

$$0 < \omega < \frac{2}{\lambda_{\max}};$$

see, *e.g.*, Saad (2003, page 114). Furthermore, the spectral radius of the iteration matrix $I - \omega BA^{-1}B^T$ of (8.3) is minimized by taking

$$\omega_* = \frac{2}{\lambda_{\min} + \lambda_{\max}}.$$

In some special cases, the optimal value of ω can be estimated analytically (Langer and Queck 1986). An important example is that of (LBB-stable) discretizations of the steady-state Stokes system (2.10)–(2.12), for which the Schur complement is *spectrally equivalent* to the identity (Verfürth 1984a). This means that the eigenvalues of $BA^{-1}B^T$ are bounded below and above by positive constants, *i.e.*, by numbers that do not depend on the mesh size h. As a result, Uzawa's iteration converges at a rate independent of h. We note that this is not the case for the so-called *generalized Stokes* problem arising from the solution of the unsteady Stokes problems using implicit methods; see Cahouet and Chabard (1988). The convergence of Uzawa's method in this case is rather slow, as it is for most other problems, particularly for the Oseen problem at high Reynolds numbers (small ν) (Fortin and Fortin 1985). Improved convergence can be achieved by suitably preconditioning the Uzawa iteration: see Cahouet and Chabard (1988) and Elman and Golub (1994). Uzawa-type algorithms for the stabilized case ($C \neq O$) were first studied by Vincent and Boyer (1992). An Uzawa-type method with variable relaxation parameters was proposed by Hu and Zou (2001). Uzawa's method is still being actively developed by many researchers: recent papers discussing various extensions and improvements of Uzawa's classical algorithm include Bertrand and Tanguy (2002), Bramble, Pasciak and Vassilev (2000), Cao (2004b), Chen (1998), Cui (2004), Hu and Zou (2002), Liu and Xu (2001), Maday, Meiron, Patera and Ronquist (1993), Nochetto and Pyo (2004), Zsaki, Rixen and Paraschivoiu (2003). Not all applications of Uzawa's method are to fluid flow problems: see Ito and Kunisch (1999) for a recent application to image restoration.

The bulk of the computational effort in Uzawa's method is spent in the solution of linear systems involving A. These systems can be solved by direct methods or, more often, by an inner iterative scheme. For instance, in the case of the Stokes problem A is a direct sum of discrete Laplace operators, and multigrid methods are a natural choice. The case of inexact inner solves has been studied in detail in Bramble, Pasciak and Vassilev (1997), Chen (1998), Cheng (2000), Cheng and Zou (2003), Cui (2002), Elman and Golub (1994), Peters, Reichelt and Reusken (2004), Robichaud, Tanguy and Fortin (1990) and Zulehner (2002).

The Arrow–Hurwicz method (Arrow and Hurwicz 1958) may be regarded as an inexpensive alternative to Uzawa's method, useful when solves with A are too expensive. Here we follow the derivation given in Saad (2003, Section 8.4). By noting that iterate x_{k+1} given by the first of (8.1) is the minimizer of the objective function

$$\phi(x) = \frac{1}{2}x^T A x - x^T(f - B^T y_k),$$

we can derive a less expensive method by taking one step in the direction of the (negative) gradient of $\phi(x)$, with fixed step length α. The resulting method is the Arrow–Hurwicz iteration:

$$\begin{cases} x_{k+1} = x_k + \alpha(f - A x_k - B^T y_k), \\ y_{k+1} = y_k + \omega(B x_{k+1} - g). \end{cases} \tag{8.5}$$

As in the case of Uzawa's method, the Arrow–Hurwicz method can be cast as a fixed-point iteration induced by the splitting

$$\mathcal{A} = \mathcal{P} - \mathcal{Q} \quad \text{where} \quad \mathcal{P} = \begin{bmatrix} \frac{1}{\alpha}I & O \\ B & -\frac{1}{\omega}I \end{bmatrix}, \quad \mathcal{Q} = \begin{bmatrix} \frac{1}{\alpha}I - A & -B^T \\ O & -\frac{1}{\omega}I \end{bmatrix}. \tag{8.6}$$

The convergence of this algorithm depends on the two relaxation parameters, α and ω. Convergence conditions and theoretical estimates for the optimal choice of parameters have been given in Fortin and Glowinski (1983), Polyak (1970), Queck (1989) and, more recently, in Astrakhantsev (2001) and Bychenkov (2002). Because the convergence of the Arrow–Hurwicz method is usually rather slow, various improvements have been proposed, including preconditioned variants of the form

$$\begin{cases} x_{k+1} = x_k + \alpha\, Q_A^{-1}(f - A x_k - B^T y_k), \\ y_{k+1} = y_k + \omega\, Q_B^{-1}(B x_{k+1} - g), \end{cases} \tag{8.7}$$

where Q_A and Q_B are appropriately chosen 'preconditioning matrices': see Astrakhantsev (2001), Queck (1989) and Robichaud et al. (1990). Obviously, the line between preconditioned versions of the Arrow–Hurwicz algorithm and inexact/preconditioned variants of Uzawa's method is blurred. Additional hybrids can be obtained by combining the Arrow–Hurwicz and CG algorithms: see, e.g., Aboulaich and Fortin (1989) and Stoyan (2001).

8.2. Penalty and multiplier methods

Here we assume that $A = A^T$ is positive semidefinite, $B_1 = B_2 = B$ is of full rank, and $C = O$. We further assume that $\ker(A) \cap \ker(B) = \{0\}$, so that the saddle point system has a unique solution. As we have noted in Section 1, the saddle point problem is then equivalent to the constrained minimization problem

$$\min\ J(x) = \frac{1}{2}x^T A x - f^T x, \tag{8.8}$$

$$\text{subject to }\ Bx = g. \tag{8.9}$$

A very old method for finding the solution x_* of (8.8)–(8.9) is based on the observation that

$$x_* = \lim_{\gamma \to \infty} x(\gamma),$$

where $x(\gamma)$ is the unique solution of the unconstrained minimization problem

$$\min\ \hat{J}(x) \equiv J(x) + \frac{\gamma}{2}\,\|Bx - g\|_2^2.$$

In mechanical terms, rigid constraints can be thought of as limiting cases of very large restoring forces, *i.e.*, in this case, forces with potential energy of the form $U(x) = \frac{\gamma}{2}(Bx - g)^T(Bx - g)$; see Courant (1943). The minimizer $x(\gamma)$ can be found by setting the gradient of $J(x) + \frac{\gamma}{2}(Bx - g)^T(Bx - g)$ to zero, leading to the linear system

$$(A + \gamma B^T B)\, x = f + \gamma B^T g. \tag{8.10}$$

If we let $y(\gamma) = \gamma(Bx(\gamma) - g)$, where $x = x(\gamma)$ is the solution of (8.10), then it is possible to prove that

$$\|x_* - x(\gamma)\|_2 = O(\gamma^{-1}) \quad \text{and} \quad \|y_* - y(\gamma)\|_2 = O(\gamma^{-1}) \quad \text{for}\ \ \gamma \to \infty;$$

see Glowinski (1984, pages 21–22). Therefore, provided that γ is taken large enough, $x(\gamma)$ and $y(\gamma)$ are approximate solutions of the original saddle point problem. The penalty method can be thought of as an *approximate* direct method. Since a monotonically increasing sequence $\gamma_1, \gamma_2, \dots$ of values of γ may be used to compute better and better approximations to x_*, it may also be regarded as a stationary iterative method. In some cases, the choice of γ may be made on the basis of physical considerations: see, *e.g.*, Destuynder and Nevers (1990), Nour-Omid and Wriggers (1986).

Since the matrix $A + \gamma B^T B$ in (8.10) is symmetric and positive definite for $\gamma > 0$, one can in principle use a Cholesky factorization or the CG method to compute $x(\gamma)$. Unfortunately, such an approach cannot be recommended, since the condition number of $A + \gamma B^T B$ grows like a (possibly large) multiple of γ; see Glowinski (1984, pages 22–23) and Van Loan (1985) for an analysis of the penalty method applied to equality-constrained least squares problems. In practice, for large values of γ the coefficient matrix $A + \gamma B^T B$ is dominated by the (highly singular) term $\gamma B^T B$, and accurate solutions of (8.10) are difficult to obtain.

This drawback of the penalty method can be overcome in two ways. One way is to observe that $(x(\gamma), y(\gamma))$ is the unique solution of the *regularized* saddle point system

$$\begin{bmatrix} A & B^T \\ B & -\varepsilon I \end{bmatrix} \begin{bmatrix} x \\ y \end{bmatrix} = \begin{bmatrix} f \\ g \end{bmatrix}, \quad \varepsilon = \gamma^{-1}. \tag{8.11}$$

Stable solution of this linear system is now possible, for instance using a sparse direct solver; this approach is popular in optimization. However, using a direct solver is not always feasible.

The other option is to modify the penalty method so as to avoid ill-conditioning. This leads to the *method of multipliers*, developed independently by Arrow and Solow (1958, page 172), Hestenes (1969) and Powell (1969). A further advantage of this method, which combines the use of penalty with Lagrange multipliers, is that it produces the *exact* solution (x_*, y_*) rather than an approximate one. The method of multipliers can be described as follows. Select $\gamma > 0$ and consider the *augmented Lagrangian*

$$\mathcal{L}(x, y) = J(x) + (Bx - g)^T y + \frac{\gamma}{2} \|Bx - g\|_2^2. \tag{8.12}$$

Given an approximation y_k for the Lagrange multiplier vector y, we compute the minimum x_{k+1} of the function $\psi(x) \equiv \mathcal{L}(x, y_k)$. This requires solving the linear system

$$(A + \gamma B^T B)x = f - B^T y_k + \gamma B^T g. \tag{8.13}$$

Now we use the computed solution x_{k+1} to obtain the new Lagrange multiplier approximation y_{k+1} according to

$$y_{k+1} = y_k + \gamma (Bx_{k+1} - g),$$

and so on. Clearly, the method of multipliers is precisely Uzawa's iteration applied to the saddle point system

$$\begin{bmatrix} A + \gamma B^T B & B^T \\ B & O \end{bmatrix} \begin{bmatrix} x \\ y \end{bmatrix} = \begin{bmatrix} f + \gamma B^T g \\ g \end{bmatrix}, \tag{8.14}$$

which has exactly the same solution (x_*, y_*) as the original one. Note that the parameter γ does double duty here, in that it appears both in the definition of the augmented Lagrangian and as the relaxation parameter for the Uzawa iteration. As we know from our discussion of Uzawa's method, the iteration converges for $\gamma \in (0, 2/\rho)$ where ρ denotes the largest eigenvalue of the Schur complement $B(A + \gamma B^T B)^{-1} B^T$. This interval becomes unbounded, and the rate of convergence arbitrarily large, as $\gamma \to \infty$. Again, taking too large a value of γ results in extreme ill-conditioning of the coefficient matrix in (8.13). It is necessary to strike a balance between the rate of convergence of the method and the conditioning properties of (8.13). The choice of γ and many other aspects of the multiplier method have been discussed by a number of authors, including Bertsekas (1982), Chen and Zou (1999), Fortin and Glowinski (1983), Greif et al. (2005), Hestenes (1975), Luenberger (1984), Zienkiewicz, Vilotte, Toyoshima and Nakazawa (1985). See further Awanou and Lai (2005) for a study of the nonsymmetric case. Another possibility is to combine the augmented Lagrangian method with a (preconditioned) Arrow–Hurwicz scheme; see Fortin and Glowinski (1983, page 26) and Kouhia and Menken (1995) for an application of this idea to problems in structural mechanics.

8.3. Other stationary iterations

In addition to the foregoing algorithms, a number of other stationary iterations based on matrix splittings $\mathcal{A} = \mathcal{P} - \mathcal{Q}$ can be found in the literature. In particular, SOR- and block-SOR-type schemes have been proposed in Strikwerda (1984) for the Stokes problem, in Barlow et al. (1988), Benzi (1993) and Plemmons (1986) for structural analysis computations, and in Chen (1998), Golub, Wu and Yuan (2001) and Li, Li, Evans and Zhang (2003) for general saddle point systems. Some of these schemes can be interpreted as preconditioned or inexact variants of the classical Uzawa algorithm. Alternating-direction iterative methods for saddle point problems have been studied in Brown (1982), Douglas Jr., Durán and Pietra (1986, 1987). Other stationary iterative methods for saddle point problems have been studied in Bank, Welfert and Yserentant (1990), Benzi and Golub (2004), Dyn and Ferguson (1983), Golub and Wathen (1998) and Tong and Sameh (1998); since these methods are most often used as preconditioners for Krylov subspace methods, we defer their description to Section 10.

9. Krylov subspace methods

In this section we discuss Krylov subspace methods for solving (preconditioned) saddle point problems. Our goal is not to survey all existing methods and implementations (more complete surveys can be found, e.g., in the monographs by Greenbaum (1997), Saad (2003) and van der Vorst (2003) or in the papers by Eiermann and Ernst (2001) and Freund et al. (1992)), but to describe the main properties of the most commonly used methods. We discuss the general theory, the main convergence results, and implementation details. For simplicity, we describe the basics of Krylov subspace methods for the unpreconditioned and nonsingular system (1.1)–(1.2). The later sections will describe the general ideas of preconditioning (Section 10), and different preconditioning techniques specifically constructed for (generalized) saddle point systems (Sections 10.1–10.4).

9.1. General theory

Suppose that u_0 is an initial guess for the solution u of (1.1)–(1.2), and define the initial residual $r_0 = b - \mathcal{A}u_0$. Krylov subspace methods are iterative methods whose kth iterate u_k satisfies

$$u_k \in u_0 + \mathcal{K}_k(\mathcal{A}, r_0), \quad k = 1, 2, \ldots, \tag{9.1}$$

where

$$\mathcal{K}_k(\mathcal{A}, r_0) \equiv \text{span} \{r_0, \mathcal{A}r_0, \ldots, \mathcal{A}^{k-1}r_0\} \tag{9.2}$$

denotes the kth Krylov subspace generated by $\mathcal{A}$ and r_0. It is well known that the Krylov subspaces form a nested sequence that ends with dimension $d \equiv \dim \mathcal{K}_{n+m}(\mathcal{A}, r_0) \leq n + m$, i.e.,

$$\mathcal{K}_1(\mathcal{A}, r_0) \subset \cdots \subset \mathcal{K}_d(\mathcal{A}, r_0) = \cdots = \mathcal{K}_{n+m}(\mathcal{A}, r_0).$$

In particular, for each $k \leq d$, the Krylov subspace $\mathcal{K}_k(\mathcal{A}, r_0)$ has dimension k. Because of the k degrees of freedom in the choice of the iterate u_k, k constraints are required to make u_k unique. In Krylov subspace methods this is achieved by requiring that the kth residual $r_k = b - \mathcal{A}u_k$ is orthogonal to a k-dimensional space $\mathcal{C}_k$, called the constraints space:

$$r_k = b - \mathcal{A}u_k \in r_0 + \mathcal{A}\mathcal{K}_k(\mathcal{A}, r_0), \qquad r_k \perp \mathcal{C}_k. \tag{9.3}$$

Orthogonality here is meant with respect to the Euclidean inner product. The relations (9.1)–(9.3) show that Krylov subspace methods are based on a general type of *projection process* that can be found in many areas of mathematics. For example, in the language of the finite element method, we may consider $\mathcal{K}_k(\mathcal{A}, r_0)$ the test and $\mathcal{C}_k$ the trial space for constructing the approximate solution u_k. In this sense the projection process (9.1)–(9.3) corresponds to the Petrov–Galerkin framework; see, *e.g.*, Quarteroni and Valli (1994, Chapter 5). The interpretation of Krylov subspace methods as projection processes was popularized by Saad in a series of papers in the early 1980s (Saad 1981, 1982). A survey of his approach can be found in his book, Saad (2003). For additional analyses of Krylov subspace methods in terms of projections see, *e.g.*, Barth and Manteuffel (1994) and Eiermann and Ernst (2001).

Knowing the properties of the system matrix $\mathcal{A}$ it is possible to determine constraint spaces $\mathcal{C}_k$ that lead to uniquely defined iterates u_k, $k = 1, 2, \ldots$, in (9.1)–(9.3). Examples for such spaces are given in the following theorem.

Theorem 9.1. Suppose that the Krylov subspace $\mathcal{K}_k(\mathcal{A}, r_0)$ has dimension k. If

(C) $\mathcal{A}$ is symmetric positive definite and $\mathcal{C}_k = \mathcal{K}_k(\mathcal{A}, r_0)$, or

(M) $\mathcal{A}$ is nonsingular and $\mathcal{C}_k = \mathcal{A}\mathcal{K}_k(\mathcal{A}, r_0)$, then there exists a uniquely defined iterate u_k of the form (9.1) for which the residual $r_k = b - \mathcal{A}u_k$ satisfies (9.3).

Proof. See Saad (2003, Proposition 5.1). □

Items (C) and (M) in Theorem 9.1 represent mathematical characterizations of the projection properties of well-known Krylov subspace methods. Item (C) characterizes the conjugate gradient (CG) method of Hestenes and Stiefel for symmetric positive definite matrices (Hestenes and Stiefel 1952). Note that if $\mathcal{A}$ is *not* symmetric positive definite, an approximate solution u_k satisfying both (9.1)–(9.2) and (9.3) with $\mathcal{C}_k = \mathcal{K}_k(\mathcal{A}, r_0)$ may not exist; *cf.*, *e.g.*, Brown (1991) and Cullum and Greenbaum (1996). Nevertheless, there are several implementations of this projection process, in particular the full orthogonalization method (FOM) of Saad (1981). Implementations of the projection process characterized by item (M) are the minimal residual (MINRES) method of Paige and Saunders (1975) for nonsingular symmetric (possibly indefinite) matrices, and the generalized minimal residual (GMRES) method of Saad and Schultz (1986) for general nonsingular matrices. Further mathematically equivalent implementations are discussed in Liesen, Rozložník and Strakoš (2002).

Numerous other choices of constraint spaces for constructing Krylov subspace methods exist. For example, in case of a nonsymmetric matrix $\mathcal{A}$ one may choose $\mathcal{C}_k = \mathcal{K}_k(\mathcal{A}^T, r_0)$, which represents a generalization of the projection process characterized in item (C). Specific implementations based on this choice include the method of Lanczos (1950) and the biconjugate gradient (BiCG) method of Fletcher (1976a). However, for a general nonsymmetric matrix $\mathcal{A}$ the process based on $\mathcal{C}_k = \mathcal{K}_k(\mathcal{A}^T, r_0)$ is not well defined, because it may happen that no iterate u_k satisfying both (9.1)–(9.2) and (9.3) exists. In an actual implementation such as BiCG this will lead to a breakdown. In practice such breakdowns are unlikely to occur, but near-breakdowns may cause irregular convergence and serious build-up of rounding errors, see Greenbaum (1997, Chapter 5) for further discussion. Such instabilities are often overcome by the stabilized BiCG (BiCGStab) method of van der Vorst (1992), which combines the BiCG projection principle with an additional minimization step in order to 'stabilize' the convergence behaviour. It is important to note that although BiCGStab is based on BiCG, it avoids using the transposed matrix $\mathcal{A}^T$. Closely related is the transpose-free quasi-minimal residual method (TFQMR) for general nonsymmetric matrices developed by Freund (1993). While both BiCGStab and TFQMR typically produce smoother convergence curves than BiCG, none of these methods is guaranteed to be free of breakdowns. A related BiCG-like method, which overcomes some of the numerical instabilities of BiCG, is the quasi-minimal residual (QMR) method for general nonsymmetric matrices of Freund and Nachtigal (1991). We point out that, despite the naming similarities, QMR and TFQMR are not mathematically equivalent, and hence may produce completely different approximate solutions even in exact arithmetic. In case of a symmetric (possibly indefinite) matrix $\mathcal{A}$ the Lanczos process underlying the QMR method can be simplified, leading to a mathematically equivalent but numerically more efficient implementation. This is exploited in the symmetric QMR (SQMR) method of Freund and Nachtigal (1994).

Additional Krylov subspace methods can be constructed by using different search spaces in (9.1). For example, the projection process

$$u_k \in u_0 + \mathcal{A}\mathcal{K}_k(\mathcal{A}, r_0), \qquad r_k \in r_0 + \mathcal{A}\mathcal{K}_k(\mathcal{A}, \mathcal{A}r_0) \perp \mathcal{K}_k(\mathcal{A}, r_0), \qquad (9.4)$$

yields a uniquely defined iterate u_k whenever $\mathcal{A}$ is nonsingular and symmetric. An implementation of this mathematical principle is the SYMMLQ method of Paige and Saunders (1975) for nonsingular symmetric (possibly indefinite) matrices.

9.2. Convergence analysis

In exact arithmetic, the methods that are mathematically described by items (C) and (M) in Theorem 9.1 as well as by (9.4) for symmetric nonsingular $\mathcal{A}$ terminate with the exact solution in step $d \equiv \dim \mathcal{K}_{n+m}(\mathcal{A}, r_0)$, i.e., they yield $u_d = u$. This feature is called the *finite termination property*. In practice, however, one is typically not interested in computing the exact solution. For example, when the linear system represents a discretized partial differential equation, then an *approximate solution* u_k of the linear system with an error norm on the level of the discretization error is often sufficient. Once this error level is reached, the iterative method can be stopped. In this way iterative methods such as Krylov subspace methods may significantly outperform any direct method which cannot be stopped prematurely, because their intermediate results do not represent approximate solutions. The question is how fast can a given Krylov subspace method reach a given accuracy level.

To analyse this question we use the fact that the geometric *orthogonality condition* expressed in (9.3) is often equivalent to an algebraic *optimality condition* for a certain norm of either the error $e_k = u - u_k$ or the residual $r_k = b - \mathcal{A}u_k$. This optimality condition can be derived using the following well-known theorem for best approximations in Hilbert spaces; see, *e.g.*, Conway (1990, Chapter 1).

Theorem 9.2. Suppose that $\mathcal{H}$ is a Hilbert space with inner product $\langle \cdot, \cdot \rangle$ and associated norm $\| \cdot \|$. If $\mathcal{M} \subset \mathcal{H}$ is a closed linear subspace, then for each fixed $h \in \mathcal{H}$ there exists a unique element $m_0 \in \mathcal{M}$ with $\|h - m_0\| = \inf_{m \in \mathcal{M}} \|h - m\|$. Moreover, $h - m_0$ is orthogonal to $\mathcal{M}$ with respect to

$\langle \cdot, \cdot \rangle$. Conversely, if $m_0 \in \mathcal{M}$ is such that $h - m_0$ is orthogonal to $\mathcal{M}$ with respect to $\langle \cdot, \cdot \rangle$, then $\| h - m_0 \| = \inf_{m \in \mathcal{M}} \| h - m \|$.

To apply this result in our context, consider a symmetric positive definite matrix $\mathcal{A}$ and the projection process (9.1)–(9.3) with $\mathcal{C}_k = \mathcal{K}_k(\mathcal{A}, r_0)$. According to item (C) in Theorem 9.1, this process leads to a uniquely defined iterate $u_k \in u_0 + \mathcal{K}_k(\mathcal{A}, r_0)$. The residual $r_k = \mathcal{A}e_k$ is orthogonal to $\mathcal{K}_k(\mathcal{A}, r_0)$ with respect to the Euclidean inner product, or, equivalently, the error $e_k = u - u_k = (u - u_0) - (u_k - u_0)$ is orthogonal to $\mathcal{K}_k(\mathcal{A}, r_0)$ with respect to the $\mathcal{A}$-inner product defined by $\langle v, w \rangle_{\mathcal{A}} \equiv w^T \mathcal{A} v$. With $\mathbb{R}^{n+m}$, $\mathcal{K}_k(\mathcal{A}, r_0)$, $u - u_0 \in \mathbb{R}^{n+m}$, and $u_k - u_0 \in \mathcal{K}_k(\mathcal{A}, r_0)$ taking the roles of $\mathcal{H}$, $\mathcal{M}$, h, and m_0 in Theorem 9.2, respectively, we see that the orthogonality condition (9.3) is equivalent to

$$\| e_k \|_{\mathcal{A}} = \min_{z \in u_0 + \mathcal{K}_k(\mathcal{A}, r_0)} \| u - z \|_{\mathcal{A}},$$

where $\| \cdot \|_{\mathcal{A}}$ denotes the $\mathcal{A}$-norm (sometimes called energy norm) associated with the $\mathcal{A}$-inner product. As mentioned above, item (C) represents a mathematical characterization of the CG method; what we have just derived is its well-known optimality property (error minimization in the energy norm). Analogously, we may show that the methods characterized by item (M) in Theorem 9.1 minimize the Euclidean norm of the residual $r_k = b - \mathcal{A}u_k$ over the affine subspace $u_0 + \mathcal{K}_k(\mathcal{A}, r_0)$. These results are summarized in the following theorem.

Theorem 9.3. Suppose that the Krylov subspace $\mathcal{K}_k(\mathcal{A}, r_0)$ in the projection process (9.1)–(9.3) has dimension k. Then the iterate u_k satisfies the following optimality properties.

(C) If $\mathcal{A}$ is symmetric positive definite and $\mathcal{C}_k = \mathcal{K}_k(\mathcal{A}, r_0)$, then

$$\| u - u_k \|_{\mathcal{A}} = \| e_k \|_{\mathcal{A}} = \min_{z \in u_0 + \mathcal{K}_k(\mathcal{A}, r_0)} \| u - z \|_{\mathcal{A}} = \min_{p \in \Pi_k} \| p(\mathcal{A})e_0 \|_{\mathcal{A}}. \quad (9.5)$$

(M) If $\mathcal{A}$ is nonsingular and $\mathcal{C}_k = \mathcal{A}\mathcal{K}_k(\mathcal{A}, r_0)$, then

$$\| b - \mathcal{A}u_k \|_2 = \| r_k \|_2 = \min_{z \in u_0 + \mathcal{K}_k(\mathcal{A}, r_0)} \| b - \mathcal{A}z \|_2 = \min_{p \in \Pi_k} \| p(\mathcal{A})r_0 \|_2.$$

$$(9.6)$$

Here Π_k denotes the set of polynomials of degree at most k with value 1 at the origin.

For reasons apparent from (9.6) we refer to methods characterized by item (M) as *minimal residual methods*. Note that the Euclidean residual norm in these methods gives a lower bound for the residual norm of all Krylov subspace methods based on (9.1)–(9.3). Therefore significant research efforts have been made to understand the convergence behaviour of these methods. The interpretation of the kth error and residual in (9.5) and (9.6) in terms of the initial error and residual multiplied by a certain polynomial in the matrix $\mathcal{A}$, respectively, is the typical starting point for the convergence analysis of the Krylov subspace methods characterized by items (C) and (M).

First consider item (C) (as implemented by the CG method). Since $\mathcal{A}$ is assumed to be symmetric positive definite, it is orthogonally diagonalizable, $\mathcal{A} = \mathcal{V} \mathcal{D} \mathcal{V}^T$, with $\mathcal{V}^T \mathcal{V} = I$ and $\mathcal{D} = \operatorname{diag}(\lambda_j)$. The $\mathcal{A}$-norm of the error at step k satisfies

$$
\begin{aligned}
\|e_k\|_{\mathcal{A}} = \min_{p \in \Pi_k} \|p(\mathcal{A}) e_0\|_{\mathcal{A}} &= \min_{p \in \Pi_k} \|\mathcal{A}^{1/2} p(\mathcal{A}) e_0\|_2 = \min_{p \in \Pi_k} \|p(\mathcal{A}) \mathcal{A}^{1/2} e_0\|_2 \\
&\leq \min_{p \in \Pi_k} \|p(\mathcal{A})\|_2 \, \|\mathcal{A}^{1/2} e_0\|_2 \\
&= \|e_0\|_{\mathcal{A}} \min_{p \in \Pi_k} \|p(\mathcal{D})\|_2 \\
&= \|e_0\|_{\mathcal{A}} \min_{p \in \Pi_k} \max_{\lambda_j} |p(\lambda_j)|.
\end{aligned}
\tag{9.7}
$$

Hence $\|e_k\|_{\mathcal{A}} / \|e_0\|_{\mathcal{A}}$, the kth relative $\mathcal{A}$-norm of the error, is bounded by the value of a polynomial approximation problem on the eigenvalues of $\mathcal{A}$. This bound is sharp in the sense that for each (symmetric positive definite) matrix $\mathcal{A}$ and each iteration step k there exists an initial error $e_0^{(k)}$ for which equality holds; see Greenbaum (1979). Consequently, the *worst-case* behaviour of the CG method is completely determined by the eigenvalues of the matrix. An immediate question is how small can a kth degree polynomial with value 1 at the origin become on a given set of matrix eigenvalues. While the polynomial that solves the min-max problem is explicitly known (see Greenbaum (1979) and Liesen and Tichý (2004b) for different derivations), no simple expression for the min-max value itself exists. The sharp bound (9.7), however, provides some intuition of how the eigenvalue distribution influences the worst-case convergence behaviour. For example, if all eigenvalues are tightly clustered around a *single* point that is far away from the origin, one may expect fast convergence. Widely spread eigenvalues, on the other hand, will potentially lead to slow convergence. The standard approach for estimating the right-hand side of (9.7) is to replace the min-max problem on the discrete set of eigenvalues by a min-max approximation problem on its convex hull (*i.e.*, on an interval from the smallest eigenvalue $\lambda_{\min}$ to the largest eigenvalue $\lambda_{\max}$ of $\mathcal{A}$). The latter is solved by scaled and shifted Chebyshev polynomials of the first kind, giving the well-known bound

$$
\min_{p \in \Pi_k} \max_{\lambda_j} |p(\lambda_j)| \leq 2 \left(\frac{\sqrt{\kappa(\mathcal{A})} - 1}{\sqrt{\kappa(\mathcal{A})} + 1} \right)^k, \quad \text{where} \quad \kappa(\mathcal{A}) = \frac{\lambda_{\max}}{\lambda_{\min}};
\tag{9.8}
$$

see, *e.g.*, Greenbaum (1997, Theorem 3.1.1). The bounds (9.7)–(9.8) show that a small condition number of $\mathcal{A}$ is *sufficient* (but not necessary) for a fast decrease of the relative $\mathcal{A}$-norm of the error. This fact motivates the classical goal of *preconditioning*, which is to modify the given linear system in order to reduce the condition number of the system matrix. However, while (9.7) is sharp, the right-hand side of (9.8) often overestimates the left-hand side and thus the worst-case relative $\mathcal{A}$-norm of the error. Moreover, the actual behaviour for a specific right-hand side vector b depends not only on the eigenvalue distribution but also on the coefficients of b in the eigenvectors of $\mathcal{A}$. Several case studies on model problems have recently been performed; see, *e.g.*, Beckermann and Kuijlaars (2001, 2002), Liesen and Tichý (2004a) and Naiman, Babuška and Elman (1997) for more details.

Next consider the minimal residual methods. The resulting projection process (9.1)–(9.3) is well defined for each nonsingular matrix $\mathcal{A}$. For simplicity, suppose that $\mathcal{A}$ is diagonalizable, $\mathcal{A} = \mathcal{X}\mathcal{D}\mathcal{X}^{-1}$, with $\mathcal{D} = \mathrm{diag}(\lambda_j)$. Then the kth Euclidean norm of the residual satisfies

$$
\begin{aligned}
\|r_k\|_2 &= \min_{p \in \Pi_k} \|p(\mathcal{A})r_0\|_2 \\
&\leq \min_{p \in \Pi_k} \|\mathcal{X}p(\mathcal{D})\mathcal{X}^{-1}\|_2 \, \|r_0\|_2 \\
&\leq \|\mathcal{X}\|_2 \min_{p \in \Pi_k} \|p(\mathcal{D})\|_2 \, \|\mathcal{X}^{-1}\|_2 \, \|r_0\|_2 \\
&= \|r_0\|_2 \, \kappa(\mathcal{X}) \min_{p \in \Pi_k} \max_{\lambda_j} |p(\lambda_j)|.
\end{aligned}
\tag{9.9}
$$

Clearly,
$$
\|r_k\|_2 / \|r_0\|_2,
$$
the kth relative Euclidean residual norm, is bounded by the value of the same type of polynomial approximation problem as in (9.7), multiplied by the condition number of the eigenvector matrix $\mathcal{X}$ of $\mathcal{A}$.

If $\mathcal{A}$ is *normal*, then $\kappa(\mathcal{X}) = 1$, and it can be shown that the bound (9.9) is sharp in the same sense as the bound (9.7); see Greenbaum and Gurvits (1994) and Joubert (1994). In this case the same intuition as described above for the worst-case behaviour of CG also applies to the worst-case behaviour of minimal residual methods. In particular, a single eigenvalue cluster far away from the origin implies fast convergence (here measured by the relative Euclidean residual norm). Additionally, (9.8) can be used if $\mathcal{A}$ is symmetric positive definite, which shows that in this case a small condition number of $\mathcal{A}$ is *sufficient* (but not necessary) for fast convergence. As in the case of CG, reducing the condition number of a symmetric positive definite system matrix $\mathcal{A}$ also represents a reasonable goal of preconditioning for minimal residual methods.

In the case of a nonsingular symmetric indefinite matrix $\mathcal{A}$, the min-max approximation problem on the matrix eigenvalues in (9.9) cannot be replaced by the min-max problem on their convex hull, as eigenvalues lie on both sides of the origin. Here one may replace the discrete set of eigenvalues by the union of two intervals containing all of them and excluding the origin, say $I^- \cup I^+ \equiv [\lambda_{\min}, \lambda_s] \cup [\lambda_{s+1}, \lambda_{\max}]$ with $\lambda_{\min} \leq \lambda_s < 0 < \lambda_{s+1} \leq \lambda_{\max}$.

When both intervals are of the same length, *i.e.*, $\lambda_{\max} - \lambda_{s+1} = \lambda_s - \lambda_{\min}$, the solution of the corresponding min-max approximation problem

$$\min_{p \in \Pi_k} \max_{\lambda \in I^- \cup I^+} |p(\lambda)|, \tag{9.10}$$

is characterized by a result of de Boor and Rice (1982). This leads to the bound

$$\min_{p \in \Pi_k} \max_{\lambda_j} |p(\lambda_j)| \leq 2 \left(\frac{\sqrt{|\lambda_{\min} \lambda_{\max}|} - \sqrt{|\lambda_s \lambda_{s+1}|}}{\sqrt{|\lambda_{\min} \lambda_{\max}|} + \sqrt{|\lambda_s \lambda_{s+1}|}} \right)^{[k/2]}, \tag{9.11}$$

where $[k/2]$ denotes the integer part of $k/2$; see Greenbaum (1997, Chap. 3). For an illustration of this bound suppose that $|\lambda_{\min}| = \lambda_{\max} = 1$ and $|\lambda_s| = \lambda_{s+1}$. Then $\kappa(\mathcal{A}) = \lambda_{s+1}^{-1}$, and the right-hand side of (9.11) reduces to

$$2 \left(\frac{1/\lambda_{s+1} - 1}{1/\lambda_{s+1} + 1} \right)^{[k/2]}. \tag{9.12}$$

Note that (9.12) corresponds to the value of the right-hand side of (9.8) at step $[k/2]$ for a symmetric positive definite matrix having all its eigenvalues in the interval $[\lambda_{s+1}^2, 1]$, and thus a condition number of λ_{s+1}^{-2}. Hence the convergence bound for an indefinite matrix with condition number κ needs twice as many steps to decrease to the value of the bound for a definite matrix with condition number κ^2. Although neither of the two bounds is sharp, this clearly indicates that solving indefinite problems represents a significant challenge.

In the general case when the two intervals I^- and I^+ are not of the same length, the explicit solution of (9.10) becomes quite complicated (see, *e.g.*, Fischer (1996, Chapter 3)), and no simple and explicit bound on the min-max value is known. One may of course extend the smaller interval to match the length of the larger one, and still apply (9.11). But this usually results in a significantly weaker convergence bound, which fails to give relevant information about the actual convergence behaviour. An alternative is to consider the asymptotic behaviour of the min-max value (9.10), and in particular the *asymptotic convergence factor*

$$\rho(I^- \cup I^+) \equiv \lim_{k \to \infty} \left(\min_{p \in \Pi_k} \max_{\lambda \in I^- \cup I^+} |p(\lambda)| \right)^{1/k}. \tag{9.13}$$

Obviously, $\rho(I^- \cup I^+)$ may be estimated even if the value (9.10) for each step k is unknown. Asymptotic convergence results are common in the theory of semi-iterative methods (Eiermann and Niethammer 1983, Eiermann, Niethammer and Varga 1985) and of classical iterative methods such as

SOR (Varga 1962). Because of the finite termination property, asymptotic convergence results for Krylov subspace methods have to be put into proper perspective. They certainly can be useful in the convergence analysis of minimal residual methods for sequences of linear systems of growing dimension, *e.g.*, when studying the dependence of the convergence behaviour on the mesh size in a discretized differential equation. An example is discussed in Section 10.1 below.

If $\mathcal{A}$ is *nonnormal*, then $\kappa(\mathcal{X})$ may be very large, and (9.9) may be a very large overestimate even of the worst-case relative residual norm of minimal residual methods. In particular, the matrix eigenvalues may in the nonnormal case give misleading information about the convergence behaviour. In fact, it has been shown that any nonincreasing convergence curve of relative residual norms can be obtained by a (nonnormal) matrix $\mathcal{A}$ having any prescribed set of eigenvalues (Greenbaum, Pták and Strakoš 1996). While the examples constructed by this theory may be artificial, misleading eigenvalue information was also demonstrated and analysed in the more practical context of discretized convection-diffusion problems; see Ernst (2000) and Liesen and Strakoš (2005) for further details. Several other sets associated with the matrix $\mathcal{A}$ have been used in the convergence analysis for nonnormal problems, among them the field of values (Ernst 2000, Starke 1997), pseudospectra (Nachtigal, Reddy and Trefethen 1992), and the polynomial numerical hull (Greenbaum 2002). In the context of saddle point problems, the field of values seems to be quite useful. In particular, rate of convergence estimates obtained from this set for preconditioned saddle point systems arising from mixed finite element discretizations of PDEs are sometimes optimal in the sense that they are independent of the mesh size parameter; see, *e.g.*, Klawonn and Starke (1999) and Loghin and Wathen (2004) for details. However, the first inequality in (9.9) has been shown to be strict for some nonnormal matrices $\mathcal{A}$; see Faber, Joubert, Knill and Manteuffel (1996) and Toh (1997). Hence no convergence analysis based solely on the matrix $\mathcal{A}$ can in the nonnormal case give a sharp bound on the worst-case residual norm, so that the initial residual should be included in the convergence analysis whenever possible. This is sometimes only possible in the context of the specific application; see Liesen and Strakoš (2005) for an example. An important and often overlooked fact is that the convergence of minimal residual methods is *not* slower for nonnormal than for normal matrices. In particular, it has been shown that for each nonnormal matrix there exists a normal matrix for which the same convergence behaviour can be observed (for the same initial residual); see Greenbaum and Strakoš (1994). Yet, as described above, the convergence behaviour in the nonnormal case is significantly more difficult to analyse than in the normal case. Sharp convergence results can usually only be obtained by considering the specific properties of $\mathcal{A}$ (*e.g.*, its eigenvalue-eigenvector structure) in relation to the given initial residual.

9.3. Implementation details

When implementing Krylov subspace methods, one needs to generate (at least implicitly) a basis of the Krylov subspace $\mathcal{K}_k(\mathcal{A}, r_0)$. For reasons of numerical stability this basis should preferably be *orthogonal*, and for reasons of computational efficiency, the basis should be generated by a *short-term recurrence*.

In the case of a *symmetric* matrix $\mathcal{A}$, these two goals are achieved simultaneously by the symmetric Lanczos method (Lanczos 1950), which generates an orthogonal basis $v_1, \ldots, v_k$ of $\mathcal{K}_k(\mathcal{A}, r_0)$ using a three-term recurrence, meaning that in step k only the vectors v_{k-1} and v_k are required to compute v_{k+1}. This highly efficient method is the basic ingredient of many Krylov subspace methods for symmetric matrices, among them CG, MINRES and SYMMLQ. Symmetry of the system matrix (even positive definiteness in case of CG) turns out to be a strong restriction in the case of generalized saddle point systems (1.1)–(1.2). First of all, $\mathcal{A}$ needs to be symmetric, which requires $A = A^T$, $C = C^T$, and $B_1 = B_2$. These conditions are satisfied in many applications: see Section 2. However, if the system is preconditioned, the preconditioned system matrix needs to be symmetric (positive definite) as well. This in general requires that the preconditioner $\mathcal{P}$ be symmetric positive definite. For example, consider left preconditioning by $\mathcal{P}$, *i.e.*, the preconditioned system

$$\mathcal{P}^{-1} \mathcal{A} u = \mathcal{P}^{-1} b.$$

If $\mathcal{P}$ is symmetric positive definite, its Cholesky decomposition exists, $\mathcal{P} = \mathcal{L}\mathcal{L}^T$, and hence the preconditioned system is equivalent to

$$(\mathcal{L}^{-1} \mathcal{A} \mathcal{L}^{-T})\,(\mathcal{L}^T u) = \mathcal{L}^{-1} b.$$

The matrix $\mathcal{L}^{-1}\mathcal{A}\mathcal{L}^{-T}$ is again symmetric (positive definite), and one can apply the same method to solve the unpreconditioned as well as the preconditioned system. If $\mathcal{P}$ is symmetric but *not* positive definite, then the preconditioned system matrix $\mathcal{P}^{-1}\mathcal{A}$ is in general nonsymmetric, regardless of $\mathcal{A}$ being symmetric or not (unless $\mathcal{P}$ and $\mathcal{A}$ commute, *e.g.*, when $\mathcal{P}$ is a polynomial in $\mathcal{A}$). If no good symmetric positive definite preconditioner is known, or if a very good nonsymmetric preconditioner is available, the possible advantage of $\mathcal{A}$ being symmetric is lost, and one usually must use a solution method for nonsymmetric matrices. See Simoncini (2004*a*) and Section 10 below.

For a general, *nonsymmetric* matrix $\mathcal{A}$, the two goals mentioned above cannot be achieved at once (with few exceptions possessing a particular structure; see Faber and Manteuffel (1984) and Liesen and Saylor (2005)). Here one has to choose between a full recurrence and an orthogonal Krylov subspace basis, or a short-term recurrence and a nonorthogonal basis. The former approach, implemented by the Arnoldi method (Arnoldi 1951), is

Table 9.1. Summary of Krylov subspace methods discussed in Section 9.

Method	Required $\mathcal{A}$	Type	Recurrence	Required $\mathcal{P}$
CG	symm. def.	optimal	three-term	symm. def.
MINRES, SYMMLQ	symm.	optimal	three-term	symm. def.
SQMR	symm.	non-optimal	three-term	symm.
GMRES	general	optimal	full	general
QMR, BiCGStab, TFQMR	general	non-optimal	three-term	general

used in the GMRES algorithm. The latter approach is implemented in the nonsymmetric (or two-sided) Lanczos method (Lanczos 1950), which forms a main ingredient of methods like BiCG, BiCGStab, and QMR. Methods based on orthogonal bases are sometimes called optimal methods, while the other class is referred to as non-optimal methods. The non-optimal methods also include truncated and restarted versions of the optimal methods. Recent research shows that when an optimal method such as GMRES converges quickly, the related non-optimal methods based on (9.1)–(9.2) often also converge quickly (Simoncini and Szyld 2005). The nonsymmetric Lanczos method is also used in SQMR, which represents a non-optimal method for symmetric indefinite matrices. However, unlike the optimal methods for this class of problems discussed above (in particular MINRES and SYMMLQ), the SQMR method does not require a symmetric positive definite preconditioner, since its underlying Lanczos method works for nonsymmetric matrices as well. A summary of the methods discussed in this section is given in Table 9.1.

10. Preconditioners

The use of preconditioning has been already referred to several times in previous sections. As we saw, preconditioning may be used in the iterative solution of reduced systems arising from Schur complement reduction or from the application of various null space methods. In this section we discuss preconditioners in greater detail, and provide a description of some of the most widely used or promising techniques. In view of the fact that preconditioning has been and remains a most active area of research, accounting for the vast majority of papers on the numerical solution of saddle point problems (more generally, on linear solvers) in the last several years, a completely exhaustive survey is impossible. In some cases we limit ourselves to very brief notes and to pointers to the literature.

As is well known, the term *preconditioning* refers to transforming the linear system $\mathcal{A}u = b$ into another system with more favourable properties for iterative solution. A *preconditioner* is a matrix $\mathcal{P}$ (or $\mathcal{P}^{-1}$) that effects such a transformation. Generally speaking, preconditioning attempts to improve the spectral properties of the system matrix. For symmetric problems, the (worst-case) rate of convergence of Krylov subspace methods like CG or MINRES depends on the distribution of the eigenvalues of $\mathcal{A}$. Ideally, the preconditioned matrix $\mathcal{M} = \mathcal{P}^{-1}\mathcal{A}$ (or $\mathcal{M} = \mathcal{A}\mathcal{P}^{-1}$) will have a smaller spectral condition number, and/or eigenvalues clustered around 1. Another favourable situation is when the preconditioned matrix has a minimum polynomial of small degree.

For nonsymmetric (nonnormal) problems the situation is more complicated, and the eigenvalues may not describe the convergence of nonsymmetric matrix iterations like GMRES; see the discussion in Section 9. Nevertheless, a clustered spectrum (away from 0) often results in rapid convergence, especially if the departure from normality of the preconditioned matrix is not too high.

Generally speaking, there are two approaches to constructing preconditioners. One is based on purely algebraic techniques, like incomplete factorizations, sparse approximate inverses, and algebraic multilevel methods. These preconditioners require little knowledge of the problem at hand besides the entries of $\mathcal{A}$, and can be applied – at least in principle – in a more or less black-box fashion. This type of preconditioning has proved quite effective in the solution of linear systems arising from the discretization of *scalar* partial differential equations of elliptic type, and is widely used in many areas of computational science and engineering; see Benzi (2002), Meurant (1999), Saad (2003), van der Vorst (2003) for recent treatments. When applied to saddle point systems, on the other hand, standard algebraic preconditioners are often found to perform poorly. Because of the indefiniteness and lack of diagonal dominance, these preconditioners are often unstable. Even when the computation of the preconditioner does not suffer from some type of breakdown (*e.g.*, zero pivots in an incomplete factorization), the quality of the resulting preconditioner is often not very satisfactory, and slow convergence is observed. Also, because of the absence of decay in $\mathcal{A}^{-1}$, it is difficult to construct good sparse approximate inverse preconditioners for saddle point matrices.

The second approach develops preconditioners that are tailored to the particular application at hand. This approach requires knowledge of the origin of the problem, including (for PDEs) details about the discretization used, the underlying geometry, properties of the coefficients, and so forth. Of course, the more information one can use, the better the quality of the resulting preconditioner. The drawback of this approach is that the range of problems that can be treated with a particular preconditioner will necessarily be narrow, but this may not be a problem from the user's viewpoint.

For saddle point problems, the construction of high-quality preconditioners necessitates exploiting the block structure of the problem, together with detailed knowledge about the origin and structure of the various blocks. Because the latter varies greatly from application to application, there is no such thing as the 'best' preconditioner for saddle point problems. The choice of a preconditioner is strongly problem-dependent. For instance, techniques that give excellent results for the time-dependent Stokes problem may be completely inadequate for the steady-state case, or with the Oseen equations. Preconditioners that have been successfully used in optimization may be useless in fluid dynamics, and conversely. The good news is that powerful preconditioning techniques have been developed for many problems of practical interest.

We review several such techniques below. We begin with block preconditioners, especially popular in fluid dynamics, and continue with constraint preconditioners, especially popular in optimization. As it turns out, the first class of preconditioners has close ties with Schur complement reduction, while the second is related to the null space approach. Next we describe a more recent, promising class of methods based on the Hermitian and skew-Hermitian splitting of $\hat{A}$ in (3.10). This approach is rather general and has already been applied to a fairly wide range of problems, from fluid dynamics to weighted least squares. We conclude this section with a brief discussion of recent attempts to develop reliable incomplete factorization techniques for symmetric indefinite systems. It is also worth noting that both the stationary iterative methods described in Section 8 and the multilevel methods described in Section 11 can be accelerated by (used as preconditioners for) Krylov subspace methods.

The survey of preconditioners in this section is by no means exhaustive; see Axelsson and Neytcheva (2003) and Zulehner (2002) for other overviews of preconditioning techniques for saddle point problems.

10.1. Block preconditioners

In this section we consider block diagonal and block triangular preconditioners for Krylov subspace methods applied to the coupled system $\mathcal{A}\,u = b$. As we shall see, the performance of such preconditioners depends on whether fast, approximate solvers for linear systems involving A and the Schur complement S are available. Therefore, the preconditioners considered in this section are related to the 'segregated' solvers considered in Section 5.

10.1.1. Block diagonal preconditioners

Here we consider block diagonal preconditioning for the case of an invertible but possibly nonsymmetric saddle point matrix $\mathcal{A}$ with $C = O$. Then the basic block diagonal preconditioner is given by

$$\mathcal{P}_d = \begin{bmatrix} A & O \\ O & -S \end{bmatrix}, \tag{10.1}$$

where $S = -B_2 A^{-1} B_1^T$ is the Schur complement. Left preconditioning of $\mathcal{A}$ with $\mathcal{P}_d$ results in the matrix

$$\mathcal{M} = \mathcal{P}_d^{-1}\mathcal{A} = \begin{bmatrix} I & A^{-1}B_1^T \\ -S^{-1}B_2 & O \end{bmatrix}. \tag{10.2}$$

The matrix $\mathcal{M}$ is nonsingular by assumption and, as pointed out in Kuznetsov (1995) and Murphy, Golub and Wathen (2000), it satisfies

$$(\mathcal{M} - I)\left(\mathcal{M} - \frac{1}{2}(1 + \sqrt{5})I\right)\left(\mathcal{M} - \frac{1}{2}(1 - \sqrt{5})I\right) = O.$$

Hence $\mathcal{M}$ is diagonalizable and has only three distinct eigenvalues, namely $1, \frac{1}{2}(1 + \sqrt{5})$, and $\frac{1}{2}(1 - \sqrt{5})$ (see de Sturler and Liesen (2005) for the complete form of the eigendecomposition of $\mathcal{M}$). Hence, for each initial residual r_0, $\dim \mathcal{K}_{n+m}(\mathcal{M}, r_0) \leq 3$, which means that GMRES applied to the preconditioned system with matrix $\mathcal{M}$ will terminate after at most 3 steps. The same can be shown for right preconditioning with $\mathcal{P}_d$, or any centred preconditioning of the form $\mathcal{P}_1^{-1}\mathcal{A}\mathcal{P}_2^{-1}$ with $\mathcal{P}_1\mathcal{P}_2 = \mathcal{P}_d$. Furthermore, these results generalize to the case of a nonzero (2,2) block C of the matrix $\mathcal{A}$ (Ipsen 2001).

At first sight this looks promising. However, a simple calculation using the formula (3.4) for the inverse of $\mathcal{A}$ shows that

$$\mathcal{A}^{-1} = \left(\mathcal{M} + \begin{bmatrix} A^{-1}B_1^T S^{-1}B_2 & O \\ O & -I \end{bmatrix}\right)\mathcal{P}_d^{-1}.$$

We see that forming the preconditioned system $\mathcal{M}u = \mathcal{P}_d^{-1}b$ out of the given saddle point system $\mathcal{A}u = b$ using the block diagonal preconditioner (10.1) is essentially as expensive as computing the inverse of $\mathcal{A}$ directly using (3.4). In practice, the exact preconditioner (10.1) needs to be replaced by an approximation,

$$\hat{\mathcal{P}}_d = \begin{bmatrix} \hat{A} & O \\ O & -\hat{S} \end{bmatrix}, \tag{10.3}$$

where both $\hat{A}$ and $\hat{S}$ are approximations of A and S, respectively.

Several different approximations have been considered in the literature. Examples from a few specific applications are given in Section 10.1.3. Here we describe a fairly general framework developed in de Sturler and Liesen (2005). There the preconditioner (10.3) is obtained by considering a splitting of A into

$$A = D - E,$$

where D is invertible. Then $\hat{A} = D$ and $\hat{S} = -B_2 D^{-1} B_1^T$ are chosen, and bounds for the eigenvalues of the resulting preconditioned matrix

$$\hat{\mathcal{M}} = \hat{\mathcal{P}}_d^{-1} \mathcal{A} = \begin{bmatrix} I - D^{-1}E & D^{-1}B_1^T \\ -\hat{S}^{-1}B_2 & O \end{bmatrix} = \begin{bmatrix} I & D^{-1}B_1^T \\ -\hat{S}^{-1}B_2 & O \end{bmatrix} - \begin{bmatrix} D^{-1}E & O \\ O & O \end{bmatrix}$$
(10.4)

are given in terms of the eigenvalues of $\mathcal{M} = \mathcal{P}_d^{-1} \mathcal{A}$. These bounds show that the distance of the eigenvalues of $\hat{\mathcal{M}}$ to the three distinct eigenvalues of $\mathcal{M}$ depends on several factors, including the norm of the matrix $D^{-1}E$. In particular, when A is diagonally dominant, and $D = \text{diag}(A)$, then a good clustering of the eigenvalues of $\hat{\mathcal{M}}$ around 1, $\frac{1}{2}(1+\sqrt{5})$, and $\frac{1}{2}(1-\sqrt{5})$ can be expected. However, the preconditioned matrix $\hat{\mathcal{M}}$ still has (now potentially $n+m$ distinct) eigenvalues that may be on both sides of the imaginary axis.

To overcome this drawback, de Sturler and Liesen (2005) propose combining the approximate block diagonal preconditioner (10.3) with an Uzawa-type fixed point iteration (see Section 8.1 above) that is based on a splitting of $\hat{\mathcal{M}}$ as in the rightmost expression in (10.4) into $\hat{\mathcal{M}} = \mathcal{P}_1 - \mathcal{P}_2$. The inverse of $\mathcal{P}_1$ is known explicitly, and if D is simple enough (e.g., diagonal), it can be efficiently computed. Then the preconditioned system $\hat{\mathcal{M}}u = \hat{\mathcal{P}}_d^{-1}b = \hat{b}$ can be written as

$$(\mathcal{P}_1 - \mathcal{P}_2)u = \hat{b} \quad \Leftrightarrow \quad u = \mathcal{P}_1^{-1}\mathcal{P}_2 u + \mathcal{P}_1^{-1}\hat{b},$$

which yields the fixed point iteration

$$u_{k+1} = \mathcal{P}_1^{-1}\mathcal{P}_2 u_k + \mathcal{P}_1^{-1}\hat{b},$$
(10.5)

$$\text{where} \quad \mathcal{P}_1^{-1}\mathcal{P}_2 = \begin{bmatrix} (I + D^{-1}B_1^T \hat{S}^{-1}B_2)D^{-1}E & O \\ -\hat{S}^{-1}B_2 D^{-1}E & O \end{bmatrix}.$$

Because of the form of the iteration matrix $\mathcal{P}_1^{-1}\mathcal{P}_2$, it is easy to see that this fixed point iteration depends only on the first n components of u, i.e., it is sufficient to consider the iteration

$$x_{k+1} = (I + D^{-1}B_1^T \hat{S}^{-1}B_2)D^{-1}E x_k + [\mathcal{P}_1^{-1}\hat{b}]_{1:n},$$

where $[\mathcal{P}_1^{-1}\hat{b}]_{1:n}$ denotes the first n components of the vector $\mathcal{P}_1^{-1}\hat{b}$. Hence this approach can be considered a segregated solution method, in which (an approximation to) x_* is computed first, and is then used to compute (an approximation to) y_*. The *reduced system* of order n here is given by

$$\left(I - (I + D^{-1}B_1^T \hat{S}^{-1}B_2)D^{-1}E\right) x = [\mathcal{P}_1^{-1}\hat{b}]_{1:n}.$$
(10.6)

Obviously, this solution technique reduces the number of unknowns compared to the preconditioned system with the matrix $\hat{\mathcal{M}}$ from $n+m$ to n. This might result in a significant saving of computational resources, especially when m is relatively large. Moreover, and possibly even more importantly, the eigenvalues of the system matrix in (10.6) tend to be clustered around 1, with tighter clustering corresponding to a smaller norm of the matrix $D^{-1}E$. In particular, the spectrum has only one instead of the three clusters of $\hat{\mathcal{M}}$; see de Sturler and Liesen (2005). Because of this, the performance of a Krylov subspace method such as GMRES applied to (10.6) is typically better than the performance for the block diagonally preconditioned system with the matrix $\hat{\mathcal{M}}$. It is important to note that the approach through (10.6) is closely related to (left) *constraint preconditioning* described below in Section 10.2. For simplicity, consider the case with $B_1 = B_2 = B$, and the constraint preconditioner $\mathcal{P}_c$ in (10.15) with $G = D$ resulting from the splitting $A = D - E$. A straightforward calculation shows that the system matrix in (10.6) is exactly the same as the (1,1) block of the preconditioned matrix $\mathcal{M}$ in (10.17). Also note that the block diagonal preconditioner $\hat{\mathcal{P}}_d$ in this case is equal to the block diagonal matrix in the factorization of the constraint preconditioner $\mathcal{P}_c$ in (10.15). Further details about the relation between these two approaches can be found in de Sturler and Liesen (2005).

Next, consider the case of a symmetric saddle point matrix $\mathcal{A}$ with A positive definite and $C = O$. Then Fischer *et al.* (1998) consider block diagonal preconditioners of the form

$$\hat{\mathcal{P}}_d^\pm = \begin{bmatrix} \eta^{-1}A & O \\ O & \pm\hat{S} \end{bmatrix}, \tag{10.7}$$

where $\eta > 0$ is a scaling parameter, and $\hat{S}$ is a symmetric positive definite approximation of the (negative) Schur complement. Note that $\hat{\mathcal{P}}_d^+$ is positive definite, while $\hat{\mathcal{P}}_d^-$ is indefinite. Since both A and $\hat{S}$ are symmetric positive definite, their Cholesky decompositions $A = LL^T$ and $\hat{S} = MM^T$ can be used to transform the system $\mathcal{A}u = b$ into the equivalent system

$$\begin{bmatrix} \eta I & (M^{-1}BL^{-T})^T \\ \pm M^{-1}BL^{-T} & O \end{bmatrix} \begin{bmatrix} \eta^{-1}L^T x \\ M^T y \end{bmatrix} = \begin{bmatrix} L^{-1}f \\ (\eta M)^{-1}g \end{bmatrix}. \tag{10.8}$$

Clearly, the system matrix in (10.8) is of the same form as $\mathcal{A}_\eta^\pm$ in (3.13), and hence its eigenvalues depending on the choice of $+$ or $-$ and of η are characterized in Theorem 3.8. There it is shown that the eigenvalue distributions vary greatly from each other. In addition, the positive definite preconditioner $\hat{\mathcal{P}}_d^+$ yields a symmetric system matrix $\mathcal{A}_\eta^+$ in (10.8), while the indefinite preconditioner $\hat{\mathcal{P}}_d^-$ leads to the nonsymmetric matrix $\mathcal{A}_\eta^-$. For a fixed η it is therefore interesting to understand which choice yields the better performance of a Krylov subspace method applied to the preconditioned system (10.8). Curiously, there may be no difference at all. In fact, assuming that $g = 0$ in the saddle point system, Fischer *et al.* (1998) show that the preconditioned matrices $\mathcal{A}_\eta^\pm$ in (10.8) both generate the same Krylov subspace iterates u_k when special initial residuals are used. In this case the

residual norms of MINRES applied to the symmetric system coincide with
the residual norms of GMRES applied to the nonsymmetric system in every
step of the iteration. This exact equivalence is lost when random initial
residuals are used, but as is shown numerically in Fischer *et al.* (1998), the
convergence curves are very close in such cases as well. Since the MINRES
method is based on three-term recurrences due to symmetry of the system
matrix (*cf.* Section 9), the positive definite preconditioner clearly represents
the superior strategy.

Block diagonal preconditioners for saddle point problems arising from
(stabilized) discretized Stokes problems (2.10)–(2.12) have been studied, for
example, in Silvester and Wathen (1994) and Wathen and Silvester (1993).
The system matrix of the discrete problem is a generalized saddle point
matrix of the form

$$\mathcal{A} = \begin{bmatrix} A & B^T \\ B & -\beta C \end{bmatrix}, \tag{10.9}$$

where A is block diagonal and symmetric positive definite, C is symmetric
positive semidefinite, and $\beta > 0$ is a stabilization parameter. Silvester and
Wathen (Silvester and Wathen 1994, Wathen and Silvester 1993) provide
several results on the eigenvalue distribution of the preconditioned matrix
$\hat{\mathcal{P}}_d^{-1}\mathcal{A}$ for different symmetric positive definite preconditioners $\hat{\mathcal{P}}_d$ of the
form (10.3).

For example, consider the case $\hat{A} = \operatorname{diag}(A)$, and $-\hat{S} = \beta\operatorname{diag}(C)$ if
$C \neq O$ or $-\hat{S} = \beta h^d I$ if $C = O$ (here d is the spatial dimension of the
Stokes problem, and h is the mesh size parameter), *i.e.*, a positive definite
diagonal preconditioner $\hat{\mathcal{P}}_d$. Then it can be shown that the eigenvalues of
the symmetric indefinite preconditioned matrix $\hat{\mathcal{P}}_d^{-1}\mathcal{A}$ are contained in the
union of two real intervals of the form

$$(-a, -bh) \cup (ch^2, d), \tag{10.10}$$

where a, b, c, d are positive constants that are independent of h (Wathen
and Silvester 1993, Theorems 1 and 2). As discussed in Section 9, the
(worst-case) convergence of MINRES applied to the preconditioned system
depends on the eigenvalue distribution of $\hat{\mathcal{P}}_d^{-1}\mathcal{A}$, and an upper convergence
bound can be found by using (the smallest) two intervals containing the
eigenvalues. If these intervals are of the same length, then such a convergence bound is given by (9.9)–(9.11). But in the case of (10.10), the two
intervals are in fact not of the same length, since the negative eigenvalues
spread out less rapidly under mesh refinement than the positive eigenvalues.
Extending the two intervals in (10.10) to make their lengths equal, say by
replacing (10.10) with $(-d, -ch^2) \cup (ch^2, d)$ (where we assume, for simplicity, $d > a$ and $b > c$), would predict an asymptotic convergence rate of
$O(1 - h^2 c/d)$ for MINRES. However, with some careful analysis it can be

shown that the asymptotic convergence factor of (10.10) is actually of order $O(1 - h^{3/2}\sqrt{bc/ad})$; see Wathen *et al.* (1995) for details. More sophistic-ated positive definite block diagonal preconditioners in this context leading to even better asymptotic convergence rates are studied in Silvester and Wathen (1994); see also Section 10.1.3.

Apart from the work in Silvester and Wathen (1994) and Wathen and Silvester (1993), block diagonal preconditioners have been studied in the context of numerous additional applications. The upshot is that for suitably chosen approximations to A and S, block diagonal preconditioners result-ing in mesh-independent rates of convergence exist for mixed finite element formulations of many problems. For example, an analysis of the eigen-value distribution of block diagonally preconditioned saddle point systems arising from mixed finite element discretizations of magnetostatics prob-lems is given in Perugia and Simoncini (2000, Section 4). Other analyses of block diagonal preconditioners in specific applications can be found in Battermann and Heinkenschloss (1998), Chizhonkov (2001, 2002), Klawonn (1998*a*), Krzyżanowski (2001), Kuznetsov (1995, 2004), Lukšan and Vlček (1998), Lyche *et al.* (2002), Mardal and Winther (2004), Pavarino (1997, 1998), Peters *et al.* (2004), Powell and Silvester (2004), Toh, Phoon and Chan (2004) and Vassilevski and Lazarov (1996).

10.1.2. *Block triangular preconditioners*
Block triangular preconditioners of the form

$$\mathcal{P}_t = \begin{bmatrix} A & B^T \\ O & S \end{bmatrix}$$

were first considered in Bramble and Pasciak (1988) and extensively de-veloped in the last few years. Note that Uzawa's method may also be re-garded as a block (lower) triangular preconditioner; *cf.* (8.2). This class of preconditioners includes some of the most effective solvers for saddle point problems, both symmetric and nonsymmetric. Note that the first block row of $\mathcal{P}_t$ coincides with the first block row of $\mathcal{A}$; hence, for the initial guess $u_0 = 0$, the solution of the initial preconditioning equation $\mathcal{P}_t z_0 = r_0 = b$ must in particular satisfy the first of the two equations in the saddle point system (1.6). In Section 10.2 we will examine preconditioning strategies that require satisfying the *second* equation in the saddle point system (constraint preconditioning). In the setting of fluid flow problems, the first approach re-quires the preconditioner to respect the conservation of momentum, whereas the second one imposes a mass balance condition.

Recalling the block factorization (3.3), we immediately see that the spec-trum of $\mathcal{P}_t^{-1}\mathcal{A}$ is $\sigma(\mathcal{P}_t^{-1}\mathcal{A}) = \{1\}$ and moreover, the preconditioned matrix has minimum polynomial of degree 2, so that a method like GMRES would converge in at most two steps. In practice, of course, approximations $\hat{A}$ and

$\hat{S}$ to A and S have to be used. The computation of $z_k = \mathcal{P}_t^{-1} r_k$ at each step of a nonsymmetric Krylov subspace method can be implemented on the basis of the factorization

$$\mathcal{P}_t^{-1} = \begin{bmatrix} \hat{A}^{-1} & O \\ O & I \end{bmatrix} \begin{bmatrix} I & B^T \\ O & -I \end{bmatrix} \begin{bmatrix} I & O \\ O & -\hat{S}^{-1} \end{bmatrix}. \tag{10.11}$$

Note that the cost of applying the preconditioner is only slightly higher than in the block diagonal case: besides the solves with $\hat{A}$ and $\hat{S}$, there is an additional multiplication by B^T. Once again, the choice of the approximations $\hat{A}$ and $\hat{S}$ is problem-dependent; see Section 10.1.3. Generally speaking, the better the approximations, the faster the convergence. However, since the preconditioned matrix $\mathcal{P}_t^{-1} \mathcal{A}$ (or $\mathcal{A} \mathcal{P}_t^{-1}$) is nonnormal, convergence estimates are not easy to obtain. Field of values analysis can be used in some instances to obtain eigenvalue bounds and convergence estimates for GMRES. These analyses and numerical experiments indicate that h-independent convergence rates can be achieved for some important problems, including the Oseen equations; see Klawonn and Starke (1999), Loghin and Wathen (2004) and the discussion in the next section. Other analyses of block triangular preconditioners for specific applications can be found in Battermann and Heinkenschloss (1998), Cao (2004a), Elman, Silvester and Wathen (2002a), Kanschat (2003), Klawonn (1998b), Krzyżanowski (2001) and Pavarino (1998); see also Bramble and Pasciak (1988), Simoncini (2004a) and Zulehner (2002) for analyses of the inexact case. An apparent disadvantage in the symmetric case is that block triangular preconditioning destroys symmetry. A symmetrized version of block triangular preconditioning has been proposed in Bramble and Pasciak (1988); see also Axelsson and Neytcheva (2003), D'Yakonov (1987) and Zulehner (2002). However, symmetrization is seldom necessary in practice: if good approximations to A and S are available, a method like GMRES with block triangular preconditioning will converge quickly and the overhead incurred from the use of a nonsymmetric solver will be negligible.

We also mention preconditioners based on incomplete block triangular factorizations of the form

$$\mathcal{P} = \begin{bmatrix} \hat{A} & O \\ B & \hat{S} \end{bmatrix} \begin{bmatrix} I & \hat{A}^{-1} B^T \\ O & I \end{bmatrix} \approx \mathcal{A},$$

where $\hat{A}$ and $\hat{S}$ are easily invertible approximations of A and the Schur complement S. Note that application of this preconditioner requires two solves with $\hat{A}$ rather than one, in addition to the solve with $\hat{S}$. Multiplying out the factors, we obtain

$$\mathcal{P} = \begin{bmatrix} \hat{A} & B^T \\ B & -\hat{C} \end{bmatrix}, \qquad \hat{C} = -(B\hat{A}^{-1}B^T + \hat{S}).$$

Hence, generally speaking, $\mathcal{P}$ is an indefinite preconditioner of the type considered in Section 10.2. Note that $\mathcal{P}$ is symmetric if $\hat{A}$ and $\hat{S}$ are. Again, the key issue here is the choice of the approximations $\hat{A} \approx A$ and $\hat{S} \approx S$. Such incomplete block LU factorization preconditioners correspond to classical solution methods known in computational fluid dynamics as SIMPLE schemes (for 'Semi-Implicit Method for Pressure-Linked Equations'): see Patankar (1980) and Patankar and Spalding (1972). In the original SIMPLE scheme $\hat{A} = D_A$ and $\hat{S} = -BD_A^{-1}B^T$, respectively, where D_A denotes the main diagonal of A. Different choices of the approximations involved lead to different preconditioners. Variants of SIMPLE are also often used as smoothers for multigrid; see the discussion in Section 11. A spectral analysis of the SIMPLE preconditioner applied to the Oseen problem can be found in Li and Vuik (2004). See also Benzi and Liu (2005) for numerical experiments and comparisons with other preconditioners. These experiments confirm the intuition that SIMPLE preconditioning is effective when A is strongly diagonally dominant, as is the case for generalized Stokes and Oseen problems with sufficiently small time steps. On the other hand, SIMPLE is not competitive when the time steps are large or 'infinite' (steady case), for in this case the approximations to A and especially to S are poor. It should be mentioned that for unsteady problems classical (Chorin-style (Chorin 1968)) pressure-correction schemes are also very efficient; see Wesseling (2001, page 303). Further references on incomplete block factorizations for the treatment of time-dependent Navier–Stokes equations include Gauthier, Saleri and Quarteroni (2004), Perot (1993), Quarteroni, Saleri and Veneziani (2000), Saleri and Veneziani (2005) and Veneziani (2003).

10.1.3. Approximating A and S

As we have seen, many algorithms for solving saddle point systems depend on the availability of good approximations for the (1,1) block A and for the Schur complement S. Such approximations are needed in segregated approaches and in preconditioned Uzawa schemes where the Schur complement system (5.1) is solved by a preconditioned iteration, and in the construction of block diagonal and block triangular preconditioners for coupled Krylov iterations applied to $\mathcal{A}u = b$.

The construction of such approximations is a strongly problem-dependent process, and a large body of literature exists on this topic. The problem has been especially well-studied for mixed finite element formulations of elliptic PDEs, for the Stokes and Oseen problems, and for linear elasticity problems. Furthermore, some work has been done towards the construction of preconditioners for the Schur complement ('normal equations') systems arising from interior point methods in optimization.

In the case of mixed finite element approximations of elliptic boundary value problems and of Stokes- and Oseen-type equations, the submatrix A usually corresponds either to a zeroth-order differential operator (multiplication by a function or tensor K, as in (2.18)), or to a direct sum of second-order differential operators of diffusion or convection-diffusion type. When A represents multiplication by a function (a mass matrix), it can be well approximated by either a (scaled) identity matrix or by diagonal lumping of A (that is, the ith diagonal entry of $\hat{A}$ is equal to the sum of the entries in the ith row of A). If the coefficients in K are sufficiently smooth, the resulting approximation $\hat{A}$ is then spectrally equivalent to A, in the sense that the eigenvalues of $\hat{A}^{-1}A$ are positive and contained in an interval whose endpoints do not depend on the mesh size; see Rusten and Winther (1992, 1993) and Wathen (1987). When A represents a discrete (vector) diffusion or convection-diffusion operator, a spectrally equivalent approximation $\hat{A}$ can be obtained in many cases by performing one or more multigrid sweeps on the linear system $A\,x = v$. Thus, the approximation $\hat{A}$ of A is not constructed explicitly, but is defined implicitly by the action $\hat{A}^{-1}v$ on a given vector v. We refer the reader to Elman *et al.* (2005*c*) for details. Here we note that for the generalized (unsteady) Stokes and Oseen problems the matrix A is usually well-conditioned because of the presence of an additional term inversely proportional to the time step Δt. This makes the (1,1) block strongly diagonally dominant, and a single multigrid V-cycle with an appropriate Gauss–Seidel smoother is often sufficient: see Mardal and Winther (2004) and Turek (1999).

When A does not arise from the discretization of a differential operator, approximate solves, if necessary, may be obtained by either incomplete factorizations or by a few iterations of a Krylov subspace method, possibly with an appropriate preconditioner.

The approximation $\hat{S}$ of the Schur complement matrix S is usually more delicate. In the context of saddle point systems arising from the discretization of PDEs, good approximations to the Schur complement are critical to the performance of block preconditioners. In many cases, excellent approximations can be constructed by thinking in terms of the underlying differential operators. We recommend Loghin and Wathen (2003) for a highly readable discussion of this problem in terms of pseudodifferential operators.

For mixed finite element discretizations of elliptic problems of the type (2.18)–(2.20), the Schur complement $S = -BA^{-1}B^T$ can be interpreted as a discretization of the second-order diffusion operator $\nabla K^{-1} \cdot \nabla$ acting on the unknown function corresponding to the y variable (*e.g.*, on the pressure for incompressible flow problems). In this case, a number of approximations $\hat{S}$ are possible. Provided that K is sufficiently well-behaved, the action of S^{-1} can be approximated by an iteration of multigrid. For higher-order methods, a low-order discretization of the same diffusion operator may also be expected to give a very good approximation. The literature in this area is vast; see, *e.g.*, Kanschat (2003), Kuznetsov (1995, 2004), Loghin and Wathen (2003), Martikainen (2003), Moulton, Morel and Ascher (1998), Perugia and Simoncini (2000), Powell and Silvester (2004), Rusten, Vassilevski and Winther (1996), Rusten and Winther (1992, 1993) and Warsa, Benzi, Wareing and Morel (2004).

For LBB-stable discretizations of the linear elasticity and steady-state Stokes problem, the Schur complement is spectrally equivalent to a mass matrix (Verfürth 1984*a*). This is not surprising if one thinks of S as a discrete counterpart of a pseudodifferential operator of the form $\operatorname{div} \Delta^{-1} \operatorname{grad}$ and keeping in mind the identity $\Delta = \operatorname{div} \operatorname{grad}$. Hence, under appropriate discretization the Schur complement has condition number bounded independently of mesh size and can be approximated effectively and cheaply in a number of ways; see the previous discussion on approximating zeroth-order differential operators, as well as the references Elman (1996), Elman *et al.* (2005*c*), Klawonn (1998*a*, 1998*b*), Loghin and Wathen (2003, 2004), Pavarino (1997), Silvester and Wathen (1994), Wathen and Silvester (1993) and Yang (2002).

For mixed finite element discretizations of the generalized Stokes problem that arises from the implicit treatment of the time-dependent Stokes problem, on the other hand, the Schur complement is of the form $S = -B(A + \beta I)^{-1}B^T$, where $\beta > 0$ is inversely proportional to the time step. In the 2D case, A is the block diagonal matrix

$$A = \begin{bmatrix} \nu H & O \\ O & \nu H \end{bmatrix},$$

where H is a discrete Laplace operator and $\nu > 0$ denotes the viscosity. In the 3D case, there will be three blocks along the diagonal instead of two. Clearly, for very large time steps (β small) the matrix $S = -B(A + \beta I)^{-1}B^T$ is close to the Schur complement of the steady Stokes problem and is well-conditioned independent of mesh size. In this case a scaled pressure mass matrix will be a good approximation. On the other hand, for small step sizes (β large) this operator has a condition number that grows as $h \to 0$. Note that in the limit as $\beta \to \infty$, the Schur complement behaves essentially as a discrete diffusion operator. The same is true for the case of small viscosity ν (high Reynolds numbers Re), regardless of the size of the time step, for in this case the matrix A, which contains terms proportional to $\nu = \mathrm{Re}^{-1}$, will be negligible. In these cases the SIMPLE approximation for the Schur complement can be quite good for sufficiently small β. Robust preconditioners that work well for small as well as large time steps and over a wide range of Reynolds numbers were first proposed and analysed in Cahouet and Chabard (1988); see also Bramble and Pasciak (1997) and Kobelkov and Olshanskii (2000). The idea is to build preconditioners that are discrete counterparts of the pseudodifferential operator L implicitly defined by

$$L^{-1} = \nu I - \beta \Delta^{-1},$$

where I denotes the identity operator and Δ is a pressure Laplacian. Here Δ^{-1} stands for the data-to-solution mapping for a pressure-Poisson equation with suitable boundary conditions. Thus, the action of the approximate Schur complement $\hat{S}$ on a vector v consists of

$$\hat{S}^{-1} v = \nu M_p v - \beta (B M_u^{-1} B^T)^{-1} v,$$

where M_p is a pressure mass matrix and M_u a velocity mass matrix. Note that $B M_u^{-1} B^T$ is simply the so-called *compatible discretization* of the Laplacian (Cahouet and Chabard 1988). Again, inexact solves can be used to approximate $(B M_u^{-1} B^T)^{-1} v$. For further details see Turek (1999) and the recent paper by Mardal and Winther (2004).

The situation is more complicated for the nonsymmetric saddle point problems arising from discretizations of the (steady and unsteady) Oseen equations. For steady problems with large viscosity (small Reynolds numbers), using a pressure mass matrix to approximate the Schur complement is usually sufficient and results in rates of convergence independent of mesh size when block diagonal or block triangular preconditioners are used. However, the rate of convergence tends to deteriorate quickly as the viscosity gets smaller. The problem is that as $\nu \to 0$, the exact Schur complement becomes increasingly nonsymmetric and cannot be well approximated by a (necessarily symmetric) mass matrix; see, *e.g.*, Cao (2004a), Elman and Silvester (1996) and Krzyżanowski (2001). Much progress has been made in the last few years towards developing robust preconditioners of the block triangular type for the Oseen equations; see Elman (1999, 2002), Elman, Howle, Shadid, Shuttleworth and Tuminaro (2005a), Elman, Howle, Shadid and Tuminaro (2003a), Elman, Silvester and Wathen (2002a, 2002b, 2005c), Kay, Loghin and Wathen (2002), Loghin and Wathen (2002) and Silvester, Elman, Kay and Wathen (2001). Out of this effort, two especially interesting techniques have emerged. The first idea is to approximate the inverse of the Schur complement matrix $S = B A^{-1} B^T$ with

$$\hat{S}^{-1} = (BB^T)^{-1} B A B^T (BB^T)^{-1} \qquad (10.12)$$

(here and in the remainder of this section we dispense with the negative sign for ease of notation). This approximation, originally proposed by Elman (1999) (see also Saleri and Veneziani (2005)), has been given various justifications. A simple one is to observe that if B were square and invertible, then the inverse of $B A^{-1} B^T$ would be $B^{-T} A B^{-1}$. However, B is rectangular. Therefore it makes sense to replace the inverse of B with the Moore–Penrose pseudo-inverse $B^\dagger$; see Campbell and Meyer Jr. (1979).

If we assume B to have full row rank, the pseudo-inverse of B is given by $B^\dagger = B^T(BB^T)^{-1}$, thus justifying the approximation (10.12). Now the approximate Schur complement is (strongly) nonsymmetric if A is. Note that evaluating the action of $\hat{S}^{-1}$ on a given vector v requires the solution of two pressure-Poisson-type problems, plus matrix-vector multiplies involving B^T, A and B. It follows from (10.11) that the block triangular precondi-tioner can be applied by performing matrix-vector products involving B^T, A and B, and a few iterations of multigrid. For 3D problems, multigrid is needed to approximately solve three convection-diffusion problems for the three components of the velocity field (approximate inversion of A), and to solve the two Poisson-type problems approximately on the pressure space. Numerical experiments for simple model problems indicate that the per-formance of this preconditioner depends only mildly on the mesh size and viscosity coefficient; see Elman (1999). However, it was found to perform poorly (occasionally failing to converge) on some difficult problems; see Vainikko and Graham (2004). As noted in Elman, Howle, Shadid, Shuttle-worth and Tuminaro (2005 a), for finite element problems the performance of this preconditioner can be greatly improved by the use of appropriate scalings. Denoting by M_1 and M_2 suitable diagonal matrices, Elman $et\ al.$ (2005 a) derive the more accurate approximation

$$\hat{S}^{-1} = (BM_2^{-2}B^T)^{-1}BM_2^{-2}AM_1^{-1}B^T(BM_1^{-2}B^T)^{-1}. \tag{10.13}$$

The matrix M_1 is taken to be the diagonal of the velocity mass matrix M_u; taking $M_2 = M_1^{1/2}$ makes the two variable-coefficient discrete diffusion operators in (10.13) identical. Although the cost of the two variants (10.12) and (10.13) is almost the same, the improvement in performance obtained by the use of scaling can be quite dramatic; see the numerical experiments in Elman $et\ al.$ (2005 a).

Another effective approximation of the Schur complement for the Os-een problem was introduced in Kay, Loghin and Wathen (2002) and ana-lysed in Elman, Silvester and Wathen (2002 b). Again, several justifica-tions have been given for this approach, the original one being based on the Green's tensor of the Oseen operator on the continuous level. An-other elegant derivation in terms of pseudodifferential operators can be found in Loghin and Wathen (2003). A simpler, heuristic argument is to first consider the case where A and B^T are both square and commute: $AB^T = B^TA$. If A is nonsingular, its inverse must also commute with B^T, hence $A^{-1}B^T = B^TA^{-1}$. If we think of A as representing a second-order differential operator and B a first-order one then, apart from possible problems near the boundary, assuming commutativity is reasonable. It follows from these assumptions that $S^{-1} = (BA^{-1}B^T)^{-1} = A(BB^T)^{-1}$. In practice, however, A and B^T have different dimensions and represent operators acting on different function spaces. Note that BB^T is, again, a (scalar) pressure-Poisson-type operator; on the other hand, A is a (vector) convection-diffusion operator acting on the velocities. This suggests introducing a discrete convection-diffusion operator A_p acting on the pressures. Including the necessary mass matrix scalings, the resulting approximation to the inverse of the Schur complement is thus

$$\hat{S}^{-1} = M_p^{-1} A_p (BM_u^{-1}B^T)^{-1}. \qquad (10.14)$$

The block triangular preconditioner based on this approximate Schur complement has been extensively tested on a variety of steady and unsteady model problems, in both 2D and 3D; see, e.g., Kay, Loghin and Wathen (2002), Elman, Howle, Shadid and Tuminaro (2003a), Elman, Silvester and Wathen (2002a, 2002b), Hemmingsson-Fdrndén and Wathen (2001) and Loghin and Wathen (2002). This method, which is referred to as the *pressure convection-diffusion* preconditioner in Elman *et al.* (2005c), appears to be very robust with respect to mesh refinement, and fairly robust with respect to the viscosity, exhibiting only a mild deterioration as $\nu \to 0$ on a wide range of problems and discretizations. Note that this approximation is less expensive than the one based on (10.13), since computing the action of $\hat{S}^{-1}$ only requires solving a single pressure-Poisson problem and a matrix-vector product with A_p.

One drawback of this approach is the need to construct an additional operator, A_p, which is not required by the formulation of the Oseen problem but is needed only to form the preconditioner. This discrete convection-diffusion operator is not usually available in standard fluid dynamics codes. This has motivated new, fully algebraic approaches to compute approximations to A_p from the existing blocks A, B and B^T based on the notion of *sparse approximate commutators* (SPAC); see Elman *et al.* (2005a). The approximate A_p is the matrix that minimizes the cost functional

$$F(X) = \|AB^T - B^T X\|_F$$

(where $\| \cdot \|_F$ denotes the Frobenius norm) over a space of matrices with prescribed sparsity pattern. Note that the solution of the *unconstrained* minimization problem is $A_p = (BB^T)^{-1}BAB^T$, corresponding to Elman's approximation $\hat{S}^{-1} = (BB^T)^{-1}BAB^T(BB^T)^{-1}$.

The block triangular preconditioner based on the pressure convection-diffusion operator was also used by Syamsudhuha and Silvester (2003) with good results as a smoother for multigrid applied to the Oseen problem. It was also used by Elman, Loghin and Wathen (2003b) as a preconditioner for the linear systems arising from Newton's method applied to the Navier–Stokes equations. In this case the method was again found to result in h-independent convergence, but some growth in the iteration number was observed for small ν.

The pressure convection-diffusion operator, on the other hand, does not seem to be applicable to the rotation form (2.15)–(2.17) of the Oseen equations. Different approximations need to be used in this case; see Olshanskii (1999) and Section 10.3 below.

Approximations for the Schur complement matrices arising in interior point methods for constrained optimization have been studied by several authors. Here we give a brief discussion based on Frangioni and Gentile (2004). As already pointed out in Section 3.5, such linear systems can be extremely ill-conditioned. An interesting technique is the *tree preconditioner* described in Resende and Veiga (1993), based on ideas first put forth in Vaidya (1990). The approximation to $S = BA^{-1}B^T$ is of the type $\hat{S} = \hat{B}M\hat{B}^T$ where $\hat{B}$ and M are now both square matrices of order m. The actual construction of these matrices depends on the application. In network flow problems (the motivating application in Resende and Veiga (1993)), the constraint matrix B is the node–edge incidence matrix of a directed graph G, and $\hat{B}$ is defined as the node–edge incidence matrix of a spanning tree of G. Furthermore, A is diagonal and M is obtained by restricting A^{-1} to the arcs that comprise the tree. (Note that when $A = I$, the matrix $S = BB^T$ is closely related to the Laplacian matrix of the undirected version of G.) More precisely, $\hat{B}$ is chosen as the node–edge incidence matrix of an (approximate) maximum weight spanning tree, where the weight of each edge (i, j) is given by the corresponding entry of M. Such spanning trees can be constructed in $O(m)$ work using a variant of the classical Kruskal algorithm. The resulting linear systems involving $\hat{S}$ can be solved in $O(n)$ work by solving linear systems with matrices $\hat{B}^T$, M and $\hat{B}$. These linear systems can be solved by visiting the spanning tree, without fill-in. This preconditioner can be expected to be especially effective in the final stages of the interior point iteration, since in this case the weights m_{ij} tend to zero on all arcs, except on those corresponding to the basic optimal solution that form a spanning tree. The preconditioner is less effective in the initial stages of the interior point algorithm. This observation has motivated the use of hybrid strategies where a diagonal preconditioner is used initially and the tree preconditioner in later stages of the algorithm: see Resende and Veiga (1993) for details. Improvements of the tree preconditioner were proposed in Mehrotra and Wang (1996). A different strategy, based on an incomplete QR factorization, was proposed in Júdice, Patricio, Portugal, Resende and Veiga (2003), and found to be superior to the tree preconditioner for the special case of transportation problems.

New preconditioners based on combinatorial principles have recently been introduced by Frangioni and Gentile (2004). Motivated by the tree preconditioner idea, these authors construct approximations of S of the form $\hat{S} = \hat{B}M\hat{B}^T$, where now $\hat{B}$ corresponds to a *subgraph* of the graph G of B; this subgraph, which may contain (strictly) a spanning tree of G, should be such that linear systems associated with the matrix $\hat{S}$ can be easily solved. For example, if the chosen subgraph happens to be *chordal*, then there is an ordering of $\hat{S}$ such that no fill-in occurs when factoring $\hat{S}$; see Rose (1970). Fast algorithms exist to compute approximate maximum weight chordal subgraphs. We refer the reader to Frangioni and Gentile (2004) for further details and for numerical experiments showing that in many cases the new subgraph-based preconditioners improve on existing techniques in terms both of iteration counts and CPU time. An interesting open question is to what extent these ideas can be exploited to construct approximate Schur complements for saddle point problems arising from mixed finite element discretizations of PDE problems, where the matrix B is a discrete divergence operator.

We conclude this section with a brief discussion of purely algebraic methods that have been used to construct approximations $\hat{S}$ or $\hat{S}^{-1}$ to the Schur complement matrix or to its inverse. See also the general treatment in Axelsson (1994, Chapter 9) and the discussion in Siefert and de Sturler (2004, Section 4). For simplicity, we confine ourselves to the case where $B_1 = B_2$ and $C = O$. Extension to cases with $B_1 \neq B_2$ is usually straightforward; the case $C \neq O$, on the other hand, may be less obvious, especially when seeking explicit approximations to S^{-1}.

The simplest approximation is perhaps to set $\hat{S} = BD_A^{-1}B^T$ where D_A is either diagonal or possibly block diagonal with blocks of small size. Note that in this case $\hat{S}$ will be sparse if B is, except for special situations (*e.g.*, when B contains one or more dense columns). The diagonal approximation $D_A = \mathrm{diag}(A)$ is used, for instance, in the already-mentioned SIMPLE-type schemes popular in computational fluid dynamics. Better approximations can be obtained by means of incomplete factorizations of the form $A \approx LU \equiv \hat{A}$. Then $\hat{S} = XY$ where $X = BU^{-1}$ and $Y = L^{-1}B^T$. Matrices X and Y may be computed by solving two lower triangular linear systems involving U^T and L and with multiple right-hand sides given by the columns of B^T. *A posteriori* dropping of small entries may be necessary to keep X and Y sparse. Alternatively, one can use (explicit) sparse approximate inverse techniques directly applied to A. For example, the approximate inverse M_A may be computed as the minimizer of the matrix functional

$$F(X) = \|I - AX\|_F$$

over a prescribed set of sparse matrices. Many other approximate inverse techniques have been developed; see Benzi (2002) for an overview. Whatever the technique employed, the resulting sparse approximate inverse $M_A \approx A^{-1}$ can then be used to form an approximation $\hat{S} = BM_AB^T$. See Chow and Saad (1997a), Little and Saad (2003) and Little, Saad and Smoch (2003) for implementation details and for experimental comparisons of various strategies. These techniques may be expected to work well when the entries in A^{-1} decay rapidly away from the main diagonal. This can be expected to hold, for instance, for implicit time discretizations of time-dependent Stokes and Oseen problems. Provided that sufficiently small time steps are taken, the diagonal entries of A will be large compared to the off-diagonals and will result in fast decay in A^{-1}; see Demko, Moss and Smith (1984).

Other approximations that have been used in the literature for specific applications can be used, at least in principle, in more general settings. Juvigny (1997) proposed the simple approximation $(BA^{-1}B^T)^{-1} \approx BAB^T$ as an approximate inverse of the Schur complement for saddle point matrices arising from the use of Lagrange multipliers in domain decomposition schemes. In this case B is often a very simple matrix representing a restriction operator, and the approximation is not unreasonable. Although this approximation is not very accurate, it has the advantage of having zero construction cost; furthermore, computing the action of $\hat{S}^{-1}$ on a vector only requires matrix vector products with sparse matrices. See also Lyche *et al.* (2002) and Tyrtyshnikov and Vassilevski (2003) for more sophisticated approximations to the Schur complements arising in similar contexts.

Another possible approximation is the already-mentioned Elman preconditioner of the form

$$\hat{S}^{-1} = (BB^T)^{-1} BAB^T (BB^T)^{-1}.$$

Although it was developed specifically for fluid flow problems, this approximation should, at least in principle, be feasible whenever the solution of linear systems with coefficient matrix BB^T (or the equivalent least squares problems) can be computed efficiently. Note that accommodating a nonzero C in this and the previous case is not easy.

Yet another possibility is to approximate the action of S^{-1} on a vector v by performing a few steps of an iterative method on the Schur complement system $Sz = v$. Krylov subspace methods only require S in the form of matrix-vector products. These in turn necessitate multiplications with B and its transpose and solves with A. The latter operation may in turn be performed by an (inner) iterative process. If a variable number of inner iterations is performed, then a generalized conjugate gradient method (Axelsson and Vassilevski 1991) or a flexible solver like FGMRES (Saad 1993) should be used for the outer iteration. These inner solves need not be performed to high accuracy. Even if an accurate approximation $z \approx S^{-1}v$ is sought, the accuracy of the inner solves with A can be relaxed in the course of the outer iteration; see Simoncini and Szyld (2003). We further mention that matrix-vector products involving S and selected vectors are used in the *probing* technique to compute banded approximations to S; see, *e.g.*, Axelsson (1994, Chapter 8.5.1). Finally, Bomhof and van der Vorst (2000) have proposed a parallel direct-iterative approximate Schur complement algorithm for linear systems arising in circuit simulation.

10.2. Constraint and indefinite preconditioning

Constraint preconditioning is based on the principle that the preconditioning matrix should have the same 2×2 block structure as the original saddle point matrix. In other words, the saddle point matrix is preconditioned by another, 'easier-to-invert' saddle point matrix. Constraint preconditioning has been extensively used in the solution of saddle point systems arising from mixed finite element formulations of elliptic partial differential equations (Axelsson and Neytcheva 2003, Bank *et al.* 1990, Ewing, Lazarov, Lu and Vassilevski 1990, Mihajlović and Silvester 2004, Perugia and Simoncini 2000, Rozložník and Simoncini 2002, Tong and Sameh 1998). The method is also popular in the solution of saddle point ('KKT') systems in optimization (Bergamaschi *et al.* 2004, Dyn and Ferguson 1983, Freund and Nachtigal 1995, Gould, Hribar and Nocedal 2001, Keller, Gould and Wathen 2000, Lukšan and Vlček 1998, Shi 1995). Komzsik, Poschmann and Sharapov (1997) applied the preconditioner to problems arising in the finite element analysis of structures.

We describe constraint preconditioning in the case where $B_1 = B_2 = B$ has full row rank and $C = O$. The preconditioning matrix is then

$$\mathcal{P}_c = \begin{bmatrix} G & B^T \\ B & O \end{bmatrix}, \tag{10.15}$$

where G is some approximation of A, $G \neq A$. Of course, G should be chosen so that $\mathcal{P}_c$ is invertible and solving linear systems involving $\mathcal{P}_c$ is significantly easier than solving the original linear system $\mathcal{A} u = b$. The name *constraint preconditioning* comes from the fact that $\mathcal{P}_c$ is the coefficient matrix for a saddle point problem with the same constraints as the original problem. Moreover, the constraint preconditioner projects the problem onto the null space of the constraint matrix B, as noted in Lukšan and Vlček (1998) and Perugia *et al.* (1999). With an appropriate choice of the initial guess u_0, the component x_k of every iterate u_k generated by a Krylov method preconditioned with $\mathcal{P}_c$ satisfies the constraint $Bx_k = g$; see, *e.g.*, Rozložník and Simoncini (2002).

When A has positive diagonal entries, G is often taken to be the diagonal part of A, and if $A = (a_{ij})$ has been scaled so that $a_{ii} = 1$ for $1 \leq i \leq n$, then $G = I$. In this case the preconditioning matrix is just

$$\mathcal{P}_c = \begin{bmatrix} I & B^T \\ B & O \end{bmatrix},$$

i.e., the augmented matrix for a least squares problem of the form

$$\min_z \|h - B^T z\|_2,$$

and application of the preconditioner within a Krylov subspace method involves repeated solution of such problems, either explicitly or implicitly.

More generally, solution of linear systems with coefficient matrix $\mathcal{P}_c$ given by (10.15) can be accomplished based on the identity

$$\begin{bmatrix} G & B^T \\ B & O \end{bmatrix}^{-1} = \begin{bmatrix} I & -G^{-1}B^T \\ O & I \end{bmatrix} \begin{bmatrix} G^{-1} & O \\ O & -(BG^{-1}B^T)^{-1} \end{bmatrix} \begin{bmatrix} I & O \\ -BG^{-1} & I \end{bmatrix}.$$

(10.16)

If G is sufficiently simple (e.g., diagonal) and linear systems with coefficient matrix $BG^{-1}B^T$ can be solved efficiently, then the action of the preconditioner on a vector can be computed at a reasonable cost. In many cases, however, the 'exact' solution of linear systems involving $BG^{-1}B^T$ is too expensive, and approximations must be used. For instance, incomplete factorizations or sparse approximate inverse techniques can be used to approximately form and invert $BG^{-1}B^T$; see Chow and Saad (1997a), Haws (2002), Perugia and Simoncini (2000) and Toh et al. (2004). These strategies are often more efficient than methods based on a direct LDL^T factorization of $\mathcal{P}_c$. However, a sparse factorization of $\mathcal{P}_c$ would be advantageous when B contains a dense column, as sometimes happens in optimization problems (Bergamaschi et al. 2004), since in this case $BG^{-1}B^T$ is completely dense. Recently, an interesting new factorization (alternative to (10.16)) has been introduced by Schilders; see Dollar and Wathen (2004) and Section 10.4 below. This can also be used to compute the action of $\mathcal{P}_c^{-1}$ efficiently.

For a symmetric positive definite G, the (left) preconditioned matrix is given by

$$\mathcal{M} = \begin{bmatrix} G & B^T \\ B & O \end{bmatrix}^{-1} \begin{bmatrix} A & B^T \\ B & O \end{bmatrix} = \begin{bmatrix} (I-\Pi)\,G^{-1}A + \Pi & O \\ X & I \end{bmatrix}$$

(10.17)

where

$$\Pi = G^{-1}B^T(BG^{-1}B^T)^{-1}B$$

is the G-orthogonal projector onto range$(G^{-1}B^T)$ and

$$X = (BG^{-1}B^T)^{-1}B(G^{-1}A - I).$$

Note that in the special case $m = n$, we have $\Pi = I$ and all the eigenvalues of the preconditioned matrix are equal to 1; furthermore,

$$\mathcal{M} - I = \begin{bmatrix} O & O \\ X & O \end{bmatrix} \Rightarrow (\mathcal{M} - I)^2 = O.$$

Therefore the minimal polynomial of the preconditioned matrix has degree 2, and GMRES applied to the preconditioned system delivers the solution in at most two steps, independently of A and G. Of course, if $m = n$ and B is nonsingular then the solution of the saddle point system is simply given by $Bx_* = g$ and $B^T y_* = f - Ax_*$, and the augmented system formulation is neither necessary nor recommended.

In the following, we deal with the more interesting case $m < n$. The eigenvalues and eigenvectors of the preconditioned matrix $\mathcal{M}$ in this case have been studied in several publications, *e.g.*, Durazzi and Ruggiero (2003*a*), Lukšan and Vlček (1998), Perugia and Simoncini (2000) and Rozložník and Simoncini (2002). The following result was given in Keller *et al.* (2000, Theorems 2.1 and 2.3).

Theorem 10.1. Let $A \in \mathbb{R}^{n \times n}$ be symmetric, let $B \in \mathbb{R}^{m \times n}$ have full rank $(= m < n)$, and suppose $\mathcal{A}$ is nonsingular. Let $Z \in \mathbb{R}^{n \times (n-m)}$ be a matrix whose columns form a basis for $\ker(B)$. Furthermore, let $G \in \mathbb{R}^{n \times n}$, with $G = G^T \neq A$, be such that $\mathcal{P}_c$ in (10.15) is nonsingular. Then the preconditioned matrix $\mathcal{M}$ in (10.17) has:

(1) the eigenvalue 1 with multiplicity $2m$;
(2) $n - m$ eigenvalues that are defined by the generalized eigenproblem $Z^T A Z x_z = \lambda Z^T G Z x_z$.

Assume, in addition, that $Z^T G Z$ is positive definite. Then $\mathcal{M}$ has the following $m + i + j$ linearly independent eigenvectors:

(1) m eigenvectors of the form $[0^T, y^T]^T$ corresponding to the eigenvalue 1 of $\mathcal{M}$;
(2) i ($0 \leq i \leq n$) eigenvectors of the form $[w^T, y^T]^T$ corresponding to the eigenvalue 1 of $\mathcal{M}$, where the components w arise from the generalized eigenproblem $Aw = Gw$;
(3) j ($0 \leq j \leq n - m$) eigenvectors of the form $[x_z^T, 0^T, y^T]^T$ corresponding to the eigenvalues of $\mathcal{M}$ not equal to 1, where the components x_z arise from the generalized eigenproblem $Z^T A Z x_z = \lambda Z^T G Z x_z$ with $\lambda \neq 1$.

Proof. See Keller *et al.* (2000). □

As pointed out in Keller *et al.* (2000), if either $Z^T A Z$ or $Z^T G Z$ are positive definite, then the indefinite preconditioner $\mathcal{P}_c$ (10.15) applied to the indefinite saddle point matrix $\mathcal{A}$ yields a preconditioned matrix $\mathcal{M}$ (10.17) that has real eigenvalues. Clearly, the eigenvalues will be more clustered near 1 the better G approximates A. Spectral properties of the preconditioned matrices in the presence of inexact solves with G and $BG^{-1}B^T$ have been studied by Perugia and Simoncini (2000) and Toh *et al.* (2004). We further mention the approach proposed by Coleman and Verma (2001), where the blocks B and B^T in the constraint preconditioner $\mathcal{P}_c$ are replaced by approximations $\bar{B}$, $\bar{B}^T$ in order to simplify solves involving the preconditioning matrix. In Coleman and Verma (2001), $\bar{B}$ is obtained from B by sparsification; in this case, care must be taken to avoid a (nearly) rank deficient approximation. See also Arbenz *et al.* (2001), Biros and Ghattas (2005*a*, 2005*b*) and Haber and Ascher (2001) for similar approaches in different application contexts.

If, in the notation of Theorem 10.1, $Z^T G Z$ is positive definite, then for each initial residual r_0, $\dim \mathcal{K}_{n+m}(\mathcal{M}, r_0) \leq n - m + 2$: see, $e.g.$, Lukšan and Vlček (1998, Theorem 3.4) and Keller et $al.$ (2000, Theorem 3.5). Hence GMRES applied to the preconditioned system with matrix $\mathcal{M}$ will terminate after at most $n - m + 2$ steps. This is a very attractive feature of constraint preconditioning when $n - m$ is small.

In many applications $n - m + 2$ is quite large, so that a total number of $n - m + 2$ iterations may be prohibitive. In addition, one is typically interested only in an approximate rather than the exact solution, which means that the iteration can be stopped after a prescribed tolerance is achieved. Then a convergence analysis of the type described in Section 9 is required. Unfortunately, the matrix $\mathcal{M}$ is not diagonalizable, so that standard convergence estimates like (9.9) are not applicable. In some cases, however, it is still possible to derive convergence bounds similar to (9.9) for GMRES applied to the preconditioned system. For example, assume that $A = A^T$ is positive definite, B has full rank, and $G = I$. Then, for a special initial residual of the form $r_0 = [s_0^T, 0^T]^T$, it can be shown that the convergence of GMRES for the $right$ constraint preconditioned system only depends on the (1,1) block of the preconditioned matrix $\mathcal{AP}_c^{-1}$. In our notation this (1,1) block is given by the (diagonalizable) matrix $A(I - \Pi) + \Pi$; see Rozložník and Simoncini (2002, Section 6) for details.

It should be pointed out that GMRES might not be the best method for solving the constraint preconditioned linear system. The fact that constraint preconditioning incorporates a projection onto $\ker(B)$ suggests that this preconditioning technique is closely related to the null space method discussed in Section 7. The two methods are, in fact, mathematically equivalent (Gould et $al.$ 2001, Lukšan and Vlček 1998, Perugia et $al.$ 1999). Exploitation of this equivalence allows the construction of a CG method for a constraint preconditioned positive definite reduced system (Gould et $al.$ 2001). Unlike GMRES (when applied to a system with the nonsymmetric matrix $\mathcal{M}$), this conjugate gradient method is based on efficient three-term recurrences. In the numerical experiments in Keller et $al.$ (2000), this preconditioned conjugate gradient method represents a viable alternative to a direct solver (here MA27 from the Harwell subroutine library (Duff and Reid 1983)), even for systems of relatively small order, and it typically outperforms GMRES applied to $\mathcal{M}$.

All numerical experiments in Keller et $al.$ (2000) are performed using test problems arising in linear and nonlinear optimization taken from the $constrained$ and $unconstrained$ $testing$ $environment$ (CUTE) (Bongartz, Conn, Gould and Toint 1995). Another situation where constraint preconditioning performs well is when the saddle point system arises from a mixed finite element formulation of a second-order, elliptic PDE. In this case it is often possible to find an easily invertible ($e.g.$, diagonal) G that is spectrally equivalent to A. In many cases of practical interest, $G = I$ will do. The resulting constraint preconditioner (10.15) leads to asymptotic rates of convergence for the *right* preconditioned system that are independent of h, the mesh discretization parameter, when the initial guess $u_0 = [0^T, g^T]^T$ (meaning that r_0 is of the form $r_0 = [s_0^T, 0^T]^T$) is used. As in the analysis in Rozložník and Simoncini (2002) mentioned above, the convergence of GMRES then only depends on the (diagonalizable) matrix $A(I - \Pi) + \Pi$. The eigenvalues of this matrix are either 1, or contained in an interval $[\alpha_0, \alpha_1]$, $0 < \alpha_0 \le \alpha_1 < 1$, where both α_0 and α_1 are independent of h (Perugia and Simoncini 2000, Theorem 1) (*cf.* also Ewing *et al.* (1990)). Furthermore, the condition number of the eigenvector matrix of $A(I - \Pi) + \Pi$ is bounded independently of h (Rozložník and Simoncini 2002, Proposition 6.3). The asymptotic optimality of these methods, however, does not always translate in computational efficiency. As shown in Perugia and Simoncini (2000), it may be much more efficient to replace the exact (but costly) preconditioner solves with inexact ones, even if this introduces a dependency on h in the number of preconditioned iterations. This is especially true of saddle point systems that arise from the solution of 3D PDE problems. See also Mihajlović and Silvester (2004) for a discussion of inexact constraint preconditioning for mixed formulations of biharmonic problems in 2D.

Some of these results have been extended to the case where both A and G are allowed to be nonsymmetric; see Cao (2002). The case where $B_1 \ne B_2$ was investigated in Ikramov (2003) and Wei and Zhang (2004). Dyn and Ferguson (1983) studied the case where A is symmetric but G may be nonsymmetric, and $B_1 = B_2 = B$. Assuming that A is symmetric positive semidefinite with $\ker(A) \cap \ker(B) = \{0\}$ and that G is nonsingular and such that $G + G^T - A$ is positive definite, they showed that the matrix $\mathcal{P}_c$ in (10.15) is nonsingular, and that the stationary iteration

$$\begin{bmatrix} G & B^T \\ B & O \end{bmatrix} \begin{bmatrix} x_{k+1} \\ y_{k+1} \end{bmatrix} = \begin{bmatrix} G - A & O \\ O & O \end{bmatrix} \begin{bmatrix} x_k \\ y_k \end{bmatrix} + \begin{bmatrix} f \\ g \end{bmatrix}, \quad k = 0, 1, \dots \quad (10.18)$$

is convergent for any choice of the initial guess. Therefore, the spectrum of $\mathcal{P}_c^{-1} \mathcal{A}$ is contained in the disk $\{z \in \mathbb{C} ; |z - 1| < 1\}$. A matrix splitting of the form $A = G - (G - A)$ such that $G + G^T - A$ is positive definite is called a *P-regular splitting*; see Ortega (1972, page 122). Examples of such splittings when A is symmetric positive semidefinite with a nonzero diagonal include the damped Jacobi, SOR, and SSOR splittings (Dyn and Ferguson 1983). We mention that variants of the coupled iteration (10.18) have been used as smoothers for multigrid methods; see Braess and Sarazin (1997), Leem, Oliveira and Stewart (2004), Wabro (2004), Zulehner (2000) and Section 11 below.

The important case where A is nonsymmetric but has positive definite symmetric part has been studied in Golub and Wathen (1998), motivated by the Oseen (linearized Navier–Stokes) equations, and more recently in Baggag and Sameh (2004), motivated by particulate flow problems. In both of these papers, the idea is to use $G = H \equiv \frac{1}{2}(A + A^T)$, the symmetric part of A. For the nonsymmetric linear system arising from the discretization of the Oseen equations (2.7)–(2.9), the preconditioner becomes the (symmetric) matrix of the corresponding Stokes system (2.10)–(2.12). When the kinematic viscosity coefficient ν in (2.7) is not too small, A is not too far from being symmetric and this preconditioning strategy is very effective, yielding convergence rates that are independent of the mesh size. In practice the exact solution of the preconditioning equation

$$\mathcal{P}_c z = r, \quad \mathcal{P}_c = \begin{bmatrix} H & B^T \\ B & O \end{bmatrix} \tag{10.19}$$

may be too expensive, and inexact solves may be required. For example, (10.19) may be solved approximately by an iterative method, leading to a nested (or inner-outer) iterative scheme. As shown in Baggag and Sameh (2004), it is often possible in this way to retain the excellent convergence properties of the outer scheme, while at the same time reducing the work per iteration significantly. On the other hand, this approach does not work well when A is far from symmetric (large Reynolds number, or small ν in (2.7)); see the numerical experiments reported in Benzi and Liu (2005) and Botchev and Golub (2004). In this case, better results can be obtained with preconditioners of the form

$$\mathcal{P}_c = \begin{bmatrix} T & B^T \\ B & O \end{bmatrix},$$

where the submatrix T is nonsymmetric positive definite with a 'large' skew-symmetric part that incorporates information from K, the skew-symmetric part of A; see Botchev and Golub (2004).

The case where $A = A^T$, $B_1 = B_2 = B$ and $C = C^T \neq O$ has been considered in Axelsson and Neytcheva (2003), Bergamaschi et al. (2004), Dollar (2005), Perugia and Simoncini (2000), Toh et al. (2004) and Zulehner (2002). Here A and C are positive semidefinite and G is chosen so that the preconditioning matrix

$$\mathcal{P}_c = \begin{bmatrix} G & B^T \\ B & -C \end{bmatrix} \tag{10.20}$$

is nonsingular. Furthermore, linear systems involving G and $C + BG^{-1}B^T$ should be easy to (approximately) solve. The spectral properties of this preconditioner have been studied in Axelsson and Neytcheva (2003), Perugia and Simoncini (2000) and Toh et al. (2004). We mention the following results from Axelsson and Neytcheva (2003).

Lemma 10.2. Let B, C, E be real matrices of order $m \times n$, $m \times m$ and $n \times n$, respectively, where B has full rank, $C = C^T$ is positive semidefinite, and $E = E^T$. Then the eigenvalues of the generalized eigenproblem

$$\gamma \begin{bmatrix} I & B^T \\ B & -C \end{bmatrix} \begin{bmatrix} x \\ y \end{bmatrix} = \begin{bmatrix} E & O \\ O & O \end{bmatrix} \begin{bmatrix} x \\ y \end{bmatrix}, \quad \|x\| + \|y\| \neq 0, \tag{10.21}$$

where $x \in \mathbb{C}^n$ and $y \in \mathbb{C}^m$, satisfy

(i) $\gamma = \frac{x^H E x}{x^H (I + B^T C^{-1} B) x} \neq 0$ if $Ex \neq 0$ and C is positive definite,

(ii) $\gamma = 0$ if and only if $Ex = 0$, $y \neq 0$ and the dimension of the eigenspace corresponding to the zero eigenvalue is $m + q$, where $q = \dim \ker(E)$,

(iii) the nonzero eigenvalues are contained in the interval $\min\{0, \lambda_{\min}(E)\} \leq \gamma \leq \lambda_{\max}(E)$.

Consider now a symmetric saddle point matrix $\mathcal{A}$ and the preconditioner $\mathcal{P}_c$ given by (10.20), with G symmetric and positive definite. The eigenvalues of the preconditioned matrix $\mathcal{P}_c^{-1} \mathcal{A}$ are those of the generalized eigenproblem $\mathcal{A} u = \lambda \mathcal{P}_c u$; but these are of the form $\lambda = \gamma + 1$, where γ are the eigenvalues of the generalized eigenproblem

$$\gamma \begin{bmatrix} G & B^T \\ B & -C \end{bmatrix} \begin{bmatrix} x \\ y \end{bmatrix} = \begin{bmatrix} A - G & O \\ O & O \end{bmatrix} \begin{bmatrix} x \\ y \end{bmatrix}.$$

The latter is equivalent to

$$\gamma \begin{bmatrix} I & \hat{B}^T \\ \hat{B} & -C \end{bmatrix} \begin{bmatrix} \hat{x} \\ y \end{bmatrix} = \begin{bmatrix} \hat{E} & O \\ O & O \end{bmatrix} \begin{bmatrix} \hat{x} \\ y \end{bmatrix}, \tag{10.22}$$

where $\hat{B} = BG^{-1/2}$, $\hat{E} = G^{-1/2} A G^{-1/2} - I$ and $\hat{x} = G^{1/2} x$. This generalized eigenproblem is of the form (10.21) and therefore Lemma 10.2 can be applied. In particular, the eigenvalues of $\mathcal{P}_c^{-1} \mathcal{A}$ are real. Clearly, the better G approximates A, the more clustered around 1 are the eigenvalues of $\mathcal{P}_c^{-1} \mathcal{A}$.

Another form of indefinite preconditioning is based on the observation that 'stabilized' saddle point systems (*i.e.*, systems of the type (1.6) with $C \neq O$) are generally easier to solve than standard saddle point systems (*i.e.*, with $C = O$), in the sense that iterative methods tend to converge faster when $C \neq O$. It is also easier to construct effective block preconditioners when $C \neq O$, since the (negative) Schur complement $C + BA^{-1} B^T$ is typically better conditioned than $BA^{-1} B^T$. This observation justifies the use of preconditioners of the form (10.20), with a suitably chosen C, even if the $(2, 2)$ block is zero in the original problem. This approach, referred to as *regularized preconditioning*, was first considered in Axelsson (1979) in the special case $G = A$ and $C = \varepsilon I$, where $\varepsilon > 0$. The preconditioning equation itself, $\mathcal{P}_c z = r$, is solved by an iterative method, leading again to an inner-outer process.

The case where $G = A$ is symmetric positive definite, $B_1 = B_2 = B$ has full rank, and C is symmetric positive semidefinite is easily analysed. In this case we have

$$\mathcal{A} - \mathcal{P}_c = \begin{bmatrix} O & O \\ O & C \end{bmatrix} \Rightarrow I - \mathcal{A}^{-1}\mathcal{P}_c = \mathcal{A}^{-1}(\mathcal{A} - \mathcal{P}_c) = \begin{bmatrix} O & Y \\ O & S^{-1}C \end{bmatrix},$$

where $S = -BA^{-1}B^T$ and Y is a certain nonzero matrix. It follows that

$$\mathcal{P}_c^{-1}\mathcal{A} = \begin{bmatrix} I & \hat{Y} \\ O & (I - S^{-1}C)^{-1} \end{bmatrix},$$

where $\hat{Y}$ is again a nonzero matrix. Since S is negative definite and C is positive semidefinite, $I - S^{-1}C$ has positive real eigenvalues. Therefore the spectrum of the preconditioned matrix $\mathcal{P}_c^{-1}\mathcal{A}$ is real and contained in the interval $(\beta, 1]$, where

$$\beta = \frac{1}{1 + \lambda_{\max}(-S^{-1}C)} \geq \frac{1}{1 + \|S^{-1}\|_2 \|C\|_2}.$$

Also, $\mathcal{P}_c^{-1}\mathcal{A}$ has eigenvalue 1 with multiplicity $n + p$, where $p = \dim \ker(C)$. This approach is especially efficient for the Stokes problem, for which S is spectrally equivalent to the identity. Then $\|S^{-1}\|_2 \approx 1$ and as long as $\|C\|_2$ is bounded above independently of h, the spectrum of $\mathcal{P}_c^{-1}\mathcal{A}$ also remains bounded as $h \to 0$ and fast convergence of the preconditioned iteration can be expected, independent of mesh size. In practice there is no need to solve with $\mathcal{P}_c$ exactly: an approximate solution will suffice and will be more efficient. Spectral bounds for the inexact case can be found in Gorelova and Chizhonkov (2004) for $C = O$ and in Dohrmann and Lehoucq (2004) for $C = C^T \neq O$. It should be mentioned that regularization of indefinite (constraint) preconditioners has also been used with good results in optimization problems; see Bergamaschi *et al.* (2004) as well as Durazzi and Ruggiero (2003a, 2003b); the latter papers also include results for a parallel implementation.

Concerning the choice of a Krylov subspace method to use with constrained preconditioning, we distinguish between the symmetric case ($A = A^T$, $B_1 = B_2$, $C = C^T$) and the nonsymmetric case. In the nonsymmetric case it is generally necessary to use a nonsymmetric Krylov subspace method, such as GMRES or BiCGStab, whether the constraint preconditioner used is symmetric or not. In the case of a symmetric A with a symmetric constraint preconditioner $\mathcal{P}_c$, on the other hand, there are alternatives to the use of nonsymmetric Krylov subspace methods. Although the preconditioned matrix $\mathcal{M}$ itself is nonsymmetric, the SQMR method can be used; see the discussion at the end of Section 9, in particular Table 9.1.

Furthermore, as described above, in some cases a combination of the null space method and the CG method may be very effective. As noted in Bramble and Pasciak (1988) and Lukšan and Vlček (1998), in the symmetric case it is sometimes even possible to use the CG method for the preconditioned system with $\mathcal{M}$. However, the indefiniteness of both $\mathcal{P}_c$ and $\mathcal{A}$ does not in general lead to a robust algorithm, and safeguard strategies have to be applied to overcome potential breakdowns. For the case of right constraint preconditioning and assuming $g = 0$ in (1.3), such strategies are discussed in detail in Lukšan and Vlček (1998) and Rozložník and Simoncini (2002). When such strategies are applied, the preconditioned conjugate gradient method can be competitive with direct solution methods, particularly for problems from nonlinear equality-constrained optimization (Lukšan and Vlček 1998, 2001). Finally, use of the CG method with an inexact form of regularized constraint preconditioning has been rigorously justified in Dohrmann and Lehoucq (2004).

10.3. Hermitian/skew-Hermitian preconditioning

The Hermitian and skew-Hermitian splitting (HSS) was introduced as a stationary iterative method in Bai, Golub and Ng (2003), where it was shown to converge for non-Hermitian positive definite systems, *i.e.*, linear systems $Ax = b$ with $A + A^*$ positive definite. In the real case (which is the only one we consider here), such systems are said to be *positive real*. Problems of this type arise, for instance, in the numerical solution of convection-diffusion equations. The use of HSS as a preconditioner for rather general saddle point problems has been studied in Benzi, Gander and Golub (2003), Benzi and Golub (2004) and Simoncini and Benzi (2004).

The HSS preconditioner is based on the nonsymmetric formulation (3.10). Here we are under the assumptions of Theorem 3.6. In particular, $B_1 = B_2 = B$ and $C = C^T$. Letting $H \equiv \frac{1}{2}(A + A^T)$ and $K \equiv \frac{1}{2}(A - A^T)$ we have the following splitting of $\hat{A}$ into its symmetric and skew-symmetric part:

$$\hat{A} = \begin{bmatrix} A & B^T \\ -B & C \end{bmatrix} = \begin{bmatrix} H & O \\ O & C \end{bmatrix} + \begin{bmatrix} K & B^T \\ -B & O \end{bmatrix} = \mathcal{H} + \mathcal{K}. \qquad (10.23)$$

Note that $\mathcal{H}$, the symmetric part of $\hat{A}$, is symmetric positive semidefinite since both H and C are. Let $\alpha > 0$ be a parameter. In the same spirit as the classical ADI (Alternating-Direction Implicit) method (Varga 1962), we consider the following two splittings of $\hat{A}$:

$$\hat{A} = (\mathcal{H} + \alpha \mathcal{I}) - (\alpha \mathcal{I} - \mathcal{K}) \quad \text{and} \quad \hat{A} = (\mathcal{K} + \alpha \mathcal{I}) - (\alpha \mathcal{I} - \mathcal{H}).$$

Here $\mathcal{I}$ denotes the identity matrix of order $n + m$. The stationary HSS iteration is obtained by alternating between these two splittings. Given an initial guess u_0, the HSS iteration computes a sequence $\{u_k\}$ as follows:

$$\begin{cases} (\mathcal{H} + \alpha\mathcal{I})\, u_{k+\frac{1}{2}} = (\alpha\mathcal{I} - \mathcal{K})\, u_k + \hat{b}, \\ (\mathcal{K} + \alpha\mathcal{I})\, u_{k+1} = (\alpha\mathcal{I} - \mathcal{H})\, u_{k+\frac{1}{2}} + \hat{b}. \end{cases} \qquad (10.24)$$

Note that both $\mathcal{H} + \alpha\mathcal{I}$ and $\mathcal{K} + \alpha\mathcal{I}$ are nonsingular. The first matrix is symmetric positive definite while the second one is a shifted skew-symmetric matrix with eigenvalues of the form $\alpha + i\nu_j$, where $\nu_j \in \mathbb{R}$ for all $j = 1, \ldots, n + m$.

The two-step process (10.24) can be written as a fixed point iteration by eliminating the intermediate vector $u_{k+\frac{1}{2}}$, yielding

$$u_{k+1} = \mathcal{T}_\alpha u_k + c. \qquad (10.25)$$

Here

$$\mathcal{T}_\alpha := (\mathcal{K} + \alpha\mathcal{I})^{-1}(\alpha\mathcal{I} - \mathcal{H})(\mathcal{H} + \alpha\mathcal{I})^{-1}(\alpha\mathcal{I} - \mathcal{K})$$

is the iteration matrix of the method, and

$$c := (\mathcal{K} + \alpha\mathcal{I})^{-1}[\mathcal{I} + (\alpha\mathcal{I} - \mathcal{H})(\mathcal{H} + \alpha\mathcal{I})^{-1}]\hat{b}.$$

The fixed point iteration (10.25) converges for arbitrary initial guesses u_0 and right-hand sides $\hat{b}$ to the solution $u = \hat{A}^{-1}\hat{b}$ if and only if $\rho(\mathcal{T}_\alpha) < 1$, where $\rho(\mathcal{T}_\alpha)$ denotes the spectral radius of $\mathcal{T}_\alpha$. It follows from the results in Bai *et al.* (2003) that when $\mathcal{H}$ is positive definite, the stationary iteration (10.24) converges for all $\alpha > 0$ to the solution of $\hat{A}u = \hat{b}$. For saddle point problems, $\mathcal{H}$ is positive definite if and only if H and C are. These conditions are rather restrictive in practice, especially the one on C. However, it was shown in Benzi and Golub (2004) that the HSS iteration converges for all $\alpha > 0$ if H is positive definite, B is of full rank and C is positive semidefinite (possibly zero). The method can be made to converge even when H is positive semidefinite and singular, provided that $\ker(H) \cap \ker(B) = \{0\}$; see Benzi and Golub (2004).

It was shown in Bai *et al.* (2003) that when $\mathcal{H}$ is positive definite, the choice

$$\alpha = \sqrt{\lambda_{\min}(\mathcal{H})\lambda_{\max}(\mathcal{H})}$$

minimizes an upper bound on the spectral radius of the iteration matrix $\mathcal{T}_\alpha$. Unfortunately, in most saddle point problems $\lambda_{\min}(\mathcal{H}) = 0$, so the result does not apply. Furthermore, the rate of convergence of the HSS iteration is rather slow, even with the 'optimal' choice of α. For these reasons Benzi and Golub (2004) proposed that GMRES or other Krylov subspace methods should be used to accelerate the convergence of the HSS method. In other words, the HSS method is best used as a preconditioner for (say) GMRES rather than as a stationary iterative method. As observed in Benzi and Golub (2004), there is a unique splitting $\hat{A} = \mathcal{P} - \mathcal{Q}$ with $\mathcal{P}$ nonsingular such that the iteration matrix $\mathcal{T}_\alpha$ is the matrix induced by that splitting, i.e., $\mathcal{T}_\alpha = \mathcal{P}^{-1}\mathcal{Q} = \mathcal{I} - \mathcal{P}^{-1}\mathcal{A}$. An easy calculation shows that the HSS iteration (10.24) can be written in *correction form* as

$$u_{k+1} = u_k + \mathcal{P}_\alpha^{-1}r_k, \qquad r_k = \hat{b} - \hat{\mathcal{A}}u_k,$$

where the preconditioner $\mathcal{P}$ is given by

$$\mathcal{P} \equiv \mathcal{P}_\alpha = \tfrac{1}{2\alpha}(\mathcal{H} + \alpha\mathcal{I})(\mathcal{K} + \alpha\mathcal{I}). \tag{10.26}$$

Note that as a preconditioner we can use $\mathcal{P}_\alpha = (\mathcal{H} + \alpha\mathcal{I})(\mathcal{K} + \alpha\mathcal{I})$ instead of the expression given in (10.26), since the factor $\frac{1}{2\alpha}$ has no effect on the preconditioned system. It is just a normalization factor that allows us to conclude that the eigenvalues of the preconditioned matrix $\mathcal{P}_\alpha^{-1}\hat{\mathcal{A}}$ (or $\hat{\mathcal{A}}\mathcal{P}_\alpha^{-1}$, which has the same spectrum) are all contained in the disk $\{z \in \mathbb{C} \,;\, |z - 1| < 1\}$. In particular, the spectrum of the preconditioned matrix, like that of $\hat{\mathcal{A}}$, lies entirely in the right half-plane: the preconditioned matrix is positive stable.

The rate of convergence of nonsymmetric Krylov iterations (like GMRES) preconditioned by $\mathcal{P}_\alpha$ depends on the particular choice of α. Finding the value of α that optimizes the rate of convergence is a very difficult problem in general. Note that the value of α that minimizes the number of GMRES iterations may be quite different from the one that minimizes the spectral radius of $\mathcal{T}_\alpha$; see Benzi *et al.* (2003). Numerical experiments show that with an appropriate scaling of the system (such that the nonzero diagonal entries of $\hat{\mathcal{A}}$ are equal to 1), there is a unique value α_* of α for which the number of preconditioned iterations is minimized, and this α_* is usually a small number, between 0 and 1. In some cases, but not always, the optimal α can be determined by trial and error on a small example (*e.g.*, a coarse discretization of the continuous problem to be solved) and then used with good results on larger problems corresponding to finer discretizations; see Benzi and Ng (2004), where HSS preconditioning was used to solve weighted least squares problems arising in image processing.

Spectral properties of the preconditioned matrix as a function of α have been studied, under different sets of assumptions, in Benzi *et al.* (2003), Benzi and Ng (2004) and Simoncini and Benzi (2004). A Fourier analysis of HSS preconditioning for saddle point formulations of Poisson's equation (including the anisotropic case) was given in Benzi *et al.* (2003). The analysis showed that using a sufficiently small value of α results in h-independent convergence. Furthermore, as $\alpha \to 0$ the eigenvalues of the preconditioned matrix are all real and fall within two small intervals $(0, \varepsilon_1)$ and $(2 - \varepsilon_2, 2)$, with $\varepsilon_1, \varepsilon_2 > 0$ and $\varepsilon_1, \varepsilon_2 \to 0$ as $\alpha \to 0$. This clustering result was generalized in Simoncini and Benzi (2004) to general saddle point systems with $A = A^T$ positive definite and $C = O$, using purely algebraic arguments. However, h-independent convergence is not always guaranteed: for example,

it does not occur for the Stokes problem. Nevertheless, good results have
been obtained in the solution of the Oseen problem, where HSS preconditioning appears to be competitive with other preconditioners for large
Reynolds number (small ν). Numerical experiments indicate that the optimal value of α for steady 2D problems is largely independent of ν and
is approximately given by $h^{1/2}$, where h denotes the mesh size. Furthermore, rapid convergence independently of h and ν is observed for unsteady
problems; see the numerical experiments in Benzi (2004) and Benzi and Liu
(2005).

Application of the HSS preconditioner within GMRES requires solving a
linear system of the form $\mathcal{P}_\alpha z_k = r_k$ at each iteration. This is done by first
solving the system

$$(\mathcal{H} + \alpha \mathcal{I}) \, w_k = r_k \qquad (10.27)$$

for w_k, followed by

$$(\mathcal{K} + \alpha \mathcal{I}) \, z_k = w_k. \qquad (10.28)$$

Recalling the form of $\mathcal{H}$, see (10.23), equation (10.27) consists of two decoupled systems with coefficient matrices $H + \alpha I$ and $C + \alpha I$, respectively.
Both matrices are symmetric positive definite, and a number of efficient
methods can be applied, including Cholesky factorization, preconditioned
CG (PCG) schemes, or multigrid, either geometric or algebraic. Multigrid
methods are the solvers of choice for a number of problems arising from
the discretization of partial differential equations, particularly in the case
of the Oseen problem (2.10)–(2.11) in $\Omega \subset \mathbb{R}^3$, where H is a direct sum of
three discrete Laplace operators. The solution of the system with matrix
$C + \alpha I$ is often much easier and reduces to a scaling by α when $C = O$.
When solving regularized weighted least squares problems, both H and C
are diagonal and the cost of solving (10.27) is negligible; see Benzi and Ng
(2004).

Note that the addition of a positive term α to the main diagonal of H
(and C) improves the condition number. This, in turn, tends to improve the
rate of convergence of iterative methods applied to (10.27). More precisely,
if H is normalized so that its largest eigenvalue is equal to 1, then for the
spectral condition number of $H + \alpha I$ we have

$$\kappa \left(H + \alpha I \right) = \frac{1 + \alpha}{\lambda_{\min}(H) + \alpha} \leq 1 + \frac{1}{\alpha},$$

independent of the size of the problem. Note that even a fairly small value of
α, such as $\alpha = 0.1$, yields a small condition number ($\kappa \left(H + \alpha I \right) \leq 11$). For
most problems, both multigrid and the CG method applied to (10.27) can
be expected to converge rapidly, independent of the number n of unknowns.

Solving (10.28) is usually more involved, especially when $K \neq O$. It requires the solution of a linear system of the form

$$\begin{cases} (\alpha I + K)\, x_{k+1} + B^T y_{k+1} = f_k, \\ -B x_{k+1} + \alpha y_{k+1} = g_k. \end{cases} \tag{10.29}$$

This system can be solved in several ways. One approach is to eliminate x_{k+1} from the second equation using the first one (Schur complement reduction), leading to a smaller (order m) linear system of the form

$$[B(I + \alpha^{-1}K)^{-1}B^T + \alpha^2 I]y_{k+1} = B(I + \alpha^{-1}K)^{-1}f_k + \alpha g_k. \tag{10.30}$$

Once the solution y_{k+1} to (10.30) has been computed, the vector x_{k+1} is given by

$$(\alpha I + K)x_{k+1} = f_k - B^T y_{k+1}.$$

When $K = O$, system (10.30) simplifies to

$$(BB^T + \alpha^2 I)\, y_{k+1} = B f_k + \alpha g_k, \tag{10.31}$$

and $x_{k+1} = \alpha^{-1}(f_k - B^T y_{k+1})$. If BB^T is sufficiently sparse, system (10.31) could be formed explicitly and solved by a sparse Cholesky factorization. Otherwise, an iterative method like LSQR with a simple preconditioner could be used. If B represents a discrete divergence operator, then $BB^T + \alpha^2 I$ is just a shifted discrete Laplace operator, and many fast solvers for system (10.31) are available. When $K \neq O$ the coefficient matrix in (10.30) is generally dense. An important exception is the Oseen problem in rotation form (2.15)–(2.17). In the 2D case the coefficient matrix has the form

$$\hat{\mathcal{A}} = \begin{bmatrix} A_1 & D & B_1^T \\ -D & A_2 & B_2^T \\ -B_1 & -B_2 & C \end{bmatrix} = \begin{bmatrix} A_1 & O & O \\ O & A_2 & O \\ O & O & C \end{bmatrix} + \begin{bmatrix} O & D & B_1^T \\ -D & O & B_2^T \\ -B_1 & -B_2 & O \end{bmatrix} = \mathcal{H} + \mathcal{K}. \tag{10.32}$$

Here A_1, A_2 are discrete Laplace operators with appropriate boundary conditions, $B = \begin{bmatrix} B_1 & B_2 \end{bmatrix}$, $C = C^T$ is a stabilization term that can be assumed to be zero if the discretization used is already stable, and $D = D^T$ is a matrix that corresponds to multiplication by $w = \nabla \times \mathbf{v}$; see Section 2.1 for details. For finite difference schemes (for instance, MAC (Harlow and Welch 1965)) D is a diagonal matrix; for finite elements, it will be a scaled mass matrix. When D is diagonal, the Schur complement $B(I + \alpha^{-1}K)^{-1}B^T + \alpha^2 I$ is a sparse matrix and can be formed explicitly. This follows from the fact that

$$(I + \alpha^{-1}K)^{-1} = \begin{bmatrix} I & \alpha^{-1}D \\ -\alpha^{-1}D & I \end{bmatrix}^{-1} = \begin{bmatrix} E_1 & -E_2 \\ E_2 & E_3 \end{bmatrix},$$

where E_1, E_2 and E_3 are diagonal matrices given by

$$E_1 = I - \alpha^{-2} D (I + \alpha^{-2} D^2)^{-1} D,$$
$$E_2 = \alpha^{-1} D (I + \alpha^{-2} D^2)^{-1},$$
$$E_3 = (I + \alpha^{-2} D^2)^{-1}.$$

When D is not diagonal, we can replace it with a spectrally equivalent diagonal approximation and still have a sparse Schur complement; since we are constructing a preconditioner, the action of $(\mathcal{K} + \alpha \mathcal{I})^{-1}$ need not be computed exactly. Hence, the Schur complement $B(I + \alpha^{-1} K)^{-1} B^T + \alpha^2 I$ (or the approximation of it obtained by replacing D with a diagonal matrix) is sparse, and system (10.28) can be efficiently solved via (10.30) using sparse matrix techniques. It is also possible to use multigrid, since the Schur complement can be interpreted as a discretization of a second-order, elliptic operator with variable coefficients. Alternatively, an ILU-preconditioned GMRES can be used. While we have focused here on the 2D case, the 3D case can be treated along the same lines; see Benzi and Liu (2005).

Besides the Schur complement reduction, there are other approaches that can be used to solve linear systems with matrix $\mathcal{K} + \alpha \mathcal{I}$. Note that this is a normal matrix of the form 'identity-plus-skew-symmetric'. Various Lanczos-type methods can be applied to systems of this kind; see Concus and Golub (1976), Widlund (1978) and, more generally, Huhtanen (2002). Other approaches to the solution of shifted skew-symmetric systems are studied in Golub and Vanderstraeten (2000). Yet another possibility is to regard (10.29) as a general nonsymmetric system and to use preconditioned GMRES (say). Many of these schemes can benefit from the fact that for even moderate values of $\alpha > 0$, the condition number of $\mathcal{K} + \alpha \mathcal{I}$ is often rather small.

It is important to stress that the linear systems in (10.24) need not be solved exactly. The use of inexact solves was considered in Bai *et al.* (2003) for the positive real case. The upshot is that inexact solves can greatly reduce the cost of each iteration, at the expense of somewhat slower convergence. Typically, in practical implementations, inexact solves result in a much more competitive algorithm. Here we observe that when the alternating scheme is used as a preconditioner for a Krylov method, inexact solves are a natural choice, and there is no theoretical restriction on the accuracy of the inner solves. Inexact solutions are often obtained by iterative methods, leading to an inner-outer scheme; in this case, a flexible method like FGMRES (Saad 1993) should be used for the outer iteration. Inexact solves may also be performed by means of matrix splittings or incomplete factorizations; see Benzi and Golub (2004). In this case, standard GMRES can be used for the outer iteration.

Other iterative solvers for the Navier–Stokes equations in rotation form have been introduced and studied in Olshanskii (1999) and Olshanskii and Reusken (2002). Although the rotation form does not appear to be widely used in practice, it has some advantages over the (standard) convective form (2.7)–(2.9). As mentioned in Olshanskii and Reusken (2002, page 1685), the numerical solution of the Navier–Stokes equations in rotation form is a topic that deserves further study.

10.4. Approximate and incomplete factorization methods

We conclude this section on preconditioners with a brief discussion of recent attempts to develop effective and robust approximate and incomplete factorization methods for saddle point matrices.

Approximate factorizations for symmetric saddle point systems have been used in Gill *et al.* (1992) in the context of interior point methods for constrained optimization. Here we assume that $A = A^T$ is positive definite, $B_1 = B_2 = B$ is of full rank, and $C = C^T$ is positive semidefinite (possibly zero). The approach taken in Gill *et al.* (1992) is to define the preconditioner in terms of an *exact* LDL^T factorization of an approximation $\mathcal{P} \approx \mathcal{A}$. More precisely, the idea is to compute a sparse Bunch–Parlett factorization

$$\mathcal{P} = \mathcal{Q}^T \mathcal{L} \mathcal{D} \mathcal{L}^T \mathcal{Q},$$

where $\mathcal{P}$ is symmetric indefinite, $\mathcal{Q}$ is a permutation matrix, $\mathcal{L}$ is unit lower triangular, and $\mathcal{D}$ a block diagonal matrix with blocks of dimension 1 and 2. The resulting factorization can be used as a preconditioner with SQMR or with a nonsymmetric Krylov subspace solver. Gill *et al.* (1992) modify $\mathcal{P}$ to guarantee it is positive definite; the resulting preconditioner can then be used with a symmetric Krylov subspace solver, such as SYMMLQ or MINRES. In order to do so, it is sufficient to compute the eigendecomposition of $\mathcal{D}$ (which can be done very easily), and then to change the sign of the negative eigenvalues of $\mathcal{D}$. If $\bar{\mathcal{D}}$ denotes the resulting block diagonal matrix, the symmetric positive definite preconditioner is

$$\bar{\mathcal{P}} = \mathcal{Q}^T \mathcal{L} \bar{\mathcal{D}} \mathcal{L}^T \mathcal{Q}.$$

The main issue is the choice of the approximation $\mathcal{P} \approx \mathcal{A}$. As usual, a trade-off is involved. On one hand, $\mathcal{P}$ must be sufficiently close to $\mathcal{A}$ so that convergence of the preconditioned iteration will be rapid; on the other, $\mathcal{P}$ must be such that an LDL^T factorization can be computed rapidly and without too much fill-in. The choice of such $\mathcal{P}$ is clearly problem-dependent. One possibility is to introduce a re-partitioning of $\mathcal{A}$ in the form

$$\mathcal{A} = \begin{bmatrix} A_{11} & A_{12}^T \\ A_{12} & A_{22} \end{bmatrix}, \tag{10.33}$$

where the order of A_{11} does not exceed $n - m$. This condition is necessary in order to have a nonsingular block A_{22}. If A_{11} is strongly diagonally dominant, then a reasonable approximation could be

$$\mathcal{P} = \begin{bmatrix} D_{11} & O \\ O & A_{22} \end{bmatrix},$$

where D_{11} denotes the main diagonal of A_{11}. For linear systems arising in interior point methods, it is often possible to find a permutation of the (1,1) block A of $\mathcal{A}$ so that $\mathcal{A}$ can be cast in the form (10.33) with A_{11} having very large entries on the main diagonal. Indeed, when a variable x_i is approaching the boundary of the feasible set, the ith diagonal entry of the Hessian A becomes very large; by numbering these variables first, the saddle point matrix $\mathcal{A}$ can be given the desired form. The block A_{22} is now a smaller saddle point-type matrix. If $m \ll n$, the submatrix A_{22} may be much smaller than $\mathcal{A}$, and a sparse Bunch–Parlett factorization of A_{22} can be computed efficiently. In the context of interior point methods, this preconditioner can be expected to be especially effective in later stages of Newton's iteration, when many of the diagonal entries in A_{11} are large. See Gill *et al.* (1992) for additional discussion and alternative approaches.

A different approximate factorization method has been presented by Dollar and Wathen (2004) in the context of constraint preconditioning. Consider a constraint preconditioner (10.15) partitioned as

$$\mathcal{P}_c = \begin{bmatrix} G_{11} & G_{12} & B_1^T \\ G_{21} & G_{22} & B_2^T \\ B_1 & B_2 & O \end{bmatrix},$$

where the block B_1 is nonsingular. Assume we choose matrices $L_1 \in \mathbb{R}^{m \times m}$ and $L_2 \in \mathbb{R}^{(n-m) \times (n-m)}$, with L_2 nonsingular; for instance, $L_2 = I$. Then $\mathcal{P}_c$ can be factorized as follows:

$$\mathcal{P}_c = \begin{bmatrix} B_1^T & O & L_1 \\ B_2^T & L_2 & E \\ O & O & I \end{bmatrix} \begin{bmatrix} D_1 & O & I \\ O & D_2 & O \\ I & O & O \end{bmatrix} \begin{bmatrix} B_1 & B_2 & O \\ O & L_2^T & O \\ L_1^T & E^T & I \end{bmatrix}, \qquad (10.34)$$

where

$$D_1 = B_1^{-T} G_{11} B_1^{-1} - L_1^T B_1^{-1} - B_1^{-T} L_1,$$

$$D_2 = L_2^{-1}(G_{22} - B_2^T D_1 B_2 - E B_2 - B_2^T E^T) L_2^{-T},$$

$$E = G_{21} B_1^{-1} - B_2^T D_1 - B_2^T L_1^T B_1^{-1}.$$

A decomposition of the form (10.34) is known as a *Schilders factorization*. Dollar and Wathen (2004) show how such a factorization can be used to solve linear systems of the form $\mathcal{P}_c z_k = r_k$ efficiently at each iteration of an iterative method. Since (10.34) is an exact factorization of an approximation to $\mathcal{A}$, we can think of it as a rather special approximate factorization of the original coefficient matrix $\mathcal{A}$. An alternative factorization (for symmetric G), related to the null space method, is the following (Saunders 2005):

$$\mathcal{P}_c = \begin{bmatrix} I & O & O \\ B_2^T B_1^{-T} & I & O \\ O & O & I \end{bmatrix} \begin{bmatrix} G_{11} & X^T & B_1^T \\ X & Z^T G Z & O \\ B_1 & O & O \end{bmatrix} \begin{bmatrix} I & B_1^{-1} B_2 & O \\ O & I & O \\ O & O & I \end{bmatrix}, \quad (10.35)$$

where $Z = \begin{bmatrix} -B_1^{-1} B_2 \\ I \end{bmatrix}$ and $X^T = G_{12} - G_{11} B_1^{-1} B_2$. Approximate variants of (10.35) may be used to construct factorized preconditioners.

A more standard approach is to compute an *incomplete* factorization of the *exact* matrix $\mathcal{A}$:

$$\mathcal{A} = \mathcal{Q}^T \mathcal{L} \mathcal{D} \mathcal{L}^T \mathcal{Q} + \mathcal{R} \equiv \mathcal{P} + \mathcal{R},$$

where $\mathcal{R}$ represents a remainder matrix that contains terms that have been discarded in the course of the incomplete factorization. The size of $\mathcal{R}$ depends on the permutation matrix $\mathcal{Q}$ and on the dropping strategy used: although levels of fill can be used, threshold-based dropping is likely to be far more effective for indefinite systems. Unfortunately, the development of reliable and effective incomplete LDL^T factorizations for highly indefinite systems has turned out to be a very difficult problem, and until recently not much progress has been made in this area. This is in contrast with the symmetric positive definite case and even with the general, nonsymmetric case, for which many successful techniques exist (Axelsson 1994, Benzi 2002, Meurant 1999, Saad 2003, van der Vorst 2003).

A possible explanation for the difficulties encountered in developing reliable incomplete factorizations for indefinite systems has been offered by van der Vorst (Dongarra, Duff, Sorensen and van der Vorst 1998, pages 198–199). When the matrix $\mathcal{A}$ is highly indefinite, it has many eigenvalues on both sides of the imaginary axis. The eigenvalues of the preconditioned matrix $\mathcal{P}^{-1} \mathcal{A}$ depend continuously on the entries of $\mathcal{P}$. For $\mathcal{P} = I$, they coincide with the eigenvalues of $\mathcal{A}$; for $\mathcal{P} = \mathcal{A}$, they are all equal to 1. As the preconditioner $\mathcal{P}$ approaches $\mathcal{A}$, for instance by allowing more and more fill-in in the incomplete factors, the eigenvalues of $\mathcal{P}^{-1} \mathcal{A}$ approach the value 1; in particular, the negative eigenvalues, *en route* to 1, have to cross the imaginary axis. There is a high risk that some (perhaps many) of these eigenvalues will come arbitrarily close to the origin. Hence, 'improving' the preconditioner (by allowing additional fill-in) may actually cause the preconditioned matrix to become very close to singular, which in turn may cause the preconditioned iteration to converge more slowly or even fail.

This non-monotonic behaviour of incomplete factorization preconditioners with respect to fill-in has been often observed in numerical experiments with indefinite matrices. Moreover, serious difficulties are often met in the course of computing an incomplete factorization, owing to various types of instabilities originating from the highly indefinite and non-diagonally dominant nature of saddle point matrices (Chow and Saad 1997b).

It is plausible that using the alternative, nonsymmetric positive definite form (3.10) of the saddle point system may alleviate this problem, particularly when $C \neq O$; however, we are not aware of any experimental studies in this direction. Another possibility would be to ignore the structure or symmetry of $\mathcal{A}$ altogether and apply one of the numerous, time-tested incomplete LU (ILU) factorization algorithms that have been developed for general sparse matrices, combined with some form of pivoting to promote stability. However, since fill-in tends to be very heavy with the original ordering of $\mathcal{A}$, large numbers of fill-ins have to be discarded, often resulting in preconditioners of low quality. Band-reducing and sparsity-preserving symmetric orderings (such as reverse Cuthill–McKee, minimum degree or nested dissection; see Duff $et\ al.$ (1986)) are of limited use here and often produce unstable pivot sequences.

Some degree of success has been achieved through the use of nonsymmetric permutations and scalings aimed at increasing the diagonal dominance of the coefficient matrix; see Benzi, Haws and Tůma (2000), Haws (2002) and Haws and Meyer (2003), where the HSL MC64 preprocessing subroutines (Duff and Koster 1999, 2001) have been used in combination with various ILU and sparse approximate inverse techniques. When applied to saddle point matrices with $C = O$, these permutations produce a matrix with a zero-free diagonal; moreover, the zero (or small) diagonal entries are replaced by (relatively) large nonzeros. The net effect of this preprocessing is that stable ILU-type factorizations can now be computed in most cases. Unfortunately, this preprocessing destroys the symmetry and other structural properties of $\mathcal{A}$, and may lead to unnecessarily high storage demands in some cases.

Very recently, some new approaches have been introduced for constructing sparse incomplete LDL^T factorizations of symmetric indefinite matrices (Fish and Qu 2000, Freund 1997, Hagemann and Schenk 2004, Li and Saad 2004, Qu 2000, Qu and Fish 2000, Qu and Fish 2002). While not necessarily targeted to indefinite matrices in saddle point form, these techniques may be quite effective when applied to such systems; see the results reported in Hagemann and Schenk (2004) and Li and Saad (2004). In particular, maximum weighted matching techniques have been used in Hagemann and Schenk (2004) to minimize the need for dynamic (Bunch-type (Bunch 1974)) pivoting in the course of the factorization (Duff and Pralet 2004). Such matching techniques can be regarded as symmetric analogues of the nonsymmetric permutations used in Duff and Koster (1999, 2001) to permute large entries to the diagonal. Here the purpose is to preprocess the matrix in a symmetry-preserving fashion with the goal of creating 'large' 2×2 diagonal blocks, thus enabling a more stable (incomplete) LDL^T factorization. Li and Saad (2004) instead promote stability by a different kind of preprocessing, referred to as *Bunch equilibration*: see Bunch (1971). These new preconditioners appear to be reasonably efficient and robust. In particular, one of the approaches presented in Hagemann and Schenk (2004) appears to be especially well suited to saddle point matrices. Since no comparison between these general-purpose methods and more specialized preconditioners for saddle point problems has been carried out yet, it is not yet clear how competitive this class of preconditioners really is. Nevertheless, in the engineering community there is certainly strong interest in general-purpose techniques (and software) that can be applied with little or no modification to a wide range of problems, even though such preconditioners may not be the best for any particular problem.

Some work has been done on developing incomplete factorization preconditioners specifically for saddle point matrices. Unlike general-purpose methods, these approaches take into account the block structure of $\mathcal{A}$. One approach is based on the following observation (Zhao 1998, Ren and Zhao 1999). If $A = A^T$ is positive definite, $B_1 = B_2 = B$ is of full rank and $C = C^T$ is positive semidefinite (possibly zero), then

$$\mathcal{A} = \begin{bmatrix} A & B^T \\ B & -C \end{bmatrix} = \begin{bmatrix} L_{11} & O \\ L_{21} & L_{22} \end{bmatrix} \begin{bmatrix} L_{11}^T & L_{21}^T \\ O & -L_{22}^T \end{bmatrix} = \mathcal{L}\mathcal{U}, \qquad (10.36)$$

where $A = L_{11}L_{11}^T$ is the Cholesky factorization of A, $L_{21} = BL_{11}^{-T}$, and $-S = C + L_{21}L_{21}^T = L_{22}L_{22}^T$ is the Cholesky factorization of the (negative) Schur complement. Note that (10.36) is a triangular, Cholesky-like factorization of $\mathcal{A}$; such a factorization always exists, without the need for pivoting (*cf.* Section 7). If we could use $\mathcal{P} = \mathcal{L}\mathcal{L}^T$ as a (split) preconditioner, it is easily verified that

$$\mathcal{L}^{-1}\mathcal{A}\mathcal{L}^{-T} = \begin{bmatrix} I & O \\ O & -I \end{bmatrix},$$

and a symmetric Krylov subspace algorithm like MINRES or SYMMLQ would deliver the solution in at most two iterations. In practice, the exact factor $\mathcal{L}$ is replaced by an incomplete one, as follows. First, an incomplete Cholesky factorization $A \approx \bar{L}_{11}\bar{L}_{11}^T$ is computed; several stable and efficient algorithms exist for this task. Next, we compute a sparse approximation $\bar{L}_{21}$ to L_{21} by solving a matrix equation of the form $\bar{L}_{11}X = B^T$ using backsubstitutions; some form of dropping may be needed to preserve sparsity in $\bar{L}_{21}$. Finally, an incomplete Cholesky factorization $C + \bar{L}_{21}\bar{L}_{21}^T \approx \bar{L}_2\bar{L}_2^T$ is computed. The resulting incomplete factor

$$\bar{\mathcal{L}} = \begin{bmatrix} \bar{L}_{11} & O \\ \bar{L}_{21} & \bar{L}_{22} \end{bmatrix}$$

and its transpose can be used to define a positive definite, factorized pre-conditioner $\mathcal{P} = \bar{\mathcal{L}}\bar{\mathcal{L}}^T$ for a method like SYMMLQ or MINRES. Numerical experiments in Ren and Zhao (1999) indicate that this can be an effective preconditioning strategy. Moreover, the resulting incomplete Cholesky fac-torization of the Schur complement $(-S \approx \bar{L}_{22}\bar{L}_{22}^T)$ can be used as a precon-ditioner for Uzawa's method, resulting in a significant acceleration. Clearly, this approach will only work well when the Schur complement S is sparse or can be well approximated by a sparse matrix; see the discussion in Sec-tion 10.1.3. This approach may be extended to nonsymmetric saddle point systems; in this case, however, the existence of the corresponding incom-plete ILU factorizations may not be guaranteed. Using an ordering different from the standard one may help; see Vasconcelos and D'Almeida (1998) for a study of incomplete LU preconditioning of the discretized Navier–Stokes equations with a nodal ordering of the unknowns.

Finally, we mention that incomplete factorizations for special classes of nonsymmetric saddle point problems have been developed by Wille and collaborators for solving the linear systems arising from Newton's method applied to mixed finite element discretizations of the steady Navier–Stokes equations; see Dahl and Wille (1992), Wille and Loula (2002) and Wille, Staff and Loula (2003). Here A is nonsymmetric (but positive definite), B_1 may or may not be equal to B_2, and $C = O$. In this work, fill-in is allowed only in predetermined locations within the factors, corresponding to the pressure block – *i.e.*, in the (2,2) block of the factors. More precisely, the guiding principle is that fill-in is accepted at locations in the global matrix where the nodes belong to the same finite element. The resulting incomplete factorizations appear to be numerically stable, and satisfactory convergence rates are observed for problems in both 2D and 3D. In Wille *et al.* (2003), parallel implementation aspects are also discussed.

11. Multilevel methods

A survey on the numerical solution of saddle point systems would not be complete without some discussion of multilevel solvers. In this section we give a very brief overview of multigrid and domain decomposition methods for saddle point systems arising from PDE problems. The literature on this topic is vast and highly technical, the construction of multilevel solvers usu-ally being tied to a particular discretization and to analytical properties of the coefficients at the continuous (infinite-dimensional) level. Since our main emphasis is on the linear algebra aspects of finite-dimensional saddle point problems, we refrain from going into details and mostly restrict ourselves to providing pointers to the literature.

11.1. Multigrid methods

As already mentioned repeatedly in the course of this survey, multigrid and, more generally, multilevel methods can be used in segregated approaches to solve reduced systems, as in Uzawa-type or block preconditioning schemes. In the incompressible flow setting these subproblems correspond to discretizations of scalar elliptic PDEs for the velocities and/or the pressure field, such as diffusion- or convection-diffusion-type problems. Multigrid methods are ideally suited for such problems, often achieving optimal computational complexity, in the sense that the number of operations scales linearly with the number of unknowns. In the context of preconditioning inexact solves are usually sufficient, and it may be enough to perform just one iteration (V-, W- or F-cycle). The choice of multigrid components (smoothers, coarse grid operators and intergrid transfer operators) is well understood for this type of PDEs, and a plethora of algorithmic and software tools is available to deal with a wide range of discretizations, properties of coefficients, and problem geometries. Standard treatments of multigrid methods can be found in Hackbusch (1985), Trottenberg, Oosterlee and Schüller (2001) and Wesseling (1992). For detailed discussions of the use of multigrid components in the solution of saddle point problems from finite element discretizations, the reader is referred to Elman *et al.* (2005*c*) and Turek (1999); see also the recent survey by Wesseling and Oosterlee (2001).

More challenging is the construction of *coupled* multilevel methods, *i.e.*, multilevel methods that are applied to the entire system $\mathcal{A}u = b$. This topic has seen extensive development (especially in the last 15 years) and there are now a number of coupled multilevel methods for incompressible flow problems (Braess and Dahmen 1999, Braess and Sarazin 1997, Brandt and Dinar 1979, Elman 1996, Pernice 2000, Vanka 1986, Verfürth 1984*b*, Wesseling 1992, Wesseling 2001, Wittum 1989, Wittum 1990), mixed formulations of second- and fourth-order elliptic equations and of Maxwell's equations (Arnold, Falk and Winther 1997, Braess, Dahmen and Wieners 1999, Hiptmair 1996, Hiptmair 1997, Hiptmair and Hoppe 1999, Hiptmair, Shiekofer and Wohlmuth 1996, Trottenberg *et al.* 2001, Vassilevski and Wang 1992), optimization (Dreyer, Maar and Schulz 2000), and parameter estimation problems (Ascher and Haber 2003). These multigrid schemes may be used alone or as preconditioners for Krylov subspace methods; see Trottenberg *et al.* (2001, pages 287–288) for a discussion of the *solver versus preconditioner* question for multigrid.

For simplicity we consider here the two-grid case only. Our discussion, although fairly general, is geared towards Stokes- and Oseen-type problems and is based on Deng, Piquet, Queutey and Visonneau (1996, Section 4) and Wesseling (2001, Chapter 7.6). Introducing subscripts h and H for the fine and the coarse grid discretization parameters (*e.g.*, $H = 2h$), we rewrite $\mathcal{A}u = b$ as $\mathcal{A}_h u_h = b_h$, and we use the H subscripts for the coarse grid problem. The restriction and prolongation operators are denoted $\mathcal{I}_H^h$ and $\mathcal{I}_h^H$, respectively. The so-called 'full approximation storage' (FAS) two-grid method consists of the following steps:

(i) perform ν_1 pre-smoothing steps on $\mathcal{A}_h u_h = b_h$;
(ii) compute the fine grid residual $r_h = b_h - \mathcal{A}_h u_h$;
(iii) apply the restriction to the fine grid residual, $r_H = \mathcal{I}_H^h r_h$;
(iv) solve the coarse grid problem $\mathcal{A}_H u_H = r_H$;
(v) add the correction, $u_h = u_h + \alpha \mathcal{I}_h^H u_H$;
(vi) perform ν_2 post-smoothing steps.

At step (v), the correction control parameter α is either held fixed (possibly $\alpha = 1$) or chosen adaptively so as to minimize an appropriate measure of the error. A truly multilevel algorithm (V-cycle) is obtained by recursively applying the above procedure in step (iv). There are many variants of the basic scheme just described.

The choice of the restriction and prolongation operators has been discussed at length in the above-mentioned monographs (*e.g.*, in Turek (1999, Chapter 3.4.2)) and will not be pursued here. More critical are the construction of the coarse grid operator $\mathcal{A}_H$ and the choice of the coarse grid solver in step (iv), and the choice of the smoother in steps (i) and (vi). The matrix $\mathcal{A}_H$ can be constructed in at least two ways. One possibility is to simply rediscretize the problem on the coarse grid: this strategy, which is referred to as DCGA (discretization coarse grid approximation), may also be applied to nonlinear problems and is frequently used for the Navier–Stokes equations. An alternative is to compute $\mathcal{A}_H$ as $\mathcal{A}_H = \mathcal{I}_H^h \mathcal{A}_h \mathcal{I}_h^H$. This choice, which is referred to as GCGA (Galerkin coarse grid approximation) is restricted to linear problems, but has the advantage that it can be computed without knowledge of the underlying discretization. Hence, it is the preferred choice for algebraic multigrid (AMG) algorithms, which only exploit information contained in the coefficient matrix (Stüben 2001, Trottenberg *et al.* 2001). If the prolongation operator $\mathcal{I}_h^H$ is taken to be the transpose of the restriction operator $\mathcal{I}_H^h$, then the coarse grid operator $\mathcal{A}_H$ is symmetric if the fine grid one is. Preserving symmetry may be important if the multigrid iteration is to be accelerated by a symmetric Krylov subspace method. In this case the pre- and post-smoothing steps must also be performed in such a way as to preserve symmetry. The restriction operator is usually of the form

$$\mathcal{I}_H^h = \begin{bmatrix} I_H^h & O \\ O & \hat{I}_H^h \end{bmatrix},$$

where I_H^h and $\hat{I}_H^h$ are restriction operators for the x and y unknowns, respectively. If we assume for simplicity that $B_1 = B_2 = B_h$ and $C_h = O$ and take $\mathcal{I}_h^H = (\mathcal{I}_H^h)^T$, the coarse grid matrix has the same form as the fine one,

$$\mathcal{A}_H = \begin{bmatrix} I_H^h & O \\ O & \hat{I}_H^h \end{bmatrix} \begin{bmatrix} A_h & B_h^T \\ B_h & O \end{bmatrix} \begin{bmatrix} I_h^H & O \\ O & \hat{I}_h^H \end{bmatrix} = \begin{bmatrix} A_H & B_H^T \\ B_H & O \end{bmatrix},$$

where $A_H = I_H^h A_h I_h^H$ and $B_H = \hat{I}_H^h B_h I_h^H$. Note that A_H is symmetric if A_h is.

The choice of the solver for the coarse grid problem is not as simple as it may appear at first. Only seldom can the grid be refined to the point that a direct method can be used to solve the coarse problem. This is because grid geometries can be highly irregular, thus allowing only a modest amount of coarsening. It is therefore necessary to resort to iterative methods and inexact coarse grid solves. In this case the multigrid method should be used as a preconditioner for an (outer) flexible Krylov subspace iteration in order to guarantee convergence. Alternatively, a hybrid nonlinear multigrid-inexact Newton method like the one proposed in Pernice (2000) can be used.

We now come to the choice of the smoother. To appreciate the importance of this component, it should be kept in mind that the smoothing steps is where most of the computational effort is usually spent, at least for geometric multigrid. Moreover, using the wrong smoother will destroy the efficiency of the entire multigrid algorithm. The definition of appropriate smoothing operations for saddle point problems is highly problem-dependent, and far from obvious. The smoothers used for scalar elliptic PDEs, such as diffusion- or convection-diffusion-type problems, are typically Gauss–Seidel or damped Jacobi relaxation, possibly combined with an appropriate renumbering of the grid points. These smoothers are not appropriate for saddle point problems, and they are not even defined when $C = O$. One approach is to use Richardson or Kaczmarz-type iterations, which is equivalent to using Jacobi or Gauss–Seidel smoothing on the normal equations (or squared system for symmetric problems): see Hackbusch (1994) and Verfürth (1984b). A better approach is to use some of the stationary iterations (or preconditioners) for saddle point problems discussed in Sections 8–10, but even this is not entirely straightforward.

The so-called *distributive relaxation schemes* (or *transforming iterations*) provide a unified framework for the description of smoothers for saddle point problems: see Brandt and Dinar (1979) and Wittum (1989, 1990). The idea is to transform the original problem $\mathcal{A}u = b$ by right preconditioning (often referred to as *postconditioning* in the multigrid literature) so that the transformed system

$$\mathcal{A}\mathcal{B}v = b, \qquad u = \mathcal{B}v$$

is more easily solved. A splitting

$$\mathcal{A}\mathcal{B} = \mathcal{S} - \mathcal{T}$$

of the transformed matrix induces the following splitting of the original matrix $\mathcal{A}$:

$$\mathcal{A} = \mathcal{S}\mathcal{B}^{-1} - \mathcal{T}\mathcal{B}^{-1}.$$

This splitting defines a stationary iterative method for the original system,

$$\mathcal{S}\mathcal{B}^{-1}u_{k+1} = \mathcal{T}\mathcal{B}^{-1}u_k + b$$

or, in correction form,

$$u_{k+1} = u_k + \mathcal{B}\mathcal{S}^{-1}(b - \mathcal{A}u_k). \qquad (11.1)$$

As long as $\mathcal{B}$ and $\mathcal{S}$ are nonsingular, the iteration (11.1) is consistent with $\mathcal{A}u = b$, in the sense that if it converges, it converges to the solution $u_* = \mathcal{A}^{-1}b$. The scheme (11.1) is called a *distributive relaxation* in the multigrid literature. The reason is that the correction $\mathcal{S}^{-1}(b - \mathcal{A}u_k)$ corresponding to non-distributive ($\mathcal{B} = I$) iterations is distributed over the entries of u_{k+1}; see Brandt and Dinar (1979) and Wesseling (2001, page 295).

A number of different distributive iterations can be obtained by special choices of $\mathcal{B}$ and $\mathcal{S}$. One possibility is to pick $\mathcal{B}$ such that $\mathcal{A}\mathcal{B}$ is block lower triangular. For instance, the block factorization (3.2) suggests the choice

$$\mathcal{B} = \begin{bmatrix} I & -A^{-1}B^T \\ O & I \end{bmatrix} \quad \Rightarrow \quad \mathcal{A}\mathcal{B} = \begin{bmatrix} A & O \\ B & S \end{bmatrix}.$$

This leads to a decoupling (segregation) of the x and y variables, and various choices of $\mathcal{S}$ lead essentially to preconditioned versions of Uzawa's method. A slightly more general form of $\mathcal{B}$ is

$$\mathcal{B} = \begin{bmatrix} I & -A^{-1}B^T Q \\ O & Q \end{bmatrix},$$

which leads to the transformed matrix

$$\mathcal{A}\mathcal{B} = \begin{bmatrix} A & O \\ B & M \end{bmatrix}, \qquad M = -BA^{-1}B^T Q.$$

A variety of methods result from the choice of Q and $\mathcal{S}$. For example, taking $Q = I$ and

$$\mathcal{S} = \begin{bmatrix} \hat{A} & O \\ B & \hat{S} \end{bmatrix},$$

where $\hat{A}$ and $\hat{S}$ are easily invertible approximations of A and of the Schur complement S, results in the SIMPLE scheme (Patankar and Spalding 1972, Patankar 1980) already described in the context of block preconditioning in Section 10.1.2. In the original SIMPLE scheme, $\hat{A} = D_A$ and $\hat{S} =$

$-BD_A^{-1}B^T$, where D_A denotes the main diagonal of A. The matrix $\hat{\mathcal{B}}$ used for the distribution step is further approximated by

$$\hat{\mathcal{B}} = \begin{bmatrix} I & -\hat{A}^{-1}B^T \\ O & I \end{bmatrix}.$$

Different choices of the approximations involved lead to variations on the original SIMPLE scheme (Patankar 1980, 1981, Shaw and Sivaloganathan 1988). Other options include the distributive Gauss–Seidel iteration and distributive ILU methods; see Brandt and Dinar (1979), Wittum (1989, 1990) and the nice description in Wesseling (2001, Chapter 7.6). For the Stokes problem, good results have been reported in Braess and Sarazin (1997) using as a smoother an iteration of the form (10.18), *i.e.*, a constraint-type preconditioner. This paper also discusses several variants of the SIMPLE scheme and shows that not all variants result in good smoothers. See also Zulehner (2000) for a generalization to the case $C \neq O$. Smoothers of the form (10.18) have also been used in connection with AMG methods for the Oseen problem in Wabro (2004), and in Leem *et al.* (2004) in the context of meshless discretizations. Recently, block triangular preconditioners of the kind discussed in Section 10.1 (so-called pressure convection-diffusion preconditioners) have been used with good results as smoothers for multigrid methods applied to the Oseen problem in Syamsudhuha and Silvester (2003). See Brandt (1998) for a discussion of the barriers that remain to be overcome before *textbook multigrid efficiency* (*i.e.*, solution of the governing equations in a number of operations not exceeding a small multiple of the nonzeros in the discretized equations) can be achieved in realistic computational fluid dynamics simulations. See also Thomas, Diskin and Brandt (2001) for recent progress in this direction.

All the smoothers discussed so far can be implemented in a segregated fashion, with each relaxation involving solution of decoupled linear systems for x and y. A strongly coupled smoothing procedure, not of the distributive type, has been introduced by Vanka (1986). The main principle of Vanka smoothing as applied to fluid flow problems is to visit each cell in some order and to apply relaxation simultaneously to each of the variables associated with that cell. This is equivalent to a reordering of the unknowns so that all the velocities and the pressures associated with each cell are numbered consecutively. Both multiplicative (Gauss–Seidel-type) and additive (Jacobi-type) variants have been developed. Vanka-type smoothing requires solving, at each step, a number of tiny saddle point problems, one for each cell. Various approximations and suitable damping (under-relaxation) are used to improve efficiency and to ensure the smoothing property. We refer to the original paper (Vanka 1986) as well as to Wesseling (2001, pages 300–301) for additional details. We note that this type of smoothing can also be interpreted as a form of domain decomposition, each subdomain reducing to a single grid cell.

Problems involving unstructured meshes on complicated geometries have motivated the development of AMG methods. AMG algorithms for scalar problems have been around for almost 20 years and have been used for some time in the solution of subproblems from segregated approaches; see, *e.g.*, Griebel, Neunhoeffer and Regler (1998), Stüben (2001) and Wagner, Kinzelbach and Wittum (1997). Coupled AMG methods for saddle point systems are barely 10 years old and are currently under development. We mention Raw (1995), Wabro (2004) and Webster (1994) for applications to incompressible flow problems, Adams (2004) for the development of AMG solvers in solid mechanics, and Leem *et al.* (2004) for AMG methods for saddle point systems arising from mesh-free discretizations. The reported results on difficult problems are promising. However, many open questions concerning coarsening strategies, the choice of appropriate smoothers, and parallel implementation issues remain to be addressed before coupled AMG solvers for saddle point problems can be considered fully mature.

Multilevel preconditioners for discontinuous Galerkin mixed finite element discretizations of radiation-diffusion problems on unstructured meshes have been developed in Warsa *et al.* (2004). Here the main idea is to use a continuous finite element approximation of the same problem to precondition the discontinuous one. Although only one mesh needs to be generated, suitable restriction and prolongation operators are needed to transfer information between the two discretizations. The continuous approximation can be solved efficiently by standard algorithms like AMG or preconditioned CG. This approach, which results in nearly optimal rates of convergence, is akin to an approximate Schur complement method. It is worth noting that in both Leem *et al.* (2004) and Warsa *et al.* (2004) the nonsymmetric formulation (3.10) was found to be advantageous, even though the original problems were symmetric.

Finally we mention some recent work on wavelet-based, *multiscale* methods. Originally developed for positive definite elliptic operators, these techniques have recently been extended to symmetric indefinite saddle point problems; see Dahlke, Hochmuth and Urban (2000), Hochmuth (1998) and Kunoth (2002). These are discretization schemes that produce saddle point systems that are well-conditioned, and can be efficiently solved by either Uzawa-type schemes or by diagonally preconditioned Krylov methods, resulting in convergence rates independent of discretization parameters.

11.2. Domain decomposition methods

Just as in the case of multigrid, domain decomposition methods can be applied straightforwardly to the (elliptic, scalar) subproblems arising from segregated approaches; see, *e.g.*, Fischer and Lottes (2004) for a recent example. Coupled algorithms have recently been developed by several authors, in particular for fluid flow and linear elasticity problems. Such methods are seldom (if ever) used as solvers: rather, they are used either as preconditioners for Krylov subspace methods, or as smoothers for multigrid, often motivated by parallel processing. We refer the reader to Chan and Mathew (1994) and Quarteroni and Valli (1999) for general treatments of domain decomposition methods, and to Toselli and Widlund (2004, Chapter 9) for an excellent discussion of methods designed specifically for saddle point problems. Because an up-to-date overview is already available, and also because of space limitations, we do not go into details and we largely limit ourselves to providing pointers to the literature.

Substructuring and additive Schwarz-type preconditioners for the symmetric saddle point problems arising from mixed finite element discretizations of second-order elliptic PDEs have been proposed and analysed in Rusten *et al.* (1996) and Rusten and Winther (1993). Different boundary conditions (Dirichlet or Neumann) on the interior interfaces are considered. The preconditioned systems are symmetrizable (so that MINRES can be used), and have condition numbers that are independent of mesh size.

An algebraic additive Schwarz domain decomposition preconditioner for solving saddle point problems arising from mixed finite element simulations of stochastic PDEs modelling flow in heterogeneous media has been described in Cliffe, Graham, Scheichl and Stals (2000). In this paper the preconditioner is not applied to to the coupled saddle point problem, but rather to a reduced (SPD) system of the type (6.1), where Z is a suitably constructed solenoidal basis. The resulting solver exhibits good robustness with respect to problem parameters and almost optimal levels of parallel efficiency. See also Oswald (1998) for a multilevel preconditioner for systems of the form (6.1) arising in the context of Stokes problems.

Several substructuring and domain decomposition preconditioners exist for linear elasticity and Stokes problems. Overlapping Schwarz preconditioners for saddle point systems containing a nonzero (2,2) block that depends on a penalty parameter (as in (10.9)) were proposed by Klawonn and Pavarino (1998). These preconditioners require the solution of local saddle point problems on overlapping subdomains, plus the solution of a coarse-level saddle point problem. Numerical experiments show that these preconditioners result in convergence rates that are independent of the mesh size, the number of subdomains, and the penalty parameter. The additive preconditioners are scalable also in the sense of parallel efficiency. In Klawonn and Pavarino (2000), the additive Schwarz preconditioner proposed in Klawonn and Pavarino (1998) is compared experimentally with a block diagonal and with a block triangular preconditioner (both with inexact solves). The results indicate that the additive Schwarz preconditioner (used with

GMRES) is superior to the other methods in terms of iteration counts and very often in terms of operation counts, with the block triangular preconditioner being occasionally slightly better in terms of operation count. A convergence analysis for the additive Schwarz preconditioner, however, is still lacking. The preconditioned matrices have complex eigenvalues and an indefinite symmetric part. On the other hand, they appear to be positive stable with eigenvalues contained in a rectangular region of the right half-plane bounded away from zero, which explains in part the good convergence properties of GMRES. In Pavarino (2000) the overlapping Schwarz preconditioner is applied to mixed spectral or finite element discretizations of time-dependent Stokes problems, again with excellent results.

A different approach, based on iterative substructuring, was introduced and analysed by Pavarino and Widlund (2000) in the context of spectral element mixed approximations. A related balancing Neumann–Neumann approach for the Stokes problem is proposed in Pavarino and Widlund (2002), and extended to heterogeneous problems from linear elasticity in Goldfeld, Pavarino and Widlund (2003). This is a non-overlapping hybrid domain decomposition method in which the local problems are treated additively while the coarse problem is treated multiplicatively. Theoretical and numerical evidence show that these methods have excellent convergence and scalability properties; see in particular Goldfeld (2000) for numerical experiments with problems with up to 100 million unknowns using parallel machines with up to 2000 processors. We further mention recent work by Li on FETI methods for incompressible flow problems; see Li (2001, 2002a, 2002b).

Another domain decomposition method for linear elasticity problems was introduced by Klawonn and Widlund (2000). This preconditioner uses inexact subdomain solves and Lagrange multipliers and is based on a reformulation of the popular FETI method as a saddle point problem with both primal and dual variables as unknowns. It is shown in Klawonn and Widlund (2000) that the condition number of the preconditioned matrices is bounded independently of the number of subdomains and grows only polylogarithmically with the number of unknowns in each subdomain. See also Klawonn and Widlund (2001), and, for recent developments, Klawonn and Widlund (2004). We also mention the overlapping additive and multiplicative two-level Schwarz methods proposed in Wang (2005) for plane elasticity problems, which exhibit convergence rates independent of mesh size.

In Ainsworth and Sherwin (1999), p and h-p finite element discretizations of the Stokes problem are considered. Both segregated (block diagonal) and coupled (indefinite) preconditioners of the additive Schwarz type are studied. For each type of scheme, eigenvalue bounds for the preconditioned matrices are derived in terms of p (the polynomial degree), the fine and coarse mesh sizes, and the inf-sup constant for the method. These estimates and the results of actual computations on representative problems show that the rate of convergence for both methods does not deteriorate significantly as the mesh is refined and the polynomial degree is increased.

We note that while there is a rich literature on domain decomposition methods for symmetric saddle point problems and in particular for mixed discretizations of the Stokes equations, there seems to be very few papers concerned with coupled domain decomposition methods for nonsymmetric saddle point systems and in particular for the Oseen equations. One such paper is Vainikko and Graham (2004), where domain decomposition methods are implemented and experimentally compared with (and found to be superior to) block triangular preconditioners for the Oseen problem. Theoretical understanding, however, is still largely lacking.

Substructuring preconditioners for saddle point problems arising from edge element discretizations of Maxwell's equation in 3D have been introduced and analysed by Hu and Zou (2003). These preconditioners are shown to result in nearly optimal convergence rates, with the condition number of the preconditioned matrices growing as the logarithm of the ratio between the subdomain diameter and the finite element mesh size.

Domain decomposition methods for saddle point systems arising from PDE-constrained optimal control problems have been studied in Heinkenschloss and Nguyen (2004) and Nguyen (2004). Numerical experiments indicate that the performance of these (overlapping and non-overlapping) preconditioners with respect to mesh size and number of subdomains is close to that of the corresponding domain decomposition preconditioners for scalar PDEs. Furthermore, the preconditioners proposed in Nguyen (2004) appear to be rather insensitive to control regularization parameters.

Finally, the application of additive Schwarz-type iterations as smoothers in coupled multigrid methods for saddle point problems has been studied in Schöberl and Zulehner (2003). These are additive variants of the already-mentioned (multiplicative) Vanka smoother (Vanka 1986). The additive smoothers are shown to be related to an inexact version of Uzawa's method. An analysis of the smoothing property in the symmetric case is also given. Numerical experiments indicate that the rates of convergence for the multiplicative Vanka-type smoothers are significantly better than for the additive smoothers, as one would expect; however, a theoretical analysis of the convergence and smoothing properties in the multiplicative case is still missing.

12. Available software

In spite of vigorous algorithmic and theoretical developments, the production of high-quality, widely accessible software for solving linear systems in saddle point form has been somewhat lagging. This may be in part a reflection of the fact that many saddle point systems arising in the applications require a customized approach – for example, a particular Schur complement approximation. Another explanation is that when segregated approaches are used, standard sparse linear equations packages are often sufficient to solve the reduced linear systems. For example, many of the existing algorithms for saddle point problems from incompressible flow calculations require the availability of efficient solvers for diffusion- or convection-diffusion-type problems, and can utilize a number of 'standard' multigrid codes. In other situations, sparse direct solvers may be used to solve the subproblems arising from segregated approaches. Because many general-purpose linear solver packages exist and are readily accessible, there has been no great demand for software specifically designed to solve linear systems in saddle point form. An exception is represented by the field of optimization, where the interior point revolution has sparked a strong need for reliable and efficient solvers for augmented systems.

As usual, the software situation tends to be better for direct solvers than for iterative ones. Professional implementations of sparse direct solvers for symmetric indefinite systems include the MA27/MA47/MA57 suite of codes from the Harwell Subroutine Library (HSL) and the PARDISO package from the University of Basel (Schenk and Gärtner 2004). Of the HSL codes, the one that is best suited for saddle point matrices is perhaps MA47, although MA57 appears to be the most efficient overall of the symmetric solvers among the HSL codes (Gould and Scott 2004). For further information on the HSL symmetric indefinite solvers, including licensing information, the reader is referred to the web page

www.cse.clrc.ac.uk/nag/hsl/contents.shtml

For information about the PARDISO package, see

www.computational.unibas.ch/cs/scicomp/software/pardiso/

Although not specifically designed with saddle point systems in mind, the out-of-core sparse symmetric indefinite code developed as part of the TAUCS package at Tel Aviv University is also of interest; see

www.tau.ac.il/~stoledo/taucs/

and the accompanying paper by Meshar and Toledo (2005).

Software tools for incorporating block preconditioners geared towards incompressible flow problems into Krylov subspace methods are now available in the Meros package, which is part of the Trilinos project from Sandia National Laboratories (Heroux, Bartlett, Howle, Hoekstra, Hu, Kolda, Lehoucq, Long, Pawlowski, Phipps, Salinger, Thornquist, Tuminaro, Willenbring, Williams and Stanley 2005); see

http://software.sandia.gov/trilinos/

Many of the solvers within Trilinos may also be used as part of segregated approaches; for instance, multilevel solvers are part of the ML package, which is part of Trilinos.

Algebraic multigrid codes for incompressible flow problems, including Markus Wabro's AMuSE (Algebraic Multigrid for Stokes-type Equations) package, can be accessed from the Institute of Computational Mathematics at the Johannes Kepler University (Linz, Austria) web site

www.numa.uni-linz.ac.at/Research/Projects/P14953.html

A number of other freely available codes implementing a host of multigrid and domain decomposition methods can be downloaded from the MGNET web site; see

www.mgnet.org/mgnet-codes.html

We conclude this brief – and by no means complete – overview of available software with the IFISS ('Incompressible Flow Iterative Solvers Software') package developed by Howard Elman, Alison Ramage, David Silvester and Andy Wathen: see

www.cs.umd.edu/~elman/ifiss/

See also the book by Elman *et al.* (2005*c*). This is a finite element package that can be used to generate a variety of incompressible flow problems on both structured and unstructured 2D meshes. The code, written in MAT-LAB, allows the user to experiment with several preconditioner and solver options.

13. Concluding remarks

In this paper we have surveyed a large number of solution methods for solving linear systems in saddle point form, with a focus on iterative methods suitable for large and sparse problems. We have discussed classical algorithms based on Schur complement reduction, null space methods, triangular factorization and stationary iterations like the Arrow–Hurwicz and Uzawa schemes and their variants. We have further discussed the use of Krylov subspace methods and described a number of preconditioning techniques including block diagonal, block triangular, constraint and incomplete factorization preconditioners. We have reviewed a promising new approach based on the Hermitian and skew-Hermitian splitting which appears to be especially well suited to the Navier–Stokes equations in rotation form. We saw that many preconditioning techniques rely (either explicitly or implicitly) on the availability of good and inexpensive approximations of a Schur complement matrix; similar comments apply to other solvers, such as multigrid or domain decomposition methods. We have seen that very effective solvers exist for some classes of problems, including mixed finite element formulations of elliptic PDEs and Stokes problems. Furthermore, great strides have been made in recent years towards the development of effective preconditioners for strongly nonsymmetric problems, such as the Oseen equations with low viscosity.

In spite of this, many challenges remain. Effective preconditioners are yet to be developed for large classes of linear systems arising from interior point methods in constrained optimization. Some degree of success has been achieved in this area with various types of approximate Schur complement and constraint preconditioners, but there is plenty of room for improvement. For many saddle point systems arising from optimal control problems, the (1,1) block A is often symmetric indefinite. Although these systems may sometimes be transformed to reduced saddle point systems with a definite (1,1) block, this reduction is not always easy to accomplish and it may be necessary to deal with the unreduced system. Hence, there is a need for effective solvers that are able to handle saddle point matrices with an indefinite (1,1) block. While some of the techniques described in this paper may be adapted to this situation, there is a need for new approaches. Also needed is a thorough study of stopping criteria for iterative methods applied to saddle point problems. In particular, criteria that take into account the different nature of the unknowns x and y should be developed.

Space and time limitations did not allow us to give very detailed treatments of many interesting techniques. In particular, important subjects such as direct solvers and multilevel methods have been touched upon only very briefly. Sparse direct solvers have been around for a long time and are widely used in some areas (such as optimization), but there are still some difficulties to overcome before these methods can be considered fully mature for solving saddle point systems: see for instance Gould and Scott (2004), where it is mentioned that the symmetric HSL sparse direct solvers could not solve four out of sixty-one of the test problems considered there. The same holds true, of course (even more so!), for iterative solvers. The ever-increasing complexity and size of the linear systems to be solved is already making the use of iterative methods absolutely mandatory for many applications, thus requiring continually improving solvers and preconditioners, with scalability more than ever a central concern. In line with current trends, it is quite likely that the future will see many new and improved multilevel algorithms, both geometric and algebraic, for solving large-scale saddle point problems from PDEs and PDE-constrained optimization.

Acknowledgements

During the preparation of this paper we have benefited from the help and advice of numerous colleagues and friends. Many thanks to Peter Arbenz,

Mario Arioli, Howard Elman, Chen Greif, Eldad Haber, Amnon Meir, Luca Pavarino, Miro Rozložník, Wil Schilders, Daniel Szyld and Andy Wathen for informative discussions and pointers to the literature. We are especially thankful to Jacek Gondzio, Jim Nagy, Maxim Olshanskii, Michael Saunders and Valeria Simoncini for their careful reading of the manuscript and for their helpful comments.

REFERENCES

R. Aboulaich and M. Fortin (1989), 'Iterative methods for the solution of Stokes equations', *Comput. Methods Appl. Mech. Engrg.* **75**, 317–324.

M. F. Adams (2004), 'Algebraic multigrid methods for constrained linear systems with applications to contact problems in solid mechanics', *Numer. Linear Algebra Appl.* **11**, 141–153.

M. Ainsworth and S. Sherwin (1999), 'Domain decomposition preconditioners for p and hp finite element approximation of Stokes equations', *Comput. Methods Appl. Mech. Engrg.* **175**, 243–266.

P. Alotto and I. Perugia (1999), 'Mixed finite element methods and tree–cotree implicit condensation', *Calcolo* **36**, 233–248.

P. Amestoy, I. S. Duff and C. Puglisi (1996), 'Multifrontal QR factorization in a multiprocessor environment', *Numer. Linear Algebra Appl.* **3**, 275–300.

R. Amit, C. A. Hall and T. A. Porsching (1981), 'An application of network theory to the solution of implicit Navier–Stokes difference equations', *J. Comput. Phys.* **40**, 183–201.

G. Ammar, C. Mehl and V. Mehrmann (1999), 'Schur-like forms for matrix Lie groups, Lie algebras and Jordan algebras', *Linear Algebra Appl.* **287**, 11–39.

P. Arbenz and R. Geus (2005), 'Multilevel preconditioned iterative eigensolvers for Maxwell eigenvalue problems', *Appl. Numer. Math.*, to appear.

P. Arbenz, R. Geus and S. Adam (2001), 'Solving Maxwell eigenvalue problems for accelerating cavities', *Phys. Rev. ST Accel. Beams* **4**, # 022001.

M. Arioli (2000), 'The use of QR factorization in sparse quadratic programming', *SIAM J. Matrix Anal. Appl.* **21**, 825–839.

M. Arioli and L. Baldini (2001), 'A backward error analysis of a null space algorithm in sparse quadratic programming', *SIAM J. Matrix Anal. Appl.* **23**, 425–442.

M. Arioli and G. Manzini (2002), 'A null space algorithm for mixed finite-element approximations of Darcy's equation', *Comm. Numer. Meth. Engng.* **18**, 645–657.

M. Arioli and G. Manzini (2003), 'Null space algorithm and spanning trees in solving Darcy's equation', *BIT* **43**, 839–848.

M. Arioli, I. S. Duff and P. P. M. De Rijk (1989), 'On the augmented system approach to sparse least-squares problems', *Numer. Math.* **55**, 667–684.

M. Arioli, J. Maryška, M. Rozložník and Tůma (2001), Dual variable methods for mixed-hybrid finite element approximation of the potential fluid flow problem in porous media, Technical Report RAL-TR-2001-023, Rutherford Appleton Laboratory, RAL-CLRC, Oxfordshire, UK.

D. N. Arnold, R. S. Falk and R. Winther (1997), 'Preconditioning in H(div) and applications', *Math. Comp.* **66**, 957–984.

W. E. Arnoldi (1951), 'The principle of minimized iteration in the solution of the matrix eigenvalue problem', *Quart. Appl. Math.* **9**, 17–29.

K. J. Arrow and L. Hurwicz (1958), Gradient method for concave programming I: Local results, in *Studies in Linear and Nonlinear Programming* (K. J. Arrow, L. Hurwicz and H. Uzawa, eds), Stanford University Press, Stanford, CA, pp. 117–126.

K. J. Arrow and R. M. Solow (1958), Gradient methods for constrained maxima, with weakened assumptions, in *Studies in Linear and Nonlinear Programming* (K. J. Arrow, L. Hurwicz and H. Uzawa, eds), Stanford University Press, Stanford, CA, pp. 166–176.

K. J. Arrow, L. Hurwicz and H. Uzawa (1958), *Studies in Linear and Nonlinear Programming*, Stanford University Press, Stanford, CA.

U. M. Ascher and E. Haber (2003), 'A multigrid method for distributed parameter estimation problems', *Electron. Trans. Numer. Anal.* **15**, 1–17.

C. Ashcraft, R. Grimes and J. G. Lewis (1998), 'Accurate symmetric indefinite linear equation solvers', *SIAM J. Matrix Anal. Appl.* **20**, 513–561.

G. P. Astrakhantsev (2001), 'Analysis of algorithms of the Arrow–Hurwicz type', *Comput. Math. Math. Phys.* **41**, 15–26.

G. M. Awanou and M. J. Lai (2005), 'On convergence rate of the augmented Lagrangian algorithm for nonsymmetric saddle point problems', *Appl. Numer. Math.*, to appear.

O. Axelsson (1979), 'Preconditioning of indefinite problems by regularization', *SIAM J. Numer. Anal.* **16**, 58–69.

O. Axelsson (1994), *Iterative Solution Methods*, Cambridge University Press, Cambridge.

O. Axelsson and M. Neytcheva (2003), 'Preconditioning methods for linear systems arising in constrained optimization problems', *Numer. Linear Algebra Appl.* **10**, 3–31.

O. Axelsson and P. S. Vassilevski (1991), 'A black-box generalized conjugate gradient solver with inner iterations and variable-step preconditioning', *SIAM J. Matrix Anal. Appl.* **12**, 625–644.

A. Baggag and A. Sameh (2004), 'A nested iterative scheme for indefinite linear systems in particulate flows', *Comput. Methods Appl. Mech. Engrg.* **193**, 1923–1957.

Z.-Z. Bai, G. H. Golub and M. K. Ng (2003), 'Hermitian and skew-Hermitian splitting methods for non-Hermitian positive definite linear systems', *SIAM J. Matrix Anal. Appl.* **24**, 603–626.

R. E. Bank, B. D. Welfert and H. Yserentant (1990), 'A class of iterative methods for solving saddle point problems', *Numer. Math.* **56**, 645–666.

J. L. Barlow (1988), 'Error analysis and implementation aspects of deferred correction for equality-constrained least squares problems', *SIAM J. Numer. Anal.* **25**, 1340–1358.

J. L. Barlow and S. L. Handy (1988), 'The direct solution of weighted and equality constrained least squares problems', *SIAM J. Sci. Statist. Comput.* **9**, 704–716.

J. L. Barlow, N. K. Nichols and R. J. Plemmons (1988), 'Iterative methods for equality-constrained least squares problems', *SIAM J. Sci. Statist. Comput.* **9**, 892–906.

T. Barth and T. Manteuffel (1994), Variable metric conjugate gradient methods, in *Advances in Numerical Methods for Large Sparse Sets of Linear Equations: Proc. 10th International Symposium on Matrix Analysis and Parallel Computing, PCG 94* (M. Natori and T. Nodera, eds), pp. 165–188.

J. R. Batt and S. Gellin (1985), 'Rapid reanalysis by the force method', *Comput. Methods Appl. Mech. Engrg.* **53**, 105–117.

A. Battermann and M. Heinkenschloss (1998), Preconditioners for Karush–Kuhn–Tucker systems arising in the optimal control of distributed systems, in *Optimal Control of Partial Differential Equations* (W. Desch, F. Kappel and K. Kunisch, eds), Birkhäuser, pp. 15–32.

A. Battermann and E. W. Sachs (2001), Block preconditioners for KKT systems arising in PDE-governed optimal control problems, in *Fast Solution of Discretized Optimization Problems, Berlin 2000* (K.-H. Hoffmann, R. H. W. Hoppe and W. Schulz, eds), Birkhäuser, pp. 1–18.

J. Bear (1972), *Dynamics of Fluids in Porous Media*, American Elsevier, New York.

B. Beckermann and A. B. J. Kuijlaars (2001), 'Superlinear convergence of conjugate gradients', *SIAM J. Numer. Anal.* **39**, 300–329.

B. Beckermann and A. B. J. Kuijlaars (2002), 'Superlinear CG convergence for special right-hand sides', *Electron. Trans. Numer. Anal.* **14**, 1–19.

S. J. Benbow (1999), 'Solving generalized least-squares problems with LSQR', *SIAM J. Matrix Anal. Appl.* **21**, 166–177.

M. Benzi (1993), 'Solution of equality-constrained quadratic programming problems by a projection iterative method', *Rend. Mat. Appl.* (7) **13**, 275–296.

M. Benzi (2002), 'Preconditioning techniques for large linear systems: a survey', *J. Comput. Phys* **182**, 418–477.

M. Benzi (2004), HSS preconditioning for the Oseen problem, in *Proc. European Congress on Computational Methods in Applied Sciences and Engineering: ECCOMAS 2004, Jyväskylä, Finland* (P. Neittaanmäki, T. Rossi, S. Korotov, E. Oñate, J. Périaux and D. Knörzer, eds), CD-ROM.

M. Benzi and G. H. Golub (2004), 'A preconditioner for generalized saddle point problems', *SIAM J. Matrix Anal. Appl.* **26**, 20–41.

M. Benzi and J. Liu (2005), 'An efficient solver for the incompressible Navier–Stokes equations in rotation form', *J. Comput. Phys.*, submitted.

M. Benzi and M. K. Ng (2004), 'Preconditioned iterative methods for weighted Toeplitz least squares', *SIAM J. Matrix Anal. Appl.*, submitted.

M. Benzi, M. J. Gander and G. H. Golub (2003), 'Optimization of the Hermitian and skew-Hermitian splitting iteration for saddle-point problems', *BIT* **43**, 881–900.

M. Benzi, J. C. Haws and M. Tůma (2000), 'Preconditioning highly indefinite and nonsymmetric matrices', *SIAM J. Sci. Comput.* **22**, 1333–1353.

L. Bergamaschi, J. Gondzio and G. Zilli (2004), 'Preconditioning indefinite systems in interior point methods for optimization', *Comput. Optim. Appl.* **28**, 149–171.

A. R. Bergen (1986), *Power Systems Analysis*, Prentice-Hall, Englewood Cliffs, NJ.

M. W. Berry and R. J. Plemmons (1987), 'Algorithms and experiments for structural mechanics on high-performance architectures', *Comput. Methods Appl. Mech. Engrg.* **64**, 487–507.

M. W. Berry, M. T. Heath, I. Kaneko, M. Lawo, R. J. Plemmons and R. C. Ward (1985), 'An algorithm for computing a sparse basis of the null space', *Numer. Math.* **47**, 483–504.

F. Bertrand and P. A. Tanguy (2002), 'Krylov-based Uzawa algorithms for the solution of the Stokes equations using discontinuous-pressure tetrahedral finite elements', *J. Comput. Phys.* **181**, 617–638.

D. P. Bertsekas (1982), *Constrained Optimization and Lagrange Multiplier Methods*, Academic Press, New York.

J. T. Betts (2001), *Practical Methods for Optimal Control Using Nonlinear Programming*, Advances in Design and Control, SIAM, Philadelphia, PA.

G. Biros and O. Ghattas (2000), Parallel preconditioners for KKT systems arising in optimal control of viscous incompressible flows, in *Proceedings of Parallel CFD '99, Williamsburg, VA, May 23–26, 1999* (D. Keyes, A. Ecer, J. Periaux and N. Satofuka, eds), North-Holland, pp. 1–8.

G. Biros and O. Ghattas (2005*a*), 'Parallel Lagrange–Newton–Krylov methods for PDE-constrained optimization, Part I: The Krylov-Schur solver', *SIAM J. Sci. Comput.*, to appear.

G. Biros and O. Ghattas (2005*b*), 'Parallel Lagrange–Newton–Krylov methods for PDE-constrained optimization, Part II: The Lagrange–Newton solver and its application to optimal control of steady viscous flows', *SIAM J. Sci. Comput.*, to appear.

L. Bitar and C. Vincent (2000), 'Eigenvalue upper bounds for the discretized Stokes operator', *Commun. Numer. Meth. Engng.* **16**, 449–457.

Å. Björck (1996), *Numerical Methods for Least Squares Problems*, SIAM, Philadelphia, PA.

Å. Björck and C. C. Paige (1994), 'Solution of augmented linear systems using orthogonal factorizations', *BIT* **34**, 1–24.

E. Y. Bobrovnikova and S. A. Vavasis (2000), 'A norm bound for projections with complex weights', *Linear Algebra Appl.* **307**, 69–75.

E. Y. Bobrovnikova and S. A. Vavasis (2001), 'Accurate solution of weighted least squares by iterative methods', *SIAM J. Matrix Anal. Appl.* **22**, 1153–1174.

P. Bochev and R. B. Lehoucq (2005), 'On finite element solution of the pure Neumann problem', *SIAM Rev.* **47**, 50–66.

A. Bojanczyk, N. J. Higham and H. Patel (2003), 'Solving the indefinite least squares problem by hyperbolic QR factorization', *SIAM J. Matrix Anal. Appl.* **24**, 914–931.

C. W. Bomhof and H. A. van der Vorst (2000), 'A parallel linear system solver for circuit-simulation problems', *Numer. Linear Algebra Appl.* **7**, 649–665.

I. Bongartz, A. R. Conn, N. I. M. Gould and P. L. Toint (1995), 'CUTE: Constrained and unconstrained testing environment', *ACM Trans. Math. Software* **21**, 123–160.

A. Bossavit (1998), ' "Mixed" systems of algebraic equations in computational electromagnetics', *COMPEL* **17**, 59–63.

M. A. Botchev and G. H. Golub (2004), A class of nonsymmetric preconditioners for saddle point problems, Technical Report SCCM-04-14, Scientific Computing and Computational Mathematics, Stanford University, Stanford, California.

D. Braess (2001), *Finite Elements. Theory, Fast Solvers, and Applications in Solid Mechanics*, second edn, Cambridge University Press, Cambridge.

D. Braess and W. Dahmen (1999), 'A cascadic multigrid algorithm for the Stokes equations', *Numer. Math.* **82**, 179–191.

D. Braess and R. Sarazin (1997), 'An efficient smoother for the Stokes problem', *Appl. Numer. Math.* **23**, 3–19.

D. Braess, W. Dahmen and C. Wieners (1999), 'A multigrid algorithm for the mortar finite element method', *SIAM J. Numer. Anal.* **37**, 48–69.

J. H. Bramble and J. E. Pasciak (1988), 'A preconditioning technique for indefinite systems resulting from mixed approximations of elliptic problems', *Math. Comp.* **50**, 1–17.

J. H. Bramble and J. E. Pasciak (1997), 'Iterative techniques for time-dependent Stokes problems', *Comput. Math. Appl.* **33**, 13–30.

J. H. Bramble, J. E. Pasciak and A. T. Vassilev (1997), 'Analysis of the inexact Uzawa algorithm for saddle point problems', *SIAM J. Numer. Anal.* **34**, 1072–1092.

J. H. Bramble, J. E. Pasciak and A. T. Vassilev (2000), 'Uzawa type algorithms for nonsymmetric saddle point problems', *Math. Comp.* **69**, 667–689.

A. Brandt (1998), Barriers to achieving textbook multigrid efficiency (TME) in CFD, Gauss Center Technical Report WI-GC 10, Weizmann Institute of Science, Rehovot, Israel.

A. Brandt and D. Dinar (1979), Multigrid solutions to elliptic flow problems, in *Numerical Methods for Partial Differential Equations* (S. Parter, ed.), Academic Press, New York. pp. 53–147.

F. Brezzi (1974), 'On the existence, uniqueness, and approximation of saddle-point problems arising from Lagrangian multipliers', *RAIRO Anal. Numér.* **8**, 129–151.

F. Brezzi (2002), Stability of saddle-points in finite dimensions, in *Frontiers in Numerical Analysis* (J. F. Blowey, A. W. Craig and T. Shardlow, eds), Springer, Berlin, pp. 17–61.

F. Brezzi and M. Fortin (1991), *Mixed and Hybrid Finite Element Methods*, Vol. 15 of *Springer Series in Computational Mathematics*, Springer, New York.

D. C. Brown (1982), Alternating-direction iterative schemes for mixed finite element methods for second order elliptic problems, PhD thesis, Department of Mathematics, University of Chicago, Chicago, IL.

P. N. Brown (1991), 'A theoretical comparison of the Arnoldi and GMRES algorithms', *SIAM J. Sci. Statist. Comput.* **12**, 58–78.

J. R. Bunch (1971), 'Equilibration of symmetric matrices in the Max-norm', *J. Assoc. Comput. Mach.* **18**, 566–572.

J. R. Bunch (1974), 'Partial pivoting strategies for symmetric matrices', *SIAM J. Numer. Anal.* **11**, 521–528.

J. R. Bunch and L. Kaufman (1977), 'Some stable methods for calculating inertia and solving symmetric linear systems', *Math. Comp.* **31**, 163–179.

J. R. Bunch and B. N. Parlett (1971), 'Direct methods for solving symmetric indefinite systems of linear equations', *SIAM J. Numer. Anal.* **8**, 639–655.

M. Burger and W. Mühlhuber (2002), 'Iterative regularization of parameter identification problems by sequential quadratic programming methods', *Inverse Problems* **18**, 943–969.

J. Burkardt, C. Hall and T. Porsching (1986), 'The dual variable method for the solution of compressible fluid flow problems', *SIAM J. Alg. Disc. Meth.* **7**, 476–483.

Y. V. Bychenkov (2002), 'Optimization of one class of nonsymmetrizable algorithms for saddle point problems', *Russian J. Numer. Anal. Math. Modelling* **17**, 521–546.

J. Cahouet and J. P. Chabard (1988), 'Some fast 3-D finite element solvers for the generalized Stokes problem', *Internat. J. Numer. Meth. Fluids* **8**, 869–895.

S. L. Campbell and C. D. Meyer Jr. (1979), *Generalized Inverses of Linear Transformations*, Pitman, London. Reprinted by Dover, New York (1991).

Z.-H. Cao (2002), 'A note on constraint preconditioning for nonsymmetric indefinite matrices', *SIAM J. Matrix Anal. Appl.* **24**, 121–125.

Z.-H. Cao (2004a), 'Fast iterative solution of stabilized Navier–Stokes systems', *Appl. Math. Comput.* **157**, 219–241.

Z.-H. Cao (2004b), 'Fast Uzawa algorithms for solving nonsymmetric stabilized saddle point problems', *Numer. Linear Algebra Appl.* **11**, 1–24.

A. Cassell, J. C. Henderson and A. Kaveh (1974), 'Cycle bases for the flexibility analysis of structures', *Internat. J. Numer. Methods Engng.* **8**, 521–528.

T. Chan and T. P. Mathew (1994), 'Domain decomposition algorithms', *Acta Numerica*, Vol. 3, Cambridge University Press, pp. 61–143.

S. Chandrasekaran, M. Gu and A. H. Sayed (1998), 'A stable and efficient algorithm for the indefinite linear least-squares problem', *SIAM J. Matrix Anal. Appl.* **20**, 354–362.

W. Chang, F. Giraldo and J. B. Perot (2002), 'Analysis of an exact fractional step method', *J. Comput. Phys.* **180**, 183–199.

X. Chen (1998), 'On preconditioned Uzawa methods and SOR methods for saddle-point problems', *J. Comput. Appl. Math.* **100**, 207–224.

X. Chen and K. Hashimoto (2003), 'Numerical validation of solutions of saddle point matrix equations', *Numer. Linear Algebra Appl.* **10**, 661–672.

Z. M. Chen and J. Zou (1999), 'An augmented Lagrangian method for identifying discontinuous parameters in elliptic systems', *SIAM J. Contr. Optim.* **37**, 892–910.

X. L. Cheng (2000), 'On the nonlinear inexact Uzawa algorithm for saddle-point problems', *SIAM J. Numer. Anal.* **37**, 1930–1934.

X. L. Cheng and J. Zou (2003), 'An inexact Uzawa-type iterative method for solving saddle point problems', *Internat. J. Computer Math.* **80**, 55–64.

E. V. Chizhonkov (2001), 'On solving an algebraic system of Stokes type under block diagonal preconditioning', *Comput. Math. Math. Phys.* **41**, 514–521.

E. V. Chizhonkov (2002), 'Improving the convergence of the Lanczos method in solving algebraic saddle point problems', *Comput. Math. Math. Phys.* **42**, 483–491.

A. J. Chorin (1968), 'Numerical solution of the Navier–Stokes equations', *Math. Comp.* **22**, 745–762.

A. J. Chorin and J. E. Marsden (1990), *A Mathematical Introduction to Fluid Mechanics*, Vol. 4 of *Texts in Applied Mathematics*, Springer, New York.

E. Chow and Y. Saad (1997*a*), 'Approximate inverse techniques for block-partitioned matrices', *SIAM J. Sci. Comput.* **18**, 1657–1675.

E. Chow and Y. Saad (1997*b*), 'Experimental study of ILU preconditioners for indefinite matrices', *J. Comput. Appl. Math.* **86**, 387–414.

E. Chow, T. Manteuffel, C. Tong and B. K. Wallin (2003), 'Algebraic elimination of slide surface constraints in implicit structural analysis', *Internat. J. Numer. Methods Engng.* **57**, 1129–1144.

L. O. Chua, C. A. Desoer and E. S. Kuh (1987), *Linear and Nonlinear Circuits*, McGraw-Hill, New York.

P. Ciarlet Jr., J. Huang and J. Zou (2003), 'Some observations on generalized saddle-point problems', *SIAM J. Matrix Anal. Appl.* **25**, 224–236.

P. Ciarlet (1988), *Mathematical Elasticity, Vol. I: Three Dimensional Elasticity*, Vol. 20 of *Studies in Mathematics and its Applications*, North-Holland, Amsterdam.

K. A. Cliffe, T. J. Garratt and A. Spence (1994), 'Eigenvalues of block matrices arising from problems in fluid mechanics', *SIAM J. Matrix Anal. Appl.* **15**, 1310–1318.

K. A. Cliffe, I. G. Graham, R. Scheichl and L. Stals (2000), 'Parallel computation of flow in heterogeneous media modelled by mixed finite elements', *J. Comput. Phys.* **164**, 258–282.

T. F. Coleman (1984), *Large Sparse Numerical Optimization*, Springer, New York.

T. F. Coleman and A. Pothen (1986), 'The null space problem I: complexity', *SIAM J. Alg. Disc. Meth.* **7**, 527–537.

T. F. Coleman and A. Pothen (1987), 'The null space problem II: algorithms', *SIAM J. Alg. Disc. Meth.* **8**, 544–563.

T. F. Coleman and A. Verma (2001), 'A preconditioned conjugate gradient approach to linear equality constrained minimization', *Comput. Optim. Appl.* **20**, 61–72.

P. Concus and G. H. Golub (1976), A generalized conjugate gradient method for nonsymmetric systems of linear equations, in *Computing Methods in Applied Sciences and Engineering* (R. Glowinski and J. L. Lions, eds), Springer, pp. 56–65.

J. B. Conway (1990), *A Course in Functional Analysis*, Vol. 96 of *Graduate Texts in Mathematics*, second edn, Springer, New York.

R. Courant (1943), 'Variational methods for the solution of problems of equilibrium and vibration', *Bull. Amer. Math. Soc.* **69**, 1–23.

A. J. Cox and N. J. Higham (1999*a*), 'Accuracy and stability of the null space method for solving the equality constrained least squares problem', *BIT* **39**, 34–50.

A. J. Cox and N. J. Higham (1999*b*), 'Row-wise backward stable elimination methods for the equality constrained least squares problem', *SIAM J. Matrix Anal. Appl.* **21**, 313–326.

M.-R. Cui (2002), 'A sufficient condition for the convergence of the inexact Uzawa algorithm for saddle point problems', *J. Comput. Appl. Math.* **139**, 189–196.

M.-R. Cui (2004), 'Analysis of iterative algorithms of Uzawa type for saddle point problems', *Appl. Numer. Math.* **50**, 133–146.

J. Cullum and A. Greenbaum (1996), 'Relations between Galerkin and norm-minimizing iterative methods for solving linear systems', *SIAM J. Matrix Anal. Appl.* **17**, 223–247.

J. Czyzyk, R. Fourer and S. Mehrotra (1998), 'Using a massively parallel processor to solve large sparse linear programs by an interior-point method', *SIAM J. Sci. Comput.* **19**, 553–565.

O. Dahl and S. Ø. Wille (1992), 'An ILU preconditioner with coupled node fill-in for iterative solution of the mixed finite element formulation of the 2D and 3D Navier–Stokes equations', *Internat. J. Numer. Meth. Fluids* **15**, 525–544.

S. Dahlke, R. Hochmuth and K. Urban (2000), 'Adaptive wavelet methods for saddle point problems', *Math. Model. Numer. Anal.* **34**, 1003–1022.

C. de Boor and J. R. Rice (1982), 'Extremal polynomials with application to Richardson iteration for indefinite linear systems', *SIAM J. Sci. Statist. Comput.* **3**, 47–57.

E. de Sturler and J. Liesen (2005), 'Block-diagonal and constraint preconditioners for nonsymmetric indefinite linear systems, Part I: Theory', *SIAM J. Sci. Comput.*, to appear.

S. Demko, W. F. Moss and P. W. Smith (1984), 'Decay rates for inverses of band matrices', *Math. Comp.* **44**, 491–499.

G. B. Deng, J. Piquet, P. Queutey and M. Visonneau (1996), Navier–Stokes equations for incompressible flows: finite-difference and finite volume methods, in *Handbook of Computational Fluid Mechanics* (R. Peyret, ed.), Academic Press, London, pp. 25–97.

P. Destuynder and T. Nevers (1990), 'Some numerical aspects of mixed finite elements for bending plates', *Comput. Methods Appl. Mech. Engrg.* **78**, 73–87.

M. Discacciati and A. Quarteroni (2004), 'Convergence analysis of a subdomain iterative method for the finite element approximation of the coupling of Stokes and Darcy equations', *Comput. Visual. Sci.* **6**, 93–103.

M. Discacciati, E. Miglio and A. Quarteroni (2002), 'Mathematical and numerical models for coupling surface and groundwater flows', *Appl. Numer. Math.* **43**, 57–74.

C. Dohrmann and R. B. Lehoucq (2004), A primal based penalty preconditioner for elliptic saddle point systems, Research Report SAND-2004-59644, Sandia National Laboratories, Albuquerque, New Mexico.

H. S. Dollar (2005), Extending constraint preconditioners for saddle point problems, Research Report NA-05/02, Numerical Analysis Group, Oxford University.

H. S. Dollar and A. J. Wathen (2004), Incomplete factorization constraint preconditioners for saddle-point matrices, Research Report NA-04/01, Numerical Analysis Group, Oxford University.

J. J. Dongarra, I. S. Duff, D. C. Sorensen and H. A. van der Vorst (1998), *Numerical Linear Algebra for High-Performance Computers*, Vol. 11 of *Software, Environments and Tools*, SIAM, Philadelphia, PA.

J. Douglas Jr., R. Durán and P. Pietra (1986), Alternating-direction iteration for mixed finite element methods, in *Computing Methods in Applied Sciences and Engineering VII: Versailles, France, 1985*, Springer, New York, pp. 181–196.

J. Douglas Jr., R. Durán and P. Pietra (1987), Formulation of alternating-direction iteration for mixed methods in three-space, in *Numerical Approximation of Partial Differential Equations*, Vol. 133 of *Studies in Mathematics*, North-Holland, Amsterdam, pp. 21–30.

T. Dreyer, B. Maar and V. Schulz (2000), 'Multigrid optimization in applications', *J. Comput. Appl. Math.* **120**, 67–84.

F. Duchin and D. B. Szyld (1979), 'Application of sparse matrix techniques to inter-regional input-output analysis', *Economics of Planning* **15**, 142–167.

I. S. Duff (1994), The solution of augmented systems, in *Numerical Analysis 1993: Proc. 15th Dundee Conf., Dundee University, June–July 1993*, Vol. 303 of *Pitman Research Notes in Mathematics*, Longman Scientific & Technical, Harlow, UK, pp. 40–55.

I. S. Duff and J. Koster (1999), 'The design and use of algorithms for permuting large entries to the diagonal of sparse matrices', *SIAM J. Matrix Anal. Appl.* **20**. 889–901.

I. S. Duff and J. Koster (2001), 'An algorithm for permuting large entries to the diagonal of a sparse matrix', *SIAM J. Matrix Anal. Appl.* **22**, 973–996.

I. S. Duff and S. Pralet (2004), Strategies for scaling and pivoting for sparse symmetric indefinite problems, Research Report TR/PA/04/59, CERFACS, Toulouse, France.

I. S. Duff and J. K. Reid (1983), 'The multifrontal solution of indefinite sparse symmetric linear systems', *ACM Trans. Math. Software* **9**, 302–325.

I. S. Duff and J. K. Reid (1995), MA47, a FORTRAN code for direct solution of indefinite sparse symmetric linear systems, Research Report RAL-95-001, Rutherford Appleton Laboratory.

I. S. Duff and J. K. Reid (1996), 'Exploiting zeros on the diagonal in the direct solution of indefinite sparse symmetric linear systems', *ACM Trans. Math. Software* **22**, 227–257.

I. S. Duff, A. M. Erisman and J. K. Reid (1986), *Direct Methods for Sparse Matrices*, Monographs on Numerical Analysis, Oxford University Press, Oxford.

I. S. Duff, N. I. M. Gould, J. K. Reid, J. A. Scott and K. Turner (1991), 'The factorization of sparse symmetric indefinite matrices', *IMA J. Numer. Anal.* **11**, 181–204.

C. Durazzi and V. Ruggiero (2003a), 'Indefinitely preconditioned conjugate gradient method for large sparse equality and inequality constrained quadratic problems', *Numer. Linear Algebra Appl.* **10**, 673–688.

C. Durazzi and V. Ruggiero (2003b), 'Numerical solution of special linear and quadratic programs via a parallel interior-point method', *Parallel Comput.* **29**, 485–503.

E. D'Yakonov (1987), 'On iterative methods with saddle operators', *Soviet Math. Dokl.* **35**, 166–170.

N. Dyn and W. E. Ferguson, Jr. (1983), 'The numerical solution of equality constrained quadratic programming problems', *Math. Comp.* **41**, 165–170.

M. Eiermann and O. G. Ernst (2001), 'Geometric aspects of the theory of Krylov subspace methods', *Acta Numerica*, Vol. 10, Cambridge University Press, pp. 251–312.

M. Eiermann and W. Niethammer (1983), 'On the construction of semi-iterative methods', *SIAM J. Numer. Anal.* **20**, 1153–1160.

M. Eiermann, W. Niethammer and R. S. Varga (1985), 'A study of semi-iterative methods for nonsymmetric systems of linear equations', *Numer. Math.* **47**, 505–533.

H. C. Elman (1996), 'Multigrid and Krylov subspace methods for the discrete Stokes equations', *Internat. J. Numer. Meth. Fluids* **22**, 755–770.

H. C. Elman (1999), 'Preconditioning for the steady-state Navier–Stokes equations with low viscosity', *SIAM J. Sci. Comput.* **20**, 1299–1316.

H. C. Elman (2002), 'Preconditioners for saddle point problems arising in computational fluid dynamics', *Appl. Numer. Math.* **43**, 75–89.

H. C. Elman and G. H. Golub (1994), 'Inexact and preconditioned Uzawa algorithms for saddle point problems', *SIAM J. Numer. Anal.* **31**, 1645–1661.

H. C. Elman and D. J. Silvester (1996), 'Fast nonsymmetric iterations and preconditioning for Navier–Stokes equations', *SIAM J. Sci. Comput.* **17**, 33–46.

H. C. Elman, V. E. Howle, J. Shadid and R. Tuminaro (2003*a*), 'A parallel block multi-level preconditioner for the 3D incompressible Navier–Stokes equations', *J. Comput. Phys.* **187**, 504–523.

H. C. Elman, V. E. Howle, J. Shadid, R. Shuttleworth and R. Tuminaro (2005*a*), 'Block preconditioners based on approximate commutators', *SIAM J. Sci. Comput.*, to appear.

H. C. Elman, D. Loghin and A. J. Wathen (2003*b*), 'Preconditioning techniques for Newton's method for the incompressible Navier–Stokes equations', *BIT* **43**, 961–974.

H. C. Elman, A. R. Ramage, D. J. Silvester and A. J. Wathen (2005*b*). http://www.cs.umd.edu/~elman/ifiss/.

H. C. Elman, D. J. Silvester and A. J. Wathen (2002*a*), 'Block preconditioners for the discrete incompressible Navier–Stokes equations', *Internat. J. Numer. Methods Fluids* **40**, 333–344.

H. C. Elman, D. J. Silvester and A. J. Wathen (2002*b*), 'Performance and analysis of saddle point preconditioners for the discrete steady-state Navier–Stokes equations', *Numer. Math.* **90**, 665–688.

H. C. Elman, D. J. Silvester and A. J. Wathen (2005*c*), *Finite Elements and Fast Iterative Solvers*, Numerical Mathematics and Scientific Computation, Oxford University Press, Oxford.

O. G. Ernst (2000), 'Residual-minimizing Krylov subspace methods for stabilized discretizations of convection-diffusion equations', *SIAM J. Matrix Anal. Appl.* **21**, 1079–1101.

R. E. Ewing, R. D. Lazarov, P. Lu and P. S. Vassilevski (1990), Preconditioning indefinite systems arising from mixed finite element discretization of second-order elliptic problems, in *Preconditioned Conjugate Gradient Methods: Nijmegen 1989*, Vol. 1457 of *Lecture Notes in Mathematics*, Springer, Berlin, pp. 28–43.

V. Faber and T. Manteuffel (1984), 'Necessary and sufficient conditions for the existence of a conjugate gradient method', *SIAM J. Numer. Anal.* **21**, 352–362.

V. Faber, W. Joubert, E. Knill and T. Manteuffel (1996), 'Minimal residual method stronger than polynomial preconditioning', *SIAM J. Matrix Anal. Appl.* **17**, 707–729.

C. Farhat and F.-X. Roux (1991), 'A method of finite element tearing and interconnecting and its parallel solution algorithm', *Internat. J. Numer. Methods Engng.* **32**, 1205–1227.

B. Fischer (1996), *Polynomial Based Iteration Methods for Symmetric Linear Systems*, Wiley–Teubner Series Advances in Numerical Mathematics, Wiley, Chichester, UK.

B. Fischer and F. Peherstorfer (2001), 'Chebyshev approximation via polynomial mappings and the convergence behaviour of Krylov subspace methods', *Electr. Trans. Numer. Anal.* **12**, 205–215.

B. Fischer, A. Ramage, D. J. Silvester and A. J. Wathen (1998), 'Minimum residual methods for augmented systems', *BIT* **38**, 527–543.

P. F. Fischer and J. W. Lottes (2004), Hybrid Schwarz-multigrid methods for the spectral element method: Extensions to Navier–Stokes, in *Domain Decomposition Methods in Science and Engineering* (R. Kornhuber, R. H. W. Hoppe, J. Périaux, O. Pironneau, O. B. Widlund and J. Xu, eds), Vol. 40 of *Lecture Notes in Computational Science and Engineering*, Springer, pp. 35–49.

J. Fish and Y. Qu (2000), 'Global basis two-level method for indefinite systems. Part 1: Convergence studies', *Internat. J. Numer. Methods Engng.* **49**, 439–460.

R. Fletcher (1976a), Conjugate gradient methods for indefinite systems, in *Numerical Analysis 1975: Proc. 6th Biennial Dundee Conf., Univ. Dundee, Dundee, 1975*, Vol. 506 of *Lecture Notes in Mathematics*, Springer, Berlin, pp. 73–89.

R. Fletcher (1976b), 'Factorizing symmetric indefinite matrices', *Linear Algebra Appl.* **14**, 257–272.

R. Fletcher (1987), *Practical Methods of Optimization*, 2nd edn, Wiley, Chichester, UK.

R. Fletcher and T. Johnson (1997), 'On the stability of null-space methods for KKT systems', *SIAM J. Matrix Anal. Appl.* **18**, 938–958.

A. Forsgren (1996), 'On linear least-squares problems with diagonally dominant weight matrices', *SIAM J. Matrix Anal. Appl.* **17**, 763–788.

A. Forsgren (2002), 'Inertia-controlling factorizations for optimization algorithms', *Appl. Numer. Math.* **43**, 91–107.

A. Forsgren and W. Murray (1993), 'Newton methods for large-scale linear equality-constrained minimization', *SIAM J. Matrix Anal. Appl.* **14**, 560–587.

A. Forsgren, P. E. Gill and J. R. Shinnerl (1996), 'Stability of symmetric ill-conditioned systems arising in interior methods for constrained optimization', *SIAM J. Matrix Anal. Appl.* **17**, 187–211.

M. Fortin (1993), 'Finite element solution of the Navier–Stokes equations', *Acta Numerica*, Vol. 2, Cambridge University Press, pp. 239–284.

M. Fortin and A. Fortin (1985), 'A generalization of Uzawa's algorithm for the solution of the Navier–Stokes equations', *Comm. Appl. Numer. Meth.* **1**, 205–208.

M. Fortin and R. Glowinski (1983), *Augmented Lagrangian Methods: Applications to the Numerical Solution of Boundary Value Problems*, Vol. 15 of *Studies in Mathematics and its Applications*, North-Holland, Amsterdam. Translated from the French by B. Hunt and D. C. Spicer.

R. Fourer and S. Mehrotra (1993), 'Solving symmetric indefinite systems in an interior-point method for linear programming', *Math. Program.* **62**, 15–39.

A. Frangioni and C. Gentile (2004), 'New preconditioners for KKT systems of network flow problems', *SIAM J. Optim.* **14**, 894–913.

R. W. Freund (1993), 'A transpose-free quasi-minimal residual algorithm for non-Hermitian linear systems', *SIAM J. Sci. Comput.* **14**, 470–482.

R. W. Freund (1997), Preconditioning of symmetric, but highly indefinite linear systems, in *Proc. 15th IMACS World Congress on Scientific Computation, Modeling and Applied Mathematics* (A. Sydow, ed.), Wissenschaft und Technik Verlag, pp. 551–556.

R. W. Freund (2003), 'Model reduction methods based on Krylov subspaces', *Acta Numerica*, Vol. 12, Cambridge University Press, pp. 267–319.

R. W. Freund and F. Jarre (1996), 'A QMR-based interior-point algorithm for solving linear programs', *Math. Program.* **76**, 183–210.

R. W. Freund and N. M. Nachtigal (1991), 'QMR: a quasi-minimal residual method for non-Hermitian linear systems', *Numer. Math.* **60**, 315–339.

R. W. Freund and N. M. Nachtigal (1994), A new Krylov-subspace method for symmetric indefinite linear systems, in *Proc. 14th IMACS World Congress on Computational and Applied Mathematics* (W. F. Ames, ed.), IMACS, pp. 1253–1256.

R. W. Freund and N. M. Nachtigal (1995), 'Software for simplified Lanczos and QMR algorithms', *Appl. Numer. Math.* **19**, 319–341.

R. W. Freund, G. H. Golub and N. M. Nachtigal (1992), 'Iterative solution of linear systems', *Acta Numerica*, Vol. 1, Cambridge University Press, pp. 57–100.

W. N. Gansterer, J. Schneid and C. W. Ueberhuber (2003), Mathematical properties of equilibrium systems, Technical Report AURORA TR2003-13, University of Vienna and Vienna University of Technology.

A. Gauthier, F. Saleri and A. Quarteroni (2004), 'A fast preconditioner for the incompressible Navier–Stokes equations', *Comput. Visual. Sci.* **6**, 105–112.

A. George and K. Ikramov (2000), 'On the condition of symmetric quasi-definite matrices', *SIAM J. Matrix Anal. Appl.* **21**, 970–977.

A. George and K. Ikramov (2002), 'The closedness of certain classes of matrices with respect to pseudoinversion', *Comput. Math. Math. Phys.* **42**, 1242–1246.

A. George, K. Ikramov and B. Kucherov (2000), 'Some properties of symmetric quasi-definite matrices', *SIAM J. Matrix Anal. Appl.* **21**, 1318–1323.

J. Gilbert and M. T. Heath (1987), 'Computing a sparse basis for the null space', *SIAM J. Alg. Disc. Meth.* **8**, 446–459.

P. E. Gill, W. Murray and M. H. Wright (1981), *Practical Optimization*, Academic Press Inc. [Harcourt Brace Jovanovich Publishers], London.

P. E. Gill, W. Murray, D. B. Ponceleón and M. A. Saunders (1992), 'Preconditioners for indefinite systems arising in optimization', *SIAM J. Matrix Anal. Appl.* **13**, 292–311.

P. E. Gill, W. Murray, M. A. Saunders and M. H. Wright (1991), 'Inertia-controlling methods for general quadratic programming', *SIAM Rev.* **33**, 1–36.

P. E. Gill, M. A. Saunders and J. R. Shinnerl (1996), 'On the stability of Cholesky factorization for symmetric quasidefinite systems', *SIAM J. Matrix Anal. Appl.* **17**, 35–46.

V. Girault and P. A. Raviart (1986), *Finite Element Approximation of the Navier–Stokes Equations*, Vol. 749 of *Lecture Notes in Mathematics*, Springer, New York.

R. Glowinski (1984), *Numerical Methods for Nonlinear Variational Problems*, Springer Series in Computational Physics, Springer, New York.

R. Glowinski (2003), *Finite Element Methods for Incompressible Viscous Flow*, Vol. IX of *Handbook of Numerical Analysis, Part 3: Numerical Methods for Fluids*, North-Holland, Amsterdam.

R. Glowinski and P. Le Tallec (1989), *Augmented Lagrangian and Operator Splitting Methods in Nonlinear Mechanics*, SIAM, Philadelphia, PA.

R. Glowinski and O. Pironneau (1979), 'Numerical methods for the first biharmonic equation and for the two-dimensional Stokes problem', *SIAM Rev.* **21**, 167–212.

I. Gohberg, P. Lancaster and L. Rodman (1983), *Matrices and Indefinite Scalar Products*, Vol. 8 of *Operator Theory: Advances and Applications*, Birkhäuser, Basel.

P. Goldfeld (2000), Balancing Neumann–Neumann preconditioners for mixed formulation of almost-incompressible linear elasticity, PhD thesis, Courant Institute of Mathematical Sciences, New York University, New York.

P. Goldfeld, L. F. Pavarino and O. B. Widlund (2003), 'Balancing Neumann–Neumann preconditioners for mixed approximations of heterogeneous problems in linear elasticity', *Numer. Math.* **95**, 283–324.

G. H. Golub and C. Greif (2003), 'On solving block-structured indefinite linear systems', *SIAM J. Sci. Comput.* **24**, 2076–2092.

G. H. Golub and M. L. Overton (1988), 'The convergence of inexact Chebyshev and Richardson iterative methods for solving linear systems', *Numer. Math.* **53**, 571–593.

G. H. Golub and C. F. Van Loan (1979), 'Unsymmetric positive definite linear systems', *Linear Algebra Appl.* **28**, 85–97.

G. H. Golub and C. F. Van Loan (1996), *Matrix Computations*, Johns Hopkins Studies in the Mathematical Sciences, third edn, Johns Hopkins University Press, Baltimore, MD.

G. H. Golub and D. Vanderstraeten (2000), 'On the preconditioning of matrices with skew-symmetric splittings', *Numer. Algorithms* **25**, 223–239.

G. H. Golub and A. J. Wathen (1998), 'An iteration for indefinite systems and its application to the Navier–Stokes equations', *SIAM J. Sci. Comput.* **19**, 530–539.

G. H. Golub, X. Wu and J.-Y. Yuan (2001), 'SOR-like methods for augmented systems', *BIT* **41**, 71–85.

R. Gonzales and R. Woods (1992), *Digital Image Processing*, Addison-Wesley, New York.

M. V. Gorelova and E. V. Chizhonkov (2004), 'Preconditioning saddle point problems with the help of saddle point operators', *Comput. Math. Math. Phys.* **44**, 1445–1455.

N. I. M. Gould (1985), 'On practical conditions for the existence and uniqueness of solutions to the general equality quadratic programming problem', *Math. Program.* **32**, 90–99.

N. I. M. Gould and J. A. Scott (2004), 'A numerical evaluation of HSL packages for the direct solution of large sparse, symmetric linear systems of equations', *ACM Trans. Math. Software* **30**, 300–325.

N. I. M. Gould, M. E. Hribar and J. Nocedal (2001), 'On the solution of equality constrained quadratic programming problems arising in optimization', *SIAM J. Sci. Comput.* **23**, 1376–1395.

I. G. Graham, A. Spence and E. Vainikko (2003), 'Parallel iterative methods for Navier–Stokes equations and application to eigenvalue calculation', *Concurrency and Computation: Practice and Experience* **15**, 1151–1168.

A. Greenbaum (1979), 'Comparison of splittings used with the conjugate gradient algorithm', *Numer. Math.* **33**, 181–194.

A. Greenbaum (1997), *Iterative Methods for Solving Linear Systems*, Vol. 17 of *Frontiers in Applied Mathematics*, SIAM, Philadelphia, PA.

A. Greenbaum (2002), 'Generalizations of the field of values useful in the study of polynomial functions of a matrix', *Linear Algebra Appl.* **347**, 233–249.

A. Greenbaum and L. Gurvits (1994), 'Max-min properties of matrix factor norms', *SIAM J. Sci. Comput.* **15**, 348–358.

A. Greenbaum and Z. Strakoš (1994), Matrices that generate the same Krylov residual spaces, in *Recent Advances in Iterative Methods*, Vol. 60 of *IMA Vol. Math. Appl.*, Springer, New York, pp. 95–118.

A. Greenbaum, V. Pták and Z. Strakoš (1996), 'Any nonincreasing convergence curve is possible for GMRES', *SIAM J. Matrix Anal. Appl.* **17**, 465–469.

C. Greif, G. H. Golub and J. M. Varah (2005), 'Augmented Lagrangian techniques for solving saddle point linear systems', *SIAM J. Matrix Anal. Appl.*, to appear.

P. M. Gresho (1991), 'Incompressible fluid dynamics: some fundamental formulation issues', *Annual Rev. Fluid Mech.* **23**, 413–453.

P. M. Gresho and R. L. Sani (1998), *Incompressible Flow and the Finite Element Method. Advection-Diffusion and Isothermal Laminar Flow*, Wiley, Chichester, UK.

M. Griebel, T. Neunhoeffer and H. Regler (1998), 'Algebraic multigrid methods for the solution of the Navier–Stokes equations in complicated geometries', *Internat. J. Numer. Meth. Fluids* **26**, 281–301.

M. Gulliksson (1994), 'Iterative refinement for constrained and weighted linear least squares', *BIT* **34**, 239–253.

M. Gulliksson and P.-Å. Wedin (1992), 'Modifying the QR-decomposition to constrained and weighted linear least squares', *SIAM J. Matrix Anal. Appl.* **13**, 1298–1313.

M. Gulliksson, X.-Q. Jin and Y. Wei (2002), 'Perturbation bounds for constrained and weighted least squares problems', *Linear Algebra Appl.* **349**, 221–232.

M. Gunzburger (1989), *Finite Element Methods for Viscous Incompressible Flows*, Academic Press, San Diego.

K. Gustafson and R. Hartmann (1983), 'Divergence-free bases for finite element schemes in hydrodynamics', *SIAM J. Numer. Anal.* **20**, 697–721.

E. Haber and U. M. Ascher (2001), 'Preconditioned all-at-once methods for large, sparse parameter estimation problems', *Inverse Problems* **17**, 1847–1864.

E. Haber and J. Modersitzki (2004), 'Numerical methods for volume-preserving image registration', *Inverse Problems* **20**, 1621–1638.

E. Haber, U. M. Ascher and D. Oldenburg (2000), 'On optimization techniques for solving nonlinear inverse problems', *Inverse Problems* **16**, 1263–1280.

W. Hackbusch (1985), *Multi-grid Methods and Applications*, Springer, Berlin.

W. Hackbusch (1994), *Iterative Solution of Large Sparse Systems of Equations*, Springer, Berlin.

G. Hadley (1964), *Nonlinear and Dynamic Programming*, Addison-Wesley, Reading, MA.

M. Hagemann and O. Schenk (2004), Weighted matchings for the preconditioning of symmetric indefinite linear systems, Technical Report CS-2004-005, Department of Computer Science, University of Basel.

C. A. Hall (1985), 'Numerical solution of Navier–Stokes problems by the dual variable method', *SIAM J. Alg. Disc. Meth.* **6**, 220–236.

C. A. Hall, J. C. Cavendish and W. H. Frey (1991), 'The dual variable method for solving fluid flow difference equations on Delaunay triangulations', *Comput. & Fluids* **20**, 145–164.

E. L. Hall (1979), *Computer Image Processing and Recognition*, Academic Press, New York.

F. H. Harlow and J. E. Welch (1965), 'Numerical calculation of time-dependent viscous incompressible flow of fluid with free surface', *Phys. Fluids* **8**, 2182–2189.

B. Hassibi, A. H. Sayed and T. Kailath (1996), 'Linear estimation in Krein spaces, Part I: Theory', *IEEE Trans. Automat. Control* **41**, 18–33.

J. C. Haws (2002), Preconditioning KKT systems, PhD thesis, Department of Mathematics, North Carolina State University, Raleigh, NC.

J. C. Haws and C. D. Meyer (2003), Preconditioning KKT systems, Technical Report M&CT-Tech-01-021, The Boeing Company.

M. T. Heath, R. J. Plemmons and R. C. Ward (1984), 'Sparse orthogonal schemes for structural optimization using the force method', *SIAM J. Sci. Statist. Comput.* **5**, 514–532.

M. Heinkenschloss and H. Nguyen (2004), Domain decomposition preconditioners for linear quadratic elliptic optimal control problems, Technical Report TR04–20, Department of Computational and Applied Mathematics, Rice University.

L. Hemmingsson-Frändén and A. Wathen (2001), 'A nearly optimal preconditioner for the Navier–Stokes equations', *Numer. Linear Algebra Appl.* **8**, 229–243.

J. C. Henderson and E. A. W. Maunder (1969), 'A problem in applied topology: on the selection of cycles for the flexibility analysis of skeletal structures', *J. Inst. Math. Appl.* **5**, 254–269.

P. Heres and W. Schilders (2005), Practical issues of model order reduction with Krylov-subspace methods, in *Progress in Electromagnetic Research Symposium: PIERS 2004*. To appear.

M. A. Heroux, R. A. Bartlett, V. E. Howle, R. J. Hoekstra, J. J. Hu, T. G. Kolda, R. B. Lehoucq, K. R. Long, R. P. Pawlowski, E. T. Phipps, A. G. Salinger, H. K. Thornquist, R. S. Tuminaro, J. S. Willenbring, A. Williams and K. S. Stanley (2005), 'An overview of the Trilinos project', *ACM Trans. Math. Software*, to appear.

M. R. Hestenes (1969), 'Multiplier and gradient methods', *J. Optim. Theory Appl.* **4**, 303–320.

M. R. Hestenes (1975), *Optimization Theory: The Finite Dimensional Case*, Wiley, New York.

M. R. Hestenes and E. Stiefel (1952), 'Methods of conjugate gradients for solving linear systems', *J. Research Nat. Bur. Standards* **49**, 409–436.

R. Hiptmair (1996), *Multilevel Preconditioning for Mixed Problems in Three Dimensions*, Vol. 8 of *Augsburger Mathematisch-Naturwissenschaftliche Schriften*, Wissner, Augsburg, Germany.

R. Hiptmair (1997), 'Multigrid method for H(div) in three dimensions', *Electron. Trans. Numer. Anal.* **6**, 133–152.

R. Hiptmair and R. H. W. Hoppe (1999), 'Multilevel methods for mixed finite elements in three dimensions', *Numer. Math.* **82**, 253–279.

R. Hiptmair, T. Shiekofer and B. Wohlmuth (1996), 'Multilevel preconditioned augmented Lagrangian techniques for 2nd order mixed problems', *Comput.* **57**, 25–48.

R. Hochmuth (1998), 'Stable multiscale discretization for saddle point problems and preconditioning', *Numer. Funct. Anal. Optim.* **19**, 789–806.

R. A. Horn and C. R. Johnson (1985), *Matrix Analysis*, Cambridge University Press, Cambridge.

P. D. Hough and S. D. Vavasis (1997), 'Complete orthogonal decomposition for weighted least squares', *SIAM J. Matrix Anal. Appl.* **18**, 369–392.

Q. Hu and J. Zou (2001), 'An iterative method with variable relaxation parameters for saddle-point problems', *SIAM J. Matrix Anal. Appl.* **23**, 317–338.

Q. Hu and J. Zou (2002), 'Two new variants of nonlinear inexact Uzawa algorithms for saddle-point problems', *Numer. Math.* **93**, 333–359.

Q. Hu and J. Zou (2003), 'Substructuring preconditioners for saddle-point problems arising from Maxwell's equations in three dimensions', *Math. Comp.* **73**, 35–61.

Q. Hu, Z. Shi and D. Yu (2004), 'Efficient solvers for saddle-point problems arising from domain decompositions with Lagrange multipliers', *SIAM J. Numer. Anal.* **42**, 905–933.

M. Huhtanen (2002), 'A Hermitian Lanczos method for normal matrices', *SIAM J. Matrix Anal. Appl.* **23**, 1092–1108.

K. D. Ikramov (2003), 'On a certain type of preconditioning for KKT-matrices', *Moscow Univ. Comput. Math. Cybernet.* **2**, 1–3.

I. C. F. Ipsen (2001), 'A note on preconditioning nonsymmetric matrices', *SIAM J. Sci. Comput.* **23**, 1050–1051.

K. Ito and K. Kunisch (1999), 'An active set strategy based on the augmented Lagrangian formulation for image restoration', *Math. Model. Numer. Anal.* **33**, 1–21.

D. James (1992), 'Implicit null space iterative methods for constrained least squares problems', *SIAM J. Matrix Anal. Appl.* **13**, 962–978.

D. James and R. J. Plemmons (1990), 'An iterative substructuring algorithm for equilibrium equations', *Numer. Math.* **57**, 625–633.

X.-Q. Jin (1996), 'A preconditioner for constrained and weighted least squares problems with Toeplitz structure', *BIT* **36**, 101–109.

W. Joubert (1994), 'A robust GMRES-based adaptive polynomial preconditioning algorithm for nonsymmetric linear systems', *SIAM J. Sci. Comput.* **15**, 427–439.

J. Júdice, J. Patricio, L. Portugal, M. Resende and G. Veiga (2003), 'A study of preconditioners for network interior point methods', *Comput. Optim. Appl.* **24**, 5–35.

X. Juvigny (1997), Solution of large linear systems on massively parallel machines, PhD thesis, Applied Mathematics, University of Paris VI, Paris, France. In French.

I. Kaneko and R. J. Plemmons (1984), 'Minimum norm solutions to linear elastic analysis problems', *Internat. J. Numer. Methods Engng.* **20**, 983–998.

I. Kaneko, M. Lawo and G. Thierauf (1982), 'On computational procedures for the force method', *Internat. J. Numer. Methods Engng.* **18**, 1469–1495.

G. Kanschat (2003), 'Preconditioning methods for local discontinuous Galerkin discretizations', *SIAM J. Sci. Comput.* **25**, 815–831.

O. A. Karakashian (1982), 'On a Galerkin–Lagrange multiplier method for the stationary Navier–Stokes equations', *SIAM J. Numer. Anal.* **19**, 909–923.

A. Kaveh (1974), Application of topology and matroid theory to the analysis of structures, PhD thesis, Imperial College of Science and Technology, London.

A. Kaveh (1979), 'A combinatorial optimization problem: optimal generalized cycle bases', *Comput. Methods Appl. Mech. Engrg.* **20**, 39–51.

A. Kaveh (1992), 'Recent developments in the force method of structural analysis', *Appl. Mech. Rev.* **45**, 401–418.

A. Kaveh (2004), *Structural Mechanics: Graph and Matrix Methods*, third edn, Research Studies Press, UK.

D. Kay, D. Loghin and A. J. Wathen (2002), 'A preconditioner for the steady-state Navier–Stokes equations', *SIAM J. Sci. Comput.* **24**, 237–256.

C. Keller, N. I. M. Gould and A. J. Wathen (2000), 'Constraint preconditioning for indefinite linear systems', *SIAM J. Matrix Anal. Appl.* **21**, 1300–1317.

G. R. Kirchhoff (1847), 'Über die Auflösung der Gleichungen, auf welche man bei der Untersuchung der linearen Verteilung galvanischer Ströme geführt wird', *Ann. Phys. Chem.* **72**, 497–508.

T. H. Kjeldsen (2000), 'A contextualized historical analysis of the Kuhn–Tucker theorem in nonlinear programming: the impact of World War II', *Historia Mathematica* **27**, 331–361.

A. Klawonn (1998a), 'Block-triangular preconditioners for saddle point problems with a penalty term', *SIAM J. Sci. Comput.* **19**, 172–184.

A. Klawonn (1998b), 'An optimal preconditioner for a class of saddle point problems with a penalty term', *SIAM J. Sci. Comput.* **19**, 540–552.

A. Klawonn and L. F. Pavarino (1998), 'Overlapping Schwarz methods for mixed linear elasticity and Stokes problems', *Comput. Methods Appl. Mech. Engrg.* **165**, 233–245.

A. Klawonn and L. F. Pavarino (2000), 'A comparison of overlapping Schwarz methods and block preconditioners for saddle point problems', *Numer. Linear Algebra Appl.* **7**, 1–25.

A. Klawonn and G. Starke (1999), 'Block triangular preconditioners for nonsymmetric saddle point problems: field-of-values analysis', *Numer. Math.* **81**, 577–594.

A. Klawonn and O. B. Widlund (2000), 'A domain decomposition method with Lagrange multipliers and inexact solves for linear elasticity', *SIAM J. Sci. Comput.* **22**, 1199–1219.

A. Klawonn and O. B. Widlund (2001), 'FETI and Neumann–Neumann iterative substructuring methods: Connections and new results', *Comm. Pure Appl. Math.* **54**, 57–90.

A. Klawonn and O. B. Widlund (2004), Dual–primal FETI methods for linear elasticity, Technical Report TR2004-855, Courant Institute of Mathematical Sciences, New York University.

G. M. Kobelkov and M. A. Olshanskii (2000), 'Effective preconditioning of Uzawa type schemes for a generalized Stokes problem', *Numer. Math.* **86**, 443–470.

L. Komzsik, P. Poschmann and I. Sharapov (1997), 'A preconditioning technique for indefinite linear systems', *Finite Elements Anal. Des.* **26**, 253–258.

J. Korzak (1999), 'Eigenvalue relations and conditions of matrices arising in linear programming', *Comput.* **62**, 45–54.

R. Kouhia and C. M. Menken (1995), 'On the solution of second-order post-buckling fields', *Commun. Numer. Meth. Engng.* **11**, 443–453.

P. Krzyżanowski (2001), 'On block preconditioners for nonsymmetric saddle point problems', *SIAM J. Sci. Comput.* **23**, 157–169.

A. Kunoth (2002), 'Fast iterative solution of saddle point problems in optimal control based on wavelets', *Comput. Optim. Appl.* **22**, 225–259.

Y. A. Kuznetsov (1995), 'Efficient iterative solvers for elliptic finite element problems on nonmatching grids', *Russian J. Numer. Anal. Math. Modelling* **10**, 187–211.

Y. A. Kuznetsov (2004), 'Efficient preconditioner for mixed finite element methods on nonmatching meshes', *Russian J. Numer. Anal. Math. Modelling* **19**, 163–172.

C. Lanczos (1950), 'An iteration method for the solution of the eigenvalue problem of linear differential and integral operators', *J. Research Nat. Bur. Standards* **45**, 255–282.

L. D. Landau and E. M. Lifschitz (1959), *Fluid Mechanics*, Pergamon Press, London.

U. Langer and W. Queck (1986), 'On the convergence factor of Uzawa's algorithm', *J. Comput. Appl. Math.* **15**, 191–202.

K. H. Leem, S. Oliveira and D. E. Stewart (2004), 'Algebraic multigrid (AMG) for saddle point systems from meshfree discretizations', *Numer. Linear Algebra Appl.* **11**, 293–308.

R. B. Lehoucq and A. G. Salinger (2001), 'Large-scale eigenvalue calculations for stability analysis of steady flows on massively parallel computers', *Internat. J. Numer. Meth. Fluids* **36**, 309–327.

W. Leontief, F. Duchin and D. B. Szyld (1985), 'New approaches in economic analysis', *Science* **228**, 419–422.

C. Li and C. Vuik (2004), 'Eigenvalue analysis of the SIMPLE preconditioning for incompressible flow', *Numer. Linear Algebra Appl.* **11**, 511–523.

C. Li, Z. Li, D. J. Evans and T. Zhang (2003), 'A note on an SOR-like method for augmented systems', *IMA J. Numer. Anal.* **23**, 581–592.

J. Li (2001), A dual–primal FETI method for incompressible Stokes equations, Technical Report TR2001-816, Courant Institute of Mathematical Sciences, New York University.

J. Li (2002*a*), A dual–primal FETI method for incompressible Stokes and linearized Navier–Stokes equations, Technical Report TR2002-828, Courant Institute of Mathematical Sciences, New York University.

J. Li (2002*b*), Dual–primal FETI methods for stationary Stokes and Navier–Stokes equations, PhD thesis, Courant Institute of Mathematical Sciences, New York. TR-830, Department of Computer Science, New York University.

N. Li and Y. Saad (2004), Crout versions of ILU factorization with pivoting for sparse symmetric matrices, Technical Report UMSI-2004-044, Minnesota Supercomputer Institute.

J. Liesen and P. E. Saylor (2005), 'Orthogonal Hessenberg reduction and orthogonal Krylov subspace bases', *SIAM J. Numer. Anal.* **42**, 2148–2158.

J. Liesen and Z. Strakoš (2005), 'GMRES convergence analysis for a convection-diffusion model problem', *SIAM J. Sci. Comput.*, to appear.

J. Liesen and P. Tichý (2004*a*), Behavior of CG and MINRES for symmetric tridiagonal Toeplitz matrices, Technical Report 34-2004, Technical University of Berlin, Department of Mathematics.

J. Liesen and P. Tichý (2004*b*), 'The worst-case GMRES for normal matrices', *BIT* **44**, 79–98.

J. Liesen, E. de Sturler, A. Sheffer, Y. Aydin and C. Siefert (2001), Preconditioners for indefinite linear systems arising in surface parameterization, in *Proc. 10th International Meshing Round Table*, pp. 71–81.

J. Liesen, M. Rozložník and Z. Strakoš (2002), 'Least squares residuals and minimal residual methods', *SIAM J. Sci. Comput.* **23**, 1503–1525.

L. Little and Y. Saad (2003), 'Block preconditioners for saddle point problems', *Numer. Algorithms* **33**, 367–379.

L. Little, Y. Saad and L. Smoch (2003), 'Block LU preconditioners for symmetric and nonsymmetric saddle point problems', *SIAM J. Sci. Comput.* **25**, 729–748.

J. W.-H. Liu (1987), 'A partial pivoting strategy for sparse symmetric matrix decomposition', *ACM Trans. Math. Software* **13**, 173–182.

W. M. Liu and S. Xu (2001), 'A new improved Uzawa method for finite element solution of Stokes' problem', *Comput. Mech.* **27**, 305–310.

D. Loghin and A. J. Wathen (2002), 'Schur complement preconditioners for the Navier–Stokes equations', *Internat. J. Numer. Methods Fluids* **40**, 403–412.

D. Loghin and A. J. Wathen (2003), 'Schur complement preconditioning for elliptic systems of partial differential equations', *Numer. Linear Algebra Appl.* **10**, 423–443.

D. Loghin and A. J. Wathen (2004), 'Analysis of preconditioners for saddle-point problems', *SIAM J. Sci. Comput.* **25**, 2029–2049.

T.-T. Lu and S.-H. Shiou (2002), 'Inverses of 2×2 block matrices', *Comput. Math. Appl.* **43**, 119–129.

G. Lube and M. A. Olshanskii (2002), 'Stable finite element calculation of incompressible flows using the rotation form of convection', *IMA J. Numer. Anal.* **22**, 437–461.

G. Luenberger (1984), *Linear and Nonlinear Programming*, Addison-Wesley, Reading, MA.

L. Lukšan and J. Vlček (1998), 'Indefinitely preconditioned inexact Newton method for large sparse equality constrained non-linear programming problems', *Numer. Linear Algebra Appl.* **5**, 219–247.

L. Lukšan and J. Vlček (2001), 'Numerical experience with iterative methods for equality constrained non-linear programming problems', *Optim. Methods Software* **16**, 257–287.

T. Lyche, T. K. Nilssen and R. Winther (2002), 'Preconditioned iterative methods for scattered data interpolation', *Adv. Comput. Math.* **17**, 237–256.

D. S. Mackey, N. Mackey and F. Tisseur (2003), 'Structured tools for structured matrices', *Electr. J. Linear Algebra* **10**, 106–145.

Y. Maday, D. Meiron, A. T. Patera and E. M. Ronquist (1993), 'Analysis of iterative methods for the steady and unsteady Stokes problem: Application to spectral element discretizations', *SIAM J. Sci. Comput.* **14**, 310–337.

H. Mahawar and V. Sarin (2003), 'Parallel iterative methods for dense linear systems in inductance extraction', *Parallel Comput.* **29**, 1219–1235.

D. S. Malkus (1981), 'Eigenproblems associated with the discrete LBB condition for incompressible finite elements', *Internat. J. Engng. Sci.* **19**, 1299–1310.

K.-A. Mardal and R. Winther (2004), 'Uniform preconditioners for the time dependent Stokes problem', *Numer. Math.* **98**, 305–327.

H. M. Markowitz (1959), *Portfolio Selection: Efficient Diversification of Investments*, Wiley, New York.

H. M. Markowitz and A. F. Perold (1981), Sparsity and piecewise linearity in large portfolio optimization problems, in *Sparse Matrices and Their Uses* (I. S. Duff, ed.), Academic Press, London, pp. 89–108.

J. Martikainen (2003), 'Numerical study of two sparse AMG methods', *Math. Model. Numer. Anal.* **37**, 133–142.

J. Maryška, M. Rozložník and M. Tůma (1995), 'Mixed-hybrid finite element approximation of the potential fluid flow problem', *J. Comput. Appl. Math.* **63**, 383–392.

J. Maryška, M. Rozložník and M. Tůma (1996), 'The potential fluid flow problem and the convergence rate of the minimal residual method', *Numer. Linear Algebra Appl.* **3**, 525–542.

J. Maryška, M. Rozložník and M. Tůma (2000), 'Schur complement systems in the mixed-hybrid finite element approximation of the potential fluid flow problem', *SIAM J. Sci. Comput.* **22**, 704–723.

P. Matstoms (1994), Sparse QR factorization with applications to linear least squares problems, PhD thesis, Linköping University, Linköping, Sweden.

W. McGuire and R. H. Gallagher (1979), *Matrix Structural Analysis*, Wiley, New York.

K. Meerbergen and A. Spence (1997), 'Implicitly restarted Arnoldi with purification for the shift-invert transformation', *Math. Comp.* **66**, 667–689.

S. Mehrotra and J. Wang (1996), Conjugate gradient based implementation of interior point methods for network flow problems, in *Linear and Nonlinear Conjugate Gradient-Related Methods* (L. Adams and L. Nazareth, eds), SIAM, Philadelphia, PA, pp. 124–142.

O. Meshar and S. Toledo (2005), 'An out-of-core sparse symmetric indefinite factorization method', *ACM Trans. Math. Software*, to appear.

G. Meurant (1999), *Computer Solution of Large Linear Systems*, Vol. 28 of *Studies in Mathematics and its Applications*, North-Holland, Amsterdam.

M. D. Mihajlović and D. J. Silvester (2004), 'Efficient parallel solvers for the biharmonic equation', *Parallel Comput.* **30**, 35–55.

J. Modersitzki (2003), *Numerical Methods for Image Registration*, Numerical Mathematics and Scientific Computation, Oxford University Press, Oxford.

J. D. Moulton, J. E. Morel and U. M. Ascher (1998), 'Approximate Schur complement preconditioning of the lowest-order nodal discretization', *SIAM J. Sci. Comput.* **19**, 185–205.

M. F. Murphy, G. H. Golub and A. J. Wathen (2000), 'A note on preconditioning for indefinite linear systems', *SIAM J. Sci. Comput.* **21**, 1969–1972.

N. M. Nachtigal, S. C. Reddy and L. N. Trefethen (1992), 'How fast are nonsymmetric matrix iterations?', *SIAM J. Matrix Anal. Appl.* **13**, 778–795.

A. E. Naiman, I. M. Babuška and H. C. Elman (1997), 'A note on conjugate gradient convergence', *Numer. Math.* **76**, 209–230.

S. G. Nash and A. Sofer (1996), 'Preconditioning reduced matrices', *SIAM J. Matrix Anal. Appl.* **17**, 47–68.

E. G. Ng and B. W. Peyton (1993), 'Block sparse Cholesky algorithms on advanced uniprocessor computers', *SIAM J. Sci. Comput.* **14**, 1034–1056.

H. Q. Nguyen (2004), Domain decomposition methods for linear-quadratic elliptic optimal control problems, PhD thesis, Department of Computational and Applied Mathematics, Rice University, Houston, TX.

R. A. Nicolaides (1982), 'Existence, uniqueness and approximation for generalized saddle point problems', *SIAM J. Numer. Anal.* **19**, 349–357.

J. Nocedal and S. J. Wright (1999), *Numerical Optimization*, Springer Series in Operations Research, Springer, Berlin.

R. H. Nochetto and J.-H. Pyo (2004), 'Optimal relaxation parameter for the Uzawa method', *Numer. Math.* **98**, 695–702.

B. Nour-Omid and P. Wriggers (1986), 'A two-level iteration method for solution of contact problems', *Comput. Methods Appl. Mech. Engrg.* **54**, 131–144.

A. R. L. Oliveira and D. C. Sorensen (2005), 'A new class of preconditioners for large-scale linear systems for interior point methods for linear programming', *Linear Algebra Appl.* **394**, 1–24.

M. A. Olshanskii (1999), 'An iterative solver for the Oseen problem and numerical solution of incompressible Navier–Stokes equations', *Numer. Linear Algebra Appl.* **6**, 353–378.

M. A. Olshanskii and A. Reusken (2002), 'Navier–Stokes equations in rotation form: a robust multigrid solver for the velocity problem', *SIAM J. Sci. Comput.* **23**, 1683–1706.

M. A. Olshanskii and A. Reusken (2004), Analysis of a Stokes interface problem, Technical Report 237, RWTH Aachen, Aachen, Germany.

J. M. Ortega (1972), *Numerical Analysis: A Second Course*, Computer Science and Applied Mathematics, Academic Press, New York and London.

P. Oswald (1998), 'An optimal multilevel preconditioner for solenoidal approximations of the two-dimensional Stokes problem', *IMA J. Numer. Anal.* **18**, 207–228.

C. C. Paige and M. A. Saunders (1975), 'Solution of sparse indefinite systems of linear equations', *SIAM J. Numer. Anal.* **12**, 617–629.

C. C. Paige and M. A. Saunders (1982), 'LSQR: An algorithm for sparse linear equations and sparse least squares', *ACM Trans. Math. Software* **8**, 43–71.

S. V. Patankar (1980), *Numerical Heat Transfer and Fluid Flow*, McGraw-Hill, New York.

S. V. Patankar (1981), 'A calculation procedure for two-dimensional elliptic situations', *Numer. Heat Transfer* **4**, 409–425.

S. V. Patankar and D. B. Spalding (1972), 'A calculation procedure for heat, mass and momentum transfer in three-dimensional parabolic flows', *Internat. J. Heat Mass Transfer* **15**, 1787–1806.

L. F. Pavarino (1997), 'Preconditioned conjugate residual methods for mixed spectral discretizations of elasticity and Stokes problems', *Comput. Methods Appl. Mech. Engrg.* **146**, 19–30.

L. F. Pavarino (1998), 'Preconditioned mixed spectral element methods for elasticity and Stokes problems', *SIAM J. Sci. Comput.* **19**, 1941–1957.

L. F. Pavarino (2000), 'Indefinite overlapping Schwarz methods for time-dependent Stokes problems', *Comput. Methods Appl. Mech. Engrg.* **187**, 35–51.

L. F. Pavarino and O. B. Widlund (2000), 'Iterative substructuring methods for spectral element discretizations of elliptic systems II: Mixed methods for linear elasticity and Stokes flow', *SIAM J. Numer. Anal.* **37**, 375–402.

L. F. Pavarino and O. B. Widlund (2002), 'Balancing Neumann–Neumann methods for incompressible Stokes equations', *Comm. Pure Appl. Math.* **55**, 302–335.

M. Pernice (2000), 'A hybrid multigrid method for the steady-state incompressible Navier–Stokes equations', *Electr. Trans. Numer. Anal.* **10**, 74–91.

J. B. Perot (1993), 'An analysis of the fractional step method', *J. Comput. Phys.* **108**, 51–58.

I. Perugia (1997), 'A field-based mixed formulation for the two-dimensional magnetostatic problem', *SIAM J. Numer. Anal.* **34**, 2382–2391.

I. Perugia and V. Simoncini (2000), 'Block-diagonal and indefinite symmetric preconditioners for mixed finite element formulations', *Numer. Linear Algebra Appl.* **7**, 585–616.

I. Perugia, V. Simoncini and M. Arioli (1999), 'Linear algebra methods in a mixed approximation of magnetostatic problems', *SIAM J. Sci. Comput.* **21**, 1085–1101.

J. Peters, V. Reichelt and A. Reusken (2004), Fast iterative solvers for discrete Stokes equations, Technical Report 241, RWTH Aachen, Aachen, Germany.

R. Peyret and T. D. Taylor (1983), *Computational Methods for Fluid Flow*, Springer Series in Computational Physics, Springer, Berlin.

R. J. Plemmons (1986), 'A parallel block iterative scheme applied to computations in structural analysis', *SIAM J. Algebraic Discrete Methods* **7**, 337–347.

R. J. Plemmons and R. E. White (1990), 'Substructuring methods for computing the nullspace of equilibrium matrices', *SIAM J. Matrix Anal. Appl.* **11**, 1–22.

B. T. Polyak (1970), 'Iterative methods using Lagrange multipliers for solving extremal problems with constraints of the equation type', *USSR Comput. Math. Math. Phys.* **10**, 42–52.

A. Pothen (1984), Sparse null bases and marriage theorems, PhD thesis, Department of Computer Science, Cornell University, Ithaca, NY.

A. Pothen (1989), 'Sparse null basis computations in structural optimization', *Numer. Math.* **55**, 501–519.

C. E. Powell and D. Silvester (2004), 'Optimal preconditioning for Raviart-Thomas mixed formulation of second-order elliptic problems', *SIAM J. Matrix Anal. Appl.* **25**, 718–738.

M. J. D. Powell (1969), A method for nonlinear constraints in minimization problems, in *Optimization* (R. Fletcher, ed.), Academic Press, London, pp. 283–298.

M. J. D. Powell (2004), On updating the inverse of a KKT matrix, Technical Report DAMTP2004/NA01, Department of Applied Mathematics and Theoretical Physics, Cambridge University.

M. L. Psiaki and K. Park (1995), 'Parallel orthogonal factorization null-space method for dynamic-quadratic programming', *J. Optim. Theory Appl.* **85**, 409–434.

Y. Qu (2000), Multilevel methods for indefinite systems, PhD thesis, Department of Civil Engineering, Renssealer Polytechnic Institute, Troy, NY.

Y. Qu and J. Fish (2000), 'Global basis two-level method for indefinite systems, Part 2: Computational issues', *Internat. J. Numer. Methods Engng.* **49**, 461–478.

Y. Qu and J. Fish (2002), 'Multifrontal incomplete factorization for indefinite and complex symmetric systems', *Internat. J. Numer. Methods Engng.* **53**, 1433–1459.

A. Quarteroni and A. Valli (1994), *Numerical Approximation of Partial Differential Equations*, Vol. 23 of *Springer Series in Computational Mathematics*, Springer, Berlin.

A. Quarteroni and A. Valli (1999), *Domain Decomposition Methods for Partial Differential Equations*, Numerical Mathematics and Scientific Computation, Oxford University Press, Oxford.

A. Quarteroni, F. Saleri and A. Veneziani (2000), 'Factorization methods for the numerical approximation of Navier–Stokes equations', *Comput. Methods Appl. Mech. Engrg.* **188**, 505–526.

W. Queck (1989), 'The convergence factor of preconditioned algorithms of the Arrow–Hurwicz type', *SIAM J. Numer. Anal.* **26**, 1016–1030.

M. J. Raw (1995), A coupled algebraic multigrid method for the 3D Navier–Stokes equations, in *Fast Solvers for Flow Problems, Proc. 10th GAMM-Seminar* (W. Hackbusch and G. Wittum, eds), Notes on Numerical Fluid Mechanics, Vieweg, Wiesbaden, Germany, pp. 204–215.

W.-Q. Ren and J.-X. Zhao (1999), 'Iterative methods with preconditioners for indefinite systems', *J. Comput. Math.* **17**, 89–96.

M. Resende and G. Veiga (1993), 'An implementation of the dual affine scaling algorithm for minimum-cost flow on bipartite uncapacitated networks', *SIAM J. Optim.* **3**, 516–537.

M. P. Robichaud, P. A. Tanguy and M. Fortin (1990), 'An iterative implementation of the Uzawa algorithm for 3-D fluid flow problems', *Internat. J. Numer. Meth. Fluids* **10**, 429–442.

J. Robinson (1973), *Integrated Theory of Finite Element Methods*, Wiley, New York.

D. Rose (1970), 'Triangulated graphs and the elimination process', *J. Math. Anal. Appl.* **32**, 597–609.

M. Rozložník and V. Simoncini (2002), 'Krylov subspace methods for saddle point problems with indefinite preconditioning', *SIAM J. Matrix Anal. Appl.* **24**, 368–391.

T. Rusten and R. Winther (1992), 'A preconditioned iterative method for saddle-point problems', *SIAM J. Matrix Anal. Appl.* **13**, 887–904.

T. Rusten and R. Winther (1993), 'Substructure preconditioners for elliptic saddle-point problems', *Math. Comp.* **60**, 23–48.

T. Rusten, P. S. Vassilevski and R. Winther (1996), 'Interior penalty preconditioners for mixed finite element approximations of elliptic problems', *Math. Comp.* **65**, 447–466.

Y. Saad (1981), 'Krylov subspace methods for solving large unsymmetric linear systems', *Math. Comp.* **37**, 105–126.

Y. Saad (1982), 'The Lanczos biorthogonalization algorithm and other oblique projection methods for solving large unsymmetric systems', *SIAM J. Numer. Anal.* **19**, 485–506.

Y. Saad (1993), 'A flexible inner-outer preconditioned GMRES algorithm', *SIAM J. Sci. Comput.* **14**, 461–469.

Y. Saad (2003), *Iterative Methods for Sparse Linear Systems*, second edn, SIAM, Philadelphia, PA.

Y. Saad and M. H. Schultz (1986), 'GMRES: a generalized minimal residual algorithm for solving nonsymmetric linear systems', *SIAM J. Sci. Statist. Comput.* **7**, 856–869.

P. Saint-Georges, Y. Notay and G. Warzée (1998), 'Efficient iterative solution of constrained finite element analyses', *Comput. Methods Appl. Mech. Engrg.* **160**, 101–114.

F. Saleri and A. Veneziani (2005), 'Pressure-correction algebraic splitting methods for the incompressible Navier–Stokes equations', *SIAM J. Numer. Anal.*, to appear.

A. Sameh and V. Sarin (2002), 'Parallel algorithms for indefinite linear systems', *Parallel Comput.* **28**, 285–299.

V. Sarin (1997), Efficient iterative methods for saddle point problems, PhD thesis, Department of Computer Science, University of Illinois, Urbana-Champaign, IL.

V. Sarin and A. Sameh (1998), 'An efficient iterative method for the generalized Stokes problem', *SIAM J. Sci. Comput.* **19**, 206–226.

V. Sarin and A. Sameh (2003), 'Hierarchical divergence-free bases and their application to particulate flows', *Trans. ASME* **70**, 44–49.

M. A. Saunders (2005), private communication.

O. Schenk and K. Gärtner (2004), On fast factorization pivoting methods for sparse symmetric indefinite systems, Technical Report CS-2004-004, Department of Computer Science, University of Basel.

J. Schöberl and W. Zulehner (2003), 'On Schwarz-type smoothers for saddle point problems', *Numer. Math.* **95**, 377–399.

G. J. Shaw and S. Sivaloganathan (1988), 'On the smoothing properties of the SIMPLE pressure correction algorithm', *Internat. J. Numer. Meth. Fluids* **8**, 441–461.

Y. Shi (1995), 'Solving linear systems involved in constrained optimization', *Linear Algebra Appl.* **229**, 175–189.

R. Sibson and G. Stone (1991), 'Computation of thin-plate splines', *SIAM J. Sci. Stat. Comput.* **12**, 1304–1313.

A. Sidi (2003), 'A zero-cost preconditioning for a class of indefinite linear systems', *WSEAS Trans. Math.* **2**, 142–150.

C. Siefert and E. de Sturler (2004), Preconditioners for generalized saddle-point problems, Technical Report UIUCDCS-R-2004-2448, Department of Computer Science, University of Illinois.

D. J. Silvester and A. J. Wathen (1994), 'Fast iterative solution of stabilised Stokes systems II: Using general block preconditioners', *SIAM J. Numer. Anal.* **31**, 1352–1367.

D. J. Silvester, H. C. Elman, D. Kay and A. J. Wathen (2001), 'Efficient preconditioning of the linearized Navier–Stokes equations for incompressible flow', *J. Comput. Appl. Math.* **128**, 261–279.

V. Simoncini (2004a), 'Block triangular preconditioners for symmetric saddle point problems', *Appl. Numer. Math.* **49**, 63–80.

V. Simoncini (2004b), private communication.

V. Simoncini and M. Benzi (2004), 'Spectral properties of the Hermitian and skew-Hermitian splitting preconditioner for saddle point problems', *SIAM J. Matrix Anal. Appl.* **26**, 377–389.

V. Simoncini and M. Benzi (2005), 'On the eigenvalues of a class of saddle point matrices'. In preparation.

V. Simoncini and D. B. Szyld (2003), 'Theory of inexact Krylov subspace methods and applications to scientific computing', *SIAM J. Sci. Comput.* **25**, 454–477.

V. Simoncini and D. B. Szyld (2005), 'The effect of non-optimal bases on the convergence of Krylov subspace methods', *Numer. Math.*, to appear.

G. Starke (1997), 'Field-of-values analysis of preconditioned iterative methods for nonsymmetric elliptic problems', *Numer. Math.* **78**, 103–117.

G. W. Stewart (1989), 'On scaled projections and pseudoinverses', *Linear Algebra Appl.* **112**, 189–193.

G. Stoyan (2001), 'Iterative Stokes solvers in the harmonic Velte subspace', *Comput.* **67**, 13–33.

G. Strang (1986), *Introduction to Applied Mathematics*, Wellesley–Cambridge Press, Wellesley, MA.

G. Strang (1988), 'A framework for equilibrium equations', *SIAM Rev.* **30**, 283–297.

J. C. Strikwerda (1984), 'An iterative method for solving finite difference approximations to the Stokes equations', *SIAM J. Numer. Anal.* **21**, 447–458.

K. Stüben (2001), 'A review of algebraic multigrid', *J. Comput. Appl. Math.* **128**, 281–309.

T. Stykel (2005), 'Balanced truncation model reduction for semidiscretized Stokes equations', *Linear Algebra Appl.*, to appear.

J. G. Sun (1999), 'Structured backward errors for KKT systems', *Linear Algebra Appl.* **288**, 75–88.

Syamsudhuha and D. J. Silvester (2003), 'Efficient solution of the steady-state Navier–Stokes equations using a multigrid preconditioned Newton–Krylov method', *Internat. J. Numer. Meth. Fluids* **43**, 1407–1427.

D. B. Szyld (1981), Using sparse matrix techniques to solve a model of the world economy, in *Sparse Matrices and Their Uses* (I. S. Duff, ed.), Academic Press, pp. 357–365.

R. E. Tarjan (1983), *Data Structures and Network Algorithms*, SIAM, Philadelphia, PA.

R. Temam (1984), *Navier–Stokes Equations: Theory and Numerical Analysis*, Vol. 2 of *Studies in Mathematics and its Applications*, third edn, North-Holland, Amsterdam.

J. L. Thomas, B. Diskin and A. Brandt (2001), 'Textbook multigrid efficiency for the incompressible Navier–Stokes equations: high Reynolds number wakes and boundary layers', *Comput. & Fluids* **30**, 853–874.

M. J. Todd (1990), 'A Dantzig–Wolfe-like variant of Karmarkar's interior point linear programming algorithm', *Oper. Res.* **38**, 1006–1018.

K.-C. Toh (1997), 'GMRES vs. ideal GMRES', *SIAM J. Matrix Anal. Appl.* **18**, 30–36.

K.-C. Toh, K.-K. Phoon and S.-H. Chan (2004), 'Block preconditioners for symmetric indefinite linear systems', *Internat. J. Numer. Methods Engng.* **60**, 1361–1381.

Z. Tong and A. Sameh (1998), 'On an iterative method for saddle point problems', *Numer. Math.* **79**, 643–646.

A. Topcu (1979), A contribution to the systematic analysis of finite element structures using the force method, PhD thesis, University of Essen, Germany. In German.

A. Toselli and O. B. Widlund (2004), *Domain Decomposition Methods: Algorithms and Theory*, Vol. 34 of *Springer Series in Computational Mathematics*, Springer.

A. M. Tropper (1962), *Matrix Theory for Electrical Engineers*, Vol. 1 of *International Series in the Engineering Sciences*, Addison-Wesley, New York.

U. Trottenberg, C. W. Oosterlee and A. Schüller (2001), *Multigrid*, Academic Press, London.

M. Tůma (2002), 'A note on the LDL^T decomposition of matrices from saddle-point problems', *SIAM J. Matrix Anal. Appl.* **23**, 903–915.

S. Turek (1999), *Efficient Solvers for Incompressible Flow Problems: An Algorithmic and Computational Approach*, Vol. 6 of *Lecture Notes in Computational Science and Engineering*, Springer, Berlin.

E. Tyrtyshnikov and Y. Vassilevski (2003), A mosaic preconditioner for a dual Schur complement, in *Numerical Mathematics and Advanced Applications* (F. Brezzi, A. Buffa, S. Corsaro and A. Murli, eds), Springer, pp. 867–877.

H. Uzawa (1958), Iterative methods for concave programming, in *Studies in Linear and Nonlinear Programming* (K. J. Arrow, L. Hurwicz and H. Uzawa, eds), Stanford University Press, Stanford, CA, pp. 154–165.

P. Vaidya (1990), Solving linear equations with diagonally dominant matrices by constructing good preconditioners, Technical report, Department of Computer Science, University of Illinois.

E. Vainikko and I. G. Graham (2004), 'A parallel solver for PDE systems and application to the incompressible Navier–Stokes equations', *Appl. Numer. Math.* **49**, 97–116.

H. A. van der Vorst (1992), 'Bi-CGSTAB: a fast and smoothly converging variant of Bi-CG for the solution of nonsymmetric linear systems', *SIAM J. Sci. Statist. Comput.* **13**, 631–644.

H. A. van der Vorst (2003), *Iterative Krylov Methods for Large Linear Systems*, Vol. 13 of *Cambridge Monographs on Applied and Computational Mathematics*, Cambridge University Press, Cambridge.

C. F. Van Loan (1985), 'On the method of weighting for equality constrained least squares problems', *SIAM J. Numer. Anal.* **22**, 851–864.

R. J. Vanderbei (1995), 'Symmetric quasidefinite matrices', *SIAM J. Optim.* **5**, 100–113.

S. P. Vanka (1986), 'Block-implicit multigrid solution of Navier–Stokes equations in primitive variables', *J. Comput. Phys.* **65**, 138–158.

R. S. Varga (1962), *Matrix Iterative Analysis*, Prentice-Hall, Englewood Cliffs, NJ.

P. B. Vasconcelos and F. D. D'Almeida (1998), 'Preconditioned iterative methods for coupled discretizations of fluid flow problems', *IMA J. Numer. Anal.* **18**, 385–397.

P. S. Vassilevski and R. D. Lazarov (1996), 'Preconditioning mixed finite element saddle-point elliptic problems', *Numer. Linear Algebra Appl.* **3**, 1–20.

P. S. Vassilevski and J. Wang (1992), 'Multilevel iterative methods for mixed finite element discretizations of elliptic probems', *Numer. Math.* **63**, 503–520.

S. A. Vavasis (1994), 'Stable numerical algorithms for equilibrium systems', *SIAM J. Matrix Anal. Appl.* **15**, 1108–1131.

S. A. Vavasis (1996), 'Stable finite elements for problems with wild coefficients', *SIAM J. Numer. Anal.* **33**, 890–916.

A. Veneziani (2003), 'Block factorized preconditioners for high-order accurate in time approximation of the Navier–Stokes equations', *Numer. Methods Partial Differential Equations* **19**, 487–510.

R. Verfürth (1984a), 'A combined conjugate gradient-multigrid algorithm for the numerical solution of the Stokes problem', *IMA J. Numer. Anal.* **4**, 441–455.

R. Verfürth (1984b), 'A multilevel algorithm for mixed problems', *SIAM J. Numer. Anal.* **21**, 264–271.

C. Vincent and R. Boyer (1992), 'A preconditioned conjugate gradient Uzawa-type method for the solution of the Stokes problem by mixed Q_1-P_0 stabilized finite elements', *Internat. J. Numer. Meth. Fluids* **14**, 289–298.

M. Wabro (2004), 'Coupled algebraic multigrid methods for the Oseen problem', *Comput. Visual. Sci.* **7**, 141–151.

C. Wagner, W. Kinzelbach and G. Wittum (1997), 'Schur-complement multigrid: a robust method for groundwater flow and transport problems', *Numer. Math.* **75**, 523–545.

Y. Wang (2005), 'Overlapping Schwarz preconditioner for the mixed formulation of plane elasticity', *Appl. Numer. Math.*, to appear.

J. S. Warsa, M. Benzi, T. A. Wareing and J. E. Morel (2004), 'Preconditioning a mixed discontinuous finite element method for radiation diffusion', *Numer. Linear Algebra Appl.* **11**, 795–811.

A. J. Wathen (1987), 'Realistic eigenvalue bounds for the Galerkin mass matrix', *IMA J. Numer. Anal.* **7**, 449–457.

A. J. Wathen and D. J. Silvester (1993), 'Fast iterative solution of stabilised Stokes systems I: Using simple diagonal preconditioners', *SIAM J. Numer. Anal.* **30**, 630–649.

A. J. Wathen, B. Fischer and D. J. Silvester (1995), 'The convergence rate of the minimal residual method for the Stokes problem', *Numer. Math.* **71**, 121–134.

R. Webster (1994), 'An algebraic multigrid solver for Navier–Stokes problems', *Internat. J. Numer. Meth. Fluids* **18**, 761–780.

M. Wei (1992a), 'Algebraic properties of the rank-deficient equality constrained and weighted least squares problems', *Linear Algebra Appl.* **161**, 27–43.

M. Wei (1992b), 'Perturbation theory for the rank-deficient equality constrained least squares problem', *SIAM J. Numer. Anal.* **29**, 1462–1481.

Y. Wei and N. Zhang (2004), 'Further note on constraint preconditioning for nonsymmetric indefinite matrices', *Appl. Math. Comput.* **152**, 43–46.

P. Wesseling (1992), *An Introduction to Multigrid Methods*, Wiley, Chichester, UK.

P. Wesseling (2001), *Principles of Computational Fluid Dynamics*, Vol. 29 of *Springer Series in Computational Mathematics*, Springer, Berlin.

P. Wesseling and C. W. Oosterlee (2001), 'Geometric multigrid with applications to fluid dynamics', *J. Comput. Appl. Math.* **128**, 311–334.

O. B. Widlund (1978), 'A Lanczos method for a class of nonsymmetric systems of linear equations', *SIAM J. Numer. Anal.* **15**, 801–812.

S. Ø. Wille and A. F. D. Loula (2002), 'A priori pivoting in solving the Navier–Stokes equations', *Commun. Numer. Meth. Engng.* **18**, 691–698.

S. Ø. Wille, O. Staff and A. F. D. Loula (2003), 'Parallel ILU preconditioning, a priori pivoting and segregation of variables for iterative solution of the mixed finite element formulation of the Navier–Stokes equations', *Internat. J. Numer. Meth. Fluids* **41**, 977–996.

G. Wittum (1989), 'Multi-grid methods for Stokes and Navier–Stokes equations with transforming smoothers: Algorithms and numerical results', *Numer. Math.* **54**, 543–563.

G. Wittum (1990), 'On the convergence of multi-grid methods with transforming smoothers', *Numer. Math.* **57**, 15–38.

P. Wolfe (1962), The reduced gradient method, Technical report, The RAND Corporation, Santa Monica, CA. Unpublished.

M. H. Wright (1992), 'Interior methods for constrained optimization', *Acta Numerica*, Vol. 1, Cambridge University Press, pp. 341–407.

S. J. Wright (1997), *Primal–Dual Interior Point Methods*, SIAM, Philadelphia, PA.

X. Wu, B. P. B. Silva and J. Y. Yuan (2004), 'Conjugate gradient method for rank deficient saddle point problems', *Numer. Algorithms* **35**, 139–154.

D. Yang (2002), 'Iterative schemes for mixed finite element methods with applications to elasticity and compressible flow problems', *Numer. Math.* **93**, 177–200.

N. Zhang and Y. Wei (2004), 'Solving EP singular linear systems', *Internat. J. Computer Math.* **81**, 1395–1405.

J.-X. Zhao (1998), 'The generalized Cholesky factorization method for saddle point problems', *Appl. Math. Comput.* **92**, 49–58.

O. C. Zienkiewicz, J. P. Vilotte, S. Toyoshima and S. Nakazawa (1985), 'Iterative method for constrained and mixed approximation: An inexpensive improvement of FEM performance', *Comput. Methods Appl. Mech. Engrg.* **51**, 3–29.

A. Zsaki, D. Rixen and M. Paraschivoiu (2003), 'A substructure-based iterative inner solver coupled with Uzawa's algorithm for the Stokes problem', *Internat. J. Numer. Meth. Fluids* **43**, 215–230.

W. Zulehner (2000), 'A class of smoothers for saddle point problems', *Comput.* **65**, 227–246.

W. Zulehner (2002), 'Analysis of iterative methods for saddle point problems: a unified approach', *Math. Comp.* **71**, 479–505.

Acta Numerica (2005), pp. 139–232
DOI: 10.1017/S0962492904000224

Computation of geometric partial differential equations and mean curvature flow

Klaus Deckelnick

Institut für Analysis und Numerik,
Otto-von-Guericke-Universität Magdeburg, Universitätsplatz 2,
D–39106 Magdeburg, Germany
E-mail: `Klaus.Deckelnick@Mathematik.Uni-Magdeburg.de`

Gerhard Dziuk

Abteilung für Angewandte Mathematik,
Albert-Ludwigs-Universität Freiburg i. Br., Hermann-Herder-Straße 10,
D–79104 Freiburg i. Br., Germany
E-mail: `gerd.dziuk@mathematik.uni-freiburg.de`

Charles M. Elliott

Department of Mathematics,
University of Sussex, Mantell Building,
Falmer, Brighton, BN1 9RF, UK
E-mail: `C.M.Elliott@sussex.ac.uk`

This review concerns the computation of curvature-dependent interface motion governed by geometric partial differential equations. The canonical problem of mean curvature flow is that of finding a surface which evolves so that, at every point on the surface, the normal velocity is given by the mean curvature. In recent years the interest in geometric PDEs involving curvature has burgeoned. Examples of applications are, amongst others, the motion of grain boundaries in alloys, phase transitions and image processing. The methods of analysis, discretization and numerical analysis depend on how the surface is represented. The simplest approach is when the surface is a graph over a base domain. This is an example of a *sharp interface* approach which, in the general *parametric approach*, involves seeking a parametrization of the surface over a base surface, such as a sphere. On the other hand an interface can be represented implicitly as a level surface of a function, and this idea gives rise to the so-called *level set method*. Another implicit approach is the *phase field method*, which approximates the interface by a zero level set of a phase field satisfying a PDE depending on a new parameter. Each approach has its own advantages and disadvantages. In the article we describe the mathematical formulations of these approaches and their discretizations. Algorithms are set out for each approach, convergence results are given and are supported by computational results and numerous graphical figures. Besides mean curvature flow, the topics of anisotropy and the higher order geometric PDEs for Willmore flow and surface diffusion are covered.

CONTENTS

1. Introduction

A geometric evolution equation defines the motion of a hypersurface by prescribing the normal velocity of the surface in terms of geometric quantities. As well as being of striking mathematical interest, geometric evolution problems occur in a wide variety of scientific and technological applications. A traditional source of problems is materials science, where the understanding of the strength and properties of materials requires the mathematical modelling of the morphology of microstructure. Evolving surfaces might be grain boundaries, which separate differing orientations of the same crystalline phase, or solid–liquid interfaces exhibiting dendritic structures in under-cooled solidification. On the other hand newer applications are associated with image processing. For example, in order to identify a dark shape in a light background in a two-dimensional image a so-called snake contour is evolved so that it wraps around the shape.

In this article we survey numerical methods for the evolution of surfaces whose normal velocity is strongly dependent on the mean curvature of the surface. The objective is to find a family $\{\Gamma(t)\}_{t\in[0,T]}$ of closed compact and orientable hypersurfaces in $\mathbb{R}^{n+1}$ whose evolution is defined by specifying the velocity V of $\Gamma(t)$ in the normal direction ν. An example of a general geometric evolution equation is

$$V = f(x, \nu, H) \quad \text{on } \Gamma(t), \tag{1.1}$$

where f depends on the application and the x dependence might arise from evaluating on the surface $\Gamma(t)$ field variables which satisfy their own system of nonlinear partial differential equations in $\mathbb{R}^{n+1}$ away from the surface. It is important to note that, in order to specify the evolution of the surface, it is sufficient to define the normal velocity.

The prototype problem is *motion by mean curvature*, for which

$$V = -H \quad \text{on } \Gamma(t), \tag{1.2}$$

where H is the sum of the n principal curvatures of $\Gamma(t)$. We call H the mean curvature rather than the arithmetic mean of the principal curvatures. Our sign convention is that H is positive for spheres, with ν being the outward normal. It is well known that, starting from an initial surface Γ_0, this equation is a gradient flow for the area functional,

$$E(\Gamma) = \int_{\Gamma} 1 \, dA. \tag{1.3}$$

In applications the area functional is an *interfacial energy* with a constant energy density 1. Equation (1.2) may be viewed as an analogue for surfaces of the parabolic heat equation

$$u_t - \Delta u = 0.$$

On the other hand, another geometric equation is

$$V = \Delta_{\Gamma(t)} H \quad \text{on } \Gamma(t), \tag{1.4}$$

where $\Delta_{\Gamma(t)}$ is the Laplace–Beltrami or surface Laplacian operator on $\Gamma(t)$. This can be viewed as an analogue of the spatially fourth order parabolic equation

$$u_t + \Delta^2 u = 0.$$

1.1. Approaches

In order to solve a surface evolution equation analytically or numerically, we need a description of $\Gamma(t)$. Each choice of description leads to a particular nonlinear partial differential equation defining the evolution. Thus the computational method depends strongly on the way we choose to describe the surface. For this article we shall focus on four possible approaches.

Parametric approach. The hypersurfaces $\Gamma(t)$ are given as

$$\Gamma(t) = X(\cdot, t)(M),$$

where M is a suitable reference manifold (fixing the topological type of $\Gamma(t)$) and $X : M \times [0, T] \to \mathbb{R}^{n+1}$ has to be determined. Here $X(p, t)$, for $p \in M$, is

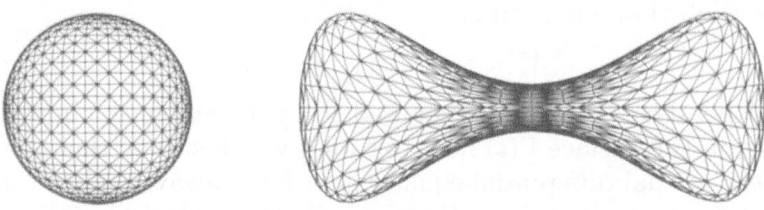

Figure 1.1. A dumbbell-shaped two-dimensional surface parametrized over the unit sphere.

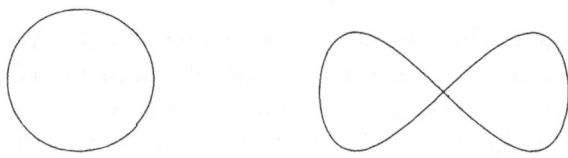

Figure 1.2. A lemniscate, parametrized over the unit circle.

the position vector at time t of a point on $\Gamma(t)$. If we are interested in closed curves in the plane then M can be the unit circle S^1, whereas if $\Gamma(t)$ is a two-dimensional surface then M could be the unit sphere S^2 (see Figures 1.1 and 1.2). Geometrical quantities are easily expressed as derivatives of the parametrization so that evolution laws such as (1.2) may be translated into nonlinear parabolic systems of PDEs for the vector X. With this approach there is no notion of the surface being the boundary of an open set and having an inside and outside, so *self-intersection* is perfectly natural for smooth parametrizations and is not necessarily associated with singularities. For example in the plane a figure of eight curve can be smoothly mapped onto the unit circle one to one (Figure 1.2). At the crossing point the curve has two smoothly evaluated normals and curvatures which depend on the parametrization. A parametrized curve evolving by mean curvature can evolve smoothly from this configuration.

Graphs. We assume that $\Gamma(t)$ can be written in the form

$$\Gamma(t) = \{(x, u(x, t)) \mid x \in \Omega\},$$

where $\Omega \subset \mathbb{R}^n$ and the height function $u : \Omega \times [0, T] \to \mathbb{R}$ has to be found. We shall see that the law (1.2) leads to a nonlinear parabolic PDE for u. Clearly, the assumption that $\Gamma(t)$ is a graph is rather restrictive; however, techniques developed for this case have turned out to be very useful in understanding more general situations. Since the height is a smooth function we can view $\Gamma(t)$ as dividing $\Omega \times \mathbb{R}$ into two sets, namely the regions above and below the graph.

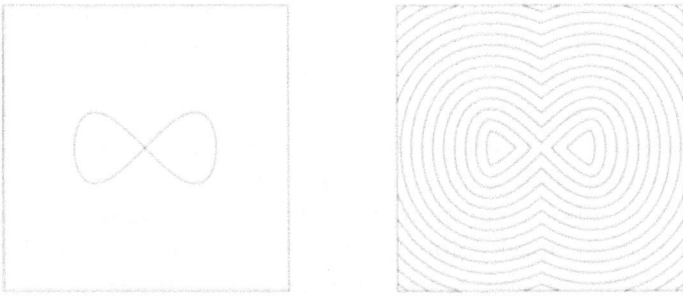

Figure 1.3. Level lines of a level set function (right) for the figure of eight curve (left).

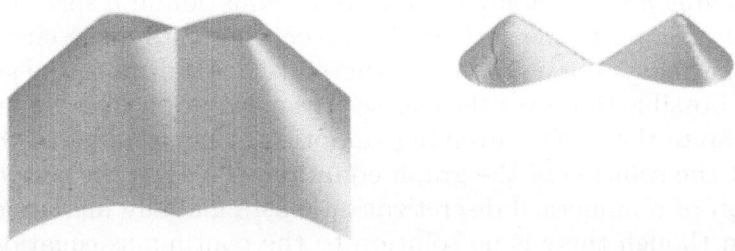

Figure 1.4. Graph of a level set function for the figure of eight curve, cut at the zero level. Negative part left and positive part (graphically enlarged) right.

Level set method. We look for $\Gamma(t)$ as the zero level set of an auxiliary function $u : \mathbb{R}^{n+1} \times [0, \infty) \to \mathbb{R}$, that is,

$$\Gamma(t) = \{x \in \mathbb{R}^{n+1} \mid u(x, t) = 0\}.$$

The law (1.2) now translates into a nonlinear, degenerate and singular PDE for u. Clearly intrinsic to this approach is the notion of $\Gamma(t)$ being a dividing surface between the two regions where the level set function is positive and negative. Thus we have the notion of inside and outside. In order to describe a figure of eight by a level set function it is necessary to have the level set function positive and negative, as shown in Figures 1.3 and 1.4.

Phase field approach. The phase field approach is based on an approximation of the sharp interface by a diffuse interface

$$\Gamma_\epsilon(t) = \{x \in \mathbb{R}^{n+1} \mid -1 + C\epsilon \leq u_\epsilon(x, t) \leq 1 - C\epsilon\}$$

of width $O(\epsilon)$, across which the phase field function u_ϵ has a transition from approximately one bulk negative value -1 to approximately a second positive bulk value $+1$. The zero level set of the phase field function approximates the surface. Just as in the level set method there is the notion of a material interface separating an inside and outside and in the basic implementation interface self-intersection and topological change are handled automatically. The bulk values of the phase field function correspond to the minima of a homogeneous energy function with two equal double wells. Interfacial energy is assigned to the diffuse interface via the gradient of the phase field function. For motion by mean curvature the evolution is defined as a semilinear parabolic equation of reaction–diffusion or Ginzburg–Landau type. Frequently in applications mathematical models are derived which, from the beginning, involve diffuse interfaces and phase field functions.

Comments

Conceptually the graph formulation is the simplest and most efficient. It involves solving a scalar nonlinear parabolic equation in n space dimensions and directly computes the surface. However, there are many circumstances where the surface is not a graph. Furthermore, even if the initial surface is a graph it is possible that over the course of the evolution that property might be lost, despite the surface evolving smoothly. This would lead to gradient blow-up of the solution of the graph equation. There is the possibility that the solution of a numerical discretization exists globally and appears to be stable even though there is no solution to the continuous equation.

The parametric approach is also direct. It is conceptually more advanced than the graph approach and one has to solve for an n-dimensional surface a system of $n + 1$ parabolic equations. If the surface is a graph then the parametric approach is less efficient than solving for the height of the surface. On the other hand it is more widely applicable. In the case of a closed curve one can use periodic boundary conditions on the unit interval in order to solve over the circle. A closed two-dimensional surface can be approximated by a polyhedral surface. A parametrized surface does not 'see' an inside or outside. From the point of view of differential geometry this may not be an issue. However, when the surface separates two phases, or two materials, or two colours, there are significant issues. For example, consider using two colours in Figure 1.2 in order to define the curve as the interface between the coloured regions. Black may be used to colour the inside of both loops and white to colour the the rest of the plane, but if black is used inside just one loop then the other loop is lost. Thus, in order to use the parametric approach with this initial condition, one either thinks of a parametrization which traverses the curve without a crossing, but with a single self-intersection, or regards them as being two separate closed curves which touch at one point. These choices lead to differing evolutions for mean curvature flow.

Contrary to the parametric approach, the level set method has the capability of tracking topological changes (like pinching-off or merging) of $\Gamma(t)$ in an automatic way. In the basic implementation of the method topological change is nothing special and is observed in post-processing the computational output. This is because, in principle, zero level sets of continuous functions can exhibit these features. However, from the mathematical point of view there are issues of existence of solutions of the degenerate partial differential equations that the level set approach generates. In the case of motion by mean curvature there is the notion of a *viscosity* solution which yields a unique evolution from any continuous function. The example of the lemniscate discussed in the context of the parametric approach introduces a new idea in the level set approach of *fattening* of the interface. The level set for this example develops an interior whose boundary yields both of the described solutions. Self-intersection, merger and pinch-off can all be simulated by this approach. This advantage, however, needs to be offset against the fact that the problem now becomes $(n + 1)$-dimensional in space.

The phase field approach can also handle topological change, self-intersection, merger and pinch-off without doing anything special. It is the one approach which in its conception involves an approximation. The fact that it involves a new parameter ϵ is both an advantage and a disadvantage. The parabolic equations are in principle easy to solve but possess a certain computational *stiffness* due to the thickness of the diffuse interface. However, in many applications phase field models arise naturally and the ϵ parameter allows us to resolve singularities in a way which may be viewed as being physically motivated. From both the mathematical and physical points of view it is widely applicable in a rational way, whereas the use of the level set method is frequently *ad hoc*.

In general, the choice of one or the other approach will depend on whether one expects topological changes in the flow.

1.2. Applications

In what follows we list some problems in which a law of the form (1.1) or generalizations of it arise.

Grain boundary motion

Grain boundaries in alloys are interfaces which separate bulk crystalline regions of the same phase but with differing orientations. Associated with the grain boundary is a surface energy which gives rise to a thermodynamic restoring force. For a constant surface energy density this is simply the surface tension force proportional to the mean curvature and the resulting evolution law is just (1.2). Frequently there is also a driving force causing motion of the grain boundary.

Surface growth

The growth of thin films on substrates is technologically important. For example, epitaxy is a method for growing single crystals by the deposition of atoms and molecules onto a growing film surface. There are numerous physical mechanisms operating at differing time and length scales which affect the growth process. A simple model would have a driving force representing the deposition flux of atoms onto the surface which might be in the normal direction or in a fixed vertical direction parallel to a beam of arriving atoms.

Image processing

One of the most important problems in image processing is to automatically detect contours of objects. We essentially follow the exposition of Aubert and Kornprobst (2002). Suppose that $M \subset \mathbb{R}^{n+1}$ ($n = 1$ or 2) is a given object and let $I(x) = \chi_{\Omega \setminus M}(x)$ be the characteristic function of $\Omega \setminus M$. The function

$$g(x) = \frac{1}{1 + |\nabla I_\sigma(x)|^2},$$

where I_σ is a mollification of I, will be small near the contour of M. It is therefore natural to look for minimizers of the functional

$$J(\Gamma) = \int_\Gamma g(x) \, \mathrm{d}A$$

where Γ is a curve in $\mathbb{R}^2$ or a surface in $\mathbb{R}^3$. The corresponding L^2-gradient flow leads to the following evolution law: find curves/surfaces (moving 'snakes') $\Gamma(t)$ such that

$$V = -\nabla \cdot (g\,\nu) = -g\,H - \nabla g \cdot \nu \quad \text{on } \Gamma(t).$$

Here, t plays the role of an artificial time; clearly this law fits into the framework (1.1).

Stefan problem for undercooled solidification

Consider a container $\Omega \subset \mathbb{R}^{n+1}$ ($n = 1$ or 2) filled with an undercooled liquid. Solidification of the liquid follows the nucleation of initial solid seed with characteristic diameter larger than the critical radius. The seed will then grow into the liquid. A mathematical model for this situation is the Stefan problem with kinetic undercooling, in which the solid–liquid interface is described by a curve/surface $\Gamma(t)$ and has to be determined together with the temperature distribution. Here the interior of $\Gamma(t)$ is the solid region $\Omega_S(t)$ and the exterior is the liquid region $\Omega_L(t)$. Using a suitable non-dimensionalization the problem then reads: for a given initial phase boundary Γ_0 and initial temperature distribution $\Theta_0 = \Theta_0(x)$ ($x \in \overline{\Omega}$), find the non-dimensional temperature $\Theta = \Theta(x, t)$ and the phase boundary $\Gamma(t)$ ($t > 0$), such that the heat equation is satisfied in the bulk, that is,

$$\Theta_t - \Delta\Theta = 0 \quad \text{in } \Omega \setminus \Gamma(t),$$

together with the initial value $\Theta(\cdot, 0) = \Theta_0$ in $\overline{\Omega}$. On the moving boundary the following two conditions are satisfied:

$$V = -\frac{1}{\varepsilon_l}\left[\frac{\partial\Theta}{\partial\nu}\right] \quad \text{on } \Gamma(t), \tag{1.5}$$

$$\Theta + \varepsilon_V \beta(\nu) V + \sigma H_\gamma = 0 \qquad \text{on } \Gamma(t). \tag{1.6}$$

Here, $[\partial\Theta/\partial\nu]$ denotes the jump in the normal derivative of the temperature field across the interface and ε_l is the constant measuring the latent heat of solidification. Equation (1.6) is the Gibbs–Thomson law; ε_V, σ are non-dimensional positive constants measuring the strength of the kinetic undercooling and surface tension which depress the temperature on the solid–liquid interface from the scaled equilibrium zero melting temperature. Furthermore, H_γ is an anisotropic mean curvature associated with a surface energy density, $\gamma(\nu)$, depending on the orientation of the normal. There may also be anisotropy, $\beta(\nu)$, in the kinetic undercooling. Note that (1.6) can be rewritten as

$$\frac{\varepsilon_V}{\sigma}\beta(\nu)V = -H_\gamma - \frac{1}{\sigma}\Theta \quad \text{on } \Gamma(t).$$

If we consider Θ as being given, this equation again fits into our general framework (1.1) provided we allow for a coefficient in front of V and a generalized notion of mean curvature.

Figure 1.5 from Schmidt (1996) shows a simulation in which the free boundary was described by the parametric approach resulting in a sharp interface model. One can see the free boundary forming a dendrite. For

Figure 1.5. Evolution of a dendrite with sixfold anisotropy. Time-steps of the free boundary (left) and adapted grid for the temperature at one time-step (right).

results concerning three-dimensional dendrites and more information about the algorithm we refer to Schmidt (1996).

Figure 1.6 from Fried (1999) illustrates a possible effect of using a level set method for the free boundary in this problem. Dendrites may seem to merge. But if a smaller time-step is used the dendrites stay apart. For more information about a level set algorithm for dendritic growth we refer to Fried (1999, 2004).

Surface diffusion and Willmore flow
The following laws do not fit into (1.1), but we list them as examples of important geometric evolution equations in which the normal velocity depends on higher derivatives of mean curvature.

The surface diffusion equation

$$V = \Delta_\Gamma H \tag{1.7}$$

models the diffusion of mass within the bounding surface of a solid body. At the atomistic level atoms on the surface move along the surface owing to a driving force consisting of a chemical potential difference. For a surface with constant surface energy density the appropriate chemical potential in this setting is the mean curvature H. This leads to the flux law

$$\rho V = -\mathrm{div}_\Gamma \mathbf{j},$$

where ρ is the mass density and $\mathbf{j}$ is the mass flux in the surface, with the constitutive flux law (Herring 1951, Mullins 1957)

$$\mathbf{j} = -D\nabla_\Gamma H.$$

Here, D is the diffusion constant. From these equations we obtain the law (1.7) after an appropriate non-dimensionalization. In order to model the

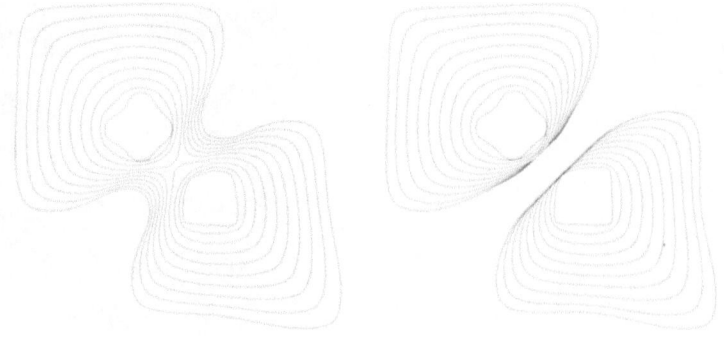

Figure 1.6. A possible effect of the use of a level set method.
Growing dendrites: merging (left) for large time-step size
and staying apart (right) for smaller time-step size.

underlying structure of the solid body bounded by Γ, anisotropic surface diffusion is important, that is,

$$V = \Delta_\Gamma H_\gamma, \qquad (1.8)$$

with H_γ denoting the anisotropic mean curvature of the surface Γ as it is introduced in (8.15).

A similar evolution law is Willmore flow,

$$V = \Delta_\Gamma H + H|\nabla_\Gamma \nu|^2 - \frac{1}{2}H^3 \quad \text{on } \Gamma(t), \qquad (1.9)$$

which arises as the L^2-gradient flow for the classical bending energy $E(\Gamma) = \frac{1}{2}\int_\Gamma H^2 \, dA$. Apart from applications in mechanics and membrane physics this flow has recently been used for surface restoration and inpainting.

1.3. Outline of article

This article is organized as follows. In Section 2 we present some useful geometric analysis, in particular the notion of mean curvature. The basic mean curvature flow is defined in Section 3 and some elementary properties are described. The next four sections consider in turn basic approaches for numerical approximation. In Section 4 we consider the parametric approach. We start with the classical curve shortening flow and present a semidiscrete numerical scheme as well as error estimates. Next, we show how to apply the above ideas to the approximation of higher-dimensional surfaces. A crucial point is to construct numerical schemes which reflect the intrinsic nature of the flow. Section 5 is concerned with graphs. We prove an error bound for a semidiscrete finite element scheme thereby showing the virtue of working with geometric quantities. A fully discrete scheme along with stability issues is discussed afterwards. In Section 6 we introduce the level set equation as a way of handling topological changes. We briefly discuss the framework of viscosity solutions which allows a satisfactory existence and uniqueness theory. For numerical purposes it is convenient to regularize the level set equation. We collect some properties of the regularized problem and clarify its formal similarity to the graph setting. The approximation of mean curvature flow by phase field methods is considered in Section 7. Even before numerical discretization there is the notion of approximation of a sharp interface by a diffuse interface of width $O(\epsilon)$. The phase field approach depends on the notion of a diffuse interfacial energy composed of quadratic gradient and homogeneous free energy terms involving a phase field function. The choice of double well energy potential is discussed. We recall some analytical results as well as a convergence analysis for a discretization in space by linear finite elements. We finish this section by discussing the discretization in time together with the question of stability. In Section 8 we introduce the concept of the anisotropy γ together with its relevant properties and subsequently generalize the ideas of the previous sections to this setting. Finally, Section 9 is concerned with fourth order flows: we present discretization techniques for both surface diffusion and Willmore flow.

For the convenience of the reader we have included a long list of references, which are related to the subject of these notes, but not all of which are cited in the text.

2. Some geometric analysis

The aim of this section is to collect some useful definitions and results from differential geometry. We refer to Gilbarg and Trudinger (1998) and Giga (2002) for a more detailed exposition of this material.

2.1. Hypersurfaces

A subset $\Gamma \subset \mathbb{R}^{n+1}$ is called a C^2-hypersurface if for each point $x_0 \in \Gamma$ there exists an open set $U \subset \mathbb{R}^{n+1}$ containing x_0 and a function $u \in C^2(U)$ such that

$$U \cap \Gamma = \{x \in U \mid u(x) = 0\}, \quad \text{and} \quad \nabla u(x) \neq 0 \quad \text{for all } x \in U \cap \Gamma. \quad (2.1)$$

The tangent space $T_x\Gamma$ is then the n-dimensional linear subspace of $\mathbb{R}^{n+1}$ that is orthogonal to $\nabla u(x)$. It is independent of the particular choice of function u which is used to describe Γ. A C^2-hypersurface $\Gamma \subset \mathbb{R}^{n+1}$ is called orientable if there exists a vectorfield $\nu \in C^1(\Gamma, \mathbb{R}^{n+1})$ (*i.e.*, $\nu \in C^1$ in an open neighbourhood of Γ) such that $\nu(x) \perp T_x\Gamma$ and $|\nu(x)| = 1$ for all $x \in \Gamma$. In what follows, we shall assume that $\Gamma \subset \mathbb{R}^{n+1}$ is an oriented C^2-hypersurface.

We define the tangential gradient of a function f, which is differentiable in an open neighbourhood of Γ by

$$\nabla_\Gamma f(x) = \nabla f(x) - \nabla f(x) \cdot \nu(x)\, \nu(x), \quad x \in \Gamma.$$

Here ∇ denotes the usual gradient in $\mathbb{R}^{n+1}$. Note also that $\nabla_\Gamma f(x)$ is the orthogonal projection of $\nabla f(x)$ onto $T_x\Gamma$. It is straightforward to show that $\nabla_\Gamma f$ only depends on the values of f on Γ. We use the notation

$$\nabla_\Gamma f(x) = (\underline{D}_1 f(x), \dots, \underline{D}_{n+1} f(x)) \quad (2.2)$$

for the $n + 1$ components of the tangential gradient. Obviously

$$\nabla_\Gamma f(x) \cdot \nu(x) = 0, \quad x \in \Gamma.$$

If f is twice differentiable in an open neighbourhood of Γ, then we define the *Laplace–Beltrami operator* of f as

$$\Delta_\Gamma f(x) = \nabla_\Gamma \cdot \nabla_\Gamma f(x) = \sum_{i=1}^{n+1} \underline{D}_i \underline{D}_i f(x), \quad x \in \Gamma. \tag{2.3}$$

2.2. Oriented distance function

A useful level set representation of a hypersurface can be obtained with the help of the distance function. Let Γ be as above and assume in addition that Γ is compact. The Jordan–Brouwer decomposition theorem then implies that there exists an open bounded set $\Omega \subset \mathbb{R}^{n+1}$ such that $\Gamma = \partial\Omega$. We assume that the unit normal field to Γ points away from Ω and define the oriented (signed) distance function d by

$$d(x) = \begin{cases} \operatorname{dist}(x, \Gamma), & x \in \mathbb{R}^{n+1} \setminus \bar{\Omega} \\ 0, & x \in \Gamma \\ -\operatorname{dist}(x, \Gamma), & x \in \Omega. \end{cases}$$

It is well known that d is globally Lipschitz-continuous and that there exists $\delta > 0$ such that

$$d \in C^2(\Gamma_\delta), \quad \text{where } \Gamma_\delta = \{x \in \mathbb{R}^{n+1} \mid |d(x)| < \delta\}. \tag{2.4}$$

Every point $x \in \Gamma_\delta$ can be uniquely written as

$$x = a(x) + d(x)\nu(a(x)), \quad x \in \Gamma_\delta, \tag{2.5}$$

where $a(x) \in \Gamma$. Furthermore, $\nabla d(x) = \nu(a(x)), x \in \Gamma_\delta$, which implies in particular that

$$|\nabla d(x)| \equiv 1 \quad \text{in } \Gamma_\delta. \tag{2.6}$$

Figure 2.1. Graph (right) of the oriented distance function for the curve (left).

2.3. Mean curvature

Let us next turn to the notion of mean curvature. By assumption, ν is C^1 in a neighbourhood of Γ so that we may introduce the matrix

$$H_{jk}(x) = \underline{D}_j \nu_k(x), \quad j, k = 1, \ldots, n+1, \ x \in \Gamma. \tag{2.7}$$

It is not difficult to show that $(H_{jk}(x))$ is symmetric. Furthermore,

$$\sum_{k=1}^{n+1} H_{jk}(x)\nu_k(x) = \sum_{k=1}^{n+1} \underline{D}_j \nu_k(x)\nu_k(x) = \frac{1}{2}\underline{D}_j|\nu|^2(x) = 0,$$

since $|\nu| = 1$ on Γ. Thus, $(H_{jk}(x))$ has one eigenvalue which is equal to zero with corresponding eigenvector $\nu(x)$. The remaining n eigenvalues $\kappa_1(x), \ldots, \kappa_n(x)$ are called the principal curvatures of Γ at the point x. We now define the mean curvature of Γ at x as the trace of the matrix $(H_{jk}(x))$, that is,

$$H(x) = \sum_{j=1}^{n+1} H_{jj}(x) = \sum_{j=1}^{n} \kappa_j(x). \tag{2.8}$$

Note that (2.8) differs from the more common definition $H = \frac{1}{n}\sum_{j=1}^{n+1} H_{jj}$. From (2.7) we derive the following expression for mean curvature,

$$H(x) = \nabla_\Gamma \cdot \nu(x), \quad x \in \Gamma, \tag{2.9}$$

where $\nabla_\Gamma \cdot f = \sum_{j=1}^{n+1} \underline{D}_j f_j$ denotes the tangential divergence of a vectorfield f. In particular we see that $H > 0$ if $\Gamma = S^n$ and the unit normal field is chosen to point away from S^n, i.e., $\nu(x) = x$.

While the sign of H depends on the choice of the normal ν, the mean curvature vector $H\nu$ is an invariant. A useful formula for this quantity can be obtained by choosing $f(x) = x_j$, $j \in \{1, \ldots, n+1\}$ in (2.3) and observing that $\underline{D}_i x_j = \delta_{ij} - \nu_j \nu_i$. We then deduce with the help of (2.9) that

$$\Delta_\Gamma x_j = -\sum_{i=1}^{n+1} \underline{D}_i(\nu_j \nu_i) = -(\nabla_\Gamma \cdot \nu)\nu_j - \nabla_\Gamma \nu_j \cdot \nu = -H\nu_j,$$

so that

$$-\Delta_\Gamma x = H\nu \quad \text{on } \Gamma. \tag{2.10}$$

Let us next fix a point $\bar{x} \in \Gamma$ and calculate $H(\bar{x})$ for various representations of the surface Γ near $\bar{x}$.

Level set representation. Suppose that Γ is given as in (2.1) near $\bar{x}$. Clearly, we then have

$$\nu(x) = \pm\frac{\nabla u(x)}{|\nabla u(x)|}$$

for $x \in U \cap \Gamma$. If the plus sign applies we obtain

$$H = \nabla_\Gamma \cdot \frac{\nabla u}{|\nabla u|} = \nabla \cdot \frac{\nabla u}{|\nabla u|} = \frac{1}{|\nabla u|} \sum_{i,j=1}^{n+1} \left(\delta_{ij} - \frac{u_{x_i} u_{x_j}}{|\nabla u|^2} \right) u_{x_i x_j}. \qquad (2.11)$$

In the special case that $u(x) = d(x)$, where d is the oriented distance function to Γ, we obtain in view of (2.6)

$$H(x) = \Delta d(x), \quad x \in \Gamma. \qquad (2.12)$$

Graph representation. Suppose that

$$U \cap \Gamma = \{(x, v(x)) \mid x \in \Omega\},$$

where $\Omega \subset \mathbb{R}^n$ is open, $x = (x_1, \ldots, x_n)$ and $v \in C^2(\Omega)$. Defining $u(x, x_{n+1}) = v(x) - x_{n+1}$ we see that $U \cap \Gamma$ is the zero level set of u and the above considerations imply that

$$H(x, v(x)) = \nabla \cdot \left(\frac{\nabla v(x)}{\sqrt{1 + |\nabla v(x)|^2}} \right), \quad (x, v(x)) \in U \cap \Gamma, \qquad (2.13)$$

where ∇ is the gradient in $\mathbb{R}^n$ and the unit normal is chosen as $\nu = \frac{(\nabla v, -1)}{\sqrt{1+|\nabla v|^2}}$.

Parametric representation. Suppose that there exists an open set $V \subset \mathbb{R}^n$ and a mapping $X \in C^2(V, \mathbb{R}^{n+1})$ such that

$$U \cap \Gamma = X(V), \quad \text{rank } DX(\theta) = n \quad \text{for all } \theta \in V.$$

The vectors $\frac{\partial X}{\partial \theta_1}(\theta), \ldots, \frac{\partial X}{\partial \theta_n}(\theta)$ then form a basis of $T_x\Gamma$ at $x = X(\theta)$. We define the metric on Γ by

$$g_{ij}(\theta) = \frac{\partial X}{\partial \theta_i}(\theta) \cdot \frac{\partial X}{\partial \theta_j}(\theta), \quad i,j = 1, \ldots, n$$

and let g^{ij} be the components of the inverse matrix of (g_{ij}). We then have the following formulae for the tangential gradient of a function f (defined in a neighbourhood of Γ) and the mean curvature vector $H\nu$:

$$\nabla_\Gamma f = \sum_{i,j=1}^n g^{ij} \frac{\partial(f \circ X)}{\partial \theta_j} \frac{\partial X}{\partial \theta_i}, \qquad (2.14)$$

$$H\nu = -\frac{1}{\sqrt{g}} \sum_{i,j=1}^n \frac{\partial}{\partial \theta_i} \left(g^{ij} \sqrt{g} \frac{\partial X}{\partial \theta_j} \right) \qquad (2.15)$$

where $g = \det(g_{ij})$.

2.4. Integration by parts

Let us assume in this section that Γ is in addition compact. The formula for integration by parts on Γ is (*cf.* Gilbarg and Trudinger (1998))

$$\int_\Gamma \underline{D}_i f \, \mathrm{d}A = \int_\Gamma f H \nu_i \, \mathrm{d}A \quad i = 1, \dots, n+1, \tag{2.16}$$

where $\mathrm{d}A$ denotes the area element on Γ and f is continuously differentiable in a neighbourhood of Γ. Applying (2.16) with $h = f\underline{D}_i g$, summing from $i = 1, \dots, n+1$ and taking into account that $\nabla_\Gamma \nu_i \cdot \nu = 0$, we obtain Green's formula,

$$\int_\Gamma \nabla_\Gamma f \cdot \nabla_\Gamma g \, \mathrm{d}A = -\int_\Gamma f \Delta_\Gamma g \, \mathrm{d}A. \tag{2.17}$$

In particular, we deduce from (2.10)

$$\int_\Gamma H\nu \cdot \phi \, \mathrm{d}A = \int_\Gamma \nabla_\Gamma x \cdot \nabla_\Gamma \phi \, \mathrm{d}A, \tag{2.18}$$

where ϕ is continuously differentiable in a neighbourhood of Γ with values in $\mathbb{R}^{n+1}$ and $\nabla_\Gamma x \cdot \nabla_\Gamma \phi = \sum_{i=1}^{n+1} \nabla_\Gamma x_i \cdot \nabla_\Gamma \phi_i$. This relation will be very important for the numerical treatment of mean curvature flow. The above formulae can be generalized to surfaces with boundaries by including an appropriate integral over $\partial\Gamma$.

2.5. Moving surfaces

In this section we shall be concerned with surfaces that evolve in time. A family $(\Gamma(t))_{t \in (0,T)}$ is called a $C^{2,1}$-family of hypersurfaces if, for each point $(x_0, t_0) \in \mathbb{R}^{n+1} \times (0, T)$ with $x_0 \in \Gamma(t_0)$, there exists an open set $U \subset \mathbb{R}^{n+1}$, $\delta > 0$ and a function $u \in C^{2,1}(U \times (t_0 - \delta, t_0 + \delta))$ such that

$$U \cap \Gamma(t) = \{x \in U \mid u(x, t) = 0\} \quad \text{and} \quad \nabla u(x, t) \neq 0, \ x \in U \cap \Gamma(t). \tag{2.19}$$

Suppose in addition that each $\Gamma(t)$ is oriented by a unit normal field $\nu(\cdot, t) \in C^1(\Gamma(t), \mathbb{R}^{n+1})$ and that $\nu \in C^0(\bigcup_{0 < t < T} \Gamma(t) \times \{t\}, \mathbb{R}^{n+1})$. The normal velocity at a point (x_0, t_0) $(x_0 \in \Gamma(t_0))$ is then defined as

$$V(x_0, t_0) = \phi'(t_0) \cdot \nu(x_0, t_0),$$

where $\phi \in C^1((t_0 - \epsilon, t_0 + \epsilon), \mathbb{R}^{n+1})$ satisfies $\phi(t_0) = x_0$ and $\phi(t) \in \Gamma(t)$ for $|t - t_0| < \epsilon$. It can be shown that $V(x_0, t_0)$ is independent of the particular choice of ϕ. Let us calculate $V(x_0, t_0)$ for various representations of $\Gamma(t)$.

Level set representation. Let u be as in (2.19); as above we then have $\nu = \pm\frac{\nabla u}{|\nabla u|}$. If the plus sign applies and $\phi \in C^1((t_0 - \epsilon, t_0 + \epsilon), \mathbb{R}^{n+1})$ satisfies $\phi(t_0) = x_0$ as well as $\phi(t) \in \Gamma(t)$ for $|t - t_0| < \epsilon$, we have

$$0 = \frac{\mathrm{d}}{\mathrm{d}t} u(\phi(t), t) = \nabla u(\phi(t), t) \cdot \phi'(t) + u_t(\phi(t), t),$$

and hence

$$V(x_0, t_0) = -\frac{u_t(x_0, t_0)}{|\nabla u(x_0, t_0)|}. \tag{2.20}$$

Graph representation. Suppose that

$$U \cap \Gamma(t) = \{(x, v(x, t)) \mid x \in \Omega\},$$

where $\Omega \subset \mathbb{R}^n$ is open and $v \in C^{2,1}(\Omega \times (t_0 - \delta, t_0 + \delta))$. Applying the formula for the level set case to $u(x, x_{n+1}, t) = v(x, t) - x_{n+1}$, we obtain

$$V = -\frac{v_t}{\sqrt{1 + |\nabla v|^2}} \tag{2.21}$$

for the unit normal field $\nu = \frac{(\nabla v, -1)}{\sqrt{1 + |\nabla v|^2}}$.

2.6. Transport theorem for integrals

Consider a family $(\Gamma(t))_{t \in (0,T)}$ of evolving hypersurfaces which satisfies the assumptions made above and suppose in addition that each surface $\Gamma(t)$ is compact. We are interested in the time derivative of certain volume and area integrals.

Lemma 2.1. Let $g \in C^1(Q)$, where Q is an open set containing

$$\bigcup_{0 < t < T} \Gamma(t) \times \{t\}.$$

Suppose in addition that each surface $\Gamma(t)$ is the boundary of an open bounded subset $\Omega(t) \subset \mathbb{R}^{n+1}$. Then

$$\frac{d}{dt} \int_{\Omega(t)} g \, dx = \int_{\Omega(t)} \frac{\partial g}{\partial t} \, dx + \int_{\Gamma(t)} g V \, dA, \tag{2.22}$$

$$\frac{d}{dt} \int_{\Gamma(t)} g \, dA = \int_{\Gamma(t)} \frac{\partial g}{\partial t} \, dA + \int_{\Gamma(t)} g V H \, dA + \int_{\Gamma(t)} \frac{\partial g}{\partial \nu} V \, dA. \tag{2.23}$$

Proof. See the Appendix. □

3. Definition and elementary properties of mean curvature flow

The purpose of this section is to introduce motion by mean curvature and to describe some basic features of this flow. Consider a $C^{2,1}$-family of hypersurfaces $(\Gamma(t))_{t \in [0,T]} \subset \mathbb{R}^{n+1}$ together with a choice ν of a unit normal.

Definition 1. We say that $(\Gamma(t))_{t \in [0,T]}$ *moves by mean curvature* if

$$V = -H \quad \text{on } \Gamma(t). \tag{3.1}$$

Here, V denotes the velocity of $\Gamma(t)$ in the direction of ν and H is mean curvature.

As we shall see later, the above equation gives rise to a parabolic equation, or a parabolic system, for the function(s) describing the surfaces $\Gamma(t)$, to which an initial condition

$$\Gamma(0) = \Gamma_0 \tag{3.2}$$

has to be added. If $\Gamma(t)$ has a boundary, then also suitable boundary conditions need to be specified.

In order to give a first idea of this flow we look at the well-known example of the shrinking sphere. Let $\Gamma(t) = \partial B_{r(t)}(x_0) \subset \mathbb{R}^{n+1}$, oriented by the unit outer normal $\nu(x) = \frac{x - x_0}{r(t)}$. Then, $V = r'(t)$, $H = \frac{n}{r(t)}$ on $\Gamma(t)$, so that $\Gamma(t)$ moves by mean curvature provided that $r'(t) = -\frac{n}{r(t)}$. The solution of this ODE is given by $r(t) = \sqrt{r_0^2 - 2nt}$, $0 \le t < \frac{r_0^2}{2n}$, where $\Gamma_0 = \partial B_{r_0}(x_0)$. Note that $\Gamma(t)$ shrinks to a point as $t \nearrow \frac{r_0^2}{2n}$.

The main feature of mean curvature flow is its area-decreasing property, which is a consequence of the following result.

Lemma 3.1. Let $\Gamma(t)$ be a family of evolving hypersurfaces satisfying $V = -H$ on $\Gamma(t)$ and assume that each $\Gamma(t)$ is compact. Then

$$\int_{\Gamma(t)} V^2 \, \mathrm{d}A + \frac{\mathrm{d}}{\mathrm{d}t} |\Gamma(t)| = 0,$$

where $|\Gamma|$ is the area of Γ.

Proof. This follows immediately from choosing $g \equiv 1$ in (2.23) and the evolution law (3.1). $\square$

Since the law (3.1) gives rise to a second order parabolic problem we expect existence of a smooth solution locally in time for a smooth initial hypersurface Γ_0. Furthermore, maximum and comparison principles are available which can be used to show that two smooth compact solutions which are initially disjoint will stay disjoint (see, *e.g.*, Ecker (2002)). Using the shrinking sphere as a comparison solution, it follows in particular that if $\Gamma(t), 0 \le t < T$ is a smooth solution with $\Gamma_0 \subset B_{r_0}(x_0)$, then $\Gamma(t) \subset B_{\sqrt{r_0^2 - 2nt}}(x_0)$ for $0 \le t < \min(T, \frac{r_0^2}{2n})$. In general, solutions will develop singularities in finite time before they disappear, but there are certain initial configurations for which they stay smooth until they shrink to a point.

Theorem 3.2. Let $n \ge 2$ and assume that $\Gamma_0 \subset \mathbb{R}^{n+1}$ is a smooth, compact and uniformly convex hypersurface. Then (3.1) and (3.2) have a smooth solution on a finite time interval $[0, T)$ and the $\Gamma(t)$ converge to a point as $t \nearrow T$. If one rescales the surfaces in such a way that the enclosed volume remains fixed, one has convergence against a sphere as $t \nearrow T$.

Proof. See Huisken (1984). □

The case $n = 1$ is usually referred to as curve shortening flow.

Theorem 3.3. Assume that $\Gamma_0 \subset \mathbb{R}^2$ is a smooth embedded closed curve. Then (3.1) and (3.2) have a smooth embedded solution on a finite time interval $[0, T)$, which shrinks to a 'round' point as $t \nearrow T$.

Proof. Gage and Hamilton (1986) proved this result for convex Γ_0; subsequently Grayson (1987) showed that a smooth embedded closed curve remains smooth and embedded and becomes convex in finite time. □

If the initial curve is not embedded, cusp-like singularities may develop (see Figures 4.2 and 4.3). The papers of Angenent (1991), Altschuler and Grayson (1992) and Deckelnick (1997) propose various methods of how to continue the solution past such a singularity. The analogue of Theorem 3.3 for surfaces does not hold, as can be seen by choosing a suitable dumbbell-shaped initial surface which develops a pinch-off singularity before it shrinks to a point (see Figure 4.5 and Grayson (1989)). This pinch-off leads to a change of the topological type of $\Gamma(t)$, so that the parametric approach – in which the topological type is fixed – will develop a singularity that is difficult to handle. Thus the question arises whether it is possible to introduce a notion of solution that is capable of following the flow through a singularity. Several such notions have been proposed and analysed starting with the pioneering work of Brakke (1978) on varifold solutions, which uses tools from geometric measure theory. In this context we also mention the surface evolver program of Brakke (1992). Level set and phase field methods constitute two completely different approaches which take an Eulerian point of view. We shall discuss these in more detail in Sections 6 and 7.

4. Parametric mean curvature flow

As is mentioned above, in the parametric approach one chooses a suitable reference manifold $M \subset \mathbb{R}^{n+1}$ (of the topological type of the evolving hypersurfaces $\Gamma(t)$) and then looks for maps $X(\cdot, t) : M \to \mathbb{R}^{n+1}$ ($0 \le t < T$) such that $\Gamma(t) = X(\cdot, t)(M)$. To fix ideas, let us assume that M is a compact hypersurface without boundary. If we can find X in such a way that

$$\frac{\partial X}{\partial t}(p, t) = -H(X(p, t))\nu(X(p, t)) \quad (p, t) \in M \times (0, T), \qquad (4.1)$$

then $V = -H$ on $\Gamma(t)$ follows by taking the dot product with the normal ν. In order to understand (4.1) let $F : \Omega \to \mathbb{R}^{n+1}$ be a local parametrization of M defined on an open set $\Omega \subset \mathbb{R}^n$ and set

$$\hat{X}(\theta, t) = X(F(\theta), t), \quad (\theta, t) \in \Omega \times (0, T).$$

Recalling (2.15), the equation (4.1) then turns into

$$\frac{\partial \hat{X}}{\partial t}(\theta, t) - \frac{1}{\sqrt{\hat{g}}} \sum_{i,j=1}^{n} \frac{\partial}{\partial \theta_i} \left(\hat{g}^{ij} \sqrt{\hat{g}} \frac{\partial \hat{X}}{\partial \theta_j} \right) = 0 \qquad (4.2)$$

where $\hat{g}_{ij}(\theta, t) = \frac{\partial \hat{X}}{\partial \theta_i} \cdot \frac{\partial \hat{X}}{\partial \theta_j}$ and $\hat{g}^{ij}, \hat{g}$ are as above. Thus (4.2), and hence (4.1), is a nonlinear parabolic system, which is not defined at points (θ, t) where $\hat{g}(\theta, t) = 0$. In order to close this system, an initial condition $X(p, 0) = X_0(p), p \in M$ needs to be prescribed, where $X_0 : M \to \mathbb{R}^{n+1}$ is a parametrization of the initial surface Γ_0.

4.1. Curve shortening flow

Mean curvature evolution in the one-dimensional case is usually referred to as curve shortening flow. In the case of closed curves, a convenient choice of a reference manifold is $M = S^1$, which can be parametrized globally by $F(\theta) = (\cos \theta, \sin \theta), \theta \in [0, 2\pi]$. In the following, for simplicity, let us identify $\hat{X}(\theta, t)$ and $X((\cos \theta, \sin \theta), t)$. Thus, (4.2) becomes

$$X_t - \frac{1}{|X_\theta|} \left(\frac{X_\theta}{|X_\theta|} \right)_\theta = 0 \qquad \text{in } I \times (0, T), \qquad (4.3)$$

$$X(\cdot, 0) = X_0 \quad \text{in } I, \qquad (4.4)$$

where $I = [0, 2\pi]$. In addition, X has to satisfy the periodicity condition

$$X(\theta, t) = X(\theta + 2\pi, t) \quad 0 \le t < T, \quad \theta \in \mathbb{R}. \qquad (4.5)$$

Suppose that $X : \mathbb{R} \times [0, T] \to \mathbb{R}^2$ is a smooth solution of (4.3)–(4.5), in particular $|X_\theta| > 0$ in $I \times [0, T]$. If we multiply (4.3) by $|X_\theta|$, take the dot product with a test function $\varphi \in H^1_{\text{per}}(I; \mathbb{R}^2) = \{\varphi \in H^1(I; \mathbb{R}^2) \mid \varphi(0) = \varphi(2\pi)\}$ and integrate over I, we obtain

$$\int_I X_t \cdot \varphi |X_\theta| + \int_I \frac{X_\theta \cdot \varphi_\theta}{|X_\theta|} = 0 \quad \text{for all } \varphi \in H^1_{\text{per}}(I; \mathbb{R}^2). \qquad (4.6)$$

We use (4.6) in order to discretize in space. For simplicity let $\theta_j = jh$ $(j = 0, \ldots, N)$ be a uniform grid with grid size $h = 2\pi/N$ and let

$$S_h = \left\{ \varphi_h \in C^0(I; \mathbb{R}^2) \mid \varphi_h|_{[\theta_{j-1}, \theta_j]} \in P_1^2, j = 1, \ldots, N; \varphi_h(0) = \varphi_h(2\pi) \right\}$$

be the space of piecewise linear continuous functions with values in $\mathbb{R}^2$. The spatial discretization of (4.3) is then given by

$$\int_I X_{ht} \cdot \varphi_h |X_{h\theta}| + \int_I \frac{X_{h\theta} \cdot \varphi_{h\theta}}{|X_{h\theta}|} = 0 \quad \text{for all } \varphi_h \in S_h. \qquad (4.7)$$

Denoting the common (scalar) nodal basis by $\{\phi_1, \ldots, \phi_N\}$, we can expand $X_h(\theta, t) = \sum_{j=1}^{N} X_j(t)\phi_j(\theta)$ with vectors $X_j(t) \in \mathbb{R}^2$. This one-dimensional finite element formulation can be rewritten as a difference scheme. To see this, insert $\varphi_h = \phi_j e^k$, $(k = 1, 2; j = 1, \ldots, N)$ into (4.7) and calculate

$$\int_I X_{ht} \cdot \varphi_h |X_{h\theta}| \, d\theta = \frac{1}{6}|X_j - X_{j-1}|\dot{X}_{j-1} \cdot e^k$$

$$+ \frac{1}{3}(|X_j - X_{j-1}| + |X_{j+1} - X_j|)\dot{X}_j \cdot e^k + \frac{1}{6}|X_{j+1} - X_j|\dot{X}_{j+1} \cdot e^k$$

as well as

$$\int_I \frac{X_{h\theta} \cdot \varphi_{h\theta}}{|X_{h\theta}|} \, d\theta = \left(-\frac{X_{j+1} - X_j}{q_{j+1}} + \frac{X_j - X_{j-1}}{q_j}\right) \cdot e^k.$$

Here, $q_j = |X_j - X_{j-1}|$ and the dot stands for the time derivative. Thus, (4.7) can be written as

$$\frac{1}{6}q_j \dot{X}_{j-1} + \frac{1}{3}(q_j + q_{j+1})\dot{X}_j + \frac{1}{6}q_{j+1}\dot{X}_{j+1} = \frac{X_{j+1} - X_j}{q_{j+1}} - \frac{X_j - X_{j-1}}{q_j} \quad (4.8)$$

$(j = 1, \ldots, N)$. If we use mass lumping in (4.8) we get the difference scheme

$$\frac{1}{2}(q_j + q_{j+1})\dot{X}_j = \frac{X_{j+1} - X_j}{q_{j+1}} - \frac{X_j - X_{j-1}}{q_j}. \quad (4.9)$$

As initial values for X_j we choose

$$X_j(0) = X_0(\theta_j), \quad j = 0, \ldots, N, \quad (4.10)$$

so that $X_h(\cdot, 0)$ is the linear interpolant of X_0. Furthermore we require the periodicity condition

$$X_j = X_{j+N}, \quad j = -1, 0, 1. \quad (4.11)$$

The following proposition shows that the lumped scheme reflects the curve shortening property of the exact solution.

Proposition 4.1. Consider solutions X of (4.3) and $X_h = \sum_{j=1}^{N} X_j(t)\phi_j(\theta)$ of (4.9) respectively. Then we have for $t \in [0, T]$

$$|X_\theta(\cdot, t)|_t = -|X_t(\cdot, t)|^2|X_\theta(\cdot, t)| \quad \text{in } I$$

$$\dot{q}_j = -\frac{1}{4}(q_{j-1} + q_j)|\dot{X}_{j-1}|^2 - \frac{1}{4}(q_j + q_{j+1})|\dot{X}_j|^2, \quad j = 1, \ldots, N$$

as long as $q_j > 0, j = 1, \ldots, N$. Thus, the faces of the polygon with vertices $X_1, \ldots, X_N$ decrease in length during time evolution.

Proof. For the proof of the first assertion we differentiate $|\frac{X_\theta}{|X_\theta|}| \equiv 1$ twice with respect to θ and get

$$\frac{X_\theta}{|X_\theta|} \cdot \left(\frac{X_\theta}{|X_\theta|}\right)_\theta = 0, \quad \left|\left(\frac{X_\theta}{|X_\theta|}\right)_\theta\right|^2 = -\frac{X_\theta}{|X_\theta|} \cdot \left(\frac{X_\theta}{|X_\theta|}\right)_{\theta\theta} \quad \text{in } I,$$

which combined with (4.3) gives

$$|X_\theta|_t = \frac{X_\theta}{|X_\theta|} \cdot X_{\theta t} = \frac{X_\theta}{|X_\theta|} \cdot \frac{1}{|X_\theta|} \left(\frac{X_\theta}{|X_\theta|} \right)_{\theta\theta}$$

$$= -\frac{1}{|X_\theta|} \left| \left(\frac{X_\theta}{|X_\theta|} \right)_\theta \right|^2 = -|X_t|^2 |X_\theta|.$$

For the discrete solution we observe that by (4.9) with the unit vectors $T_j = \frac{X_j - X_{j-1}}{q_j}$ we have

$$\dot{q}_j = T_j \cdot (\dot{X}_j - \dot{X}_{j-1})$$

$$= T_j \cdot \left(\frac{2}{q_j + q_{j+1}} (T_{j+1} - T_j) - \frac{2}{q_{j-1} + q_j} (T_j - T_{j-1}) \right)$$

$$= -\frac{2}{q_j + q_{j+1}} (1 - T_j \cdot T_{j+1}) - \frac{2}{q_{j-1} + q_j} (1 - T_{j-1} \cdot T_j)$$

$$= -\frac{1}{q_j + q_{j+1}} |T_j - T_{j+1}|^2 - \frac{1}{q_{j-1} + q_j} |T_{j-1} - T_j|^2$$

$$= -\frac{1}{4} (q_{j-1} + q_j) |\dot{X}_{j-1}|^2 - \frac{1}{4} (q_j + q_{j+1}) |\dot{X}_j|^2.$$

For this we have used the discrete equation (4.9) twice.   □

Under the assumption that a smooth and regular solution of the curve shortening flow (4.3)–(4.5) exists, one obtains the following convergence result together with error estimates for the position vector X and the velocity vector X_t, which by (4.1) is equal to the curvature vector. The proof follows from Dziuk (1994) and is a special case of Theorem 8.4.

Theorem 4.2. Let $X : I \times [0, T] \to \mathbb{R}^2$ be a periodic smooth solution of the curve shortening flow (4.3)–(4.5) with $|X_\theta| \geq c_0 > 0$ in $I \times [0, T]$. Then there exists an $h_0 > 0$ depending on X and T such that for every $0 < h \leq h_0$ there exists a unique solution $X_h(\theta, t) = \sum_{j=1}^{N} X_j(t)\phi_j(\theta)$ of the difference scheme (4.9), (4.10) and

$$\max_{[0,T]} \|X - X_h\|_{L^2(I)} + \left(\int_0^T \|X_\theta - X_{h\theta}\|_{L^2(I)}^2 \, dt \right)^{1/2} \leq ch, \qquad (4.12)$$

$$\max_{[0,T]} \|X_t - X_{ht}\|_{L^2(I)} + \left(\int_0^T \|X_{t\theta} - X_{ht\theta}\|_{L^2(I)}^2 \, dt \right)^{1/2} \leq ch, \qquad (4.13)$$

where c depends on X and T.

This algorithm can be generalized without changes to curves evolving in higher codimension, i.e., $X : I \times [0, T] \to \mathbb{R}^m$ and $m > 2$. The curve solving (4.3) has a velocity only in the normal direction. It is also possible to use the parametric equation

$$X_t = \frac{X_{\theta\theta}}{|X_\theta|^2}$$

instead, which defines the same curve evolving in the normal direction with a normal velocity being given by the curvature. However, the parametrization is different, with the points on the curve now having a tangential velocity. A finite element error analysis for the motion of a closed curve is given in Deckelnick and Dziuk (1994), while error bounds for the evolution of a curve attached to a fixed boundary with a normal contact condition are proved in Deckelnick and Elliott (1998).

In order to obtain a practical method we still have to discretize in time. Choose a time-step $\tau > 0$ and let $t_m = m\tau$, $m = 0, \ldots, M$, $M \le [\frac{T}{\tau}]$. We let $X_h^m \in S_h$ denote the approximation to $X(\cdot, t_m)$. On the basis of (4.7) we suggest the following scheme:

$$\frac{1}{\tau} \int_I (X_h^{m+1} - X_h^m) \cdot \varphi_h |X_{h\theta}^m| + \int_I \frac{X_{h\theta}^{m+1} \cdot \varphi_{h\theta}}{|X_{h\theta}^m|} = 0 \quad \text{for all } \varphi_h \in S_h. \quad (4.14)$$

Calculations similar to those above yield a time discrete analogue of (4.9), which we formulate as the following algorithm.

Algorithm 1. (Curve shortening flow)

(1) Let $X_j^0 = X_0(\theta_j)$ $(j = 0, \ldots, N)$.

(2) Compute X_j^{m+1} $(j = 0, \ldots, N)$ from the tridiagonal systems

$$\frac{1}{2\tau}(q_j^m + q_{j+1}^m)(X_j^{m+1} - X_j^m) - \left(\frac{X_{j+1}^{m+1} - X_j^{m+1}}{q_{j+1}^m} - \frac{X_j^{m+1} - X_{j-1}^{m+1}}{q_j^m} \right) = 0.$$

(3) If $\min_{j=1,\ldots,N+1} q_j^{m+1} > 0$ then replace m by $m + 1$ and GOTO 2.

Thus, in each time-step a positive definite and symmetric linear system has to be solved for each component of X_h^{m+1}. Each of these linear systems is of tridiagonal form with two additional entries reflecting the periodicity condition. The system decouples with respect to the dimension of the space in which the curve moves. For practical purposes a redistribution of nodes according to arc length on the curve is sometimes convenient.

Let us go back to the more precise notation $\hat{X}(\theta, t) = X((\cos \theta, \sin \theta), t)$. For later purposes it is convenient to look at (4.14) from a slightly different angle. We introduce the polygon $\Gamma_h^m = \hat{X}_h^m(I)$ along with the space

$$S_h^m = \{\phi_h : \Gamma_h^m \to \mathbb{R}^2 \mid \phi_h \text{ is affine on each face of } \Gamma_h^m\}. \quad (4.15)$$

Thus, if $\phi_h \in S_h^m$, then ϕ_h is the restriction of an affine function on $\mathbb{R}^2$ on each face of the polygon and therefore

$$\varphi_h(\theta) = \phi_h(\hat{X}_h^m(\theta)), \quad \theta \in I,$$

Figure 4.1. Curve shortening flow applied to a star-shaped
curve. Time-steps 0, 100, 200, 300, 500, 700, 5000, 7000
(time-step size $= 8.5586 \times 10^{-5}$), 480 nodes.

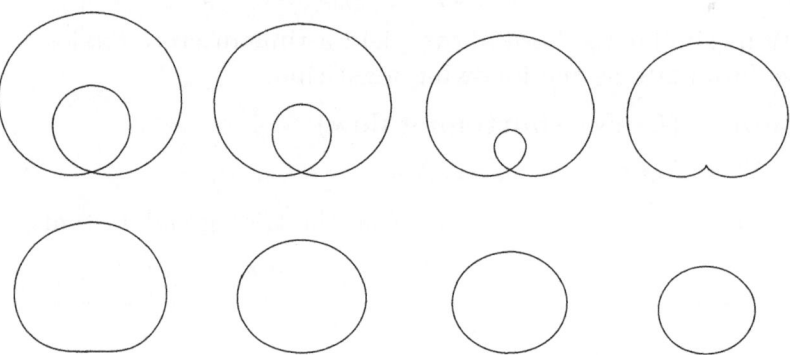

Figure 4.2. Curve shortening flow applied to a curve with a
self-intersection. A singularity (cusp) appears. The effect
is that the algorithm jumps across the singularity. See
Figure 4.3 for a magnified image. Time-steps 0, 1000, 2000,
2500, 3000, 5000, 6000, 7000 (time-step size $= 8.5586 \times 10^{-5}$),
480 nodes.

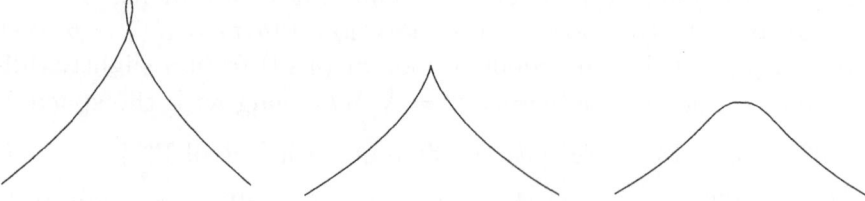

Figure 4.3. Close-up of Figure 4.2. Time-steps 3498 and 3499
and 3505. The parametric theory breaks down.

belongs to $\hat{S}_h$. Recalling (2.14) we have

$$\nabla_{\Gamma_h^m}\phi_h = \frac{1}{|\hat{X}_{h\theta}^m|}\varphi_{h\theta}\otimes\frac{\hat{X}_{h\theta}^m}{|\hat{X}_{h\theta}^m|}, \quad p = \hat{X}_h^m(\theta),$$

where $(u\otimes v)_{ij} = u_i v_j$ $(u,v\in\mathbb{R}^2)$ and $\nabla_{\Gamma_h^m}\phi_h$ is given piecewise on each face of Γ_h^m. Let us define $X_h^{m+1}\in S_h^m$ by $X_h^{m+1}(p) = \hat{X}_h^{m+1}(\theta)$, $p = \hat{X}_h^m(\theta)$. Observing that

$$\nabla_{\Gamma_h^m}X_h^{m+1}\cdot\nabla_{\Gamma_h^m}\phi_h\,|\hat{X}_{h\theta}^m| = \frac{\hat{X}_{h\theta}^{m+1}\cdot\varphi_{h\theta}}{|\hat{X}_{h\theta}^m|}\quad\text{for all }\varphi_h\in\hat{S}_h$$

we can rewrite (4.14) as

$$\frac{1}{\tau}\int_{\Gamma_h^m}(X_h^{m+1}-\text{id})\cdot\phi_h\,\mathrm{d}A + \int_{\Gamma_h^m}\nabla_{\Gamma_h^m}X_h^{m+1}\cdot\nabla_{\Gamma_h^m}\phi_h\,\mathrm{d}A = 0\quad\text{for all }\phi_h\in S_h^m.$$
$$(4.16)$$

Note that the dot between the matrices $\nabla_{\Gamma_h^m}X_h^{m+1}$ and $\nabla_{\Gamma_h^m}\phi_h$ is the standard scalar product in $\mathbb{R}^4$. The key point about the formulation (4.16) is that Γ_h^{m+1} is now parametrized with the help of the polygon Γ_h^m from the previous time-step, so that the reference manifold M is no longer needed. We can interpret the second integral on the left-hand side of (4.16) as an approximation to

$$\int_{\Gamma(t_{m+1})}\nabla_{\Gamma(t_{m+1})}x\cdot\nabla_{\Gamma(t_{m+1})}\phi\,\mathrm{d}A,$$

which equals $-\int_{\Gamma(t_{m+1})}H\nu\cdot\phi\,\mathrm{d}A$ by (2.17) and (2.10). Here, H is just the usual curvature of the curve $\Gamma(t_{m+1})$, but of course it is now natural to also use (4.16) for approximating surfaces evolving by mean curvature. We will discuss this issue in the next section.

4.2. Mean curvature flow of surfaces

In this section we shall use a higher-dimensional version of (4.16) in order to approximate parametric surfaces $\Gamma(t) = X(M,t)$, which satisfy (4.1). To begin, we need an analogue of the polygons used in the previous section.

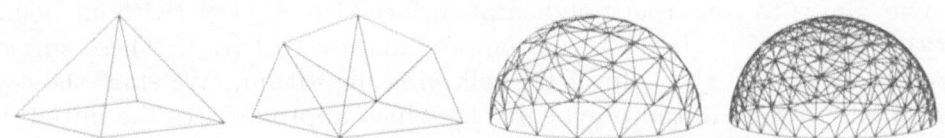

Figure 4.4. Polyhedral surfaces: successively refined grids approximating a half sphere. Macro triangulation (left) and triangulation levels 1, 5 and 7.

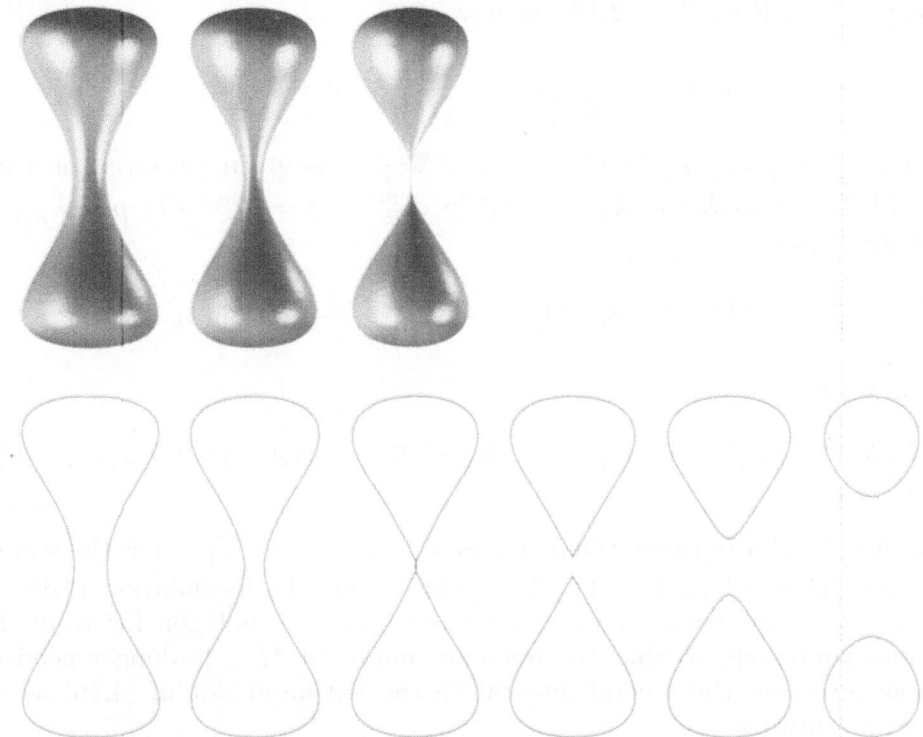

Figure 4.5. First row: Parametric mean curvature flow of a
dumbbell-shaped surface. Development of a singularity.
Second row: Axially symmetric level set computation of the
same flow going beyond the topological change of the surface.

Definition 2. We call a set $\Gamma \subset \mathbb{R}^{n+1}$ a *polyhedral surface* if

$$\Gamma = \bigcup_{T \in \mathcal{T}_h} T,$$

where the triangulation $\mathcal{T}_h$ consists of closed, nondegenerate, n-dimensional simplices. The intersection of two adjacent simplices is an $(n - k)$-dimensional subsimplex of these simplices ($k \in \{1, \ldots, n\}$).

Our aim is to construct polyhedral surfaces $\Gamma_h^0, \ldots, \Gamma_h^M$ (without boundary) in such a way that Γ_h^m is an approximation to $\Gamma(t_m)$. These surfaces are obtained with the help of the following algorithm. We start the computations with an initial polyhedral Γ_h^0 which approximates the initial surface Γ_0. In practice there are several ways to construct the initial discrete surface. One way is to map triangulations of charts onto the continuous surface and to glue them together. A much better way is to construct a macro triangulation, that is, a coarse approximation $\tilde{\Gamma}_h^0$ of Γ_0 such that

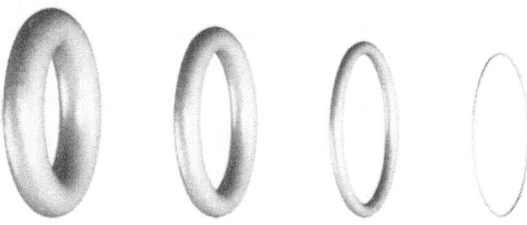

Figure 4.6. A thin two-dimensional torus shrinks under parametric mean curvature flow to a circle.

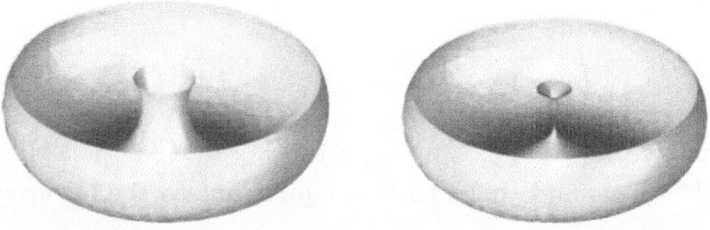

Figure 4.7. A thick two-dimensional torus (cut open) shrinks under parametric mean curvature flow to a sphere developing a singularity.

$\tilde{\Gamma}_h^0 \subset \Gamma_\delta$ (see (2.4), (2.5)) and then to refine this triangulation in $\mathbb{R}^{n+1}$ and project the new nodes x orthogonally onto the smooth surface according to $x' = x - d(x)\nu(x)$ to obtain the new nodes x' of the next-finer triangulation for Γ_h^0 (see Figure 4.4).

Algorithm 2. (Mean curvature flow of surfaces)

Let Γ_h^0 be a polyhedral approximation of Γ_0.

For $m = 0, 1, \ldots, M - 1$ define

$$S_h^m = \{\phi_h \in C^0(\Gamma_h^m) \mid \phi_{h|T} \text{ is affine for each } T \subset \Gamma_h^m\},$$

and find $X_h^{m+1} \in S_h^m$ with

$$\frac{1}{\tau} \int_{\Gamma_h^m} (X_h^{m+1} - \text{id})\phi_h \, dA + \int_{\Gamma_h^m} \nabla_{\Gamma_h^m} X_h^{m+1} \cdot \nabla_{\Gamma_h^m} \phi_h \, dA = 0 \quad \text{for all } \phi_h \in S_h^m$$

(4.17)

Generate the new surface $\Gamma_h^{m+1} = X_h^{m+1}(\Gamma_h^m)$, and if it is a polyhedral surface then GOTO to the next m.

This algorithm is based on a finite element method for partial differential equations on surfaces, developed in Dziuk (1988). Let us have a look at the implementation of the above algorithm. Fix $m \in \{0, \dots, M-1\}$ and denote by $a_1, \dots, a_N \in \mathbb{R}^{n+1}$ the nodes of the polyhedral surface Γ_h^m. The functions $\phi_i : \Gamma_h^m \to \mathbb{R}$, $i = 1, \dots, N$ are uniquely defined by the requirements

$$\phi_i \in S_h^m, \quad \phi_i(a_j) = \delta_{ij}, \quad i, j = 1, \dots, N.$$

It is not difficult to verify that $\phi_1, \dots, \phi_N$ actually form a basis of S_h^m. Now, stiffness and mass matrix are defined by

$$S_{ij} = \int_{\Gamma_h^m} \nabla_{\Gamma_h^m} \phi_i \cdot \nabla_{\Gamma_h^m} \phi_j \, \mathrm{d}A, \quad i, j = 1, \dots, N$$

$$M_{ij} = \int_{\Gamma_h^m} \phi_i \phi_j \, \mathrm{d}A, \quad\quad\quad i, j = 1, \dots, N.$$

Expanding $(X_h^{m+1})_k(p) = \sum_{j=1}^N \alpha_j^{(k)} \phi_j(p)$ (where $(X_h^{m+1})_k$ is the kth component of X_h^{m+1}), we find that (4.17) is equivalent to the linear systems

$$M\alpha^{(k)} + \tau S\alpha^{(k)} = b^{(k)}, \quad k = 1, \dots, n+1. \tag{4.18}$$

Here, $\alpha^{(k)} = (\alpha_1^{(k)}, \dots, \alpha_N^{(k)})$ and $b^{(k)} \in \mathbb{R}^N$ is given by

$$b_j^{(k)} = \int_{\Gamma_h^m} x_k \phi_j \, \mathrm{d}A, \quad j = 1, \dots, N.$$

Since the matrix $M + \tau S$ is symmetric and positive definite, the systems (4.18) can be solved with a conjugate gradient method. The only difference to a 'Cartesian' FEM is that the nodes have one more coordinate.

5. Mean curvature flow of graphs

We turn our attention to the mean curvature evolution of surfaces $\Gamma(t)$, which can be written as graphs over some base domain $\Omega \subset \mathbb{R}^n$, that is,

$$\Gamma(t) = \{(x, u(x, t)) \mid x \in \Omega\}.$$

In order to find the differential equation to be satisfied by the height function u, we recall (2.13) and (2.21) to see that the mean curvature H and the velocity V in the direction of $\nu = \frac{(\nabla u, -1)}{\sqrt{1+|\nabla u|^2}}$ are given by

$$H = \nabla \cdot \left(\frac{\nabla u}{\sqrt{1 + |\nabla u|^2}} \right), \quad V = -\frac{u_t}{\sqrt{1 + |\nabla u|^2}}. \tag{5.1}$$

Thus, the evolution law $V = -H$ on $\Gamma(t)$ translates into the nonlinear parabolic partial differential equation

$$u_t - \sqrt{1 + |\nabla u|^2}\, \nabla \cdot \left(\frac{\nabla u}{\sqrt{1 + |\nabla u|^2}} \right) = 0 \quad \text{in } \Omega \times (0, T), \tag{5.2}$$

to which we add the following boundary and initial conditions

$$u = g \quad \text{on } \partial\Omega \times (0, T), \tag{5.3}$$
$$u(\cdot, 0) = u_0 \quad \text{in } \Omega, \tag{5.4}$$

where $g : \partial\Omega \to \mathbb{R}$ and $u_0 : \bar{\Omega} \to \mathbb{R}$ are given functions. The boundary condition (5.3) implies that the boundaries of the surfaces $\Gamma(t)$ are kept fixed during the evolution. It would also be possible to replace (5.3) by

$$\frac{\partial u}{\partial n} = 0 \quad \text{on } \partial\Omega \times (0, T), \tag{5.5}$$

in which case the surfaces $\Gamma(t)$ would meet the boundary of the cylinder $\Omega \times \mathbb{R}$ at a right angle.

5.1. Analytical results

The main difficulties for the mathematical analysis are due to the fact that the operator

$$A(u) = \sqrt{1 + |\nabla u|^2}\, \nabla \cdot \left(\frac{\nabla u}{\sqrt{1 + |\nabla u|^2}} \right)$$

is not uniformly parabolic and not in divergence form. Only in one space dimension the equation is in divergence form, since $A(u) = (\arctan u_x)_x$.

Theorem 5.1. Let Ω be a bounded domain in $\mathbb{R}^n$ with $\partial\Omega \in C^{2+\alpha}$ and $u_0 \in C^{2,\alpha}(\bar{\Omega})$.

(a) Suppose that $g \in C^{2,\alpha}(\bar{\Omega})$ and that the compatibility conditions

$$u_0 = g \quad \text{and} \quad \sqrt{1 + |\nabla u_0|^2}\, \nabla \cdot \left(\frac{\nabla u_0}{\sqrt{1 + |\nabla u_0|^2}} \right) = 0 \quad \text{on } \partial\Omega$$

are satisfied. If $\partial\Omega$ has nonnegative mean curvature, the initial-boundary value problem (5.2), (5.3), (5.4) has a unique smooth solution which converges to the solution of the minimal surface equation with boundary data g as $t \to \infty$.

(b) Suppose that the compatibility condition $\frac{\partial u_0}{\partial n} = 0$ on $\partial\Omega$ holds. Then the initial-boundary value problem (5.2), (5.5), (5.4) has a unique smooth solution which converges to a constant function as $t \to \infty$.

Proof. See Lieberman (1986) and also Huisken (1989) for (a); (b) is proved in Huisken (1989). ☐

The assumption that the boundary of the domain has nonnegative mean curvature is a necessary condition. If it is dropped, the gradient of the solution will become infinite on the boundary: see Oliker and Uraltseva (1993). The main tool in the proof of the previous theorem is the derivation of an evolution equation for the surface element. Our numerical algorithms will be based on a variational formulation of (5.2), (5.3). To derive it, divide (5.2) by

$$Q = \sqrt{1 + |\nabla u|^2}, \tag{5.6}$$

multiply by a test function $\phi \in H_0^1(\Omega)$ and integrate. Integration by parts implies

$$\int_\Omega \frac{u_t \phi}{Q} + \int_\Omega \frac{\nabla u \cdot \nabla \phi}{Q} = 0, \quad \phi \in H_0^1(\Omega),\ 0 < t < T. \tag{5.7}$$

It is straightforward to derive from (5.7) the decrease in area.

Lemma 5.2. Suppose that u is a smooth solution of (5.2). Then

$$\int_\Omega \frac{u_t^2}{Q} + \frac{\mathrm{d}}{\mathrm{d}t} \int_\Omega Q = 0. \tag{5.8}$$

Proof. Since $u(\cdot, t) = g$ on $\partial\Omega \times (0, T)$ we have $u_t(\cdot, t) = 0$ on $\partial\Omega$ for $0 < t < T$. The relation (5.8) now follows by inserting $\phi = u_t(\cdot, t)$ in (5.7) and observing that $Q_t = \frac{\nabla u \cdot \nabla u_t}{Q}$. $\qquad\square$

Recalling that $V = -\frac{u_t}{Q}$ we may rewrite the relation (5.8) in the more geometric form of Lemma 3.1.

5.2. Spatial discretization

Let $\mathcal{T}_h$ be an admissible nondegenerate triangulation of the domain Ω with mesh size bounded by h, simplices S and $\Omega_h = \bigcup_{S \in \mathcal{T}_h} S$ the corresponding discrete domain. We assume that vertices on $\partial\Omega_h$ are contained in $\partial\Omega$. The space of finite elements of order $s \in \mathbb{N}$ is chosen to be

$$X_h = \{v_h \in C^0(\overline{\Omega}_h) \mid v_h \text{ is a polynomial of order } s \text{ on each } S \in \mathcal{T}_h\}. \tag{5.9}$$

The subspace containing functions with zero boundary values will be denoted by X_{h0}.

We assume that for $s \in \mathbb{N}, p \in [1, \infty]$ there exists an interpolation operator $I_h : H^{s+1,p}(\Omega) \to X_h$ which satisfies $I_h v \in X_{h0}$ for $v \in H^{s+1,p}(\Omega) \cap H_0^1(\Omega)$, as well as

$$\|v - I_h v\|_{L^p(\Omega \cap \Omega_h)} + h \|\nabla(v - I_h v)\|_{L^p(\Omega \cap \Omega_h)} \le ch^{s+1} \|v\|_{H^{s+1,p}(\Omega)} \tag{5.10}$$

for all $v \in H^{s+1,p}(\Omega)$. For dimensions $n < p(s + 1)$, we can, for instance, choose the usual Lagrange interpolation operator; in higher dimensions a possible choice is the Clément operator. For what follows we choose piece-wise linear finite elements: $s = 1$.

We now use (5.7) in order to define a semidiscrete approximation to the solution of (5.2)–(5.4) as follows: find $u_h(\cdot, t) \in X_h$ with $u_h(\cdot, t) - I_h g \in X_{h0}$ and $u_h(\cdot, 0) = u_{h0} = I_h u_0$ such that

$$\int_{\Omega_h} \frac{u_{ht}\phi_h}{Q_h} + \int_{\Omega_h} \frac{\nabla u_h \cdot \nabla \phi_h}{Q_h} = 0, \quad \text{for all } \phi_h \in X_{h0} \tag{5.11}$$

and all $t \in (0, T)$. Here, we have abbreviated $Q_h = \sqrt{1 + |\nabla u_h|^2}$. The following lemma establishes the global existence of the discrete solution.

Lemma 5.3. The semidiscrete problem has a unique solution u_h which exists globally in time.

Proof. We denote by $a_i, i = 1, \ldots, N$ the nodes of the triangulation $\mathcal{T}_h$ and by χ_i the corresponding nodal basis functions. We assume that $a_1, \ldots, a_{N_1}$ are the interior nodes, while $a_{N_1+1}, \ldots, a_N$ lie on $\partial\Omega_h$. We expand $u_h(\cdot . t) = \sum_{i=1}^{N_1} \alpha_i(t)\chi_i + \sum_{i=N_1+1}^{N} g(a_i)\chi_i$ and the relation (5.11) then amounts to a nonlinear system of ODEs for $\alpha = (\alpha_1, \ldots, \alpha_{N_1})$. Existence of a unique local solution follows from standard ODE theory, while the analogue of (5.8) implies a uniform bound on u_h and therefore on α since X_h is finite-dimensional. This allows us to continue the solution for all times. □

In order to prove error estimates for the semidiscrete problem we need to make regularity assumptions on the solution of the continuous problem. Let us suppose that u satisfies

$$\int_0^T \|u_t\|_{H^{1,\infty}(\Omega)}^2 \, dt + \int_0^T \|u_t\|_{H^2(\Omega)}^2 \, dt \leq N \tag{5.12}$$

for some $N > 0$ (see Deckelnick and Dziuk (1999) for sufficient conditions which imply (5.12)). In the following we shall assume that we have a solution of this kind until the time T. We shall formulate our error estimates in terms of geometric quantities, more specifically in terms of the normals $\nu = \frac{(\nabla u, -1)}{Q}$, $\nu_h = \frac{(\nabla u_h, -1)}{Q_h}$ and the normal velocities $V = -\frac{u_t}{Q}$, $V_h = -\frac{u_{ht}}{Q_h}$ reflecting the form of the *a priori* estimate (5.8).

Theorem 5.4. Let u be a solution of the continuous problem (5.2)–(5.4), which satisfies (5.12). Then

$$\int_0^T \int_{\Omega \cap \Omega_h} (V - V_h)^2 Q_h + \sup_{(0,T)} \int_{\Omega \cap \Omega_h} |\nu - \nu_h|^2 Q_h \leq ch^2.$$

The constant c depends on N.

Proof. Let us give the proof of this theorem for polygonal domains, $\Omega = \Omega_h$. The proof shows how important it is to work with the geometric quantities. The difference of the discrete weak form (5.11) and the corresponding continuous weak form of equation (5.2) reads

$$\int_\Omega \left(\frac{u_t}{Q} - \frac{u_{ht}}{Q_h} \right) \phi_h + \int_\Omega \left(\frac{\nabla u}{Q} - \frac{\nabla u_h}{Q_h} \right) \cdot \nabla \phi_h = 0 \qquad (5.13)$$

for all discrete test functions $\phi_h \in X_{h0}$. As a test function we choose

$$\phi_h = I_h u_t - u_{ht} = (u_t - u_{ht}) - (u_t - I_h u_t).$$

We observe that

$$\left(\frac{u_t}{Q} - \frac{u_{ht}}{Q_h} \right)(u_t - u_{ht}) = (V - V_h)(VQ - V_h Q_h) \qquad (5.14)$$

$$= (V - V_h)^2 Q_h + (V - V_h)V(Q - Q_h)$$

$$\geq (V - V_h)^2 Q_h - |V - V_h||V||Q| \left| \frac{1}{Q} - \frac{1}{Q_h} \right| Q_h$$

$$\geq \frac{1}{2}(V - V_h)^2 Q_h - \frac{1}{2}|u_t|^2|\nu - \nu_h|^2 Q_h.$$

Here we have used the fact that

$$\left| \frac{1}{Q} - \frac{1}{Q_h} \right| \leq |\nu - \nu_h|. \qquad (5.15)$$

For the gradient term in (5.13) we exploit the fact that the last component of the vector $\nu Q - \nu_h Q_h$ is zero, and get

$$\left(\frac{\nabla u}{Q} - \frac{\nabla u_h}{Q_h} \right) \cdot (\nabla u_t - \nabla u_{ht}) = (\nu - \nu_h) \cdot (\nabla u_t - \nabla u_{ht}, 0) \qquad (5.16)$$

$$= (\nu - \nu_h) \cdot (\nu Q - \nu_h Q_h)_t.$$

With the elementary relation

$$(\nu - \nu_h) \cdot \nu = -(\nu - \nu_h) \cdot \nu_h = \frac{1}{2}|\nu - \nu_h|^2,$$

the right-hand side in (5.16) can be estimated as follows:

$(\nu - \nu_h) \cdot (\nu Q - \nu_h Q_h)_t$

$\quad = (\nu - \nu_h) \cdot (\nu_t Q - \nu_{ht} Q_h + \nu Q_t - \nu_h Q_{ht})$

$\quad = \frac{1}{2}|\nu - \nu_h|^2(Q_t + Q_{ht}) + (\nu - \nu_h) \cdot (\nu - \nu_h)_t Q_h + (\nu - \nu_h) \cdot \nu_t(Q - Q_h)$

$\quad = \frac{1}{2}(|\nu - \nu_h|^2 Q_h)_t + \frac{1}{2}|\nu - \nu_h|^2 Q_t + (\nu - \nu_h) \cdot \nu_t(Q - Q_h)$

$\quad \geq \frac{1}{2}(|\nu - \nu_h|^2 Q_h)_t - \frac{1}{2}|Q_t||\nu - \nu_h|^2 - |\nu_t|Q\,|\nu - \nu_h|^2 Q_h,$

where again we have used (5.15). With this estimate, (5.14) and (5.16) the error relation (5.13) implies the bound

$$\frac{1}{2} \int_\Omega (V - V_h)^2 Q_h + \frac{1}{2} \frac{\mathrm{d}}{\mathrm{d}t} \int_\Omega |\nu - \nu_h|^2 Q_h$$

$$\leq \frac{1}{2} \big(\|u_t\|_{L^\infty(\Omega)}^2 + 3\|\nabla u_t\|_{L^\infty(\Omega)}^2 \big) \int_\Omega |\nu - \nu_h|^2 Q_h$$

$$+ \int_\Omega |V - V_h| \, |u_t - I_h u_t| + \int_\Omega |\nu - \nu_h| \, |\nabla(u_t - I_h u_t)|.$$

We estimate the interpolation terms with the help of (5.10), that is,

$$\int_\Omega |V - V_h| \, |u_t - I_h u_t| + \int_\Omega |\nu - \nu_h| \, |\nabla(u_t - I_h u_t)|$$

$$\leq c \|u_t\|_{H^2(\Omega)} \left(h^2 \left(\int_\Omega (V - V_h)^2 \right)^{\frac{1}{2}} + h \left(\int_\Omega |\nu - \nu_h|^2 \right)^{\frac{1}{2}} \right)$$

$$\leq \delta \int_\Omega (V - V_h)^2 Q_h + \delta \int_\Omega |\nu - \nu_h|^2 Q_h + \frac{c}{\delta} \|u_t\|_{H^2(\Omega)}^2 h^2$$

for every $\delta > 0$, since $Q_h \geq 1$. After a suitable choice of δ we arrive at

$$\frac{1}{2} \int_\Omega (V - V_h)^2 Q_h + \frac{\mathrm{d}}{\mathrm{d}t} \int_\Omega |\nu - \nu_h|^2 Q_h$$

$$\leq c \big(1 + \|u_t\|_{H^{1,\infty}(\Omega)}^2\big) \int_\Omega |\nu - \nu_h|^2 Q_h + c \|u_t\|_{H^2(\Omega)}^2 h^2.$$

A Gronwall argument and the choice $u_h(\cdot, 0) = I_h u_0$ then finally proves the theorem. $\qquad \square$

Remark 1. It is possible to show that in the two-dimensional case the above error bounds imply that $\sup_{\bar\Omega \times [0,T]} Q_h \leq C$ uniformly in h. As a consequence the error estimate can be written down with the help of the usual norms, namely

$$\int_0^T \|u_t - u_{h,t}\|_{L^2(\Omega \cap \Omega_h)}^2 \, dt + \sup_{(0,T)} \|\nabla(u - u_h)\|_{L^2(\Omega \cap \Omega_h)}^2 \leq c h^2.$$

5.3. Time discretization

Let us choose a time-step $\tau > 0$ and let $t_m = m\tau, m = 0, \ldots, M, M \leq [\frac{T}{\tau}]$ as well as $v^m = v(\cdot, m\tau)$ for $m = 0, \ldots, M$. Based on (5.11) we suggest the following algorithm.

Algorithm 3. (Mean curvature flow of graphs) Let $u_h^0 = I_h u_0$. For $m = 0, \ldots, M - 1$, compute $u_h^{m+1} \in X_h$ such that $u_h^{m+1} - I_h g \in X_{h0}$ and, for every $\varphi_h \in X_{h0}$,

$$\frac{1}{\tau}\int_{\Omega_h}\frac{u_h^{m+1}\varphi_h}{Q_h^m} + \int_{\Omega_h}\frac{\nabla u_h^{m+1}\cdot\nabla\varphi_h}{Q_h^m} = \frac{1}{\tau}\int_{\Omega_h}\frac{u_h^m\varphi_h}{Q_h^m}. \qquad (5.17)$$

with $Q_h^m = \sqrt{1+|\nabla u_h^m|^2}$.

The above scheme is semi-implicit in time and has the property that in each time-step a linear Laplace-type equation with stiffness matrix weighted by Q_h^m has to be solved. In order to analyse its stability we go back to the basic energy norms introduced in (5.8).

Theorem 5.5. The solution $u_h^m, 0 \le m \le M$ of (5.17) satisfies, for every $m \in \{1, \dots, M\}$,

$$\tau\sum_{k=0}^{m-1}\int_{\Omega_h}|V_h^k|^2 Q_h^k + \int_{\Omega_h}Q_h^m \le \int_{\Omega_h}Q_h^0 \qquad (5.18)$$

where $V_h^k = -\frac{(u_h^{k+1}-u_h^k)}{\tau Q_h^k}$ is the discrete normal velocity.

Proof. We choose $\varphi_h = u_h^{k+1} - u_h^k$ as a test function in (5.17) for $m = k$ and get

$$\frac{1}{\tau}\int_{\Omega_h}\frac{(u_h^{k+1}-u_h^k)^2}{Q_h^k} + \int_{\Omega_h}\frac{\nabla u_h^{k+1}\cdot\nabla(u_h^{k+1}-u_h^k)}{Q_h^k} = 0. \qquad (5.19)$$

Let us use the notation $\nu_h^k = \frac{(\nabla u_h^k, -1)}{Q_h^k}$. The integrand in the second term can be rewritten as

$$\frac{\nabla u_h^{k+1}\cdot\nabla(u_h^{k+1}-u_h^k)}{Q_h^k} = \frac{(Q_h^{k+1})^2-1}{Q_h^k} - \frac{\nabla u_h^{k+1}}{Q_h^{k+1}}\cdot\frac{\nabla u_h^k}{Q_h^k}Q_h^{k+1}$$

$$= \frac{(Q_h^{k+1})^2}{Q_h^k} + \frac{1}{2}|\nu_h^{k+1}-\nu_h^k|^2 Q_h^{k+1} - Q_h^{k+1}$$

$$= \frac{1}{2}|\nu_h^{k+1}-\nu_h^k|^2 Q_h^{k+1} + Q_h^{k+1} - Q_h^k + \frac{(Q_h^{k+1}-Q_h^k)^2}{Q_h^k}.$$

We insert this result into (5.19), sum over $k = 0, \dots, m-1$ and obtain the equation

$$\tau\sum_{k=0}^{m-1}\int_{\Omega_h}|V_h^k|^2 Q_h^k + \sum_{k=0}^{m-1}\int_{\Omega_h}\frac{(Q_h^{k+1}-Q_h^k)^2}{Q_h^k} + \frac{1}{2}\sum_{k=0}^{m-1}\int_{\Omega_h}|\nu_h^{k+1}-\nu_h^k|^2 Q_h^{k+1}$$

$$+ \int_{\Omega_h}Q_h^m = \int_{\Omega_h}Q_h^0$$

which implies the stability estimate (5.18). $\qquad \square$

Let us emphasize that our scheme is unconditionally stable even though the nonlinear expressions are treated explicitly. Other schemes, such as fully explicit and fully implicit variants are discussed in Dziuk (1999a). It is natural to follow the ideas of the semidiscrete case in order to analyse the above algorithm. For the analysis of the fully discrete scheme we need the following regularity assumptions:

$$\sup_{t\in(0,T)} \left(\|u(\cdot,t)\|_{H^{2,\infty}(\Omega)} + \|u_t(\cdot,t)\|_{H^{1,\infty}(\Omega)} \right)$$

$$+ \int_0^T \left(\|u_t\|_{H^2(\Omega)}^2 + \|u_{tt}\|^2 \right) \mathrm{d}s \leq N. \tag{5.20}$$

This leads to the following result.

Theorem 5.6. Assume that there exists a solution of (5.2)–(5.4) on $[0,T]$, which satisfies (5.20) and let u_h^m, $(m = 1, \ldots, M = [\frac{T}{\tau}])$ be the solution of Algorithm 3. Then there exists a $\tau_0 > 0$ such that, for all $0 < \tau \leq \tau_0$,

$$\tau \sum_{m=0}^{M-1} \int_{\Omega \cap \Omega_h} (V^m - V_h^m)^2 Q_h^m \leq c(\tau^2 + h^2), \tag{5.21}$$

$$\sup_{m=0,\ldots,M} \int_{\Omega \cap \Omega_h} |\nu^m - \nu_h^m|^2 Q_h^m \leq c(\tau^2 + h^2). \tag{5.22}$$

Proof. This is a special case of the results obtained in Deckelnick and Dziuk (2002a). $\square$

For computational tests we refer to the anisotropic case; see Table 8.2. Here we give some test results for the usual norms. Error estimates in these norms for the two-dimensional case are contained in Deckelnick and Dziuk (2000). For the tests we have solved the partial differential equation

Table 5.1. Absolute errors in $L^\infty((0,T);L^2(\Omega))$, $L^\infty((0,T);H^1(\Omega))$ and experimental orders of convergence (EOC) for the test problem.

h	E_1	EOC	E_2	EOC
2.0	1.1932	–	0.9428	–
1.0	0.6649	0.84	0.9453	0.00
0.7368	0.2878	2.74	0.5873	1.56
0.4203	0.1067	1.77	0.2919	1.25
0.2219	0.04211	1.46	0.1375	1.18
0.1137	0.01775	1.29	0.06536	1.11
0.05754	0.007986	1.17	0.03168	1.06

(5.2) with a given additional right-hand side. We have chosen $u(x,t) = \sin(|x|^2 - t) - \sin(1-t)$ and calculated a right-hand side from this function. The computational domain was $\Omega = \{x \in \mathbb{R}^2 | \, |x| < 1\}$ and we used the boundary condition $u = 0$ on $\partial\Omega$. The time interval was $[0,4]$ and as time-step size we have chosen $\tau = 0.125h$. For two successive grids with grid sizes h_1 and h_2 we computed the absolute errors $E(h_j)$, $(j = 1,2)$ between discrete solution and exact solution for certain norms. The experimental order of convergence was then defined by $\mathrm{EOC} = \log(E(h_1)/E(h_2))/\log(h_1/h_2)$. In Table 5.1 the errors in the norms $E_1 = \sup_{0 \le m \le M} \|u^m - u_h^m\|$ with $M\tau = T$ and $E_2 = \sup_{0 \le m \le M} \|\nabla(u^m - u_h^m)\|$ are shown. The results confirm the theoretical estimates. Note that the $L^\infty((0,T), L^2(\Omega))$-error behaves linearly in the grid size h because we have chosen the time-step proportional to the spatial grid size.

6. Mean curvature flow of level sets

If we want to compute topological changes of free boundaries then it is necessary to leave the parametric world, because this fixes the topological type of the interface. One method to do this is to define the interface as the level set of a scalar function:

$$\Gamma(t) = \{x \in \mathbb{R}^{n+1} | \, u(x,t) = 0\}.$$

Let us assume for the moment that $u \in C^{2,1}(\mathbb{R}^{n+1} \times (0,T))$ with $\nabla u \ne 0$ in a neighbourhood of $\bigcup_{t \in (0,T)} \Gamma(t) \times \{t\}$. Recalling (2.11) and (2.20), the relation $V = -H$ on $\Gamma(t)$ would hold if

$$u_t - \sum_{i,j=1}^{n+1} \left(\delta_{ij} - \frac{u_{x_i} u_{x_j}}{|\nabla u|^2} \right) u_{x_i x_j} = 0 \tag{6.1}$$

in a neighbourhood of $\bigcup_{t \in (0,T)} \Gamma(t) \times \{t\}$. This partial differential equation is highly nonlinear, degenerate parabolic and not defined where the gradient of u vanishes. Therefore, standard methods for parabolic equations fail, but it is possible to develop an existence and uniqueness theory for (6.1) within the framework of viscosity solutions. The corresponding notion involves a pointwise relation and the analysis relies mainly on the maximum principle. It is therefore not straightforward to use finite element methods, which are typically L^2-methods and normally do not allow a maximum principle. This difficulty will be reflected in the numerical analysis. An example of the evolution of level sets under mean curvature flow is shown in Figure 6.1 (Deckelnick and Dziuk 2001).

Crandall, Ishii and Lions (1992) give a concise introduction to the theory of viscosity solutions, while Giga (2002) describes in detail the application of level set techniques to a large class of geometric evolution equations.

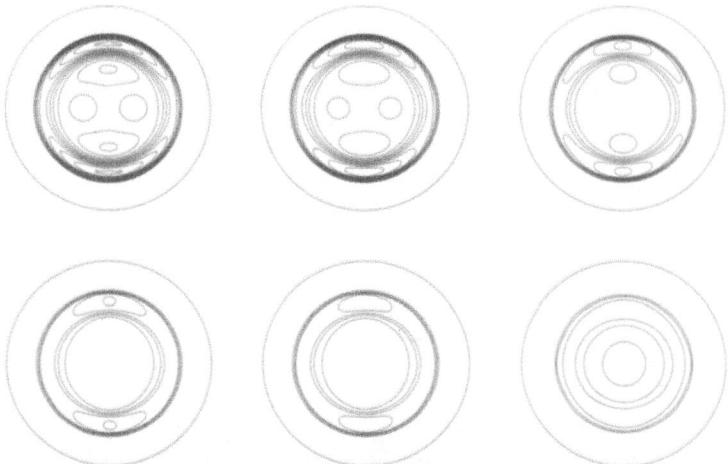

Figure 6.1. Evolution of level lines under mean curvature flow.

Detailed descriptions of computational techniques for level set methods along with a host of applications can be found in the monographs by Sethian (1999) and Osher and Fedkiw (2003).

6.1. Analytical results

Starting from (6.1), we are interested in the following problem:

$$u_t - \sum_{i,j=1}^{n+1} \left(\delta_{ij} - \frac{u_{x_i} u_{x_j}}{|\nabla u|^2} \right) u_{x_i x_j} = 0 \quad \text{in } \mathbb{R}^{n+1} \times (0, \infty) \qquad (6.2)$$

$$u(\cdot, 0) = u_0 \quad \text{in } \mathbb{R}^{n+1}. \qquad (6.3)$$

An existence and uniqueness theory for (6.2), (6.3) can be carried out within the framework of viscosity solutions.

Definition 3. A function $u \in C^0(\mathbb{R}^{n+1} \times [0, \infty))$ is called a *viscosity sub-solution* of (6.2) provided that for each $\phi \in C^\infty(\mathbb{R}^{n+2})$, if $u - \phi$ has a local maximum at $(x_0, t_0) \in \mathbb{R}^{n+1} \times (0, \infty)$, then

$$\phi_t - \sum_{i,j=1}^{n+1} \left(\delta_{ij} - \frac{\phi_{x_i} \phi_{x_j}}{|\nabla \phi|^2} \right) \phi_{x_i x_j} \leq 0 \qquad \text{at } (x_0, t_0), \text{ if } \nabla\phi(x_0, t_0) \neq 0,$$

$$\qquad (6.4)$$

$$\phi_t - \sum_{i,j=1}^{n+1} (\delta_{ij} - p_i p_j) \phi_{x_i x_j} \leq 0 \qquad \text{at } (x_0, t_0) \text{ for some } |p| \leq 1,$$
$$\text{if } \nabla\phi(x_0, t_0) = 0.$$

Figure 6.2. Evolution of a lemniscate under level set mean curvature flow: the zero level.

A viscosity supersolution is defined analogously: maximum is replaced by minimum and $\leq$ by $\geq$. A viscosity solution of (6.2) is a function $u \in C^0(\mathbb{R}^{n+1} \times [0, \infty))$ that is both a subsolution and a supersolution.

We shall assume that the initial function u_0 is smooth and satisfies

$$u_0(x) = 1 \quad \text{for } |x| \geq S \tag{6.5}$$

for some $S > 0$. The following existence and uniqueness theorem is a special case of results proved independently by Evans and Spruck (1991) and Chen, Giga and Goto (1991).

Theorem 6.1. Assume $u_0 : \mathbb{R}^{n+1} \to \mathbb{R}$ satisfies (6.5). Then there exists a unique viscosity solution of (6.2), (6.3), such that

$$u(x, t) = 1 \quad \text{for } |x| + t \geq R$$

for some $R > 0$ depending only on S.

The level set approach can now be described as follows: given a compact hypersurface Γ_0, choose a continuous function $u_0 : \mathbb{R}^{n+1} \to \mathbb{R}$ such that $\Gamma_0 = \{x \in \mathbb{R}^{n+1} \,|\, u_0(x) = 0\}$. If $u : \mathbb{R}^{n+1} \times [0, \infty) \to \mathbb{R}$ is the unique viscosity solution of (6.2), (6.3), we then call

$$\Gamma(t) = \{x \in \mathbb{R}^{n+1} \,|\, u(x, t) = 0\}, \quad t \geq 0$$

Figure 6.3. Evolution of the oriented distance function
of a lemniscate: level lines.

a generalized solution of the mean curvature flow problem. We remark that
Evans and Spruck (1991) and Chen, Giga and Goto (1991) also established
that the sets $\Gamma(t) = \{x \in \mathbb{R}^{n+1} \mid u(x,t) = 0\}, t > 0$ are independent of
the particular choice of u_0 which has Γ_0 as its zero level set, so that the
generalized evolution $(\Gamma(t))_{t \geq 0}$ is well defined for a given Γ_0. As $\Gamma(t)$ exists
for all times, it provides a notion of solution beyond singularities in the flow.
For this reason, the level set approach has also become very important in
the numerical approximation of mean curvature flow and related problems.
Note however that it is possible that the set $\Gamma(t)$ may develop an interior for
$t > 0$, even if Γ_0 had none, a phenomenon which is referred to as *fattening*.
The level set solution has been investigated further in several papers: in
particular we mention Evans and Spruck (1992a, 1992b, 1995) and Soner
(1993).

6.2. Regularization

Evans and Spruck (1991) proved that the (smooth) solutions u^ϵ of

$$u_t^\epsilon - \sum_{i,j=1}^{n+1} \left(\delta_{ij} - \frac{u_{x_i}^\epsilon u_{x_j}^\epsilon}{\epsilon^2 + |\nabla u^\epsilon|^2} \right) u_{x_i x_j}^\epsilon = 0 \quad \text{in } \mathbb{R}^{n+1} \times (0, \infty), \qquad (6.6)$$

$$u^\epsilon(\cdot, 0) = u_0 \quad \text{in } \mathbb{R}^{n+1} \qquad (6.7)$$

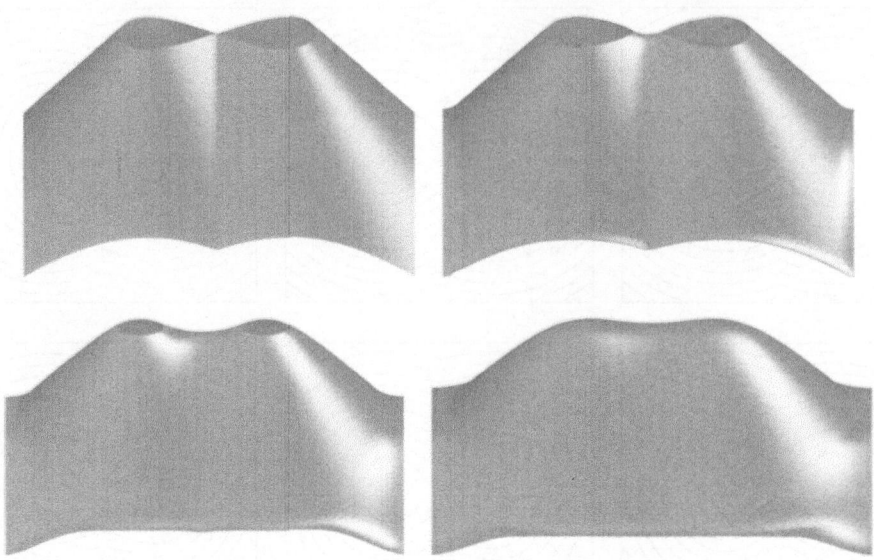

Figure 6.4. Evolution of the oriented distance
function of a lemniscate: graph.

converge locally uniformly as $\epsilon \to 0$ to the unique viscosity solution of (6.2),
(6.3). For numerical purposes it is important to know the asymptotic error
between the viscosity solution and the solution of the regularized problem
quantitatively as $\epsilon \to 0$. In Deckelnick (2000) there is a proof of the following
theorem together with several *a priori* estimates and their dependence on
the regularization parameter ϵ.

Theorem 6.2. For every $\alpha \in (0, \frac{1}{2}), 0 < T < \infty$ there is a constant
$C = C(u_0, T, \alpha)$ such that

$$\sup_{0 \leq t \leq T} \|u - u^\epsilon\|_{L^\infty(\mathbb{R}^{n+1})} \leq C\epsilon^\alpha \quad \text{for all } \epsilon > 0.$$

If one wants to calculate approximations to the viscosity solution u of
(6.2), (6.3) then, according to Theorem 6.2, it is sufficient to solve the
regularized problem (6.6), (6.7), which we have to study for computational
purposes, on a bounded domain. For simplicity we choose $\Omega = B_{\tilde{S}}(0)$ with
$\tilde{S} > R = R(S)$, where R is the radius from Theorem 6.1, and consider

$$u_{\epsilon t} - \sum_{i,j=1}^{n+1} \left(\delta_{ij} - \frac{u_{\epsilon x_i} u_{\epsilon x_j}}{\epsilon^2 + |\nabla u_\epsilon|^2} \right) u_{\epsilon x_i x_j} = 0 \quad \text{in } \Omega \times (0, \infty), \qquad (6.8)$$

$$u_\epsilon = 1 \quad \text{on } \partial\Omega \times (0, \infty), \qquad (6.9)$$

$$u_\epsilon(\cdot, 0) = u_0 \quad \text{in } \Omega. \qquad (6.10)$$

An application of the parabolic comparison theorem yields the following corollary of Theorem 6.2.

Corollary 6.3. For every $\alpha \in (0, \frac{1}{2}), 0 < T < \infty$ there is a constant $C = C(u_0, T, \alpha)$ such that

$$\|u - u_\epsilon\|_{L^\infty(\Omega \times (0,T))} \leq C\epsilon^\alpha. \tag{6.11}$$

We are now in position to look at the regularized level set mean curvature flow problem as a problem for graphs. If we scale

$$U = \frac{u_\epsilon}{\epsilon} \tag{6.12}$$

then U is a solution of the mean curvature flow problem for graphs (see (5.2)), that is,

$$U_t - \sqrt{1 + |\nabla U|^2} \, \nabla \cdot \frac{\nabla U}{\sqrt{1 + |\nabla U|^2}} = 0 \quad \text{in } \Omega \times (0,T). \tag{6.13}$$

This is a theoretical observation and implies that we can apply techniques developed for the mean curvature flow of graphs to the mean curvature flow of level sets. But for computations we shall not use (6.13) but the unscaled version for u_ϵ.

6.3. The approximation of viscosity solutions

Numerical schemes based on the level set approach were first introduced in Osher and Sethian (1988); see also Sethian (1990). Chen, Giga, Hitaka and Honma (1994) proposed a finite difference scheme for which they proved stability with respect to the L^∞-norm. Walkington (1996) used a finite element approach on the dual mesh to construct a discretization that is stable both with respect to L^∞ and to $W^{1,1}$. Evans (1993) analysed a scheme based on the solution of the usual heat equation, continually re-initialized after short time-steps, and which was proposed in Merriman, Bence and Osher (1994). Crandall and Lions (1996) constructed a finite difference scheme that is both monotone and consistent, and obtained the first convergence result for an approximation of (6.2), (6.3). An error analysis for this scheme can be found in Deckelnick (2000).

Here we want to consider a different finite element scheme which exploits the above-described formal similarity to the graph case. This will also allow us to carry out some basic numerical analysis. In the following we use the abbreviations

$$\nu_\epsilon(v) = \frac{(\nabla v, -\epsilon)}{Q_\epsilon(v)}, \quad Q_\epsilon(v) = \sqrt{\epsilon^2 + |\nabla v|^2}, \quad V_\epsilon(v) = -\frac{v_t}{Q_\epsilon(v)}.$$

Our results for the mean curvature flow of a graph can directly be transformed into a convergence result for the regularized level set problem.

Theorem 6.4. Let u_ϵ be the solution of (6.8), (6.10) and let $u_{\epsilon h}$ be the solution of the semidiscrete problem $u_{\epsilon h}(\cdot, t) \in X_h$ with $u_{\epsilon h}(\cdot, t) - 1 \in X_{h0}$, $u_{\epsilon h}(\cdot, 0) = u_{h0} = I_h u_0$ and

$$\int_{\Omega_h} \frac{u_{\epsilon ht} \phi_h}{Q_\epsilon(u_{\epsilon h})} + \int_{\Omega_h} \frac{\nabla u_{\epsilon h} \cdot \nabla \phi_h}{Q_\epsilon(u_{\epsilon h})} = 0 \qquad (6.14)$$

for all $t \in (0, T)$ and all discrete test functions $\phi_h \in X_{h0}$. Then

$$\int_0^T \int_{\Omega \cap \Omega_h} (V_\epsilon(u_\epsilon) - V_\epsilon(u_{\epsilon h}))^2 Q_\epsilon(u_{\epsilon h}) \le c_\epsilon h^2,$$

$$\sup_{(0,T)} \int_{\Omega \cap \Omega_h} |\nu_\epsilon(u_\epsilon) - \nu_\epsilon(u_{\epsilon h})|^2 Q_\epsilon(u_{\epsilon h}) \le c_\epsilon h^2.$$

We omit the proof as it is based on the scaling argument (6.12). Unfortunately, the constants c_ϵ contain a term that depends exponentially on $\frac{1}{\epsilon}$, which is due to an application of Gronwall's lemma. Numerical tests, however, suggest that the resulting bound overestimates the error.

In two space dimensions we can prove that the computed solutions $u_{\epsilon h}$ converge in L^∞ to the viscosity solution. The proof is contained in Deckelnick and Dziuk (2001).

Theorem 6.5. Let u be the viscosity solution of (6.2), (6.3) and let $u_{\epsilon h}$ be the solution of the problem (6.14) with $\Omega \subset \mathbb{R}^2$ as in Corollary 6.3. Then there exists a function $h = h(\epsilon) \to 0$ as $\epsilon \to 0$ such that

$$\lim_{\epsilon \to 0} \|u - u_{\epsilon h(\epsilon)}\|_{L^\infty(\Omega \times (0,T))} = 0.$$

Finally, the fully discrete numerical scheme for (regularized) isotropic mean curvature flow of level sets is now a straightforward adaption of Algorithm 3.

Algorithm 4. (Mean curvature flow of level sets)   Let $u_{\epsilon h}^0 = I_h u_0$. For $m = 0, \ldots, M - 1$, compute $u_{\epsilon h}^{m+1} \in X_h$ such that $u_{\epsilon h}^{m+1} - 1 \in X_{h0}$ and, for every $\phi_h \in X_{h0}$,

$$\frac{1}{\tau} \int_{\Omega_h} \frac{u_{\epsilon h}^{m+1} \phi_h}{Q_\epsilon(u_{\epsilon h}^m)} + \int_{\Omega_h} \frac{\nabla u_{\epsilon h}^{m+1} \cdot \nabla \phi_h}{Q_\epsilon(u_{\epsilon h}^m)} = \frac{1}{\tau} \int_{\Omega_h} \frac{u_{\epsilon h}^m \phi_h}{Q_\epsilon(u_{\epsilon h}^m)}. \qquad (6.15)$$

For this scheme we have the following convergence result.

Theorem 6.6. Let u_ϵ be the solution of (6.8)–(6.10) and let $u_{\epsilon h}^m$, $(m = 1, \ldots, M)$ be the solution from Algorithm 4. Then there exists a $\tau_0 > 0$

such that, for all $0 < \tau \leq \tau_0$,

$$\tau \sum_{m=0}^{M-1} \int_{\Omega \cap \Omega_h} (V_\epsilon(u_\epsilon^m) - V_{\epsilon h}^m)^2 Q_\epsilon(u_{\epsilon h}^m) \leq c_\epsilon(\tau^2 + h^2), \qquad (6.16)$$

$$\sup_{m=0,\dots,M} \int_{\Omega \cap \Omega_h} |\nu_\epsilon(u_\epsilon^m) - \nu_\epsilon(u_{\epsilon h}^m)|^2 Q_\epsilon(u_{\epsilon h}^m) \leq c_\epsilon(\tau^2 + h^2), \qquad (6.17)$$

with $M = [\frac{T}{\tau}]$. Here $V_{\epsilon h}^m = -(u_{\epsilon h}^{m+1} - u_{\epsilon h}^m)/(\tau\, Q_\epsilon(u_{\epsilon h}^m))$ is the regularized discrete normal velocity.

This result implies the convergence of the fully discrete regularized solution to the viscosity solution.

Theorem 6.7.   Let u be the viscosity solution from Theorem 6.1 and let Ω be the domain from Corollary 6.3 in $\mathbb{R}^2$. Let $u_{\epsilon h \tau}$ denote the time-interpolated solution of the fully discrete scheme (6.15). Then there exist functions $h = h(\epsilon) \to 0$ and $\tau = \tau(\epsilon) \to 0$ as $\epsilon \to 0$ such that

$$\lim_{\epsilon \to 0} \|u - u_{\epsilon h(\epsilon) \tau(\epsilon)}\|_{L^\infty(\Omega_h \times (0,T))} = 0.$$

7. Phase field approach to mean curvature flow

7.1. Introduction

The phase field approach to interface evolution is based on physical models for problems involving phase transitions. In this section Ω is a bounded domain in $\mathbb{R}^{n+1}$ and $\Gamma(t)$ is a hypersurface moving through Ω. In the case of two phases one has the notion of an order parameter or phase field function $\varphi : \Omega \times (0,T) \to \mathbb{R}$ which indicates the phase of a material by associating with the phases the minima of a C^2 double well bulk energy function $W(\cdot) : \mathbb{R} \to \mathbb{R}$. For simplicity we suppose that $W(r) = W(-r)$ and the minima of $W(\cdot)$ are at ± 1. The canonical example is

$$W(r) = \frac{1}{4}(r^2 - 1)^2. \qquad (7.1)$$

Consider the gradient energy functional

$$E(\varphi) = \int_\Omega \left(\frac{\epsilon}{2}|\nabla\varphi|^2 + \frac{W(\varphi)}{\epsilon} \right) \mathrm{d}x, \qquad (7.2)$$

where ϵ is a small parameter. Steepest descent or gradient flow for this functional leads to the parabolic Allen–Cahn equation (Allen and Cahn 1979)

$$\epsilon \varphi_t - \epsilon \Delta\varphi + \frac{1}{\epsilon}W'(\varphi) = 0 \quad \text{in } \Omega \times (0,T) \qquad (7.3)$$

with Neumann boundary conditions. In order to understand the behaviour of this evolution equation for an initial function $\varphi_0 : \Omega \to \mathbb{R}$, observe that the flow of the ordinary differential equation $\varphi_t = -\frac{W'(\varphi)}{\epsilon^2}$ drives positive values of φ_0 to 1 and negative values to -1. On the other hand the Laplacian term in the equation (7.3) has a smoothing effect which will diffuse large gradients of the solution. Thus, on the basis of these heuristics, after a short time the solution of (7.3) will develop a structure consisting of bulk regions in which φ is smooth and takes the values ± 1, and separating these regions there will be interfacial transition layers across which φ changes rapidly from one bulk value to the other. These transition layers are due to the interaction between the regularizing effect of the gradient energy term and the flow associated with the bi-stable potential term W'. It turns out that the motion of these interfacial transition layers approximates mean curvature flow.

We can argue informally to support this in the following way. Let, for $t \in (0, T)$, $\Gamma(t)$ be a smoothly evolving closed and compact hypersurface satisfying $V = -H$. Suppose that $\Gamma(t)$ is the boundary of an open set $\Omega(t) \subset \Omega$ and denote by $d(\cdot, t)$ the signed distance function to $\Gamma(t)$. We consider the semilinear parabolic operator

$$P(v) = \epsilon v_t - \epsilon \Delta v + \frac{1}{\epsilon} W'(v).$$

A calculation yields for $v(x, t) = \psi\left(\frac{d(x,t)}{\epsilon}\right)$, where $\psi : \mathbb{R} \to \mathbb{R}$, that

$$P(v) = (d_t - \Delta d)\psi'\left(\frac{d}{\epsilon}\right) - \frac{1}{\epsilon}\left(\psi''\left(\frac{d}{\epsilon}\right) - W'\left(\psi\left(\frac{d}{\epsilon}\right)\right)\right).$$

Hence it is natural to define $\psi = \psi(z)$ to be the unique solution of

$$-\psi''(z) + W'(\psi(z)) = 0, \quad z \in \mathbb{R}, \tag{7.4}$$

$$\psi(z) \to \pm 1, \quad z \to \pm\infty, \quad \psi(0) = 0, \quad \psi'(z) > 0. \tag{7.5}$$

If W is given by (7.1) we have that $\psi(z) = \tanh\left(\frac{z}{\sqrt{2}}\right)$ and therefore

$$P(v) = (d_t - \Delta d)\psi'\left(\frac{d}{\epsilon}\right).$$

Recalling (2.12) and (2.20) we obtain $d_t - \Delta d = -V - H = 0$ on $\Gamma(t)$, so that the smoothness of d implies

$$|d_t - \Delta d| \le C|d|$$

in a neighbourhood U of $\bigcup_{0 < t < T} \Gamma(t) \times \{t\}$. Hence

$$|P(v)| \le C\epsilon|\frac{d}{\epsilon}\psi'\left(\frac{d}{\epsilon}\right)| \le C\epsilon \quad \text{in } U$$

and it follows that $v = \psi\left(\frac{d}{\epsilon}\right)$ is close to being a solution of (7.3) with initial data $\varphi_0 = \psi\left(\frac{d(\cdot,0)}{\epsilon}\right)$. That (7.3) is gradient flow for (7.2) is easily shown by testing the equation with φ_t and integrating by parts, leading to

$$\epsilon \int_\Omega |\varphi_t|^2 \, \mathrm{d}x + \frac{\mathrm{d}E(\varphi)}{\mathrm{d}t} = 0,$$

which is the analogue of the energy equation in Lemma 3.1.

A more general isotropic phase field equation is

$$\epsilon\varphi_t = \epsilon\Delta\varphi - \frac{1}{\epsilon}W'(\varphi) + c_W g, \tag{7.6}$$

where g is a forcing term. The constant c_W is a scaling constant dependent on the precise definition of the double well potential W and is given by the formula

$$c_W = \frac{1}{\sqrt{2}} \int_{-1}^{1} \sqrt{W(r)} \, \mathrm{d}r. \tag{7.7}$$

The equation of motion that this phase field model approximates is

$$V = -H - g. \tag{7.8}$$

We refer to Rubinstein, Sternberg and Keller (1989) and de Mottoni and Schatzman (1995) for formal and rigorous interface asymptotics relating the Allen–Cahn equation to mean curvature flow. Error bounds for the Hausdorff distance between the zero level set of the phase field function and the interface have been derived (Chen 1992, Bellettini and Paolini 1996). In particular, a convergence rate of $\mathcal{O}(\varepsilon^2|\log\varepsilon|^2)$ was established by Bellettini and Paolini (1996). These bounds are proved using comparison theorems for the phase field equation and this can be extended to prove convergence to the viscosity solution of the level set equation in the case of nonsmooth evolution and without the interface thickening (fattening) (Evans, Soner and Souganidis 1992).

7.2. The double obstacle phase field model

We consider the phase field model

$$\epsilon\varphi_t - \varepsilon\Delta\varphi + \frac{1}{\epsilon}W'(\varphi) = c_W g. \tag{7.9}$$

The potential W is taken to be of double obstacle form

$$W(r) = \frac{1}{2}(1 - r^2) + I_{[-1,1]}(r), \tag{7.10}$$

where

$$I_{[-1,1]}(r) = \begin{cases} +\infty & \text{for } |r| > 1, \\ 0 & \text{for } |r| \leq 1, \end{cases} \tag{7.11}$$

introduced in the gradient phase field models by Bellettini, Paolini and Verdi

(1990), Blowey and Elliott (1991, 1993b), Chen and Elliott (1994), Paolini and Verdi (1992).

Properly we should interpret $W'(r)$ in the following way:

$$W'(r) = \begin{cases} (-\infty, 1] & \text{if } r = -1, \\ -r & \text{if } |r| < 1, \\ [-1, \infty) & \text{if } r = 1. \end{cases}$$

For this potential, a calculation reveals that the profile of the phase variable in the transition layer given by the solution of (7.4), (7.5) is

$$\psi(r) = \begin{cases} -1 & \text{if } r \leq -\frac{\pi}{2}, \\ \sin(r) & \text{if } |r| < \frac{\pi}{2}, \\ 1 & \text{if } r \geq \frac{\pi}{2}. \end{cases}$$

Furthermore, $c_W = \frac{\pi}{4}$. The double obstacle problem can be written in an equivalent variational inequality formulation. Let $\mathcal{K}$ be the convex set

$$\mathcal{K} = \{\eta \in H^1(\Omega) : |\eta| \leq 1\}.$$

Then the problem is to seek $\varphi \in L^\infty(0, T; \mathcal{K}) \cap H^1(0, T; L^2(\Omega))$ such that $\varphi(\cdot, 0) = \varphi_0$ and

$$\varepsilon \int_\Omega \varphi_t(\eta - \varphi) + \varepsilon \int_\Omega \nabla\varphi \cdot \nabla(\eta - \varphi) - \frac{1}{\varepsilon} \int_\Omega \varphi(\eta - \varphi) \geq \frac{\pi}{4} \int_\Omega g(\eta - \varphi) \quad (7.12)$$

for all $\eta \in \mathcal{K}$ and for almost every $t \in (0, T)$. It is well known that this problem has a unique solution.

Theorem 7.1. Suppose that the smooth hypersurfaces $\Gamma(t) \subset \mathbb{R}^{n+1}$ satisfy:

(i) $\Gamma(t) = \partial\Omega(t)$ for open sets $\Omega(t) \subset \mathbb{R}^{n+1}$;

(ii) there exists $\delta > 0$ such that $\text{dist}(\Gamma(t), \partial\Omega) \geq \delta$ for $t \in [0, T]$;

(iii) $|d_t - \Delta d| \leq D_0 |d|$ for $|d| \leq \delta, t \in [0, T]$, where $d(\cdot, t)$ is the signed distance function to $\Gamma(t)$;

(iv) $V = -H$ on $\Gamma(t)$ for $t \in [0, T]$.

Let ε be sufficiently small such that $\frac{1}{2}\pi\varepsilon \leq \delta(1 + 2e^{2D_0 T})^{-1}$ and let $\varphi = \varphi_\varepsilon$ be the unique solution of (7.12) with $g = 0$ and initial data $\varphi_0 = \psi\left(\frac{d(\cdot, 0)}{\varepsilon}\right)$. Then, for all $t \in [0, T]$,

$$d(x, t) \geq \frac{1}{2}\pi\varepsilon(1 + 2e^{2D_0 t}) \Rightarrow \varphi_\varepsilon(x, t) = 1,$$

$$d(x, t) \leq -\frac{1}{2}\pi\varepsilon(1 + 2e^{2D_0 t}) \Rightarrow \varphi_\varepsilon(x, t) = -1.$$

Proof. See Chen and Elliott (1994). ☐

A consequence of this theorem is that the diffuse interfacial region

$$\big\{(x, t) : |\varphi_\varepsilon(x, t)| < 1\big\}$$

is sharply defined with finite width bounded by $c(t)\pi\varepsilon$ and that both the zero level set of $\varphi_\varepsilon(\cdot, t)$ and $\Gamma(t)$ are in a narrow strip of width $c(t)\pi\varepsilon$. Here $c(t) = \frac{1}{2}(1 + e^{2D_0 t})$; but in practice it is observed that this is pessimistic and the growth of the interface width is not usually an issue. A more refined analysis by Nochetto, Paolini and Verdi (1994) revealed in the case of a smooth evolution of the forced mean curvature flow that the Hausdorff distance between the zero level set of φ_ε and the interface of the flow (7.8) is of order $\mathcal{O}(\varepsilon^2)$. Furthermore, there is convergence to the unique viscosity solution of the level set formulation (Nochetto and Verdi 1996a).

7.3. Discretization of the Allen–Cahn equation

We use the same notation for the discrete spaces as in Section 5.2. We will identify any function $\Phi \in X_h$ with the vector $\{\Phi_j\}_{j=1}^N$ of its nodal values, so that $\Phi = \sum_{j=1}^N \Phi_j \chi_j$. By $(\cdot, \cdot)$ we denote the $L^2(\Omega)$ inner product.

For computational convenience we use a discrete inner product $(\cdot, \cdot)_h$ on $C^0(\overline{\Omega})$ defined by

$$(\chi, \eta)_h = \int_\Omega I_h(\chi\eta)\, dx \quad \text{for all } \chi,\ \eta \in \mathrm{C}^0(\overline{\Omega}), \tag{7.13}$$

where I_h is the usual Lagrange interpolation operator for X_h. Furthermore, let $\tau = T/M > 0$ be the uniform time-step and $t_m = m\tau$. For any $\{\Phi^m\}_{m=0}^M$, we set $\partial\Phi^m = \tau^{-1}(\Phi^{m+1} - \Phi^m)$. The fully discrete approximation using explicit ($\theta = 0$) and implicit ($\theta = 1$) time-stepping reads as follows.

Algorithm 5. (Allen–Cahn equation)   Let $\Phi^0 = I_h\varphi_0$. For $m = 0, \dots,$ $M - 1$, find $\Phi^{m+1} \in X_h$, $1 \le m \le M - 1$, such that, for all $\chi \in X_h$,

$$(\partial\Phi^m, \chi)_h + (\nabla\Phi^{m+\theta}, \nabla\chi) - \frac{1}{\varepsilon^2}(W'(\Phi^{m+\theta}), \chi)_h = \frac{c_W}{\epsilon}(g, \chi)_h. \tag{7.14}$$

For initial data we choose the finite element interpolant of the transition layer profile

$$\Phi^0 = I_h\psi\left(\frac{d_0(x)}{\epsilon}\right),$$

where d_0 is the signed distance function to the initial interface.

The explicit scheme requires the usual time-step constraint for parabolic equations,

$$\tau \le Ch^2, \tag{7.15}$$

where the constant C depends on the mesh and the L^∞ norm of the initial data through the magnitude of $|W''|$. On the other hand the implicit scheme requires a time-step constraint in order for the nonlinear equations defining Φ^{m+1} to have a unique solution. This constraint is

$$\tau \leq \alpha \epsilon^2, \tag{7.16}$$

where α is the minimum value of W''. See Elliott and Stuart (1993) and Chen, Elliott, Gardiner and Zhao (1998).

The analysis of convergence to mean curvature flow requires consideration of the three approximation parameters ϵ, h, τ tending to zero. Standard *a priori* finite element error analysis for fixed ϵ would lead to, for the difference between the finite element solution and the solution of the Allen–Cahn equation, optimal order error bounds in terms of the mesh sizes τ, h but with constants depending on the Gronwall-induced factor $\exp(\frac{T}{\epsilon^2})$. Feng and Prohl (2003) have improved the finite element error analysis of the Allen–Cahn equation using the special structure of the solution. Indeed, they exploit spectral estimates of Chen (1994) which lead to error bounds whose constants show just polynomial growth in $\frac{1}{\epsilon}$. They specifically consider the implicit scheme without numerical integration. As a consequence they derive an error bound of order ϵ^2 between the zero level set of the solution of the Allen–Cahn equation and the limiting surface.

7.4. Discretization of the double obstacle phase field model

We use the finite element setting of Section 7.3. Let $\mathcal{K}^h = \{\chi \in X_h : |\chi| \leq 1\}$. The double obstacle version of Algorithm 5 is as follows.

Algorithm 6. (Double obstacle phase field) Let $\Phi^0 = I_h \varphi_0$. For $m = 0, \ldots, M - 1$, find $\Phi^{m+1} \in \mathcal{K}^h$ such that, for all $\chi \in \mathcal{K}^h$,

$$(\partial \Phi^m, \chi - \Phi^{m+1})_h + (\nabla \Phi^{m+\theta}, \nabla \chi - \Phi^{m+1}) \tag{7.17}$$

$$- \frac{1}{\varepsilon^2}(\Phi^{m+\theta} + \epsilon c_W g^{m+\theta}, \chi - \Phi^{m+1})_h \geq 0.$$

For initial data we choose the finite element interpolant of the transition layer profile. The explicit scheme is a discrete obstacle variational inequality associated with the mass matrix. Without mass lumping the solution of this nonlinear algebraic problem would require quadratic programming or linear complementarity methods. However, with the mass lumping quadrature rule the explicit scheme is as simple as the explicit scheme for a semilinear parabolic equation. It can be simply written as

$$\Phi^{m+1/2} = \left(\left(1 + \frac{\tau}{\epsilon^2} \right) I - \tau A \right) \Phi^m + c_W \frac{\tau}{\epsilon} g^m, \tag{7.18}$$

$$\Phi^{m+1} = \mathcal{P} \Phi^{m+1/2}. \tag{7.19}$$

Here $A = M^{-1} K$, where M and K are defined by

$$M_{ij} = (\chi_i, \chi_j)_h, \qquad K_{ij} = (\nabla \chi_i, \nabla \chi_j),$$

for $1 \leq i,\ j \leq N$. Furthermore, $\mathcal{P} : \mathbb{R}^N \to \mathbb{R}^N$ is the component-wise projection onto $[-1,1]^N$ defined by

$$(\mathcal{P}V)_j = \max(-1, \min(1, V_j)).$$

On the other hand, in linear algebraic form the implicit scheme leads to the discrete variational inequality: find $\Phi^{m+1} \in \mathbb{R}^N$ such that $|\Phi_j| \leq 1$ and

$$\left(\left(1 - \frac{\tau}{\epsilon^2}\right)I + \tau A \right)\Phi^{m+1} \cdot (\chi - \Phi^{m+1}) \geq \left(\Phi^m + c_W \frac{\tau}{\epsilon} g^{m+1} \right) \cdot (\chi - \Phi^{m+1}) \quad (7.20)$$

for all $\chi \in \mathbb{R}^N$ with $|\chi_j| \leq 1$. Because A is symmetric this is equivalent to minimizing a quadratic function subject to bound constraints and can easily be solved by projected SOR (Elliott and Ockendon 1982). Such a system can also be solved by nonlinear multigrid (Kornhuber and Krause 2003).

As for the continuous parabolic variational inequality, a discrete comparison principle holds for these schemes if the triangulation is acute. This provides the basis for a convergence analysis (Nochetto and Verdi 1996b, 1997). For the implicit scheme without numerical integration an $O(\epsilon)$ error bound for the interface is obtained when $\tau = O(h^2) = O(\epsilon^4)$. For the explicit scheme without numerical integration in the potential term an $O(\epsilon^2)$ is proved for $\tau = O(h^2) = O(\epsilon^5)$.

7.5. *Implementation*

One expects there to be a relationship between ϵ and h in order that the discrete phase field model can approximate the sharp interface motion. Since the convergence analysis in the continuous case relies heavily on understanding the profile of the phase field function across the transition layer, one would expect that for any ϵ the mesh size h should be sufficiently small in order to resolve the interface. Indeed the existing convergence analysis described above indicates that h should tend to zero faster than ϵ. In practice this implies that across the discrete interfacial layer in the normal direction there should be a sufficient number of elements.

In the case of the double obstacle potential, at the mth time-step, the finite elements may be divided into three sets:

$$\mathcal{J}^h_-(m) = \{\Phi_j = -1 \ \text{ for each element vertex}\},$$
$$\mathcal{J}^h_+(m) = \{\Phi_j = 1 \ \text{ for each element vertex}\},$$
$$\mathcal{I}^h(m) = \mathcal{T}^h \setminus (\mathcal{J}^h_-(m) \cup \mathcal{J}^h_+(m)).$$

Clearly the approximation to the interface is the zero level set of Φ^m which lies inside the discrete interfacial region $\mathcal{I}^h(m)$. We view $\mathcal{I}^h(m)$ as a *sharp diffuse interface*, as opposed to the interfacial region associated with the smooth double well, which is not sharply defined. The computational work in evolving the interface is then associated with the small number of

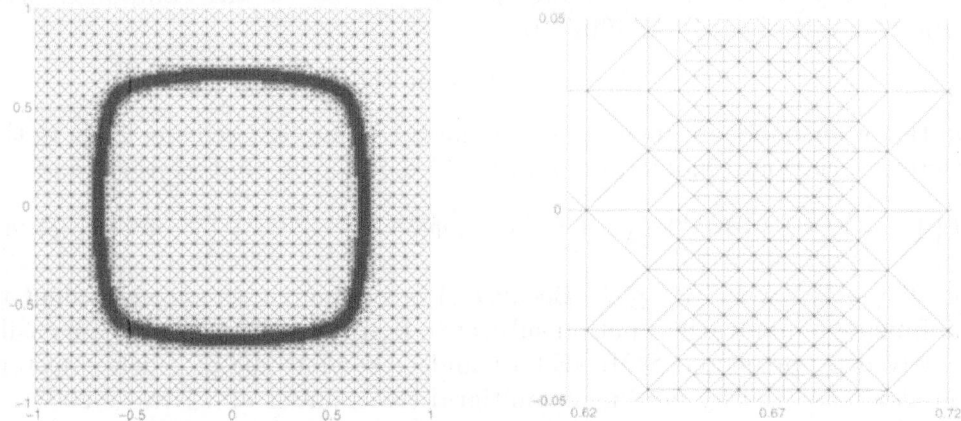

Figure 7.1. Meshes.

elements in this region. As observed above, the time-step, τ, in the phase field calculations is substantially smaller than the mesh size, h. Thus, in a numerical simulation one would expect that, for finite normal velocity of the interface, the sharp diffuse interface should only move by at most the addition or subtraction of a single layer of elements. In the case of the explicit scheme this can be made precise. For nodes in $\mathcal{J}_+^h(m)$ (or $\mathcal{J}_-^h(m)$) whose nearest neighbours are also in $\mathcal{J}_+^h(m)$ (or $\mathcal{J}_-^h(m)$), we find

$$\Phi_j^{m+1/2} = \pm 1 + \frac{\tau}{\epsilon^2}(\pm 1 + c_W \epsilon g^m(a_j))$$

which, provided $|g^m(a_j)| \leq \frac{1}{c_W \epsilon}$, implies that $\Phi_j^{m+1} = \pm 1$. It follows that the *sharp diffuse interface* can not move more than one element per time-step. It also implies that it is only necessary to compute Φ^{m+1} on the closure of the transition layer. This can be exploited in a number of ways.

The two-dimensional *dynamic mesh algorithm* (Nochetto, Paolini and Verdi 1996) is based on the explicit scheme and carries a mesh only in the sharp diffuse interface; it adds and removes triangles where necessary.

The *mask* method (Elliott and Gardiner 1996) keeps an underlying fixed mesh and computes in the sharp diffuse interface only. It is possible to store nodal values only in this region.

An amalgam of the above is an adaptive procedure which uses a fine mesh within the diffuse interface and a coarse mesh outside. In Figure 7.1 a typical mesh is shown for a phase field calculation of anisotropic mean curvature flow. The global mesh is shown together with a zoom. This approach requires a fine mesh slightly larger than the diffuse interface. As the interface region moves the mesh is refined and coarsened appropriately.

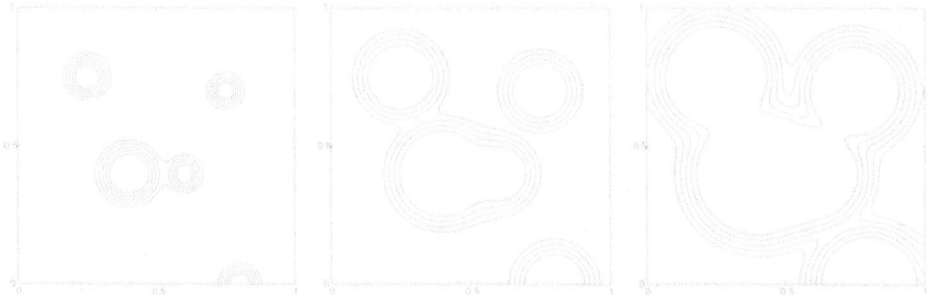

Figure 7.2. Topological change.

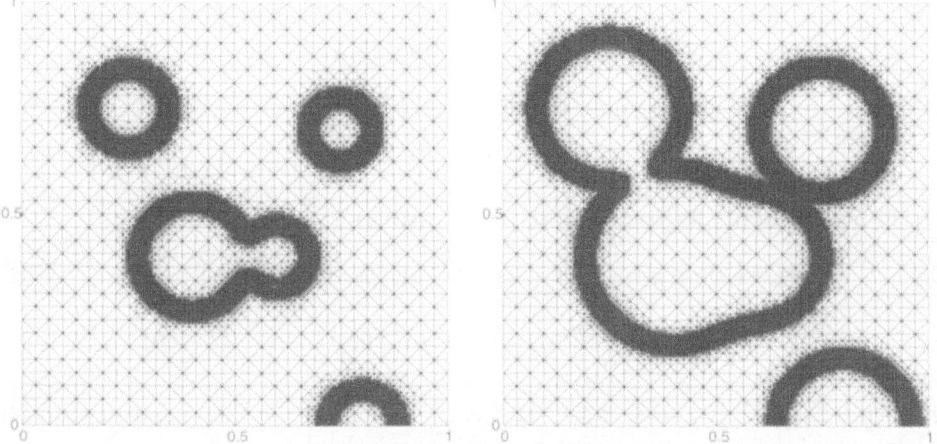

Figure 7.3. Diffuse interfaces with topological change.

Sharp diffuse interface front tracking

Using the double obstacle phase field method and only computing within a sharp diffuse interface as described above can be viewed as a front tracking method, which has the advantage of being able to handle topological change. In Figure 7.2 the interfaces at various times are displayed for a forced mean curvature flow starting from initial circles. Eventually the circles intersect. Meshes associated with these computations are shown in Figure 7.3.

8. Anisotropic mean curvature flow

8.1. *The concept of anisotropy*

In free boundary problems such as phase transition problems it is often necessary to treat interfaces which are driven by anisotropic curvature. This is induced by modelling an anisotropic surface energy, which generalizes area in the isotropic case to weighted area in the anisotropic case. Anisotropic

surface energy has the form

$$E_\gamma(\Gamma) = \int_\Gamma \gamma(\nu)\, \mathrm{d}A, \qquad (8.1)$$

where Γ is a surface with normal ν and γ is a given anisotropy function. For $\gamma(p) = |p|$ this energy is the area of Γ. For our purposes it will be necessary to restrict the admissible anisotropies to a certain class.

Definition 4.   An anisotropy function $\gamma : \mathbb{R}^{n+1} \to \mathbb{R}$ is called *admissible* if

(1) $\gamma \in C^3(\mathbb{R}^{n+1} \setminus \{0\})$, $\gamma(p) > 0$ for $p \in \mathbb{R}^{n+1} \setminus \{0\}$;

(2) γ is positively homogeneous of degree one, *i.e.*,

$$\gamma(\lambda p) = |\lambda| \gamma(p) \quad \text{for all } \lambda \neq 0, p \neq 0; \qquad (8.2)$$

(3) there exists $\gamma_0 > 0$ such that

$$D^2\gamma(p)q \cdot q \geq \gamma_0 |q|^2 \quad \text{for all } p, q \in \mathbb{R}^{n+1},\ |p| = 1,\ p \cdot q = 0. \qquad (8.3)$$

It is not difficult to verify that (8.2) implies

$$D\gamma(p) \cdot p = \gamma(p), \qquad D^2\gamma(p)p \cdot q = 0, \qquad (8.4)$$

$$D\gamma(\lambda p) = \frac{\lambda}{|\lambda|} D\gamma(p), \qquad D^2\gamma(\lambda p) = \frac{1}{|\lambda|} D^2\gamma(p) \qquad (8.5)$$

for all $p \in \mathbb{R}^{n+1} \setminus \{0\}$, $q \in \mathbb{R}^{n+1}$ and $\lambda \neq 0$. The convexity assumption (8.3) will be crucial for analysis and numerical methods.

Anisotropy is normally visualized by using the Frank diagram $\mathcal{F}$ and the Wulff shape $\mathcal{W}$:

$$\mathcal{F} = \{p \in \mathbb{R}^{n+1} \mid \gamma(p) \leq 1\},$$
$$\mathcal{W} = \{q \in \mathbb{R}^{n+1} \mid \gamma^*(q) \leq 1\}.$$

 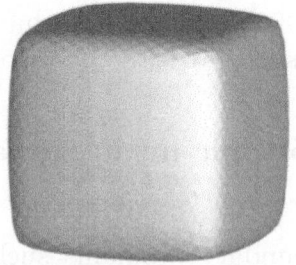

Figure 8.1. Frank diagram (left) and Wulff shape (right) for the regularized l^1-anisotropy $\gamma(p) = \sum_{j=1}^{3} \sqrt{\varepsilon^2 |p|^2 + p_j^2}$.

Here γ^* is the dual of γ, which is given by

$$\gamma^*(q) = \sup_{p \in \mathbb{R}^{n+1} \setminus \{0\}} \frac{p \cdot q}{\gamma(p)}. \tag{8.6}$$

Let us consider some examples. Note that not all of them are admissible.

The choice $\gamma(p) = |p|$ is called the isotropic case; in particular we have that $\mathcal{F} = \mathcal{W} = \{p \in \mathbb{R}^{n+1} \mid |p| \leq 1\}$ is the closed unit ball.

A typical choice for anisotropy is the discrete l^r-norm for $1 \leq r \leq \infty$,

$$\gamma(p) = \|p\|_{l^r} = \left(\sum_{k=1}^{n+1} |p_k|^r \right)^{\frac{1}{r}}, \quad 1 \leq r < \infty, \tag{8.7}$$

with the obvious modification for $r = \infty$.

For a given positive definite $(n+1) \times (n+1)$ matrix G, the anisotropy function

$$\gamma(p) = \sqrt{Gp \cdot p} \tag{8.8}$$

models an anisotropy which is defined by a (constant) Riemannian metric. In Figure 8.2 we show the Frank diagram and Wulff shape for the anisotropy

$$\gamma(p) = \sqrt{(5.5 + 4.5 \, \mathrm{sign}(p_1)) p_1^2 + p_2^2 + p_3^2}. \tag{8.9}$$

One anisotropy function often used in a physical context is

$$\gamma(p) = \left(1 - A \left(1 - \frac{\|p\|_{l^4}^4}{\|p\|_{l^2}^4} \right) \right) \|p\|_{l^2} \tag{8.10}$$

where A is a parameter. For $A < 0.25$ the Frank diagram is convex.

For more information on this subject, including anisotropies that may depend on space, see Bellettini and Paolini (1996).

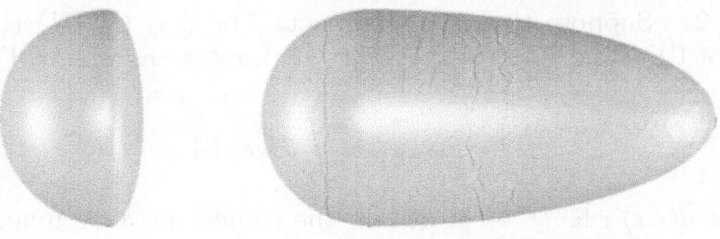

Figure 8.2. Frank diagram $\mathcal{F}$ (left) and Wulff shape $\mathcal{W}$ (right) for the anisotropy (8.9).

8.2. Anisotropic distance function

Let γ be an admissible anisotropy function. We can associate with γ a nonsymmetric metric Υ on $\mathbb{R}^{n+1}$ by setting

$$\Upsilon(x,y) = \gamma^*(x-y), \quad x,y \in \mathbb{R}^{n+1}. \tag{8.11}$$

It is possible to prove that Υ is equivalent to the standard Euclidean metric. Suppose next that $\Omega \subset \mathbb{R}^{n+1}$ is a bounded open set with smooth boundary Γ. Using Υ we now define an anisotropic signed distance function $d_\gamma : \mathbb{R}^{n+1} \to \mathbb{R}$ by

$$d_\gamma(x) = \begin{cases} \inf_{y \in \Gamma} \Upsilon(x,y), & x \in \mathbb{R}^{n+1} \setminus \bar{\Omega}, \\ 0, & x \in \Gamma, \\ -\inf_{y \in \Gamma} \Upsilon(x,y), & x \in \Omega. \end{cases}$$

Lemma 8.1. There exists an open neighbourhood U of Γ such that $d_\gamma \in C^2(U)$ and

$$\gamma(\nabla d_\gamma) = 1, \tag{8.12}$$
$$D^2 d_\gamma D\gamma(\nabla d_\gamma) = 0. \tag{8.13}$$

8.3. Anisotropic mean curvature

Our goal is to generalize the notion of mean curvature to the anisotropic setting. Suppose that γ is an admissible anisotropy function and that $\Gamma \subset \mathbb{R}^{n+1}$ is an oriented hypersurface with normal ν. We define the Cahn–Hoffmann vector ν_γ on Γ by

$$\nu_\gamma(x) = D\gamma(\nu(x)), \quad x \in \Gamma, \tag{8.14}$$

and the anisotropic mean curvature by

$$H_\gamma(x) = \nabla_\Gamma \cdot \nu_\gamma(x), \quad x \in \Gamma. \tag{8.15}$$

Note that $H_\gamma = H$ in the isotropic case $\gamma(p) = |p|$. The following lemma shows that H_γ is a natural generalization of mean curvature as the first variation of the area functional with respect to normal variations.

Lemma 8.2. Suppose that Γ is compact. For $\phi \in C_0^\infty(U)$ (U a neighbourhood of Γ) define $F_\epsilon(x) = x + \epsilon\phi(x)\nu(x), x \in U$ as well as $\Gamma_\epsilon = F_\epsilon(\Gamma)$. Then,

$$\frac{d}{d\epsilon} E_\gamma(\Gamma_\epsilon)_{|\epsilon=0} = \int_\Gamma H_\gamma \phi \, dA.$$

Proof. Let $d(\cdot, \epsilon) : \mathbb{R}^{n+1} \to \mathbb{R}$ denote the signed distance function to Γ_ϵ. Consider $g : U \times (-\epsilon_0, \epsilon_0) \to \mathbb{R}$, defined by

$$g(x,\epsilon) = \gamma(\nu_\epsilon(x)) = \gamma(\nabla d(x,\epsilon)),$$

where ∇ acts on the x variables only. Now (2.23), (2.20) and (2.6) imply

$$\frac{\mathrm{d}}{\mathrm{d}\epsilon}E_\gamma(\Gamma_\epsilon)_{|\epsilon=0} = \frac{\mathrm{d}}{\mathrm{d}\epsilon}\int_{\Gamma_\epsilon} g(\cdot,\epsilon)\,\mathrm{d}A_{|\epsilon=0}$$

$$= \int_\Gamma \frac{\partial g}{\partial \epsilon}(\cdot,0)\,\mathrm{d}A - \int_\Gamma g(\cdot,0)\frac{\partial d}{\partial \epsilon}(\cdot,0)H\,\mathrm{d}A - \int_\Gamma \frac{\partial g}{\partial \nu}(\cdot,0)\frac{\partial d}{\partial \epsilon}(\cdot,0)\,\mathrm{d}A.$$

It is not difficult to see that $\frac{\partial d}{\partial \epsilon}(\cdot,0) = -\phi(x), x \in \Gamma$, which also implies that

$$\frac{\partial g}{\partial \epsilon}(\cdot,0) = D\gamma(\nu)\cdot\nabla\frac{\partial d}{\partial \epsilon}(\cdot,0) = D\gamma(\nu)\cdot\nabla_\Gamma\frac{\partial d}{\partial \epsilon}(\cdot,0) = -\nu_\gamma\cdot\nabla_\Gamma\phi.$$

Here we have used the definition of ν_γ and the fact that $\nabla\frac{\partial d}{\partial \epsilon}(\cdot,0)\cdot\nu = 0$ on Γ. Thus,

$$\frac{\mathrm{d}}{\mathrm{d}\epsilon}E_\gamma(\Gamma_\epsilon)_{|\epsilon=0} = -\int_\Gamma \nu_\gamma\cdot\nabla_\Gamma\phi\,\mathrm{d}A + \int_\Gamma \gamma(\nu)\phi H\,\mathrm{d}A + \int_\Gamma \frac{\partial g}{\partial \nu}(\cdot,0)\phi\,\mathrm{d}A$$

$$= \int_\Gamma \nabla_\Gamma\cdot\nu_\gamma\phi\,\mathrm{d}A + \int_\Gamma \frac{\partial g}{\partial \nu}(\cdot,0)\phi\,\mathrm{d}A,$$

where the last identity follows from (2.16). Finally, observing that $\frac{\partial g}{\partial \nu}(\cdot,0) = \gamma_{p_i}(\nu)d_{x_ix_j}(\cdot,0)d_{x_j}(\cdot,0) = 0$, and recalling the definition of H_γ, the claim follows. $\square$

Let us next calculate H_γ for various descriptions of Γ.

Level set representation. Suppose that Γ is given as in (2.1) and oriented by $\nu = \frac{\nabla u}{|\nabla u|}$. Since γ_{p_i} is homogeneous of degree 0, we have (see also (2.2))

$$H_\gamma = \nabla_\Gamma\cdot\nu_\gamma = \sum_{i=1}^{n+1}D_i\left(\gamma_{p_i}\left(\frac{\nabla u}{|\nabla u|}\right)\right) = \sum_{i=1}^{n+1}D_i(\gamma_{p_i}(\nabla u))$$

$$= \sum_{i,j=1}^{n+1}\gamma_{p_ip_j}(\nabla u)u_{x_ix_j} - \sum_{i,k,l=1}^{n+1}\gamma_{p_ip_l}(\nabla u)u_{x_lx_k}\frac{u_{x_k}}{|\nabla u|}\frac{u_{x_i}}{|\nabla u|}.$$

Recalling (8.4) we therefore deduce

$$H_\gamma = \sum_{i,j=1}^{n+1}\gamma_{p_ip_j}(\nabla u)u_{x_ix_j}. \tag{8.16}$$

Graph representation. If Γ is locally given as the graph of the function $x' \mapsto v(x'), x' = (x_1,\ldots,x_n)$ with normal $\nu = \frac{(\nabla_{x'}v,-1)}{\sqrt{1+|\nabla_{x'}v|^2}}$, formula (8.16) applied to $u(x',x_{n+1}) = v(x') - x_{n+1}$ gives

$$H_\gamma = \sum_{i,j=1}^{n}\gamma_{p_ip_j}(\nabla_{x'}v,-1)v_{x_ix_j}. \tag{8.17}$$

Let us next derive an analogue of (2.16) with H replaced by H_γ. Observing that $\underline{D}_k \nu_l = \underline{D}_l \nu_k$ and recalling that $\underline{D}_k x_l = \delta_{kl} - \nu_k \nu_l$, we obtain

$$
\begin{aligned}
H_\gamma \nu_l &= \underline{D}_k \big(\gamma_{p_k}(\nu)\big)\nu_l = \underline{D}_k \big(\gamma_{p_k}(\nu)\nu_l\big) - \gamma_{p_k}(\nu)\underline{D}_k \nu_l \\
&= \underline{D}_k \big(\gamma_{p_k}(\nu)\nu_l\big) - \gamma_{p_k}(\nu)\underline{D}_l \nu_k \\
&= \underline{D}_k \big(\gamma_{p_k}(\nu)\nu_l\big) - \underline{D}_l \big(\gamma(\nu)\big) \\
&= \underline{D}_k \big(\gamma_{p_k}(\nu)\nu_l\big) - \underline{D}_k \big(\gamma(\nu)(\delta_{kl} - \nu_k \nu_l)\big) - \gamma(\nu)\nu_l \underline{D}_k \nu_k \\
&= \underline{D}_k \big(\gamma_{p_k}(\nu)\nu_l\big) - \underline{D}_k \big(\gamma(\nu)\underline{D}_k x_l\big) - \gamma(\nu)H\nu_l,
\end{aligned}
$$

where summation over k is from 1 to $n+1$. For a smooth test function $\phi = (\phi_1, \ldots, \phi_{n+1})$ we multiply the above relation by ϕ_l, sum over l and integrate over Γ. Using (2.16) we infer

$$
\int_\Gamma H_\gamma \nu \cdot \phi = -\sum_{k,l=1}^{n+1} \int_\Gamma \gamma_{p_k}(\nu)\nu_l \underline{D}_k \phi_l + \sum_{k,l=1}^{n+1} \int_\Gamma \gamma_{p_k}(\nu)\nu_l H\nu_k \phi_l
$$
$$
+ \sum_{k,l=1}^{n+1} \int_\Gamma \gamma(\nu)\underline{D}_k x_l \underline{D}_k \phi_l - \sum_{l=1}^{n+1} \int_\Gamma \gamma(\nu)H\nu_l \phi_l
$$

and (8.4) yields

$$
\int_\Gamma H_\gamma \nu \cdot \phi = -\sum_{k,l=1}^{n+1} \int_\Gamma \gamma_{p_k}(\nu)\nu_l \underline{D}_k \phi_l + \sum_{k,l=1}^{n+1} \int_\Gamma \gamma(\nu)\underline{D}_k x_l \underline{D}_k \phi_l. \tag{8.18}
$$

This relation will be at the heart of the numerical methods in the parametric case. For additional information on the subject of weighted mean curvature including the crystalline case, see Taylor (1992).

8.4. Motion by anisotropic mean curvature with mobility

Having introduced the notion of anisotropic mean curvature we can now formulate the following generalization of (3.1):

$$
\beta(\nu)V = -H_\gamma + g \quad \text{on } \Gamma(t). \tag{8.19}
$$

Here, $\beta : S^n \to \mathbb{R}$ is a given positive and smooth function of degree zero. In applications where $\Gamma(t)$ models a sharp phase-interface, the coefficient β measures the drag opposing interfacial motion and the function $\frac{1}{\beta}$ is called mobility. The function g represents the energy difference in the bulk phases. A detailed derivation of (8.19) from the force balances and the second law of thermodynamics can be found in Angenent and Gurtin (1989) and Gurtin (1993). Taylor, Cahn and Handwerker (1992) give an overview of various mathematical approaches to (8.19).

In what follows we shall consider the simpler problem

$$\beta(\nu)V = -H_\gamma \quad \text{on } \Gamma(t), \tag{8.20}$$

even though all our techniques can be generalized to (8.19). It can be shown (see Bellettini and Paolini (1996)) that for the choice $\beta(\nu) = \frac{1}{\gamma(\nu)}$ there is an explicit solution of (8.20) consisting of shrinking boundaries of Wulff shapes; the sets

$$\Gamma(t) = \{p \in \mathbb{R}^{n+1} \mid \gamma^*(p) = \sqrt{r(0)^2 - 2nt}\}$$

satisfy $\frac{1}{\gamma(\nu)}V = -H_\gamma$ and are therefore a generalization of the shrinking circles from the isotropic case. We also have the following analogue of Lemma 3.1.

Lemma 8.3. Let $\Gamma(t)$ be a family of evolving hypersurfaces satisfying (8.20) on $\Gamma(t)$, and assume that each $\Gamma(t)$ is compact. Then

$$\int_{\Gamma(t)} \beta(\nu)V^2 \, \mathrm{d}A + \frac{\mathrm{d}}{\mathrm{d}t} \int_{\Gamma(t)} \gamma(\nu) = 0.$$

Proof. In the same way as in the proof of Lemma 8.2, we derive

$$\frac{\mathrm{d}}{\mathrm{d}t} \int_{\Gamma(t)} \gamma(\nu) = \int_{\Gamma(t)} H_\gamma V,$$

and the claim follows from the evolution law (8.20). $\qquad \square$

8.5. Anisotropic curve shortening flow

Let us consider a family $\Gamma(t)$ of closed curves in $\mathbb{R}^2$ which move according to (8.20). As in Section 4.1 we describe the evolution by means of a mapping $X: \mathbb{R} \times [0, T) \to \mathbb{R}^2$ which satisfies $X(\theta, t) = X(\theta + 2\pi, t)$ for $t \in [0, T)$, $\theta \in \mathbb{R}$. The curves $\Gamma(t) = X(\cdot, t)$ will move by (8.20) provided that

$$\beta(\nu)X_t = -H_\gamma \nu. \tag{8.21}$$

Using the notation $(a_1, a_2)^\perp = (-a_2, a_1)$ we may write $\nu = \tau^\perp$, where $\tau = \frac{X_\theta}{|X_\theta|}$ is the unit tangent to the curve $\Gamma(t)$. Equation (8.21) amounts to a system of partial differential equations for the vector function X. In order to write down this system, let $\varphi \in H^1_{\text{per}}(I; \mathbb{R}^2)$, $I = [0, 2\pi]$, be a test function, which we can think of as being defined on $\Gamma(t)$ via $\phi(X(\theta, t)) = \varphi(\theta)$. It follows from (8.18) that

$$\int_{\Gamma(t)} H_\gamma \nu \cdot \phi = -\sum_{k,l=1}^{2} \int_{\Gamma(t)} \gamma_{p_k}(\nu)\nu_l \underline{D}_k \phi_l + \sum_{k,l=1}^{2} \int_{\Gamma(t)} \gamma(\nu)\underline{D}_k x_l \underline{D}_k \phi_l$$

$$= -\sum_{k,l=1}^{2} \int_{\Gamma(t)} \big(\gamma_{p_k}(\nu)\nu_l - \gamma(\nu)\delta_{kl}\big)\underline{D}_k \phi_l,$$

since $\underline{D}_k x_l = \delta_{kl} - \nu_k \nu_l$. Using $\nabla_\Gamma \phi_l = \frac{\varphi_{l,\theta}}{|X_\theta|}\tau$ and recalling that $\gamma(p) = D\gamma(p) \cdot p$, we obtain after some calculations

$$\sum_{k,l=1}^{2} \left(\gamma_{p_k}(\nu)\nu_l - \gamma(\nu)\delta_{kl}\right)\underline{D}_k\phi_l = -D\gamma(\nu) \cdot \frac{\varphi_\theta^\perp}{|X_\theta|}.$$

In conclusion we have

$$\int_{\Gamma(t)} H_\gamma \nu \cdot \varphi \, \mathrm{d}A = \int_0^{2\pi} D\gamma(X_\theta^\perp) \cdot \varphi_\theta^\perp \, \mathrm{d}\theta,$$

so that we obtain the following weak form of (8.21):

$$\int_0^{2\pi} \beta\left(\frac{X_\theta^\perp}{|X_\theta|}\right) X_t \cdot \varphi \, |X_\theta| \, \mathrm{d}\theta + \int_0^{2\pi} D\gamma(X_\theta^\perp) \cdot \varphi_\theta^\perp \, \mathrm{d}\theta = 0 \quad \text{for all } \varphi \in H^1_{\text{per}}(I; \mathbb{R}^2). \tag{8.22}$$

We shall base our numerical scheme on this formulation. The classical form of (8.22) is

$$\beta\left(\frac{X_\theta^\perp}{|X_\theta|}\right) X_t + \frac{1}{|X_\theta|}\frac{\partial}{\partial\theta}\left(D\gamma(X_\theta^\perp)^\perp\right) = 0 \quad \text{in } I \times (0, T). \tag{8.23}$$

For the convenience of the reader we explicitly write down the two equations of this system:

$$\beta\left(\frac{X_\theta^\perp}{|X_\theta|}\right) X_{1t}|X_\theta| - \gamma_{p_2 p_2}(-X_{2\theta}, X_{1\theta})X_{1\theta\theta} + \gamma_{p_2 p_1}(-X_{2\theta}, X_{1\theta})X_{2\theta\theta} = 0,$$

$$\beta\left(\frac{X_\theta^\perp}{|X_\theta|}\right) X_{2t}|X_\theta| - \gamma_{p_1 p_1}(-X_{2\theta}, X_{1\theta})X_{2\theta\theta} + \gamma_{p_1 p_2}(-X_{2\theta}, X_{1\theta})X_{1\theta\theta} = 0.$$

It is easy to see that this system can be written as

$$\beta\left(\frac{X_\theta^\perp}{|X_\theta|}\right) X_t - a\left(\frac{X_\theta^\perp}{|X_\theta|}\right)\frac{1}{|X_\theta|}\frac{\partial}{\partial\theta}\left(\frac{X_\theta}{|X_\theta|}\right) = 0,$$

where

$$a(p) = \gamma_{pp}(p)\, p^\perp \cdot p^\perp, \quad p \in \mathbb{R}^2 \setminus \{0\}.$$

Note that (8.3) implies $a(p) \geq \gamma_0 > 0$ for all $p \in \mathbb{R}^2, |p| = 1$. Analytical results for this problem which generalize the theory for the isotropic case ($a = 1$) have been obtained by Gage (1993). We shall continue to use the form (8.22) because this equation only contains first derivatives of the anisotropy function γ. Recall the definition of S_h from Section 4.1. A discrete solution of (8.22) will be a function $X_h : [0, T] \to S_h$, such that

$$X_h(\cdot, 0) = X_{h0} = I_h X_0 = \sum_{j=1}^{N} X_0(\theta_j)\phi_j,$$

and for all discrete test functions $\varphi_h \in S_h$

$$\int_0^{2\pi} \beta\left(\frac{X_{h\theta}}{|X_{h\theta}|}\right) X_{ht} \cdot \varphi_h \, |X_{h\theta}| \, \mathrm{d}\theta + \int_0^{2\pi} D\gamma(X_{h\theta}^\perp) \cdot \varphi_{h\theta}^\perp \, \mathrm{d}\theta = 0. \qquad (8.24)$$

In the same way as in the isotropic case we can write

$$X_h(\theta, t) = \sum_{j=1}^{N} X_j(t) \phi_j(\theta)$$

with $X_j(t) \in \mathbb{R}^2$, and find that the discrete weak equation (8.24) is equivalent to the following system of $2N$ ordinary differential equations:

$$\frac{1}{6}\beta_j q_j \dot{X}_{j-1} + \frac{1}{3}(\beta_j q_j + \beta_{j+1} q_{j+1})\dot{X}_j + \frac{1}{6}\beta_{j+1} q_{j+1} \dot{X}_{j+1}$$
$$+ D\gamma(X_{j+1}^\perp - X_j^\perp)^\perp - D\gamma(X_j^\perp - X_{j-1}^\perp)^\perp = 0,$$

for $j = 1, \ldots, N$, where $X_0(t) = X_N(t)$, $X_{N+1} = X_1(t)$, and

$$q_j = |X_j - X_{j-1}|, \qquad \beta_j = \beta\left(\frac{X_j - X_{j-1}}{q_j}\right).$$

Furthermore, the initial values are given by

$$X_j(0) = X_0(\theta_j), \ j = 1, \ldots, N.$$

We again use mass lumping, which is equivalent to a quadrature formula. Thus we replace this system by the lumped scheme

$$\frac{1}{2}(\beta_j q_j + \beta_{j+1} q_{j+1})\dot{X}_j + D\gamma(X_{j+1}^\perp - X_j^\perp)^\perp - D\gamma(X_j^\perp - X_{j-1}^\perp)^\perp = 0 \quad (8.25)$$

together with the initial conditions $X_j(0) = X_0(\theta_j)$ for $j = 1, \ldots, N$. We are now ready to say what we mean by a discrete solution of anisotropic curve shortening flow. The above system is equivalent to the one we use in the following definition of discrete anisotropic curve shortening flow.

Definition 5. A solution of the discrete anisotropic curve shortening flow for the initial curve $\Gamma_{h0} = X_{h0}([0, 2\pi])$ is a polygon $\Gamma_h(t) = X_h([0, 2\pi], t)$, which is parametrized by a piecewise linear mapping $X_h(\cdot, t) \in S_h$, $t \in [0, T]$, such that $X_h(\cdot, 0) = X_{h0}$ and for all $\varphi_h \in S_h$

$$\int_0^{2\pi} \beta\left(\frac{X_{h\theta}}{|X_{h\theta}|}\right) X_{ht} \cdot \varphi_h \, |X_{h\theta}| \, \mathrm{d}\theta + \int_0^{2\pi} D\gamma(X_{h\theta}^\perp) \cdot \varphi_{h\theta}^\perp \, \mathrm{d}\theta$$
$$+ \frac{1}{6}h^2 \int_0^{2\pi} \beta\left(\frac{X_{h\theta}}{|X_{h\theta}|}\right) X_{h\theta t} \cdot \varphi_{h\theta} \, |X_{h\theta}| \, \mathrm{d}\theta = 0. \qquad (8.26)$$

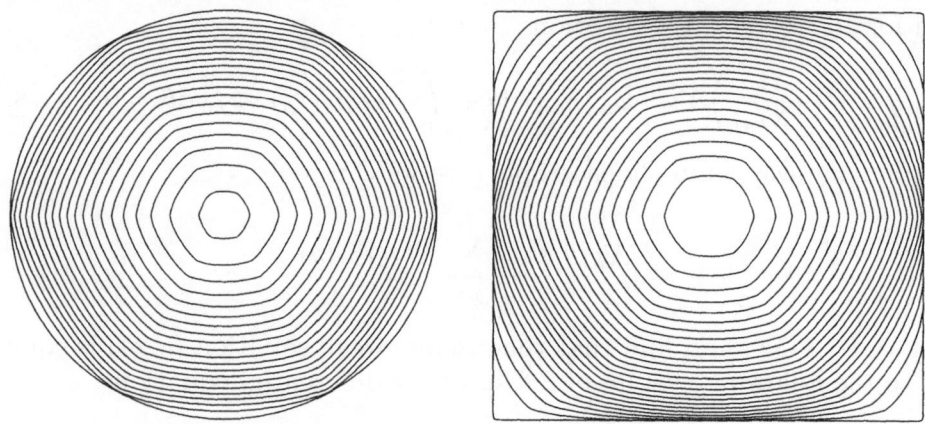

Figure 8.3. Anisotropic curve shortening flow with a
sixfold anisotropy function applied to a circle (left) and
to a square (right).

Here h is the constant grid size of the uniform grid in $[0, 2\pi]$. The last
term of (8.26) is introduced by mass lumping. One could also define the
discrete curve shortening flow without this quantity, but then the geometric
property of length shortening would not be true for the discrete problem.

Dziuk (1999b) proved the following convergence result for $\beta = 1$. It
is easily extended to the case of general β. We formulate the result for
the geometric quantities normal, length and normal velocity. The error
estimates in standard norms then follow easily.

Theorem 8.4. Suppose that $\beta : S^2 \to \mathbb{R}$ is a smooth positive function.
Let X be a solution of the anisotropic curve shortening flow (8.23) on the
interval $[0, T]$ with $X(\cdot, 0) = X_0$, $\min_{[0,2\pi] \times [0,T]} |X_\theta| \geq c_0 > 0$ and $X_t \in
L^2((0, T), H^2(0, 2\pi))$. Then there is an $h_0 > 0$ such that, for all $0 < h \leq h_0$,
there exists a unique solution X_h of the discrete anisotropic curve shortening
flow (8.26) on $[0, T]$ with $X_h(\cdot, 0) = X_{h0} = I_h X_0$, and the error between
smooth and discrete solution can be estimated as follows:

$$\sup_{(0,T)} \int_0^{2\pi} |\nu - \nu_h|^2 |X_{h\theta}| \, \mathrm{d}\theta + \sup_{(0,T)} \int_0^{2\pi} (|X_\theta| - |X_{h\theta}|)^2 \, \mathrm{d}\theta \leq ch^2,$$

$$\int_0^T \int_0^{2\pi} |X_t - X_{ht}|^2 |X_{h\theta}| \, \mathrm{d}\theta \, \mathrm{d}t \leq ch^2.$$

The constants depend on c_0, T and $\|X_t\|_{L^2((0,T),H^2(0,2\pi))}$.

Table 8.1. Convergence test for anisotropic curve shortening flow.

h	E_1	EOC_1	E_2	EOC_2	E_3	EOC_3	E_4	EOC_4
0.3927	0.4929		1.049		0.1236		1.042	
0.1964	0.2544	0.954	0.5467	0.940	0.04703	1.39	0.5500	0.922
0.09818	0.1327	0.939	0.2762	0.985	0.02060	1.19	0.2787	0.981
0.04909	0.06698	0.986	0.1345	0.996	0.009875	1.06	0.1398	0.995
0.02454	0.03354	0.998	0.06928	0.999	0.004882	1.02	0.06996	0.999
0.01227	0.01680	0.998	0.03465	1.0	0.002434	1.0	0.03499	1.0

We tested the algorithm with an exact solution,

$$X(\theta, t) = \sqrt{1 - 2t}\,(\cos g(\theta), \sin g(\theta)),$$

where we have chosen $g(\theta) = \theta + 0.1 \sin \theta$. The anisotropy function is $\gamma(p) = |p| - 0.25 p_1$. We compute the errors

$$E_1 = \|X_t - X_{ht}\|_{L^2((0,T),L^2(\Gamma_h))}, \qquad E_2 = \|\nu - \nu_h\|_{L^\infty((0,T),L^2(\Gamma_h))},$$

$$E_3 = \||X_\theta| - |X_{h\theta}|\|_{L^2((0,T),L^2(S^1))}, \qquad E_4 = \|X_\theta - X_{h\theta}\|_{L^\infty((0,T),L^2(S^1))}$$

with $\Gamma_h = X_h([0, 2\pi], \cdot)$. For two successive grid sizes h_1 and h_2 with corresponding errors $E(h_1)$ and $E(h_2)$, the experimental order of convergence $EOC = \ln(E(h_1)/E(h_2))/\ln(h_1/h_2)$ is calculated and shown in Table 8.1 from Dziuk (1999b). The time-step τ was chosen $\tau = 0.5\,h^2$ for these computations. We emphasize that the algorithm for anisotropic curve shortening flow does not use the second derivatives of the anisotropy function γ.

The system (8.25) can be formally written in complex tridiagonal form. For details and a suitable time discretization we refer to Dziuk (1999b). Let us finally mention that in Girao (1995) simple closed convex curves evolving by (8.20) are computed by approximating the smooth anisotropy by a crystalline one. Also, an error analysis for the resulting method is given.

8.6. Anisotropic curvature flow of graphs

Let us next turn to the evolution of hypersurfaces which are given as graphs, *i.e.*, $\Gamma(t) = \{(x, u(x, t)) \mid x \in \Omega\}$. In order to translate (8.20) into an evolution equation for u we recall that

$$H_\gamma = \sum_{i,j=1}^{n} \gamma_{p_i p_j}(\nabla u, -1) u_{x_i x_j} = \sum_{i=1}^{n} \frac{\partial}{\partial x_i}\left(\gamma_{p_i}(\nabla u, -1)\right).$$

Furthermore, since $V = -\frac{u_t}{Q}$ with $Q = \sqrt{1 + |\nabla u|^2}$ we see that (8.20) leads to the following nonlinear partial differential equation,

$$\beta\left(\frac{(\nabla u, -1)}{Q}\right) u_t - Q \sum_{i=1}^{n} \frac{\partial}{\partial x_i}(\gamma_{p_i}(\nabla u, -1)) = 0 \quad \text{in } \Omega \times (0, T), \qquad (8.27)$$

to which we add the following initial and boundary conditions:

$$u = g \quad \text{on } \partial\Omega \times (0, T),$$
$$u(\cdot, 0) = u_0 \quad \text{in } \Omega.$$

In the sequel we shall again assume that this problem has a solution u which satisfies (5.12) and refer to Deckelnick and Dziuk (1999) for a corresponding existence and uniqueness result.

Discretization in space and estimate of the error
As in the isotropic case we may use a variational approach even though the differential equation is not in divergence form. Starting from (8.27) we obtain, with the abbreviation $\nu = \frac{(\nabla u, -1)}{Q}$,

$$\int_{\Omega} \frac{\beta(\nu) u_t \varphi}{Q} + \sum_{i=1}^{n} \int_{\Omega} \gamma_{p_i}(\nabla u, -1) \varphi_{x_i} = 0 \qquad (8.28)$$

for all $\varphi \in H_0^1(\Omega)$, $t \in (0, T)$ together with the above initial and boundary conditions. We now consider a semidiscrete approximation of (8.28): find $u_h(\cdot, t) \in X_h$ with $u_h(\cdot, t) - I_h g \in X_{h0}$ such that

$$\int_{\Omega_h} \frac{\beta(\nu_h) u_{h,t} \varphi_h}{Q_h} + \sum_{i=1}^{n} \int_{\Omega_h} \gamma_{p_i}(\nabla u_h, -1) \varphi_{h,x_i} = 0 \quad \text{for all } \varphi_h \in X_{h0}, \quad (8.29)$$

for all $t \in (0, T]$, where we have set

$$Q_h = \sqrt{1 + |\nabla u_h|^2}, \qquad \nu_h = \frac{(\nabla u_h, -1)}{Q_h}.$$

As an initial condition we use $u_h(\cdot, 0) = u_h^0 = I_h u_0$. Our main result gives an error bound for the important geometric quantities V and ν. The proof is contained in Deckelnick and Dziuk (1999).

Theorem 8.5. Suppose that (8.27) has a solution u that satisfies (5.12). Then (8.29) has a unique solution u_h and

$$\int_0^T \|V - V_h\|_{L^2(\Gamma_h(t))}^2 \, dt + \sup_{t \in (0,T)} \|(\nu - \nu_h)(\cdot, t)\|_{L^2(\Gamma_h(t))}^2 \le Ch^2.$$

Here, $\Gamma_h(t) = \{(x, u_h(x, t)) \mid x \in \Omega_h \cap \Omega\}$ and V, V_h are as in Theorem 5.4.

Fully discrete scheme, stability and error estimate
Let us next consider discretization in time in order to get a practical method. Compared to the isotropic case, our problem has become more complicated because of the presence of two additional nonlinearities, namely the functions β and γ. In order to keep the computational effort as small as possible it would be desirable to have a method that only requires the solution of a linear problem at each time-step. This can be achieved by treating the nonlinearities in an explicit way and guaranteeing stability via the introduction of an additional stabilizing term. We start again from the variational formulation (8.28) and choose a time-step $\tau > 0$. Using the notation from Section 5.3 our scheme reads as follows.

Algorithm 7. (Anisotropic mean curvature flow of graphs)   Given u_h^m, find $u_h^{m+1} \in X_h$ such that $u_h^{m+1} - I_h g \in X_{h0}$ and

$$
\frac{1}{\tau} \int_{\Omega_h} \frac{\beta(\nu_h^m)(u_h^{m+1} - u_h^m)}{Q_h^m} \varphi_h + \sum_{i=1}^{n} \int_{\Omega_h} \gamma_{p_i}(\nu_h^m) \varphi_{h x_i}
$$
$$
+ \lambda \int_{\Omega_h} \frac{\gamma(\nu_h^m)}{Q_h^m} \nabla(u_h^{m+1} - u_h^m) \cdot \nabla \varphi_h = 0 \qquad (8.30)
$$

for all $\varphi_h \in X_{h0}$. Here we have set $u_h^0 = I_h u_0$ as well as

$$
Q_h^m = \sqrt{1 + |\nabla u_h^m|^2}, \qquad \nu_h^m = \frac{(\nabla u_h^m, -1)}{Q_h^m}.
$$

The above scheme is semi-implicit and requires the solution of a linear system in each time-step. We shall see that it is unconditionally stable provided the parameter λ is chosen appropriately.

Theorem 8.6.   Let $\overline{\gamma} = \frac{1}{\sqrt{5}-1} \max \left\{ \sup_{|p|=1} |\nabla \gamma(p)|, \sup_{|p|=1} |D^2 \gamma(p)| \right\}$. Then we have for $0 \le M \le [\frac{T}{\tau}]$

$$
\tau \sum_{m=1}^{M-1} \int_{\Omega_h} \frac{\beta(\nu_h^m)}{Q_h^m} \left| \frac{u_h^{m+1} - u_h^m}{\tau} \right|^2 + \lambda \tau \sum_{m=1}^{M-1} \int_{\Omega_h} \frac{\gamma(\nu_h^m)}{Q_h^m} \left(\frac{Q_h^{m+1} - Q_h^m}{\sqrt{\tau}} \right)^2
$$
$$
+ \left(\lambda \inf_{|p|=1} \gamma(p) - \overline{\gamma} \right) \tau \sum_{m=1}^{M-1} \int_{\Omega_h} \left| \frac{\nu_h^{m+1} - \nu_h^m}{\sqrt{\tau}} \right|^2 Q_h^{m+1} + \int_{\Omega_h} \gamma(\nu_h^M) Q_h^M
$$
$$
\le \int_{\Omega_h} \gamma(\nu_h^0) Q_h^0.
$$

In particular, if λ is chosen in such a way that $\lambda \inf_{|p|=1} \gamma(p) > \overline{\gamma}$, then we have for $\Gamma_h^m = \{(x, u_h^m(x)) \mid x \in \Omega_h\}$

$$
E_\gamma(\Gamma_h^m) \le E_\gamma(\Gamma_h^0) \quad \text{for all } 0 \le m \le \left[\frac{T}{\tau} \right]. \qquad (8.31)
$$

Thus we have proved stability for the semi-implicit scheme without any restriction on the time-step size. An error analysis for the above scheme has been carried out in Deckelnick and Dziuk (2002*a*). The precise result is as follows.

Theorem 8.7. Suppose that $\lambda \inf_{|p|=1} \gamma(p) > \overline{\gamma}$ ($\overline{\gamma}$ as in Theorem 8.6). Then there exists $\tau_0 > 0$ such that, for all $0 < \tau \leq \tau_0$,

$$\sum_{m=0}^{[\frac{T}{\tau}]-1} \tau \int_{\Omega \cap \Omega_h} (V^m - V_h^m)^2 Q_h^m + \max_{0 \leq m \leq [\frac{T}{\tau}]} \int_{\Omega \cap \Omega_h} |\nu^m - \nu_h^m|^2 Q_h^m \leq c(\tau^2 + h^2).$$

We have run numerical tests for anisotropic mean curvature flow of graphs. The Wulff shape shrinks homothetically during the evolution. We have chosen the very strong anisotropy $\gamma(p) = \sqrt{0.01 p_1^2 + p_2^2 + p_3^2}$. The equation $\gamma^*(x, u(x, t)) = \sqrt{1 - 4t}$ defines a solution of the differential equation when the mobility is chosen as $\beta = 1/\gamma$. The exact solution is given by $u(x, t) = \sqrt{1 - 4t - 100x_1^2 - x_2^2}$. The condition on the stabilizing parameter λ (see Theorem 8.6) is satisfied for $\lambda = 81.0$. We use $\tau = 0.01h$ as a uniform time-step size. The coupling between time-step size and spatial grid size is done in order not to spoil the asymptotic orders of convergence. For a discussion with respect to the choice of λ and τ we refer to Deckelnick and Dziuk (2002*a*). Table 8.2 shows the grid size h, the errors

$$E(V) = \left(\sum_{m=0}^{M} \tau \int_{\Omega_h} |V^m - V_h^m|^2 Q_h^m \right)^{\frac{1}{2}},$$

$$E(\nu) = \left(\max_{0 \leq m \leq M} \int_{\Omega_h} |\nu^m - \nu_h^m|^2 Q_h^m \right)^{\frac{1}{2}},$$

Table 8.2. Convergence test for anisotropic mean curvature flow of graphs.

h	$E(\nu)$	EOC	$E(V)$	EOC	$L^\infty(H^1)$	EOC
7.0711e-2	9.7027e-2	–	2.3278e-2	–	1.3366e-1	–
3.5355e-2	2.3213e-2	2.06	6.0827e-3	1.94	4.4935e-2	1.57
2.6050e-2	2.4818e-2	−0.22	7.5203e-3	−0.70	4.5372e-2	−0.03
1.4861e-2	1.3868e-2	1.04	4.1117e-3	1.08	2.4163e-2	1.12
7.8462e-3	7.0232e-3	1.07	1.9806e-3	1.14	1.2256e-2	1.06
4.0210e-3	3.5368e-3	1.03	1.0176e-3	1.00	6.1725e-3	1.03
2.0342e-3	1.8103e-3	0.98	5.2675e-4	0.97	3.1225e-3	1.00
1.0229e-3	9.2938e-4	0.97	2.6988e-4	0.97	1.5799e-3	0.99

Figure 8.4. Level lines for the time-steps 0, 250, 500, 750, 3000 for a regularized crystalline anisotropy.

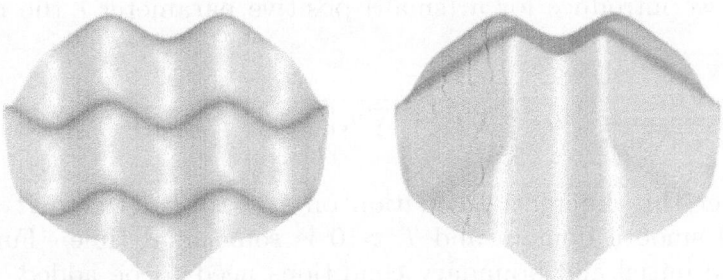

Figure 8.5. Initial value and stationary solution for a regularized crystalline anisotropy.

and the corresponding experimental orders of convergence (EOC) between two successive grid sizes. We add a column with the $L^\infty((0,T), H^1(\Omega))$ error. Obviously the results of the asymptotic error estimates of Theorem 8.7 are reproduced in our test computations. We add a long time computation from Deckelnick and Dziuk (2002a). As an anisotropy function we used an anisotropy which is a regularized form of $\gamma(p) = |p|_{l^\infty}$ (see also Figure 8.1 for the dual situation). We have chosen a nonzero constant right-hand side for the equation. In Figure 8.4 we show the level lines of the initial function, of four time-steps and of the stationary solution. The boundary values were kept fixed during the evolution. We can see that the octahedral shape develops during the evolution. In Figure 8.5 we show the initial graph and the stationary graph. The domain was the unit disk.

8.7. Anisotropic mean curvature flow of level sets

Here we briefly sketch how the level set approach can be adapted to the anisotropic case. Let us look for solutions of (8.20) in the form

$$\Gamma(t) = \{x \in \mathbb{R}^{n+1} \,|\, u(x,t) = 0\}$$

where the scalar function u has to be determined. The relations (2.20) and (8.16) lead to the following nonlinear partial differential equation

$$\beta\left(\frac{\nabla u}{|\nabla u|}\right) u_t - |\nabla u| \sum_{j,k=1}^{n+1} \gamma_{p_j p_k}(\nabla u) u_{x_j x_k} = 0 \quad \text{in } \mathbb{R}^{n+1} \times (0, \infty), \quad (8.32)$$

which is degenerate parabolic since $D^2 \gamma(p) p = 0$. We regularize the equation by using an extension of the anisotropy to $n+2$ space dimensions. Let us assume that there is an admissible weight function $\bar{\gamma} = \bar{\gamma}(p_1, \ldots, p_{n+1}, p_{n+2})$ such that

$$\bar{\gamma}(p_1, \ldots, p_{n+1}, 0) = \gamma(p_1, \ldots, p_{n+1}).$$

In the following we denote this extension again by γ. Rather than treating (8.32) we introduce for a (small) positive parameter ϵ the regularized problem

$$\beta\left(\frac{\nabla u_\epsilon}{\sqrt{\epsilon^2 + |\nabla u_\epsilon|^2}}\right) u_{\epsilon t} - \sqrt{\epsilon^2 + |\nabla u_\epsilon|^2} \sum_{j,k=1}^{n+1} \gamma_{p_j p_k}(\nabla u_\epsilon, -\epsilon) u_{\epsilon\, x_j x_k} = 0.$$

We consider this differential equation on $\Omega \times (0, T)$, where $\Omega \subset \mathbb{R}^{n+1}$ is a bounded smooth domain and $T > 0$ is some final time. Furthermore, appropriate initial and boundary conditions need to be added, which can be done similarly to the isotropic case. The numerical approximation of the resulting problem follows the ideas of the graph case. The same applies to the analysis of the schemes, where of course one has to bear in mind the dependency on the regularization parameter ϵ.

8.8. Anisotropic phase field

We turn now to the setting of Section 7. Anisotropic phase field models are based on the following anisotropic interfacial energy functional

$$E_\gamma(\varphi) = \int_\Omega \left(\epsilon A(\nabla \varphi) + \frac{1}{\epsilon} W(\varphi)\right) dx, \quad (8.33)$$

where $A : \mathbb{R}^{n+1} \to \mathbb{R}$ is smooth, convex and positively homogeneous of degree two which replaces the quadratic gradient energy used in the isotropic case. In order to relate it to the anisotropic energy density used in this section we set

$$A(p) = \frac{1}{2}\gamma(p)^2, \quad p \in \mathbb{R}^{n+1}. \quad (8.34)$$

The double well bulk energy function W may be chosen as in the isotropic situation. The L^2-gradient flow of E_γ leads to the following quasilinear parabolic equation:

$$\epsilon \varphi_t - \epsilon \nabla \cdot DA(\nabla \varphi) + \frac{1}{\epsilon} W'(\varphi) = 0 \quad \text{in } \Omega \times (0, T). \tag{8.35}$$

For small ϵ and suitable initial data, (8.35) approximates the following anisotropic mean curvature flow:

$$\frac{1}{\gamma(\nu)} V = -H_\gamma \quad \text{on } \Gamma(t). \tag{8.36}$$

This can be motivated in a similar manner to the isotropic case. For convenience we suppose that W is a smooth double well. Set

$$P(v) = \epsilon v_t - \epsilon \nabla \cdot DA(\nabla v) + \frac{1}{\epsilon} W'(v) \tag{8.37}$$

and

$$v(x, t) = \psi\left(\frac{d_\gamma(x, t)}{\epsilon}\right),$$

where ψ is the transition profile defined by (7.4) and (7.5) and $d_\gamma(\cdot, t)$ denotes the anisotropic signed distance function to the smoothly evolving interface $\Gamma(t)$ which satisfies (8.36) on $\Gamma(t)$. A calculation shows

$$v_t = \psi'\left(\frac{d_\gamma}{\epsilon}\right) \frac{d_{\gamma,t}}{\epsilon}, \qquad \nabla v = \psi'\left(\frac{d_\gamma}{\epsilon}\right) \frac{\nabla d_\gamma}{\epsilon},$$

$$D^2 v = \psi\left(\frac{d_\gamma}{\epsilon}\right) \frac{\nabla d_\gamma \otimes \nabla d_\gamma}{\epsilon^2} + \psi'\left(\frac{d_\gamma}{\epsilon}\right) \frac{D^2 d_\gamma}{\epsilon},$$

and using (8.4), (8.5) as well as Lemma 8.1, we obtain

$$P(v) = \psi'\left(\frac{d_\gamma}{\epsilon}\right) |\nabla d_\gamma| \left(\frac{d_{\gamma,t}}{|\nabla d_\gamma|} - \gamma\left(\frac{\nabla d_\gamma}{|\nabla d_\gamma|}\right) \sum_{i,j=1}^{n+1} \gamma_{p_i p_j}(\nabla d_\gamma) d_{\gamma, x_i x_j} \right)$$

$$+ \frac{1}{\epsilon}\left(-\psi''\left(\frac{d_\gamma}{\epsilon}\right) + W'\left(\psi\left(\frac{d_\gamma}{\epsilon}\right)\right)\right).$$

Observing that

$$\frac{d_{\gamma,t}}{|\nabla d_\gamma|} - \gamma\left(\frac{\nabla d_\gamma}{|\nabla d_\gamma|}\right) \sum_{i,j=1}^{n+1} \gamma_{p_i p_j}(\nabla d_\gamma) d_{\gamma, x_i x_j} = -V - \gamma(\nu) H_\gamma = 0 \quad \text{on } \Gamma(t)$$

and choosing ψ as in (7.4), (7.5) we see that v is close to being a solution of $P(v) = 0$ in a neighbourhood of $\bigcup_{0 < t < T} \Gamma(t) \times \{t\}$.

Kinetic anisotropy and a generalized double obstacle phase field model
As a generalization of (8.35) we consider the phase field model

$$\varepsilon \beta(\nabla\varphi)\varphi_t - \varepsilon \nabla \cdot DA(\nabla\varphi) + \frac{1}{\epsilon}W'(\varphi) = c_W \rho(\varphi)g. \qquad (8.38)$$

Here, the potential W is taken to be of double obstacle form,

$$W(r) = W_0(r) + I_{[-1,1]}(r), \qquad (8.39)$$

where $W_0 \in C^2[-1,1]$ and

$$I_{[-1,1]}(r) = \begin{cases} +\infty & \text{for } |r| > 1, \\ 0 & \text{for } |r| \le 1. \end{cases}$$

A possible example of W_0 is

$$W_0(r) = \frac{1}{4(1+\xi^2)}\left[(r^2 - 1 - \xi^2)^2 - \xi^4\right] \qquad (8.40)$$

with $\xi \in (0, \infty)$. For $\xi = 0$, W_0 takes the classical smooth double well Ginzburg–Landau quadratic form $\frac{1}{4}(r^2 - 1)^2$, whereas for $\xi \to \infty$ we recover the classical double obstacle potential

$$W(r) = \frac{1}{2}(1 - r^2) + I_{[-1,1]}(r). \qquad (8.41)$$

The function ρ is nonnegative and even with a positive integral across the transition region $[-1, 1]$. As above one can show that the zero level set of φ approximates an interface which evolves according to the anisotropic forced mean curvature flow:

$$\frac{\beta(\nu)}{\gamma(\nu)}V = -H_\gamma - g. \qquad (8.42)$$

Properly (8.38) should be written as the parabolic variational inequality

$$\epsilon \int_\Omega \beta(\nabla\varphi)\varphi_t(\eta - \varphi) + \epsilon \int_\Omega DA(\nabla\varphi) \cdot (\nabla\eta - \nabla\varphi) + \frac{1}{\epsilon}\int_\Omega W_0'(\varphi)(\eta - \varphi)$$

$$\ge c_W \int_\Omega \rho(\varphi)g(\eta - \varphi) \quad \text{for all } \eta \in \mathcal{K} = \{\eta \in \mathrm{H}^1(\Omega) : |\eta| \le 1\}, \qquad (8.43)$$

which is treated in the viscosity sense in Elliott and Schätzle (1997) because of the singularity in β at the origin.

Convergence
The phase field approximation of anisotropic interface motion can be established as in the isotropic case for smooth potential W (McFadden, Wheeler, Braun, Coriell and Sekerka 1993, Wheeler and McFadden 1996). Convergence of the double obstacle model to the unique viscosity solution of the anisotropic level set equation was proved in Elliott and Schätzle (1997) even in the case of kinetic anisotropy. The error bounds for smoothly

evolving flows are again $O(\epsilon^2)$ (Elliott and Schätzle 1996, Elliott, Paolini and Schätzle 1996).

Discretization of anisotropic phase field equation
The numerical approximation of (8.43) follows the approach used for the isotropic Allen–Cahn equation (Section 7.3). We use the same notation for the finite element spaces and time discretizations. In order to implement the method it is necessary to use a regularization β_ϵ of β.

Fully explicit time-stepping
The fully discrete approximation of (8.43) using explicit time-stepping reads as follows.

Algorithm 8. (Anisotropic double obstacle phase field) Let $\Phi^0 = I_h\varphi_0$. For $m = 0, \ldots, M - 1$, find $\Phi^{m+1} \in \mathcal{K}_h$ such that, for all $\chi \in \mathcal{K}_h$,

$$\epsilon(\beta_\varepsilon(\nabla\Phi^m)\partial\Phi^m, \chi - \Phi^{m+1})_h + \epsilon(DA(\nabla\Phi^m), \nabla\chi - \nabla\Phi^{m+1}) \qquad (8.44)$$

$$+ \frac{1}{\varepsilon}(W_0'(\Phi^m), \chi - \Phi^{m+1})_h - c_W(\rho(\Phi^m)g^m, \chi - \Phi^{m+1})_h \geq 0.$$

This scheme is as simple to implement as in the isotropic situation. Let M_β^m, K^m, M_ρ^m, and M_W^m be defined by

$$(M_\beta^m)_{ij} = (\beta_\varepsilon(\nabla\Phi^m)\chi_i, \chi_j)_h, \quad (K^m)_{ij} = (D^2A(\nabla\Phi^m)\nabla\chi_i, \nabla\chi_j),$$

$$(M_\rho^m)_j = c_W(\rho(\Phi^m)g^m, \chi_j)_h, \quad (M_W^m)_j = (W_0'(\Phi^m), \chi_j)_h$$

for $1 \leq i, j \leq N, 0 \leq m \leq M$. Here we made use of the fact that $DA(\nabla\Phi^m) = D^2A(\nabla\Phi^m)\nabla\Phi^m$, which follows from (8.4) and the fact that DA is homogeneous of degree 1. The variational formulation (8.44) is then equivalent to the following matrix formulation

$$M_\beta^m\Phi^{m+1/2} = \left(M_\beta^m - \tau K^m\right)\Phi^m + \frac{\tau}{\varepsilon}M_\rho^m - \frac{\tau}{\varepsilon^2}M_W^m \qquad (8.45)$$

and it remains to project $\Phi^{m+1/2}$ component-wise yielding $\Phi^{m+1} = \mathcal{P}\Phi^{m+1/2}$. The use of the mass lumping in (8.45), which diagonalizes M_β^m, is crucial to eliminate any iteration in solving (8.45).

Semi-implicit time-stepping scheme
A semi-implicit scheme is obtained by treating the gradient energy term implicitly, yielding the following method.

Algorithm 9. Let $\Phi^0 = I_h\varphi_0$. For $m = 0, \ldots, M - 1$, find $\Phi^{m+1} \in \mathcal{K}_h$ such that, for all $\chi \in \mathcal{K}_h$,

$$\epsilon(\beta_\varepsilon(\nabla\Phi^m)\partial\Phi^m, \chi - \Phi^{m+1})_h + \epsilon(DA(\nabla\Phi^{m+1}), \nabla\chi - \nabla\Phi^{m+1}) \qquad (8.46)$$

$$+ \frac{1}{\varepsilon}(W_0'(\Phi^m), \chi - \Phi^{m+1})_h - c_W(\rho(\Phi^m)g^m, \chi - \Phi^{m+1})_h \geq 0.$$

The algebraic problem is now a convex optimization problem with obstacle constraints.

These schemes are stable in the sense of satisfying energy norm bounds analogous to those enjoyed by the solution of the PDE. The stability constraints are analogous to those holding in the isotropic case. However, owing to the anisotropy in the discrete elliptic operator there is a lack of a comparison principle which has proved a barrier to proving convergence.

9. Fourth order flows

9.1. Surface diffusion

In this paragraph we study various ways to approximate surfaces which evolve according to surface diffusion, that is,

$$V = \Delta_\Gamma H_\gamma \quad \text{on } \Gamma(t). \tag{9.1}$$

Here, H_γ denotes the anisotropic mean curvature of the surface $\Gamma(t)$ as it was introduced in (8.15). This evolution has interesting geometrical properties: if $\Gamma(t)$ is a closed surface bounding a domain $\Omega(t)$, then the volume of $\Omega(t)$ is preserved and the weighted surface area of $\Gamma(t)$ decreases. At present, the existence and uniqueness theory for surface diffusion is limited to the isotropic case $\gamma(q) = |q|, q \in \mathbb{R}^{n+1}$. For example, it is known that for closed curves in the plane or closed surfaces in $\mathbb{R}^3$, balls are asymptotically stable subject to small perturbations: see Elliott and Garcke (1997) and Escher, Mayer and Simonett (1998). However, topological changes such as pinch-off are possible (Giga and Ito 1998, Mayer and Simonett 2000) and a one-dimensional graph may lose its graph property. An example of pinch-off is shown in Figure 9.1. We start with the axially symmetric initial surface given by

$$r_0(x) = 1 + 0.05(\sin(5.5x) + \sin(5x)), \quad x \in (0, 8\pi). \tag{9.2}$$

Pinch-off happens after a very long computation time. Note the different scaling of the x- and the r-axis. This example was first computed in Coleman, Falk and Moakher (1995).

9.2. Surface diffusion for axially symmetric surfaces

In applications one is interested in the stability of so-called whiskers, which are axially symmetric cylindrical bodies of small diameter with respect to their length: see Nichols and Mullins (1965) and Coleman, Falk and Moakher (1995). Let us consider an axially symmetric cylindrical body, whose boundary

$$\Gamma(t) = \{\mathbf{x} \in \mathbb{R}^3 \mid \mathbf{x} = (x, r(x,t)\cos\phi, r(x,t)\sin\phi), x \in [0,L], \phi \in [0,2\pi]\}$$

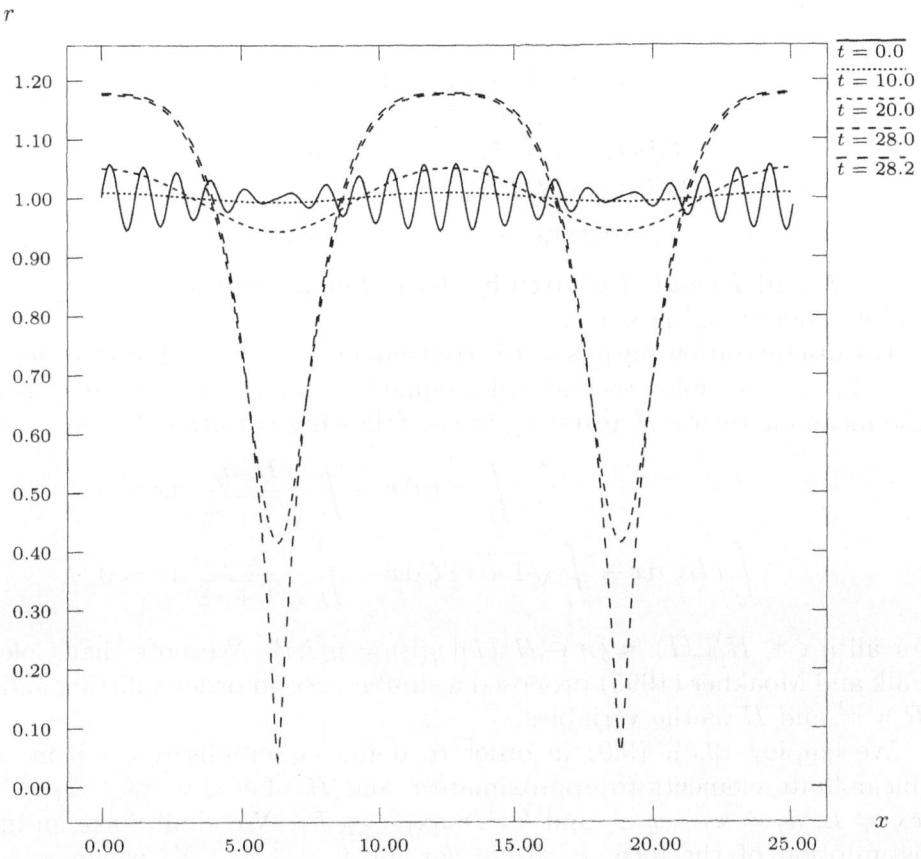

Figure 9.1. Evolution of the initial surface given by (9.2) for $t = 0.0$, 10.0, 20.0, 28.0 and 28.2.

evolves by $V = \Delta_\Gamma H$. We assume that the radius r is a smooth positive function, which is periodic in x, so that $r(0,t) = r(L,t)$. In these coordinates the mean curvature of $\Gamma(t)$ is

$$H = \frac{1}{r\sqrt{1+r_x^2}} - \frac{r_{xx}}{\sqrt{1+r_x^2}^3} = \frac{1}{r\sqrt{1+r_x^2}} - \left(\frac{r_x}{\sqrt{1+r_x^2}}\right)_x, \qquad (9.3)$$

while normal velocity and the surface Laplacian, respectively, are given by

$$V = \frac{r_t}{\sqrt{1+r_x^2}}, \qquad \Delta_\Gamma H = \frac{1}{r\sqrt{1+r_x^2}}\left(\frac{r H_x}{\sqrt{1+r_x^2}}\right)_x.$$

It follows from these two equations that r satisfies the quasilinear fourth order parabolic problem

$$r_t = \frac{1}{r}\left(\frac{rH_x}{\sqrt{1+r_x^2}}\right)_x \quad \text{in } I \times (0,T], \tag{9.4}$$

$$r(0,t) = r(L,t) \quad \text{in } (0,T], \tag{9.5}$$

$$H(0,t) = H(L,t) \quad \text{in } (0,T], \tag{9.6}$$

$$r(\cdot,0) = r_0 \quad \text{in } I, \tag{9.7}$$

where $I = (0,L)$ and H is given by (9.3). The initial function r_0 is assumed to be periodic and positive.

For discretization purposes it is convenient to split the fourth order problem into two coupled second order equations for the radial variable r and the mean curvature H resulting in the following variational form:

$$\int_I r r_t \eta \, \mathrm{d}x + \int_I \frac{r H_x \eta_x}{\sqrt{1+r_x^2}} \, \mathrm{d}x = 0 \tag{9.8}$$

$$\int_I r H \zeta \, \mathrm{d}x - \int_I \sqrt{1+r_x^2}\, \zeta \, \mathrm{d}x - \int_I \frac{r r_x \zeta_x}{\sqrt{1+r_x^2}} \, \mathrm{d}x = 0 \tag{9.9}$$

for all $\eta, \zeta \in H^1_{\mathrm{per}}(I) = \{\eta \in H^1(I) \mid \eta(0) = \eta(L)\}$. We note that Coleman, Falk and Moakher (1995) proposed a similar second order splitting and used $R = r^2$ and H as the variables.

We employ (9.8), (9.9) in order to define a semidiscrete scheme using linear finite elements to approximate r and H. Let $0 = x_0 < x_1 < \cdots < x_N = L$, $h_j = x_j - x_{j-1}$ and $h = \max_{1 \le j \le N} h_j$. We shall make an inverse assumption of the form $h \le \rho h_j$ for all $j = 1, \ldots, N$, where $\rho > 0$ is independent of h. The spatial discretization is based on piecewise linear finite elements,

$$X_h = \{\phi_h \in C^0(\bar{I}) \mid \phi_{h|[x_{j-1},x_j]} \in P^1, 1 \le j \le N, \phi_h(0) = \phi_h(L)\}.$$

Our discrete problem now reads: find $r_h, H_h : [0,T] \to X_h$ such that

$$\int_I r_h r_{h,t} \eta_h \, \mathrm{d}x + \int_I \frac{r_h H_{h,x} \eta_{h,x}}{\sqrt{1+r_{h,x}^2}} \, \mathrm{d}x = 0, \tag{9.10}$$

$$\int_I r_h H_h \zeta_h \, \mathrm{d}x - \int_I \sqrt{1+r_{h,x}^2}\, \zeta_h \, \mathrm{d}x - \int_I \frac{r_h r_{h,x} \zeta_{h,x}}{\sqrt{1+r_{h,x}^2}} \, \mathrm{d}x = 0 \tag{9.11}$$

for all $\eta_h, \zeta_h \in X_h$, $t \in [0,T]$ and with $r_h(0) = I_h r_0$, where I_h denotes the Lagrange interpolation operator. In Deckelnick, Dziuk and Elliott (2003a) a convergence analysis for the above scheme is carried out. The principal results are error bounds for position and mean curvature as described in the following theorem.

Theorem 9.1.   Let us assume that (9.4)–(9.7) has a sufficiently smooth positive solution on a maximal time interval $[0, T_{\max})$. Then the discrete solution (r_h, H_h) exists on $[0, T]$ for all $T < T_{\max}$ and there is an $h_0 > 0$ such that, for all $0 < h \leq h_0$,

$$\sup_{(0,T)} \|r - r_h\|_{H^1(I)}^2 + \int_0^T \|H - H_h\|_{H^1(I)}^2 \, dt \leq Ch^2. \tag{9.12}$$

9.3. Surface diffusion for graphs

The anisotropic surface diffusion (1.8) of a graph $\Gamma(t) = \{(x, u(x,t)) \,|\, x \in \Omega\}$ sitting above a domain $\Omega \subset \mathbb{R}^n$ leads to a highly nonlinear fourth order geometric partial differential equation. For graphs the Laplace–Beltrami operator applied to anisotropic mean curvature H_γ reads

$$\Delta_\Gamma H_\gamma = \frac{1}{Q} \nabla \cdot \left(Q \left(I - \frac{\nabla u \otimes \nabla u}{Q^2} \right) \nabla H_\gamma \right), \tag{9.13}$$

where we have again written $Q = \sqrt{1 + |\nabla u|^2}$. Recalling (8.17) as well as $V = -\frac{u_t}{Q}$, we see that (1.8) for graphs is equivalent to the partial differential equation

$$u_t + \nabla \cdot \left(Q \left(I - \frac{\nabla u}{Q} \otimes \frac{\nabla u}{Q} \right) \nabla \left(\sum_{i,j=1}^n \gamma_{p_i p_j} (\nabla u, -1) u_{x_i x_j} \right) \right) = 0. \tag{9.14}$$

As in the previous section, it is convenient to split the fourth order problem into two second order problems as follows:

$$u_t = \nabla \cdot \left(Q \left(I - \frac{\nabla u}{Q} \otimes \frac{\nabla u}{Q} \right) \nabla w \right), \tag{9.15}$$

$$w = -\sum_{i,j=1}^n \gamma_{p_i p_j} (\nabla u, -1) u_{x_i x_j}. \tag{9.16}$$

The system is closed using Neumann boundary conditions and an initial condition for u:

$$Q \left(I - \frac{\nabla u}{Q} \otimes \frac{\nabla u}{Q} \right) \nabla w \cdot \nu_{\partial \Omega} = 0, \tag{9.17}$$

$$D\gamma(\nabla u, -1) \cdot (\nu_{\partial \Omega}, 0) = 0, \tag{9.18}$$

$$u(\cdot, 0) = u_0. \tag{9.19}$$

The first equation, (9.17), is the zero mass flux condition, whereas the second equation, (9.18), is the natural variational boundary condition which defines w as the variational derivative or chemical potential for the surface energy functional. Note that an initial condition on w is not required. The problem (9.15)–(9.18) can easily be rewritten in variational form, namely

$$\int_\Omega u_t \eta + \int_\Omega Q \left(I - \frac{\nabla u}{Q} \otimes \frac{\nabla u}{Q} \right) \nabla w \cdot \nabla \eta = 0, \tag{9.20}$$

$$\int_\Omega w \psi - \sum_{j=1}^n \int_\Omega \gamma_{p_j}(\nabla u, -1) \psi_{x_j} = 0 \tag{9.21}$$

for all $\eta, \psi \in H^1(\Omega), t \in (0, T]$.

Replacing $H^1(\Omega)$ by the space X_h of piecewise linear finite elements we immediately arrive at a natural way to discretize in space. A finite element error analysis for the resulting semidiscrete scheme in the isotropic case was carried out by Bänsch, Morin and Nochetto (2004) for graphs in arbitrary space dimension. The time discretization follows the ideas of the discretization techniques introduced in Algorithm 7 and Theorem 8.7, leading to the following.

Algorithm 10. (Anisotropic surface diffusion of graphs) Let $\tau > 0$ be the time-step size with $M\tau = T$ and assume that $\lambda > 0$ is as in Theorem 8.7. Let the initial value $u_{h0} \in X_h$. For $m = 1, \ldots, M$, compute $u_h^{m+1}, w_h^{m+1} \in X_h$ such that

$$\frac{1}{\tau} \int_\Omega (u_h^{m+1} - u_h^m)\eta_h + \int_\Omega Q_h^m \left(I - \frac{\nabla u_h^m}{Q_h^m} \otimes \frac{\nabla u_h^m}{Q_h^m} \right) \nabla w_h^{m+1} \cdot \nabla \eta_h = 0,$$

$$\int_\Omega w_h^{m+1} \psi_h - \lambda \int_\Omega \frac{\gamma(\nabla u_h^m, -1)}{(Q_h^m)^2} \nabla(u_h^{m+1} - u_h^m) \cdot \nabla \psi_h$$

$$- \sum_{i=1}^n \int_\Omega \gamma_{p_i}(\nabla u_h^m, -1) \psi_{hx_i}$$

$$- \tau \int_\Omega Q_h^m \left(I - \frac{\nabla u_h^m}{Q_h^m} \otimes \frac{\nabla u_h^m}{Q_h^m} \right) \nabla(u_h^{m+1} - u_h^m) \cdot \nabla \psi_h = 0$$

for all discrete test functions $\eta_h, \psi_h \in X_h$. Here $Q_h^m = \sqrt{1 + |\nabla u_h^m|^2}$.

Note that in each time-step a linear system of equations has to be solved. An error analysis for this scheme is carried out in Deckelnick, Dziuk and Elliott (2003*b*).

Theorem 9.2.   Let u be a sufficiently smooth solution of anisotropic surface diffusion (9.14), (9.17)–(9.19) on the domain $\Omega \times (0, T)$ and set $w = -H_\gamma$. Let X_h be the space of continuous piecewise linear finite elements.

Figure 9.2. Anisotropic surface diffusion with a very strong anisotropy. Level lines are shown for the time-steps $0, 10, 200$ and a view from a position vertically above the graph for time-step 300.

Then for the discrete solution u_h^m, w_h^m $(m = 1, \ldots, M)$ we have the error estimates

$$\max_{m=1,\ldots,M} \|u^m - u_h^m\|_{L^2(\Omega)}^2 + \tau \sum_{m=1}^{M} \|w^m - w_h^m\|_{L^2(\Omega)}^2 \le c(\tau^2 + h^2),$$

$$\max_{m=1,\ldots,M} \int_{\Omega} \frac{|\nabla(u^m - u_h^m)|^2}{Q_h^m} + \tau \sum_{m=1}^{M} \int_{\Omega} \frac{|\nabla(w^m - w_h^m)|^2}{Q_h^{m-1}} \le c(\tau^2 + h^2).$$

Here $u^m = u(\cdot, m\tau)$, $w^m = w(\cdot, m\tau)$.

The proof uses ideas that were developed for the motion of graphs by anisotropic mean curvature. There is neither a restriction on the space dimension nor a coupling of time-step size and grid size. In two dimensions inverse estimates yield $L^\infty((0,T), H^1(\Omega))$-convergence for u and convergence in $L^2((0,T), H^1(\Omega))$ for w.

In Figure 9.2 we show computational results for anisotropic surface diffusion of a graph. The anisotropy is chosen to be a regularized l^1 norm (see Figure 8.1),

$$\gamma(p) = \sum_{j=1}^{3} \sqrt{p_j^2 + \varepsilon^2 |p|^2} \tag{9.22}$$

with $\varepsilon = 10^{-3}$. Thus the Frank diagram is a smoothed octahedron and the Wulff shape is a smoothed cube. The initial data were taken to depend on three random numbers $r_1, r_2, r_3 \in (0,1)$,

$$u_0(x) = 0.25 \left(\sin\left(2\pi r_1 x_1 \right) + 0.25 \sin\left(3\pi r_2 x_2 \right) \right) \times$$
$$\left(0.1 \sin\left(2\pi r_3 x_1 \right) + \sin\left(5\pi r_1 x_2 \right) \right) \sin\left(2\pi r_2 x_1 x_2 \right). \tag{9.23}$$

We used Neumann boundary conditions and the right-hand side (for the curvature equation) $f = 1 - x_1^2 - x_2^2$. The level lines for some time-steps are shown in Figure 9.2. The Wulff shape (a smooth cube) appears in the solution as a consequence of the right-hand side f. For more computational results we refer to Deckelnick, Dziuk and Elliott (2003b).

9.4. Phase field model for surface diffusion

Just as the phase field model for mean curvature flow is gradient flow for the gradient energy functional and leads to a second order parabolic equation, a phase field model for surface diffusion may also be based on the same energy functional and a suitable approximation of the Laplace–Beltrami operator leading to a nonlinear degenerate fourth order parabolic equation. The appropriate setting is in the context of the Cahn–Hilliard equation for phase separation in binary alloys. The phase function φ may be viewed as the difference in mass fractions of the two species so that the values $\varphi = \pm 1$ are associated with the pure materials. Stable phases of the alloy are then associated with the minima of a double well bulk energy W, which in the regular solution form is

$$W(\varphi) = \frac{\theta}{2}[(1+\varphi)\ln[1+\varphi] + (1-\varphi)\ln[1-\varphi]] + \frac{1}{2}(1-\varphi^2).$$

This homogeneous free energy function is non-convex with a double well, for $|\theta| < 1$, and $W'(\varphi)$ is said to be the homogeneous chemical potential.

The Cahn–Hilliard gradient energy functional is then

$$E(\varphi) = \int_\Omega \left[\frac{\varepsilon}{2} |\nabla \varphi|^2 + \frac{1}{\varepsilon} W(\varphi) \right]. \tag{9.24}$$

The functional derivative of this energy is used to define the chemical potential

$$w = -\epsilon \Delta \varphi + \frac{W'(\varphi)}{\epsilon}. \tag{9.25}$$

Mass conservation is

$$\partial_t \varphi + \nabla \cdot \mathcal{J} = 0, \tag{9.26}$$

where $\mathcal{J}$ is the mass flux and typically for diffusion

$$\mathcal{J} = -\mathcal{M}(\varphi) \nabla w \tag{9.27}$$

with the degenerate mobility $\mathcal{M}(\varphi) = 1 - \varphi^2$. The upshot is a fourth order Cahn–Hilliard equation with degenerate mobility. Interface asymptotics (Cahn, Elliott and Novick-Cohen 1996) show that, as $\theta(\epsilon)$ and ϵ tend to zero, the zero level set of φ approximates a surface evolving by surface diffusion. Computational results in a setting which includes a forcing due to an electric field may be found in Barrett, Nürnberg and Styles (2004).

9.5. Willmore flow

Our starting point is the Willmore functional

$$E(X) = \frac{1}{2} \int_\Gamma H^2 \, \mathrm{d}A, \quad \Gamma = X(M), \tag{9.28}$$

where M is an n-dimensional reference manifold and $X : M \to \mathbb{R}^{n+1}$ is a smooth immersion. Considering variations $X_\epsilon(p) = X(p) + \epsilon \phi(p), p \in M$, where $\phi : M \to \mathbb{R}^{n+1}$ is smooth and vanishes near ∂M, one obtains the formula

$$\langle E'(X), \phi \rangle = \frac{\mathrm{d}}{\mathrm{d}\epsilon} E(X_\epsilon)_{|\epsilon=0} \tag{9.29}$$

$$= \int_\Gamma \Delta_\Gamma X \cdot (\Delta_\Gamma \phi + 2\nu \nabla_\Gamma \nu \cdot \nabla_\Gamma \phi) + \frac{1}{2} \int_\Gamma H^2 \nabla_\Gamma X \cdot \nabla_\Gamma \phi$$

$$= \int_\Gamma \nabla_\Gamma (H\nu) \cdot \nabla_\Gamma \phi - 2 \int_\Gamma H \nabla_\Gamma \nu \cdot \nabla_\Gamma \phi + \frac{1}{2} \int_\Gamma H^2 \nabla_\Gamma X \cdot \nabla_\Gamma \phi,$$

where we have used (2.10). Note that $\nabla_\Gamma X \cdot \nabla_\Gamma \phi = \sum_{j,k=1}^{n+1} \underline{D}_j X_k \underline{D}_j \phi_k$. Willmore flow then arises as the L^2-gradient flow of the Willmore functional, that is,

$$\int_\Gamma X_t \cdot \phi \, \mathrm{d}A = -\langle E'(X), \phi \rangle. \tag{9.30}$$

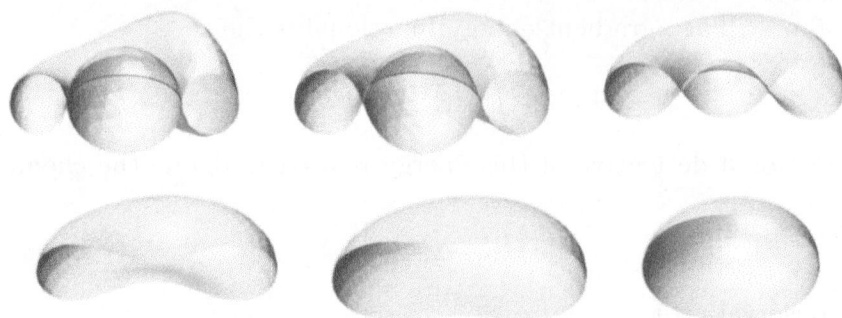

Figure 9.3. Surface relaxing under Willmore flow, cut open at $x_2 = 0$.

Using integration by parts one obtains the nonlinear evolution equation of fourth order, that is,

$$X_t = \Delta_\Gamma(H\nu) - 2\nabla_\Gamma \cdot (H\nabla_\Gamma\nu) + H\nabla_\Gamma H - \frac{1}{2}H^3\nu. \qquad (9.31)$$

If we take the scalar product of the above expression with ν and observe that $\Delta_\Gamma\nu \cdot \nu = -|\nabla_\Gamma\nu|^2$, we obtain the evolution law

$$V = \Delta_\Gamma H + H|\nabla_\Gamma\nu|^2 - \frac{1}{2}H^3 \quad \text{on } \Gamma(t). \qquad (9.32)$$

Note that from Section 2.3 we have

$$|\nabla_\Gamma\nu|^2 = \sum_{j,k=1}^{n+1} (\underline{D}_j\nu_k)^2 = \kappa_1^2 + \cdots + \kappa_n^2.$$

For two-dimensional surfaces Γ we then have

$$|\nabla_\Gamma\nu|^2 = \kappa_1^2 + \kappa_2^2 = (\kappa_1 + \kappa_2)^2 - 2\kappa_1\kappa_2 = H^2 - 2K$$

with Gauss curvature K. This leads to the evolution law

$$V = \Delta_\Gamma H + \frac{1}{2}H^3 - 2KH. \qquad (9.33)$$

Compared with the surface diffusion problem (1.7) additional dimension-dependent nonlinearities appear. Up to now analytical results for the above evolution law have been primarily obtained for the case of closed surfaces. In Simonett (2001) it is shown that a unique local solution of (9.32) satisfying $\Gamma(0) = \Gamma_0$ exists provided that Γ_0 is a compact closed immersed and orientable $C^{2,\alpha}$-surface in $\mathbb{R}^3$. The solution exists globally in time if Γ_0 is sufficiently close to a sphere in $C^{2,\alpha}$. Using different methods, Kuwert and Schätzle (2004a) obtain global existence of solutions provided that $\int_{\Gamma_0} |A^\circ|^2$ is sufficiently small, where A° denotes the trace-free part of the second fundamental form. They were subsequently able to remove the smallness assumption and to prove the existence of a global smooth solution provided that $E(X_0) \le 16\pi$, where $\Gamma_0 = X_0(S^2)$ (see Kuwert and Schätzle (2004b) and note that our definition differs from theirs by a factor of 2). There is numerical evidence (Mayer and Simonett 2002) that the above condition is optimal in the sense that the flow develops a singularity if the initial surface has energy greater that 16π. A major problem in the numerical solution of this problem is the treatment of Gauss curvature, which is a nonlinear expression of the principal curvatures and – contrary to mean curvature – is not easily accessible to variational methods.

The elastic flow of curves

Let us start with the one-dimensional parametric problem. The Bernoulli model of an elastic rod (Truesdell 1983) described by a closed curve $X :$ $S^1 \to \mathbb{R}^2$ uses the curvature integral (9.28) as elastic energy. Since this energy can be minimized by scaling, one usually adds length multiplied by a parameter $\lambda > 0$ resulting in the functional

$$E_\lambda(X) = \frac{1}{2} \int_\Gamma H^2 \, \mathrm{d}s + \lambda \int_\Gamma 1 \, \mathrm{d}s.$$

Let us introduce $Y = H\nu$, where H is just the usual curvature of a curve. We then obtain from (9.29) and the Frenet formula $\nabla_\Gamma \nu = H\tau \otimes \tau$ (with the unit tangent τ)

$$\langle E_\lambda'(X), \phi \rangle = \int_\Gamma \nabla_\Gamma Y \cdot \nabla_\Gamma \phi \, \mathrm{d}s - 2 \int_\Gamma H \nabla_\Gamma \nu \cdot \nabla_\Gamma \phi$$

$$+ \frac{1}{2} \int_\Gamma H^2 \nabla_\Gamma X \cdot \nabla_\Gamma \phi + \lambda \int_\Gamma Y \cdot \phi$$

$$= \int_\Gamma \nabla_\Gamma Y \cdot \nabla_\Gamma \phi - \frac{3}{2} \int_\Gamma H^2 \nabla_\Gamma X \cdot \nabla_\Gamma \phi + \lambda \int_\Gamma Y \cdot \phi.$$

Thus one may expect the gradient flow for E_λ to be given by the equation

$$X_t = \Delta_\Gamma Y - \frac{3}{2} \nabla_\Gamma \cdot (H^2 \nabla_\Gamma X) - \lambda Y. \tag{9.34}$$

Long time existence for this problem has been proved by Polden (1996). Just as in Section 4.1 we think of X as a mapping from $\mathbb{R} \times [0, T)$ into $\mathbb{R}^2$. We then have the following system to be satisfied by X and Y:

$$X_t - \frac{1}{|X_\theta|} \left(\frac{Y_\theta}{|X_\theta|} \right)_\theta + \frac{3}{2} \frac{1}{|X_\theta|} \left(|Y|^2 \frac{X_\theta}{|X_\theta|} \right)_\theta + \lambda Y = 0, \tag{9.35}$$

$$Y + \frac{1}{|X_\theta|} \left(\frac{X_\theta}{|X_\theta|} \right)_\theta = 0 \tag{9.36}$$

in $[0, 2\pi] \times (0, T)$. In addition, X has to satisfy the initial condition $X(\cdot, 0) = X_0$ in $I = [0, 2\pi]$ and the periodicity condition $X(\theta, t) = X(\theta + 2\pi, t)$ for

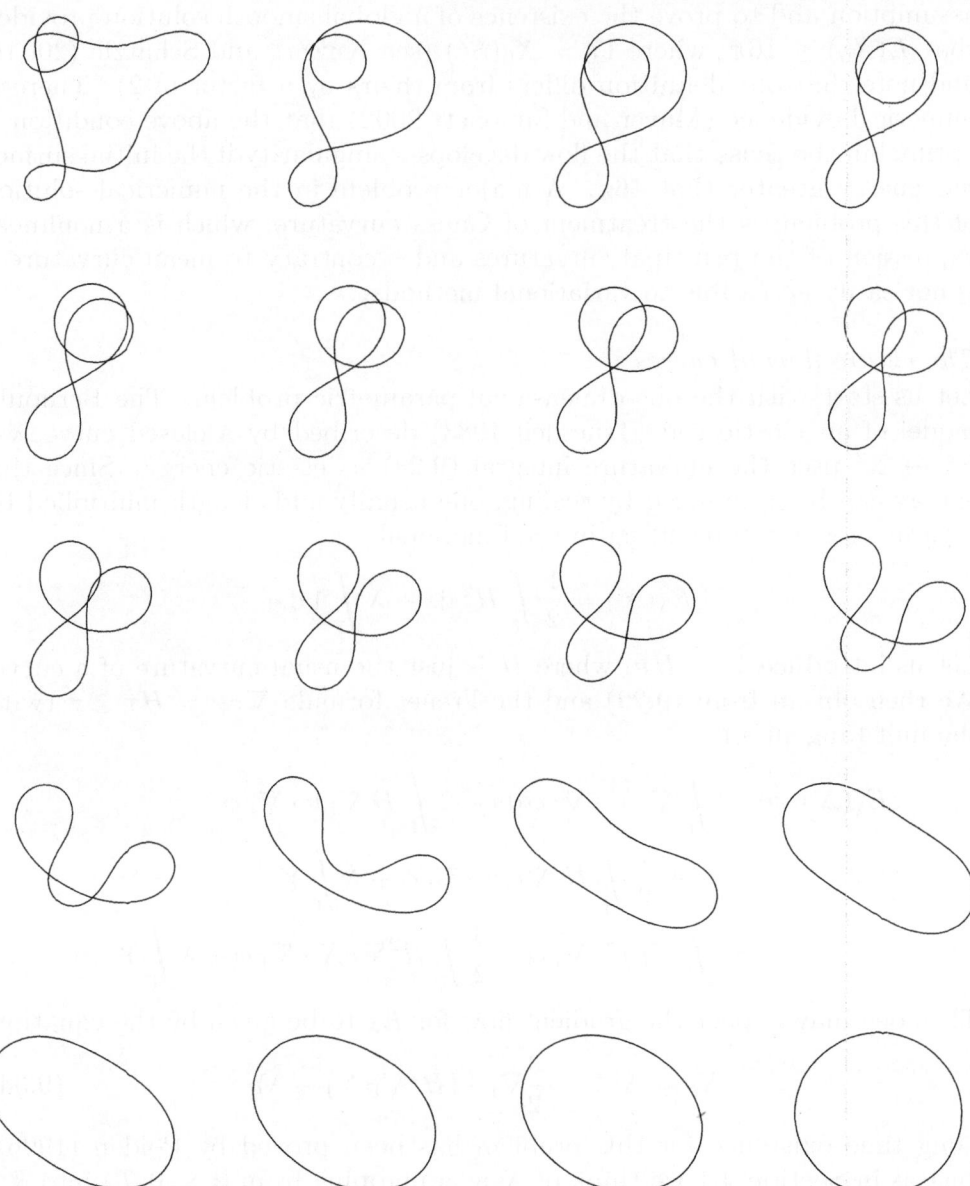

Figure 9.4. Time series of the two-dimensional
length-preserving elastic flow (graphically scaled).

$0 \leq t < T$, $\theta \in \mathbb{R}$. As in the derivation of (4.6) we obtain a variational formulation of (9.35), (9.36),

$$\int_I X_t \cdot \varphi |X_\theta| + \int_I \frac{Y_\theta \cdot \varphi_\theta}{|X_\theta|} - \frac{3}{2} \int_I |Y|^2 \frac{X_\theta \cdot \varphi_\theta}{|X_\theta|} + \lambda \int_I Y \cdot \varphi |X_\theta| = 0,$$

$$\int_I Y \cdot \psi |X_\theta| - \int_I \frac{X_\theta \cdot \psi_\theta}{|X_\theta|} = 0$$

for all test functions $\varphi, \psi \in H^1_{\text{per}}([0, 2\pi]; \mathbb{R}^2)$. We use this weak form of our problem for a finite element discretization in space, which in one space dimension leads to a suitable difference scheme. The derivation of this scheme follows the derivation of (4.9), additionally using mass lumping in both equations. Let us denote by $\phi_1, \ldots, \phi_N$ the basis of the finite element space S_h introduced in Section 4.1. Then, expanding $X_h(\theta, t) = \sum_{j=1}^N X_j(t)\phi_j(\theta)$, $Y_h(\theta, t) = \sum_{j=1}^N Y_j(t)\phi_j(\theta)$ with vectors $X_j(t), Y_j(t) \in \mathbb{R}^2$, yields the following system of $2N$ ordinary differential equations:

$$\frac{1}{2}(q_j + q_{j+1})(\dot{X}_j + \lambda Y_j) + \frac{Y_j - Y_{j-1}}{q_j} - \frac{Y_{j+1} - Y_j}{q_{j+1}} \tag{9.37}$$

$$- p_j \frac{X_j - X_{j-1}}{q_j} + p_{j+1} \frac{X_{j+1} - X_j}{q_{j+1}} = 0$$

$$\frac{1}{2}(q_j + q_{j+1})Y_j - \frac{X_j - X_{j-1}}{q_j} + \frac{X_{j+1} - X_j}{q_{j+1}} = 0 \tag{9.38}$$

$(j = 1, \ldots, N)$, where $X_0 = X_N, X_{N+1} = X_1, Y_0 = Y_N, Y_{N+1} = Y_1$, and the initial values are given by $X_j(0) = X_0(\theta_j)(j = 1, \ldots, N)$. Furthermore,

$$q_j = |X_j - X_{j-1}|, \qquad p_j = \frac{1}{2}\left(|Y_{j-1}|^2 + Y_{j-1} \cdot Y_j + |Y_j|^2\right). \tag{9.39}$$

A more detailed description can be found in Dziuk, Kuwert and Schätzle (2002). The paper actually treats curves in arbitrary codimension both showing long time existence of solutions as well as numerical examples. We include here a computation which shows the unravelling of a planar knotted curve under the length-preserving elastic flow in Figure 9.4.

Parametric Willmore flow of surfaces
The equation for Willmore flow of two-dimensional surfaces in $\mathbb{R}^3$ is much more difficult to treat. This is because Gauss curvature appears in the equation (9.33) for Willmore flow. Mean curvature H is given as a divergence expression (see (2.9)), so that in the discretization of parametric mean curvature flow, for example, we were able to formulate the mean curvature vector in a weak form, which then lead to a finite element scheme for parametric mean curvature flow. We were able to define the mean curvature vector of a polyhedron as a continuous and piecewise linear vector-valued

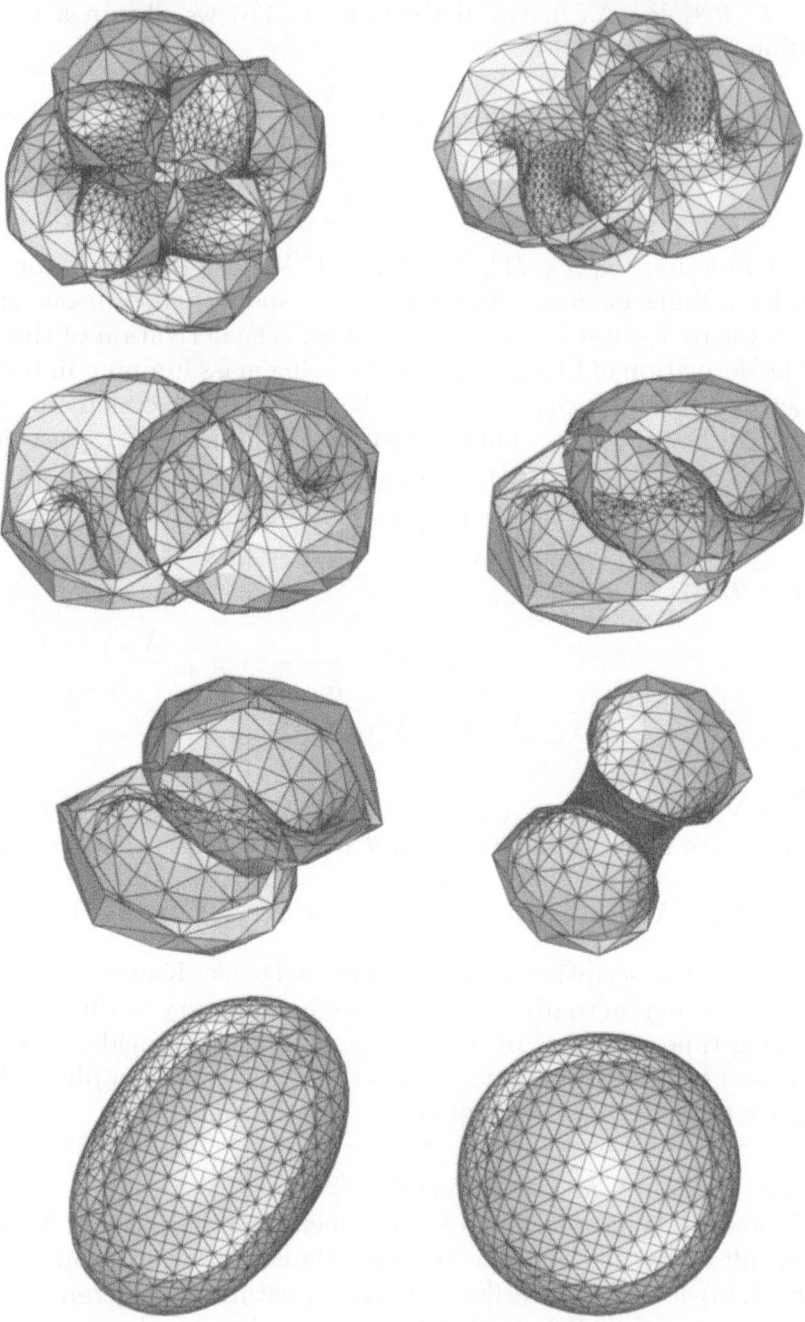

Figure 9.5. Half of a sphere eversion: the unravelling of a perturbed Willmore sphere under parametric Willmore flow (scaled graphically). Time-steps 0, 5600, 6000, 6400, 6800, 7000, 7200, 7400, 7600 and 8000.

function. Rusu (2001) employed a trick to remove Gauss curvature from the equations. Let us briefly describe the underlying idea, which we think is very important for applications of Willmore flow. For simplicity we look at closed surfaces. Going back to (9.29) and introducing the mean curvature vector $Y = H\nu$ as a new variable, we have

$$\langle E'(X), \phi \rangle = \int_\Gamma \nabla_\Gamma Y \cdot \nabla_\Gamma \phi - 2 \int_\Gamma Y \cdot \nu \, \nabla_\Gamma \nu \cdot \nabla_\Gamma \phi + \frac{1}{2} \int_\Gamma |Y|^2 \nabla_\Gamma X \cdot \nabla_\Gamma \phi.$$

Integration by parts gives

$$\int_\Gamma Y \cdot \nu \, \nabla_\Gamma \nu \cdot \nabla_\Gamma \phi = \int_\Gamma \left(\nabla_\Gamma Y \cdot \nabla_\Gamma \phi - (\nu \nabla_\Gamma Y) \cdot (\nu \nabla_\Gamma \phi) \right).$$

If we insert this identity into the above expression for $E'(X)$ we obtain

$$\langle E'(X), \phi \rangle = - \int_\Gamma R(\nu) \nabla_\Gamma Y \cdot \nabla_\Gamma \phi + \frac{1}{2} \int_\Gamma |Y|^2 \nabla_\Gamma X \cdot \nabla_\Gamma \phi$$

with the reflection matrix $R_{kl}(\nu) = \delta_{kl} - 2\nu_k \nu_l$. Starting from (9.30) we can now write down a variational formulation for parametric Willmore flow which uses position X and mean curvature vector Y as variables: find $X : M \times [0, T) \to \mathbb{R}^3$ such that

$$\int_\Gamma X_t \cdot \phi \; \mathrm{d}A - \int_\Gamma R(\nu) \nabla_\Gamma Y \cdot \nabla_\Gamma \phi \; \mathrm{d}A + \frac{1}{2} \int_\Gamma |Y|^2 \nabla_\Gamma X \cdot \nabla_\Gamma \phi \; \mathrm{d}A = 0,$$

$$\int_\Gamma Y \cdot \psi \; \mathrm{d}A - \int_\Gamma \nabla_\Gamma X \cdot \nabla_\Gamma \psi \; \mathrm{d}A = 0$$

for all test functions $\phi, \psi \in H^1(\Gamma)^3$. Here, $\Gamma = \Gamma(t) = X(M, t)$. Furthermore we require the initial condition $X(\cdot, 0) = X_0$. We observe that all quantities are well defined for a polyhedral surface Γ so that it is possible to use this formulation in order to approximate solutions by linear finite elements (see Rusu (2001) for more details and Clarenz, Diewald, Dziuk, Rumpf and Rusu (2004) for applications to problems in image restoration).

9.6. Willmore flow of graphs

If the two-dimensional surface $\Gamma = \{(x, u(x, t)) \mid x \in \Omega\}$ is a graph above some domain $\Omega \subset \mathbb{R}^2$, then we can directly derive a fourth order parabolic partial differential equation for u. We write the equation (9.33) for a graph. In order to write down this equation we note that the quantities V, H, K and $\Delta_\Gamma H$ appearing in (9.33) are expressed in terms of u as in (5.1) and

$$K = \frac{\det D^2 u}{Q^4}, \tag{9.40}$$

$$\Delta_\Gamma H = \frac{1}{Q} \nabla \cdot \left(Q \left(I - \frac{\nabla u}{Q} \otimes \frac{\nabla u}{Q} \right) \nabla H \right). \tag{9.41}$$

We can rewrite the last equation as

$$\Delta_\Gamma H = \nabla \cdot \left(\frac{1}{Q} \left(I - \frac{\nabla u \otimes \nabla u}{Q^2} \right) \nabla(QH) \right) - H \nabla \cdot \left(\frac{1}{Q} \left(I - \frac{\nabla u \otimes \nabla u}{Q^2} \right) \nabla Q \right).$$
(9.42)

With the expression (5.1) for H we conclude

$$\frac{1}{Q} \left(I - \frac{\nabla u \otimes \nabla u}{Q^2} \right) \nabla Q = \frac{1}{Q} \left(\nabla Q - \frac{\Delta u}{Q} \nabla u \right) + H \frac{\nabla u}{Q},$$
(9.43)

and a calculation shows that

$$\nabla \cdot \left(\frac{1}{Q} \left(\nabla Q - \frac{\Delta u}{Q} \nabla u \right) \right) = -2K.$$
(9.44)

Inserting (9.43) and (9.44) into (9.42), we obtain

$$\Delta_\Gamma H = \nabla \cdot \left(\frac{1}{Q} \left(I - \frac{\nabla u \otimes \nabla u}{Q^2} \right) \nabla(QH) \right) + 2HK - H \nabla \cdot \left(H \frac{\nabla u}{Q} \right)$$

$$= \nabla \cdot \left(\frac{1}{Q} \left(I - \frac{\nabla u \otimes \nabla u}{Q^2} \right) \nabla(QH) \right) + 2HK - \frac{1}{2} \nabla \cdot \left(\frac{H^2}{Q} \nabla u \right) - \frac{1}{2} H^3.$$

Comparing this expression with (9.33), we obtain a fourth order parabolic partial differential equation for u,

$$u_t + Q \nabla \cdot \left(\frac{1}{Q} \left(I - \frac{\nabla u \otimes \nabla u}{Q^2} \right) \nabla(QH) \right) - \frac{1}{2} Q \nabla \cdot \left(\frac{H^2}{Q} \nabla u \right) = 0 \quad \text{in } \Omega \times (0, T).$$
(9.45)

As before we can split the fourth order problem into two second order equations. The above equation suggests using the height u and

$$w = -QH$$

as variables which is different from the case of surface diffusion. Note that Gauss curvature no longer appears. The above ideas were introduced by Droske and Rumpf (2004) for a level set approach to Willmore flow.

The finite element approach is now based on dividing (9.45) by Q, multiplying by a test function $\varphi \in H_0^1(\Omega)$ and integrating by parts. This leads to

$$\int_\Omega \frac{u_t \varphi}{Q} + \int_\Omega \frac{1}{Q} \left(I - \frac{\nabla u \otimes \nabla u}{Q^2} \right) \nabla w \cdot \nabla \varphi + \frac{1}{2} \int_\Omega \frac{w^2}{Q^3} \nabla u \cdot \nabla \varphi = 0, \quad (9.46)$$

$$\int_\Omega \frac{w \zeta}{Q} - \int_\Omega \frac{\nabla u \cdot \nabla \zeta}{Q} = 0, \quad (9.47)$$

for all $\varphi, \zeta \in H_0^1(\Omega)$. As boundary conditions we choose

$$u = u_0 \quad \text{on } \partial\Omega \times [0, T] \cup \Omega \times \{0\}, \quad (9.48)$$

$$w = 0 \quad \text{on } \partial\Omega \times [0, T] \quad (9.49)$$

with a given function u_0, which is independent of time. For the error estimates we need the following regularity of the continuous solution:

$$\frac{\partial^k u}{\partial t^k} \in L^\infty((0,T); H^{4-2k,\infty}(\Omega)) \cap L^2((0,T); H^{5-2k}(\Omega)), \quad k = 0,1,2.$$
(9.50)

Thus we need high compatibility of initial and boundary data. The spatially discrete problem now reads as follows. Find $(u_h(t), w_h(t))$, $0 \le t \le T$, such that $u_h(t) - I_h u_0 \in X_{h0}$, $w_h(t) \in X_{h0}$, $u_h(0) = u_{0h} \in X_{h0}$ and

$$\int_\Omega \frac{u_{ht}\varphi h}{Q_h} + \int_\Omega \frac{1}{Q_h}\left(I - \frac{\nabla u_h \otimes \nabla u_h}{Q_h^2}\right)\nabla w_h \cdot \nabla \varphi_h$$

$$+ \frac{1}{2}\int_\Omega \frac{w_h^2}{Q_h^3}\nabla u_h \cdot \nabla \varphi_h = 0 \quad \text{for all } \varphi_h \in X_{h0},$$

$$\int_\Omega \frac{w_h \zeta_h}{Q_h} - \int_\Omega \frac{\nabla u_h \cdot \nabla \zeta_h}{Q_h} = 0 \quad \text{for all } \zeta_h \in X_{h0}.$$

The discrete initial value $u_h(\cdot, 0) = u_{h0} \in X_h$ is chosen as the 'minimal surface projection'

$$\int_\Omega \frac{\nabla u_{h0} \cdot \nabla \zeta_h}{\sqrt{1 + |\nabla u_{h0}|^2}} = \int_\Omega \frac{\nabla u_0 \cdot \nabla \zeta_h}{\sqrt{1 + |\nabla u_0|^2}} \quad \text{for all } \zeta_h \in X_{h0},$$
(9.51)

of the continuous initial value u_0.

Theorem 9.3. Let us assume that (9.45), (9.48), (9.49) has a unique solution u on the interval $[0, T]$, which satisfies (9.50). Also suppose that u_{0h} is defined as the projection (9.51) of u_0. Then

$$\sup_{0 \le t \le T} \|(u - u_h)(t)\| + \sup_{0 \le t \le T} \|(w - w_h)(t)\| \le ch^2 |\log h|^2,$$
(9.52)

$$\sup_{0 \le t \le T} \|\nabla(u - u_h)(t)\| \le ch,$$
(9.53)

$$\int_0^T \|u_t - u_{ht}\|^2 \, dt \le ch^4 |\log h|^4,$$
(9.54)

$$\int_0^T \|\nabla(w - w_h)\|^2 \, dt \le ch^2.$$
(9.55)

Appendix

Proof of Lemma 2.1. We prove (2.23), leaving (2.22) to the reader. Fix $t_0 \in (0, T)$. For $x \in \Gamma(t_0)$ let $U_x, \delta_x > 0$ and u be as in (2.19). By the implicit function theorem there exists an open set $\tilde{U}_x \subset U_x$, $0 < \tilde{\delta}_x \le \delta_x$ such that $\tilde{U}_x \times (t_0 - \tilde{\delta}_x, t_0 + \tilde{\delta}_x) \subset Q$ and $\tilde{U}_x \cap \Gamma(t)$ can be written as a graph over some open set $\Omega_x \subset \mathbb{R}^n$ for $|t - t_0| < \tilde{\delta}_x$. Since $\Gamma(t_0) \subset \cup_{x \in \Gamma(t_0)}\tilde{U}_x$ and

$\Gamma(t_0)$ is compact, there exist $a_1, \ldots, a_N$ with $\Gamma(t_0) \subset \cup_{j=1}^N \tilde{U}_j$, $\tilde{U}_j = \tilde{U}_{a_j}$. Let $Q_j = \tilde{U}_j \times (t_0 - \tilde{\delta}_j, t_0 + \tilde{\delta}_j)$ and let $\eta_j \in C_0^\infty(Q_j)$, $1 \leq j \leq N$, be a partition of unity which satisfies $\sum_{j=1}^N \eta_j(x, t) = 1$ for (x, t) in a neighbourhood of $\Gamma(t_0) \times \{t_0\}$. For t close to t_0 we then have

$$\frac{\mathrm{d}}{\mathrm{d}t} \int_{\Gamma(t)} g(x, t)\, \mathrm{d}A = \sum_{j=1}^N \frac{\mathrm{d}}{\mathrm{d}t} \int_{\tilde{U}_j \cap \Gamma(t)} \eta_j(x, t) g(x, t)\, \mathrm{d}A. \tag{9.56}$$

Let us fix $j \in \{1, \ldots, N\}$; by construction there exists $\Omega \subset \mathbb{R}^n$ and $v \in C^{2,1}(\Omega \times (t_0 - \tilde{\delta}_j, t_0 + \tilde{\delta}_j))$ such that w.l.o.g.

$$\tilde{U}_j \cap \Gamma(t) = \{(x', v(x', t)) \mid x' = (x_1, \ldots, x_n) \in \Omega\}.$$

Abbreviating $h = \eta_j g$, we have

$$\int_{\tilde{U}_j \cap \Gamma(t)} h(x, t)\, \mathrm{d}A = \int_\Omega h(x', v(x', t), t) \sqrt{1 + |\nabla_{x'} v|^2}\, \mathrm{d}x'$$

so that we obtain, for t close to t_0,

$$\frac{\mathrm{d}}{\mathrm{d}t} \int_{\tilde{U}_j \cap \Gamma(t)} h\, \mathrm{d}A = \int_\Omega \left(h_{x_{n+1}} v_t + h_t \right) \sqrt{1 + |\nabla_{x'} v|^2} + \int_\Omega h \frac{\nabla_{x'} v \cdot \nabla_{x'} v_t}{\sqrt{1 + |\nabla_{x'} v|^2}}$$

$$= \int_\Omega \left(h_{x_{n+1}} v_t + h_t \right) \sqrt{1 + |\nabla_{x'} v|^2} - \int_\Omega h \nabla_{x'} \cdot \left(\frac{\nabla_{x'} v}{\sqrt{1 + |\nabla_{x'} v|^2}} \right) v_t$$

$$- \int_\Omega \left(\nabla_{x'} h + h_{x_{n+1}} \nabla_{x'} v \right) \cdot \frac{\nabla_{x'} v}{\sqrt{1 + |\nabla_{x'} v|^2}} v_t$$

where we have used integration by parts observing that supp $h(\cdot, t) \subset \Omega$. Recalling that $\nu = \frac{(\nabla_{x'} v, -1)}{\sqrt{1 + |\nabla_{x'} v|^2}}$ and $H = \nabla_{x'} \cdot \left(\frac{\nabla_{x'} v}{\sqrt{1 + |\nabla_{x'} v|^2}} \right)$ we finally get

$$\frac{\mathrm{d}}{\mathrm{d}t} \int_{\tilde{U}_j \cap \Gamma(t)} h\, \mathrm{d}A = \int_{\tilde{U}_j \cap \Gamma(t)} \left(\frac{\partial h}{\partial \nu} V + h_t + h V H \right) \mathrm{d}A.$$

Note that the above identity has been derived under the implicit assumption that $\nabla_{x'} v_t$ exists; the general case can be justified with the help of an approximating argument. If we return to (9.56) and recall that $\sum_{j=1}^N \eta_j \equiv 1$ in a neighbourhood of $\Gamma(t_0) \times \{t_0\}$, we obtain (2.23) at $t = t_0$. $\square$

Acknowledgements

We would like to thank Michael Fried and Alfred Schmidt for providing figures of calculations for the Stefan problem with undercooling and Vanessa Styles for providing figures of phase field calculations. The work was supported by the Deutsche Forschungsgemeinschaft via DFG-Forschergruppe 'Nonlinear partial differential equations: Theoretical and numerical analysis' and via DFG-Graduiertenkolleg: 'Nichtlineare Differentialgleichungen: Modellierung, Theorie, Numerik, Visualisierung'. Partial support was provided by EPSRC/network research 'Computation and numerical analysis for multiscale and multiphysics modelling'. The graphical presentations were performed with the packages GRAPE and Xgraph.

REFERENCES

S. Allen and J. Cahn (1979), 'A microscopic theory for antiphase boundary motion and its application to antiphase domain coarsening', *Acta Metall.* **27**, 1084–1095.

R. Almgren (1993), 'Variational algorithms and pattern formation in dendritic solidification', *J. Comput. Phys.* **106**, 337–354.

S. J. Altschuler and M. A. Grayson (1992), 'Shortening space curves and flow through singularities', *J. Differential Geom.* **35**, 283–298.

S. Angenent (1991), 'Parabolic equations for curves on surfaces, Part II: Intersections, blow-up and generalized solutions', *Ann. of Math.* **133**, 171–215.

S. Angenent and M. Gurtin (1989), 'Multiphase thermomechanics with interfacial structure, 2: Evolution of an isothermal interface', *Arch. Rat. Mech. Anal.* **108**, 323–391.

G. Aubert and P. Kornprobst (2002), *Mathematical Problems in Image Processing*, Vol. 147 of *Applied Mathematical Sciences*, Springer.

E. Bänsch and A. Schmidt (2000), 'Simulation of dendritic crystal growth with thermal convection', *Interfaces Free Bound.* **2**, 95–115.

E. Bänsch, P. Morin and R. H. Nochetto (2004), 'Surface diffusion of graphs: Variational formulation, error analysis and simulation', *SIAM J. Numer. Anal.* **42**, 773–799.

J. W. Barrett, R. Nürnberg and V. Styles (2004), 'Finite element approximation of a phase field model for void electromigration', *SIAM J. Numer. Anal.* **42**, 738–772.

G. Bellettini and M. Paolini (1995), 'Quasi-optimal error estimates for the mean curvature flow with a forcing term', *Differential Integral Equations* **8**, 735–752.

G. Bellettini and M. Paolini (1996), 'Anisotropic motion by mean curvature in the context of Finsler geometry', *Hokkaido Math. J.* **25**, 537–566.

G. Bellettini, M. Novaga and M. Paolini (1999), 'Facet-breaking for three dimensional crystals evolving by mean curvature', *Interfaces Free Bound.* **1**, 39–55.

G. Bellettini, M. Novaga and M. Paolini (2001), 'Characterization of facet-breaking for non-smooth mean curvature flow in the convex case', *Interfaces Free Bound.* **3**, 415–446.

G. Bellettini, M. Paolini and C. Verdi (1990), 'Γ-convergence of discrete approximations to interfaces with prescribed mean curvature', *Rendconti Mat. Acc. Lincei* **9**, 317–328.

J. F. Blowey and C. M. Elliott (1991), 'The Cahn–Hilliard gradient theory for phase separation with non-smooth free energy, Part I: Mathematical Analysis', *European J. Appl. Math.* **2**, 233–280.

J. F. Blowey and C. M. Elliott (1993a), 'The Cahn–Hilliard gradient theory for phase separation with non-smooth free energy, Part II: Numerical Analysis', *European J. Appl. Math.* **3**, 147–179.

J. F. Blowey and C. M. Elliott (1993b), Curvature dependent phase boundary motion and parabolic obstacle problems, in *Proc. IMA Workshop on Degenerate Diffusion, 1991* (W. Ni, L. Peletier and J. L. Vasquez, eds) Springer, IMA **47**, pp. 19–60.

J. F. Blowey and C. M. Elliott (1994), A phase field model with a double obstacle potential, in *Motion by Mean Curvature and Related Topics* (G. Buttazzo and A. Visintin, ed.), de Gruyter, New York, pp. 1–22.

K. A. Brakke (1978), *The Motion of a Surface by its Mean Curvature*, Princeton University Press, NJ.

K. A. Brakke (1992), 'The surface evolver', *Exp. Math.* **1**, 141–165.

R. J. Braun, S. R. Coriell, G. B. McFadden and A. A. Wheeler (1993), 'Phase field models for anisotropic interfaces', *Phys. Rev. E* **48**, 2016–2024.

E. Burman and M. Picasso (2003), 'Anisotropic, adaptive finite elements for the computation of a solutal dendrite', *Interfaces Free Bound.* **5**, 103–127.

G. Caginalp (1986), 'An analysis of a phase field model of a free boundary', *Arch. Rat. Mech. Anal.* **92**, 206–245.

J. W. Cahn and J. E. Hilliard (1958), 'Free energy of a nonuniform system I: Interfacial free energy', *J. Chem. Phys.* **28**, 258–267.

J. W. Cahn and J. E. Taylor (1994), 'Surface motion by surface diffusion', *Acta Metall. Mater.* **42**, 1045–1063.

J. W. Cahn, C. M. Elliott and A. Novick-Cohen (1996), 'The Cahn–Hilliard equation with a concentration dependent mobility: motion by minus the Laplacian of the mean curvature', *European J. Appl. Math.* **7**, 287–301.

V. Caselles, R. Kimmel, G. Sapiro and C. Sbert (1997), 'Minimal surfaces: a geometric three dimensional segmentation approach', *Numer. Math.* **77**, 423–451.

A. Chambolle (2004), 'An algorithm for mean curvature motion', *Interfaces Free Bound.* **6**, 195–218.

X. Chen (1992), 'Generation and propagation of interface in reaction diffusion equations', *J. Differential Equations* **96**, 116–141.

X. Chen (1994), 'Spectrum for the Allen–Cahn, Cahn–Hilliard and phase field equations for generic interfaces', *Comm. Partial Differential Equations* **19**, 1371–1395.

X. Chen and C. M. Elliott (1994), 'Asymptotics for a parabolic double obstacle problem', *Proc. R. Soc. London A* **444**, 429–445.

X. Chen, C. M. Elliott, A. R. Gardiner and J. J. Zhao (1998), 'Convergence of numerical solutions to the Allen–Cahn equation', *Appl. Anal.* **69**, 47–56.

Y.-G. Chen, Y. Giga and S. Goto (1991), 'Uniqueness and existence of viscosity solutions of generalized mean curvature flow equations', *J. Differential Geom.* **33**, 749–786.

Y.-G. Chen, Y. Giga, Y. T. Hitaka and M. Honma (1994), A stable difference scheme for computing motion of level surfaces by the mean curvature, in *Proceedings of the Global Analysis Research Center Symposium, Seoul, Korea* (D. Kim *et al.*, eds), pp. 1–19.

D. L. Chopp (1993), 'Computing minimal surfaces via level set curvature flow', *J. Comput. Phys.* **106**, 77–91.

D. L. Chopp and J. A. Sethian (1999), 'Motion by intrinsic Laplacian of curvature', *Interfaces Free Bound.* **1**, 107–123.

U. Clarenz, U. Diewald, G. Dziuk, M. Rumpf and R. Rusu (2004), 'A finite element method for surface restoration with smooth boundary conditions', *Computer Aided Geometric Design* **21** (5), 427–445.

B. D. Coleman, R. S. Falk and M. Moakher (1995), 'Stability of cylindrical bodies in the theory of surface diffusion', *Phys. D* **89**, 123–135.

M. G. Crandall and P.-L. Lions (1996), 'Convergent difference schemes for nonlinear parabolic equations and mean curvature motion', *Numer. Math.* **75**, 17–41.

M. G. Crandall, H. Ishii and P.-L. Lions (1992), 'User's guide to viscosity solutions of second order partial differential equations', *Bull. Amer. Math. Soc.* **27**, 1–67.

K. Deckelnick (1997), 'Weak solutions of the curve shortening flow', *Calc. Var.* **5**, 489–510.

K. Deckelnick (2000), 'Error analysis for a difference scheme approximating mean curvature flow', *Interfaces Free Bound.* **2**, 117–142.

K. Deckelnick and G. Dziuk (1994), On the approximation of the curve shortening flow, in *Calculus of Variations, Applications and Computations: Pont-à-Mousson, 1994* (C. Bandle, J. Bemelmans, M. Chipot, J. Saint Jean Paulin and I. Shafrir, eds), Pitman Research Notes in Mathematics Series, pp. 100–108.

K. Deckelnick and G. Dziuk (1995), 'Convergence of a finite element method for non-parametric mean curvature flow', *Numer. Math.* **72**, 197–222.

K. Deckelnick and G. Dziuk (1999), 'Discrete anisotropic curvature flow of graphs', *Math. Model. Numer. Anal.* **33**, 1203–1222.

K. Deckelnick and G. Dziuk (2000), 'Error estimates for a semi implicit fully discrete finite element scheme for the mean curvature flow of graphs', *Interfaces Free Bound.* **2**, 341–359.

K. Deckelnick and G. Dziuk (2001), Convergence of numerical schemes for the approximation of level set solutions to mean curvature flow, in *Numerical Methods for Viscosity Solutions and Applications* (M. Falcone and C. Makridakis, eds), Vol. 59 of *Series Adv. Math. Appl. Sciences*, pp. 77–94.

K. Deckelnick and G. Dziuk (2002a), 'A fully discrete numerical scheme for weighted mean curvature flow', *Numer. Math.* **91**, 423–452.

K. Deckelnick and G. Dziuk (2002b), A finite element level set method for anisotropic mean curvature flow with space dependent weight, Abschlußband SFB 256, Bonn.

K. Deckelnick and C. M. Elliott (1998), 'Finite element error bounds for curve shrinking with prescribed normal contact to a fixed boundary', *IMA J. Numer. Anal.* **18**, 635–654.

K. Deckelnick and C. M. Elliott (2004), 'Uniqueness and error analysis for Hamilton–Jacobi equations with discontinuities', *Interfaces Free Bound.* **6**, 329–349.

K. Deckelnick, G. Dziuk and C. M. Elliott (2003a), 'Error analysis of a semidiscrete numerical scheme for diffusion in axially symmetric surfaces', *SIAM J. Numer. Anal.* **41**, 2161–2179.

K. Deckelnick, G. Dziuk and C. M. Elliott (2003b), Fully discrete semi-implicit second order splitting for anisotropic surface diffusion of graphs, Preprint No. 33/2003, Universität Magdeburg, *SIAM J. Numer. Anal.*, to appear.

K. Deckelnick, C. M. Elliott and V. M. Styles (2001), 'Numerical diffusion-induced grain boundary motion', *Interfaces Free Bound.* **3**, 393–414.

P. de Mottoni and M. Schatzman (1995), 'Geometric evolution of developed interfaces', *Trans. Amer. Math. Soc.* **347**, 1533–1589.

M. Droske and M. Rumpf (2004), 'A level set formulation for Willmore flow', *Interfaces Free Bound.* **6**, 361–378.

G. Dziuk (1988), Finite elements for the Beltrami operator on arbitrary surfaces, in *Partial Differential Equations and Calculus of Variations* (S. Hildebrandt and R. Leis, eds), Vol. 1357 of *Lecture Notes in Mathematics*, Springer, pp. 142–155.

G. Dziuk (1991), 'An algorithm for evolutionary surfaces', *Numer. Math.* **58**, 603–611.

G. Dziuk (1994), 'Convergence of a semi-discrete scheme for the curve shortening flow', *Math. Models Methods Appl. Sci.* **4**, 589–606.

G. Dziuk (1999a), Numerical schemes for the mean curvature flow of graphs, in *IUTAM Symposium on Variations of Domains and Free-Boundary Problems in Solid Mechanics* (P. Argoul, M. Frémond and Q. S. Nguyen, eds), Kluwer, Dordrecht/Boston/London, pp. 63–70.

G. Dziuk (1999b), 'Discrete anisotropic curve shortening flow', *SIAM J. Numer. Anal.* **36**, 1808–1830.

G. Dziuk, E. Kuwert and R. Schätzle (2002), 'Evolution of elastic curves in $\mathbb{R}^n$: existence and computation', *SIAM J. Math. Anal.* **33**, 1228–1245.

K. Ecker (2002) Lectures on regularity for mean curvature flow, Preprint No. 23/2002, Universität Freiburg.

K. Ecker and G. Huisken (1989), 'Mean curvature evolution of entire graphs', *Ann. of Math.* **130**, 453–471.

C. M. Elliott (1997), Approximation of curvature dependent interface motion, in *The State of the Art in Numerical Analysis* (I. S. Duff *et al.*, ed.), Vol. 63 of *Inst. Math. Appl. Conf. Ser., New Ser.*, Clarendon Press, Oxford, pp. 407–440.

C. M. Elliott and H. Garcke (1997), 'Existence results for diffusive surface motion laws', *Adv. Math. Sci. Appl.* **7**, 467–490.

C. M. Elliott and A. R. Gardiner (1996), Double obstacle phase field computations of dendritic growth, University of Sussex, CMAIA Research Report 96-15.

C. M. Elliott and J. R. Ockendon (1982), *Weak and Variational Methods for Moving Boundary Problems*, Pitman, London.

C. M. Elliott and R. Schätzle (1996), 'The limit of the anisotropic double-obstacle Allen–Cahn equation', *Proc. Roy. Soc. Edinburgh* **126**, 1217–1234.

C. M. Elliott and R. Schätzle (1997), The limit of the fully anisotropic double-obstacle Allen–Cahn equation in the non-smooth case, *SIAM J. Math. Anal.* **28**, 274–303.

C. M. Elliott and A. M. Stuart (1993), 'The global dynamics of discrete semilinear parabolic equations', *SIAM J. Numer. Anal.* **30**, 1622–1663.

C. M. Elliott and V. M. Styles (2003), 'Computations of bidirectional grain boundary dynamics in thin metallic films', *J. Comput. Phys.* **187**, 524–543.

C. M. Elliott, A. R. Gardiner and T. Kuhn (1996), Generalizes double obstacle phase field approximation of the anisotropic mean curvature flow, University of Sussex, CMAIA Research Report 96-17.

C. M. Elliott, A. R. Gardiner and R. Schätzle (1998), 'Crystalline curvature flow of a graph in a variational setting', *Adv. Math. Sci. Appl.* **8**, 425–460.

C. M. Elliott, M. Paolini and R. Schätzle (1996), 'Interface estimates for the fully anisotropic Allen–Cahn equation and anisotropic mean curvature flow', *Math. Models Methods Appl. Sci.* **6**, 1103–1118.

J. Escher, U. F. Mayer and G. Simonett (1998), 'The surface diffusion flow for immersed hypersurfaces', *SIAM J. Math. Anal.* **29**, 1419–1433.

L. C. Evans (1993), 'Convergence of an algorithm for mean curvature motion', *Indiana Univ. Math. J.* **42**, 533–557.

L. C. Evans and J. Spruck (1991), 'Motion of level sets by mean curvature I', *J. Differential Geom.* **33**, 636–681.

L. C. Evans and J. Spruck (1992a), 'Motion of level sets by mean curvature II', *Trans. Amer. Math. Soc.* **330**, 321–332.

L. C. Evans and J. Spruck (1992b), 'Motion of level sets by mean curvature III', *J. Geom. Anal.* **2**, 121–150.

L. C. Evans and J. Spruck (1995), 'Motion of level sets by mean curvature IV', *J. Geom. Anal.* **5**, 77–114.

L. C. Evans, H. M. Soner and P. E. Souganidis (1992), 'Phase transition and generalised motion by mean curvature', *Comm. Pure Appl. Math.* **45**, 1097–1123.

X. Feng and A. Prohl (2003), 'Numerical analysis of the Allen–Cahn equation and approximation for mean curvature flows', *Numer. Math.* **94**, 33–65.

F. Fierro, R. Goglione and M. Paolini (1994), 'Finite element minimization of curvature functionals with anisotropy', *Calcolo* **3–4**, 191–210.

M. Fried (1999), Niveauflächen zur Berechnung zweidimensionaler Dendrite, Dissertation Freiburg.

M. Fried (2004), 'A level set based finite element algorithm for the simulation of dendritic growth', *Comput. Vis. Sci.* **7** 97–110.

M. Gage (1993), 'Evolving plane curves by curvature in relative geometries', *Duke Math. J.* **72**, 441–466.

M. Gage and R. S. Hamilton (1996), 'The heat equation shrinking convex plane curves', *J. Differential Geom.* **23** 69–96.

H. Garcke and V. M. Styles (2004), 'Bi-directional diffusion induced grain boundary motion with triple junctions', *Interfaces Free Bound.* **6**, 271–294.

Y. Giga (2002), Surface evolution equations: a level set method, Hokkaido University Technical Report Series in Mathematics No. 71.

Y. Giga and K. Ito (1998), 'On pinching of curves moved by surface diffusion', *Commun. Appl. Anal.* **2**(3), 393–405.

D. Gilbarg and N. S. Trudinger (1998), *Elliptic Partial Differential Equations of Second Order*, Grundlehren der mathematischen Wissenschaften, Springer.

P. M. Girao (1995), 'Convergence of a crystalline algorithm for the motion of a simple closed convex curve by weighted curvature', *SIAM J. Numer. Anal.* **32**, 886–899.

M. A. Grayson (1987), 'The heat equation shrinks embedded plane curves to round points', *J. Differential Geom.* **26**, 285–314.

M. A. Grayson (1989) 'A short note on the evolution of a surface by its mean curvature', *Duke Math. J.* **58** (3), 555–558.

M. E. Gurtin (1993), *Thermomechanics of Evolving Phase Boundaries in the Plane*, Oxford Mathematical Monographs, Oxford University Press.

C. Herring (1951), Surface diffusion as a motivation for sintering, in *The Physics of Powder Metallurgy* (W. E. Kingston, ed.), McGraw-Hill, New York, pp. 143–179.

G. Huisken (1984), 'Flow by mean curvature of convex surfaces into spheres', *J. Differential Geom.* **20**, 237–266.

G. Huisken (1989), 'Non-parametric mean curvature evolution with boundary conditions', *J. Differential Equations* **77**, 369–378.

T. Ilmanen (1993), 'Convergence of the Allen–Cahn equation to Brakke's motion by mean curvature', *J. Differential Geom.* **38**, 417–461.

C. Johnson and V. Thom'ee (1975), 'Error estimates for a finite element approximation of a minimal surface', *Math. Comp.* **29**, 343–349.

M. A. Katsoulakis and A. T. Kho (2001), 'Stochastic curvature flows: asymptotic derivation, level set formulation and numerical experiments', *Interfaces Free Bound.* **3**, 265–290.

M. Kimura (1997), 'Numerical analysis of moving boundary problems using the boundary tracking method', *Japan J. Indust. Appl. Math.* **14**, 373–398.

R. Kornhuber and R. Krause (2003), On multi-grid methods for vector Allen–Cahn equations, in *Domain Decomposition Methods in Science and Engineering* (I. Herrera *et al.*, eds), UNAM, Mexico City, Mexico, pp. 307–314.

T. Kuhn (1996), Approximation of anisotropic and advected mean curvature flows by phase field models, DPhil thesis, University of Sussex.

T. Kuhn (1998), 'Convergence of a fully discrete approximation for advected mean curvature flows', *IMA J. Numer. Anal.* **18**,595–634.

E. Kuwert and R. Schätzle (2004a), 'The Willmore flow with small initial energy', *J. Differential Geom.* **57**, 409–441.

E. Kuwert and R. Schätzle (2004b), 'Removability of point singularities of Willmore surfaces', *Ann. of Math.* **159**, 1–43.

O. A. Ladyzhenskaya, V. A. Solonnikov and N. N. Ural'tseva (1968), *Linear and Quasilinear Equations of Parabolic Type*, AMS, Providence, RI.

G. Lieberman (1986), 'The first initial-boundary value problem for quasilinear second order parabolic equations', *Ann. Scuola Norm. Sup. Pisa* **13**, 347–387.

G. B. McFadden, A. A. Wheeler, R. J. Braun, S. R. Coriell and R. F. Sekerka (1993), 'Phase-field models for anisotropic interfaces', *Phys. Rev. E* **48** (3), 2016–2024.

U. F. Mayer and G. Simonett (2000), 'Self-intersections for the surface diffusion and the volume-preserving mean curvature flow', *Differential Integral Equations* **13**, 1189–1199.

U. F. Mayer and G. Simonett (2002), 'A numerical scheme for axi-symmetric solutions of curvature driven free boundary problems, with applications to the Willmore flow', *Interfaces Free Bound.* **4**, 89–109.

B. Merriman, J. K. Bence and S. Osher (1994), 'Motion of multiple junctions: A level set approach', *J. Comput. Phys.* **112**, 343–363.

K. Mikula and J. Kacur (1996), 'Evolution of convex plane curves describing anisotropic motions of phase interfaces', *SIAM J. Sci. Comput.* **17** (6), 1302–1327.

K. Mikula and D. Sevcovic (2001), 'Evolution of plane curves driven by a nonlinear function of curvature and anisotropy', *SIAM J. Appl. Math.* **61** (5), 1473–1501.

W. W. Mullins (1957), 'Theory of thermal grooving', *J. Appl. Phys.* **28**, 333–339.

F. A. Nichols and W. W. Mullins (1965), 'Surface-(interface-) and volume–diffusion contributions to morphological changes driven by capillarity', *Trans. Metall. Soc.*, AIME **233**, 1840–1847.

R. H. Nochetto and C. Verdi (1996a), 'Convergence of double obstacle problems to the geometric motions of fronts', *SIAM J. Math. Anal.* **26**, 1514–1526.

R. H. Nochetto and C. Verdi (1996b), 'Combined effect of explicit time-stepping and quadrature for curvature driven flows', *Numer. Math.* **74**, 105–136.

R. H. Nochetto and C. Verdi (1997), 'Convergence past singularities for a fully discrete approximation of curvature driven interfaces', *SIAM J. Numer. Anal.* **34**, 490–512.

R. H. Nochetto, M. Paolini and C. Verdi (1993), 'Sharp error analysis for curvature dependent evolving fronts', *Math. Models Methods Appl. Sci.* **3**, 771–723.

R. H. Nochetto, M. Paolini and C. Verdi (1994), 'Optimal interface error estimates for the mean curvature flow', *Ann. Scuola Norm. Sup. Pisa Cl. Sci.* **21**, 193–212.

R. H. Nochetto, M. Paolini and C. Verdi (1996), 'A dynamic mesh algorithm for curvature dependent evolving interfaces', *J. Comput. Phys.* **123**, 296–310.

V. I. Oliker and N. N. Uraltseva (1993), 'Evolution of nonparametric surfaces with speed depending on curvature II: The mean curvature case', *Comm. Pure Appl. Math.* **46**, No 1, 97–135.

S. Osher and R. Fedkiw (2003), *Level Set Methods and Dynamic Implicit Surfaces*, Vol. 153 of *Applied Mathematical Sciences*, Springer.

S. Osher and J. A. Sethian (1988), 'Fronts propagating with curvature dependent speed: Algorithms based on Hamilton–Jacobi formulations', *J. Comput. Phys.* **79**, 12–49.

M. Paolini (1995), An efficient algorithm for computing anisotropic evolution by mean curvature, in *Proceedings of the International Conference on Curvature Flows and Related Topics held in Levico, Italy, June 27–July 2nd, 1994* (A. Damlamian *et al.*, eds), Vol. 5 of *GAKUTO Int. Ser., Math. Sci. Appl.*, Gakkotosho, Tokyo, pp. 199–213.

M. Paolini (1997), 'A quasi-optimal error estimate for a discrete singularly perturbed approximation to the prescribed curvature problem', *Math. Comp.* **66**, 45–67.

M. Paolini and C. Verdi (1992), 'Asymptotic and numerical analysis of the mean curvature flow with a space dependent relaxation parameter', *Asymptot. Anal.* **5**, 553–574.

A. Polden (1996), Curves and surfaces of least total curvature and fourth-order flows, PhD dissertation, Universität Tübingen, Germany.

T. Preußer and M. Rumpf (2000), A level set method for anisotropic geometric diffusion in 3D image processing, Report SFB 256 Bonn, **37**.

J. Rubinstein, P. Sternberg and J. B. Keller (1989), 'Fast reaction, slow diffusion and curve shortening', *SIAM J. Appl. Math.* **49**, 116–133.

R. Rusu (2001), An algorithm for the elastic flow of surfaces, Preprint Math. Fak. Univ. Freiburg 01-35.

A. Schmidt (1996), 'Computation of three dimensional dendrites with finite elements', *J. Comput. Phys.* **125**, 293–312.

J. A. Sethian (1990), 'Numerical algorithms for propagating interfaces: Hamilton–Jacobi equations and conservation laws', *J. Differential Geom.* **31** (1), 131–161.

J. A. Sethian (1999), *Level Set Methods and Fast Marching Methods*, Vol. 3 of *Cambridge Monographs on Applied and Computational Mathematics*, Cambridge University Press.

G. Simonett (2001), 'The Willmore flow near spheres', *Differential Integral Equations* **14**, 1005–1014.

H. M. Soner (1993), 'Motion of a set by the curvature of its boundary', *J. Differential Equations* **101**, 313–372.

J. E. Taylor (1992), 'Mean curvature and weighted mean curvature', *Acta Metall. Mater.* **40**, 1475–1485.

J. E. Taylor, J. W Cahn and C. A. Handwerker (1992), 'Geometric models of crystal growth', *Acta Metall. Mater.* **40**, 1443–1474.

C. Truesdell (1983), 'The influence of elasticity on analysis: The classical heritage', *Bull. Amer. Math. Soc. (N.S.)* **9**, 293–310.

A. Veeser (1999), 'Error estimates for semi-discrete dendritic growth', *Interfaces Free Bound.* **1**, 227–254.

A. Visintin (1996), *Models of Phase Transitions*, Vol. 28 of *Progress in Nonlinear Differential Equations*, Birkhäuser, Boston.

N. J. Walkington (1996), 'Algorithms for computing motion by mean curvature', *SIAM J. Numer. Anal.* **33**, 2215–2238.

A. A. Wheeler and G. B. McFadden (1996), 'A ξ-vector formulation of anisotropic phase-field models: 3-D Asymptotics', *European J. Appl. Math.* **7**, 367–381.

T. J. Willmore (1993), *Riemannian Geometry*, Clarendon, Oxford.

E. Yokoyama and R. F. Sekerka (1992), 'A numerical study of the combined effect of anisotropic surface tension and interface kinetics on pattern formation during the growth of two-dimensional crystals', *J. Crystal Growth* **125**, 389–403.

Acta Numerica (2005), pp. 233–297
DOI: 10.1017/S0962492904000236

Random matrix theory

Alan Edelman

Department of Mathematics,
Massachusetts Institute of Technology,
Cambridge, MA 02139, USA
E-mail: edelman@math.mit.edu

N. Raj Rao

Department of Electrical Engineering and Computer Science,
Massachusetts Institute of Technology,
Cambridge, MA 02139, USA
E-mail: raj@mit.edu

Random matrix theory is now a big subject with applications in many disciplines of science, engineering and finance. This article is a survey specifically oriented towards the needs and interests of a numerical analyst. This survey includes some original material not found anywhere else. We include the important mathematics which is a very modern development, as well as the computational software that is transforming the theory into useful practice.

CONTENTS

1. Introduction

Texts on 'numerical methods' teach the computation of solutions to non-random equations. Typically we see integration, differential equations, and linear algebra among the topics. We find 'random' there too, but only in the context of random number generation.

The modern world has taught us to study stochastic problems. Already many articles exist on stochastic differential equations. This article covers topics in stochastic linear algebra (and operators). Here, the equations themselves are random. Like the physics student who has mastered the lectures and now must face the sources of randomness in the laboratory, numerical analysis is heading in this direction as well. The irony to newcomers is that often randomness imposes more structure, not less.

2. Linear systems

The limitations on solving large systems of equations are computer memory and speed. The speed of computation, however, is not only measured by clocking hardware; it also depends on numerical stability, and for iterative methods, on convergence rates. At this time, the fastest supercomputer performs Gaussian elimination, *i.e.*, solves $Ax = b$ on an n by n matrix A for $n \approx 10^6$. We can easily imagine $n \approx 10^9$ on the horizon. The standard benchmark HPL ('high-performance LINPACK') chooses A to be a random matrix with elements from a uniform distribution on $[-1/2, 1/2]$. For such large n, a question to ask would be whether a double precision computation would give a single precision answer.

Turning back the clock, in 1946 von Neumann and his associates saw $n = 100$ as the large number on the horizon. How do we pick a good test matrix A? This is where von Neumann and his colleagues first introduced the assumption of random test matrices distributed with elements from independent normals. Any analysis of this problem necessarily begins with an attempt to characterize the condition number $\kappa = \sigma_1/\sigma_n$ of the $n \times n$ matrix A. They give various 'rules of thumb' for κ when the matrices are so distributed. Sometimes these estimates are referred to as an expectation and sometimes as a bound that holds with high, though unspecified, probability. It is interesting to compare their 'rules of thumb' with what we now know about the condition numbers of such random matrices as $n \rightarrow \infty$ from Edelman (1989).

Quote. *For a 'random matrix' of order n the expectation value has been shown to be about n.*
(von Neumann 1963, p. 14)

Fact. $E[\kappa] = \infty$.

Quote. ... *we choose two different values of* κ, *namely* n *and* $\sqrt{10}n$.
(von Neumann 1963, p. 477)

Fact. $\Pr(\kappa < n) \approx 0.02, \Pr(\kappa < \sqrt{10}\,n) \approx 0.44$.

Quote. *With probability* ≈ 1, $\kappa < 10n$
(von Neumann and Goldstine 1947, p. 555)

Fact. $\Pr(\kappa < 10\,n) \approx 0.80$.

Results on the condition number have been extended recently by Edelman and Sutton (2004), and Azaïs and Wschebor (2004). Related results include the work of Viswanath and Trefethen (1998).

Analysis of Gaussian elimination of random matrices[1] began with the work of Trefethen and Schreiber (1990), and later Yeung and Chan (1997). Of specific interest is the behaviour of the 'growth factor' which influences numerical accuracy. More recently, Sankar, Spielman and Teng (2004) analysed the performance of Gaussian elimination using smoothed analysis, whose basic premise is that bad problems tend to be more like isolated spikes. Additional details can be found in Sankar (2003).

Algorithmic developers in need of guinea pigs nearly always take random matrices with standard normal entries, or perhaps close cousins, such as the uniform distribution of $[-1, 1]$. The choice is highly reasonable: these matrices are generated effortlessly and might very well catch programming errors. But are they really 'test matrices' in the sense that they can catch every type of error? It really depends on what is being tested; random matrices are not as random as the name might lead one to believe. Our suggestion to library testers is to include a carefully chosen range of matrices rather than rely on randomness. When using random matrices as test matrices, it can be of value to know the theory.

We want to convey is that random matrices are very *special* matrices. It is a mistake to link psychologically a random matrix with the intuitive notion of a 'typical' matrix or the vague concept of 'any old matrix'. In fact, the larger the size of the matrix the more predictable it becomes. This is partly because of the central limit theorem.

3. Matrix calculus

We have checked a few references on 'matrix calculus', yet somehow none were quite right for a numerical audience. Our motivation is twofold. Firstly, we believe that condition number information has not traditionally been widely available for matrix-to-matrix functions. Secondly, matrix

---

[1] On a personal note, the first author started down the path of random matrices because his adviser was studying Gaussian elimination on random matrices.

calculus allows us to compute Jacobians of familiar matrix functions and transformations.

Let $x \in \mathbb{R}^n$ and $y = f(x) \in \mathbb{R}^n$ be a differentiable vector-valued function of x. In this case, it is well known that the Jacobian matrix

$$J = \begin{pmatrix} \dfrac{\partial f_1}{\partial x_1} & \cdots & \dfrac{\partial f_1}{\partial x_n} \\ \vdots & & \vdots \\ \dfrac{\partial f_n}{\partial x_1} & \cdots & \dfrac{\partial f_n}{\partial x_n} \end{pmatrix} = \left(\frac{\partial f_i}{\partial x_j} \right)_{i,j=1,2,\dots,n} \tag{3.1}$$

evaluated at a point x approximates $f(x)$ by a linear function. Intuitively $f(x + \delta x) \approx f(x) + J\delta x$, i.e., J is the matrix that allows us to invoke first-order perturbation theory. The function f may be viewed as performing a change of variables. Often the matrix J is denoted df and 'Jacobian' refers to $\det J$. In the complex case, the Jacobian matrix is real $2n \times 2n$ in the natural way.

3.1. Condition numbers of matrix transformations

A matrix function/transformation (with no breakdown) can be viewed as a local linear change of variables. Let f be a (differentiable) function defined in the neighbourhood of a (square or rectangular) matrix A.

We think of functions such as $f(A) = A^3$ or $f(A) = \text{lu}(A)$, the LU factorization, or even the SVD or QR factorizations. The linearization of f is df which (like Kronecker products) is a linear operator on matrix space. For general A, df is $n^2 \times n^2$, but it is rarely helpful to write df explicitly in this form.

We recall the notion of condition number which we put into the context of matrix functions and factorizations. The condition number for $A \to f(A)$ is defined as

$$\begin{aligned} \kappa &= \frac{\text{relative change in } f(A)}{\text{relative change in } A} \\ &= \lim_{\substack{\epsilon \to 0 \\ \|E\|=\epsilon}} \sup \frac{\|f(A+E) - f(A)\|/\|f(A)\|}{\|E\|/\|A\|} \\ &= \|df\| \left(\frac{\|A\|}{\|f(A)\|} \right). \end{aligned}$$

Figure 3.1 illustrates the condition number to show that the key factor in the two-norm κ is related to the largest axis of an ellipsoid in the matrix factorization space, i.e., the largest singular value of df. The product of the semi-axis lengths is related to the volume of the ellipsoid and is the Jacobian determinant of f.

In summary

$$\kappa = \sigma_{\max}(\mathrm{d}f)\,\frac{\|A\|}{\|f(A)\|}, \tag{3.2}$$

$$J = \prod_i \sigma_i(\mathrm{d}f) = \det(\mathrm{d}f). \tag{3.3}$$

Example 1. Let $f(A) = A^2$ so that $\mathrm{d}f(E) = AE + EA$. This can be rewritten in terms of the Kronecker (or tensor) product operator $\otimes$ as $\mathrm{d}f = I \otimes A + A^T \otimes I$. Therefore

$$\kappa = \sigma_{\max}(I \otimes A + A^T \otimes I)\frac{\|A\|}{\|A^2\|}.$$

Recall that $A \otimes B : X \to BXA^T$ is the linear map from X to BXA^T. The Kronecker product has many wonderful properties, as described in the article by Van Loan (2000).

Example 2. Let $f(A) = A^{-1}$, so that $\mathrm{d}f(E) = -A^{-1}EA^{-1}$, or in terms of the Kronecker product operator as $\mathrm{d}f = -A^{-T} \otimes A^{-1}$.

This implies that the singular values of $\mathrm{d}f$ are $(\sigma_i(A)\sigma_j(A))^{-1}$, for $1 \le i, j \le n$.

The largest singular value $\sigma_{\max}(\mathrm{d}f)$ is thus equal to $1/\sigma_n(A)^2 = \|A^{-1}\|^2$ so that κ as defined in (3.2) is simply the familiar matrix condition number

$$\kappa = \|A\|\,\|A^{-1}\| = \frac{\sigma_1}{\sigma_n},$$

while in contrast, the Jacobian given by (3.3) is

$$\text{Jacobian} = \prod_{i,j} \frac{1}{\sigma_i(A)\sigma_j(A)} = (\det A)^{-2n}.$$

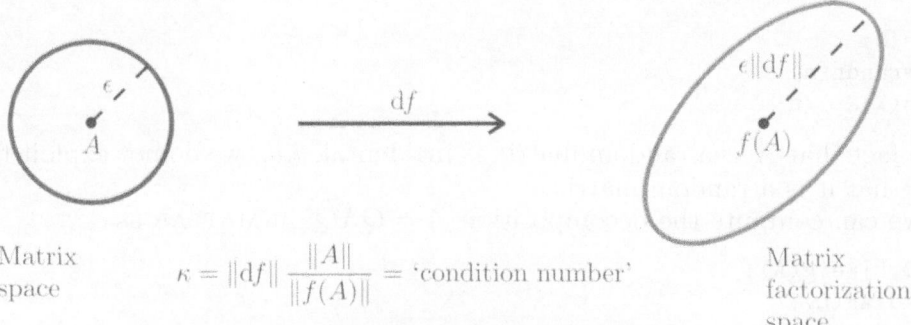

$$\kappa = \|\mathrm{d}f\|\frac{\|A\|}{\|f(A)\|} = \text{`condition number'}$$

Matrix space

Matrix factorization space

Figure 3.1. The condition number of a matrix factorization is related to the largest axis of an ellipsoid in matrix factorization space.

Without dwelling on the point, κ has a worst case built into the 'lim sup', while J contains information about an average behaviour under perturbations.

3.2. Matrix Jacobians numerically computed with finite differences

Consider the symmetric eigenvalue decomposition $A = Q\Lambda Q'$, where A is an $n \times n$ symmetric matrix. The Jacobian for this factorization is the term $\prod_{i<j} |\lambda_i - \lambda_j|$ in

$$(\mathrm{d}A) = \prod_{i<j} |\lambda_i - \lambda_j| \, (\mathrm{d}\Lambda) \, (Q' \, \mathrm{d}Q). \tag{3.4}$$

This equation is derived from the first-order perturbation $\mathrm{d}A$ in A due to perturbations $\mathrm{d}\Lambda$ and $Q' \, \mathrm{d}Q$ in the eigenvalues Λ and the eigenvectors Q. Note that since Q is orthogonal, $Q'Q = I$ so that $Q' \, \mathrm{d}Q + \mathrm{d}Q'Q = 0$ or that $Q' \, \mathrm{d}Q$ is antisymmetric with zeros along the diagonal. Restricting $Q' \, \mathrm{d}Q$ to be antisymmetric ensures that $A + \mathrm{d}A$ remains symmetric.

Numerically, we compute the perturbations in Λ and Q due to perturbations in A. As numerical analysts we always think of A as the input and Q and Λ as output, so it is natural to ask for the answer in this direction. Assuming the eigenvalue decomposition is unique after fixing the phase of the columns of Q, the first-order perturbation in Λ and Q due to perturbations in A is given by

$$\frac{(\mathrm{d}\Lambda)(Q' \, \mathrm{d}Q)}{(\mathrm{d}A)} = \frac{1}{\prod_{i<j} |\lambda_i - \lambda_j|} = \frac{1}{\Delta(\Lambda)}, \tag{3.5}$$

where $\Delta(\lambda) = \prod_{i<j} |\lambda_i - \lambda_j|$ is the absolute value of the Vandermonde determinant.

We can create an $n \times n$ symmetric matrix A by, for example, creating an $n \times n$ matrix X with independent Gaussian entries and then symmetrizing it as $A = (X + X')/n$. This can be conveniently done in MATLAB as

```
n=15;
X=randn(n);
A=(X+X')/n;
```

The fact that X is a random matrix is incidental, *i.e.*, we do not exploit the fact that it is a random matrix.

We can compute the decomposition $A = Q\Lambda Q'$ in MATLAB as

```
[Q,L]=eig(A);
L=diag(L);
```

Since A is an $n \times n$ symmetric matrix, the Jacobian matrix as in (3.1) resides in an $(n(n+1)/2)^2$-dimensional space:

```
JacMatrix=zeros(n*(n+1)/2);
```

Table 3.1. Jacobians computed numerically with finite differences.

```
n=15;                           % Size of the matrix
X=randn(n);
A=(X+X')/n;                     % Generate a symmetric matrix
[Q,L]=eig(A);                   % Compute its eigenvalues/eigenvectors
L=diag(L);
JacMatrix=zeros(n*(n+1)/2);     % Initialize Jacobian matrix
epsilon=1e-7; idx=1;
mask=triu(ones(n),1); mask=logical(mask(:)); % Upper triangular mask
for i=1:n
    for j=i:n

        %%% Perturbation Matrix
        Eij=zeros(n);              % Initialize perturbation
        Eij(i,j)=1; Eij(j,i) = 1;  % Perturbation matrix
        Ap=A+epsilon*Eij;          % Perturbed matrix

        %%% Eigenvalues and Eigenvectors

        [Qp,Lp] = eig(Ap);
        dL= (diag(Lp)-L)/epsilon;  % Eigenvalue perturbation
        QdQ = Q'*(Qp-Q)/epsilon;   % Eigenvector perturbation

        %%% The Jacobian Matrix
        JacMatrix(1:n,idx)=dL;                 % Eigenvalue part of Jacobian
        JacMatrix((n+1):end,idx) = QdQ(mask);  % Eigenvector part of Jacobian
        idx=idx+1;                             % Increment column counter
    end
end
```

Let ϵ be any small positive number, such as

```
epsilon=1e-7;
```

Generate the symmetric perturbation matrix E_{ij} for $1 \le i \le j < n$ whose entries are equal to zero except in the (i, j) and (j, i) entries, where they are equal to 1. Construct the Jacobian matrix by computing the eigenvalues and eigenvectors of the perturbed matrix $A + \epsilon E_{ij}$, and the quantities $d\Lambda$ and $Q' dQ$. This can be done in MATLAB using the code in Table 3.1. We note that this simple forward difference scheme can be replaced by a central difference scheme for added accuracy.

We can compare the numerical answer obtained by taking the determinant of the Jacobian matrix with the theoretical answer expressed in terms of the Vandermonde determinant as in (3.5). For a particular choice of A we can run the MATLAB code in Table 3.1 to get the answer:

```
format long
disp([abs(det(JacMatrix))    1/abs(det(vander(L)))]);
>> ans = 1.0e+49 *
         3.32069877679394    3.32069639128242
```

This is, in short, the 'proof by MATLAB' to show how Jacobians can be computed numerically with finite differences.

3.3. Jacobians of matrix factorizations

The computations of matrix Jacobians can be significantly more complicated than the scalar derivatives familiar in elementary calculus. Many Jacobians have been rediscovered in various communities. We recommend Olkin (1953, 2002), and the books by Muirhead (1982) and Mathai (1997). When computing Jacobians of matrix transformations or factorizations, it is important to identify the dimension of the underlying space occupied by the matrix perturbations.

Wedge products and the accompanying notation are used to facilitate the computation of matrix Jacobians. The notation also comes in handy for expressing the concept of volume on curved surfaces as in differential geometry. Mathai (1997) and Muirhead (1982) are excellent references for readers who truly wish to understand wedge products as a tool for computing the Jacobians of commonly used matrix factorizations such as those listed below.

While we expect our readers to be familiar with real and complex matrices, it is reasonable to consider quaternion matrices as well. The parameter β has been traditionally used to count the dimension of the underlying algebra as in Table 3.2. In other branches of mathematics, the parameter $\alpha = 2/\beta$ is used.

We provide, without proof, the formulas containing the Jacobians of familiar matrix factorizations. We encourage readers to notice that the vanishing

Table 3.2. Notation used to denote whether the elements of a matrix are real, complex or quaternion ($\beta = 2/\alpha$).

β	α	Division algebra
1	2	real ($\mathbb{R}$)
2	1	complex ($\mathbb{C}$)
4	1/2	quaternion ($\mathbb{H}$)

of the Jacobian is connected to difficult numerical problems. The parameter count is only valid where the Jacobian does not vanish.

QR (Gram–Schmidt) decomposition ($A = QR$). Valid for all three cases ($\beta = 1, 2, 4$). Q is orthogonal/unitary/symplectic, R is upper triangular. A and Q are $m \times n$ (assume $m \geq n$), R is $n \times n$. The parameter count for the orthogonal matrix is the dimension of the Stiefel manifold $V_{m,n}$.

Parameter count:

$$\beta mn = \beta mn - \beta \frac{n(n-1)}{2} - n + \beta \frac{n(n-1)}{2} + n.$$

Jacobian:

$$(\mathrm{d}A) = \prod_{i=1}^{n} r_{ii}^{\beta(m-i+1)-1} \, (\mathrm{d}R) \, (Q' \, \mathrm{d}Q). \tag{3.6}$$

Notation: $(\mathrm{d}A)$, $(\mathrm{d}R)$ are volumes of little boxes around A and R, while $(Q' \, \mathrm{d}Q)$ denotes the volume of a little box around the strictly upper triangular part of the antisymmetric matrix $Q' \, \mathrm{d}Q$ (see a numerical illustration in Section 3.2).

LU (Gaussian elimination) decomposition ($A = LU$). Valid for all three cases ($\beta = 1, 2, 4$). All matrices are $n \times n$, L and U are lower and upper triangular respectively, $l_{ii} = 1$ for all $1 \leq i \leq n$. Assume there is no pivoting.

Parameter count:

$$\beta n^2 = \beta \frac{n(n-1)}{2} + \beta \frac{n(n+1)}{2}.$$

Jacobian:

$$(\mathrm{d}A) = \prod_{i=1}^{n} |u_{ii}|^{\beta(n-i)} \, (\mathrm{d}L) \, (\mathrm{d}U). \tag{3.7}$$

$Q\Lambda Q'$ (symmetric eigenvalue) decomposition ($A = Q\Lambda Q'$). Valid for all three cases ($\beta = 1, 2, 4$). Here A is $n \times n$ symmetric/Hermitian/self-dual, Q is $n \times n$ and orthogonal/unitary/symplectic, Λ is $n \times n$ diagonal and real. To make the decomposition unique, we must fix the phases of the columns of Q (that eliminates $(\beta - 1)n$ parameters) and order the eigenvalues.

Parameter count:

$$\beta \frac{n(n-1)}{2} + n = \beta \frac{n(n+1)}{2} - n - (\beta - 1)n + n.$$

Jacobian:

$$(\mathrm{d}A) = \prod_{i<j} (\lambda_i - \lambda_j)^{\beta} \, (\mathrm{d}\Lambda) \, (Q' \, \mathrm{d}Q). \tag{3.8}$$

UΣV′ (singular value) decomposition (A = UΣV′). Valid for all three cases ($\beta = 1, 2, 4$). A is $m \times n$, U is $m \times n$ orthogonal/unitary/symplectic, V is $n \times n$ orthogonal/unitary/symplectic, Σ is $n \times n$ diagonal, positive, and real (suppose $m \geq n$). Again, to make the decomposition unique, we need to fix the phases on the columns of U (removing $(\beta - 1)n$ parameters) and order the singular values.

Parameter count:

$$\beta mn = \beta mn - \beta\frac{n(n-1)}{2} - n - (\beta - 1)n + n + \beta\frac{n(n+1)}{2} - n.$$

Jacobian:

$$(dA) = \prod_{i<j}(\sigma_i^2 - \sigma_j^2)^\beta \prod_{i=1}^{n} \sigma_i^{\beta(m-n+1)-1} \, (U' \, dU) \, (d\Sigma) \, (V' \, dV). \qquad (3.9)$$

References: real, Muirhead (1982), Dumitriu (2003), Shen (2001).

CS (Cosine–sine) decomposition. Valid for all three cases ($\beta = 1, 2, 4$). Q is $n \times n$ orthogonal/unitary/symplectic. Then, for any $k + j = n$, $p = k - j \geq 0$, the decomposition is

$$Q = \begin{pmatrix} U_{11} & U_{12} & 0 \\ U_{21} & U_{22} & 0 \\ 0 & 0 & U_2 \end{pmatrix} \begin{pmatrix} I_p & 0 & 0 \\ 0 & C & S \\ 0 & S & -C \end{pmatrix} \begin{pmatrix} V'_{11} & V'_{12} & 0 \\ V'_{21} & V'_{22} & 0 \\ 0 & 0 & V'_2 \end{pmatrix},$$

such that U_2, V_2 are $j \times j$ orthogonal/unitary/symplectic,

$$\begin{pmatrix} U_{11} & U_{12} \\ U_{21} & U_{22} \end{pmatrix} \quad \text{and} \quad \begin{pmatrix} V'_{11} & V'_{12} \\ V'_{21} & V'_{22} \end{pmatrix}$$

are $k \times k$ orthogonal/unitary/symplectic, with U_{11} and V_{11} being $p \times p$, and C and S are $j \times j$ real, positive, and diagonal, and $C^2 + S^2 = I_j$. Now let $\theta_i \in (0, \frac{\pi}{2})$, $q \leq i \leq j$ be the angles such that $C = \text{diag}(\cos(\theta_1), \ldots, \cos(\theta_j))$, and $S = \text{diag}(\sin(\theta_1), \ldots, \sin(\theta_j))$. To ensure uniqueness of the decomposition we order the angles, $\theta_i \geq \theta_j$, for all $i \leq j$.

This parameter count is a little special since we have to account for the choice of the cases in the decomposition.

Parameter count:

$$\beta\frac{n(n+1)}{2} - n = \big(\beta j(j+1) - (\beta - 1)j\big) + j$$

$$+ \left(\beta k(k+1) - k - \beta\frac{p(p+1)}{2} + p\right).$$

Jacobian:

$$(Q' \, dQ) = \prod_{i<j} \sin(\theta_i - \theta_j)^\beta \, \sin(\theta_i + \theta_j)^\beta \prod_{i=1}^{j} \cos(\theta_i)^{\beta-1} \sin(\theta_i) \, d\theta$$

$$\times (U'_1 \, dU_1) \, (U'_2 \, dU_2) \, (V'_1 \, dV_1) \, (V'_2 \, dV_2).$$

Tridiagonal $Q\Lambda Q'$ (eigenvalue) decomposition ($T = Q\Lambda Q'$). Valid for real matrices. T is an $n \times n$ tridiagonal symmetric matrix, Q is an orthogonal $n \times n$ matrix, and Λ is diagonal. To make the factorization unique, we impose the condition that the first row of Q is all positive. The number of independent parameters in Q is $n-1$ and they can be seen as being all in the first row q of Q. The rest of Q can be determined from the orthogonality constraints, the tridiagonal symmetric constraints on A, and from Λ.

Parameter count:
$$2n - 1 = n - 1 + n.$$

Jacobian:
$$(\mathrm{d}T) = \frac{\prod_{i=1}^{n-1} T_{i+1,i}}{\prod_{i=1}^{n} q_i} \, \mu(\mathrm{d}q) \, (\mathrm{d}\Lambda). \tag{3.10}$$

Note that the Jacobian is written as a combination of parameters from T and q, the first row of Q, and $\mu(\mathrm{d}q)$ is the surface area on the sphere.

Tridiagonal BB' (Cholesky) decomposition ($T = BB'$). Valid for real matrices. T is an $n \times n$ real positive definite tridiagonal matrix, B is an $n \times n$ real bidiagonal matrix.

Parameter count:
$$2n - 1 = 2n - 1.$$

Jacobian:
$$\mathrm{d}T = 2^n b_{11} \prod_{i=2}^{n} b_{ii}^2 \, (\mathrm{d}B). \tag{3.11}$$

4. Classical random matrix ensembles

We now turn to some of the most well-studied random matrices. They have names such as Gaussian, Wishart, MANOVA, and circular. We prefer Hermite, Laguerre, Jacobi, and perhaps Fourier. In a sense, they are to random matrix theory as Poisson's equation is to numerical methods. Of course, we are thinking in the sense of the problems that are well tested, well analysed, and well studied because of nice fundamental analytic properties.

These matrices play a prominent role because of their deep mathematical structure. They have arisen in a number of fields, often independently. The tables that follow are all keyed by the first column to the titles Hermite, Laguerre, Jacobi, and Fourier.

4.1. Classical ensembles by discipline

We connect classical random matrices to problems roughly by discipline. In each table, we list the 'buzz words' for the problems in the field; where a

Table 4.1. Matrix factorizations associated with the classical random matrix ensembles.

Ensemble	Numerical procedure	MATLAB
Hermite	symmetric eigenvalue decomposition	`eig`
Laguerre	singular value decomposition	`svd`
Jacobi	generalized singular value decomposition	`gsvd`
Fourier	unitary eigenvalue decomposition	`eig`

Table 4.2. Equilibrium measure for classical random matrix ensembles.

Ensemble	Weight function	Equilibrium measure
Hermite	$e^{-x^2/2}$	semi-circular law (Wigner 1958)
Laguerre	$x^a e^{-x}$	Marčenko and Pastur (1967)
Jacobi	$(1-x)^a (1+x)^b$	generalized McKay law
Fourier	$e^{j\theta}$	uniform

classical random matrix has not yet appeared or, as we would rather believe, is yet to be discovered, we indicate with a blank. The actual definition of the random matrices may be found in Section 4.3. Note that for every problem there is an option of considering random matrices over the reals ($\beta = 1$), complexes ($\beta = 2$), quaternions ($\beta = 4$), or there is the general β approach.

We hope the reader will begin to see a fascinating story already emerging in Tables 4.1 and 4.2, where the symmetric eigenvalue problem is connected to the Hermite weight factor e^{-x^2}, the SVD to the Laguerre weight factor $x^a e^{-x}$ and so forth.

In multivariate statistics, the problems of interest are random covariance matrices (known as Wishart matrices) and ratios of Wishart matrices that

Table 4.3. Multivariate statistics.

Ensemble	Problem solved	Univariate distribution
Hermite	–	normal
Laguerre	Wishart	chi-squared
Jacobi	MANOVA	beta
Fourier	–	–

Table 4.4. Graph theory.

Ensemble	Type of graph	Author
Hermite	undirected	Wigner (1955)
Laguerre	bipartite	Jonsson (1982)
Jacobi	d-regular	McKay (1981)
Fourier	–	–

Table 4.5. Free probability and operator algebras.

Ensemble	Terminology
Hermite	semi-circle
Laguerre	free Poisson
Jacobi	free product of projections
Fourier	–

arise in the multivariate analysis of variance (MANOVA). This is a central theme of texts such as Muirhead (1982).

The same matrices also arise elsewhere, especially in the modern physics of super-symmetry. This is evident in the works of Zirnbauer (1996), Ivanov (2002) and Caselle and Magnea (2004). More classically Dyson and Wigner worked on the Hermite and Fourier cases, known, respectively, as the Gaussian and circular ensembles. (See Mehta (1991).)

The cases that correspond to symmetric spaces are quantized perhaps unnecessarily. In mathematics, a symmetric space is a geometric object such as a sphere that can be identified as the quotient space of two Lie groups, with an involution that preserves geodesics. The Grassmann manifold is a symmetric space, while the Stiefel manifold of $m \times n$ orthogonal matrices is not, unless $m = 1$ or $m = n$, *i.e.*, the sphere and the orthogonal group respectively.

Many of the classical techniques for computing the eigenvalue distributions are ultimately related to interconnectivity of the matrix. For each case Table 4.4 shows a graph structure underlying the matrix.

'Free probability' is an important branch of operator algebra developed in 1985 by Voiculescu that has deep connections to random matrix theory. Table 4.5 uses the names found in that literature. From the random matrix viewpoint, free probability describes the eigenvalues of such operations as $A + B$ or AB in a language similar to that of the distribution of independent random variables $a + b$ or ab, respectively. There will be more on this in Section 12.

The authors would be delighted if the reader is awed by the above set of tables. Anything that manifests itself in so many ways in so many fields must be deep in the foundations of the problem. We indicate the four channels of structure lurking underneath computation (Table 4.1), multivariate statistics (Table 4.3), graph theory (Table 4.4) and operator algebras (Table 4.5).

There is a deep structure begging the dense matrix expert to forget the SVD for a moment, or the sparse expert to forget bipartite graphs, if only briefly, or the statistician to forget the chi-squared distribution and sample covariance matrices. Something ties these experts together. Probably random matrix theory is not the only way to reveal the hidden message, but it is the theory that has compelled us to see what is truly there.

A few words for the numerical analyst. The symmetric and unitary eigenvalue problems, the SVD, and the GSVD have important mathematical roles because of certain symmetries not enjoyed by LU or the asymmetric eigenvalue problem. More can be said, but this may not be the place. We plant the seed and we hope it will be picked up by many.

In the remainder of this chapter we will explore these random matrix ensembles in depth. We begin with the basic Gaussian matrices and briefly consider the joint element density and invariance properties. We then construct the classical ensembles, derive their joint element densities, and their joint eigenvalue densities, all in the context of the natural numerical procedures listed in Table 4.1.

4.2. Gaussian random matrices

$G_1(m, n)$ is an $m \times n$ matrix of independent and identically distributed (i.i.d.) standard real random normals. More simply, in MATLAB notation:

```
G1=randn(m,n);
```

Table 4.6 lists MATLAB commands that can be used to generate $G_\beta(m, n)$ for general β. Note that since quaternions do not exist in MATLAB they are 'faked' using 2×2 complex matrices.

If A is an $m \times n$ Gaussian random matrix $G_\beta(m, n)$ then its joint element density is given by

$$\frac{1}{(2\pi)^{\beta mn/2}}\exp\left(-\frac{1}{2}\|A\|_F^2\right). \tag{4.1}$$

Some authors also use the notation $\operatorname{etr}(A)$ for the exponential of the trace of a matrix.

The most important property of G_β, be it real, complex, or quaternion, is its *orthogonal invariance*. This makes the distribution impervious to multiplication by an orthogonal (unitary, symplectic) matrix, provided that

Table 4.6. Generating the Gaussian random matrix $G_\beta(m,n)$ in MATLAB.

β	MATLAB command
1	`G = randn(m,n)`
2	`G = randn(m,n) + i*randn(m,n)`
4	`X = randn(m,n) + i*randn(m,n); Y = randn(m,n) + i*randn(m,n);`
	`G = [X Y; - conj(Y) conj(X)]`

the two are independent. This can be inferred from the joint element density in (4.1) since its Frobenius norm, $\|A\|_F$, is unchanged when A is multiplied by an orthogonal (unitary, symplectic) matrix. The orthogonal invariance implies that no test can be devised that would differentiate between $Q_1 A$, A, and AQ_2, where Q_1 and Q_2 are non-random orthogonal and A is Gaussian.

4.3. Construction of the classical random matrix ensembles

The classical ensembles are constructed from G_β as follows. Since they are constructed from multivariate Gaussians, they inherit the orthogonality property as well, *i.e.*, they remain invariant under orthogonal transformations.

Gaussian orthogonal ensemble (GOE): symmetric $n \times n$ matrix obtained as $(A + A^T)/2$ where A is $G_1(n,n)$. The diagonal entries are i.i.d. with distribution $N(0,1)$, and the off-diagonal entries are i.i.d. (subject to the symmetry) with distribution $N(0,\frac{1}{2})$.

Gaussian unitary ensemble (GUE): Hermitian $n \times n$ matrix obtained as $(A + A^H)/2$, where A is $G_2(n,n)$ and H denotes the Hermitian transpose of a complex matrix. The diagonal entries are i.i.d. with distribution $N(0,1)$, while the off-diagonal entries are i.i.d. (subject to being Hermitian) with distribution $N_2(0,\frac{1}{2})$.

Gaussian symplectic ensemble (GSE): self-dual $n \times n$ matrix obtained as $(A + A^D)/2$, where A is $G_4(n,n)$ and D denotes the dual transpose of a quaternion matrix. The diagonal entries are i.i.d. with distribution $N(0,1)$, while the off-diagonal entries are i.i.d. (subject to being self-dual) with distribution $N_4(0,\frac{1}{2})$.

Similarly, the Wishart and MANOVA ensembles can be defined as follows.

Wishart ensemble ($W_\beta(m,n)$, $m \geq n$): symmetric/Hermitian/self-dual $n \times n$ matrix which can be obtained as $A'A$, where A is $G_\beta(m,n)$ and A' denotes A^T, A^H and A^D, depending on whether A is real, complex, or quaternion, respectively.

MANOVA ensemble $(J_\beta(m_1, m_2, n)$, $m_1, m_2 \geq n)$: symmetric/Hermitian/self-dual $n \times n$ matrix which can be obtained as $A/(A + B)$, where A and B are $W_\beta(m_1, n)$ and $W_\beta(m_2, n)$, respectively. See Sutton (2005) for details on a construction using the CS decomposition.

Circular ensembles: constructed as $U^T U$ and U for $\beta = 1, 2$ respectively, where U is a uniformly distributed unitary matrix (see Section 4.6). For $\beta = 4$, it is defined analogously as in Mehta (1991).

The β-Gaussian ensembles arise in physics, and were first identified by Dyson (1963) by the group over which they are invariant: Gaussian orthogonal or, for short, GOE (with real entries, $\beta = 1$), Gaussian unitary or GUE (with complex entries, $\beta = 2$), and Gaussian symplectic or GSE (with quaternion entries $\beta = 4$).

The Wishart ensembles owe their name to Wishart (1928), who studied them in the context of statistics applications as sample covariance matrices. The β-Wishart models for $\beta = 1, 2, 4$ could be named Wishart real, Wishart complex, and Wishart quaternion respectively, though the β notation is not as prevalent in the statistical community.

The MANOVA ensembles arise in statistics in the Multivariate Analysis of Variance, hence the name. They are in general more complicated to characterize, so less is known about them than the Gaussian and Wishart ensembles.

4.4. Computing the joint element densities

The joint eigenvalue densities of the classical random matrix ensembles have been computed in many different ways by different authors. Invariably, the basic prescription is as follows.

We begin with the probability distribution on the matrix elements. The next step is to pick an appropriate matrix factorization whose Jacobians are used to derive the joint densities of the elements in the matrix factorization space. The relevant variables in this joint density are then appropriately transformed and 'integrated out' to yield the joint eigenvalue densities.

This prescription is easy enough to describe, though in practice the normal distribution seems to be the best choice to allow us to continue and get analytical expressions. Almost any other distribution would stop us in our tracks, at least if our goal is some kind of exact formula.

Example. Let A be an $n \times n$ matrix from the Gaussian orthogonal ensemble ($\beta = 1$). As described earlier, this is an $n \times n$ random matrix with elements distributed as $N(0, 1)$ on the diagonal and $N(0, 1/2)$ off the diagonal, that is,

$$a_{ij} \sim \begin{cases} N(0, 1) & i = j, \\ N(0, 1/2) & i > j. \end{cases}$$

Table 4.7. Joint element densities of an $n \times n$ matrix A from a Gaussian ensemble.

Gaussian	orthogonal $\beta = 1$ unitary $\beta = 2$ symplectic $\beta = 4$	$\dfrac{1}{2^{n/2}} \dfrac{1}{\pi^{n/2 + n(n-1)\beta/4}} \exp\left(-\dfrac{1}{2}\|A\|_F^2\right)$

Recall that the normal distribution with mean μ and variance σ^2, *i.e.*, $N(\mu, \sigma^2)$, has a density given by

$$\frac{1}{\sqrt{2\pi\,\sigma^2}} \exp\left(-\frac{(x-\mu)^2}{2\sigma^2}\right),$$

from which it is fairly straightforward to verify that the joint element density of A written as

$$\frac{1}{2^{n/2}} \frac{1}{\pi^{n(n+1)/4}} \exp\left(-\|A\|_F^2/2\right) \tag{4.2}$$

can be obtained by taking products of the n normals along the diagonal having density $N(0,1)$ and $n(n-1)/2$ normals in the off-diagonals having density $N(0, 1/2)$.

Table 4.7 lists the joint element density for the three Gaussian ensembles parametrized by β.

Now that we have obtained the joint element densities, we simply have to follow the prescription discussed earlier to obtain the joint eigenvalue densities.

In the case of the Gaussian ensembles, the matrix factorization $A = Q\Lambda Q'$ directly yields the eigenvalues and the eigenvectors. Hence, applying the Jacobian for this transformation given by (3.8) allows us to derive the joint densities for the eigenvalues and the eigenvectors of A. We obtain the joint eigenvalue densities by 'integrating' out the eigenvectors.

We like to think of the notion of the 'most natural' matrix factorization that allows us to compute the joint eigenvalue densities in the easiest manner. For the Gaussian ensembles, the symmetric eigenvalue decomposition $A = Q\Lambda Q'$ is the most obvious choice. This not the case for the Wishart and the MANOVA ensembles. In this context, what makes a matrix factorization 'natural'? Allow us to elaborate.

Consider the Wishart matrix ensemble $W_\beta(m, n) = A'A$, where $A = G_\beta(m, n)$ is a multivariate Gaussian. Its joint element density can be computed rather laboriously in a two-step manner whose first step involves writing $W = QR$ and then integrating out the Q, leaving the R. The second step is the transformation $W = R'R$ which is the Cholesky factorization of a matrix in numerical analysis. The conclusion is that although we may obtain the joint density of the elements of W as listed in Table 4.8, this procedure is much more involved than it needs to be.

Table 4.8. Joint element density of the Wishart ensemble $W_\beta(m, n)$ $(m \geq n)$.

Wishart	orthogonal $\beta = 1$ unitary $\beta = 2$ symplectic $\beta = 4$	$\dfrac{\operatorname{etr}(-W/2)\,(\det W)^{\beta(m-n+1)/2-1}}{2^{mn\beta/2}\ \Gamma_n^\beta(m\beta/2)}$

This is where the notion of a 'natural' matrix factorization comes in. Though it seems statistically obvious to think of Wishart matrices as covariance matrices and compute the joint density of the eigenvalues of $A'A$ directly, it is more natural to derive the joint density of the singular values of A instead. Since A is a multivariate Gaussian, the Jacobian of the factorization $A = U\Sigma V'$ given by (3.9) can be used to directly determine the joint density of the singular values and the singular vectors of W from the joint element density of A in (4.1). We can then integrate out the singular vectors to obtain the joint density of the singular values of A and hence the eigenvalues of $W = A'A$. The technicalities of this may be found in Edelman (1989).

Similarly, the corresponding 'natural' factorization for the MANOVA ensembles is the generalized singular value decomposition. Note that the square of the generalized singular values of two matrices A and B is the same as the eigenvalues of $(BB')^{-1}(AA')$, so that the eigenvalues of the MANOVA matrix $J_\beta(m_1, m_2, n) = (I + W(m_1, n)^{-1}W(m_2, n))^{-1}$ can be obtained by a suitable transformation.

Table 4.1 summarizes the matrix factorizations associated with the classical random matrix ensembles that allow us to compute the joint eigenvalue densities in the most natural manner. Later we will discuss additional connections between these matrix factorizations, and classical orthogonal polynomials.

4.5. Joint eigenvalue densities of the classical ensembles

The three Gaussian ensembles have joint eigenvalue probability density function

$$\text{Gaussian:}\quad f_\beta(\lambda) = c_H^\beta \prod_{i<j} |\lambda_i - \lambda_j|^\beta e^{-\sum_{i=1}^n \lambda_i^2/2}, \qquad (4.3)$$

with $\beta = 1$ corresponding to the reals, $\beta = 2$ to the complexes, $\beta = 4$ to the quaternion, and with

$$c_H^\beta = (2\pi)^{-n/2} \prod_{j=1}^n \frac{\Gamma(1 + \frac{\beta}{2})}{\Gamma(1 + \frac{\beta}{2}j)}. \qquad (4.4)$$

The best references are Mehta (1991) and the original paper by Dyson (1963).

Similarly, the Wishart (or Laguerre) models have joint eigenvalue PDF

$$\text{Wishart:} \quad f_\beta(\lambda) = c_L^{\beta,a} \prod_{i<j} |\lambda_i - \lambda_j|^\beta \prod_i \lambda_i^{a-p} e^{-\sum_{i=1}^n \lambda_i/2}, \qquad (4.5)$$

with $a = \frac{\beta}{2}m$ and $p = 1 + \frac{\beta}{2}(n-1)$. Again, $\beta = 1$ for the reals, $\beta = 2$ for the complexes, and $\beta = 4$ for the quaternion. The constant

$$c_L^{\beta,a} = 2^{-na} \prod_{j=1}^n \frac{\Gamma(1 + \frac{\beta}{2})}{\Gamma(1 + \frac{\beta}{2}j)\Gamma(a - \frac{\beta}{2}(n-j)))}. \qquad (4.6)$$

Good references are Muirhead (1982), Edelman (1989), and James (1964), and for $\beta = 4$, Macdonald (1998).

To complete the triad of classical orthogonal polynomials, we will mention the β-MANOVA ensembles, which are associated to the multivariate analysis of variance (MANOVA) model. They are better known in the literature as the Jacobi ensembles, with joint eigenvalue PDF, that is,

$$\text{MANOVA:} \quad f_\beta(\lambda) = c_J^{\beta,a_1,a_2} \prod_{i<j} |\lambda_i - \lambda_j|^\beta \prod_{j=1}^n \lambda_i^{a_1-p} (1 - \lambda_i)^{a_2-p}, \qquad (4.7)$$

with $a_1 = \frac{\beta}{2}m_1$, $a_2 = \frac{\beta}{2}m_2$, and $p = 1 + \frac{\beta}{2}(n-1)$. As usual, $\beta = 1$ for real and $\beta = 2$ for complex; also

$$c_J^{\beta,a_1,a_2} = \prod_{j=1}^n \frac{\Gamma(1 + \frac{\beta}{2})\Gamma(a_1 + a_2 - \frac{\beta}{2}(n-j))}{\Gamma(1 + \frac{\beta}{2}j)\Gamma(a_1 - \frac{\beta}{2}(n-j))\Gamma(a_2 - \frac{\beta}{2}(n-j))}. \qquad (4.8)$$

Good references are the original paper by Constantine (1963), and Muirhead (1982) for $\beta = 1, 2$.

4.6. Haar-distributed orthogonal, unitary and symplectic eigenvectors

The eigenvectors of the classical random matrix ensembles are distributed with Haar measure. This is the uniform measure on orthogonal/unitary/symplectic matrices; see Chapter 1 of Milman and Schechtman (1986) for a derivation.

A measure $\mu(E)$ is a generalized volume defined on E. A measure μ, defined on a group G, is a Haar measure if $\mu(gE) = \mu(E)$, for every $g \in G$. For the example $O(n)$ of orthogonal $n \times n$ matrices, the condition that our measure be Haar is, for any continuous function f, that

$$\int_{Q \in O(n)} f(Q)\, d\mu(Q) = \int_{Q \in O(n)} f(Q_o Q)\, d\mu(Q), \qquad \text{for any } Q_o \in O(n).$$

In other words, Haar measure is symmetric: no matter how we rotate our

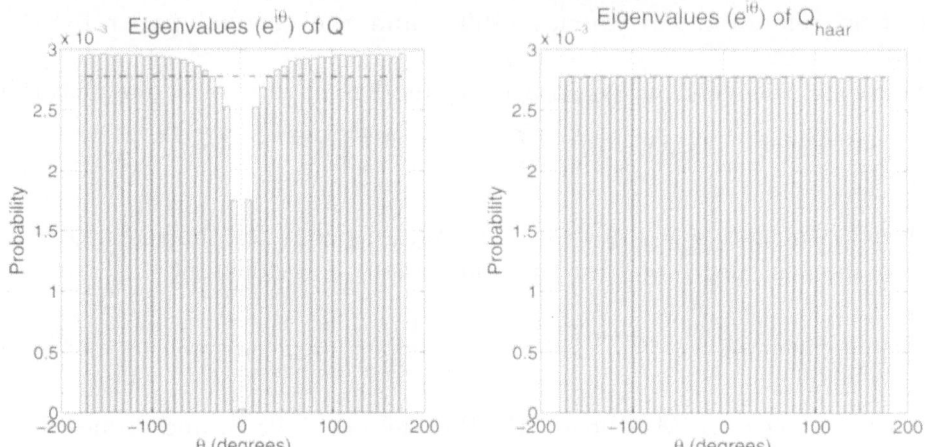

Figure 4.1. QR (Gram–Schmidt) factorization of `randn(n)`; no correction in the left panel, phase correction in the right panel.

sets, we get the same answer. In numerical terms, we can devise the following experiment to get some intuition on whether or not randomly generated unitary matrices are Haar-distributed.

Suppose we started with an $n \times n$ complex random matrix A constructed in MATLAB as

```
% Pick n
A=randn(n)+i*randn(n);
```

Compute its QR decomposition to generate a random unitary matrix Q:

```
[Q,R]=qr(A);
```

The eigenvalues of Q will be complex with a magnitude of 1, *i.e.*, they will be distributed on the unit circle in the complex plane. Compute the phases associated with these complex eigenvalues:

```
Qphase=angle(eig(Q));
```

Now, perform this experiment several times and collect the phase information in the variable `Qphase`. Plot the histogram of the phases (in degrees) normalized to have area 1. The left-hand panel of Figure 4.1 shows this histogram for $n = 50$ and $100,000$ trials. The dotted line indicates a uniform density between $[-180, 180]$. From this we conclude that since the phases of Q are not uniformly distributed, Q as constructed in this experiment is not distributed with Haar measure.

It is interesting to recognize why the experiment described above does not produce a Haar-distributed unitary matrix. This is because the QR algorithm in MATLAB does not guarantee nonnegative diagonal entries in R. A simple correction by randomly perturbing the phases as:

```
Q=Q*diag(exp(i*2*pi*rand(n,1)));
```

or even by the sign of the diagonal entries of R:

```
Q=Q*diag(sign(diag(R)));
```

would correct this problem and produce the histogram in the right-hand panel of Figure 4.1 for the experiment described above. Note that the MATLAB command **rand** generates a random number uniformly distributed between 0 and 1. While this method of generating random unitary matrices with Haar measure is useful for simplicity, it is not the most efficient. For information on the efficient numerical generation of random orthogonal matrices distributed with Haar measure, see Stewart (1980).

4.7. The longest increasing subsequence

There is an interesting link between the moments of the eigenvalues of Q and the number of permutations of length n with longest increasing subsequence k. For example, the permutation $(3\,1\,8\,4\,5\,7\,2\,6\,9\,10)$ has $(1\,4\,5\,7\,9\,10)$ or $(1\,4\,5\,6\,9\,10)$ as the longest increasing subsequences of length 6.

This problem may be recast for the numerical analyst as the parallel complexity of solving an upper triangular system whose sparsity is given by a permutation π:

$$U_{ij}(\pi) \begin{cases} \neq 0 & \text{if } \pi(i) \leq \pi(j) \text{ and } i \leq j, \\ = 0 & \text{if } \pi(i) > \pi(j) \text{ or } i > j. \end{cases}$$

The result from random matrix theory is that the number of permutations of length n with longest increasing subsequence less than or equal to length k is given by

$$E_{Q_k}\left(|\mathrm{tr}(Q_k)|^{2n}\right).$$

We can verify this numerically using what we know about generating Haar unitary random matrices from Section 4.6. We can construct a function in MATLAB that generates a Haar unitary matrix, computes the quantity $|\mathrm{tr}Q_k|^{2n}$ and averages it over many trials:

```
function L = longestsubsq(n,k,trials);
expt=[];
for idx=1:trials,
    % Generate Haar unitary matrix
    [Q,R]=qr(randn(k)+i*randn(k));
    Q=Q*diag(exp(i*2*pi*rand(k,1)));
    expt=[expt;abs(trace(Q))^(2*n)];
end

mean(exp)
```

Table 4.9. Permutations for $n = 4$.

1 2 3 4	2 1 3 4	3 1 2 4	4 1 2 3
1 2 4 3	2 1 4 3	3 1 4 2	4 1 3 2
1 3 2 4	2 3 1 4	3 2 1 4	4 2 1 3
1 3 4 2	2 3 4 1	3 2 4 1	4 2 3 1
1 4 2 3	2 4 1 3	3 4 1 2	4 3 1 2
1 4 3 2	2 4 3 1	3 4 2 1	4 3 2 1

For $n = 4$, there are 24 possible permutations listed in Table 4.9. We underline the fourteen permutations with longest increasing subsequence of length ≤ 2. Of these, one permutation (4 3 2 1) has length 1 and the other thirteen have length 2.

If we were to run the MATLAB code for $n = 4$ and $k = 2$ and 30000 trials we would get:

```
>> longestsubsq(4,2,30000)
ans = 14.1120
```

which is approximately equal to the number of permutations of length less than or equal to 2. It can be readily verified that the code gives the right answer for other combinations of n and k as well. We note that for this numerical verification, it was critically important that a Haar unitary matrix was generated. If we were to use a matrix without Haar measure, for example simply using the command [Q,R]=qr(randn(n)+i*randn(n)) without randomly perturbing the phases, as described in Section 4.6, we would not get the right answer.

The authors still find it remarkable that the answer to a question this simple (at least in terms of formulation) involves integrals over Haar unitary matrices. There is, of course, a deep mathematical reason for this that is related to the correspondence between, on the one hand, permutations and combinatorial objects known as Young tableaux, via the Schensted correspondence, and, on the other hand, representations of the symmetric group and the unitary group. The reader may wish to consult Rains (1998), Aldous and Diaconis (1999) and Odlyzko and Rains (1998) for additional details. Related works include Borodin (1999), Tracy and Widom (2001) and Borodin and Forrester (2003).

5. Numerical algorithms stochastically

Matrix factorization algorithms may be performed stochastically given Gaussian inputs. What this means is that instead of performing the matrix reductions on a computer, they can be done by mathematics. The three that are well known, though we will focus on the latter two, are:

(1) Gram–Schmidt (the `qr` decomposition)
(2) symmetric tridiagonalization (standard first step of `eig`), and
(3) bidiagonalization (standard first step of `svd`).

The bidiagonalization method is due to Golub and Kahan (1965), while the tridiagonalization method is due to Householder (1958).

These two linear algebra algorithms can be applied stochastically, and it is not very hard to compute the distributions of the resulting matrix.

The two key ideas are:

(1) the χ_r distribution, and
(2) the orthogonal invariance of Gaussians.

The χ_r is the χ-distribution with r degrees of freedom where r is any real number. It can be derived from the univariate Gaussian and is also the square root of the χ_r^2-distribution. Hence it may be generated using the MATLAB Statistics Toolbox using the command `sqrt(chi2rnd(r))`. If the parameter r is a positive integer n, one definition of χ_n is given by $\|G(n,1)\|_2$, in other words, the 2-norm of an $n \times 1$ vector of independent standard normals (or `norm(randn(n,1))` in MATLAB). The probability density function of χ_n can then be extended to any real number r so that the probability density function of χ_r is given by

$$f_r(x) = \frac{1}{2^{r/2-1}\,\Gamma(\frac{1}{2}r)}\; x^{r-1}\; e^{-x^2/2}.$$

The orthogonal invariance of Gaussians is mentioned in Section 4.3. In this form it means that

$$H \begin{pmatrix} G_1 \\ G_1 \\ \vdots \\ \vdots \\ G_1 \end{pmatrix} \overset{\mathcal{D}}{=} \begin{pmatrix} G \\ G \\ \vdots \\ \vdots \\ G \end{pmatrix},$$

if each G denotes an independent standard Gaussian and H any independent orthogonal matrix (such as a reflector).

Thus, for example, the first two steps of Gram–Schmidt applied to an $n \times n$ real Gaussian matrix ($\beta = 1$) are:

$$\begin{pmatrix} G & G & \cdots & G \\ G & G & \cdots & G \\ \vdots & \vdots & \cdots & \vdots \\ G & G & \cdots & G \end{pmatrix} \rightarrow \begin{pmatrix} \chi_n & G & \cdots & G \\ & G & \cdots & G \\ & \vdots & \cdots & \vdots \\ & G & \cdots & G \end{pmatrix} \rightarrow \begin{pmatrix} \chi_n & G & \cdots & G \\ & \chi_{n-1} & \cdots & G \\ & & \cdots & \vdots \\ & & G & G \end{pmatrix}.$$

Table 5.1. Tri- and bidiagonal models for the Gaussian and Wishart ensembles.

Gaussian Ensemble $n \in \mathbb{N}$	$H_n^\beta \sim \frac{1}{\sqrt{2}} \begin{pmatrix} N(0,2) & \chi_{(n-1)\beta} & & & & \\ \chi_{(n-1)\beta} & N(0,2) & \chi_{(n-2)\beta} & & & \\ & \ddots & \ddots & \ddots & & \\ & & & \chi_{2\beta} & N(0,2) & \chi_\beta \\ & & & & \chi_\beta & N(0,2) \end{pmatrix}$
Wishart ensemble $n \in \mathbb{N}$ $a \in \mathbb{R}$ $a > \frac{\beta}{2}(n-1)$	$L_n^\beta = B_n^\beta \, B_n^{\beta'}$, where $B_n^\beta \sim \begin{pmatrix} \chi_{2a} & & & \\ \chi_{\beta(n-1)} & \chi_{2a-\beta} & & \\ & \ddots & \ddots & \\ & & \chi_\beta & \chi_{2a-\beta(n-1)} \end{pmatrix}$

Applying the same ideas for tridiagonal or bidiagonal reduction gives the answer listed in Table 5.1, where the real case corresponds to $\beta = 1$, complex $\beta = 2$ and quaternion $\beta = 4$. For the Gaussian ensembles, before scaling the diagonal elements are i.i.d. normals with mean 0 and variance 2. The subdiagonal has independent elements that are χ variables as indicated. The superdiagonal is copied to create a symmetric tridiagonal matrix. The diagonal and the subdiagonals for the bidiagonal Wishart ensembles are independent elements that are χ-distributed with degrees of freedom having arithmetic progressions of step size β.

There is a tridiagonal matrix model for the β-Jacobi ensemble also, as described in Killip and Nenciu (2004); the correspondence between the CS decomposition and the Jacobi model is spelled out in Sutton (2005). Other models for the β-Jacobi ensemble include Lippert (2003).

There is both an important computational and theoretical implication of applying these matrix factorizations stochastically. Computationally speaking, often much of the time goes into performing these reductions for a given realization of the ensemble. Having them available analytically means that the constructions in Section 4.3 are highly inefficient for numerical simulations of the Hermite and Laguerre ensembles. Instead, we can generate then much more efficiently using MATLAB code and the Statistics Toolbox as listed in Table 5.2. The tangible savings in storage $O(n^2)$ to $O(n)$ is reflected in similar savings in computational complexity when computing their eigenvalues too. Not surprisingly, these constructions have been rediscovered independently by several authors in different contexts. Trotter (1984) used it in his alternate derivation of Wigner's semi-circular law.

Table 5.2. Generating the β-Hermite and β-Laguerre ensembles efficiently.

Ensemble	MATLAB commands

β-Hermite	```% Pick n, beta```
	 ```d=sqrt(chi2rnd(beta*[n:-1:1])))';```
	 ```H=spdiags(d,1,n,n)+spdiags(randn(n,1),0,n,n);```
	 ```H=(H+H')/sqrt(2);```

β-Laguerre	```% Pick m, n, beta```
	 ```% Pick a > beta*(n-1)/2;```
	 ```d=sqrt(chi2rnd(2*a-beta*[0:1:n-1]))';```
	 ```s=sqrt(chi2rnd(beta*[n:-1:1])))';```
	 ```B=spdiags(s,-1,n,n)+spdiags(d,0,n,n)```
	 ```L=B*B';```

Similarly, Silverstein (1985) and, more recently, Baxter and Iserles (2003) have rederived this result; probably many others have as well.

Theoretically the parameter β plays a new important role. The answers show that insisting on $\beta = 1, 2$ and 4 is no longer necessary. While these three values will always play something of a special role, like the mathematician who invents the Gamma function and forgets about counting permutations, we now have a whole continuum of possible betas available to us. While clearly simplifying the 'book-keeping' in terms of whether the elements are real, complex or quaternion, this formalism can be used to re-interpret and rederive familiar results as in Dumitriu (2003).

The general β version requires a generalization of $G_\beta(1,1)$. We have not seen any literature but formally it seems clear how to work with such an object (rules of algebra are standard, rules of addition come from the normal distribution, and the absolute value must be a χ_β distribution). From there, we might formally derive a general Q for each β.

6. Classical orthogonal polynomials

We have already seen in Section 4 that the weight function associated with classical orthogonal polynomials plays an important role in random matrix theory.

Given a weight function $w(x)$ and an interval $[a, b]$ the orthogonal polynomials satisfy the relationship

$$\int_a^b p_j(x)p_k(x)w(x)\,\mathrm{d}x = \delta_{jk}.$$

In random matrix theory there is interest in matrices with joint eigen-

Table 6.1. The classical orthogonal polynomials.

Polynomial	Interval $[a, b]$	$w(x)$
Hermite	$(-\infty, \infty)$	$e^{-x^2/2}$
Laguerre	$[0, \infty)$	$x^k e^{-x}$
Jacobi	$(-1, 1)$	$(1 - x)^a (1 + x)^b$

value density proportional to $\prod w(\lambda_i)|\Delta(\lambda)|^\beta$ where $\Delta(x) = \prod_{i<j}(x_i - x_j)$. Table 6.1 lists the weight functions and the interval of definition for the classical Hermite, Laguerre and Jacobi orthogonal polynomials as found in classical references such as Abramowitz and Stegun (1970).

Note that the Jacobi polynomial reduces to the Legendre polynomial when $\alpha = \beta = 0$, and to the Chebyshev polynomials when $\alpha, \beta = \pm 1/2$.

Classical mathematics suggests that a procedure such as Gram–Schmidt orthonormalization can be used to generate these polynomials. Numerically, however, other procedures are available, as detailed in Gautschi (1996).

There are deep connections between these classical orthogonal polynomials and three of the classical random matrix ensembles as alluded to in Section 4.

The most obvious link is between the form of the joint eigenvalue densities for these matrix ensembles and the weight functions $w(x)$ of the associated orthogonal polynomial. Specifically, the joint eigenvalue densities of the Gaussian (Hermite), Wishart (Laguerre) and MANOVA (Jacobi) ensembles given by (4.3), (4.5), and (4.7) can be written in terms of the weight functions where $\Delta(\Lambda) = \prod_{i<j}|\lambda_i - \lambda_j|$ is the absolute value of the Vandermonde determinant.

6.1. Equilibrium measure and the Lanczos method

In orthogonal polynomial theory, given a weight function $w(x)$, with integral 1, we obtain Gaussian quadrature formulas for computing

$$\int f(x)w(x)\,dx \approx \sum_{i=1}^{n} f(x_i)q_i^2.$$

In other words, we have the approximation

$$w(x) \approx \sum \delta(x - x_i)q_i^2.$$

Here the x_i are the roots of the nth orthogonal polynomial, and the q_i^2 are the related Christoffel numbers also obtainable from the nth orthogonal polynomial.

The Lanczos algorithm run on the continuous operator $w(x)$ gives a tridiagonal matrix whose eigenvalues are the x_i and the first component of the ith eigenvector is q_i.

As the q_i^2 weigh each x_i differently, the distribution of the roots

$$e_n(x) = \sum_{i=1}^{n} \delta(x - x_i)$$

converges to a different answer from $w(x)$. For example, if $w(x)$ corresponds to a Gaussian, then the limiting $e_n(x)$ is semi-circular. Other examples are listed in Table 4.2.

These limiting measures, $e(x) = \lim_{n \to \infty} e_n(x)$, are known as the equilibrium measure for $w(x)$. They are characterized by a solution to a two-dimensional force equilibrium problem on a line segment. These equilibrium measures become the characteristic densities of random matrix theory as listed in Table 4.2. They have the property that, under the right conditions,

$$\operatorname{Re} m(x) = \frac{w'(x)}{w(x)},$$

where $m(x)$ is the Cauchy transform of the equilibrium measure.

In recent work, Kuijlaars (2000) has made the connection between the equilibrium measure and how Lanczos finds eigenvalues. Under reasonable assumptions, if we start with a large matrix, and take a relatively smaller number of Lanczos steps, then Lanczos follows the equilibrium measure. This is more or less intuitively clear. What he discovered was how one interpolates between the equilibrium measure and the original measure as the algorithm proceeds. There is a beautiful combination of a cut-off equilibrium measure and the original weight that applies during the transition.

For additional details on the connection see Kuijlaars and McLaughlin (2000). For a good reference on equilibrium measure, see Deift (1999, Chapter 6).

6.2. Matrix integrals

A strand of random matrix theory that is connected to the classical orthogonal polynomials is the evaluation of matrix integrals involving the joint eigenvalue densities. One can see this in works such as Mehta (1991).

Definition. Let A be a matrix with eigenvalues $\lambda_1, \ldots, \lambda_n$. The empirical distribution function for the eigenvalues of A is the probability measure

$$\frac{1}{n} \sum_{i=1}^{n} \delta(x - \lambda_i).$$

Definition. The level density of an $n \times n$ ensemble A with real eigenvalues is the distribution of a random eigenvalue chosen from the ensemble. Equivalently, it is the average (over the ensemble) empirical density. It is denoted by ρ_n^A.

There is another way to understand the level density in terms of a matrix integral. If one integrates out all but one of the variables in the joint (unordered) eigenvalue distribution of an ensemble, what is left is the level density.

Specifically, the level density can be written in terms of the joint eigenvalue density $f_A(\lambda_1, \ldots, \lambda_n)$ as

$$\rho_{n,\beta}^A(\lambda_1) = \int_{\mathbb{R}^{n-1}} f_A(\lambda_1, \ldots, \lambda_n) \, d\lambda_2 \cdots d\lambda_n.$$

For the case of the β-Hermite ensemble, this integral can be written in terms of its joint eigenvalue density in (4.3) as

$$\rho_{n,\beta}^H(\lambda_1) = c_H^\beta \int_{\mathbb{R}^{n-1}} |\Delta(\Lambda)|^\beta e^{-\sum_{i=1}^n \lambda_i^2/2} \, d\lambda_2 \cdots d\lambda_n. \tag{6.1}$$

The integral in (6.1) certainly looks daunting. Surprisingly, it turns out that closed form expressions are available in many cases.

6.3. Matrix integrals for complex random matrices

When the underlying random matrix is complex ($\beta = 2$), some matrix integrals become particularly easy. They are an application of the Cauchy–Binet theorem that is sometimes familiar to linear algebraists from texts such as Gantmacher and Krein (2002).

Theorem 6.1. (Cauchy–Binet) Let $C = AB$ be a matrix product of any kind. Let $M\left(\begin{smallmatrix} i_1 \cdots i_p \\ j_1 \cdots j_p \end{smallmatrix}\right)$ denote the $p \times p$ minor

$$\det(M_{i_k j_l})_{1 \leq k \leq p, 1 \leq l \leq p}.$$

In other words, it is the determinant of the submatrix of M formed from rows $i_1, \ldots, i_p$ and columns $j_1, \ldots, j_p$.

The Cauchy–Binet theorem states that

$$C\begin{pmatrix} i_1, \ldots, i_p \\ k_1, \ldots, k_p \end{pmatrix} = \sum_{j_1 < j_2 < \cdots < j_p} A\begin{pmatrix} i_1, \ldots, i_p \\ j_1, \ldots, j_p \end{pmatrix} B\begin{pmatrix} j_1, \ldots, j_p \\ k_1, \ldots, k_p \end{pmatrix}.$$

Notice that when $p = 1$ this is the familiar formula for matrix multiplication. When all matrices are $p \times p$, then the formula states that

$$\det C = \det A \det B.$$

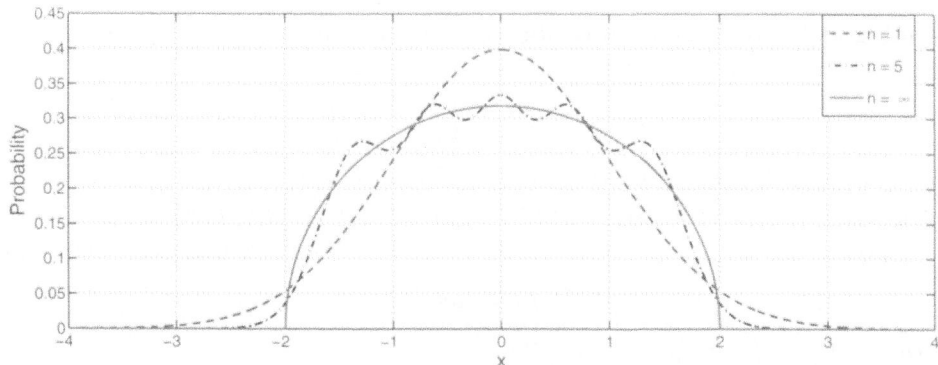

Figure 6.1. Level density of the GUE ensemble ($\beta = 2$) for different values of n. The limiting result when $n \to \infty$ is Wigner's famous semi-circular law.

Cauchy–Binet extends in the limit to matrices with infinitely many columns. If the columns are indexed by a continuous variable, we now have a vector of functions.

Replacing A_{ij} with $\varphi_i(x_j)$ and B_{jk} with $\psi_k(x_j)$, we see that Cauchy–Binet becomes

$$\det C = \int \cdots \int \det(\varphi_i(x_j))_{i,j=1,\ldots,n} \det(\psi_k(x_j))_{k,j=1,\ldots,n} \, dx_1 \, dx_2 \cdots dx_n.$$

where $C_{ik} = \int \varphi_i(x)\psi_k(x) \, dx$, $i,k = 1,\ldots,n$.

This continuous version of Cauchy–Binet may be traced back to Andréief (1883).

We assume that $\beta = 2$ so that $w_n(x) = \Delta(x)^2 \prod_{i=1}^n w(x_i)$. For classical weight function $\omega(x)$, Hermitian matrix models have been constructed. We have already seen the GUE corresponding to Hermite matrix models, and complex Wishart matrices for Laguerre. We also get the complex MANOVA matrices corresponding to Jacobi.

Notation: we define $\phi_n(x) = p_n(x)w(x)^{1/2}$. Thus the $\phi_i(x)$ are not polynomials, but they do form an orthonormal set of functions on the support of the weight function, $w(x)$.

It is a general fact that the level density of an $n \times n$ complex ($\beta = 2$) classical random matrix ensemble

$$f_w(x) = \frac{1}{n} \sum_{i=0}^{n-1} \phi_i(x)^2.$$

Figure 6.1 compares the normalized level density of the GUE for different values of n using $w(x) = \frac{1}{\sqrt{2\pi}} e^{-x^2/2}$. When $n = 1$, it is simply the normal distribution. The asymptotic result is the celebrated Wigner's semi-circular law (Wigner 1958).

Analogously to the computation of the level density, given any function $f(x)$ one can ask for

$$E(f) \equiv E_{\omega_n}\left(\prod (f(x_i))\right).$$

When we have a matrix model, this is $E(\det(f(X)))$.

It is a simple result that $E(f) = \int (\det(\phi_i(x)\phi_j(x)f(x))_{i,j=0,\ldots,n-1}\, dx$. This implies, by the continuous version of the Cauchy–Binet theorem, that

$$E(f) = \det C_n,$$

where $(C_n)_{ij} = \int \phi_i(x)\phi_j(x)f(x)\, dx$.

Some important functions to use are $f(x) = 1 + \sum z_i(\delta(x - y_i))$. The coefficients of the resulting polynomial are then the marginal density of k eigenvalues. See Tracy and Widom (1998) for more details.

Another important function is $f(x) = 1 - \chi_{[a,b]}$, where $\chi_{[a,b]}$ is the indicator function on $[a, b]$. Then we obtain the probability that no eigenvalue is in the interval $[a, b]$. If b is infinite, we obtain the probability that all eigenvalues are less than a, that is, the distribution function for the largest eigenvalue.

Research on integrable systems is a very active area within random matrix theory in conjunction with applications in statistical physics, and statistical growth processes. Some good references on this subject are van Moerbeke (2001), Tracy and Widom (2000b), Its, Tracy and Widom (2001), Deift, Its and Zhou (1997) and Deift (2000). The connection with the Riemann–Hilbert problem is explored in Deift (1999), Kuijlaars (2003) and Bleher and Its (1999).

7. Multivariate orthogonal polynomials

We feel it is safe to say that classical orthogonal polynomial theory and the theory of special functions reached prominence in numerical computation just around or before computers were becoming commonplace. The knowledge has been embodied in such handbooks as Abramowitz and Stegun (1970), Erdélyi, Magnus, Oberhettinger and Tricomi (1981a), Erdélyi, Magnus, Oberhettinger and Tricomi (1981b), Erdélyi, Magnus, Oberhettinger and Tricomi (1955), Spanier and Oldham (1987) and Weisstein (2005).

Very exciting developments linked to random matrix theory are the orthogonal polynomials and special functions of a matrix argument. These are scalar functions of a matrix argument that depend on the eigenvalues of the matrix, but in highly nontrivial ways. They are not mere trivial generalizations of the univariate objects. They are also linked to the other set of special functions that arise in random matrix theory: the Painlevé equations (see Section 9).

We refer readers to works by James (1964), Muirhead (1982), and Forrester (2005) for statistical and random matrix applications, and Hanlon,

Stanley and Stembridge (1992) for combinatorial aspects. Stanley (1989) is another good reference on the subject.

The research terrain is wide open to study fully the general multivariate orthogonal polynomial theory. Generalizations of Lanczos and other applications seem like low-hanging fruit for anyone to pick. Also, the numerical computation of these functions were long considered out of reach. As we describe in Section 8, applications of dynamic programming have suddenly now made these functions computable.

Our goal is to generalize orthogonal polynomials $p_k(x)$ with respect to a weight function $w(x)$ on $[a, b]$. The objects will be denoted $p_\kappa(X)$, where $\kappa \equiv (k_1, k_2, \ldots)$ is a partition of K, i.e., $k_1 \geq k_2 \geq \cdots$ and $K = k_1 + k_2 + \cdots$. The partition κ is the multivariate degree in the sense that the leading term of $p_\kappa(X)$ is

$$\sum_{\text{sym terms}} \lambda_1^{k_1} \lambda_2^{k_2} \cdots ,$$

where the $\lambda_1 \leq \cdots \leq \lambda_n$ are the eigenvalues of X.

We define $W(X) = \det(w(X)) = \prod_i w(\lambda_i)$ for X such that $\lambda_1 \geq a$ and $\lambda_n \leq b$. The multivariate orthogonality property is then

$$\int_{aI \leq X \leq bI} p_\kappa(X) p_\mu(X) W(X) \, \mathrm{d}X = \delta_{\kappa\mu}.$$

The multivariate orthogonal polynomials may also be defined as polynomials in n variables:

$$\int_{\substack{a \leq x_i \leq b, \\ i=1,2,\ldots,n}} p_\kappa(x_1, \ldots, x_n) p_\mu(x_1, \ldots, x_n) \prod_{i<j} |x_i - x_j|^\beta \prod_{i=1}^n w(x_i) \, \mathrm{d}x_1 \cdots \mathrm{d}x_n = \delta_{\kappa\mu},$$

where $\beta = 1, 2, 4$, according to Table 3.2, or may be arbitrary.

The simplest univariate polynomials are the monomials $p_n(x) = x^n$. They are orthogonal on the unit circle. This is Fourier analysis. Formally we take $w(x) = 1$ if $|x| = 1$ for $x \in \mathbb{C}$. The multivariate version is the famous Jack polynomial $C_\kappa^{2/\beta}(X)$ introduced in 1970 by Henry Jack as a one-parameter family of polynomials that include the Schur functions ($\beta = 2, \alpha = 1$) and (as conjectured by Jack (1970) and later proved by Macdonald (1982)) the zonal polynomials ($\beta = 1, \alpha = 2$). The Schur polynomials are well known in combinatorics, representation theory and linear algebra in their role as the determinant of a generalized Vandermonde matrix: see Koev (2002). One may also define the Jack polynomials by performing the QR factorization on the matrix that expresses the power symmetric functions $p_\kappa(X) = \prod \mathrm{tr}(X^{k_i})$ in terms of the monomial symmetric function $m_\kappa(X) = \sum x_i^{\kappa_i}$. The Q in the QR decomposition becomes a generalized character table while R defines the Jack polynomials. Additional details may be found in Knop and Sahi (1997).

Dumitriu has built a symbolic package (MOPs) written in Maple, for the evaluation of multivariate polynomials symbolically. This package allows the user to write down and compute the Hermite, Laguerre, Jacobi and Jack multivariate polynomials.

This package has been invaluable in the computation of matrix integrals and multivariate statistics for general β or a specific $\beta \neq 2$ for which traditional techniques fall short. For additional details see Dumitriu and Edelman (2004).

8. Hypergeometric functions of matrix argument

The classical univariate hypergeometric function is well known:

$$
{}_pF_q(a_1, \ldots, a_p; b_1, \ldots, b_q; x) \equiv \sum_{k=0}^{\infty} \frac{(a_1)_k \cdots (a_p)_k}{k!(b_1)_k \cdots (b_q)_k} \cdot x^k,
$$

where $(a)_k = a(a+1)\cdots(a+k-1)$.

The multivariate version is

$$
{}_pF_q^\alpha(a_1, \ldots, a_p; b_1, \ldots, b_q; x_1, \ldots, x_n) \equiv \sum_{k=0}^{\infty} \sum_{\kappa \vdash k} \frac{(a_1)_\kappa \cdots (a_p)_\kappa}{k!(b_1)_\kappa \cdots (b_q)_\kappa} C_\kappa^\alpha(x_1, \ldots, x_n),
$$

where

$$
(a)_\kappa \equiv \sum_{(i,j) \in \kappa} \left(a - \frac{i-1}{\alpha} + j - 1 \right)
$$

is the Pochhammer symbol and $C_\kappa^\alpha(x_1, x_2, \ldots, x_n)$ is the Jack polynomial.

Some random matrix statistics of the multivariate hypergeometric functions are the largest and smallest eigenvalue of a Wishart matrix. As in Section 5, the Wishart matrix can be written as $L = BB^T$, where

$$
B = \begin{bmatrix} \chi_{2a} & & & \\ \chi_{\beta(n-1)} & \chi_{2a-\beta} & & \\ & \ddots & \ddots & \\ & & \chi_\beta & \chi_{2a-\beta(n-1)} \end{bmatrix},
$$

where $a = m\frac{\beta}{2}$. The probability density function of the smallest eigenvalue of the Wishart matrix is

$$
f(x) = x^{kn} \cdot e^{-\frac{nx}{2}} \cdot {}_2F_0^{2/\beta}\left(-k, \beta\frac{n}{2} + 1; ; -\frac{2}{x} I_{n-1} \right),
$$

where $k = a - (n-1)\frac{\beta}{2} - 1$ is a nonnegative integer. Figure 8.1 shows

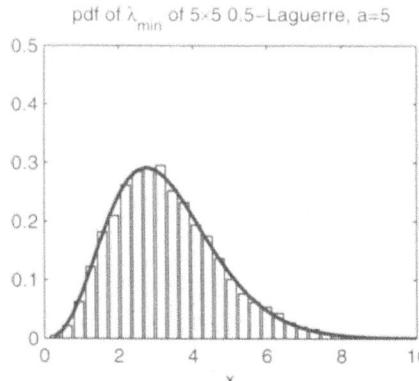 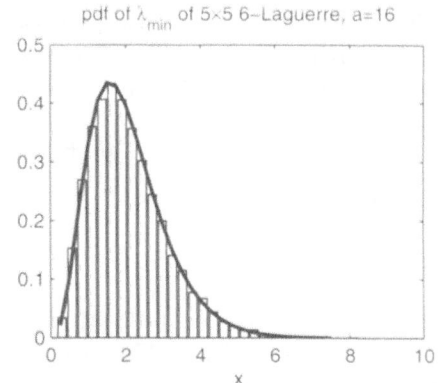

Figure 8.1. The probability density function
of $\lambda_{\min}$ of the β-Laguerre ensemble.

this distribution against a Monte Carlo simulation for 5×5 matrices with $\beta = 0.5$ and $a = 5$ and $\beta = 6$ and $a = 16$.

Hypergeometrics of matrix argument also solve the random hyperplane angle problem. One formulation picks two random p-hyperplanes through the origin in n dimensions and asks for the distribution of the angle between them. For numerical applications and the formulae see Absil, Edelman and Koev (2004).

A word on the computation of these multivariate objects. The numerical computation of the classical function is itself difficult if the user desires accuracy over a large range of parameters. Many articles and books on multivariate statistics consider the multivariate function difficult.

In recent work Koev has found an algorithm for computing matrix hypergeometrics based on exploiting the combinatorial properties of the Pochhammer symbol, dynamic programming, and the algorithm for computing the Jack function. For a specific computation, this replaces an algorithm in 2000 that took 8 days to one that requires 0.01 seconds. See Koev and Edelman (2004) for more details.

9. Painlevé equations

The Painlevé equations, already favourites of those who numerically study solitons, now appear in random matrix theory and in the statistics of zeros of the Riemann zeta function. In this section we introduce the equations, show the connection to random matrix theory, and consider numerical solutions matched against theory and random matrix simulations.

We think of the Painlevé equations as the most famous equations not found in the current standard handbooks. This will change rapidly. They are often introduced in connection to the problem of identifying second-order differential equations whose singularities may be poles depending on the initial conditions ('movable poles') and other singularities that are not movable. For example, the first-order equation

$$y' + y^2 = 0, \quad y(0) = \alpha$$

has solution

$$y(x) = \frac{\alpha}{\alpha x + 1},$$

which has a movable pole at $x = -1/\alpha$. (To repeat, the pole moves with the initial condition.) The equation

$$y'' + (y')^2 = 0, \quad y(0) = \alpha, \quad y'(0) = \beta$$

has solution

$$y(x) = \log(1 + x\beta) + \alpha.$$

This function has a movable log singularity $(x = -1/\beta)$ and hence would not be of the type considered by Painlevé.

Precisely, Painlevé allowed equations of the form $y'' = R(x, y, y')$, where R is analytic in x and rational in y and y'. He proved that the equations whose only movable singularities are poles can be transformed into either a linear equation, an elliptic equation, a Riccati equation or one of the six families of equations below:

(I) $y'' = 6y^2 + t,$

(II) $y'' = 2y^3 + ty + \alpha$

(III) $y'' = \frac{1}{y}y'^2 - \frac{y'}{t} + \frac{\alpha y^2 + \beta}{t} + \gamma y^3 + \frac{\delta}{y},$

(IV) $y'' = \frac{1}{2y}y'^2 + \frac{3}{2}y^3 + 4ty^2 + 2(t^2 - \alpha)y + \frac{\beta}{y},$

(V) $y'' = \left(\frac{1}{2y} + \frac{1}{y-1}\right)y'^2 - \frac{1}{t}y' + \frac{(y-1)^2}{t}\left(\alpha y + \frac{\beta}{y}\right) + \gamma\frac{y}{t} + \delta\frac{y(y+1)}{y-1},$

(VI) $y'' = \frac{1}{2}\left(\frac{1}{y} + \frac{1}{y-1} + \frac{1}{y-t}\right)y'^2 - \left(\frac{1}{t} + \frac{1}{t-1} + \frac{1}{y-t}\right)y'$

$\qquad\qquad + \frac{y(y-1)(y-t)}{t^2(t-1)^2}\left[\alpha - \beta\frac{t}{y^2} + \gamma\frac{t-1}{(y-1)^2} + \left(\frac{1}{2} - \delta\right)\frac{t(t-1)}{(y-t)^2}\right].$

A nice history of the Painlevé equation may be found in Takasaki (2000). Deift (2000) has a good exposition on this as well, where the connection to Riemann–Hilbert problems, explored in greater detail in Deift *et al.* (1997), is explained nicely. (A Riemann–Hilbert problem prescribes the jump condition across a contour and asks which problems satisfy this condition.)

In random matrix theory, distributions of some statistics related to the eigenvalues of the classical random matrix ensembles are obtainable from solutions to a Painlevé equation. The Painlevé II, III, V equations have been well studied, but others arise as well. More specifically, it turns out that integral operator discriminants related to the eigenvalue distributions satisfy differential equations, which involve the Painlevé equations in the large n limit. Connections between Painlevé theory and the multivariate hypergeometric theory of Section 7 are discussed in Forrester and Witte (2004) though more remains to be explored.

9.1. Eigenvalue distributions for large random matrices

In the study of eigenvalue distributions, two general areas can be distinguished. These are, respectively, the *bulk*, which refers to the properties of all of the eigenvalues and the *edges*, which (generally) addresses the largest and smallest eigenvalues.

A kernel $K(x, y)$ defines an operator K on functions f via

$$K[f](x) = \int K(x, y) f(y) \, dy. \tag{9.1}$$

With appropriate integration limits, this operator is well defined if $K(x, y)$ is chosen as in Table 9.1. Discretized versions of these operators are famous 'test matrices' in numerical analysis as in the case of the sine-kernel which discretizes to the prolate matrix (Varah 1993).

The determinant becomes a 'Fredholm determinant' in the limit of large random matrices. This is the first step in the connection to Painlevé theory. The full story may be found in the Tracy–Widom papers (Tracy and Widom 1993, 1994a, 1994b) and in the paper by Forrester (2000). The term 'soft

Table 9.1. Operator kernels associated with the different eigenvalue distributions.

Painlevé	Statistic	Interval $(s > 0)$	Kernel	$K(x, y)$
V	'bulk'	$[-s, s]$	sine	$\dfrac{\sin(\pi(x - y))}{\pi(x - y)}$
III	'hard edge'	$(0, s]$	Bessel	$\dfrac{\sqrt{y} J_\alpha(\sqrt{x}) J_\alpha'(\sqrt{y}) - \sqrt{x} J_\alpha(\sqrt{y}) J_\alpha'(\sqrt{x})}{2(x - y)}$
II	'soft edge'	$[s, \infty)$	Airy	$\dfrac{\mathrm{Ai}(x)\,\mathrm{Ai}'(y) - \mathrm{Ai}'(x)\,\mathrm{Ai}(y)}{x - y}$

A. EDELMAN AND N. R. RAO

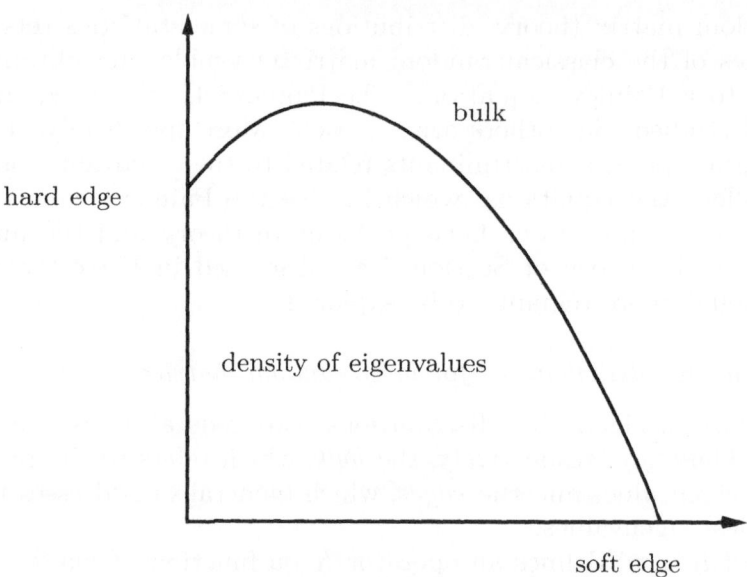

Figure 9.1. Regions corresponding to eigenvalue
distributions that are of interest in random matrix theory.

edge' applies (because there is still 'wiggle room') when the density hits
the horizontal axis, while the 'hard edge' applies when the density hits the
vertical axis (no further room on the left because of positivity constraints
on the eigenvalues, for example as is the case for the smallest eigenvalue of
the Laguerre and Jacobian ensembles). This is illustrated in Figure 9.1 and
is reflected in the choice of the integration intervals in Table 9.1 as well.

The distributions arising here are becoming increasingly important as
they are showing up in many places. Authors have imagined a world (perhaps in the past) where the normal distribution might be found experimentally or mathematically but without the central limit theorem to explain why.
This is happening here with these distributions as in the connection to the
zeros of the Riemann zeta function (discussed in Section 9.3), combinatorial problems (Deift 2000), and growth processes (Johansson 2000*a*). The
relevance of β in this context has not been fully explored.

9.2. The largest eigenvalue distribution and Painlevé II

The distribution of the appropriately normalized largest eigenvalues of the
Hermite ($\beta = 1, 2, 4$) and Laguerre ($\beta = 1, 2$) ensembles can be computed
from the solution of the Painlevé II equation:

$$q'' = sq + 2q^3 \tag{9.2}$$

with the boundary condition

$$q(s) \sim \text{Ai}(s), \qquad \text{as } s \to \infty. \tag{9.3}$$

The probability distributions thus obtained are the famous Tracy–Widom distributions.

The probability distribution $f_2(s)$, corresponding to $\beta = 2$, is given by

$$f_2(s) = \frac{\mathrm{d}}{\mathrm{d}s} F_2(s), \tag{9.4}$$

where

$$F_2(s) = \exp\left(-\int_s^\infty (x - s)q(x)^2 \, \mathrm{d}x\right). \tag{9.5}$$

The distributions $f_1(s)$ and $f_4(s)$ for $\beta = 1$ and $\beta = 4$ are the derivatives of $F_1(s)$ and $F_4(s)$ respectively, which are given by

$$F_1(s)^2 = F_2(s) \exp\left(-\int_s^\infty q(x) \, \mathrm{d}x\right) \tag{9.6}$$

and

$$F_4\left(\frac{s}{2^{\frac{2}{3}}}\right)^2 = F_2(s)\left(\cosh\left(\int_s^\infty q(x) \, \mathrm{d}x\right)\right)^2. \tag{9.7}$$

These distributions can be readily computed numerically. To solve using MATLAB, first rewrite (9.2) as a first-order system:

$$\frac{\mathrm{d}}{\mathrm{d}s}\begin{pmatrix} q \\ q' \end{pmatrix} = \begin{pmatrix} q' \\ sq + 2q^3 \end{pmatrix}. \tag{9.8}$$

This can be solved as an initial value problem starting at $s = s_0 = $ sufficiently large positive number, and integrating backwards along the s-axis. The boundary condition (9.3) then becomes the initial values

$$\begin{cases} q(s_0) &= \text{Ai}(s_0), \\ q'(s_0) &= \text{Ai}'(s_0). \end{cases} \tag{9.9}$$

This problem can be solved in just a few lines of MATLAB using the built-in Runge–Kutta-based ODE solver ode45. First define the system of equations as an inline function

```
deq=inline('[y(2); s*y(1)+2*y(1)^3]','s','y');
```

Next specify the integration interval and the desired output times:

```
s0=5;
sn=-8;
sspan=linspace(s0,sn,1000);
```

The initial values can be computed as

```
y0=[airy(s0); airy(1,s0)]
```

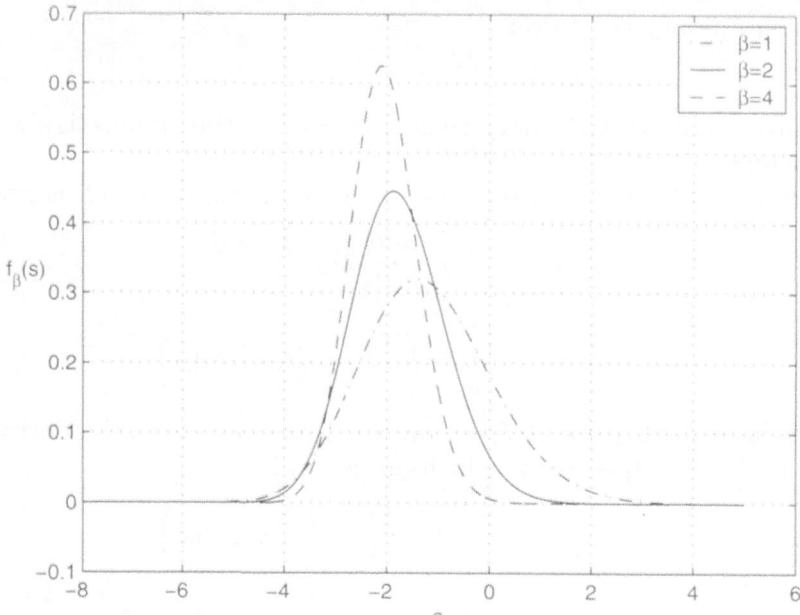

Figure 9.2. The Tracy–Widom distributions for $\beta = 1, 2, 4$.

Now, the integration tolerances can be set and the system integrated:

```
opts=odeset('reltol',1e-13,'abstol',1e-15);
[s,y]=ode45(deq,sspan,y0,opts);
q=y(:,1);
```

The first entry of the MATLAB variable y is the function $q(s)$. The distributions $F_2(s)$, $F_1(s)$ and $F_4(s)$ can be obtained from $q(s)$ by first setting the initial values:

```
dI0=I0=J0-0;
```

then numerically integrating to obtain:

```
dI=-[0;cumsum((q(1:end-1).^2+q(2:end).^2)/2.*diff(s))]+dI0;
I=-[0;cumsum((dI(1:end-1)+dI(2:end))/2.*diff(s))]+I0;
J=-[0;cumsum((q(1:end-1)+q(2:end))/2.*diff(s))]+J0;
```

Finally, using equations (9.5), (9.6), and (9.7) we obtain the desired distributions as:

```
F2=exp(-I);
F1=sqrt(F2.*exp(-J));
F4=sqrt(F2).*(exp(J/2)+exp(-J/2))/2;
s4=s/2^(2/3);
```

Note that the trapezoidal rule (cumsum function in MATLAB) is used to approximate numerically the integrals in (9.5), (9.6) and (9.7) respectively.

The probability distributions $f_2(s)$, $f_1(s)$, and $f_4(s)$ can then computed by numerical differentiation:

```
f2=gradient(F2,s);
f1=gradient(F1,s);
f4=gradient(F4,s4);
```

The result is shown in Figure 9.2. Note that more accurate techniques for computing the Tracy–Widom distributions are known and have been implemented as in Edelman and Persson (2002). Dieng (2004) discusses the numerics of another such implementation.

These distributions are connected to random matrix theory by the following theorems.

Theorem 9.1. (Tracy and Widom 2000a)   Let $\lambda_{\max}$ be the largest eigenvalue of $G_\beta(n,n)$, the β-Hermite ensemble, for $\beta = 1, 2, 4$. The normalized largest eigenvalue $\lambda'_{\max}$ is calculated as

$$\lambda'_{\max} = n^{\frac{1}{6}}(\lambda_{\max} - 2\sqrt{n}).$$

Then, as $n \to \infty$,

$$\lambda'_{\max} \xrightarrow{\mathcal{D}} F_\beta(s).$$

Theorem 9.2. (Johnstone 2001)   Let $\lambda_{\max}$ be the largest eigenvalue of $W_1(m,n)$, the real Laguerre ensemble ($\beta = 1$). The normalized largest eigenvalue $\lambda'_{\max}$ is calculated as

$$\lambda'_{\max} = \frac{\lambda_{\max} - \mu_{mn}}{\sigma_{mn}},$$

where μ_{mn} and σ_{mn} are given by

$$\mu_{mn} = (\sqrt{m-1} + \sqrt{n})^2, \sigma_{mn} = (\sqrt{m-1} + \sqrt{n})\left(\frac{1}{\sqrt{m-1}} + \frac{1}{n}\right)^{\frac{1}{3}}.$$

Then, if $m/n \to \gamma \geq 1$ as $n \to \infty$,

$$\lambda'_{\max} \xrightarrow{\mathcal{D}} F_1(s).$$

Theorem 9.3. (Johansson 2000b)   Let $\lambda_{\max}$ be the largest eigenvalue of $W_2(m,n)$, the complex Laguerre ensemble ($\beta = 2$). The normalized largest eigenvalue $\lambda'_{\max}$ is calculated as

$$\lambda'_{\max} = \frac{\lambda_{\max} - \mu_{mn}}{\sigma_{mn}},$$

where μ_{mn} and σ_{mn} are given by

$$\mu_{mn} = (\sqrt{m} + \sqrt{n})^2, \sigma_{mn} = (\sqrt{m} + \sqrt{n})\left(\frac{1}{\sqrt{m}} + \frac{1}{n}\right)^{\frac{1}{3}}.$$

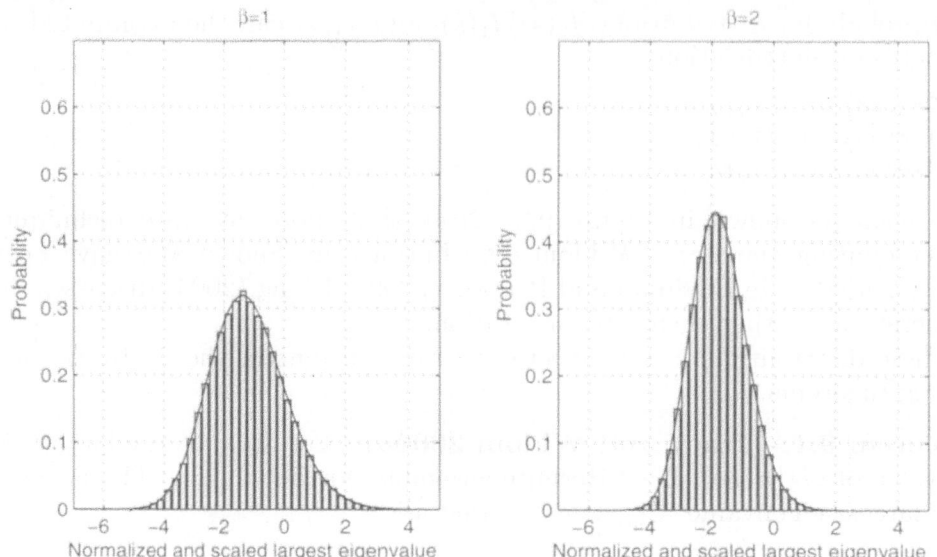

Figure 9.3. Probability distribution of scaled largest eigenvalue
of the Hermite ensembles (10^5 repetitions, $n = 10^9$).

Then, if $m/n \to \gamma \geq 1$ as $n \to \infty$,

$$\lambda'_{\max} \xrightarrow{\mathcal{D}} F_2(s).$$

Figure 9.3 compares the probability distribution of the scaled large eigenvalue of the GOE, and GUE with the numerical results for a billion by
billion matrix over 10^5 trials. We talk about how we generate data points
for a billion by billion matrix later in this article. Related results include
Soshnikov (1999). Dieng (2004) derives Painlevé-type expressions for the
distribution of the kth-largest eigenvalue in the GOE and GSE in the edge
scaling limit.

9.3. The GUE level spacings distribution and Painlevé V

Another quantity with an interesting probability distribution is the spacings of the eigenvalues of the Gaussian unitary ensemble, $G_2(n,n)$. The
normalized spacings of the eigenvalues $\lambda_1 \leq \lambda_2 \leq \cdots \leq \lambda_m$ are computed
according to

$$\delta'_k = \frac{\lambda_{k+1} - \lambda_k}{\pi \beta} \sqrt{2\beta n - \lambda_k^2}, \qquad k \approx n/2. \tag{9.10}$$

The distribution of the eigenvalues is almost uniform, with a slight deviation
at the two ends of the spectrum. Therefore, only half of the eigenvalues are
used, and one quarter of the eigenvalues at each end is discarded.

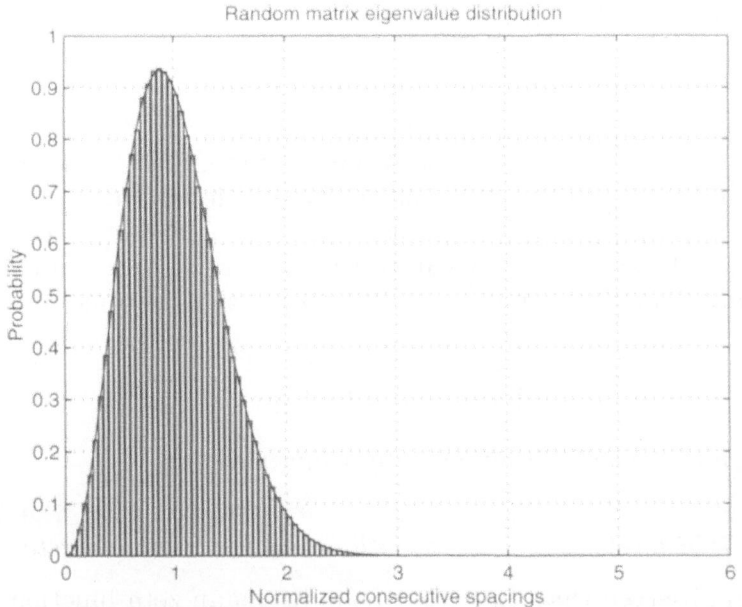

Figure 9.4. Probability distribution of consecutive spacings of
the eigenvalues of a GUE ensemble (1000 repetitions, $n = 1000$).

The probability distribution $p(s)$ for the eigenvalue spacings when $\beta = 2$
can be computed with the solution to the Painlevé V nonlinear differential
equation:

$$(t\sigma'')^2 + 4(t\sigma' - \sigma)\big(t\sigma' - \sigma + (\sigma')^2\big) = 0 \tag{9.11}$$

with the boundary condition

$$\sigma(t) \approx -\frac{t}{\pi} - \left(\frac{t}{\pi}\right)^2, \qquad \text{as } t \to 0^+. \tag{9.12}$$

Then $p(s)$ is given by

$$p(s) = \frac{\mathrm{d}^2}{\mathrm{d}s^2} E(s), \tag{9.13}$$

where

$$E(s) = \exp\left(\int_0^{\pi s} \frac{\sigma(t)}{t}\, \mathrm{d}t\right). \tag{9.14}$$

Explicit differentiation gives

$$p(s) = \frac{1}{s^2}\big(\pi s\sigma'(\pi s) - \sigma(\pi s) + \sigma(\pi s)^2\big) E(s). \tag{9.15}$$

The second-order differential equation (9.11) can be written as a first-order system of differential equations:

$$\frac{\mathrm{d}}{\mathrm{d}t}\begin{pmatrix} \sigma \\ \sigma' \end{pmatrix} = \begin{pmatrix} \sigma' \\ -\frac{2}{t}\sqrt{(\sigma - t\sigma')(t\sigma' - \sigma + (\sigma')^2)} \end{pmatrix}. \tag{9.16}$$

This is solved as an initial value problem starting at $t = t_0 =$ very small positive number. The value $t = 0$ has to be avoided because of the division by t in the system of equations. This is not a problem, since the boundary condition (9.12) provides an accurate value for $\sigma(t_0)$ (as well as $E(t_0/\pi)$). The boundary conditions for the system (9.16) then become

$$\begin{cases} \sigma(t_0) &= -\frac{t_0}{\pi} - (\frac{t_0}{\pi})^2, \\ \sigma'(t_0) &= -\frac{1}{\pi} - \frac{2t_0}{\pi}. \end{cases} \tag{9.17}$$

This system can be solved numerically using MATLAB.

9.4. The GUE level spacings distribution and the Riemann zeta zeros

It has been observed that the zeros of the Riemann zeta function along the critical line $\mathrm{Re}(z) = \frac{1}{2}$ (for z large) behave similarly to the eigenvalues of random matrices in the GUE. Here, the distribution of the scaled spacings of the zeros is compared to the corresponding level spacing distribution computed using the Painlevé V equation.

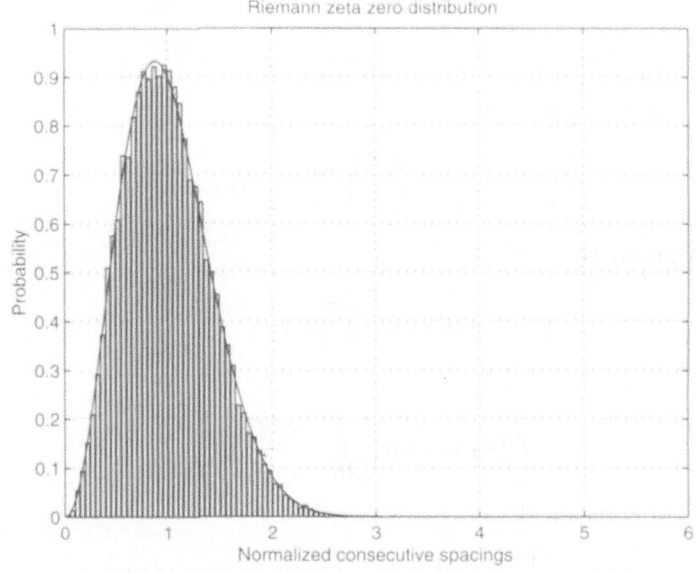

Figure 9.5. Probability distribution of consecutive spacings of Riemann zeta zeros ($30,000$ zeros, $n \approx 10^{12}, 10^{21}, 10^{22}$).

Define the nth zero γ_n as

$$\zeta\left(\frac{1}{2} + i\gamma_n\right) = 0, \qquad 0 < \gamma_1 < \gamma_2 < \cdots. \tag{9.18}$$

Compute a normalized spacing:

$$\tilde{\gamma}_n = \frac{\gamma_n}{\text{av spacing near } \gamma_n} = \gamma_n \cdot \left[\frac{\log \gamma_n/2\pi}{2\pi}\right]. \tag{9.19}$$

Zeros of the Riemann zeta function can be downloaded from Odlyzko (2001). Assuming that the MATLAB variable gamma contains the zeros, and the variable offset the offset, these two lines compute the consecutive spacings $\tilde{\gamma}_{n+1} - \tilde{\gamma}_n$ and plot the histogram:

```
delta=diff(gamma)/2/pi.*log((gamma(1:end-1)+offset)/2/pi);

% Normalize and plot the histogram of the spacings
```

The result can be found in Figure 9.5, along with the Painlevé V distribution.

10. Eigenvalues of a billion by billion matrix

We discuss how knowledge of numerical algorithms and software allow us to perform random matrix simulations very efficiently. In this case study, we illustrate an improvement rarely seen in computation. We succeeded in going from $n = 100$ to $n = 10^9$, i.e., we can compute the largest eigenvalue of a billion by billion matrix in the time required by naive methods for a hundred by hundred matrix. Pushing to $n = 10^{12}$ is within reach.

We devise a numerical experiment to verify that the distribution of the appropriately normalized and scaled largest eigenvalue of the GOE ensemble is given by the Tracy–Widom distribution $F_2(s)$ in (9.5).

Recall that an instance of the GOE ensemble ($\beta = 1$) can be created conveniently in MATLAB as:

```
A=randn(n);
A=(A+A')/2;
```

It is now straightforward to compute the distribution for $\lambda'_{\max}$ by simulation:

```
for idx=1:trials
  A=randn(n);
  A=(A+A')/2;
  lmax=max(eig(A));
  lmaxscaled=n^(1/6)*(lmax-2*sqrt(n));
  % Store lmax
end

% Create and plot histogram
```

The problem with this technique is that the computational requirements and the memory requirements grow rapidly with n. Storing the matrix A requires n^2 double-precision numbers, so on most computers today n has to be less than 10^4. Furthermore, computing all the eigenvalues of a full Hermitian matrix requires computing time proportional to n^3. This means that it will take many days to create a smooth histogram by simulation, even for relatively small values of n.

To improve upon this situation, we can instead study the β-Hermite tridiagonal ensemble as in Table 5.1:

$$
H_n^\beta \sim \frac{1}{\sqrt{2}} \begin{pmatrix} N(0,2) & \chi_{(n-1)\beta} & & & & \\ \chi_{(n-1)\beta} & N(0,2) & \chi_{(n-2)\beta} & & & \\ & \ddots & \ddots & \ddots & & \\ & & \chi_{2\beta} & N(0,2) & \chi_\beta \\ & & & \chi_\beta & N(0,2) \end{pmatrix}. \tag{10.1}
$$

Recall that $N(0,2)$ is a zero-mean Gaussian with variance 2, and χ_r is the square-root of a χ^2-distributed number with r degrees of freedom. Note that the matrix is symmetric, so the subdiagonal and the superdiagonal are always equal.

This matrix has a tridiagonal sparsity structure, and only $2n - 1$ double-precision numbers are required to store an instance of it. The time for computing the largest eigenvalue is proportional to n, either using Krylov subspace-based methods or the method of bisection (Trefethen and Bau 1997). This is certainly an improvement, though not substantial enough to do a simulation of a billion by billion GOE as in Figure 9.3.

The following code can, however, be used to compute the largest eigenvalue of a billion by billion GOE ($\beta = 1$):

```
beta=1; n=1e9; opts.disp=0; opts;issym=1;
alpha=10;k=round(alpha*n^(1/3)); % cutoff parameters
d=sqrt(chi2rnd(beta*n:-1:(n-k-1)))';
H=spdiags(d,1,k,k)+spdiags(randn(k,1),0,k,k);
H=(H+H')/sqrt(4*n*beta); % Scale so largest eigenvalue is near 1
eigs(H,1,1,opts);
```

The technology underlying this code is remarkable and deserves to be widely known. A number of interesting tricks are combined together.

- The observation that if $k = 10n^{1/3}$, then the largest eigenvalue is determined numerically by the top $k \times k$ segment of n. (This is related to the decay of the Airy function that arises in the kernel whose eigenvalues determine the largest eigenvalue distribution. The 'magic number' 10 here is not meant to be precise. It approximates the index k such that $\frac{v(k)}{v(1)} \approx \epsilon$, where $\epsilon = 2^{-52}$ for double precision arithmetic, and v is the eigenvector corresponding to the largest eigenvalue. For small β, it may be necessary to crank up the number 10 to a larger value.)

- Sparse matrix storage. (Only $O(n)$ storage is used.)

- Tridiagonal ensemble formulas. (Any beta is available due to the tridiagonal ensemble.)

- The Lanczos algorithm for eigenvalue computation. (This allows the computation of the largest eigenvalue faster than typical general purpose eigensolvers.)

- The shift-and-invert accelerator to Lanczos and Arnoldi. (Since we know the eigenvalues are near 1, we can accelerate the convergence of the largest eigenvalue.)

- The ARPACK software package as made available seamlessly in MATLAB. (The Arnoldi package contains state of the art data structures and numerical choices.)

Two of these tricks are mathematical. The first one is the ability to use tridiagonal ensembles to generate matrices whose eigenvalues match the GOE distribution. This allows us to avoid using dense matrices again for random matrix experiments. The second mathematical trick is the ability to cut off the top segment of the matrix to obtain accurately the largest eigenvalue.

It would be all too easy to take for granted the technology available for the remaining tricks. It was not so many years ago that the user would have to code up the sparse matrix storage made available by the 'spdiags' command or the ability to peel off the largest eigenvalue and give a starting guess that is made available in 'eigs'. Though numerical analysts are well versed in such numerical techniques, we would probably still not have bothered to implement the shift-and-invert Arnoldi-style algorithms ourselves. It has been said that technology advances by the number of operations that we do *not* have to think about. This is a great example.

Incidentally, for users interested in all of the eigenvalues of the tridiagonal matrix (Hermite ensembles such as the GOE, GUE, GSE) or all the singular values of a bidiagonal matrix (Laguerre ensembles such as Wishart matrices), then the LAPACK routines DSTEQR and DBDSQR can compute the eigenvalues with linear storage and in quadratic time. Users who simply use MATLAB's eig, Mathematica's Eigenvalues, or Maple's linalg[eigenvalues] are at a severe disadvantage.

We remark that further improvements are possible (and have been implemented!) if we use the approximation $\chi_n \approx \sqrt{n} + \frac{1}{\sqrt{2}} G$. This approximation forms the basis of the ideas in the next section. There are further tricks available, such as using the method of bisection (Trefethen and Bau 1997) and approximating χ_n with simply $\sqrt{n}$. See Edelman and Persson (2002) for more details.

11. Stochastic operators

For years, the first author was mystified by the notation $\sqrt{dt}$ often found in integrals connected with the Black–Scholes model of options pricing in finance. The simple fact that he was missing is that, if one has Gaussian random variables, the natural quantity that adds (thus, the linear function) is the variance, which is connected to the square of the variable.

There is some mathematics to be completed to understand fully how well-defined is the notion of the eigenvalues of a stochastic operator. Careful analysis will tell whether different discretizations give the same limiting eigenvalue distributions. Nonetheless, as we will outline, there is an idea here that we feel is sufficiently important that we can not afford to wait for this sort of analysis.

We define a Wiener process differentially as

$$dW = (\text{standard normal}) \cdot \sqrt{dt}.$$

The integral of such a process $W(t)$ (Brownian motion) is

$$W(t) = \int dW.$$

This is probably best graphed in MATLAB with the command:

```
t = [dt:dt:1]';
W = cumsum(randn(length(t),1))*sqrt(dt);
plot([0;t],[0;W])
```

where $dt =$ very small number not equal to zero and $W(0) = 0$. A good reference on Brownian motion is Karatzas and Shreve (1991).

Every time we 'roll the dice' we get a new W, but it is always the case that $W(t)$ is a Gaussian with variance t.

We are interested in operators exactly or closely related to the form

$$\frac{d^2}{dx^2} \quad + \quad V(x) \quad + \quad \sigma\, dW.$$

	$\uparrow$	$\uparrow$	$\uparrow$
Discretization:	tridiagonal	diagonal	diagonal or tridiagonal

When discretized each term can be thought of as a tridiagonal or diagonal matrix. The last part requires Gaussians.

11.1. From random matrices to stochastic operators

Consider the β-Hermite ensemble. The eigenvalue distribution of this ensemble is shared by a tridiagonal matrix with real elements that could be constructed as

$$
H_n^\beta = \frac{1}{\sqrt{2}}
\begin{bmatrix}
\sqrt{2}G & \chi_{\beta(n-1)} & & & \\
\chi_{\beta(n-1)} & \sqrt{2}G & \chi_{\beta(n-2)} & & \\
& \ddots & \ddots & \ddots & \\
& & \chi_{\beta\cdot2} & \sqrt{2}G & \chi_\beta \\
& & & \chi_\beta & \sqrt{2}G
\end{bmatrix}.
$$

This matrix is symmetric with independent entries in the upper triangular part. G represents an element taken from the standard Gaussian distribution, and χ_r represents an element taken from the χ-distribution with r degrees of freedom.

We are interested in the distribution of the largest eigenvalue, which is related to the solution of the Painlevé II transcendent.

Consider the β-Hermite ensemble from a numerical linear algebra point of view. The tridiagonal form suggests that H_n^β may be a finite difference approximation to some differential operator. We proposed that the β-Hermite ensemble is a finite difference approximation of the *stochastic Airy operator*:

$$
\frac{\mathrm{d}^2}{\mathrm{d}x^2} - x + \sigma\, \mathrm{d}W, \tag{11.1}
$$

in which $\mathrm{d}W$ represents a Wiener process. Recall that the Airy kernel in Table 9.1 plays an important role.

Hence, the random matrix model *itself* has a large n limit, and the eigenvalues should converge in distribution to the eigenvalues of the stochastic Airy operator as $n \to \infty$.

When $\sigma = 0$, the stochastic Airy operator in (11.1) specializes to the well-known, non-noisy, Airy operator on $[0, \infty)$ with boundary condition $u(0) = 0$. It has a particularly simple eigendecomposition in terms of the Airy special function,

$$
\left(\frac{\mathrm{d}^2}{\mathrm{d}x^2} - x\right) u_i(x) = u_i''(x) - x u_i(x) = \lambda_i u_i(x),
$$

which has solutions

$$
u_i(x) = \frac{1}{\mathrm{Ai}'(\lambda_i)^2}\, \mathrm{Ai}(x + \lambda_i),
$$

where λ_i is the ith root of the Airy function, $\mathrm{Ai}(x)$.

We can discretize the non-noisy Airy operator using finite differences. Taking some mesh size $h = h(n) \to 0$ and defining $x_i = hi$, the matrix

$$A_n = \frac{1}{h^2} \begin{bmatrix} -2 & 1 & & & & \\ 1 & -2 & 1 & & & \\ & \ddots & \ddots & \ddots & & \\ & & 1 & -2 & 1 & \\ & & & 1 & -2 \end{bmatrix} - \begin{bmatrix} x_1 & & & & \\ & x_2 & & & \\ & & \ddots & & \\ & & & x_{n-1} & \\ & & & & x_n \end{bmatrix}$$

$$= \frac{1}{h^2} D_n^2 - h\, \mathrm{diag}(1, 2, \ldots, n)$$

is a natural finite difference approximation to the non-noisy Airy operator, *i.e.*, the stochastic Airy operator in (11.1) with $\sigma = 0$. We expect the eigenvalues nearest 0 and the corresponding eigenvectors to converge to the eigenvalues and eigenfunctions of the Airy operator as $n \to \infty$.

The β-Hermite ensemble H_n^β, which is clearly 'noisy', admits a similar representation. There are some manipulations that need to be done to get to that form.

The first step is to obtain the right scaling, focusing on the largest eigenvalue. From Tracy and Widom's result on the distribution of the largest eigenvalue, we know that the largest eigenvalue of

$$\tilde{H}_n^\beta = \frac{\sqrt{2}}{\sqrt{\beta}} n^{1/6} (H_m^\beta - \sqrt{2\beta n}\, I)$$

converges in distribution as $n \to \infty$ for $\beta = 1, 2, 4$.

Using the approximation $\chi_r \approx \sqrt{r} + \frac{1}{\sqrt{2}} G$, valid for large r, and breaking the matrix into a sum of a non-random part and a random part, it follows that

$$\tilde{H}_n^\beta \approx \begin{bmatrix} -2n^{2/3} & n^{1/6}\sqrt{n-1} & & & & \\ n^{1/6}\sqrt{n-1} & -2n^{2/3} & n^{1/6}\sqrt{n-2} & & & \\ & \ddots & \ddots & \ddots & & \\ & & & n^{1/6}\sqrt{2} & -2n^{2/3} & n^{1/6}\sqrt{1} \\ & & & & n^{1/6}\sqrt{1} & -2n^{2/3} \end{bmatrix}$$

$$+ \frac{1}{\sqrt{2\beta}} n^{1/6} \begin{bmatrix} 2G & G & & & \\ G & 2G & G & & \\ & \ddots & \ddots & \ddots & \\ & & G & 2G & G \\ & & & G & 2G \end{bmatrix}.$$

Next, replacing $\sqrt{n-i}$ with the first-order Taylor series expansion $\sqrt{n} - \frac{1}{2}n^{-1/2}i$, the following approximation is obtained:

$$
\tilde{H}_n^\beta \approx n^{2/3}
\begin{bmatrix}
-2 & 1 & & & & \\
1 & -2 & 1 & & & \\
& \ddots & \ddots & \ddots & & \\
& & 1 & -2 & 1 & \\
& & & 1 & -2 \\
\end{bmatrix}
+ \frac{1}{\sqrt{2\beta}} n^{1/6}
\begin{bmatrix}
2G & G & & & & \\
G & 2G & G & & & \\
& \ddots & \ddots & \ddots & & \\
& & & G & 2G & G \\
& & & & G & 2G \\
\end{bmatrix}
$$

$$
- \frac{1}{2} n^{-1/3}
\begin{bmatrix}
& 1 & & & & \\
1 & & 2 & & & \\
& \ddots & & \ddots & & \ddots \\
& & n-2 & & n-1 \\
& & & n-1 & \\
\end{bmatrix}.
$$

The first term is a second difference operator, the second term injects noise, and the third term resembles a diagonal multiplication. Introducing $h = n^{-1/3}$ and replacing the second and third terms with analogous diagonal matrices, preserving total variance, the final approximation obtained is:

$$
\tilde{H}_n^\beta \approx \frac{1}{h^2} D_n^2 - h\,\mathrm{diag}(1,2,\ldots,n) + \frac{2}{\sqrt{\beta}} \frac{1}{\sqrt{h}}\,\mathrm{diag}(G,G,\ldots,G)
$$

$$
\approx A_n + \frac{2}{\sqrt{\beta}} \frac{1}{\sqrt{h}}\,\mathrm{diag}(G,G,\ldots,G),
$$

$$
h = n^{-1/3}.
$$

This final approximation appears to be a reasonable discretization of the stochastic Airy operator

$$
L^\beta = \frac{\mathrm{d}^2}{\mathrm{d}x^2} - x + \frac{2}{\sqrt{\beta}}\,\mathrm{d}W, \tag{11.2}
$$

with the boundary conditions $f(0) = f(+\infty) = 0$, in which W is Gaussian white noise.

Therefore, the largest eigenvalue distribution of L^β should follow the Tracy–Widom distribution in the cases $\beta = 1, 2, 4$. Figure 11.1 plots the distribution for $\beta = 1, 2, 4$ and compares it to simulation results for $\beta = 1$.

The stochastic operator approach is also advantageous when dealing with 'general β'. The traditional approaches are limited to the cases $\beta = 1, 2, 4$. In the stochastic operator approach, β is related to the variance of the noise; specifically, $\sigma = 2/\sqrt{\beta}$ in the case of the stochastic Airy operator as in (11.2). Instead of working with three discrete values of β, the stochastic operators vary continuously with β. Numerical simulations, as in Figure 11.1, indicate

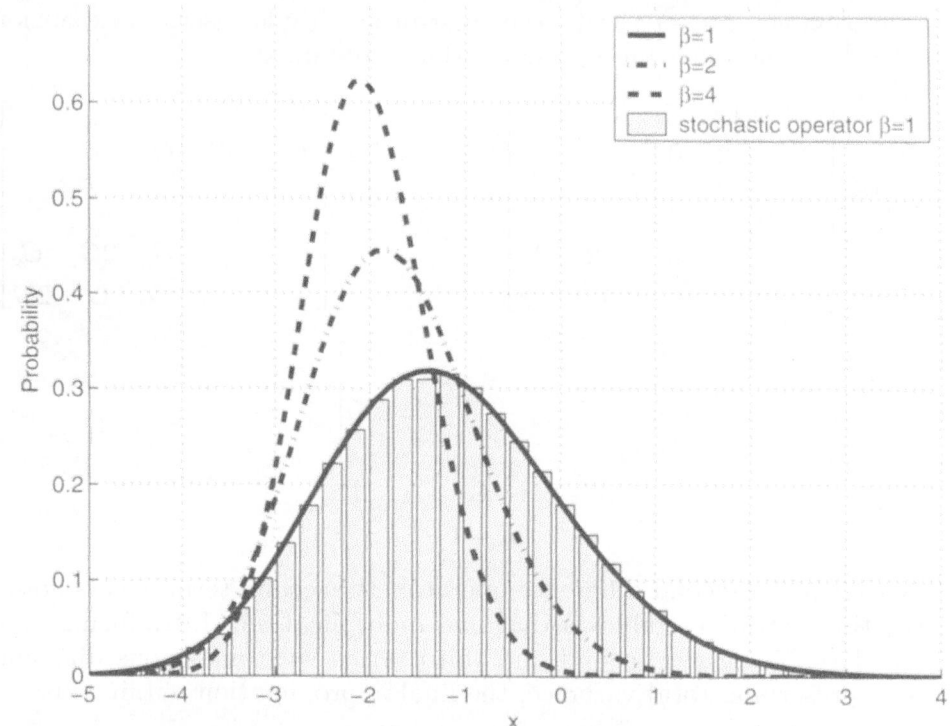

Figure 11.1. The largest eigenvalue distribution: comparison
of discretized stochastic Airy operator with the Tracy–Widom
law ($\beta = 1$). Monte Carlo simulations involved 10^5 trials of
500-by-500 matrices.

some sort of convection-diffusion process that can be explained in general
terms.

The diffusion comes from the high noise associated with small β. Increase
the volatility (decrease β) and we increase the range. The convection comes
from the repulsion of eigenvalues seen by any perturbation.

The reader can play with a simple experiment to observe the same phenomenon. Consider the 2×2 symmetric random matrix

$$\begin{bmatrix} x & z \\ z & y \end{bmatrix} + \frac{2}{\sqrt{\beta}} \begin{bmatrix} G & 0 \\ 0 & G \end{bmatrix},$$

where the G are independent standard normals. As $\beta \to 0$ the largest
eigenvalue will have a larger mean and a larger variance no matter what
matrix you start with, *i.e.*, for any choice of x, y, and z.

Similar stochastic operators corresponding to the discretization of the sine
and Bessel kernels in Table 9.1 can also be readily derived, as detailed in
Sutton (2005).

12. Free probability and infinite random matrices

There is a new mathematical field of 'free probability' emerging as a counterpart to classical probability. Some good references are Voiculescu, Dykema and Nica (1992), Hiai and Petz (2000) and Biane (2003). These references and even the name 'free probability' are worthy of some introduction. The forthcoming book by Speicher and Nica (2005) promises to serve as invaluable resource for making this subject more accessible.

We begin with a viewpoint on classical probability. If we are given probability densities f and g for random variables X and Y respectively, and if we know that X and Y are independent, we can compute the moments of $X + Y$, and XY, for example, from the moments of X and Y.

Our viewpoint on free probability is similar. Given two random matrices, A and B with eigenvalue density f and g, we would like to compute the eigenvalue densities for $A + B$ and AB in terms of the moments of f and g. Of course, A and B do not commute so we are in the realm of non-commutative algebra. Since all possible products of A and B are allowed, we have the 'free' product, *i.e.*, all words in A and B are allowed. (We recall that this is precisely the definition of the free product in algebra.) The theory of free probability allows us to compute the moments of these products in the limit of large matrices, as long as at least one of A or B has what amounts to eigenvectors that are essentially uniformly distributed with Haar measure. Speicher (2003) places these moment computations in an elegant combinatorial context.

We like to think of the difference between classical and free probability as being illustrated by the following maxim:

---

sum of the eigenvalues of random matrices   (*classical probability*)

versus

eigenvalues of the sum of random matrices   (*free probability*)

We take a closer look with an example.

Suppose A_i is an $m \times m$ matrix from the Gaussian orthogonal ensemble (GOE). Let λ_i be a random eigenvalue chosen uniformly from the m eigenvalues of A_i.

The classical central limit theorem states that if we form

$$\lambda = \frac{\lambda_1 + \lambda_2 + \cdots + \lambda_n}{\sqrt{n}},$$

no matter what m is, for large n, we obtain a normal distribution. The central limit theorem does not care at all that these λ_is were eigenvalues of random matrices.

However, if rather λ is a random eigenvalue of $A_1 + \cdots + A_n$ (eigenvalue of the sum), then λ is no longer normal. Free probability tells us that as $m, n \to \infty$, the λ follows Wigner's semi-circular density. This is the analogous 'free' central limit theorem for asymptotically large random matrices.

In a broader sense, free probability is studied in the context of non-commutative operator algebras. The synergy between random matrices and free probability arises because matrices are a natural model for a non-commutative algebra. The general theory of free probability is, however, more than just infinite random matrix theory.

In this sense, we find it remarkable that free probabilists were able to derive many of the well-known results in infinite random matrix theory by abstracting away the matrix in question. In special cases, techniques first used by Marčenko and Pastur (1967) and later perfected by Silverstein (1986) and Girko (1998) yield the same results as well. More details on these techniques can be found in Bai (1999) and the references therein.

12.1. Finite free probability

We propose that there is a finite counterpart, which we might call finite free probability. This is an area that is yet to be fully explored but some of the formulas for the moments of AB may be computed using the Jack polynomial theory mentioned in Section 7. There would be a beta dependence that is not necessary when $n = 1$ or $n = \infty$, but otherwise the theory is sensible.

In Figure 12.1, we illustrate (what we call) the finite free central limit theorem for a case when $n = 5$ and $\beta = 2$ (complex random matrices). The answer is neither a semi-circle as in standard free probability or a normal distribution as in classical probability. Here we took 5×5 complex Wishart

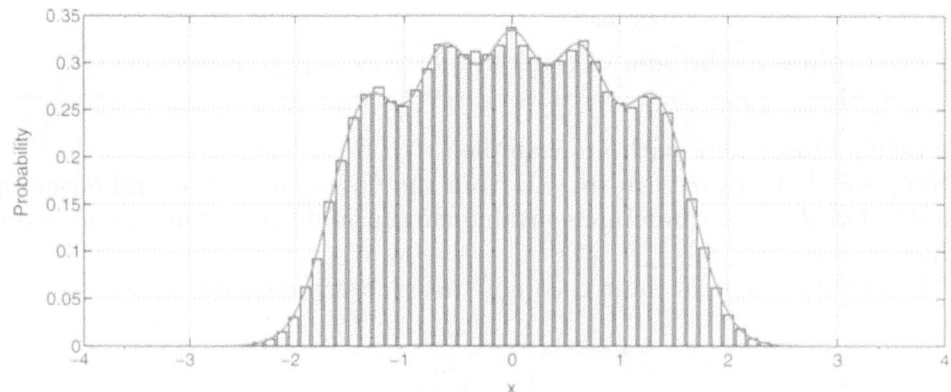

Figure 12.1. Finite free probability: the level density of the $\beta = 2$, $n = 5$ Hermite ensemble obtained by summing a large number of independent realizations of the $\beta = 2$, $n = 5$ Laguerre ensemble.

matrices, subtracted the mean and added them. There is a sensible notion of finite free probability, though it is not clear if finite free cumulants can or do exist. The details have yet to be worked out, though efforts by many authors on different fronts are underway. We invite readers to work in this area.

13. A random matrix calculator

In principle, the formulas from free probability allow us to combine very general combinations of random matrices and still compute the eigenvalue densities. In practice, however, researchers have been constrained from doing so because the relevant theorems are expressed explicitly in terms of transforms that are difficult to compute beyond some simple 'toy examples'.

It turns out that these theorems can be described implicitly as well. The key object is not the transform itself but the algebraic equation that the transform satisfies. The practical implication of this is that we can actually compute the limiting level density and moments for an infinitely large class of random matrices. We label such random matrices as 'characterizable'. Figure 13.1 uses a calculator analogy to describe how one characterizable matrix can be constructed from another.

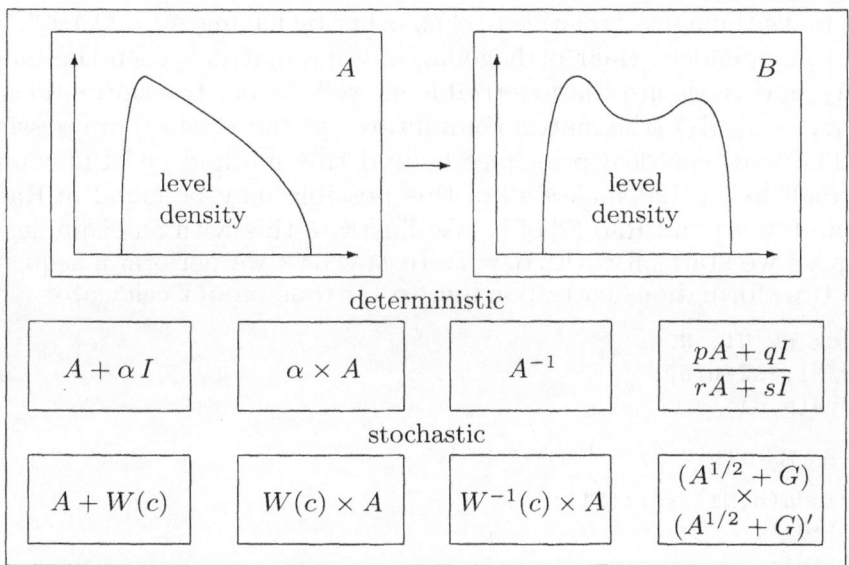

Figure 13.1. A random matrix calculator where a sequence of deterministic and stochastic operations performed on a 'characterizable' matrix A produces a 'characterizable' matrix B. The level density and moments of a 'characterizable' matrix can be computed analytically.

The 'buttons' in the top row of Figure 13.1 represent deterministic operations that can be performed on it (α, p, q, r, s are scalars). The 'buttons' in the bottom row are stochastic operations where additional randomness is injected.

The G matrix is an $m \times n$ matrix with independent, identically distributed (i.i.d.) zero mean elements with a variance of $1/n$ and bounded higher-order moments. We could generate G of this form in MATLAB as

```
G=randn(m,n)/sqrt(n);
```

or

```
G=sign(randn(m,n))/sqrt(n);
```

as examples. The $W(c)$ matrix is a Wishart-like matrix constructed as $W(c) = GG'$ where $m/n \to c > 0$ as $m, n \to \infty$.

The idea behind the calculator is that if we start off with a characterizable matrix A and if we were to generate the matrix B by pressing any of the buttons of the calculator we generate another characterizable matrix B. We can repeat this process forever, and by virtue of it being characterizable we can compute the limiting level density and limiting moments, often in closed form.

We can extend this idea even further by using the theorems of free probability. If we are given two characterizable random matrices, A_1 and A_2, then we can make them 'free' relative to each other by letting $A_2 = QA_2Q'$, where Q is an independent Haar orthogonal/unitary matrix. Then the matrices $A_1 + A_2$, and A_1A_2 are characterizable as well. Other transformations such as $i(A_1A_2 - A_2A_1)$ (the matrix commutator in Lie algebra) are possible as well. The mathematical principles behind this method and the computational realization that makes all of this possible may be found in Rao and Edelman (2005) and Rao (2005). We illustrate this with an example.

Suppose we start off with $A_1 = I$. In MATLAB we perform a sequence of simple transformations corresponding to buttons on our calculator:

```
% Pick n, N1, N2
c1=n/N1;  c2=n/N2;
A1=eye(n,n);
```

Then, we generate $A_2 = W_1(c_1) \times A_1$:

```
G1=randn(n,N1)/sqrt(N1);
W1=G1*G1';
A2=A1*W1;
```

Let $A_3 = A_2^{-1}$ and $A_4 = W_2(c_2) \times A_3$:

```
A3=inv(A2);
G2=randn(n,N2)/sqrt(N2);
W2=G2*G2';
A4=A3*W2
```

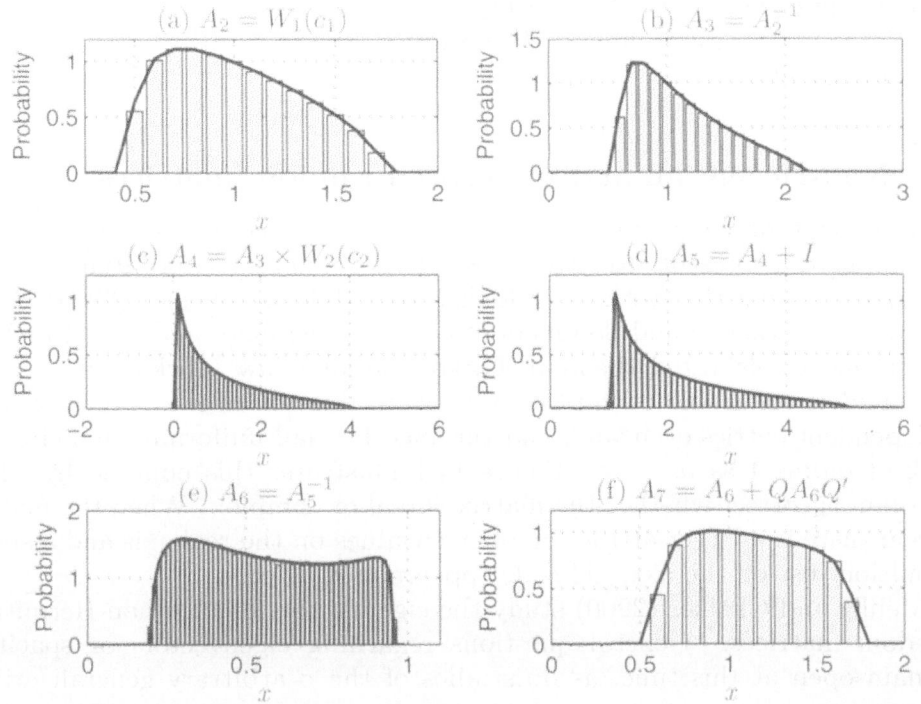

Figure 13.2. Comparison of the theoretical limiting level density (solid line) with the experimental level density for 1000 random matrix ensemble realizations with $c_1 = 0.1$, $c_2 = 0.625$, with $n = 100$, $N_1 = n/c_1 = 1000$ and $N_2 = n/c_2 = 160$.

Now, $A_5 = A_4 + I$ and $A_6 = A_5^{-1}$:

```
A5=A4+eye(n,n);
A6=inv(A5);
```

Generate a Haar unitary matrix and let $A_7 = A_6 + Q\,A_6\,Q'$:

```
[Q,R]=qr(randn(n)+i*randn(n));
Q=Q*diag(exp(2*pi*i*rand(n,1)));
A7=A6+Q*A6*Q';

% Collect eigenvalues
% Repeat over several trials
% Histogram eigenvalues
```

Since we constructed the matrices A_2 to A_7 using the 'buttons' of the random matrix calculator, they turn out to be characterizable. Figure 13.2 shows the limiting level density of these matrix ensembles compared with the experimental version. It is clear that although the predictions were asymptotic in nature (with respect to large n, N_1, N_2) the agreement with experimental data is excellent. Empirical evidence suggests that a 10×10 matrix is often 'good enough' to corroborate the limiting predictions of free probability.

14. Non-Hermitian and structured random matrices

Our understanding of non-Hermitian and structured random matrices is very limited at present. Relatively recent results on non-Hermitian random matrices include the works by Goldsheid and Khoruzhenko (2000), Fyodorov, Khoruzhenko and Sommers (1997), and Feinberg and Zee (1997).

The most celebrated theorem, Girko's circular law (Girko 1994) states that under reasonable conditions, the eigenvalues of an $n \times n$ matrix with independent entries of mean 0 and variance $1/n$ fall uniformly on a circular disk of radius 1 as $n \to \infty$. Figure 14.1 illustrates this numerically. The theorem is correct whether the matrix is real or complex. When the matrix is real there is a larger attraction of eigenvalues on the real axis and a small repulsion just off the axis. This disappears as $n \to \infty$.

Mehlig and Chalker (2000) study the eigenvectors of such non-Hermitian random matrices. General questions regarding eigenvectors or spacings remain open at this time, as do studies of the β-arbitrary generalization.

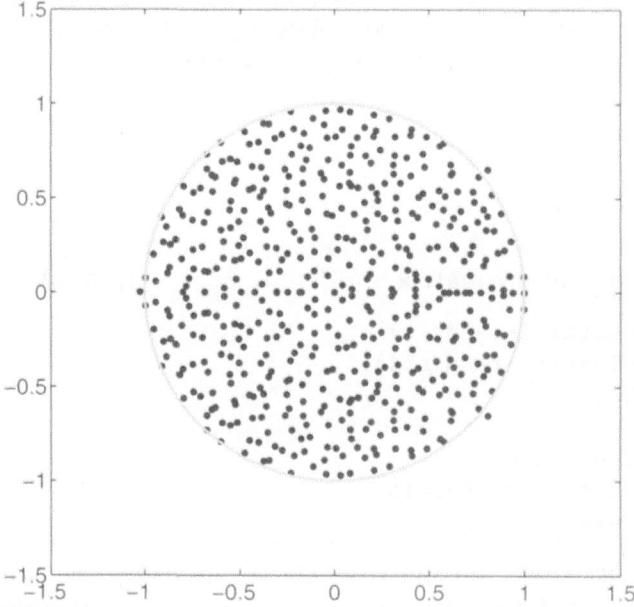

Figure 14.1. The eigenvalues of a 500×500
Gaussian random matrix (`randn(500)/sqrt(500)`
in MATLAB) in the complex plane.

The theory of pseudospectra is a rich area that allows for the study of non-Hermitian matrices, specifically those that are highly non-normal. Many tools for drawing pseudospectra are available, such as EigTool by Wright (2000). Figure 14.2 shows the pseudospectra for the same random matrix whose eigenvalues were plotted in Figure 14.1. The Pseudospectra Gateway compiled by Embree and Trefethen (2000) and their well-researched book, Trefethen and Embree (2005), contain discussions of nonsymmetric random matrices.

Random matrix theory is relevant in two distinct ways. An instance of a random matrix itself becomes a valuable object to study, as in Girko's circular law or in the Hatano and Nelson's non-Hermitian Anderson model as described by Trefethen, Contedini and Embree (2001). Also, perturbing a matrix randomly allows for insights into the pseudospectra and has been elevated to the level of a tool, as in the book by Chaitin-Chatelin and Fraysse (1996), where, for example, the Jordan structure is studied.

Another interesting result concerns the probability $p_{n,k}$ that $G_1(n,n)$ has k real eigenvalues. A formula for this may be found in Edelman (1997). Numerical analysts might be interested in the use of the real Schur decomposition in the computation $p_{n,k}$. This is the decomposition used in standard

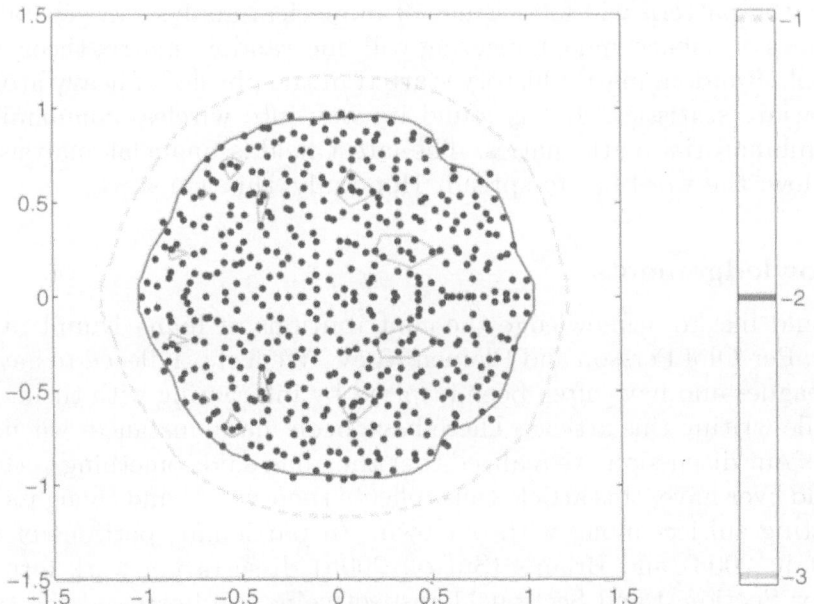

Figure 14.2. The pseudospectra of a 500×500 Gaussian random matrix (`randn(500)/sqrt(500)` in MATLAB). The illustration was produced with the `eigtool` pseudospectra plotter from Oxford University. The values on the colour bar show $10 \log_{10} \epsilon$.

eigenvalue computations. For example, to compute the probability that a matrix has all real eigenvalues, one integrates the measure on $G_1(n, n)$ over matrices of the form $A = QRQ^T$, where Q is orthogonal and R is upper triangular with ordered diagonal elements. This is the Schur form for real matrices with all real eigenvalues.

For random sparse matrices we refer the reader to Rodgers and Bray (1988) and Semerjian and Cugliandolo (2002), and the general theory of random graphs (Bollobás 1985). In Spiridonov (2005) one finds an interesting fractal pattern in the histogram of the eigenvalues of a sparse random matrix depending on the degree of sparsity.

The classical reference on deterministic Toeplitz matrices is Grenander and Szegő (1958). Recent work by Byrc, Dembo and Jiang (2005) provides a free probability-like characterization of the limiting spectral measure of Toeplitz, Hankel and Markov random matrices. Anderson and Zeitouni (2005) discuss central limit theorems related to generalized 'banded' random matrix models.

15. A segue

We make some final predictions about the application of random matrix theory: the pattern will follow that of numerical analysis in general. Most disciplines of science and engineering will find random matrix theory a valuable tool. Random matrix history started in the physics of heavy atoms and multivariate statistics. It has found its way into wireless communications and combinatorial mathematics. The latest field is financial analysis. More will follow; the word has to spread. Hopefully, this is a start.

Acknowledgements

We would like to acknowledge the contributions of Ioana Dumitriu, Brian Sutton, Per-Olof Persson and Plamen Koev. We feel privileged to have them as colleagues and have often been inspired by interacting with them, including while writing this article. There have been many instances when, in the midst of our discussion, we realized that they had said something better than we could ever have; this article thus reflects their words and thoughts on this fascinating subject along with our own. In particular, portions of Ioana's (Dumitriu 2003) and Brian's (Sutton 2005) dissertation work formed the basis for Section 4 and Section 11 respectively. We borrowed the MATLAB code in Section 9 from Persson's beautifully written documentation of the same in Edelman and Persson (2002). We thank Pierre-Antoine Absil, Peter Forrester, Matthew Harding, Nick Higham, Eric Kostlan, Julius Kusuma, Gil Strang, Craig Tracy, and Nick Trefethen for their valuable feedback. Especially, we thank Arieh Iserles and Brad Baxter for their comments and encouragement, Glennis Starling for being incredibly patient with us in the face of severe time constraints, and Brett Coonley for his typesetting wizardry. We thank the National Science Foundation (DMS-0411962) and the Singapore-MIT Alliance for supporting our work.

REFERENCES

M. Abramowitz and I. Stegun, eds (1970), *Handbook of Mathematical Functions*, Dover Publications, New York.

P.-A. Absil, A. Edelman and P. Koev (2004), 'On the largest principal angle between random subspaces'. Submitted.

D. Aldous and P. Diaconis (1999), 'Longest increasing subsequences: from patience sorting to the Baik–Deift–Johansson theorem', *Bull. Amer. Math. Soc.* **36**, 413–432.

G. W. Anderson and O. Zeitouni (2005), A CLT for a band matrix model: `arXiV.org/math.PR/0412040`.

C. Andréief (1883), 'Note sur une relation les intégrales définies des produits des fonctions', *Mém. de la Soc. Sci. Bordeaux* **2**, 1–14.

J.-M. Azaïs and M. Wschebor (2004), Upper and lower bounds for the tails of the distribution of the condition number of a Gaussian matrix: `www.lsp.ups-tlse.fr/Azais/publi/upper.ps`.

Z. D. Bai (1999), 'Methodologies in spectral analysis of large-dimensional random matrices: a review', *Statist. Sinica* **9**(3), 611–677. With comments by G. J. Rodgers and J. W. Silverstein, and a rejoinder by the author.

B. J. C. Baxter and A. Iserles (2003), On the foundations of computational mathematics, in *Handbook of Numerical Analysis*, Vol. XI, North-Holland, Amsterdam, pp. 3–34.

P. Biane (2003), Free probability for probabilists, in *Quantum probability communications, Vol. XI: Grenoble, 1998*, World Scientific, River Edge, NJ, pp. 55–71.

P. Bleher and A. Its (1999), 'Semiclassical asymptotics of orthogonal polynomials, Riemann–Hilbert problem, and the universality in the matrix model', *Ann. Math.* **150**, 185–266.

B. Bollobás (1985), *Random Graphs*, Academic Press, London.

A. Borodin (1999), 'Longest increasing subsequences of random colored permutations', *Electron. J. Combin.* **6**, # 13 (electronic).

A. Borodin and P. J. Forrester (2003), 'Increasing subsequences and the hard-to-soft edge transition in matrix ensembles', *J. Phys. A* **36**(12), 2963–2981.

W. Byrc, A. Dembo and T. Jiang (2005), Spectral measure of large random Hankel, Markov, and Toeplitz matrices: `arXiV.org/abs/math.PR/0307330`.

M. Caselle and U. Magnea (2004), 'Random matrix theory and symmetric spaces', *Phys. Rep.* **394**(2–3), 41–156.

F. Chaitin-Chatelin and V. Frayssé (1996), *Lectures on Finite Precision Computations: Software, Environments, and Tools*, SIAM, Philadelphia, PA.

A. Constantine (1963), 'Some noncentral distribution problems in multivariate analysis', *Ann. Math. Statist.* **34**, 1270–1285.

P. A. Deift (1999), *Orthogonal Polynomials and Random Matrices: A Riemann–Hilbert Approach*, Vol. 3 of *Courant Lecture Notes in Mathematics*, New York University Courant Institute of Mathematical Sciences, New York.

P. Deift (2000), 'Integrable systems and combinatorial theory', *Notices Amer. Math. Soc.* **47**(6), 631–640.

P. Deift, A. Its and X. Zhou (1997), 'A Riemann–Hilbert problem approach to asymptotic problems arising in the theory of random matrix models, and also in the theory of integrable statistical mechanics', *Ann. Math.* **146**(1), 149–235.

M. Dieng (2004), 'Distribution functions for edge eigenvalues in orthogonal and symplectic ensembles: Painlevé representations': arXiV.org/math.PR/0411421.

I. Dumitriu (2003), Eigenvalue statistics for beta-ensembles, PhD thesis, Department of Mathematics, Massachusetts Institute of Technology, Cambridge, MA.

I. Dumitriu and A. Edelman (2004), MOPS: Multivariate Orthogonal Polynomials (symbolically): arXiV.org/abs/math-ph/0409066.

F. J. Dyson (1963), 'The threefold way: algebraic structures of symmetry groups and ensembles in quantum mechanics', *J. Math. Phys.* **3**, 1199–1215.

A. Edelman (1989), Eigenvalues and condition numbers of random matrices, PhD thesis, Department of Mathematics, Massachusetts Institute of Technology, Cambridge, MA.

A. Edelman (1997), 'The probability that a random real Gaussian matrix has k real eigenvalues, related distributions, and the circular law', *J. Multivar. Anal.* **60**, 203–232.

A. Edelman and P.-O. Persson (2002), Numerical methods for random matrices. Technical report, Massachusetts Institute of Technology: arXiV.org/abs/math-ph/0501068.

A. Edelman and B. Sutton (2004), 'Tails of condition number distributions'. Submitted.

M. Embree and L. N. Trefethen (2000), 'Pseudospectra gateway': www.comlab.ox.ac.uk/pseudospectra.

A. Erdélyi, W. Magnus, F. Oberhettinger and F. G. Tricomi (1955), *Higher Transcendental Functions*, Vol. III, McGraw-Hill, New York/Toronto/London.

A. Erdélyi, W. Magnus, F. Oberhettinger and F. G. Tricomi (1981a), *Higher Transcendental Functions*, Vol. I, Robert E. Krieger, Melbourne, FL. Reprint of the 1953 original.

A. Erdélyi, W. Magnus, F. Oberhettinger and F. G. Tricomi (1981b), *Higher Transcendental Functions*, Vol. II, Robert E. Krieger, Melbourne, FL. Reprint of the 1953 original.

J. Feinberg and A. Zee (1997), 'Non-Hermitian random matrix theory: method of Hermitian reduction', *Nuclear Phys. B* **504**(3), 579–608.

P. J. Forrester (2000), Painlevé transcendent evaluation of the scaled distribution of the smallest eigenvalue in the Laguerre orthogonal and symplectic ensembles. Technical report: www.lanl.gov, arXiV.org/nlin.SI/0005064.

P. J. Forrester (2005), *Log-Gases and Random Matrices*, book in progress.

P. J. Forrester and N. S. Witte (2004), 'Application of the τ-function theory of Painlevé equations to random matrices: P_{VI}, the JUE, CyUE, cJUE and scaled limits', *Nagoya Math. J.* **174**, 29–114.

Y. Fyodorov, B. A. Khoruzhenko and H.-J. Sommers (1997), 'Almost-Hermitian random matrices: Crossover from Wigner–Dyson to Ginibre eigenvalue statistics', *Phys. Rev. Lett.* **79**, 557–560.

F. P. Gantmacher and M. G. Krein (2002), *Oscillation Matrices and Kernels and Small Vibrations of Mechanical Systems*, revised edn, AMS Chelsea Publishing, Providence, RI. Translation based on the 1941 Russian original.

W. Gautschi (1996), Orthogonal polynomials: Applications and computation, in *Acta Numerica*, Vol. 5, Cambridge University Press, pp. 45–119.

V. L. Girko (1994), 'The circular law: ten years later', *Random Oper. Stoch. Equations* **2**, 235–276, 377–398.

V. L. Girko (1998), *An Introduction to Statistical Analysis of Random Arrays*, VSP, Utrecht. Translated from the Russian.

I. Y. Goldsheid and B. A. Khoruzhenko (2000), 'Eigenvalue curves of asymmetric tridiagonal random matrices', *Electron. J. Probab.* **5**, # 16 (electronic).

G. Golub and W. Kahan (1965), 'Calculating the singular values and pseudo-inverse of a matrix', *SIAM J. Numer. Anal.* **2**, 205–224.

U. Grenander and G. Szegő (1958), *Toeplitz Forms and their Applications*, California Monographs in Mathematical Sciences, University of California Press, Berkeley.

P. J. Hanlon, R. P. Stanley and J. R. Stembridge (1992), Some combinatorial aspects of the spectra of normally distributed random matrices, in *Hypergeometric Functions on Domains of Positivity, Jack Polynomials, and Applications: Tampa, FL, 1991*, Vol. 138 of *Contemp. Math.*, AMS, Providence, RI, pp. 151–174.

F. Hiai and D. Petz (2000), *The Semicircle Law, Free Random Variables and Entropy*, Mathematical Surveys and Monographs, AMS.

A. S. Householder (1958), 'Unitary triangularization of a nonsymmetric matrix', *J. Assoc. Comput. Mach.* **5**, 339–342.

A. Its, C. A. Tracy and H. Widom (2001), 'Random words, Toeplitz determinants and integrable systems II', *Physica* **152–153D**, 199–224.

D. A. Ivanov (2002), Random-matrix ensembles in p-wave vortices, in *Proc. Dresden Workshop 'Vortices in Unconventional Superconductors and Superfluids'*, Springer, Heidelberg: arXiV.org/abs/cond-mat/0103089.

H. Jack (1970), 'A class of symmetric polynomials with a parameter', *Proc. R. Soc. Edinburgh* **69**, 1–18.

A. T. James (1964), 'Distributions of matrix variates and latent roots derived from normal samples', *Ann. Math. Stat.* **35**, 475–501.

K. Johansson (2000a), 'Random growth and random matrices', *European Congress of Mathematics* **I**, 445–456.

K. Johansson (2000b), 'Shape fluctuations and random matrices', *Comm. Math. Phys.* **209**, 437–476.

I. M. Johnstone (2001), 'On the distribution of the largest eigenvalue in principal components analysis', *Ann. Statist.* **29**(2), 295–327.

D. Jonsson (1982), 'Some limit theorems for the eigenvalues of a sample covariance matrix', *J. Multivar. Anal.* **12**, 1–38.

I. Karatzas and S. E. Shreve (1991), *Brownian Motion and Stochastic Calculus*, Vol. 113 of *Graduate Texts in Mathematics*, second edn, Springer, New York.

R. Killip and I. Nenciu (2004), 'Matrix models for circular ensembles', *Int. Math. Research Notices* **50**, 2665–2701.

F. Knop and S. Sahi (1997), 'A recursion and a combinatorial formula for Jack polynomials', *Invent. Math.* **128**(1), 9–22.

P. Koev (2002), Accurate and efficient matrix computation with structured matrices, PhD thesis, University of California, Berkeley.

P. Koev and A. Edelman (2004), The efficient evaluation of the hypergeometric function of a matrix argument. Preprint.

A. B. J. Kuijlaars (2000), 'Which eigenvalues are found by the Lanczos method?', *SIAM J. Matrix Anal. Appl.* **22**(1), 306–321 (electronic).

A. B. J. Kuijlaars (2003), Riemann–Hilbert analysis for orthogonal polynomials, in *Orthogonal Polynomials and Special Functions: Leuven, 2002*, Vol. 1817 of *Lecture Notes in Mathematics*, Springer, Berlin, pp. 167–210.

A. Kuijlaars and K. T.-R. McLaughlin (2000), 'Generic behavior of the density of states in random matrix theory and equilibrium problems in the presence of real analytic external fields', *Comm. Pure Appl. Math.* **53**, 736–785.

R. A. Lippert (2003), 'A matrix model for the β-Jacobi ensemble', *J. Math. Phys.* **44**(10), 4807–4816.

I. Macdonald (1982), 'Some conjectures for root systems', *SIAM J. Math. Anal.* **13**, 998–1004.

I. Macdonald (1998), *Symmetric Functions and Hall Polynomials*, Oxford Mathematical Monographs, 2nd edn, Oxford University Press.

B. D. McKay (1981), 'The expected eigenvalue distribution of a large regular graph', *Linear Algebra Appl.* **40**, 203–216.

V. Marčenko and L. Pastur (1967), 'Distribution of eigenvalues for some sets of random matrices', *Math USSR Sbornik* **1**, 457–483.

A. Mathai (1997), *Jacobians of Matrix Transformations and Functions of Matrix Arguments*, World Scientific, Singapore.

B. Mehlig and J. T. Chalker (2000), 'Statistical properties of eigenvectors in non-Hermitian Gaussian random matrix ensembles', *J. Math. Phys.* **41**(5), 3233–3256.

M. L. Mehta (1991), *Random Matrices*, second edn, Academic Press, Boston.

V. D. Milman and G. Schechtman (1986), *Asymptotic Theory of Finite Dimensional Normed Spaces*, Vol. 1200 of *Lecture Notes in Mathematics*, Springer.

R. J. Muirhead (1982), *Aspects of Multivariate Statistical Theory*, Wiley, New York.

A. M. Odlyzko (2001), 'Tables of zeros of the Riemann zeta function':
www.dtc.umn.edu/~odlyzko/zeta_tables/index.html.

A. Odlyzko and E. Rains (1998), On longest increasing subsequences in random permutations. Technical report, AT&T Laboratories.

I. Olkin (1953), 'Note on "The Jacobians of certain matrix transformations useful in multivariate analysis"', *Biometrika* **40**, 43–46.

I. Olkin (2002), 'The 70th anniversary of the distribution of random matrices: a survey', *Linear Algebra Appl.* **354**, 231–243.

E. M. Rains (1998), 'Increasing subsequences and the classical groups', *Electron. J. Combin.* **5**, # 12 (electronic).

N. R. Rao (2005), Infinite random matrix theory for multi-channel signal processing, PhD thesis, Massachusetts Institute of Technology.

N. R. Rao and A. Edelman (2005), The polynomial method for random matrices. Preprint.

G. J. Rodgers and A. J. Bray (1988), 'Density of states of a sparse random matrix', *Phys. Rev. B* (3) **37**(7), 3557–3562.

A. Sankar (2003), Smoothed analysis of Gaussian elimination, PhD thesis, Massachusetts Institute of Technology.

A. Sankar, D. A. Spielman and S.-H. Teng (2004), Smoothed analysis of the condition numbers and growth factors of matrices:
arXiV.org/abs/cs.NA/0310022.

G. Semerjian and L. F. Cugliandolo (2002), 'Sparse random matrices: the eigenvalue spectrum revisited', *J. Phys. A* **35**(23), 4837–4851.

J. Shen (2001), 'On the singular values of Gaussian random matrices', *Linear Algebra Appl.* **326**(1–3), 1–14.

J. W. Silverstein (1985), 'The smallest eigenvalue of a large-dimensional Wishart matrix', *Ann. Probab.* **13**(4), 1364–1368.

J. W. Silverstein (1986), Eigenvalues and eigenvectors of large dimensional sample covariance matrices, chapter in *Random Matrices and their Applications*, AMS, Providence, RI, pp. 153–159.

A. Soshnikov (1999), 'Universality at the edge of the spectrum in Wigner random matrices', *Comm. Math. Phys.* **207**, 697–733.

J. Spanier and K. B. Oldham (1987), *An Atlas of Functions*, Taylor & Francis/ Hemisphere.

R. Speicher (2003), Free probability theory and random matrices, in *Asymptotic Combinatorics with Applications to Mathematical Physics: St. Petersburg, 2001*, Vol. 1815 of *Lecture Notes in Mathematics*, Springer, Berlin, pp. 53–73.

R. Speicher and A. Nica (2005), Combinatorics of free probability. In preparation.

A. N. Spiridonov (2005), Spectra of sparse graphs and matrices. Preprint.

R. P. Stanley (1989), 'Some combinatorial properties of Jack symmetric functions', *Adv. Math.* **77**, 76–115.

G. W. Stewart (1980), 'The efficient generation of random orthogonal matrices with an application to condition estimators', *SIAM J. Numer. Anal.* **17**(3), 403–409 (loose microfiche suppl).

B. D. Sutton (2005), The stochastic operator approach to random matrix theory, PhD thesis, Massachusetts Institute of Technology, Department of Mathematics.

K. Takasaki (2000), 'Painlevé equations', Soliton Laboratory, Chronology of Mathematics:
www.math.h.kyoto-u.ac.jp/~takasaki/soliton-lab/chron/painleve.html.

C. A. Tracy and H. Widom (1993), Introduction to random matrices, in *Geometric and Quantum Aspects of Integrable Systems* (G. Helminck, ed.), Vol. 424 of *Lecture Notes in Physics*, Springer, Berlin, pp. 103–130.

C. A. Tracy and H. Widom (1994a), 'Fredholm determinants, differential equations and matrix models', *Comm. Math. Phys.* **163**, 33–72.

C. A. Tracy and H. Widom (1994*b*), 'Level-spacing distributions and the Bessel kernel', *Comm. Math. Phys.* **161**, 289–310.

C. A. Tracy and H. Widom (1998), 'Correlation functions, cluster functions and spacing distributions for random matrices', *J. Statist. Phys.* **92**, 809–835.

C. A. Tracy and H. Widom (2000*a*), The distribution of the largest eigenvalue in the Gaussian ensembles, in *Calogero–Moser–Sutherland Models*, Vol. 4 of *CRM Series in Mathematical Physics* (J. van Diejen and L. Vinet, eds), Springer, Berlin, pp. 461–472.

C. A. Tracy and H. Widom (2000*b*), Universality of the distribution functions of random matrix theory II, in *Integrable Systems: From Classical to Quantum* (J. Harnad, G. Sabidussi and P. Winternitz, eds), Vol. 26 of *CRM Proceedings & Lecture Notes*, AMS, Providence, pp. 251–264.

C. A. Tracy and H. Widom (2001), 'On the distributions of the lengths of the longest monotone subsequences in random words', *Probab. Theory Rel. Fields* **119**, 350–380.

L. N. Trefethen and D. Bau, III (1997), *Numerical Linear Algebra*, SIAM, Philadelphia, PA.

L. N. Trefethen and A. Embree (2005), *Spectra and Pseudospectra: The Behavior of Nonnormal Matrices and Operators*, Princeton.

L. N. Trefethen and R. S. Schreiber (1990), 'Average-case stability of Gaussian elimination', *SIAM J. Matrix Anal. Appl.* **11**(3), 335–360.

L. N. Trefethen, M. Contedini and M. Embree (2001), 'Spectra, pseudospectra, and localization for random bidiagonal matrices', *Comm. Pure Appl. Math.* **54**(5), 595–623.

H. F. Trotter (1984), 'Eigenvalue distributions of large Hermitian matrices; Wigner's semi-circle law and a theorem of Kac, Murdock, and Szegő', *Adv. Math.* **54**, 67–82.

C. F. Van Loan (2000), 'The ubiquitous Kronecker product', *J. Comput. Appl. Math.* **123**(1–2), 85–100.

P. van Moerbeke (2001), Integrable lattices: random matrices and random permutations, in *Random Matrices and their Applications*, MSRI Publications, Cambridge University Press, Cambridge.

J. M. Varah (1993), 'The prolate matrix', *Linear Algebra Appl.* **187**, 269–278.

D. Viswanath and L. N. Trefethen (1998), 'Condition numbers of random triangular matrices', *SIAM J. Matrix Anal. Appl.* **19**(2), 564–581 (electronic).

D. V. Voiculescu, K. J. Dykema and A. Nica (1992), *Free Random Variables*, AMS, Providence, RI.

J. von Neumann (1963), *Collected Works*, Vol. V: *Design of Computers, Theory of Automata and Numerical Analysis* (A. H. Taub, ed.), Pergamon Press, Macmillan, New York.

J. von Neumann and H. H. Goldstine (1947), 'Numerical inverting of matrices of high order', *Bull. Amer. Math. Soc.* **53**, 1021–1099.

E. W. Weisstein (2005), 'Special function', From MathWorld, A Wolfram Web Resource: `http://mathworld.wolfram.com/SpecialFunction.html`

E. P. Wigner (1955), 'Characteristic vectors of bordered matrices with infinite dimensions', *Ann. Math.* **62**, 548–564.

E. P. Wigner (1958), 'On the distribution of the roots of certain symmetric matrices', *Ann. Math.* **67**, 325–327.

J. Wishart (1928), 'The generalized product moment distribution in samples from a normal multivariate population', *Biometrika* **20A**, 32–52.

T. G. Wright (2000), 'EigTool':
`www.comlab.ox.ac.uk/projects/pseudospectra/eigtool`.

M.-C. Yeung and T. F. Chan (1997), 'Probabilistic analysis of Gaussian elimination without pivoting', *SIAM J. Matrix Anal. Appl.* **18** (2), 499–517.

M. R. Zirnbauer (1996), 'Riemannian symmetric superspaces and their origin in random-matrix theory', *J. Math. Phys.* **37**(10), 4986–5018.

Acta Numerica (2005), pp. 299–361
DOI: 10.1017/S0962492904000248

Numerical methods for large-scale nonlinear optimization

Nick Gould

Computational Science and Engineering Department,
Rutherford Appleton Laboratory,
Chilton, Oxfordshire, England
E-mail: `n.i.m.gould@rl.ac.uk`

Dominique Orban

Department of Mathematics and Industrial Engineering,
Ecole Polytechnique de Montréal,
2900, Bd E. Montpetit, H3T 1J4 Montréal, Canada
E-mail: `dominique.orban@polymtl.ca`

Philippe Toint

Department of Mathematics,
University of Namur,
61, rue de Bruxelles, B-5000 Namur, Belgium
E-mail: `philippe.toint@fundp.ac.be`

Recent developments in numerical methods for solving large differentiable nonlinear optimization problems are reviewed. State-of-the-art algorithms for solving unconstrained, bound-constrained, linearly constrained and non-linearly constrained problems are discussed. As well as important conceptual advances and theoretical aspects, emphasis is also placed on more practical issues, such as software availability.

CONTENTS

1. Introduction

Large-scale nonlinear optimization is concerned with the numerical solution of continuous problems expressed in the form

$$\underset{x \in \mathbb{R}^n}{\text{minimize}} \; f(x) \; \text{subject to} \; c_{\mathcal{E}}(x) = 0 \; \text{and} \; c_{\mathcal{I}}(x) \geq 0, \qquad (1.1)$$

where $f \colon \mathbb{R}^n \to \mathbb{R}$, $c_{\mathcal{E}} \colon \mathbb{R}^n \to \mathbb{R}^{n_{\mathcal{E}}}$ and $c_{\mathcal{I}} \colon \mathbb{R}^n \to \mathbb{R}^{n_{\mathcal{I}}}$ are smooth and n, and possibly $n_{\mathcal{E}}$ and/or $n_{\mathcal{I}}$, are large. Here, the components of the vector x are the *variables*, $f(x)$ is the *objective function* and the components of the vectors $c_{\mathcal{E}}(x)$ and $c_{\mathcal{I}}(x)$ are the *constraint* functions. Such problems arise throughout science, engineering, planning and economics. Fortunately algorithmic development and theoretical understanding generally continue to keep apace with the needs of such applications.

Our purpose in this paper is to review recent developments, with an emphasis on discussing state-of-the-art methods for various problem types fitting within the broad definition (1.1). As the title indicates, we will focus on nonlinear problems, that is, on problems for which at least one of the functions involved is nonlinear – although many of the methods for linear programming are variants of those in the nonlinear case, extra efficiencies are generally possible in this first case, and the general state of the art is to be able to solve linear problems perhaps ten times larger than nonlinear ones (Bixby, Fenlon, Gu, Rothberg and Wunderling 2000). We shall also mostly be concerned with large problems, that is, at the time of writing, those involving of the order of 100,000 variables and perhaps a similar number of constraints. However, we accept that this estimate may be too conservative for some problem classes – for instance, larger quadratic programs can certainly be solved today. Moreover, structure plays an important role in the size of problems that can be tackled: large sparse or partially separable cases are easier to handle than dense ones. Finally, the definition of a large problem may also depend on the hardware used, although this effect is less visible than in the past because of the remarkable evolution of personal computers in terms of memory processing power.

We will not review the history of the field here, but refer the interested reader to Gould and Toint (2004a) for a brief perspective and a discussion of the reasons why this mature research domain remains so active and why this is likely to continue for some time. The field has acquired a vast literature, and there have been numerous attempts to synthesize various aspects of it in books, such as those by Bertsekas (1995), Bonnans, Gilbert, Lemaréchal and Sagastizábal (1997), Dennis and Schnabel (1983), Fletcher (1981), Gill, Murray and Wright (1981), Moré and Wright (1993), Nash and Sofer (1990), Nocedal and Wright (1999), Conn, Gould and Toint (2000a), in volumes of conference proceedings, such as those edited by Coleman and Li (1990), Leone, Murli, Pardalos and Toraldo (1998), Di Pillo and Gianessi (1996, 1999), Di Pillo and Murli (2003), Hager, Hearn and Pardalos (1994), Spedicato (1994), Yuan (1998), in survey articles, like those given by Conn, Gould and Toint (1994, 1996), Fletcher (1987b), Forsgren, Gill and Wright (2002), Gould (2003), Marazzi and Nocedal (2001), Nash (2000b) and, in this series, by Boggs and Tolle (1995), Lewis and Overton (1996), Nocedal (1992), Powell (1998), Todd (2001), and Wright (1992).

The paper is structured as follows. Sections of the paper deal with problem classes: Section 2 covers unconstrained problems, while bound-constrained and linearly constrained problems are reviewed in Sections 3 and 4, respectively, and Section 5 considers general nonlinearly constrained cases. In each of these sections, subsections refer to method classes, allowing the interested reader to focus on these across different problem types. In particular, we discuss linesearch and trust region methods successively. We conclude most sections with a paragraph on practicalities and a paragraph on software. Final comments are made in Section 6.

2. Large-scale unconstrained optimization

2.1. General problems

Although general unconstrained optimization problems (that is, problems where $\mathcal{E}$ and $\mathcal{I}$ are empty in (1.1)) arise relatively infrequently in practice – nonlinear least-squares problems (see Section 2.2) being a notable exception – a brief discussion of methods for unconstrained optimization is useful if only for understanding those for problems involving constraints. For a fuller discussion see Nocedal (1992, 1997). While hybrids are possible, the essential distinction over the past 35 years has been between the linesearch and trust region approaches.

Given an estimate x_k of an unconstrained minimizer of $f(x)$, both paradigms rely on simple (differentiable) models $m_k(d)$ of the objective function $f(x_k + d)$. For *linesearch* methods m_k will normally be convex while this is not required in the trust region case; for both it is usually important that $m_k(0) = f(x_k)$ and $\nabla_x m_k(0) = \nabla_x f(x_k)$. Given a suitable model, a model-improving approximate minimizer d_k is computed. In the trust region case, possible unboundedness of the model is naturally handled by the trust region constraint $\|d\| \leq \Delta_k$ for some $\Delta_k > 0$. Since the model is only a local representation of the objective function, it is possible that predicted improvements in f may not actually be realized. Linesearch methods account for this by retracting the step along d_k so that $x_k + \alpha_k d_k$ gives an improvement in f. In contrast, *trust region* methods reject steps for which there is poor agreement between the decrease in m_k and f, and rely on a reduction of the radius Δ_{k+1}, and thus a re-computation of d_{k+1}, to ensure improvement. The mechanics of finding the step-size α_k for linesearch methods (Hager and Zhang 2003, Moré and Thuente 1994) and adjusting the radius Δ_k in trust region methods (Conn *et al.* 2000*a*, §17.1) has been much studied, and can have a significant effect on the performance of an algorithm. But overall the dominant computational cost of both classes of algorithms is in evaluating the values and required derivatives of f and in computing the step d_k; the cost of evaluating f often dominates in simulation-based applications or industry problems, but quite rarely in problems defined in commonly occurring modelling languages such as AMPL (Fourer, Gay and Kernighan 2003) or GAMS (Brooke, Kendrick and Meeraus 1988).

Computation of derivatives
In the early days, researchers invested much effort in finding methods with modest derivative requirements. Typically function values and, sometimes, gradients were available, but second derivatives frowned upon. The advent of automatic differentiation (Griewank 2000) and (group) partial separability (Griewank and Toint 1982*b*, Conn, Gould and Toint 1990) has somewhat altered this position at least amongst researchers, and now methods that are designed to exploit second derivatives (or good approximations thereof) are commonplace. But it is arguable that such new methods have not been as widely used by practitioners as might have been hoped, often because application codes capable of computing function values are unnameable to automatic differentiation for a variety of reasons, size and unavailability of the source-code being two common complaints. Indeed, there are still many practitioners who prefer methods that avoid derivatives at all (Powell 1998), although such methods are usually only appropriate for small-scale problems (but see Colson and Toint (2003) or Price and Toint (2004) for recent attempts to extend these techniques to large-scale cases).

Automatic differentiation offers the possibility of computing gradients and Hessian-vector products at a few times the cost of a function value (Griewank 2000). Tools for automatic differentiation are available both as stand-alone software or as part of modelling languages (AMPL and GAMS being good examples). Partial separability allows the computation of finite-difference gradients at a similar cost if only function values are available, and the same for Hessians if (structured) gradients can be found (Conn *et al.* 1990). Moreover, accurate structured secant approximations to second derivatives can be computed (Griewank and Toint 1982*b*), and this allows one to approximate gradients (by finite-differences) and Hessians (by secant formulae) just given function values if the problem functions are partially separable and the structure specified (Conn, Gould and Toint 1996).

Note that these comments on evaluating derivatives are of interest not only for unconstrained problems, but also for most of the other problems that are discussed in this paper. In the constrained case, the derivative of the constraint and Lagrangian functions will also be concerned, and the techniques to compute them are similar to what we have just mentioned.

Computation of the step

Even if function and derivative values are available, in general the cost of computing the step d_k may be significant if the problem involves a large number of variables. This computation often follows the following line: if H_k is a symmetric positive definite approximation to $\nabla_{xx} f(x_k)$, if the quasi-Newton (QN) model

$$m_k(d) = f(x_k) + d^T \nabla_x f(x_k) + \tfrac{1}{2} d^T H_k d \qquad (2.1)$$

is used, and if the minimizer of this model is sought, the resulting step d_k satisfies the QN equations

$$H_k d_k = -\nabla_x f(x_k). \qquad (2.2)$$

Since H_k is positive definite, realistic solution options include a (sparse) Cholesky factorization of H_k or application of the (preconditioned) conjugate gradient (CG) method (Hestenes and Stiefel 1952). The former may not be viable if the factors fill in significantly, but is capable of giving a numerical solution with small relative error. The latter is more flexible – rather than needing H_k, it merely requires a series of products $H_k p$ for given vectors p (and possibly preconditioned residuals $r = P_k^{-1} g$ for some suitable symmetric preconditioner P_k), and thus is better equipped for automatic differentiation or finite-difference gradient approximations ($\nabla_x f(x_k + \epsilon p) - \nabla_x f(x_k))/\epsilon$ for small ϵ – but less likely to be able to compute highly accurate numerical solutions of (2.2). When the approximation H_k is indefinite, it may be modified during factorization (Schlick 1993) or as the CG process proceeds (Nash 1984) to restore definiteness. Alternatively, the CG method may be terminated appropriately as soon as one of the products $H_k p$ in the CG method reveals negative curvature (Dembo and Steihaug 1983) or even continued in the subspace of positive curvature whilst gathering negative curvature information (Gould, Lucidi, Roma and Toint 2000).

A significant breakthrough for large-scale unconstrained optimization occurred in the early 1980s with the advent of truncated-QN methods (Dembo,

Eisenstat and Steihaug 1982). Here, rather than requiring that d_k satisfies (2.2), instead d_k is asked to satisfy

$$\|H_k d_k + \nabla_x f(x_k)\| \leq \eta_k \|\nabla_x f(x_k)\|, \tag{2.3}$$

where $0 < \eta_k < 1$ and $\eta_k \to 0$ if $\nabla_x f(x_k) \to 0$. This is helpful for use in conjunction with CG methods, since one could anticipate being able to satisfy (2.3) after few CG iterations for modest values of η_k. But more significantly – and perhaps overlooked by those who view CG as simply a method for solving linear systems – the iterates $\{d_{k,j}\}_{j \geq 0}$ generated by the CG method from x_k have two further fundamental properties. Firstly, by construction each successive CG step further reduces the model, that is, $m_k(d_{k,j+1}) < m_k(d_{k,j})$ for $j \geq 0$. Secondly, an appropriate norm of the CG iterates increases at each step, that is, $\|d_{k,j+1}\| > \|d_{k,j}\|$ for $j \geq 0$ (Steihaug 1983). This enables one to construct globally convergent linesearch (Dembo and Steihaug 1983) and trust region (Steihaug 1983, Toint 1981) truncated Newton methods, *i.e.*, methods that converge to local solutions from arbitrary starting points. In the linesearch case, d_k is chosen as the first $d_{k,j}$ for which (2.3) is satisfied, unless negative curvature is discovered when computing the required product $H_k p$ at CG iteration j, in which case either the steepest descent direction $-\nabla_x f(x_k)$ (when $j = 0$) or the current CG approximation $d_{k,j-1}$ (when $j > 0$) may be used instead (Dembo and Steihaug 1983). For the trust region case, such methods should be stopped on the trust region boundary if $\|d_{k,j}\| > \Delta_k$ or negative curvature is discovered, since once the CG iterates leave the trust region they will not return (Steihaug 1983). By judicious control of η_k in (2.3), such methods may also be shown to be superlinearly convergent under reasonable conditions on the approximation H_k to $\nabla_{xx} f(x_k)$.

In the trust region case, an accurate solution of the model problem needs to account for the trust region constraint $\|d\| \leq \Delta_k$. When H_k is positive semi-definite, the strategy of truncating the CG iteration on the trust region boundary (Steihaug 1983, Toint 1981) ensures a model decrease which is at least half as good as the optimal decrease (Yuan 2000). For indefinite H_k this is not so. Although there are excellent methods for solving the problem in the small-scale case (Moré and Sorensen 1983), these rely on being able to solve a (small) sequence of linear systems with coefficient matrices $H_k + \sigma_{k,l} I$ for given $\sigma_{k,l} \geq 0$, and thus implicitly on being able to factorize each coefficient matrix. Since this may be expensive or even impossible in the large-scale case, an alternative is to note that the CG and Lanczos methods compute different bases for the same Krylov space, and that after j steps of the Lanczos method, $Q_{k,j}^T H_k Q_{k,j} = T_{k,j}$ where the columns of the n by j matrix $Q_{k,j}$ are orthonormal and $T_{k,j}$ is tridiagonal. Thus if we seek an approximation to the solution of the trust region problem in the range of the expanding matrix $Q_{k,j}$, we may compute

$$d_{k,j} = Q_{k,j}h_{k,j}, \quad \text{where} \quad h_{k,j} = \underset{\|h\|\leq\Delta_k}{\arg\min} \; e_1^T Q_{k,j}^T \nabla_x f(x_k) e_1^T h + \tfrac{1}{2}h^T T_{k,j}h,$$

where $e_1 = [1,0,0,\ldots,0]^T$. Since $T_{k,j}$ is tridiagonal, we may reasonably factorize $T_{k,j} + \sigma_{k,j,l}I$, and thus the earlier Moré–Sorensen method is now applicable (Gould, Lucidi, Roma and Toint 1999). The Lanczos iteration may be truncated in a similar way to (2.3), preconditioning may be readily incorporated, and the resulting so-called GLTR method has been used as a subproblem solver in a number of large-scale optimization packages (Byrd, Gould, Nocedal and Waltz 2004a, Gould, Orban and Toint 2003a). Other iterative methods for the exact minimization of (2.1) within the trust region have been proposed (Hager 2001, Rendl and Wolkowicz 1997, Sorensen 1997), but as far as we are aware they have not been used in truncated form.

Another popular and effective method is the limited-memory secant approach (Gilbert and Lemaréchal 1989, Liu and Nocedal 1989, Nocedal 1980). Secant methods maintain Hessian approximations by sequences of low-rank updates, each using a pair of vectors (d_k, y_k), where $y_k = \nabla_x f(x_{k+1}) - \nabla_x f(x_k)$, to satisfy the secant condition $H_k d_k = y_k$ (Nocedal and Wright 1999, §2.2). Noting the success of (particularly) the BFGS secant method for small-scale computation, and recognizing that such methods are generally inappropriate for large problems because the generated matrices are almost invariably dense, the idea of limited memory methods is simply to use no more than m pairs $\{(d_j, y_j)\}_{j=k-m+1}^k$ to generate a secant approximation from a given, easily invertible initial matrix. If m is small, application of the resulting limited-memory approximation H_k or its inverse to a given vector may be performed extremely efficiently (Byrd, Nocedal and Schnabel 1994). Although this approach is perhaps most natural in a linesearch framework – because the QN direction $-H_k^{-1}\nabla_x f(x_k)$ is easy to obtain – it may also be used in a trust region one (Burke and Weigmann 1997, Kaufman 1999).

Since estimating H_k directly by secant methods is likely to be out of the question for large problems, an alternative we have already briefly mentioned is to exploit problem structure, and most especially partial separability, to obtain good Hessian approximations. By definition, a partially separable function has the form $f(x) = \sum_i f^{(i)}(x)$, where each element $f^{(i)}$ has a large invariant subspace. Thus it is reasonable to approximate $\nabla_{xx} f(x)$ by $\sum_i H^{(i)}$, where each $H^{(i)}$ approximates the low-rank element Hessian $\nabla_{xx} f^{(i)}(x)$. So-called partitioned QN methods (Griewank and Toint 1982c) use suitable secant formulae to build (often highly accurate) approximations $H^{(i)}$. Although the resulting $H_k = \sum_i H_k^{(i)}$ may not be as easily inverted as, say, that from a limited-memory method, it often gives more accurate approximations, and has been used with great success within a truncated CG framework (Conn $et\ al.$ 1990).

The final major class of methods are nonlinear variants of the CG method. Briefly, these methods aim to mimic the linear CG approach, and the step d_k is updated every iteration so that

$$d_{k+1} = -\nabla_x f(x_k) + \beta_k d_k$$

for some appropriate scalar β_k. Such methods have a long pedigree (Fletcher and Reeves 1964, Gilbert and Nocedal 1992, Polak and Ribière 1969, Powell 1977). Early methods chose β_k using formulae derived from the linear CG method, but sometimes subsequent steps tended to be closely dependent. A number of modifications have been proposed to avoid this defect, many of them resorting to steps in, or close to, the steepest-descent direction. The most successful recent methods (Dai and Yuan 2000, Hager and Zhang 2003) achieve this seamlessly, and additionally use linesearches with weak step-size acceptance criteria.

Practicalities

Despite the large number of papers devoted to large-scale unconstrained optimization, it is quite difficult to find comparisons between the various approaches proposed. A 1991 survey by Nash and Nocedal (1991) compares the limited-memory L-BFGS method (Liu and Nocedal 1989) with both the (early) Polak–Ribière nonlinear CG method (Polak and Ribière 1969) and a truncated-Newton method in which Hessian-vector products are obtained by differences. Although the results are mixed, the truncated-Newton approach seems preferable for problems well-approximated by a quadratic while L-BFGS appears best for more nonlinear problems. The nonlinear CG method is often best in terms of time, but requires more function evaluations. A contemporary survey by Gilbert and Nocedal (1992) which compares various nonlinear CG methods indicates there is little to choose between variants on the Polak–Ribière theme. However, while the test problems might have been large by 1990 standards, they are certainly not by today's. The only recent comparison we are aware of is that by Hager and Zhang (2003), in which their modern nonlinear CG method is compared with L-BFGS and Gilbert and Nocedal's (1992) improvement to Polak–Ribière. At least on the basis of these tests, modern nonlinear CG appears to be the method of choice if second derivatives are unavailable. However, we should exercise some caution as again the problems were not really large by today's standard, nor do we know how second-derivative-based truncated-Newton fits into the picture.

Two other issues are vital for good performance of many of the methods we have discussed. The first is preconditioning, where beyond very simple ideas such as diagonal or band scaling using Hessian terms (Conn *et al.* 1990), little has been done except for using standard incomplete factorization ideas from numerical linear algebra – Lin and Moré's (1999*a*) memory-conserving incomplete factorization is widely used in optimization circles. One interesting idea is to use a limited-memory approximation to H_k to precondition the next subproblem H_{k+1} (Morales and Nocedal 2000), although more experience is needed to see if this is generally applicable.

The second important advance is based on the observation that while there should be some overall monotonically reducing trend of function values in algorithms for minimization, this is not necessary for every iteration (Grippo, Lampariello and Lucidi 1986). Non-monotonic methods for unconstrained problems were first proposed in a linesearch framework (Grippo, Lampariello and Lucidi 1989), and have been observed to offer significant gains when compared with their monotone counterparts (Toint 1996). The same is true in a trust region context (Deng, Xiao and Zhou 1993, Toint 1997), and many algorithms now offer non-monotonic variants (Gould *et al.* 2003*a*).

Another technique that exploits the potential benefits of non-monotonicity uses the idea of filters. Inspired by multi-objective optimization and originally intended by Fletcher and Leyffer (2002) for constrained problems (see Section 5.1 below), the aim of a *filter* is to allow conflicting abstract objectives within the design of numerical algorithms. To understand the idea, consider an abstract situation where an algorithm attempts to simultaneously reduce two potentially conflicting objectives $\theta_1(x)$ and $\theta_2(x)$. A point x is then said to dominate another point y if and only if $\theta_i(x) < \theta_i(y)$ for $i = 1$ *and* 2 (this definition can obviously be generalized to more than two conflicting objectives). Remembering a dominated y is of little interest when aiming to reduce both θ_1 and θ_2 since x is at least as good as y for each objective. Obviously, an algorithm using this selection criterion should therefore store some or all pairs (θ_1, θ_2) corresponding to successful previous iterates.

It turns out that this concept allows the design of new non-monotonic techniques for unconstrained minimization. For convex problems, we know that finding the (unique) minimizer is equivalent to finding a zero of the gradient. This in turn may be viewed as the (potentially conflicting) objective of zeroing each of the n gradient components $[\nabla_x f(x)]_i$ $(i = 1, \ldots, n)$. One may therefore decide that a new trial point $x_k + d_k$ is not acceptable as a new iterate only if it is dominated by x_p, one of (a subset of) the previous iterates, in the sense that

$$|[\nabla_x f(x_p)]_i| < |[\nabla_x f(x_k + d_k)]_i| \qquad (2.4)$$

for all $i = 1, \ldots, n$, which corresponds to the choice $\theta_i(x) = |[\nabla_x f(x_k)]_i|$ $(i = 1, \ldots, n)$. The subset of previous iterates x_p for which the values of the gradient components are remembered and this comparison conducted is called the 'filter' and is maintained dynamically. If $x_k + d_k$ is not acceptable according to (2.4), it can still be evaluated using the more usual trust region technique, which then guarantees that a step is eventually acceptable and that a new iterate can be found. Unfortunately, this technique might prevent progress away from a saddle point for nonconvex problems, in which case an increase in the gradient components is warranted. The filter mechanism is thus modified to dynamically disregard the filter in these cases. The details of the resulting algorithm are described by Gould, Sainvitu and Toint (2004), where encouraging numerical results are also reported on both small- and large-scale problems.

Software

There is a lot of easily available software for unconstrained minimization. Here, and later, we refer the reader to the on-line software guides

> http://www-fp.mcs.anl.gov/otc/Guide/SoftwareGuide/ and
>
> http://plato.asu.edu/guide.html,

by Moré and Wright, and Mittelmann and Spellucci, respectively. Of the methods discussed in this section, TN/TNBC (Nash 1984) is a truncated CG method, LBFGS (Liu and Nocedal 1989) is a limited-memory QN method, VE08 (Griewank and Toint 1982c) is a partitioned QN method, and CG+ (Gilbert and Nocedal 1992) and CG_DESCENT (Hager and Zhang 2003) are nonlinear CG methods. In addition, software designed for more general problems – for example IPOPT, KNITRO, LANCELOT, LOQO and TRON – is often more than capable when applied in the unconstrained case.

2.2. Least-squares problems

Nonlinear least-squares problems, for which

$$f(x) = \frac{1}{2} \sum_{i=1}^{m} f_i^2(x),$$

are perhaps the major source of really unconstrained problems. In particular, large sets of nonlinear equations, parameter estimation in large dynamical systems and free surface optimization often result in sizeable and difficult instances (see Gould and Toint (2004a) for examples). Methods for solving problems of this type follow the general trends of Section 2.1, but specifically exploit the special form of the objective function to select – sometimes adaptively (Dennis, Gay and Welsh 1981) – between the 'full QN' model, where the matrix H_k in (2.1) is chosen to approximate the Hessian

$$\nabla_{xx} f(x_k) = J(x_k)^T J(x_k) + \sum_{i=1}^{m} f_i(x_k) \nabla_{xx} f_i(x_k)$$

(where $J(x)$ is the $m \times n$ matrix whose rows are the gradients $\nabla_x f_i(x)$), and the cheaper 'Gauss–Newton' model, for which $H_k = J(x_k)^T J(x_k)$. Furthermore, algorithmic stopping criteria can be adapted to exploit the special structure of $\nabla_x f(x)$ and the fact that zero provides an obvious lower bound on the value of the objective function.

Apart from the contributions of Al-Baali (2003) on dedicated QN updates, the work of Lukšan (1993, 1994, 1996) on incorporating iterative linear algebra techniques in trust region algorithms for nonlinear least-squares and that of Gulliksson, Söderkvist and Wedin (1997) on handling weights (and constraints), there has been little recent research in this area. Of course, most new ideas applicable to general unconstrained optimization may also be applied in the nonlinear least-squares case.

This is in particular the case for filter methods. In this context, the idea is to associate one filter objective $\theta_i(x)$ with each residual, *i.e.*, $\theta_i(x) = f_i(x)$ $(i = 1, \ldots, m)$, or perhaps with the norm of a block of residuals, *i.e.*, $\theta_i(x) = (\sum_{j \in J_i} f_j^2(x))^{\frac{1}{2}}$ for some $J_i \subset \{1, \ldots, m\}$. Details along with proofs of convergence are given by Gould, Leyffer and Toint (2005). Such ideas may be trivially extended to incorporate inequality constraints, thus providing practical means for solving the nonlinear feasibility problems (that is, to find a solution to a set of nonlinear equality and inequality constraints in the least-squares sense). Numerical efficiency and reliability is considered by Gould and Toint (2003a).

Software

The only dedicated large-scale nonlinear least-squares packages we are aware of are the sparsity-exploiting SPRNLP (Betts and Frank 1994), VE10 (Toint 1987), which uses the obvious partially separable structure of such problems, and the filter-based code FILTRANE from the GALAHAD library (Gould *et al.* 2003a). Of course much general-purpose software is applicable to nonlinear least-squares problems.

2.3. Discretized problems

In practice, many large-scale finite-dimensional unconstrained optimization problems arise from the discretization of those in infinite dimensions, a primary example being least-squares parameter identification in systems defined in terms of either ordinary or partial differential equations. The direct solution of such problems for a given discretization yielding the desired accuracy is often possible using general packages for large-scale numerical optimization (see Section 2.1). However, such techniques rarely make use of the underlying infinite-dimensional nature of the problem, for which several levels of discretization are possible, and thus such an approach rapidly becomes cumbersome. Multi-scale (sometimes known as multi-level) optimization aims at making explicit use of the problem structure in the hope of improving efficiency and, possibly, enhancing reliability.

Using differing scales of discretization for an infinite-dimensional problem is not a new idea. An obvious simple 'mesh refinement' approach is to use coarser grids in order to compute approximate solutions which can then be used as starting points for the optimization problem on a finer grid (Griewank and Toint 1982a, Bank, Gill and Marcia 2003, Betts and Erb 2003, Benson, McInnes, Moré and Sarich 2004). However, potentially more efficient techniques are inspired by the multigrid paradigm in the solution of partial differential equations and associated systems of linear algebraic equations (Brandt 1977, Bramble 1993, Hackbusch 1995, Briggs, Henson and McCormick 2000), and have only been discussed relatively recently in the optimization community. Contributions along this direction include the 'generalized truncated Newton algorithm' presented in Fisher (1998), and those by Moré (2003), Nash (2000a) and Lewis and Nash (2002, 2005). The latter three papers present the description of MG/OPT, a linesearch-based recursive algorithm, an outline of its convergence properties and impressive numerical results. The generalized truncated Newton algorithm and MG/OPT are very similar and, like many linesearch methods, naturally suited to convex problems, but their extension to nonconvex cases is also possible. Very recently, Gratton, Sartenaer and Toint (2004) have proposed a recursive multi-scale trust region algorithm (RMTR) which fits nonconvex problems more naturally and is backed by a strong convergence theory. The main idea of all the methods mentioned here is to (recursively) exploit the cheaper optimization on a coarse mesh to produce steps that significantly decrease the objective function on a finer mesh, while of course continuing to benefit from mesh refinement for obtaining good starting points. In principle, low frequency components of the problem solution (after suitable prolongation in the original infinite-dimensional space of interest) are determined by the coarse mesh calculations, and optimizing on the fine mesh then only fixes high frequency components.

While the idea appears to be very powerful and potentially leads to the solution of very large-scale problems, the practical algorithms that implement them are still mostly experimental. Preliminary numerical results are encouraging, but the true potential of these methods will only be confirmed by continued success in the coming years.

A second interesting approach to very large problems arising from continuous applications is to look at other ways to simplify them and make them more amenable to classical optimization techniques. For instance, Arian, Fahl and Sachs (2000) and Fahl and Sachs (2003) investigate the use of reduced-order models (using proper orthogonal decomposition techniques) in the framework of trust region algorithms, and apply this technique to fluid-mechanics problems. Note that model simplification of that type can also be thought of as a recursive process, although not immediately based on discretization. The idea is thus close in spirit to the proposals described above. Again, the practical power of this approach, although promising at this stage, is still the object of ongoing evaluation.

3. Large-scale bound-constrained optimization

In the simplest of constrained optimization problems, we seek the minimizer of $f(x)$ within a feasible box, $\Omega = \{x \mid l \le x \le u\}$ for given (possibly infinite) lower and upper bounds l and u. Without loss of generality, we assume that $l_i < u_i$ for all $i = 1, \ldots, n$. It has been argued that all unconstrained problems should actually include simple bounds to prevent bad effects of computer arithmetic such as overflows, and certainly many real problems have simple bounds to prevent unreasonable or physically impossible values.

Active set methods
Early methods for this problem tended to be of the active set variety. The active set at x is $\mathcal{A}(x) = \mathcal{L}(x) \cup \mathcal{U}(x)$, where $\mathcal{L}(x) = \{i \mid x_i = l_i\}$ and $\mathcal{U}(x) = \{i \mid x_i = u_i\}$. Trivially, if x_* is a (local) minimizer of f within Ω, x_* is a (local) minimizer of $f(x)$ subject to $x_i = l_i$, $i \in \mathcal{L}(x_*)$ and $x_i = u_i$, $i \in \mathcal{U}(x_*)$. Active set methods aim to predict $\mathcal{L}(x_*)$ and $\mathcal{U}(x_*)$ using suitably chosen disjoint sets $\mathcal{L}$, $\mathcal{U} \subseteq \{1, \ldots, n\}$. Given $\mathcal{L}$ and $\mathcal{U}$, a typical method will aim to (approximately)

$$\text{minimize } f(x)$$
$$\text{subject to } x_i = l_i, \ i \in \mathcal{L} \text{ and } x_i = u_i, \ i \in \mathcal{U};$$

such a calculation is effectively an unconstrained minimization over the variables (x_i), $i \notin \mathcal{A} = \mathcal{L} \cup \mathcal{U}$, and thus any of the methods mentioned in Section 2 is appropriate. Of course the predictions $\mathcal{L}$ and $\mathcal{U}$ may be incorrect, and the 'art' of active set methods is to adjust the sets as the iteration proceeds either by adding variables which violate one of their bounds or by removing those for which further progress is predicted – the same idea is possible (and indeed used) to deal with more general inequality constraints. See Gill *et al.* (1981, §5.5) or Fletcher (1987*a*, §10.3) for more details. Especially effective methods for the quadratic programming case, for which f is quadratic, have been developed (Coleman and Hulbert 1989).

The main disadvantage of (naive) active set methods for large-scale problems is the potential worst-case complexity in which each of the possible 3^n active sets is visited before discovering the optimal one. Although it is possible to design active set methods for the simple-bound case that are capable of making rapid changes to incorrect predictions (Facchinei, Judice and Soares 1998), it is now more common to use gradient-projection methods.

Gradient-projection methods

The simplest gradient-projection algorithm (Bertsekas 1976, Dunn 1981, Levitin and Polyak 1966) is the obvious linesearch extension of the steepest-descent method to deal with convex constraints, and is based on the iteration

$$x_{k+1} = P_\Omega[x_k - \alpha_k \nabla_x f(x_k)],$$

where $P_\Omega(v)$ projects v into Ω and α_k is a suitable step-size. In the case of simple bounds, $P_\Omega[v] = \mathrm{mid}(l, v, u)$, the (componentwise) median of v with respect to the bounds l and u, is trivial to compute. The method possesses one extremely helpful feature: for non-degenerate problems (*i.e.*, those for which the removal of one or more active constraints necessarily changes the solution), the optimal 'face' of active constraints will be determined in a finite number of iterations (Bertsekas 1976). Of course, its steepest-descent ancestry hints that this is unlikely to be an effective method as it stands, and some form of acceleration is warranted.

The simplest idea exploits the finite optimal-face identification property: if the active faces visited by consecutive iterates of the gradient-projection algorithm are identical, a higher order (Newton-like) method should be used to investigate this face. This was first suggested for quadratic f (Moré and Toraldo 1991), but is now commonplace for general objectives.

A natural question is whether there are other algorithms which have the finite optimal-face identification property for non-degenerate problems. It turns out that the result is true for any algorithm for convex constraints for which the projected gradient $P_{T(x)}[-\nabla_x f(x)]$ converges to zero (Burke and Moré 1988, Calamai and Moré 1987) – here $T(x)$ is the closure of the cone of all feasible directions (the tangent cone) at x. Although the (discontinuous) projected gradient is often hard to compute, in the simple-bound case it is merely (componentwise)

$$\left(P_{T(x)}[-\nabla_x f(x)]\right)_i = \begin{cases} -\min\{0, (\nabla_x f(x))_i\} & \text{if } x_i = l_i \\ -(\nabla_x f(x))_i & \text{if } l_i < x_i < u_i \text{ and} \\ -\max\{0, (\nabla_x f(x))_i\} & \text{if } x_i = u_i. \end{cases}$$

Its continuous variant $P_\Omega[x_k - \nabla_x f(x_k)] - x_k$ is sometimes preferred, and plays a similar role in theory and algorithms.

A restricted version of the finite identification result also holds in the degenerate case, namely that the set of strongly active constraints (*i.e.*, those whose removal will change the solution) will be identified in a finite number of iterations if the projected gradient converges to zero (Lescrenier 1991, Burke and Moré 1994). These finite-identification results apply to many contemporary methods.

Trust region methods for the problem typically consider the gradient-projection arc

$$d(\alpha) = P_{\Omega \cap \{y \,\mid\, \|y - x_k\| \le \Delta_k\}}[x_k - \alpha \nabla_x f(x_k)] - x_k,$$

from x_k. Given a QN model $m_k(d)$, a so-called (generalized) Cauchy point $d(\alpha_k^C)$ is found by approximately minimizing $m_k(d)$ along $d(\alpha)$; either the first local arc minimizer (Conn, Gould and Toint 1988a) or a point satisfying sufficient-decrease linesearch conditions (Burke, Moré and Toraldo 1990, Toint 1988) is required – the computation of a suitable Cauchy point may be performed very efficiently when the Hessian is sparse (Conn, Gould and Toint 1988b, Lin and Moré 1999b). Thereafter a step d_k is computed so that

$$x_k + d_k \in \Omega, \quad \|d_k\| \le \Delta_k \text{ and } m_k(d_k) \le m_k(d_k^C), \tag{3.1}$$

and the usual trust region acceptance rules applied (e.g., Conn et al., 2000a, §6.1). Since it has been shown that the projected gradient converges to zero for these methods, the flexibility in (3.1) is typically used to accelerate convergence by allowing a truncated CG method to explore the face of active constraints at $x_k + d_k^C$. Since the CG iterates may try to leave Ω, early methods simply fixed variables to their bounds and restarted the CG iteration (Conn et al. 1988a), while more modern ones allow infeasible CG iterates by periodically projecting them back into Ω (Gould et al. 2003a, Lin and Moré 1999b).

If second derivatives are unavailable, they may be estimated by any of the methods discussed in Section 2. A particularly appealing approach is to use a limited-memory secant method to estimate the Hessian. Although this approximation is dense, it is so structured that a generalized Cauchy point may still be calculated. Moreover, one of the advantages of limited memory methods, namely that the QN step may be computed directly, is retained, despite the requirement that the QN step be restricted to the face determined by d_k^C, by judicious use of the Sherman–Morrison–Woodbury formula (Byrd, Lu, Nocedal and Zhu 1995).

Although we have only considered methods which remain feasible with respect to the bounds, there is no theoretical reason – as long as the objective function is well-defined outside Ω – to do so provided there is some mechanism for ensuring that the iterates are asymptotically feasible (Facchinei, Lucidi and Palagi 2002). It is also unsurprising that, just as in the unconstrained case, there is no need for the objective function to decrease monotonically as long as there is some overall monotonic trend (Facchinei et al. 1998, Gould et al. 2003a). Efforts have also been made to embed nonlinear CG methods within a gradient-projection framework (Pytlak 1998). Filter ideas have also been investigated (Gould and Toint 2003a) that use penalty techniques (see Section 5) to handle the bounds. Research is on-

going to merge filter methods with the projection methods discussed above
or the interior-point techniques discussed below.

Interior-point methods

Interior-point methods provide an alternative means of solving bound-constrained problems. For simplicity, consider the case where $\Omega = \{x \mid x \geq 0\}$,
suppose that μ is a positive – so-called *barrier* – parameter, and let

$$\phi(x, \mu) \stackrel{\text{def}}{=} f(x) - \mu \sum_{i=1}^{n} \log(x)_i$$

be the logarithmic barrier function for the problem, where $(x)_i$ denotes
the ith component of x. The key idea is to trace approximate minimizers
of $\phi(x, \mu)$ as μ decreases to zero. Under reasonable assumptions, and for
sufficiently small positive values of μ, (local) minimizers of $\phi(x, \mu)$ exist
and describe continuous trajectories – primal central paths – converging
to (local) solutions of the required problem (Fiacco and McCormick 1968,
Wright 1992). Likewise, if X is the diagonal matrix whose ith diagonal
element is the ith component of x and e is a vector of ones, the associated
(first-order) dual variables estimates

$$z = \mu X^{-1} e, \tag{3.2}$$

are located on trajectories enjoying similar properties and converge to Lagrange multipliers associated with the bound constraints. The cross-product
of each pair of trajectories is known as a *primal–dual central path*, and most
barrier methods attempt to follow one with increasing accuracy as μ decreases (Fiacco and McCormick 1968, Wright 1992); for this reason, interior-
point methods are sometimes also referred to as *path-following methods*.

The unconstrained minimization of ϕ can be handled using the techniques
described in Section 2 as long as care is taken to ensure that the iterates
remain within the interior of Ω. A QN model of the form (2.1) might be
used, and as such would be

$$m_k(d) = \phi(x_k, \mu) + d^T (\nabla_x f(x_k) - \mu X_k^{-1} e) + \tfrac{1}{2} d^T (H_k + \mu X_k^{-2}) d, \tag{3.3}$$

where H_k is, as before, an approximation to $\nabla_{xx} f(x_k)$. However, considerable numerical experience has shown that it is usually preferable to replace
the first-order dual variable estimates $z_k = \mu X_k^{-1} e$ in the Hessian term of
(3.3) to obtain instead

$$m_k(d) = \phi(x_k, \mu) + d^T (\nabla_x f(x_k) - \mu X_k^{-1} e) + \tfrac{1}{2} d^T (H_k + X_k^{-1} Z_k) d, \tag{3.4}$$

and to compute the dual variable z_k by other means. In this case, since the
optimal Lagrange multipliers for the problem are necessarily positive, it is
reasonable to require the same of z_k. Rather than computing z_k explicitly

from (3.2), it is better to multiply both sides of (3.2) by X, giving $Xz = \mu e$. Applying Newton's method to this last system then yields the alternative

$$z_{k+1} = \mu X_k^{-1} e - X_k^{-1} Z_k d_k, \tag{3.5}$$

involving the current step d_k from x_k. Additional safeguards need to be employed to enforce convergence of the process (Conn *et al.* 2000*a*, Chapter 13). Methods of this latter type are referred to as *primal–dual methods* because they explicitly consider both primal and dual variables, while methods based on the model (3.3) (with its implicitly computed dual variables) are called *primal methods*.

An approximate minimizer of the model (3.4) may be computed by either a direct (factorization) or iterative (CG) method. If the latter is used, it is normally essential to precondition the iteration to remove the effects of the extreme eigenvalues of $X_k^{-1} Z_k$ (Luenberger 1984, Chapter 12). A preconditioner of the form $P_k = G_k + X_k^{-1} Z_k$ for some suitable approximation G_k of H_k is usually recommended, with G_k varying from naive ($G_k = 0$) to sophisticated ($G_k = H_k$).

Both linesearch or trust region globalization of interior-point methods are possible and essentially identical to that discussed in Section 2. The major difference in both cases is the addition of a so-called fraction-to-the-boundary rule, preventing iterates from prematurely approaching the boundary of the feasible set. A trust region algorithm will accept a step d_k from x_k if

(1) $x_k + d_k \geq \gamma_k x_k$ holds componentwise, for some given $0 < \gamma_k < 1$, and
(2) there is good agreement between the changes in m_k and $\phi(x, \mu)$.

For most practical purposes, the fraction-to-the-boundary parameter γ_k is held constant, a typical value being 0.005. It may however be permitted to converge to zero, allowing for fast asymptotic convergence. Wächter and Biegler (2004) choose (in a more general context) $\gamma_k = \max(\gamma_{\min}, \mu_k)$, where $0 < \gamma_{\min} < 1$ is a prescribed minimal value. The fraction-to-the-boundary rule also applies for linesearch methods, and a (backtracking) linesearch is typically performed until it is satisfied. A corresponding backtracking may then also be applied to the dual step to ensure consistency.

Although there is no difficulty in providing a strictly interior primal–dual starting point, (x_0, z_0), in the bound-constrained case, it is generally sensible to ensure that such a point is well separated from the boundary of the feasible region; failure to do this can (and in practice does seriously) delay convergence. Given suitable primal estimates x_0, traditional choices for dual estimates z_0 include the vector of ones or those given from (3.2) although there is little reason to believe that these are more than heuristics. An initial value for μ is then typically $\mu_0 = x_0^T z_0 / n$, so as to obtain good centrality at the initial point.

It is well known that computing d_k to be a critical point of (3.4) and recovering z_{k+1} via (3.5) is equivalent to applying Newton's method from (x_k, z_k) to the perturbed optimality conditions

$$\nabla_x f(x) - z = 0 \ \text{ and } \ Xz = \mu e$$

for our problem. While it is tempting to try similar approaches directly with $\mu = 0$ – so-called affine methods – these have both theoretical and practical shortcomings for general problems (see, for example, Conn *et al.*, 2000*a*, §13.11). A more promising approach is to note that equivalent first-order optimality conditions are that

$$W(x)\nabla_x f(x) = 0, \tag{3.6}$$

where

$$W(x) = \operatorname{diag} w(x) \ \text{ and } \ w_i(x) = \begin{cases} x_i & \text{if } (\nabla_x f(x))_i \geq 0 \\ -1 & \text{otherwise.} \end{cases}$$

As long as strictly feasible iterates are generated, $W(x)\nabla_x f(x)$ is differentiable, and Newton's method may be applied to (3.6). To globalize such an iteration, combined linesearch/trust region methods have been proposed (Coleman and Li 1994, 1996) and variants which allow quadratic convergence even in the presence of degeneracy are possible (Heinkenschloss, Ulbrich and Ulbrich 1999).

Practicalities

There has been a number of comparative studies of algorithms for bound-constrained optimization (Facchinei *et al.* 2002, Gould *et al.* 2003*a*, Lin and Moré 1999*b*, Zhu, Byrd, Lu and Nocedal 1997), but we feel that none of these makes a compelling case as to the best approach(es) for the large-scale case. In practice, both gradient projection and interior-point methods appear to require a modest number of iterations.

Software

Once again there is a reasonable choice of reliable software for the bound-constrained case. Both TRON (Lin and Moré 1999*b*) and LANCELOT/SBMIN (Conn, Gould and Toint 1992) are trust region gradient-projection algorithms with subspace conjugate gradient acceleration – there is an improved version of the latter within GALAHAD (Gould *et al.* 2003*a*) – while L-BFGS-B (Zhu *et al.* 1997) is a linesearch implementation of the limited-memory approach. The MATLAB function fmincon (Branch, Coleman and Li 1999) uses an interior-point subspace method based on (3.6). FILTRANE, another algorithm of the GALAHAD library, uses a filter method combined with penalty techniques for the bounds. As before, more general codes such as KNITRO and LOQO are also highly appropriate.

4. Large-scale linearly constrained optimization

As the next level of generality, we now turn to problems involving general
linear constraints.

4.1. Equality-constrained quadratic programming

A – some might say *the* – basic subproblem in constrained optimization is to

$$\underset{x \in \mathbb{R}^n}{\text{minimize}} \ q(x) = g^T x + \tfrac{1}{2} x^T H x \ \text{subject to} \ Ax = b, \qquad (4.1)$$

where H is symmetric (but possibly indefinite) and A is m by n and (without
loss of generality) of full rank. Such equality-constrained quadratic programming (EQP) subproblems arise when computing search directions for
either general methods for equality-constrained or active set methods for
inequality-constrained optimization. It is actually often more convenient to
consider the related homogeneous problem

$$\underset{\bar{x} \in \mathbb{R}^n}{\text{minimize}} \ \bar{q}(\bar{x}) = \bar{g}^T \bar{x} + \tfrac{1}{2} \bar{x}^T H \bar{x} \ \text{subject to} \ A\bar{x} = 0; \qquad (4.2)$$

as long as there is some easy way to find x_0 satisfying $Ax_0 = b$, the solutions
of the two problems satisfy $x = \bar{x} + x_0$ provided that $\bar{g} = g + Hx_0$.

Critical points of (4.1) necessarily satisfy the augmented (KKT) system

$$\begin{pmatrix} H & A^T \\ A & 0 \end{pmatrix} \begin{pmatrix} x \\ y \end{pmatrix} = \begin{pmatrix} -g \\ b \end{pmatrix}, \qquad (4.3)$$

and solutions of (4.3) can only be solutions of (4.1) if the coefficient matrix
of (4.3) has precisely m negative eigenvalues (Chabrillac and Crouzeix 1984,
Gould 1985). Thus direct (factorization) methods for solving (4.3) must be
capable of coping with indefinite matrices; fortunately there is a growing
number of highly capable symmetric, indefinite linear solvers (for example
BCSEXT, MA27/57, Oblio or PARDISO; see Scott, Hu and Gould (2004) for a
comparison), and in particular if H is (block) diagonal a Schur-complement
decomposition involving factorizations of H and $-AH^{-1}A^T$ is often to be
recommended. Nevertheless, for very large problems direct methods may
be unviable or too expensive, and iterative methods may be the only alternative. Although non-symmetric or indefinite iterative methods may be
applied, we only consider CG-type methods here, since these have the desirable property of decreasing $q(x)$ at every iteration.

It should be apparent that CG methods can be applied explicitly to (4.2)
by computing a basis N for the null-space of A, and then using the transformation $\bar{x} = N x_n$ to derive the equivalent (unconstrained) problem of
minimizing $q_n(x_n) = x_n^T g_n + \tfrac{1}{2} x_n^T H_n x_n$, where $g_n = N^T \bar{g}$ and $H_n = N^T H N$
are known as the reduced gradient and Hessian respectively. Perhaps not
so obviously, the same may be achieved *implicitly* by using the standard

preconditioned CG (PCG) method but using a block (so-called), constraint
preconditioner of the form

$$\begin{pmatrix} G & A^T \\ A & 0 \end{pmatrix} \begin{pmatrix} r \\ w \end{pmatrix} = -\begin{pmatrix} \bar{g} + Hx \\ 0 \end{pmatrix}, \tag{4.4}$$

to obtain the 'preconditioned' residual r from the 'unpreconditioned' $\bar{g}+Hx$,
for some suitable G (Coleman 1994, Gould, Hribar and Nocedal 2001,
Lukšan and Vlček 1998, Polyak 1969). Various choices for G, ranging from
the identity matrix to H, have been suggested, and all require a suitable
(block) factorization of the coefficient matrix K of (4.4); basic requirements
are that K should be non-singular and have precisely m negative eigenvalues. A further advantage of the PCG approach is that any additional
(properly scaled) trust region constraint may easily be incorporated using
the GLTR strategy mentioned in Section 2. Nevertheless, requiring a factorization of K may still be considered a disadvantage, and methods which
avoid this are urgently needed.

4.2. General quadratic programming

Another important subproblem in constrained optimization is the general
quadratic programming (QP) problem, namely to

$$\underset{x \in \mathbb{R}^n}{\text{minimize}} \ q(x) \ \text{subject to} \ A_{\mathcal{E}}x = b_{\mathcal{E}} \ \text{and} \ A_{\mathcal{I}}x \geq b_{\mathcal{I}}. \tag{4.5}$$

Of particular interest is the non-convex case where the symmetric H may
be indefinite, although in these circumstances we must normally be content
with local solutions. The main application area we are concerned with is in
solving subproblems which arise within sequential quadratic programming
(SQP) algorithms for general nonlinear optimization (see Section 5.1, but
note the caveat there that expresses our concerns over the SQP approach),
although there are actually a large number of other (usually convex) applications of (4.5) (Gould, and Toint 2000a), including VLSI design, optimal
control, economic dispatch and financial planning, to mention only a few.
They also constitute a class apart as their necessary and sufficient optimality
conditions coincide (Contesse 1980, Mangasarian 1980, Borwein 1982).

Active set methods for general quadratic programming

As was the case for bound-constrained problems we considered in Section 3, QP methods may broadly be categorized as either active-set-based
or interior-point-based. As the name suggests, active set methods aim to
predict which of the inequality constraints $A_{\mathcal{I}}x \geq b_{\mathcal{I}}$ are active at a solution
to (4.5). At each iteration, a working set $\mathcal{W}_{\mathcal{I}} \subseteq \mathcal{I}$ is selected so that the
gradients of the constraints $A_{\mathcal{W}}$, $\mathcal{W} = \mathcal{E} \cup \mathcal{W}_{\mathcal{I}}$, are linearly independent.

For this working set, the EQP

$$\underset{x \in \mathbb{R}^n}{\text{minimize}} \; q(x) \text{ subject to } A_\mathcal{W} x = b_\mathcal{W} \qquad (4.6)$$

is solved (if possible) using one of the methods described in Section 4.1. There are a number of possibilities. If (4.6) is unbounded from below or if the solution to (4.6) violates one of the inequality constraints indexed by $\mathcal{I} \setminus \mathcal{W}_\mathcal{I}$, one (or more) constraints should be added to the working set. If (4.6) is infeasible or if the solution to (4.6) is not that of (4.5) – the latter is true if any of the Lagrange multipliers y from (4.3) are negative – one (or more) constraints should be removed from the working set. In the convex case, only when the solution to (4.6) satisfies all of the constraints indexed by $\mathcal{I} \setminus \mathcal{W}_\mathcal{I}$ and all of the Lagrange multipliers are positive can we be certain that we have solved (4.5). Unfortunately, for non-convex problems, even checking if such a critical point is a local minimizer may be (NP) hard (Murty and Kabadi 1987). While such a strategy is simple – it may be reduced to the solution of a sequence of EQPs – potentially a large number of iterations may be required. Fortunately, just as with the simplex method for linear programming (LP), the working set usually changes very gradually, and the potentially dominant cost of matrix factorization is lessened through updates to existing factors. Thus active set methods for QP usually comprise a large number of very cheap iterations – in contrast, interior-point methods require a few, more expensive ones.

There have been relatively few active set methods for large-scale QP, especially in the nonconvex case. The majority of these are based on the idea of inertia control (Fletcher 1971). Suppose that the coefficient matrix K for the optimality conditions (4.3) corresponding to the current EQP (4.6) has the 'correct' inertia, i.e., K has $|\mathcal{W}|$ negative eigenvalues. If a constraint is added to the working set, the new subproblem will inherit the correct inertia. However, if a constraint is removed from the working set, it is possible that the resulting K may have $|\mathcal{W}| + 1$ rather than $|\mathcal{W}|$ negative eigenvalues. If this happens, there must be a feasible direction of negative curvature, and an inertia-controlling method will follow this direction until it encounters a currently inactive constraint (or perhaps q is unbounded from below on the feasible set). This new constraint will be added to the working set, and once again the resulting K will have either $|\mathcal{W}|$ or $|\mathcal{W}| + 1$ negative eigenvalues. In the former case, the correct inertia has been restored, while in the latter there is again a direction of feasible negative curvature. This process of following negative curvature and adding currently inactive constraints must ultimately terminate (unless the problem is unbounded below) at a vertex of the feasible region, at which point the correct inertia will have been restored. The principal differences between the inertia-controlling methods that have been proposed are the algebraic means by which the factors are maintained and updated. These include using Schur-complement (Gill, Murray, Saunders and Wright 1990, Gill, Murray, Saunders and Wright 1991) or linear-programming basis-type (Gould 1991) updates to a factorization of an initial K, or Cholesky-factor updates of the (dense) reduced Hessian (Fletcher 2000), the latter only really being appropriate for problems with few degrees of freedom.

For problems for which a direct solution of the sequence of generated EQPs is unviable or too expensive, it is also possible to use the PCG method described in Section 4.1. Now rather than controlling the inertia of the KKT matrix, inertia control is only required for the preconditioner (4.4). Once again, factors of the preconditioner must adapt to changes in the working set, but the ability to choose G gives considerable flexibility (Gould and Toint 2002a).

Interior-point methods for general quadratic programming

It is easy to generalize the interior-point methods discussed in Section 3 to cope with the quadratic program (4.5). Denoting the ith row of $A_{\mathcal{I}}$ by a_i and the ith component of $b_{\mathcal{I}}$ by b_i, typical barrier methods for such problems aim to

$$\underset{x \in \mathbb{R}^n}{\text{minimize}} \quad \phi(x,\mu) \overset{\text{def}}{=} q(x) - \mu \sum_{i \in \mathcal{I}} \log(a_i^T x - b_i) \tag{4.7}$$
$$\text{subject to} \quad A_{\mathcal{E}} x = b_{\mathcal{E}}$$

as μ decreases to 0, while ensuring that x remains interior to $\Omega_{\mathcal{I}} = \{x \,|\, A_{\mathcal{I}} x \geq b_{\mathcal{I}}\}$. Just as in the bound-constrained case and under reasonable conditions, the minimizers of (4.7) and their dual variable (Lagrange multiplier) estimates

$$y_{\mathcal{I}} = \mu C_{\mathcal{I}}^{-1}(x) e, \quad \text{where} \quad C_{\mathcal{I}}(x) = \text{diag } c_{\mathcal{I}}(x) \quad \text{and} \quad c_{\mathcal{I}}(x) = A_{\mathcal{I}} x - b_{\mathcal{I}},$$

define continuous trajectories – primal–dual central paths – leading to (local) solutions of (4.5).

For fixed μ and feasible x_k, basic iterative methods might compute a suitable step d_k by building a primal QN model

$$m_k(d) = \phi(x_k,\mu) + d^T(g_k - \mu A_{\mathcal{I}}^T C_{\mathcal{I}k}^{-1} e) + \tfrac{1}{2} d^T (H + \mu A_{\mathcal{I}}^T C_{\mathcal{I}k}^{-2} A_{\mathcal{I}}) d, \tag{4.8}$$

where $g_k \overset{\text{def}}{=} H x_k + g$ and $C_{\mathcal{I}k} \overset{\text{def}}{=} C_{\mathcal{I}}(x_k)$, and then trying to (approximately)

$$\underset{s \in \mathbb{R}^n}{\text{minimize}} \quad m_k(d) \text{ subject to } A_{\mathcal{E}} s = 0 \text{ and (possibly) } \|s\| \leq \Delta_k, \tag{4.9}$$

involving an additional trust region constraint. To ensure feasibility of the next iterate, a step-size $0 < \alpha_k \leq \alpha_k^{\max}$ should be imposed along an approximate solution d_k to (4.9) – a fraction-to-the-boundary rule such as

$$\alpha_k^{\max} = \max\{0 < \alpha \leq 1 \mid c_{\mathcal{I}}(x_k + \alpha d_k) \geq \gamma_k c_{\mathcal{I}}(x_k)\}$$

is appropriate – and to guarantee convergence it may also be necessary to linesearch along d_k or adjust Δ_k in the usual manner. But as in Section 3, a primal–dual model

$$m_k(d) = \phi(x_k, \mu) + d^T(g_k - \mu A_{\mathcal{I}}^T C_{\mathcal{I}k}^{-1} e) + \tfrac{1}{2} d^T(H + A_{\mathcal{I}}^T C_{\mathcal{I}k}^{-1} Y_{\mathcal{I}k} A_{\mathcal{I}}) d \tag{4.10}$$

involving explicit (positive) dual variables $y_{\mathcal{I}k}$ is generally preferable to (4.8) both in theory and in practice (Conn, Gould, Orban and Toint 2000b). In particular, the analogue of the Newton update (3.5),

$$y_{\mathcal{I}k+1} = \mu C_{\mathcal{I}k}^{-1} e - C_{\mathcal{I}k}^{-1} Y_{\mathcal{I}k} d_k, \tag{4.11}$$

is appropriate, so long as appropriate precautions are taken to modify (4.11) to ensure that $y_{\mathcal{I}k+1}$ remains sufficiently positive (Conn $et\ al.$ 2000a, Chapter 13).

The key subproblem here is (4.9), but this is precisely of the form discussed in Section 4.1. The only significant extra issue when the objective function has the form (4.10) is that any preconditioner should respect the potentially ill-conditioned Hessian term (Luenberger 1984, Chapter 12), and thus that the leading block in (4.4) should be $G_k + A_{\mathcal{I}}^T C_{\mathcal{I}k}^{-1} Y_{\mathcal{I}k} A_{\mathcal{I}}$ for some suitable approximation G_k to H. Although it might first appear that such a leading block may be unacceptably dense, the preconditioning step (4.4) for the model (4.10) may be trivially rearranged to give the potentially sparser

$$\begin{pmatrix} G_k & A_{\mathcal{E}}^T & A_{\mathcal{I}}^T \\ A_{\mathcal{E}} & 0 & 0 \\ A_{\mathcal{I}} & 0 & -Y_{\mathcal{I}k}^{-1} C_{\mathcal{I}k} \end{pmatrix} \begin{pmatrix} r \\ w_{\mathcal{E}} \\ w_{\mathcal{I}} \end{pmatrix} = - \begin{pmatrix} g_k + Hs \\ 0 \\ -\mu Y_{\mathcal{I}k}^{-1} e \end{pmatrix} \tag{4.12}$$

for auxiliary variables $w_{\mathcal{I}} = C_{\mathcal{I}k}^{-1} Y_{\mathcal{I}k} A_{\mathcal{I}} s - \mu C_{\mathcal{I}k}^{-1} e$ (Gould 1986).

The method sketched above presupposes that an initial point x_0 is known within the intersection of $\Omega_{\mathcal{E}} = \{x \mid A_{\mathcal{E}} x = b_{\mathcal{E}}\}$ and the interior of $\Omega_{\mathcal{I}}$. A suitable point may be found by solving an auxiliary $phase\text{-}I$ problem such as

$$\underset{x \in \mathbb{R}^n, \, s_{\mathcal{I}} \in \mathbb{R}^{n_{\mathcal{I}}}}{\text{minimize}} - \sum_{i \in \mathcal{I}} \log(s_i) \text{ subject to } A_{\mathcal{E}} x = b_{\mathcal{E}} \text{ and } A_{\mathcal{I}} x - s_{\mathcal{I}} = b_{\mathcal{I}}, \tag{4.13}$$

where the $s_{\mathcal{I}}$ are being treated as auxiliary, positive $slack$ variables. The intention here is to find a point which is significantly interior, and in the above case will give the analytic centre of the feasible region. Fortunately, although finding feasible points for (4.13) may not be obvious, the problem is convex and may be solved using an infeasible interior-point method, such as those discussed in the next section.

With this in mind, an equivalent formulation of (4.5) is to

$$\underset{x \in \mathbb{R}^n, \, s_{\mathcal{I}} \in \mathbb{R}^{n_{\mathcal{I}}}}{\text{minimize}} \, q(x) \text{ subject to } A_{\mathcal{E}} x = b_{\mathcal{E}}, \, A_{\mathcal{I}} x - s_{\mathcal{I}} = b_{\mathcal{I}} \text{ and } s_{\mathcal{I}} \geq 0, \tag{4.14}$$

and an alternative barrier method for (4.5) might aim to

$$\operatorname*{minimize}_{x \in \mathbb{R}^n,\, s_\mathcal{I} \in \mathbb{R}^{n_\mathcal{I}}} \phi(x, s_\mathcal{I}, \mu) \overset{\text{def}}{=} q(x) - \mu \sum_{i \in \mathcal{I}} \log(s_i)$$

$$\text{subject to} \quad A_\mathcal{E} x = b_\mathcal{E} \text{ and } A_\mathcal{I} x - s_\mathcal{I} = b_\mathcal{I}$$

$$(4.15)$$

as μ decreases to 0. Although the distinction between this 'slack variable' formulation and (4.7) is actually very slight if the constraints $A_\mathcal{I} x - s_\mathcal{I} = b_\mathcal{I}$ are enforced throughout, as we shall see later the distinction is more pronounced for nonlinear constraints.

Interior-point methods for convex quadratic programming

Quadratic programs are traditionally classified as convex or nonconvex, depending on whether the Hessian matrix H is positive semidefinite or not. Simply finding a local minimizer of a nonconvex QP is an NP-hard problem (Vavasis 1990), as is proving that a first-order critical point is in fact a minimizer (Murty and Kabadi 1987) – most algorithms for general QP are consequently only designed to locate first-order critical points.

Convex QPs are provably solvable by algorithms having polynomial complexity (Nesterov and Nemirovskii 1994, Vavasis 1991). The use of a barrier function to force convergence in the convex case is usually inefficient, and the best methods are more closely allied to those for LP (Wright 1997). The basis of these primal–dual path-following methods for convex QP is to solve the perturbed optimality conditions

$$g + Hx - A_\mathcal{E}^T y_\mathcal{E} - A_\mathcal{I}^T y_\mathcal{I} = 0,$$
$$A_\mathcal{E} x - b_\mathcal{E} = 0,$$
$$\text{and } Y_\mathcal{I}(A_\mathcal{I} x - b_\mathcal{I}) - \mu e = 0$$

$$(4.16)$$

for (4.5) or, more commonly,

$$g + Hx - A_\mathcal{E}^T y_\mathcal{E} - A_\mathcal{I}^T y_\mathcal{I} = 0,$$
$$y_\mathcal{I} - z_\mathcal{I} = 0,$$
$$A_\mathcal{E} x - b_\mathcal{E} = 0,$$
$$A_\mathcal{I} x - s_\mathcal{I} - b_\mathcal{I} = 0,$$
$$\text{and } Z_\mathcal{I} s_\mathcal{I} - \mu e = 0$$

$$(4.17)$$

for (4.14), using Newton's method or a variant thereof, while maintaining strict feasibility for $s_\mathcal{I} \geq 0$ and $z_\mathcal{I} \geq 0$ (or $A_\mathcal{I} x \geq b_\mathcal{I}$ and $y_\mathcal{I}$ for (4.5)), and letting μ gradually decrease to zero. The most popular are based on the linesearch-based predictor–corrector algorithm of Mehrotra (1992), originally developed for LP.

A typical *predictor–corrector* iteration for (4.17) involves the solution of a pair of (symmetrized) linear systems of the form

$$
\begin{pmatrix}
G & 0 & A_{\mathcal{E}}^T & A_{\mathcal{I}}^T & 0 \\
0 & 0 & 0 & -I & -I \\
A_{\mathcal{E}} & 0 & 0 & 0 & 0 \\
A_{\mathcal{I}} & -I & 0 & 0 & 0 \\
0 & -I & 0 & 0 & -Z_{\mathcal{I}}^{-1}S_{\mathcal{I}}
\end{pmatrix}
\begin{pmatrix}
d_x \\
d_{s_{\mathcal{I}}} \\
-d_{y_{\mathcal{E}}} \\
-d_{y_{\mathcal{I}}} \\
d_{z_{\mathcal{I}}}
\end{pmatrix}
= -
\begin{pmatrix}
g(x) - A_{\mathcal{E}}^T y_{\mathcal{E}} - A_{\mathcal{I}}^T y_{\mathcal{I}} \\
y_{\mathcal{I}} - z_{\mathcal{I}} \\
c_{\mathcal{E}}(x) \\
c_{\mathcal{I}}(x) - s_{\mathcal{I}} \\
-Z_{\mathcal{I}}^{-1} r_{\mathrm{c}}(\mu)
\end{pmatrix}
$$

(4.18)

for different values of $r_{\mathrm{c}}(\mu)$, where (as before) $c_{\mathcal{I}}(x) = A_{\mathcal{I}}x - b_{\mathcal{I}}$, $c_{\mathcal{E}}(x) = A_{\mathcal{E}}x - b_{\mathcal{E}}$, $g(x) = g + Hx$ and $G \approx H$. Note that Newton's method for (4.17) results when $G = H$ and $r_{\mathrm{c}}(\mu) = S_{\mathcal{I}}z_{\mathcal{I}} - \mu e$ with the current value of μ. The first of the two systems uses $r_{\mathrm{c}}(\mu) = S_{\mathcal{I}}z_{\mathcal{I}}$ and defines a *predictor* step intended to reduce primal and dual feasibility. This step is often referred to as an *affine scaling* step and is denoted d^{AFF}. A steplength α^{AFF} is determined to preserve positivity of $z_{\mathcal{I}}$ and $s_{\mathcal{I}}$. On defining the duality gap after the predictor step $\mu^{\mathrm{AFF}} = (z_{\mathcal{I}} + \alpha^{\mathrm{AFF}} d_{z_{\mathcal{I}}}^{\mathrm{AFF}})^T (s_{\mathcal{I}} + \alpha^{\mathrm{AFF}} d_{s_{\mathcal{I}}}^{\mathrm{AFF}})/n_{\mathcal{I}}$ and the centering parameter $\sigma = (\mu^{\mathrm{AFF}}/\mu)^\tau$ with $2 \leq \tau \leq 4$, the second system uses $r_{\mathrm{c}}(\mu) = S_{\mathcal{I}}z_{\mathcal{I}} - \sigma\mu e + D_{s_{\mathcal{I}}}^{\mathrm{AFF}} d_{z_{\mathcal{I}}}^{\mathrm{AFF}}$ and defines a *corrector* step aiming to improve centrality. The final primal and dual common steplength (Mehrotra 1992) is determined by

$$
\alpha = \min(1, \eta\alpha_{\mathrm{MAX}}^{\mathrm{P}}, \eta\alpha_{\mathrm{MAX}}^{\mathrm{D}}),
$$

where $\eta \in [0.9, 1.0)$ converges to 1 as a solution is approached, and $\alpha_{\mathrm{MAX}}^{\mathrm{P}}$ and $\alpha_{\mathrm{MAX}}^{\mathrm{D}}$ are primal and dual steplengths enforcing a fraction-to-the-boundary rule.

The higher-order corrections scheme of Gondzio (1996), again originally developed for LP, generalizes to convex quadratic programming. Several corrector-like steps are taken, as long as substantial steplengths are acceptable and individual complementarity pairs cluster around their average value. These steps aim for dynamically computed targets (Jansen, Roos, Terlaky and Vial 1996) located in a loose neighbourhood of the central path. The number of corrector steps is computed at the first iteration by balancing the cost of the linear algebra and the expected progress towards optimality.

Just as in Section 4.1, (block) direct or iterative methods may be used to solve the indefinite system (4.18). Further savings often result from the block elimination

$$
\begin{pmatrix}
G & A_{\mathcal{E}}^T & A_{\mathcal{I}}^T \\
A_{\mathcal{E}} & 0 & 0 \\
A_{\mathcal{I}} & 0 & -Z_{\mathcal{I}}^{-1}S_{\mathcal{I}}
\end{pmatrix}
\begin{pmatrix}
d_x \\
-y_{\mathcal{E}} - d_{y_{\mathcal{E}}} \\
-y_{\mathcal{I}} - d_{y_{\mathcal{I}}}
\end{pmatrix}
= -
\begin{pmatrix}
g(x) \\
c_{\mathcal{E}}(x) \\
c_{\mathcal{I}}(x) - s_{\mathcal{I}} - Z_{\mathcal{I}}^{-1} r_{\mathrm{c}}(\mu)
\end{pmatrix}
$$

(4.19)

of (4.18), or possibly even from

$$\begin{pmatrix} G + A_{\mathcal{I}}^T S_{\mathcal{I}}^{-1} Z_{\mathcal{I}} A_{\mathcal{I}} & A_{\mathcal{E}}^T \\ A_{\mathcal{E}} & 0 \end{pmatrix} \begin{pmatrix} d_x \\ -y_{\mathcal{E}} - d_{y_{\mathcal{E}}} \end{pmatrix} = \\ - \begin{pmatrix} g(x) + A_{\mathcal{I}}^T S_{\mathcal{I}}^{-1} Z_{\mathcal{I}} [c_{\mathcal{I}}(x) - s_{\mathcal{I}} - Z_{\mathcal{I}}^{-1} r_c(\mu)] \\ c_{\mathcal{E}}(x) \end{pmatrix}, \tag{4.20}$$

which arises by further eliminating $y_{\mathcal{I}} + d_{y_{\mathcal{I}}}$ from (4.19) – of course (4.20) may be inappropriate if the term $A_{\mathcal{I}}^T S_{\mathcal{I}}^{-1} Z_{\mathcal{I}} A_{\mathcal{I}}$ is significantly denser than G, but has the virtue of being considerably smaller if there are many inequality constraints. It is also worth noting that the corresponding predictor–corrector steps for (4.16) satisfy

$$\begin{pmatrix} G & A_{\mathcal{E}}^T & A_{\mathcal{I}}^T \\ A_{\mathcal{E}} & 0 & 0 \\ A_{\mathcal{I}} & 0 & -Y_{\mathcal{I}}^{-1} C_{\mathcal{I}} \end{pmatrix} \begin{pmatrix} d_x \\ -y_{\mathcal{E}} - d_{y_{\mathcal{E}}} \\ -y_{\mathcal{I}} - d_{y_{\mathcal{I}}} \end{pmatrix} = - \begin{pmatrix} g(x) \\ c_{\mathcal{E}}(x) \\ Y_{\mathcal{I}}^{-1} r_c(\mu) \end{pmatrix} \tag{4.21}$$

which is simply (4.19) in the special case $s_{\mathcal{I}} = c_{\mathcal{I}}(x)$ and $z_{\mathcal{I}} = y_{\mathcal{I}}$ – also cf. (4.12). The coefficient matrices from (4.18), (4.19)/(4.21) and (4.20) are appropriate preconditioners for PCG as long as they have, respectively, $\operatorname{rank}(A_{\mathcal{E}}) + 2|\mathcal{I}|$, $\operatorname{rank}(A_{\mathcal{E}}) + |\mathcal{I}|$ and $\operatorname{rank}(A_{\mathcal{E}})$ negative eigenvalues (Conn et al. 2000b, with Sylvester's law of inertia); equivalently $G + A_{\mathcal{I}}^T S_{\mathcal{I}}^{-1} Z_{\mathcal{I}} A_{\mathcal{I}}$ should be positive definite on the null-space of $A_{\mathcal{E}}$, and this will always be the case if G is positive definite. Any of the factorizations mentioned in Section 4.1 are appropriate.

For large problems, it is vital to be able to exploit commonly occurring sub-structure when solving (4.19). Applications from multi-stage stochastic programming, network communications or asset liability management give rise to matrices H and A having one of a number of predefined block structure – examples include H and A being block diagonal, primal or dual block angular or bordered block diagonal. Moreover, this block structure appears recursively in the sense that the structure of the blocks is similar to that of the matrix containing them. This nestedness is fully exploitable if matrices H and A have compatible structures – i.e., have the same number of diagonal blocks with matching numbers of columns – the coefficient matrix K of (4.19) can be reordered to have similarly exploitable block structure (Gondzio and Grothey 2003a).

Frequently in practice K may be very ill-conditioned or even singular, and it is common to *regularize* K to avoid such deficiencies. Typically, diagonal blocks R will be added to K so that the resulting matrix $K + R$ is quasi-definite. *Quasi-definite* matrices (Vanderbei 1995) are strongly factorizable in the sense that, for any symmetric permutation P, there exist a unit lower triangular matrix L and a diagonal matrix D such that $P(K + R)P^T = LDL^T$ without recourse to 2×2 pivoting, as is common with other popular factorizations of indefinite matrices (see Section 4.1). If the system is block-structured as above, the quasi-definite factorization may easily be parallelized, since block structure in $P(K + R)P^T$ induces block structure in L and D (Gondzio and Grothey 2003a).

The requirement that slack variables introduced in (4.14) remain strictly feasible suggests that a steplength $0 < \alpha_k \leq \alpha_k^{\mathrm{max}}$ be chosen, where

$$\alpha_k^{\mathrm{max}} = \max\{0 < \alpha \leq 1 \mid s_{\mathcal{I},k} + \alpha d_{s_{\mathcal{I}}} \geq \gamma_k s_{\mathcal{I},k}\},$$

to enforce a fraction-to-the-boundary rule. A similar rule applies to primal variables that are subject to bounds and to dual variables. A strictly feasible initial point is any $s_{\mathcal{I},0} > 0$, but in practice it may be prudent to initialize $s_{\mathcal{I}}$ to a significantly positive value. Since the inequality constraints also need to be satisfied, a common choice is to pick $s_{\mathcal{I},0} = \max(A_{\mathcal{I}}x_0 - b_{\mathcal{I}}, \sigma e)$ componentwise, where x_0 is supplied by the user or the model, $\sigma > 0$ is a given constant, e.g., $\sigma = 1$ and e is the vector of all ones. Often, explicit bound constraints will be honoured by first moving x_0 to satisfy them, and computing $s_{\mathcal{I},0}$ from this perturbed initial point. Another possibility is to compute an affine-scaling step d^{AFF}, i.e., using $\mu = 0$, for the primal–dual system associated with (4.5). On defining $s_{\mathcal{I}}^{\mathrm{AFF}} = s_{\mathcal{I},0} + d_{s_{\mathcal{I}}}^{\mathrm{AFF}}$, an initial $s_{\mathcal{I},1}$ is computed based on the feasibility of $s_{\mathcal{I}}^{\mathrm{AFF}}$, using a rule such as

$$s_{\mathcal{I},1} = \max(\beta e, |s_{\mathcal{I}}^{\mathrm{AFF}}|) \quad \text{or} \quad s_{\mathcal{I},1} = s_{\mathcal{I}}^{\mathrm{AFF}} + \gamma \max(0, -s_{\mathcal{I}}^{\mathrm{AFF}}) + \beta e, \quad (4.22)$$

where the absolute values and maxima are understood elementwise and $\beta, \gamma > 0$ (Gertz, Nocedal and Sartenaer 2003).

Good general-purpose initial values for the Lagrange multipliers y in primal–dual interior methods are hard to find, and poor guesses may introduce unwarranted nonconvexity into the model if the problem is nonconvex. Nonetheless, they are often initialized to approximate least-squares solutions for dual feasibility, i.e., values of y for which the gradient lies on the null-space of the constraints at the starting point, and adjusted to ensure that those corresponding to inequality constraints are strictly positive.

Not all path-following interior-point methods for convex QP are of the predictor–corrector type. The simplest alternative is to solve (4.18) or (4.19) for a pre-assigned μ but to ensure ensure strict feasibility by means of a fraction-to-the-boundary rule, in which the step-size α is chosen as

$$\alpha = \min\left\{1, (1 - \epsilon) \max_{[d_s]_i < 0} \frac{s_i}{-[d_s]_i}, (1 - \epsilon) \max_{[d_z]_i < 0} \frac{z_i}{-[d_z]_i}\right\},$$

for a small $\epsilon > 0$. A merit function such as

$$\phi(x, s_{\mathcal{I}}, y, z_{\mathcal{I}}) \equiv s_{\mathcal{I}}^T z_{\mathcal{I}} + \|\nabla \mathcal{L}(x, s_{\mathcal{I}}, y_{\mathcal{E}}, y_{\mathcal{I}}, z_{\mathcal{I}})\|_2, \quad (4.23)$$

where $\mathcal{L}(x, s_{\mathcal{I}}, y_{\mathcal{E}}, y_{\mathcal{I}}, z_{\mathcal{I}})$ is the Lagrangian associated with (4.14), is used to assess suitability of such a steplength. Such a fraction-to-the-boundary condition may be implicitly ensured by Zhang's (1994) step-size rule, and in this case yields a global linear convergence rate and a polynomial algorithm.

One further interesting idea in both convex and non-convex cases is to solve (4.5) by a sequence of minimizations over the intersection of the interior of the feasible region with iteratively generated low-dimensional (typically 2- or 3-dimensional) subspaces. The advantage here is that the resulting subproblems are small, so that global optimization is possible. Clearly the choice of subspaces is crucial, and should include at least one 'descent' direction for whatever globalization mechanism is to be used, and others which are geared towards fast asymptotic convergence – solutions of (4.18) for different $r_C(\mu)$ may be used (Boggs, Domich, Rogers and Witzgall 1996).

Practicalities

The only comparison of the competing QP ideologies we are aware of is that of Gould and Toint (2002b). As perhaps one might expect, the interior-point approach seems generally to be preferable to the active set, especially for very large problems where the number of active set iterations can be enormous. For 'warm-start' problems, where a solution to a small perturbation of an existing already solved problem is required, there is some virtue in using the active set approach as it seems better able to use good estimates of the optimal active set. Whether this trend will continue is debatable, especially as current research for LP indicates promise for warm-started interior-point methods (Gondzio and Grothey 2003b, Yildirim and Wright 2002).

When carefully implemented, interior methods for QP scale almost perfectly with the number of variables, and rarely do they need more than, say, 30–35 iterations. Moreover, unlike active-set-type methods, the linear systems which arise at each iteration have identical block structure. Nonetheless, the solution of such systems may still be costly, and implementations must pay particular attention to exploiting structure – an example of a disastrous situation caused by the lack of exploitation of low-rank-corrector structure is given by Ferris and Munson (2000).

A final important idea is to simplify QPs before solution. Such 'presolve' methods have proved to be very effective for LP (Gondzio 1997), and similar gains are also possible for QP (Gould and Toint 2004b).

Software

Currently available active set non-convex QP codes include VE09 (Gould 1991), bqpd (Fletcher 2000) and QPA (Gould and Toint 2002a). The PRESOLVE package (Gould and Toint 2004b) is, as its name suggests, intended for presolving QPs.

Highly efficient commercial interior-point-based software such as CPLEX 6.0 (1998), MOSEK (Andersen and Andersen 2000) and XPRESS-MP (Guéret, Prins and Seveaux 2002) is available for convex QP. These packages implement path-following algorithms in a primal–dual setting, and are available for parallel machines as well as for personal computers. Significantly, they may be tested online on the NEOS Server for Optimization (Czyzyk, Mesnier and Moré 1998, Gropp and Moré 1997, Dolan 2001).

The object-oriented QP package OOQP (Gertz and Wright 2003) implements generalizations of both Mehrotra's (1992) predictor–corrector and Gondzio's (1996) higher-order correction methods. OOQP has the advantage of being customizable to various application domains, and has been tailored to solve problems arising from support vector machines and Huber regression. Similar features are implemented in COPLQP (Ye 1997).

Specialized structure is exploited automatically by the object-oriented parallel solver OOPS (Gondzio and Grothey 2003a). Currently OOPS has been able to solve nontrivial problems involving 52 million variables and 20 million constraints.

Although now a code for general nonlinear programming, a set of default parameter values for convex QP and a careful implementation of a tailored LDL^T factorization for the quasi-definite systems at the heart of the algorithm make LOQO (Vanderbei 1999) one of the most robust predictor–corrector, primal–dual path-following convex QP solvers. Much of this is due to the care with which the factorization is obtained. An LDL^T factorization of the regularized matrix $K + R$ from (4.19) is computed using a two-stage ordering scheme assigning priorities to pivots based on estimates of the fill-in in both AA^T and $A^T A$. Pivots corresponding to the current priority are treated using a minimum-degree ordering heuristic.

QPB from the GALAHAD library of Gould et al. (2003a) implements a primal–dual interior method for general QP – for non-convex problems, QPB is only capable of identifying a weak second-order critical point. The Phase-I relies on the package LSQP, itself a primal–dual infeasible method for convex separable QP (Zhang 1994) which is also part of GALAHAD. Numerical tests on a monoprocessor machine on small, $n + m \lesssim 10^4$, medium, $10^4 \lesssim n + m \lesssim 10^5$ and large-scale, $10^5 \lesssim n + m \lesssim 10^6$, problems illustrate how well the method scales with the dimension, and the superiority of interior-point approaches over active-set-type methods when a reliable estimate of the optimal working set is not available and when the number of variables and constraints are large (Conn et al. 2000b, Gould et al. 2003a, Gould and Toint 2002b).

4.3. General linearly constrained optimization

When the constraints are linear but the objective neither linear nor quadratic, most algorithms try to emulate the QP methods described above, by ensuring feasibility with respect to constraints and requiring a reduction in the objective function (or perhaps barrier function) at each iteration – if an interior-point method is used, the iterates will remain interior to all inequality constraints. The only significant differences occur because the Hessian of the objective function changes at each iteration and must be periodically evaluated or estimated by some means. If the objective is close to linear, solutions (and intermediate iterates) often have a high proportion of active constraints ($|W| \approx n$) and some methods (Murtagh and Saunders 1982, Gill, Murray and Saunders 2002, Friedlander and Saunders 2005) exploit this by maintaining (dense) secant approximations of the reduced Hessian.

Interior-point methods for convex problems have received extensive attention since the existence of self-concordant barriers leads to polynomial algorithms (Nesterov and Nemirovskii 1994, Renegar 2001), and specialized methods have been devised for important applications. A good example is the minimization of a nonlinear but convex, (and preferably, but not necessarily) separable objective subject to linear equalities and bounds which arise in transportation planning, knowledge management or world-wide web traffic modelling (Saunders and Tomlin 1996). The problem is stated as

$$\underset{x \in \mathbb{R}^n}{\text{minimize}} \ f(x) \ \text{subject to} \ Ax = b \ \text{and} \ l \leq x \leq u,$$

and is regularized and reformulated as

$$\underset{x,r}{\text{minimize}} \ f(x) + \tfrac{1}{2}\|D_1 x\|^2 + \tfrac{1}{2}\|r\|^2 \ \text{subject to} \ Ax + D_2 r = b \ \text{and} \ l \leq x \leq u,$$

for some diagonal positive definite regularization matrices D_1 and D_2. A primal–dual path-following method is then applied. The bulk of the computation involves solving systems of the form

$$\begin{pmatrix} H & A^T \\ A & -D_2^2 \end{pmatrix} \begin{pmatrix} d_x \\ -d_y \end{pmatrix} = -\begin{pmatrix} \nabla_x f(x) - A^T y - \mu[(X-L)^{-1} + (U-X)^{-1}]e \\ Ax + D_2^2 y - b \end{pmatrix}$$

where $H = \nabla_{xx} f(x) + D_1^2 + (X-L)^{-1} Z_l + (U-X)^{-1} Z_u$ and y, z_l and z_u are suitable Lagrange multiplier estimates. As the coefficient matrix here is quasi-definite, it admits an LDL^T factorization. Alternatively, eliminating d_y to obtain normal equations and treating them as a least-squares problem, a trial step (d_x, d_y) is computed using a least-squares method, *e.g.*, LSQR of Paige and Saunders (1982).

Not all proposed interior-point methods are of the path-following variety. For example, it is possible to generalize the affine-scaling approach of Coleman and Li (1996) to handle linear inequality constraints (Coleman and Li 2000).

Software

Although it is capable of handling general constraints, the venerable active set NLP solver MINOS (Murtagh and Saunders 1982) is perhaps best regarded for its ability to deal with linear constraints. Likewise its successors SNOPT

(Gill *et al.* 2002) and KNOSSOS (Friedlander and Saunders 2005) are both highly effective for such problems, particularly if there are relatively few degrees of freedom. As usual, other general nonlinear programming packages, such as LOQO and KNITRO may be applied and are comfortable with such problems, although we would not recommend LANCELOT in this case.

5. Large-scale nonlinearly constrained optimization

Finally, we turn our attention to our most general nonlinear programming problem (1.1) and the attendant difficulties of coping with constraint curvature.

5.1. Sequential linear and quadratic programming methods

The phrase 'sequential quadratic programming' (SQP) seems to mean different things to different people, but the central theme is undoubtedly to apply an iteration for which a new iterate is generated by trying to minimize a quadratic approximation of the appropriate Lagrangian function $\ell(x, y) \stackrel{\text{def}}{=} f(x) - y_{\mathcal{E}}^T c_{\mathcal{E}}(x) - y_{\mathcal{I}}^T c_{\mathcal{I}}(x)$ subject to linearizations of some or all of the constraints. Here we will examine several aspects of this approach. There has been a number of surveys of SQP methods over the past 10 years (Boggs and Tolle 1995, 2000, Conn, Gould and Toint 1997, Gould and Toint 2000b) and we urge readers to consult these for details since we do not have room to give them all here.

We start by considering problems only involving equality constraints – for some people, such as those who work on PDE-constrained optimization (*e.g.*, Biros and Ghattas (2000)), this *is* SQP – for which the central ideas are best understood. But it is in the context of the general problem (1.1) that we believe most people understand the term SQP, and which we consider next. There is a strong distinction between linearizing a subset of the constraints at each iteration – the EQP subproblem approach, which is strongly influenced by methods for equality constraints – and linearizing all constraints at every iteration – the IQP subproblem approach.

5.2. SQP methods for equality-constrained problems

We first consider SQP methods for the equality-constrained (EC) problem

$$\underset{x \in \mathbb{R}^n}{\text{minimize}} \; f(x) \text{ subject to } c_{\mathcal{E}}(x) = 0. \tag{5.1}$$

SQP methods for EC problems (SQPE) aim to find a correction d_k to the current solution estimate x_k so as to (approximately)

$$\underset{d}{\text{minimize}} \; q_k(d) = d^T g(x_k) + \tfrac{1}{2} d^T H_k d$$
$$\text{subject to } c_{\mathcal{E}}(x_k) + J_{\mathcal{E}}(x_k)d = 0. \tag{5.2}$$

Here $g(x) = \nabla_x f(x)$ is the gradient of the objective, $J(x) = \nabla_x c_{\mathcal{E}}(x)$ is the Jacobian of the constraints, and H_k is (an approximation to) the Hessian of the Lagrangian function $\ell_{\mathcal{E}}(x, y_{\mathcal{E}}) = f(x) - y_{\mathcal{E}}^T c_{\mathcal{E}}(x)$ for given estimates $y_{\mathcal{E}k}$ of the Lagrange multipliers $y_{\mathcal{E}}$ at x_k. If $H_k = \nabla_{xx} \ell_{\mathcal{E}}(x_k, y_{\mathcal{E}k})$ and

$$d^T H_k d > 0 \text{ for all } d \text{ for which } J_{\mathcal{E}}(x_k)d = 0, \tag{5.3}$$

the solution to (5.2) is identical to that obtained by applying Newton's method to the criticality conditions $\nabla_{(x,y_{\mathcal{E}})} \ell_{\mathcal{E}}(x, y_{\mathcal{E}}) = 0$ at $(x_k, y_{\mathcal{E}k})$. Aside from the fundamental issues of how to choose H_k and $y_{\mathcal{E}k}$, SQPE methods have a number of obvious possible shortcomings. In particular (i) the linearized constraints may be inconsistent, (ii) (5.3) may be violated, and (iii) the iteration may diverge.

Possible shortcoming (i) is best dealt with in one of two, related, ways. The first is to re-pose (1.1) as the related penalty problem

$$\underset{x \in \mathbb{R}^n}{\text{minimize}} \ \phi(x, \sigma) \overset{\text{def}}{=} f(x) + \sigma \sum_{i \in \mathcal{E}} |c_i(x)| + \sigma \sum_{i \in \mathcal{I}} \min(-c_i(x), 0) \tag{5.4}$$

for some sufficiently large $\sigma > 0$ and given norm $\|\cdot\|$, and instead to minimize some model of the (non-smooth) penalty function $\phi(x, \sigma)$. A typical model problem then might be to approximately

$$\underset{d}{\text{minimize}} \ q_k(d) + \sigma \|c_{\mathcal{E}}(x_k) + J_{\mathcal{E}}(x_k)d\|; \tag{5.5}$$

if $\|\cdot\|$ is polyhedral (e.g., the ℓ_1- or ℓ_∞-norm), (5.5) may be reformulated as a (consistent) inequality-constrained QP, while if $\|\cdot\|$ is elliptical (e.g., the ℓ_2-norm) a quadratic conic-programming reformulation is possible.

Notice that the intention here is implicitly to allow inconsistent linearized constraints by merely reducing their infeasibility as much as is possible. A second, more direct way of dealing with inconsistency is to aim for reduction in infeasibility rather than full satisfaction of the constraints. A composite step $d_k = n_k + t_k$ may be used to achieve this. The idea is simply that the '(quasi-)normal' step n_k tries to reduce $\|c_{\mathcal{E}}(x) + J_{\mathcal{E}}(x_k)n\|$ while the 'tangential' step t_k aims to reduce $q_k(d)$ while maintaining the infeasibility at the level achieved by n_k; if n_k reduces the infeasibility to zero and t_k solves

$$\underset{t \in \mathbb{R}^n}{\text{minimize}} \ q_k(n_k + t) \text{ subject to } J_{\mathcal{E}}(x_k)t = 0, \tag{5.6}$$

d_k will be the solution to (5.2). Although there is a number of composite-step methods (Conn et al. 2000a, §15.4), the most appealing is the so-called Byrd–Omojokun approach (Byrd, Hribar and Nocedal 1999, Lalee, Nocedal and Plantenga 1998, Omojokun 1989), in which the CG method is used both to reduce $\|c_{\mathcal{E}}(x_k) + J_{\mathcal{E}}(x_k)n\|_2^2$ and subsequently to approximately solve the EQP (5.6) (see Section 4.1).

If shortcoming (ii) occurs, $q_k(d)$ will be unbounded from below on the feasible region. Suitable remedies are just as in the unconstrained case (see Section 2). Linesearch-based methods cope with such an eventuality either by obtaining a direction of feasible negative curvature or by modifying H_k, although good methods for achieving the latter during matrix factorization are still in their infancy (Forsgren 2002, Forsgren and Murray 1993, Gould 1999). Trust region-based methods impose a constraint to stop steps to infinity, but there is the added complication that the trust region constraint $\|s\| \leq \Delta_k$ may be incompatible with the linearizations $J_{\mathcal{E}}(x_k)d = -c_{\mathcal{E}}(x_k)$ if Δ_k is too small. In this case, one of the remedies proposed for shortcoming (i) may be required.

Shortcoming (iii) may be overcome in the usual way, namely by requiring descent (monotonic or otherwise) with respect to a suitable merit function such as (5.4). A good choice of σ is vital if such a method is to be efficient, and we will return to this later. An unfortunate consequence – the Maratos (1978) 'effect' – is that the SQP step may not be acceptable to merit functions like (5.4), and that an auxiliary calculation (a 'second-order correction') may be required to modify the step to allow fast convergence. Other merit functions, such as the augmented Lagrangian function, avoid this defect and have been used with much success (Boggs, Kearsley and Tolle 1999a, Gill et al. 2002).

A modern alternative to merit functions, which avoids the need to compute a penalty parameter, is to use the filter idea introduced in Section 2.1. For EC problems, we consider the conflicting objectives $\theta_1(x)$ and $\theta_2(x)$ to be the objective function and the constraints violation $\|c_{\mathcal{E}}(x)\|$, respectively. A step d is thus accepted if either the objective function decreases or if the constraints violation is reduced, while it is rejected if no decrease is obtained in either. But of course many further refinements are necessary in order to devise a workable algorithm. One is the way that filter methods deal with incompatible model constraints. Rather than resorting to the remedies for shortcoming (i) given above, filter trust region algorithms switch to a 'restoration phase', i.e., to the minimization of the constraint violation alone (the objective function is momentarily forgotten) until a model with compatible constraints is found. Since this will be true for any feasible point for the original problem, this restoration phase must terminate at a suitable point as long as it is capable of finding one – or indeed if it is even possible to find one at all. This restoration phase may use any suitable algorithm, including the filter method for nonlinear least-squares mentioned in Section 2.2. It may also be triggered more frequently – the method of Gonzaga, Karas and Vanti (2003) performs the equivalent of a restoration phase at every iteration.

A drawback that is common to SQPE approaches is that they all potentially suffer from the Maratos effect and therefore may need a second-order correction step to guarantee fast convergence. In theory this may be avoided by the filter remembering Lagrangian rather than objective function values (Ulbrich 2004b), but, to our knowledge, numerical experience is not yet available to support this idea in practice.

Rival trust region SQPE filter methods impose different requirements on the step computation – Fletcher, Leyffer and Toint (2002b) require the global solution of the trust region constrained SQPE, while others (Fletcher and Leyffer 2002, Fletcher, Gould, Leyffer, Toint and Wächter 2002a, Gonzaga *et al.* 2003, Gould and Toint 2005, 2003b) permit approximate local minimizers – and on the precise technique for maintaining the filter. This technical decision is often based on the distinction between iterations whose main effect is to reduce the objective function (f-iterations), and iterations whose main effect is to reduce constraint violation (θ-iterations).

Linesearch variants of the filter idea are also possible. Despite using a different globalization technique, the proposal of Wächter and Biegler (2003a, 2003b) remains similar in structure to the trust region variants, in that it also involves restorations, second-order correction steps and similarly uses the distinction between f- and θ-iterations to manage the filter.

For all SQPE algorithms, two other issues which are of great practical importance are the choice of Hessian approximation H_k and Lagrange multiplier estimates $y_{\mathcal{E}k}$. Although exact second derivatives of the Hessian of the Lagrangian are often available, the use of approximations still persists especially for problems where $|\mathcal{E}| \approx n$. In particular, as we noted in Section 4, solving (5.6) may be reduced to minimizing $t_n^T g_n + \frac{1}{2} t_n^T H_n t_n$ and recovering $t_k = N^T t_n$, where $g_n = N^T(g(x_k) + H_k n_k)$ and $H_n = N^T H_k N$, and the columns of N form a basis for the null-space of $J_{\mathcal{E}}(x_k)$. Thus, as long as $|\mathcal{E}| \approx n$, H_n will be small and it will be feasible to maintain H_n as a dense secant approximation to $N^T \nabla_{xx} \ell_{\mathcal{E}}(x_k, y_{\mathcal{E}k}) N$ (see the survey articles mentioned at the start of this section). If $|\mathcal{E}| \not\approx n$, it may still be possible to maintain a useful limited-memory secant approximation to the same matrix (Gill *et al.* 2002). Lagrange multipliers $y_{\mathcal{E}\,k+1}$ are often taken as those from the approximate solution to (5.4), although some form of interpolation between these values and $y_{\mathcal{E}k}$ may be necessary if the merit function, the trust region or constraint inconsistency intervene; little work seems to have been performed to discover the influence of such distractions which is somewhat surprising given the influence $y_{\mathcal{E}k}$ may have on H_k. As an alternative, a direct or CG least-squares solution to $J_{\mathcal{E}}^T(x_k)y = g(x_k)$ may be appropriate (Lalee *et al.* 1998).

5.3. SQP methods for the general problem

Suffice it to say, as the name suggests, an SQP method aims to solve the general problem (1.1) by solving a sequence of (cleverly) chosen QP problems. There are essentially two classes of SQP methods.

Sequential equality-constrained quadratic programming (SEQP) methods
The first, which we call sequential equality-constrained QP (SEQP) methods are essentially SQPE methods for which the set $\mathcal{E}$ is replaced by a (changing) estimate $\mathcal{A}_k \subseteq \mathcal{E} \cup \mathcal{I}$ of (1.1)'s optimal active set. All of the salient points we made about SQPE methods apply equally here, but now the dynamic data structures necessary to accommodate changes in $\mathcal{A}_k$ and, more importantly, the choice of $\mathcal{A}_k$ itself introduce extra complications. Of paramount importance is the globalization strategy, since otherwise there will be little control over constraints not in $\mathcal{A}_k$. In particular, it is vital that all constraints are represented in whatever merit function or filter is used.

A common strategy is to use the non-differentiable penalty function

$$\underset{x \in \mathbb{R}^n}{\text{minimize}} \ \phi(x, \sigma) \overset{\text{def}}{=} f(x) + \sigma \|c_{\mathcal{E}}(x)\| + \sigma \| \min(-c_{\mathcal{I}}(x), 0)\| \qquad (5.7)$$

as a merit function, and to use an EQP model in which a second-order approximation to the (locally) differentiable part of $\phi(x, \sigma)$ is minimized subject to linearized approximations to the (locally) non-differentiable part remaining fixed (Coleman and Conn 1982); $\mathcal{A}_k$ is thus defined by those constraints with (almost) zero values.

An alternative is to use the active set at a minimizer of a 'simpler' model of (1.1) or (5.7) to predict the active set of (1.1). The most obvious models are linear, and lead to linear programming subproblems which aim to

$$\underset{d}{\text{minimize}} \ l_k(d) = d^T g(x_k) \ \text{subject to} \ c_{\mathcal{E}}(x_k) + J_{\mathcal{E}}(x_k)d = 0 \qquad (5.8)$$
$$\text{and} \ c_{\mathcal{I}}(x_k) + J_{\mathcal{I}}(x_k)d \geq 0.$$

or

$$\underset{d}{\text{minimize}} \ l_k(d) + \sigma \|c_{\mathcal{E}}(x_k) + J_{\mathcal{E}}(x_k)d\| + \sigma \| \min(-c_{\mathcal{I}}(x_k) - J_{\mathcal{I}}(x_k)d, 0)\|; \qquad (5.9)$$

the advantage here is that there are excellent (simplex and interior-point) methods for large-scale linear programming. However, since the solutions to these subproblems almost inevitably lie at vertices of their feasible regions, and as there is no reason to expect that the solution to (1.1) has n active constraints, (5.8) or (5.9) alone are not sufficient to determine $\mathcal{A}_k$.

One way of remedying this is to impose artificial constraints whose role is simply to cut off those problem constraints which are likely to be inactive at the solution to (1.1); if an artificial constraint is active at the solution of (5.8) or (5.9) it will not be included in $\mathcal{A}_k$. Care must be taken, however, to ensure that the artificial constraints do not exclude optimally active problem constraints, and the balance between these aims is quite delicate. Early sequential linear programming (SLP) methods (Griffith and Stewart 1961) imposed artificial constraints of the form $\|s\|_\infty \leq \Delta$ in which Δ was dynamically adjusted, but it was Fletcher and Sainz de la Maza

(1989) who first interpreted this as a trust region constraint. Crucially they showed that the usual trust region acceptance and adjustment rules are sufficient to correctly identify the optimal active set in a finite number of iterations. Both filter-based (Chin and Fletcher 2003) and merit-function-based (Fletcher and Sainz de la Maza 1989, Byrd *et al.* 2004*a*) SLP variants are possible.

If the non-differentiable penalty function $\phi(x, \sigma)$ in (5.7) is used, it is important that the penalty parameter σ be adjusted to ensure that ultimately feasible critical points of the latter correspond to critical points of (1.1). Although in principle one could simply adjust σ once an approximate solution of (5.7) has been found (Mayne and Polak 1976), this is wasteful. It is preferable to adjust σ as soon as there is model-based evidence that the current value is not reducing the constraints, and means for doing this while ensuring convergence to critical points of (1.1) (or perhaps finding a critical point of infeasibility) are known (Byrd, Gould, Nocedal and Waltz 2004*b*).

Sequential inequality-constrained quadratic programming (SIQP) methods
The second class of SQP methods are those we refer to as sequential inequality-constrained QP (SIQP) methods. In these, no *a priori* prediction is made about the active set, but instead a correction d_k is chosen to (approximately)

$$\text{minimize}_d \ q_k(d) \ \text{subject to} \ c_{\mathcal{E}}(x_k) + J_{\mathcal{E}}(x_k)d = 0$$
$$\text{and} \ c_{\mathcal{I}}(x_k) + J_{\mathcal{I}}(x_k)d \geq 0. \tag{5.10}$$

Now H_k is an approximation to the Hessian of the full Lagrangian, $\ell(x, y) = f(x) - y^T c(x)$, for the problem, and linearizations of all constraints are included. Although we no longer need to specify $\mathcal{A}_k$, constraint inconsistency and iterate divergence are still serious concerns, and now we have the added complication that (iv) (5.10) may have (many) local minimizers.

Given all of these potential pitfalls, why are SIQP methods so popular? One reason is obviously their potential for fast local convergence; under reasonable assumption, the iteration based on (5.10) will correctly identify the active set and thereafter converge rapidly (Robinson 1974). Another is favourable empirical evidence accumulated on small-scale problems (Hock and Schittkowski 1981). But, on this basis and given the growing number of successful codes for large-scale QP, it might be thought surprising that there are so few large-scale SIQP algorithms. We now believe that this is not a coincidence and most likely an indication of the unsuitability of the SIQP paradigm for large-scale optimization. Why do we believe this?

Our first objection to SIQP is simply that, given even the most efficient QP method, the cost of solving a large-scale inequality-constrained

QP (IQP) is usually far greater than, say, an equivalently sized EQP or interior-point subproblem. Thus a method that uses IQPs either needs to ensure that relatively few overall iterations are required, or have some mechanism for stopping short of QP optimality. Although there is anecdotal evidence that SIQP methods require few iterations for small-scale problems, we are unaware of any proof that this will always be the case. Likewise, the methods suggested in the tiny body of work on IQP truncation (Goldsmith 1999, Murray and Prieto 1995) may, in the worst case, require the solution of n (related) EQPs per IQP.

Of more serious concern are the dangers posed by allowing indefinite H_k. This possibility rarely surfaced in the small-scale case, since almost always positive definite secant Hessian approximations were used. But for large problems traditional secant approximations are rarely viable on sparsity grounds – limited-memory secant methods are possible (Gill *et al.* 2002) but may give inaccurate approximations, while the alternatives of using partitioned secant approximations or exact second derivatives often generate indefinite Hessians. Indefinite H_k may cause difficulties for a number of reasons. Firstly, the possibility of moving to an unwelcome (possibly higher) local minimizer, d_k, cannot be discounted, particularly when using an interior-point QP solver. Such a d_k may well be unsuitable for use with a globalization strategy. While this appears to be a defect specifically for interior-point QP solvers, active set methods may also be fooled. Consider the simple-bound QP

$$\operatorname*{minimize}_{x\in\mathbb{R}^2} \tfrac{1}{2}(x_1^2 + x_2^2) - 3x_1x_2 - \tfrac{5}{4}x_1 + \tfrac{7}{4}x_2 \text{ subject to } 0 \le x_1, x_2 \le 1,$$

whose contours are illustrated in Figure 5.1 – this is a simplified version of that given by Goldsmith (1999). Starting from $x = (0,0)$, many active set QP solvers would move downhill via the corner $(1,0)$ to the (global) minimizer at $(1,1)$ – both steps are along directions of positive curvature. Unfortunately the overall step $(1,1)$ is an initially uphill direction of negative curvature, and thus again unlikely to be suitable for use with a globalization strategy. Of course this does not mean that the method will fail, merely that the approach may be inefficient as extra precautions (such as reducing a trust region radius or modifying curvature) may have to be applied.

Whatever our reservations, SIQP methods remain popular. Linesearch, trust region and filter variants have been proposed. Some avoid difficulty (iv) above by insisting on positive definite (sometimes limited-memory) secant approximations to second derivatives (Gill *et al.* 2002). Others modify true second derivatives to ensure that the reduced Hessian is positive definite (Boggs, Kearsley and Tolle 1999*b*, Boggs *et al.* 1999*a*), while some use the restoration-phase of the filter approach to recover from bad steps (Fletcher and Leyffer 2002).

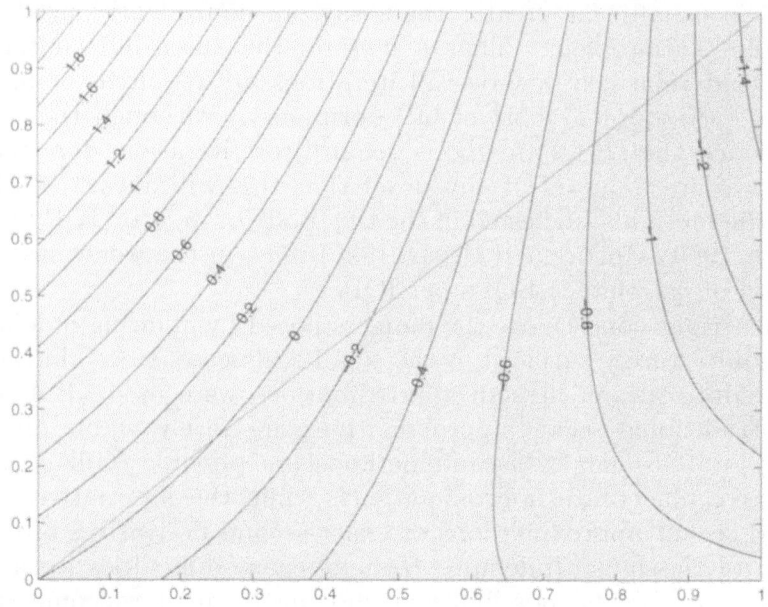

Figure 5.1. Active set QP method gives uphill step

5.4. Interior-point methods for nonlinear programs

As the reader might anticipate, the last decade has seen an explosion in interest in path-following methods for the general nonlinear program (1.1). Amongst the large number of papers devoted to the topic, two related approaches have emerged.

The first places all inequality constraints directly into a logarithmic barrier, leaving only explicit equality constraints. A sequence of barrier subproblems of the form

$$\operatorname*{minimize}_{x\in\mathbb{R}^n} \phi(x;\mu) \overset{\text{def}}{=} f(x) - \mu\sum_{i\in\mathcal{I}} \log(c_i(x))$$
$$\text{subject to } c_{\mathcal{E}}(x) = 0,$$

(5.11)

parametrized by the scalar $\mu > 0$, is solved for positive values of μ which eventually decrease to zero. This approach is particularly appealing when $c_{\mathcal{E}}(x) = 0$ are sufficiently simple to be handled directly, *e.g.*, when they are linear (Conn *et al.* 2000*b*) – indeed, this is simply a generalization of (4.7) – but does require that the inequality constraints are strictly satisfied throughout. Since this may be difficult to achieve – even finding an initial point for which this is true may be far from trivial – the second approach allows inequality constraints to be violated at intermediate stages, but for each introduces a slack variable which is treated by a barrier function. The resulting problem is thus of the form

$$\underset{x \in \mathbb{R}^n, \, s_{\mathcal{I}} \in \mathbb{R}^{n_{\mathcal{I}}}}{\text{minimize}} \phi(x, s; \mu) \overset{\text{def}}{=} f(x) - \mu \sum_{i \in \mathcal{I}} \log(s_i) \tag{5.12}$$

$$\text{subject to } c_{\mathcal{E}}(x) = 0 \text{ and } c_{\mathcal{I}}(x) - s_{\mathcal{I}} = 0.$$

Clearly, the introduction of slacks $s_{\mathcal{I}}$ is reminiscent of (4.15). Although it is vital that the slacks remain strictly feasible throughout, not all methods of this type remain infeasible right up to the solution (Byrd, Nocedal and Waltz 2003).

For both approaches, the barrier subproblems are equality-constrained, and the SQPE methods described in Section 5.2 are appropriate (Byrd, Gilbert and Nocedal 2000, Vanderbei and Shanno 1999, Wächter and Biegler 2004). Note, however, that extra precautions to ensure that the barrier terms remain finite must be taken, and it is here that (5.12) has some advantage, since in this case the barrier terms only involve (trivial) linear expressions.

Just as in the linearly constrained case, locally convergent methods may be devised by applying (variants of) Newton's method to the perturbed optimality conditions

$$\nabla f(x) - J_{\mathcal{E}}(x)^T y_{\mathcal{E}} - J_{\mathcal{I}}(x)^T y_{\mathcal{I}} = 0,$$
$$c_{\mathcal{E}}(x) = 0, \tag{5.13}$$
$$\text{and } c_{\mathcal{I}}(x) y_{\mathcal{I}} - \mu e = 0,$$

of (1.1), or

$$\nabla f(x) - J_{\mathcal{E}}(x)^T y_{\mathcal{E}} - J_{\mathcal{I}}(x)^T y_{\mathcal{I}} = 0,$$
$$y_{\mathcal{I}} - z_{\mathcal{I}} = 0,$$
$$c_{\mathcal{E}}(x) = 0, \tag{5.14}$$
$$c_{\mathcal{I}}(x) - s_{\mathcal{I}} = 0,$$
$$\text{and } Z_{\mathcal{I}} s_{\mathcal{I}} - \mu e = 0,$$

of

$$\underset{x \in \mathbb{R}^n, \, s_{\mathcal{I}} \in \mathbb{R}^{n_{\mathcal{I}}}}{\text{minimize}} f(x) \text{ subject to } c_{\mathcal{E}}(x) = 0, \, c_{\mathcal{I}}(x) - s_{\mathcal{I}} = 0 \text{ and } s_{\mathcal{I}} \geq 0, \tag{5.15}$$

(cf. (4.16) and (4.17)). The only differences between the variants on Newton's method described in Section 4.2 and those applicable here are that the Jacobians $A_{\mathcal{E}}$ and $A_{\mathcal{I}}$ in (4.18)–(4.21) are now $J_{\mathcal{E}}(x)$ and $J_{\mathcal{I}}(x)$, respectively, and that G should now be an approximation to the Hessian of the Lagrangian, $\nabla_{xx}\ell(x, y)$; (direct or iterative) methods for solving these systems are identical to those in Section 4.2.

In (5.12), when the constraint Jacobian

$$J(x) = \begin{pmatrix} J_{\mathcal{E}}(x) & 0 \\ J_{\mathcal{I}}(x) & I \end{pmatrix}$$

has full rank, composite step (reduced space) variants of (4.18)–(4.21) are also possible. Just as in Section 5.2, the step may be decomposed using the Byrd–Omojokun scheme (Omojokun 1989), and least-squares estimates of the Lagrange multipliers for the equality constraints obtained. If similar multiplier estimates for inequality constraints are found, care needs to be taken to ensure that these remain positive (Wright 1997) or that the quadratic model remains convex in the slacks (Byrd *et al.* 1999). For problems arising from, *e.g.*, dynamical systems where multiplier estimates are not available, it is remarkable that schemes to update the penalty parameter may still be derived (Wächter 2002).

As always, it is necessary to globalize Newton's method in some way, and both (smooth and non-smooth) merit function- and filter-based possibilities have been proposed. Issues that arise with the linesearch globalization of the Newton direction d_x

$$
\begin{pmatrix} G & J_{\mathcal{E}}^T(x) & J_{\mathcal{I}}^T(x) \\ J_{\mathcal{E}}(x) & 0 & 0 \\ J_{\mathcal{I}}(x) & 0 & -Z_{\mathcal{I}}^{-1}S_{\mathcal{I}} \end{pmatrix} \begin{pmatrix} d_x \\ -y_{\mathcal{E}} - d_{y_{\mathcal{E}}} \\ -y_{\mathcal{I}} - d_{y_{\mathcal{I}}} \end{pmatrix} = - \begin{pmatrix} g(x) \\ c_{\mathcal{E}}(x) \\ c_{\mathcal{I}}(x) - s_{\mathcal{I}} + \mu Z_{\mathcal{I}}^{-1}e \end{pmatrix}
$$

(5.16)

(or its Section 4.2 equivalents (4.18)–(4.21)) include the choices of step-size and other (penalty and barrier) parameters and strategies to ensure that G is chosen to guarantee that d_x gives descent for whatever merit function is used – to date, the simple expedient of adding a diagonal matrix λI to G for suitably large λ seems to be the most sophisticated strategy used in the large-scale case (Vanderbei and Shanno 1999, Wächter and Biegler 2004), although, just as in Section 4.2, all that is actually required is that $G + J_{\mathcal{I}}^T(x)S_{\mathcal{I}}^{-1}Z_{\mathcal{I}}J_{\mathcal{I}}(x)$ should be positive definite on the null-space of $J_{\mathcal{E}}(x)$. Murray and Wright (1992) devised a linesearch procedure tailored to the logarithmic barrier function, given a search direction d, by identifying the closest constraint for which d is a descent direction at the current iterate. A step-size is computed by identifying a root of an approximation to the gradient of the barrier along d, by linearizing f and the constraint in question and ignoring all other constraints. Several other interpolating functions are used as approximations of the logarithmic barrier by the same authors, so as to devise specialized linesearches. To the best of our knowledge, these have not been incorporated into large-scale interior-point codes.

Typical merit functions for (5.12) might be the non-smooth penalty-barrier function (Yamashita, Yabe and Tanabe 2004)

$$
\phi(x, s; \mu, \nu) = f(x) - \mu \sum_{i \in \mathcal{I}} \log(s_i) + \nu \|c_{\mathcal{E}}(x)\| + \nu \|c_{\mathcal{I}}(x) - s_{\mathcal{I}}\|, \quad (5.17)
$$

the smooth variant (Gay, Overton and Wright 1998, Vanderbei and Shanno 1999)

$$\psi(x, s; \mu, \nu) = f(x) - \mu \sum_{i \in \mathcal{I}} \log(s_i) + \tfrac{\nu}{2} \|c_{\mathcal{E}}(x)\|_2^2 + \tfrac{\nu}{2} \|c_{\mathcal{I}}(x) - s_{\mathcal{I}}\|_2^2, \quad (5.18)$$

or some scaled equivalents, perhaps even involving a different penalty parameter ν_i per constraint to account for poor scaling. Although these parameters should be handled globally as described in Section 5.3, care must be taken to ensure that the direction computed from (5.16) or its variants is a descent direction for the merit function. This may be guaranteed for ϕ by iteratively increasing the penalty parameter until its directional derivative is negative; only a finite number of increases are required under standard assumptions (Byrd $et\ al.$ 2000).

A disadvantage of (5.17) and (5.18) is that they really only measure suitability of the primal step d_x; other means are used to compute steps in the dual variables. One function which does not suffer from this drawback is the augmented penalty-barrier merit function (Forsgren and Gill 1998),

$$\theta(x, y; \mu, \nu) = f(x) + \frac{1}{2\mu} \sum_{i \in \mathcal{E}} \left\{ c_i(x)^2 + \nu(c_i(x) + \mu y_i)^2 \right\} \quad (5.19)$$

$$- \mu \sum_{i \in \mathcal{I}} \left\{ \log(c_i(x)) + \nu \left(\log \left(\frac{c_i(x) y_i}{\mu} \right) + 1 - \frac{c_i(x) y_i}{\mu} \right) \right\},$$

which allows simultaneous minimization in both the primal and dual variables. So long as $G + J_{\mathcal{I}}^T(x) S_{\mathcal{I}}^{-1} Z_{\mathcal{I}} J_{\mathcal{I}}(x)$ is positive definite on the null-space of $J_{\mathcal{E}}(x)$, the primal–dual Newton step (5.16) for (5.13) is a descent direction for θ. If not, negative curvature descent directions are easy to obtain.

For the most part, theoretical analyses of these techniques make relatively strong assumptions – a linear independence qualification condition (LICQ) is often required to establish global convergence, while fast local convergence analyses rely on strict complementarity. Because the objective function and the barrier objective function both decrease monotonically with μ along the exact central path (Fiacco and McCormick 1968, Wright 1992), path-following algorithms for nonlinear programming have a monotone flavour. This is at variance with other methods discussed earlier.

A most disturbing aspect of linesearch-based interior-point methods which use (5.13) to compute the search direction d_x has recently been discovered (Wächter and Biegler 2000). The issue is that if there is a mixture of equality and inequality constraints, if the former are approximated by linearizations and if feasibility of the latter are controlled by restricting the step along the search direction, the resulting iteration may converge to a worthless infeasible point. This surprising result has caused a reassessment of linesearch methods, and in some cases consideration of filter methods with appropriate acceptance measures as a replacement (Benson, Vanderbei and Shanno 2002, Wächter and Biegler 2004). The natural alternative, though, is to consider trust region-based methods, which fortunately do not suffer from this convergence failure.

In trust region interior methods for general nonlinear programming, any of the SQPE approaches discussed in Section 5.2 are appropriate, but now extra care needs to be taken to cope with the required feasibility of the inequality constraints in (5.11) or slacks in (5.12). In particular, in the latter case, it is important that the slack variables do not approach their bounds either prematurely or too rapidly. The obvious SQPE trust region subproblem would minimize a quadratic approximation to the Lagrangian of (5.12) – in which as usual a primal–dual approximation $Z_{\mathcal{I}} S_{\mathcal{I}}^{-1}$ to the Hessian of the barrier terms rather than the primal one $\mu S_{\mathcal{I}}^{-2}$ is used – subject to linearized approximations to the constraints within an appropriately scaled trust region and perhaps a suitable fraction-to-the-boundary constraint. For example, the step (d_x, d_s) may be constrained so that

$$\|(d_x, S^{-1} d_s)\|_2 \leq \Delta \quad \text{and} \quad s + d_s \geq (1 - \tau)s,$$

with $0 < \tau \lesssim 1$ (Byrd, Gilbert and Nocedal 2000, Byrd, Hribar and Nocedal 1999), or the fraction-to-the-boundary rule may be imposed after the event (Conn *et al.* 2000*b*). In general, it is especially important that the shape of the trust region mirrors that of the ill-conditioned barrier terms (Conn *et al.* 2000*b*). As before, the issue of linearized constraint incompatibility – particularly when there is a trust region – is present, and a composite-step strategy as outlined in Section 5.3 is appropriate. As in the linesearch case, the penalty parameter ν must be adjusted as the algorithm proceeds to try to ensure asymptotic feasibility of the constraints, and rules to achieve this within a trust region framework are known (Byrd *et al.* 2000). It is also possible to use the augmented penalty-barrier merit function (5.19) within such a framework (Gertz and Gill 2004).

Although primal–dual multiplier estimates $z_{\mathcal{I}}$ are usually preferred to primal ones $\mu S_{\mathcal{I}}^{-1} e$, it is important for global convergence that the former do not differ arbitrarily from the latter. To ensure this property, and to encourage fast asymptotic convergence, generated primal–dual estimates are typically projected into a box containing the primal values. This in turn guarantees proximity of the primal–dual Hessian to the pure primal Hessian, which is also required for fast convergence. An alternative is always to compute least-squares multipliers from an estimate of the optimal active set (Dussault 1995).

Some problems may not be defined when the constraints are violated, and methods based on (5.11) directly respect this requirement. Methods based on (5.12) may be modified to address it by resetting slacks to ensure that all iterates are strictly feasible. This is sometimes referred to as 'feasible mode'

(Byrd *et al.* 2003) and is often used in practice (Byrd *et al.* 2000). In a linesearch framework, as soon as an iterate x_k strictly satisfies the constraint c_i, $i \in \mathcal{I}$, *i.e.*,

$$c_i(x_k) \geq \epsilon > 0, \tag{5.20}$$

the ith component of the trial slack variables $s_i^{\mathrm{T}} = s_{ki} + d_{si}$ is reset to $c_i(x^{\mathrm{T}}) = c_i(x_k + d_x)$. Should the resulting step be rejected by the merit function, a shorter step (d_x, d_s) is attempted and the process is repeated. In trust region frameworks, the situation is more complicated since possible successive increases in the merit function caused by this reset might dominate decreases attempted by the step. It should also be kept in mind that (5.20) might very well never happen.

In practice it is common to encounter degenerate problems, that is, those for which the set of Lagrange multipliers is unbounded or, worse, does not exist – often such problems result from 'over-modelling'. For instance, it is easily seen that

$$\underset{x \in \mathbb{R}}{\text{minimize}} \ f(x) \ \text{subject to} \ x^2 = 0, \tag{5.21}$$

where $f : \mathbb{R} \to \mathbb{R}$ is such that $f'(0) \neq 0$, admits no Lagrange multiplier. To deal with this possibility (1.1) may be transformed so as to

$$\underset{x,s}{\text{minimize}} \quad \phi^{\mathrm{s}}(x, s; \nu) \overset{\text{def}}{=} f(x) + \nu \sum_{i \in \mathcal{E}} [c_i(x) + 2s_i] + \nu \sum_{i \in \mathcal{I}} s_i$$
$$\text{subject to} \ c_i(x) + s_i \geq 0 \ \text{and} \ s_i \geq 0, \ \text{for all} \ i \in \mathcal{E} \cup \mathcal{I}, \tag{5.22}$$

in terms of so-called *elastic variables* $s_{\mathcal{E}} \in \mathbb{R}^{n_{\mathcal{E}}}$ and $s_{\mathcal{I}} \in \mathbb{R}^{n_{\mathcal{I}}}$; the objective $\phi^{\mathrm{s}}(x, s; \nu)$ is simply a smooth reformulation of the exact ℓ_1-penalty function for (1.1). This new problem is not only smooth but regular – it satisfies the Mangasarian–Fromovitz constraint qualification, and thus has bounded multipliers, for all fixed $\nu > 0$. Furthermore, the problem only involves inequality constraints and is thus well suited to an interior-point approach (Gould, Orban and Toint (2003*b*); see also Tits, Wächter, Bakhtiari, Urban and Lawrence (2003) for a simplified variant).

One other possibility is to balance satisfaction of centrality and feasibility against optimality using a filter. The central idea is to compute a primal–dual step for (5.15) in a manner similar to that described in (5.16). But now, instead of defining a new iterate by a linesearch along the step, or by some classical trust region scheme, a two-dimensional filter with conflicting objectives (see Section 2.1)

$$\theta_1(x, s_{\mathcal{I}}, y, z_{\mathcal{I}}) = \|c_{\mathcal{E}}(x)\| + \|c_{\mathcal{I}}(x) - s_{\mathcal{I}}\| + \left\| Z_{\mathcal{I}} s_{\mathcal{I}} - \frac{z_{\mathcal{I}}^T s_{\mathcal{I}}}{n_{\mathcal{I}}} e \right\|$$

$$\text{and} \quad \theta_2(x, s_{\mathcal{I}}, y, z_{\mathcal{I}}) = \frac{z_{\mathcal{I}}^T s_{\mathcal{I}}}{n_{\mathcal{I}}} + \|\nabla_{(x, s_{\mathcal{I}})} \ell(x, s_{\mathcal{I}}, y, z_{\mathcal{I}})\|$$

is used to accept or reject the step; here $\ell(x, s_\mathcal{I}, y, z_\mathcal{I})$ is the Lagrangian of (5.15). The first objective, θ_1, measures feasibility and centrality of the vector $(x, s_\mathcal{I}, y, z_\mathcal{I})$ while θ_2 attempts to measure optimality. The resulting algorithm, which decomposes the primal–dual step into normal (towards the central path) and tangential (to the central path) components whose sizes are controlled by a trust region scheme, is globally convergent to first-order critical points (Ulbrich, Ulbrich and Vicente 2004).

Practicalities

Many practical issues are to be considered with extreme care when implementing path-following methods. Among those, we have already touched on the treatment of indefiniteness, degeneracy, unboundedness, poor scaling and handling of feasible sets with no strict interior. We now briefly comment on two other outstanding issues, the choice of the initial barrier parameter and its update.

The initial value of μ, although irrelevant in theory, is crucial in practice and may determine the success of a method within the allowed limits. Most algorithms set the initial barrier parameter to some prescribed constant value which seems to perform well on average over a large class of problems, *e.g.*, $\mu_0 = 0.1$.

For the formulation (5.12), if initial values for $s_\mathcal{I}$ and $z_\mathcal{I}$ are determined using (4.22), the initial value $\mu_0 = s_\mathcal{I}^T z_\mathcal{I} / n_\mathcal{I}$ is reminiscent of linear programming (Wright 1997) to obtain good centrality at the initial point. To take scaling into account and in an attempt to locate nearby points on the central path, one might set $\mu_0 = \max_i \|\nabla c_i(x_0)\|_\infty$ and perform a heuristic test by selecting the value of μ producing the smallest residual in the primal–dual system among the values $0.01\mu_0$, $0.1\mu_0$, μ_0, $10\mu_0$ and $100\mu_0$ – this rule is used by interior-point codes in the **GALAHAD** library (Gould *et al.* 2003*a*). Perhaps more usefully, if $P_{J_\mathcal{E}(x)}(v)$ denotes the orthogonal projection of v onto the null-space of $J_\mathcal{E}(x)$ and $\phi(x; \mu)$ is the objective of (5.11), Gay *et al.* (1998) suggest computing

$$\mu_{\mathrm{LS}} = \mathrm{argmin}_{\mu > 0} \|P_{J_\mathcal{E}(x_0)}(\nabla_x \phi(x; \mu))\|,$$

and subsequently setting the initial barrier parameter for (5.11) to the value

$$\mu_0 = \min(100, \max(1, \mu_{\mathrm{LS}})).$$

For (5.12), the same recipe involving $P_{J(x)}$ and the objective $\phi(x, s; \mu)$ is appropriate.

In short-step, long-step or predictor–corrector methods for linear programming (Wright 1997) and convex quadratic programming, the barrier parameter is updated using a rule similar to $\mu_{k+1} = \sigma_k s_{\mathcal{I}k}^T z_{\mathcal{I}k} / n_\mathcal{I}$, where $0 < \sigma_k < 1$ is a centering parameter. More traditional rules, such as

$\mu_{k+1} = \sigma_k \mu_k$ with $0 < \sigma_k < 1$ are commonplace in nonlinear programming, given that there is no concept of duality gap, and virtually all convergence theory has been established for such rules.

In the framework (5.11), Gay $et\ al.$ (1998) suggest the rule

$$\mu_{k+1} = \min\left(\mu_k, \sigma_k \frac{c_{\mathcal{I}}(x)^T z_{\mathcal{I}}}{n_{\mathcal{I}}}\right) \quad \text{where} \quad \sigma_k = \min\left(0.2, 100 \frac{c_{\mathcal{I}}(x)^T z_{\mathcal{I}}}{n_{\mathcal{I}}}\right),$$

where $z_{\mathcal{I}}$ are the estimates of the Lagrange multipliers associated to the inequality constraints of (1.1) at x_k. This rule is clearly reminiscent of linear programming and enforces that $\{\mu_k\}$ be decreasing. For some problems, this monotone behaviour causes difficulties and, sometimes, failure in practice, and more $dynamic$ rules are investigated, such as the laxer

$$\mu_{k+1} = \sigma \frac{c_{\mathcal{I}}(x)^T z_{\mathcal{I}}}{n_{\mathcal{I}}},$$

with $0 < \sigma < 1$, which allows the barrier parameter to increase (Bakry, Tapia, Tsuchiya and Zhang 1996). Similar rules have been used in the framework of (5.12), using $s_{\mathcal{I}k}^T z_{\mathcal{I}k}/n_{\mathcal{I}}$ instead of $c_{\mathcal{I}}(x)^T z_{\mathcal{I}}/n_{\mathcal{I}}$.

Vanderbei and Shanno (1999) note that, in practice, it is important to keep individual complementarity pairs clustered together. Using the formulation (5.12), they define

$$\xi_k = \frac{\min_i s_{ki} z_{ki}}{s_{\mathcal{I}k}^T z_{\mathcal{I}k}/n_{\mathcal{I}}}$$

to measure deviation from complementarity and use the heuristic update

$$\mu_{k+1} = 0.1 \min\left(0.05 \frac{1-\xi_k}{\xi_k}, 2\right)^3 \frac{s_{\mathcal{I}k}^T z_{\mathcal{I}k}}{n_{\mathcal{I}}}.$$

Such rules have had some success in practice but are unfortunately not covered by convergence theory and can indeed cause failure if μ becomes too small prematurely or diverges.

Problems with equilibrium constraints

Several formulations of mathematical programs with equilibrium constraints (MPECs) are given in the literature. Generalizing mathematical programs with complementarity constraints (MPCCs), their trait is the presence of a constraint of the form

$$0 \le F_1(x) \perp F_2(x) \ge 0, \tag{5.23}$$

where $F_1, F_2 : \mathbb{R}^n \to \mathbb{R}^{n_{\mathrm{CC}}}$ and, for $x, y \in \mathbb{R}^{n_{\mathrm{CC}}}$, the notation $x \perp y$ is understood componentwise as meaning $x_i y_i = 0$ for all $i = 1, \ldots, n_{\mathrm{CC}}$. Such a constraint might originate, $e.g.$, from variational inequalities, optimality conditions of the inner problem in a bilevel setting, or from an economic equilibrium requiring that either the price or the excess production for a product be zero. In game theory, F_1 and F_2 might represent the strategy of the leader and the follower, respectively. In design problems, F_1 is the design while F_2 is the response of the system. It is easily seen that problems with a constraint of the form (5.23) violate the Mangasarian–Fromovitz constraint qualification at every feasible point. Such problems thus always have unbounded sets of multipliers which typically consist in *rays*. We refer the reader to the recent overview by Leyffer (2003) for references.

Practical implementations able to reliably treat such problems remain rare and in an active development stage and there is much room left for improvement and the advent of new methods. Of particular importance is the impact of the formulation of the complementarity constraints on the performance of algorithms. An additional difficulty appears when studying interior methods for (1.1) with constraints of the form (5.23) as no central path exists. To circumvent this issue, most practical methods consider a sequence of *relaxed problems* with nonempty strict interior (Scheel and Scholtes 2000), and an interior method is applied to them. A rather simple modification of the step described in the filter linesearch interior algorithm of Wächter and Biegler (2004) is described by Raghunathan and Biegler (2003) who perform a single interior-point iteration per relaxed problem. This modification ensures nonsingularity of the step-defining augmented matrix and alleviates the need for centrality conditions. Numerical difficulties may appear, for in the limit, the strict interior of the feasible set vanishes. DeMiguel, Friedlander, Nogales and Scholtes (2004) propose an alternative where this limit is nonempty, removing the need to modify the search directions.

Anitescu (2000) reformulates MPCCs with nonempty Lagrange multiplier sets by smoothing an ℓ_∞-penalty function. The resulting nonlinear program depends on an elastic variable and has an isolated local minimizer at a solution of the MPCC which, under a quadratic growth condition, can be approached with a finite penalty parameter. This last problem may be solved, *e.g.*, using an SQP approach.

Luo, Pang and Ralph (1998) propose a disjunctive approach, in which the feasible region is decomposed in *branches*, also called *local pieces*. A single SQP step is performed on the nonlinear program defined by the current piece, and all pieces must be examined. Superlinear convergence holds under uniqueness of the multipliers.

Using elastic variables in a manner similar to Anitescu (2000), Benson, Sen, Shanno and Vanderbei (2003) reformulate the MPCC by smoothing an ℓ_∞-penalty function. Under strict complementarity, multipliers at a solution are bounded. The algorithm of Vanderbei and Shanno (1999) implemented in the LOQO package is used to solve the penalty subproblems, using an *ad hoc* rule to update the penalty parameter.

Convergence properties of algorithms for (5.23) typically rely on MPCC-specific regularity conditions, *e.g.*, strong stationarity, the so-called MPCC-LICQ, a strong constraint qualification, and the MPCC-SOSC, a specialized second-order condition. A form of strict complementarity usually ensures fast local convergence. For complete details regarding MPCCs and MPECs, we refer the reader to Luo, Pang and Ralph (1996).

Finally, filter methods can also be adapted for the solution of mixed complementarity problems. Ulbrich (2004a) uses a reformulation of the problem into semi-smooth equations, to which a filter method for least-squares (in a variant very close to that described in Section 2.2) is then applied. Although preliminary experiments are interesting, extensive numerical evidence is still missing and the effectiveness of the approach remains to be confirmed.

General convex programs

We finally consider the special case of problems of the form (1.1), in which f is convex and the constraints define a convex feasible set. Interior methods for such problems inherit many properties of those for linear and convex quadratic programming. Algorithms for the latter may therefore relatively painlessly be extended to the former. In particular, the multiple target tracking strategy of Gondzio (1996) generalizes, a key being the reduction of the Newton matrix for the primal–dual equation to a quasi-definite matrix. The method has the peculiarity of defining one barrier parameter per constraint.

The fact that there is a great deal of well-understood theory covering the convex case, and that efficient algorithms from linear programming carry over does not imply by any means that tracking the central path is an easy task. Indeed, even for infinitely differentiable convex data, the central path can exhibit an infinite number of segments of constant length and assume the shape of an 'antenna' or zigzag infinitely (Gilbert, Gonzaga and Karas 2002),

To control the step-size, linesearch-based methods for general convex programming use the ℓ_2 merit function (5.18). The rationale for this approach is that for sufficiently large values of $\nu > 0$, the direction d computed from (5.16) is a descent direction for (5.18) whenever the problem is strictly convex. The additional difficulty introduced by the use of such a merit function is the need to manage the penalty parameter. For most practical purposes, simple updating rules such as $\nu_{k+1} = 10\nu_k$ suffice. More clever rules (approximately) compute the smallest value $\nu_{\min}$ of ν which makes d a descent direction for the merit function, and set $\nu_{k+1} = 10\nu_{\min}$. The linesearch procedure next determines an appropriate step-size based on a fraction-to-the-boundary rule and an Armijo-type acceptance condition.

Software

Perhaps the most widely known SQP method is SNOPT (Gill *et al.* 2002), a worthy successor to the augmented-Lagrangian-based MINOS (Murtagh and Saunders 1982). Both methods are especially designed for the case where there are relatively few degrees of freedom – and most successful in this case – and neither requires second derivatives. The augmented-Lagrangian-based LANCELOT (Conn *et al.* 1992) operates at the other extreme, being most effective when there are relatively few general constraints, and is capable of running without gradients if necessary – (group) partial separability (Griewank and Toint 1982*b*, Conn *et al.* 1990) allows for the efficient estimation of derivatives. More modern SQP interior-point hybrids like LOQO (Vanderbei and Shanno 1999), KNITRO (Byrd *et al.* 2000) and NLPSPR (Betts and Frank 1994) are effective regardless of the relative number of (active) constraints. Of the filter-based methods, both the trust region SQP-based FilterSQP (Fletcher and Leyffer 1998) and the linesearch interior-point-based IPOPT (Wächter and Biegler 2004) have proved to be robust and efficient. The primal–dual method of Forsgren and Gill (1998) is being implemented in the object-oriented code IOTR which acts as a template for implementing interior-point algorithms. Some codes – for example, the augmented-Lagrangian-based PENNON (Kočvara and Stingl 2003) – have even wider scope, permitting semi-definite matrix constraints. Others, such as CONOPT (Drud 1994) and LSGRG2 (Smith and Lasdon 1992), use (generalized) reduced gradient methods not even covered in this survey. A welcome development has certainly been the flurry of papers – see for example those just cited – comparing and contrasting rival nonlinear programming packages. At this stage, algorithm development is still so rapid that it is impossible to identify the best method(s). We urge potential users to try the award-winning NEOS server (Dolan, Fourer, Moré and Munson 2002, Czyzyk *et al.* 1998)

$$\texttt{www-neos.mcs.anl.gov}$$

to compare many of the leading contenders.

Turning to convex programming, both MOSEK (Andersen and Andersen 2000, Andersen and Ye 1998) – which is based on a homogeneous model (Andersen and Ye 1999) – and NLPHOPDM (Epelly, Gondzio and Vial 2000) – which applies a multiple target tracking strategy (Gondzio 1996) – are designed for general problems, having evolved from linear programming beginnings. The same is true of PDCO (Saunders and Tomlin 1996), which implements the regularization scheme of Section 4.3. PDCO has been successfully used to solve large-scale entropy maximization problems using Shannon's entropy function $S(x) = -\sum_i x_i \log(x_j)$ as objective and has proved able of solving a maximum entropy model of web traffic with $662,463$ variables and $51,152$ sparse constraints in 12 iterations.

6. Conclusion

We have reviewed recent developments in algorithms for large-scale optimization, successively considering the unconstrained, bound-constrained, linearly constrained and nonlinearly constrained cases. Emphasis has been put on the underlying principles and theoretical underpinnings of the described methods as well as on practical issues and software.

We are aware that, despite our best efforts, the picture remains incomplete and biased by our experience. This is reflected, for instance, in our lack of cover of neighbouring subjects such as variational inequalities and nonsmooth problems, despite their intrinsic interest. It is nevertheless hoped that the overview presented will make the field of nonlinear programming and its application to solving large problems easier to understand, both for scholars and practitioners.

Acknowledgements

The work of the first author was supported by the EPSRC grant GR/S42170, and that of the second author by NSERC grant RGPIN299010-04 and PIED grant 131FR88. The work of the third author has been conducted in the framework of the Interuniversity Attraction Poles Programme of the Belgian Science Policy Agency. The authors are indebted to Annick Sartenaer for her comments on a draft of the manuscript.

REFERENCES

M. Al-Baali (2003), Quasi-Newton algorithms for large-scale nonlinear least-squares, in Di Pillo and Murli (2003), pp. 1–21.

E. D. Andersen and K. D. Andersen (2000), The MOSEK interior point optimizer for linear programming: An implementation of the homogeneous algorithm, in *High Performance Optimization* (T. T. H. Frenk, K. Roos and S. Zhang, eds), Kluwer, pp. 197–232.

E. D. Andersen and Y. Ye (1998), 'A computational study of the homogeneous algorithm for large-scale convex optimization', *Comput. Optim. Appl.* **10**, 243–269.

E. D. Andersen and Y. Ye (1999), 'On a homogeneous algorithm for the monotone complementarity problem', *Math. Program.* **84**(2), 375–399.

M. Anitescu (2000), On using the elastic mode in nonlinear programming approaches to mathematical programs with complementarity constraints, Preprint ANL/MCS-P864-1200, Argonne National Laboratory, IL, USA.

E. Arian, M. Fahl and E. W. Sachs (2000), Trust-region proper orthogonal decomposition for flow control, Technical Report 2000-25, Institute for Computer Applications in Science and Engineering, NASA Langley Research Center, Hampton, VA, USA.

A. S. El Bakry, R. A. Tapia, T. Tsuchiya and Y. Zhang (1996), 'On the formulation and theory of Newton interior point methods for nonlinear programming', *J. Optim. Theory Appl.* **89**(3), 507–541.

R. E. Bank, P. E. Gill and R. F. Marcia (2003), Interior point methods for a class of elliptic variational inequalities, in Biegler, Ghattas, Heinkenschloss and Van Bloemen Waanders (2003), pp. 218–235.

H. Y. Benson, A. Sen, D. F. Shanno and R. J. Vanderbei (2003), Interior-point algorithms, penalty methods and equilibrium problems, Technical Report ORFE-03-02, Operations Research and Financial Engineering, Princeton University.

H. Y. Benson, R. J. Vanderbei and D. F. Shanno (2002), 'Interior-point methods for nonconvex nonlinear programming: Filter methods and merit functions', *Comput. Optim. Appl.* **23**, 257–272.

S. J. Benson, L. C. McInnes, J. Moré and J. Sarich (2004), Scalable algorithms in optimization: Computational experiments, Preprint ANL/MCS-P1175-0604, Mathematics and Computer Science, Argonne National Laboratory, Argonne, IL, USA. To appear in *Proc. 10th AIAA/ISSMO Multidisciplinary Analysis and Optimization (MA&O) Conference, August 30–September 1, 2004.*

D. P. Bertsekas (1976), 'On the Goldstein–Levitin–Poljak gradient projection method', *IEEE Trans. Automat. Control* **AC-21**, 174–184.

D. P. Bertsekas (1995), *Nonlinear Programming*, Athena Scientific, Belmont, MA, USA.

J. T. Betts and S. O. Erb (2003), 'Optimal low thrust trajectory to the moon', *SIAM J. Appl. Dyn. Syst.* **2**(2), 144–170.

J. T. Betts and P. D. Frank (1994), 'A sparse nonlinear optimization algorithm', *J. Optim. Theory Appl.* **82**(3), 519–541.

T. Biegler, O. Ghattas, M. Heinkenschloss and B. Van Bloemen Waanders, eds (2003), *High Performance Algorithms and Software for Nonlinear Optimization*, Springer, Heidelberg/Berlin/New York.

G. Biros and O. Ghattas (2000), 'A Lagrange–Newton–Krylov–Schur method for PDE-constrained optimization', *SIAG/OPT Views-and-News* **11**(2), 12–18.

R. E. Bixby, M. Fenlon, Z. Gu, E. Rothberg and R. Wunderling (2000), MIP: theory and practice; closing the gap, in *System Modelling and Optimization: Methods, Theory and Applications* (M. J. D. Powell and S. Scholtes, eds), Kluwer, Dordrecht, Netherlands, pp. 10–49.

P. T. Boggs and J. W. Tolle (1995), Sequential quadratic programming, in *Acta Numerica*, Vol. 4, Cambridge University Press, pp. 1–51.

P. T. Boggs and J. W. Tolle (2000), 'Sequential quadratic programming for large-scale nonlinear optimization', *Comput. Appl. Math.* **124**, 123–137.

P. T. Boggs, P. D. Domich, J. E. Rogers and C. Witzgall (1996), 'An interior point method for general large scale quadratic programming problems', *Ann. Oper. Res.* **62**, 419–437.

P. T. Boggs, A. J. Kearsley and J. W. Tolle (1999a), 'A global convergence analysis of an algorithm for large scale nonlinear programming problems', *SIAM J. Optim.* **9**(4), 833–862.

P. T. Boggs, A. J. Kearsley and J. W. Tolle (1999*b*), 'A practical algorithm for general large scale nonlinear optimization problems', *SIAM J. Optim.* **9**(3), 755–778.

J. F. Bonnans, J.-Ch. Gilbert, C. Lemaréchal and C. Sagastizábal (1997), *Optimisation Numérique: Aspects Théoriques et Pratiques*, Vol. 27 of *Mathématiques & Applications*, Springer.

J. M. Borwein (1982), 'Necessary and sufficient conditions for quadratic minimality', *Numer. Funct. Anal. Optim.* **5**, 127–140.

J. H. Bramble (1993), *Multigrid Methods*, Longman, New York.

M. A. Branch, T. F. Coleman and Y. Li (1999), 'A subspace, interior and conjugate gradient method for large-scale bound-constrained minimization problems', *SIAM J. Sci. Comput.* **21**(1), 1–23.

A. Brandt (1977), 'Multi-level adaptative solutions to boundary value problems', *Math. Comp.* **31**(138), 333–390.

W. L. Briggs, V. E. Henson and S. F. McCormick (2000), *A Multigrid Tutorial*, second edn, SIAM, Philadelphia, USA.

A. Brooke, D. Kendrick and A. Meeraus (1988), *GAMS: A User's Guide*, The Scientific Press, Redwood City, USA.

J. V. Burke and J. J. Moré (1988), 'On the identification of active constraints', *SIAM J. Numer. Anal.* **25**(5), 1197–1211.

J. V. Burke and J. J. Moré (1994), 'Exposing constraints', *SIAM J. Optim.* **4**(3), 573–595.

J. V. Burke and A. Weigmann (1997), Notes on limited memory BFGS updating in a trust-region framework, Technical report, Department of Mathematics, University of Washington, Seattle, Washington, USA.

J. V. Burke, J. J. Moré and G. Toraldo (1990), 'Convergence properties of trust region methods for linear and convex constraints', *Math. Program.* **47**(3), 305–336.

R. H. Byrd, J.-Ch. Gilbert and J. Nocedal (2000), 'A trust region method based on interior point techniques for nonlinear programming', *Math. Program., Ser. A* **89**(1), 149–185.

R. H. Byrd, N. I. M. Gould, J. Nocedal and R. A. Waltz (2004*a*), 'An algorithm for nonlinear optimization using linear programming and equality constrained subproblems', *Math. Program., Ser. B* **100**(1), 27–48.

R. H. Byrd, N. I. M. Gould, J. Nocedal and R. A. Waltz (2004*b*), On the convergence of successive linear-quadratic programming algorithms, Technical Report RAL-TR-2004-032, Rutherford Appleton Laboratory, Chilton, Oxfordshire, UK.

R. H. Byrd, M. E. Hribar and J. Nocedal (1999), 'An interior point method for large scale nonlinear programming', *SIAM J. Optim.* **9**(4), 877–900.

R. H. Byrd, P. Lu, J. Nocedal and C. Zhu (1995), 'A limited memory algorithm for bound constrained optimization', *SIAM J. Sci. Comput.* **16**(5), 1190–1208.

R. H. Byrd, J. Nocedal and R. B. Schnabel (1994), 'Representations of quasi-Newton matrices and their use in limited memory methods', *Math. Program.* **63**(2), 129–156.

R. H. Byrd, J. Nocedal and R. A. Waltz (2003), 'Feasible interior methods using slacks for nonlinear optimization', *Comput. Optim. Appl.* **26**, 35–61.

P. H. Calamai and J. J. Moré (1987), 'Projected gradient methods for linearly constrained problems', *Math. Program.* **39**(1), 93–116.

Y. Chabrillac and J.-P. Crouzeix (1984), 'Definiteness and semidefiniteness of quadratic forms revisited', *Linear Algebra Appl.* **63**, 283–292.

C. M. Chin and R. Fletcher (2003), 'On the global convergence of an SLP-filter algorithm that takes EQP steps', *Math. Program.* **96**(1), 161–177.

T. F. Coleman (1994), Linearly constrained optimization and projected preconditioned conjugate gradients, in *Proc. Fifth SIAM Conference on Applied Linear Algebra* (J. Lewis, ed.), SIAM, Philadelphia, USA, pp. 118–122.

T. F. Coleman and A. R. Conn (1982), 'Nonlinear programming via an exact penalty function method: Asymptotic analysis', *Math. Program.* **24**(3), 123–136.

T. F. Coleman and L. A. Hulbert (1989), 'A direct active set algorithm for large sparse quadratic programs with simple bounds', *Math. Program., Ser. B* **45**(3), 373–406.

T. F. Coleman and Y. Li, eds (1990), *Large Scale Numerical Optimization*, SIAM, Philadelphia, USA.

T. F. Coleman and Y. Li (1994), 'On the convergence of interior-reflective Newton methods for nonlinear minimization subject to bounds', *Math. Program.* **67**(2), 189–224.

T. F. Coleman and Y. Li (1996), 'An interior trust region approach for nonlinear minimization subject to bounds', *SIAM J. Optim.* **6**(2), 418–445.

T. F. Coleman and Y. Li (2000), 'A trust region and affine scaling interior point method for nonconvex minimization with linear inequality constraints', *Math. Program., Ser. A* **88**, 1–31.

B. Colson and Ph. L. Toint (2003), Optimizing partially separable functions without derivatives, Technical Report 03/20, Department of Mathematics, University of Namur, Namur, Belgium.

A. R. Conn, N. I. M. Gould and Ph. L. Toint (1988a), 'Global convergence of a class of trust region algorithms for optimization with simple bounds', *SIAM J. Numer. Anal.* **25**(2), 433–460. See also same journal **26** (1989), 764–767.

A. R. Conn, N. I. M. Gould and Ph. L. Toint (1988b), 'Testing a class of methods for solving minimization problems with simple bounds on the variables', *Math. Comp.* **50**, 399–430.

A. R. Conn, N. I. M. Gould and Ph. L. Toint (1990), An introduction to the structure of large scale nonlinear optimization problems and the **LANCELOT** project, in *Computing Methods in Applied Sciences and Engineering* (R. Glowinski and A. Lichnewsky, eds), SIAM, Philadelphia, USA, pp. 42–51.

A. R. Conn, N. I. M. Gould and Ph. L. Toint (1992), **LANCELOT**: *A Fortran package for Large-scale Nonlinear Optimization (Release A)*, Springer Series in Computational Mathematics, Springer, Heidelberg/Berlin/New York.

A. R. Conn, N. I. M. Gould and Ph. L. Toint (1994), Large-scale nonlinear constrained optimization: a current survey, in *Algorithms for Continuous Optimization: The State of the Art* (E. Spedicato, ed.), Vol. 434 of *NATO ASI Series C: Mathematical and Physical Sciences*, Kluwer, Dordrecht, Netherlands, pp. 287–332.

A. R. Conn, N. I. M. Gould and Ph. L. Toint (1996), 'Numerical experiments with the LANCELOT package (Release A) for large-scale nonlinear optimization', *Math. Program., Ser. A* **73**(1), 73–110.

A. R. Conn, N. I. M. Gould and Ph. L. Toint (1997), Methods for nonlinear constraints in optimization calculations, in Duff and Watson (1997), pp. 363–390.

A. R. Conn, N. I. M. Gould and Ph. L. Toint (2000*a*), *Trust-Region Methods*, SIAM, Philadelphia, USA.

A. R. Conn, N. I. M. Gould, D. Orban and Ph. L. Toint (2000*b*), 'A primal-dual trust-region algorithm for non-convex nonlinear programming', *Math. Program., Ser. B* **87**(2), 215–249.

B. L. Contesse (1980), 'Une caractérisation complète des minima locaux en programmation quadratique', *Numer. Math.* **34**(3), 315–332.

CPLEX 6.0 (1998), *High-Performance Linear, Integer and Quadratic Programming Software*, ILOG SA, Gentilly, France. www.cplex.com.

J. Czyzyk, M. Mesnier and J. Moré (1998), 'The NEOS server', *IEEE J. Comput. Sci. Engr.* **5**, 68–75.

Y. H. Dai and Y. Yuan (2000), 'A nonlinear conjugate gradient method with a strong global convergence property', *SIAM J. Optim.* **10**(1), 177–182.

R. S. Dembo and T. Steihaug (1983), 'Truncated-Newton algorithms for large-scale unconstrained optimization', *Math. Program.* **26**(2), 190–212.

R. S. Dembo, S. C. Eisenstat and T. Steihaug (1982), 'Inexact-Newton methods', *SIAM J. Numer. Anal.* **19**(2), 400–408.

A.-V. DeMiguel, M. P. Friedlander, F. J. Nogales and S. Scholtes (2004), An interior-point method for MPECs based on strictly feasible relaxations, Technical Report ANL/MCS-P1150-0404, Argonne National Laboratory, IL, USA.

N. Deng, Y. Xiao and F. Zhou (1993), 'Nonmonotonic trust region algorithms', *J. Optim. Theory Appl.* **76**(2), 259–285.

J. E. Dennis and R. B. Schnabel (1983), *Numerical Methods for Unconstrained Optimization and Nonlinear Equations*, Prentice-Hall, Englewood Cliffs, USA. Reprinted as Vol. 16 of *Classics in Applied Mathematics*, SIAM, Philadelphia, USA.

J. E. Dennis, D. M. Gay and R. E. Welsh (1981), 'An adaptive nonlinear least squares algorithm', *ACM Trans. Math. Software* **7**(3), 348–368.

G. Di Pillo and F. Gianessi, eds (1996), *Nonlinear Optimization and Applications*, Plenum Publishing, New York.

G. Di Pillo and F. Gianessi, eds (1999), *Nonlinear Optimization and Related Topics*, Vol. 2, Kluwer, Dordrecht, Netherlands.

G. Di Pillo and A. Murli, eds (2003), *High Performance Algorithms and Software in Nonlinear Optimization*, Kluwer, Dordrecht, Netherlands.

E. Dolan (2001), The NEOS server 4.0 administrative guide, Technical Memorandum ANL/MCS-TM-250, Argonne National Laboratory, IL.

E. D. Dolan, R. Fourer, J. J. Moré and T. S. Munson (2002), 'Computing a trust region step', *SIAM News* **35**(5), 8–9.

A. S. Drud (1994), 'CONOPT: A large scale GRG code', *ORSA J. Comput.* **6**, 207–216.

I. Duff and A. Watson, eds (1997), *The State of the Art in Numerical Analysis*, Oxford University Press, Oxford.

J. C. Dunn (1981), 'Global and asymptotic convergence rate estimates for a class of projected gradient processes', *SIAM J. Control Optim.* **19**, 368–400.

J.-P. Dussault (1995), 'Numerical stability and efficiency of penalty algorithms', *SIAM J. Numer. Anal.* **32**(1), 296–317.

O. Epelly, J. Gondzio and J.-P. Vial (2000), An interior-point solver for smooth convex optimization with an application to environmental-energy-economic models, Technical Report 2000.08, Logilab, HEC, University of Geneva, Switzerland.

F. Facchinei, J. Judice and J. Soares (1998), 'An active set Newton algorithm for large-scale nonlinear programs with box constraints', *SIAM J. Optim.* **8**(1), 158–186.

F. Facchinei, S. Lucidi and L. Palagi (2002), 'A truncated Newton algorithm for large scale box constrained optimization', *SIAM J. Optim.* **12**(4), 1100–1125.

M. Fahl and E. Sachs (2003), Reduced order modelling approaches to PDE-constrained optimization based on proper orthogonal decomposition, in Biegler *et al.* (2003), pp. 268–281.

M. C. Ferris and T. S. Munson (2000), Interior-point methods for massive support vector machines, Data Mining Institute Technical Report 00-05, Computer Science Department, University of Wisconsin, Madison, WI, USA.

A. V. Fiacco and G. P. McCormick (1968), *Nonlinear Programming: Sequential Unconstrained Minimization Techniques*, Wiley, Chichester, UK. Reprinted as *Classics in Applied Mathematics*, SIAM, Philadelphia, USA (1990).

M. Fisher (1998), Minimization algorithms for variational data assimilation, in *Recent Developments in Numerical Methods for Atmospheric Modelling*, ECMWF, pp. 364–385.

R. Fletcher (1971), 'A general quadratic programming algorithm', *J. Inst. Math. Appl.* **7**, 76–91.

R. Fletcher (1981), *Practical Methods of Optimization: Constrained Optimization*, Wiley, Chichester, UK.

R. Fletcher (1987a), *Practical Methods of Optimization*, second edn, Wiley, Chichester, UK.

R. Fletcher (1987b), Recent developments in linear and quadratic programming, in *The State of the Art in Numerical Analysis* (A. Iserles and M. J. D. Powell, eds), Oxford University Press, Oxford, pp. 213–243.

R. Fletcher (2000), 'Stable reduced Hessian updates for indefinite quadratic programming', *Math. Program.* **87**(2), 251–264.

R. Fletcher and S. Leyffer (1998), User manual for filterSQP, Numerical Analysis Report NA/181, Department of Mathematics, University of Dundee, Dundee, UK.

R. Fletcher and S. Leyffer (2002), 'Nonlinear programming without a penalty function', *Math. Program.* **91**(2), 239–269.

R. Fletcher and C. M. Reeves (1964), 'Function minimization by conjugate gradients', *Computer Journal* **7**, 149–154.

R. Fletcher and E. Sainz de la Maza (1989), 'Nonlinear programming and nonsmooth optimization by successive linear programming', *Math. Program.* **43**(3), 235–256.

R. Fletcher, N. I. M. Gould, S. Leyffer, Ph. L. Toint and A. Wächter (2002*a*), 'Global convergence of trust-region SQP-filter algorithms for nonlinear programming', *SIAM J. Optim.* **13**(3), 635–659.

R. Fletcher, S. Leyffer and Ph. L. Toint (2002*b*), 'On the global convergence of a filter-SQP algorithm', *SIAM J. Optim.* **13**(1), 44–59.

A. Forsgren (2002), 'Inertia-controlling factorizations for optimization algorithms', *Appl. Numer. Math.* **43**(1–2), 91–107.

A. Forsgren and P. E. Gill (1998), 'Primal-dual interior methods for nonconvex nonlinear programming', *SIAM J. Optim.* **8**(4), 1132–1152.

A. Forsgren and W. Murray (1993), 'Newton methods for large-scale linear equality-constrained minimization', *SIAM J. Matrix Anal. Appl.* **14**(2), 560–587.

A. Forsgren, P. E. Gill and M. H. Wright (2002), 'Interior-point methods for nonlinear optimization', *SIAM Review* **44**, 525–597.

R. Fourer, D. M. Gay and B. W. Kernighan (2003), *AMPL: A Modeling Language for Mathematical Programming*, second edn, Brooks/Cole-Thompson Learning, Pacific Grove, CA, USA.

M. P. Friedlander and M. A. Saunders (2005), 'A globally convergent linearly constrained Lagrangian method for nonlinear optimization', *SIAM J. Optim.*, to appear.

D. M. Gay, M. L. Overton and M. H. Wright (1998), A primal-dual interior method for nonconvex nonlinear programming, in *Advances in Nonlinear Programming* (Y. Yuan, ed.), Kluwer, Dordrecht, Netherlands, pp. 31–56.

E. M. Gertz and Ph. E. Gill (2004), 'A primal-dual trust region algorithm for nonlinear optimization', *Math. Program., Ser. A* **100**(1), 49–94.

E. M. Gertz and S. J. Wright (2003), 'Object-oriented software for quadratic programming', *Trans. ACM Math. Software* **29**(1), 58–81.

E. M. Gertz, J. Nocedal and A. Sartenaer (2003), A starting-point strategy for nonlinear interior methods, Technical Report OTC 2003/4, Optimization Technology Center, Evanston, IL, USA.

J.-Ch. Gilbert and C. Lemaréchal (1989), 'Some numerical experiments with variable-storage quasi-Newton algorithms', *Math. Program., Ser. B* **45**(3), 407–435.

J.-Ch. Gilbert and J. Nocedal (1992), 'Global convergence properties of conjugate gradient methods for optimization', *SIAM J. Optim.* **2**(1), 21–42.

J.-Ch. Gilbert, C. C. Gonzaga and E. Karas (2002), Examples of ill-behaved central paths in convex optimization, Technical Report 4179, INRIA, Rocquencourt, Le Chesnay, France.

P. E. Gill, W. Murray and M. A. Saunders (2002), 'SNOPT: An SQP algorithm for large-scale constrained optimization', *SIAM J. Optim.* **12**(4), 979–1006.

P. E. Gill, W. Murray and M. H. Wright (1981), *Practical Optimization*, Academic Press, London.

P. E. Gill, W. Murray, M. A. Saunders and M. H. Wright (1990), A Schur-complement method for sparse quadratic programming, in *Reliable Scientific Computation* (M. G. Cox and S. J. Hammarling, eds), Oxford University Press, pp. 113–138.

P. E. Gill, W. Murray, M. A. Saunders and M. H. Wright (1991), 'Inertia-controlling methods for general quadratic programming', *SIAM Review* **33**(1), 1–36.

M. J. Goldsmith (1999), Sequential quadratic programming methods based on indefinite Hessian approximations, PhD thesis, Dept of Management Science and Engineering, Stanford University, CA, USA.

J. Gondzio (1996), 'Multiple centrality corrections in a primal-dual method for linear programming', *Comput. Optim. Appl.* **6**, 137–156.

J. Gondzio (1997), 'Presolve analysis of linear programs prior to applying an interior point method', *INFORMS J. Comput.* **9**(1), 73–91.

J. Gondzio and A. Grothey (2003a), Parallel interior point solver for structured quadratic programs: Application to financial planning problems, Technical Report MS-03-001, School of Mathematics, University of Edinburgh.

J. Gondzio and A. Grothey (2003b), 'Reoptimization with the primal-dual interior point method', *SIAM J. Optim.* **13**(3), 842–864.

C. C. Gonzaga, E. Karas and M. Vanti (2003), 'A globally convergent filter method for nonlinear programming', *SIAM J. Optim.* **14**(3), 646–669.

N. I. M. Gould (1985), 'On practical conditions for the existence and uniqueness of solutions to the general equality quadratic-programming problem', *Math. Program.* **32**(1), 90–99.

N. I. M. Gould (1986), 'On the accurate determination of search directions for simple differentiable penalty functions', *IMA J. Numer. Anal.* **6**, 357–372.

N. I. M. Gould (1991), 'An algorithm for large-scale quadratic programming', *IMA J. Numer. Anal.* **11**(3), 299–324.

N. I. M. Gould (1999), 'On modified factorizations for large-scale linearly-constrained optimization', *SIAM J. Optim.* **9**(4), 1041–1063.

N. I. M. Gould (2003), 'Some reflections on the current state of active-set and interior point methods for constrained optimization', *SIAG/OPT Views-and-News* **14**(1), 2–7.

N. I. M. Gould, and Ph. L. Toint (2000a), A quadratic programming bibliography, Numerical Analysis Group Internal Report 2000-1, Rutherford Appleton Laboratory, Chilton, Oxfordshire, UK.

N. I. M. Gould and Ph. L. Toint (2000b), SQP methods for large-scale nonlinear programming, in *System Modelling and Optimization: Methods, Theory and Applications* (M. J. D. Powell and S. Scholtes, eds), Kluwer, Dordrecht, Netherlands, pp. 149–178.

N. I. M. Gould and Ph. L. Toint (2002a), 'An iterative working-set method for large-scale non-convex quadratic programming', *Appl. Numer. Math.* **43**(1–2), 109–128.

N. I. M. Gould and Ph. L. Toint (2002b), Numerical methods for large-scale non-convex quadratic programming, in *Trends in Industrial and Applied Mathematics* (A. H. Siddiqi and M. Kočvara, eds), Kluwer, Dordrecht, Netherlands, pp. 149–179.

N. I. M. Gould and Ph. L. Toint (2003a), FILTRANE: A Fortran 95 filter-trust-region package for solving systems of nonlinear equalities, nonlinear inequalities and nonlinear least-squares problems, Technical Report 03/15, Rutherford Appleton Laboratory, Chilton, Oxfordshire, UK.

N. I. M. Gould and Ph. L. Toint (2003b), Global convergence of a hybrid trust-region SQP-filter algorithm for general nonlinear programming, in *System*

Modeling and Optimization XX (E. Sachs and R. Tichatschke, eds), Kluwer, Dordrecht, Netherlands, pp. 23–54.

N. I. M. Gould and Ph. L. Toint (2004*a*), How mature is nonlinear optimization?, in *Applied Mathematics Entering the 21st Century: Invited Talks from the ICIAM 2003 Congress* (J. M. Hill and R. Moore, eds), SIAM, Philadelphia, USA, pp. 141–161.

N. I. M. Gould and Ph. L. Toint (2004*b*), 'Preprocessing for quadratic programming', *Math. Program., Ser. B* **100**(1), 95–132.

N. I. M. Gould and Ph. L. Toint (2005), Global convergence of a non-monotone trust-region filter algorithm for nonlinear programming, in *Proc. 2004 Gainesville Conference on Multilevel Optimization* (W. Hager, ed.), Kluwer, Dordrecht, Netherlands, to appear.

N. I. M. Gould, M. E. Hribar and J. Nocedal (2001), 'On the solution of equality constrained quadratic problems arising in optimization', *SIAM J. Sci. Comput.* **23**(4), 1375–1394.

N. I. M. Gould, S. Leyffer and Ph. L. Toint (2005), 'A multidimensional filter algorithm for nonlinear equations and nonlinear least-squares', *SIAM J. Optim.* **15**(1), 17–38.

N. I. M. Gould, S. Lucidi, M. Roma and Ph. L. Toint (1999), 'Solving the trust-region subproblem using the Lanczos method', *SIAM J. Optim.* **9**(2), 504–525.

N. I. M. Gould, S. Lucidi, M. Roma and Ph. L. Toint (2000), 'Exploiting negative curvature directions in linesearch methods for unconstrained optimization', *Optim. Methods Software* **14**(1–2), 75–98.

N. I. M. Gould, D. Orban and Ph. L. Toint (2003*a*), 'GALAHAD: A library of thread-safe Fortran 90 packages for large-scale nonlinear optimization', *ACM Trans. Math. Software* **29**(4), 353–372.

N. I. M. Gould, D. Orban and Ph. L. Toint (2003*b*), An interior-point ℓ_1-penalty method for nonlinear optimization, Technical Report RAL-TR-2003-0xx, Rutherford Appleton Laboratory, Chilton, Oxfordshire, UK.

N. I. M. Gould, C. Sainvitu and Ph. L. Toint (2004), A filter-trust-region method for unconstrained optimization, Technical Report 04/03, Department of Mathematics, University of Namur, Belgium.

S. Gratton, A. Sartenaer and Ph. L. Toint (2004), Recursive trust-region methods for multilevel nonlinear optimization (Part I): Global convergence and complexity, Technical Report 04/06, Department of Mathematics, University of Namur, Belgium.

A. Griewank (2000), *Evaluating Derivatives: Principles and Techniques of Algorithmic Differentiation*, Vol. 19 of *Frontiers in Applied Mathematics*, SIAM, Philadelphia, USA.

A. Griewank and Ph. L. Toint (1982*a*), 'Local convergence analysis for partitioned quasi-Newton updates', *Numer. Math.* **39**, 429–448.

A. Griewank and Ph. L. Toint (1982*b*), On the unconstrained optimization of partially separable functions, in *Nonlinear Optimization 1981* (M. J. D. Powell, ed.), Academic Press, London, pp. 301–312.

A. Griewank and Ph. L. Toint (1982*c*), 'Partitioned variable metric updates for large structured optimization problems', *Numer. Math.* **39**, 119–137.

R. E. Griffith and R. A. Stewart (1961), 'A nonlinear programming technique for the optimization of continuous processing systems', *Management Science* **7**, 379–392.

L. Grippo, F. Lampariello and S. Lucidi (1986), 'A nonmonotone line search technique for Newton's method', *SIAM J. Numer. Anal.* **23**(4), 707–716.

L. Grippo, F. Lampariello and S. Lucidi (1989), 'A truncated Newton method with nonmonotone line search for unconstrained optimization', *J. Optim. Theory Appl.* **60**(3), 401–419.

W. Gropp and J. Moré (1997), Optimization environments and the NEOS server, in *Approximation Theory and Optimization* (M. D. Buhmann and A. Iserles, eds), Cambridge University Press, pp. 167–182.

C. Guéret, C. Prins and M. Seveaux (2002), *Applications of Optimization with Xpress-MP*, Dash Optimization. www.dashoptimization.com.

M. Gulliksson, I. Söderkvist and P.-A. Wedin (1997), 'Algorithms for constrained and weighted nonlinear least-squares', *SIAM J. Optim.* **7**(1), 208–224.

W. Hackbusch (1995), *Multi-Grid Methods and Applications*, Vol. 4 of *Series in Computational Mathematics*, Springer, Heidelberg/Berlin/New York.

W. W. Hager (2001), 'Minimizing a quadratic over a sphere', *SIAM J. Optim.* **12**(1), 188–208.

W. W. Hager and H. Zhang (2003), CG_DESCENT: A conjugate-gradient method with guaranteed descent, Technical report, Department of Mathematics, University of Florida, Gainesville, USA.

W. W. Hager, D. W. Hearn and P. M. Pardalos, eds (1994), *Large Scale Optimization: State of the Art*, Kluwer, Dordrecht, Netherlands.

M. Heinkenschloss, M. Ulbrich and S. Ulbrich (1999), 'Superlinear and quadratic convergence of affine-scaling interior-point Newton methods for problems with simple bounds without strict complementarity assumption', *Math. Program.* **86**(3), 615–635.

M. R. Hestenes and E. Stiefel (1952), 'Methods of conjugate gradients for solving linear systems', *J. Res. National Bureau of Standards* **49**, 409–436.

W. Hock and K. Schittkowski (1981), *Test Examples for Nonlinear Programming Codes*, Vol. 187 of *Lecture Notes in Economics and Mathematical Systems*, Springer, Heidelberg/Berlin/New York.

B. Jansen, C. Roos, T. Terlaky and J.-Ph. Vial (1996), 'Primal-dual target following algorithms for linear programming', *Ann. Oper. Res.* **62**, 197–231.

L. C. Kaufman (1999), 'Reduced storage, quasi-Newton trust region approaches to function optimization', *SIAM J. Optim.* **10**(1), 56–69.

M. Kočvara and M. Stingl (2003), 'PENNON, a code for nonconvex nonlinear and semidefinite programming', *Optim. Methods Software* **18**(3), 317–333.

M. Lalee, J. Nocedal and T. D. Plantenga (1998), 'On the implementation of an algorithm for large-scale equality constrained optimization', *SIAM J. Optim.* **8**(3), 682–706.

R. D. Leone, A. Murli, P. M. Pardalos and G. Toraldo, eds (1998), *High Performance Algorithms and Software in Nonlinear Optimization*, Kluwer, Dordrecht, Netherlands.

M. Lescrenier (1991), 'Convergence of trust region algorithms for optimization with bounds when strict complementarity does not hold', *SIAM J. Numer. Anal.* **28**(2), 476–495.

E. S. Levitin and B. T. Polyak (1966), 'Constrained minimization problems', *USSR Comput. Math. Math. Phys.* **6**, 1–50.

A. S. Lewis and M. L. Overton (1996), Eigenvalue optimization, in *Acta Numerica*, Vol. 5, Cambridge University Press, pp. 149–190.

M. Lewis and S. G. Nash (2002), Practical aspects of multiscale optimization methods for VLSICAD, in *Multiscale Optimization and VLSI/CAD* (J. Cong and J. R. Shinnerl, eds), Kluwer, Dordrecht, Netherlands, pp. 265–291.

M. Lewis and S. G. Nash (2005), 'Model problems for the multigrid optimization of systems governed by differential equations', *SIAM J. Sci. Comput.*, to appear.

S. Leyffer (2003), 'Mathematical programs with complementarity constraints', *SIAG/OPT Views-and-News* **14**(1), 15–18.

C. Lin and J. J. Moré (1999*a*), 'Incomplete Cholesky factorizations with limited memory', *SIAM J. Sci. Comput.* **21**(1), 24–45.

C. Lin and J. J. Moré (1999*b*), 'Newton's method for large bound-constrained optimization problems', *SIAM J. Optim.* **9**(4), 1100–1127.

D. C. Liu and J. Nocedal (1989), 'On the limited memory BFGS method for large-scale optimization', *Math. Program., Ser. B* **45**(3), 503–528.

D. G. Luenberger (1984), *Linear and Nonlinear Programming*, 2nd edn, Addison-Wesley, Reading, MA, USA.

L. Lukšan (1993), 'Inexact trust region method for large sparse nonlinear least-squares', *Kybernetica* **29**(4), 305–324.

L. Lukšan (1994), 'Inexact trust region method for large sparse systems of nonlinear equations', *J. Optim. Theory Appl.* **81**(3), 569–590.

L. Lukšan (1996), 'Hybrid methods for large sparse nonlinear least-squares', *J. Optim. Theory Appl.* **89**(3), 575–595.

L. Lukšan and J. Vlček (1998), 'Indefinitely preconditioned inexact Newton method for large sparse equality constrained nonlinear programming problems', *Numer. Linear Algebra Appl.* **5**(3), 219–247.

Z.-Q. Luo, J. S. Pang and D. Ralph (1996), *Mathematical Programs with Equilibrium Constraints*, Cambridge University Press, Cambridge.

Z.-Q. Luo, J. S. Pang and D. Ralph (1998), Piecewise sequential quadratic programming for mathematical programs with complementarity constraints, in *Multilevel Optimization: Complexity and Applications* (A. Migdala *et al.*, ed.), Kluwer.

O. L. Mangasarian (1980), 'Locally unique solutions of quadratic programs, linear and non-linear complementarity problems', *Math. Program., Ser. B* **19**(2), 200–212.

N. Maratos (1978), Exact penalty function algorithms for finite-dimensional and control optimization problems, PhD thesis, University of London.

M. Marazzi and J. Nocedal (2001), Feasibility control in nonlinear optimization, in *Foundations of Computational Mathematics* (A. DeVore, A. Iserles and E. Suli, eds), Vol. 284 of *London Mathematical Society Lecture Note Series*, Cambridge University Press, pp. 125–154.

D. Q. Mayne and E. Polak (1976), 'Feasible directions algorithms for optimisation problems with equality and inequality constraints', *Math. Program.* **11**(1), 67–80.

S. Mehrotra (1992), 'On the implementation of a primal-dual interior point method', *SIAM J. Optim.* **2**, 575–601.

J. L. Morales and J. Nocedal (2000), 'Automatic preconditioning by limited memory quasi-Newton updating', *SIAM J. Optim.* **10**(4), 1079–1096.

J. J. Moré (2003), 'Terascale optimal PDE solvers', Talk at the ICIAM 2003 Conference in Sydney.

J. J. Moré and D. C. Sorensen (1983), 'Computing a trust region step', *SIAM J. Sci. Statist. Comput.* **4**(3), 553–572.

J. J. Moré and D. J. Thuente (1994), 'Line search algorithms with guaranteed sufficient decrease', *ACM Trans. Math. Software* **20**(3), 286–307.

J. J. Moré and G. Toraldo (1991), 'On the solution of large quadratic programming problems with bound constraints', *SIAM J. Optim.* **1**(1), 93–113.

J. J. Moré and S. J. Wright (1993), *Optimization Software Guide*, Vol. 14 of *Frontiers in Applied Mathematics*, SIAM, Philadelphia, USA.

W. Murray and F. J. Prieto (1995), 'A sequential quadratic programming algorithm using an incomplete solution of the subproblem', *SIAM J. Optim.* **5**(3), 590–640.

W. Murray and M. H. Wright (1992), Project Lagrangian methods based on the trajectories of penalty and barrier functions, Numerical Analysis Manuscript 92-01, AT&T Bell Laboratories.

B. A. Murtagh and M. A. Saunders (1982), 'A projected Lagrangian algorithm and its implementation for sparse non-linear constraints', *Math. Program. Studies* **16**, 84–117.

K. G. Murty and S. N. Kabadi (1987), 'Some NP-complete problems in quadratic and nonlinear programming', *Math. Program.* **39**(2), 117–129.

S. G. Nash (1984), 'Newton-type minimization via the Lanczos method', *SIAM J. Numer. Anal.* **21**(4), 770–788.

S. G. Nash (2000a), 'A multigrid approach to discretized optimization problems', *Optim. Methods Software* **14**, 99–116.

S. G. Nash (2000b), 'A survey of truncated-Newton methods', *J. Comput. Appl. Math.* **124**, 45–59.

S. G. Nash and J. Nocedal (1991), 'A numerical study of the limited memory BFGS method and the truncated-Newton method for large-scale optimization', *SIAM J. Optim.* **1**(3), 358–372.

S. G. Nash and A. Sofer (1990), 'Assessing a search direction within a truncated-Newton method', *Oper. Res. Lett.* **9**(4), 219–221.

Y. Nesterov and A. Nemirovskii (1994), *Interior-Point Polynomial Algorithms in Convex Programming*, SIAM, Philadelphia, USA.

J. Nocedal (1980), 'Updating quasi-Newton matrices with limited storage', *Math. Comp.* **35**, 773–782.

J. Nocedal (1992), Theory of algorithms for unconstrained optimization, in *Acta Numerica*, Vol. 1, Cambridge University Press, pp. 199–242.

J. Nocedal (1997), Large scale unconstrained optimization, in Duff and Watson (1997), pp. 311–338.

J. Nocedal and S. J. Wright (1999), *Large Sparse Numerical Optimization*, Series in Operations Research, Springer, Heidelberg/Berlin/New York.

E. O. Omojokun (1989), Trust region algorithms for optimization with nonlinear equality and inequality constraints, PhD thesis, University of Colorado, Boulder, Colorado, USA.

C. C. Paige and M. A. Saunders (1982), 'LSQR: an algorithm for sparse linear equations and sparse least squares', *Trans. ACM Math. Software* **8**, 43–71.

E. Polak and G. Ribière (1969), 'Note sur la convergence de méthodes de directions conjuguées', *Revue Francaise d'Informatique et de Recherche Opérationelle* **16-R1**, 35–43.

B. T. Polyak (1969), 'The conjugate gradient method in extremal problems', *USSR Comput. Math. Math. Phys.* **9**, 94–112.

M. J. D. Powell (1977), 'Restart procedures for the conjugate gradient method', *Math. Program.* **12**(2), 241–254.

M. J. D. Powell (1998), Direct search algorithms for optimization calculations, in *Acta Numerica*, Vol. 7, Cambridge University Press, pp. 287–336.

C. J. Price and Ph. L. Toint (2004), Exploiting problem structure in pattern-search methods for unconstrained optimization, Technical Report November, Department of Mathematics and Statistics, University of Canterbury, Christchurch, New Zealand.

R. Pytlak (1998), 'An efficient algorithm for large-scale nonlinear programming problems with simple bounds on the variables', *SIAM J. Optim.* **8**(2), 532–560.

A. U. Raghunathan and L. T. Biegler (2003), Interior point methods for Mathematical Programs with Complementarity Constraints (MPCCs), Technical Report, Department of Chemical Engineering, Carnegie Mellon University, Pittsburgh, PA, USA.

F. Rendl and H. Wolkowicz (1997), 'A semidefinite framework for trust region subproblems with applications to large scale minimization', *Math. Program.* **77**(2), 273–299.

J. Renegar (2001), *A Mathematical View of Interior-Point Methods in Convex Optimization*, MPS/SIAM series on Optimization, SIAM, Philadelphia, PA, USA.

S. M. Robinson (1974), 'Perturbed Kuhn–Tucker points and rates of convergence for a class of nonlinear programming algorithms', *Math. Program.* **7**(1), 1–16.

M. A. Saunders and J. A. Tomlin (1996), Solving regularized linear programs using barrier methods and KKT systems, Technical Report SOL 96-4, Department of EESOR, Stanford University, Stanford, CA, USA.

H. Scheel and S. Scholtes (2000), 'Mathematical programs with complementarity constraints: Stationarity, optimality, and sensitivity', *Math. Oper. Res.* **25**, 1–22.

T. Schlick (1993), 'Modified Cholesky factorizations for sparse preconditioners', *SIAM J. Sci. Comput.* **14**(2), 424–445.

J. A. Scott, Y. Hu and N. I. M. Gould (2004), An evaluation of sparse direct symmetric solvers: An introduction and preliminary findings, Numerical Analysis Group Internal Report 2004-1, Rutherford Appleton Laboratory, Chilton, Oxfordshire, UK.

S. Smith and L. Lasdon (1992), 'Solving large sparse nonlinear programs using GRG', *ORSA J. Comput.* **4**, 1–15.

D. C. Sorensen (1997), 'Minimization of a large-scale quadratic function subject to a spherical constraint', *SIAM J. Optim.* **7**(1), 141–161.

E. Spedicato, ed. (1994), *Algorithms for Continuous Optimization: The State of the Art*, Vol. 434 of *NATO ASI Series C: Mathematical and Physical Sciences*, Kluwer, Dordrecht, Netherlands.

T. Steihaug (1983), 'The conjugate gradient method and trust regions in large scale optimization', *SIAM J. Numer. Anal.* **20**(3), 626–637.

A. Tits, A. Wächter, S. Bakhtiari, T. J. Urban and C. T. Lawrence (2003), 'A primal-dual interior-point method for nonlinear programming with strong global and local convergence properties', *SIAM J. Optim.* **14**(1), 173–199.

M. J. Todd (2001), Semidefinite optimization, in *Acta Numerica*, Vol. 10, Cambridge University Press, pp. 515–560.

Ph. L. Toint (1981), Towards an efficient sparsity exploiting Newton method for minimization, in *Sparse Matrices and Their Uses* (I. S. Duff, ed.), Academic Press, London, pp. 57–88.

Ph. L. Toint (1987), 'On large scale nonlinear least squares calculations', *SIAM J. Sci. Statist. Comput.* **8**(3), 416–435.

Ph. L. Toint (1988), 'Global convergence of a class of trust region methods for nonconvex minimization in Hilbert space', *IMA J. Numer. Anal.* **8**, 231–252.

Ph. L. Toint (1996), 'An assessment of non-monotone linesearch techniques for unconstrained optimization', *SIAM J. Sci. Statist. Comput.* **17**(3), 725–739.

Ph. L. Toint (1997), 'A non-monotone trust-region algorithm for nonlinear optimization subject to convex constraints', *Math. Program.* **77**(1), 69–94.

M. Ulbrich (2004a), 'A multidimensional filter trust-region method for mixed complementarity problems', Talk at ICCOPT 1, Troy, USA.

M. Ulbrich, S. Ulbrich and L. N. Vicente (2004), 'A globally convergence primal-dual interior-point filter method for nonlinear programming', *Math. Program., Ser. B* **100**(2), 379–410.

S. Ulbrich (2004b), 'On the superlinear local convergence of a filter-SQP method', *Math. Program., Ser. B* **100**(1), 217–245.

R. J. Vanderbei (1995), 'Symmetric quasi-definite matrices', *SIAM J. Optim.* **5**, 100–113.

R. J. Vanderbei (1999), 'LOQO: An interior point code for quadratic programming', *Optim. Methods Software* **12**, 451–484.

R. J. Vanderbei and D. F. Shanno (1999), 'An interior point algorithm for nonconvex nonlinear programming', *Comput. Optim. Appl.* **13**, 231–252.

S. A. Vavasis (1990), 'Quadratic programming is is NP', *Inform. Process. Lett.* **36**(2), 73–77.

S. A. Vavasis (1991), Convex quadratic programming, in *Nonlinear Optimization: Complexity Issues*, Oxford University Press, Oxford, pp. 36–75.

A. Wächter (2002), An interior point algorithm for large-scale nonlinear optimization with applications in process engineering, PhD thesis, Department of Chemical Engineering, Carnegie-Mellon University, Pittsburgh, PA, USA.

A. Wächter and L. T. Biegler (2000), 'Failure of global convergence for a class of interior point methods for nonlinear programming', *Math. Program.* **88**(3), 565–574.

A. Wächter and L. T. Biegler (2003a), Line search filter methods for nonlinear programming: Local convergence, Technical Report RC23033(W0312-090), T. J. Watson Research Center, Yorktown Heights, NY, USA.

A. Wächter and L. T. Biegler (2003b), Line search filter methods for nonlinear programming: Motivation and global convergence, Technical Report RC23036(W0304-181), T. J. Watson Research Center, Yorktown Heights, NY, USA.

A. Wächter and L. T. Biegler (2004), On the implementation of an interior-point filter line-search algorithm for large-scale nonlinear programming, Research report RC 23149, IBM T. J. Watson Research Center, Yorktown Heights, NY, USA.

M. H. Wright (1992), Interior methods for constrained optimization, in *Acta Numerica*, Vol. 1, Cambridge University Press, pp. 341–407.

S. J. Wright (1997), *Primal-Dual Interior-Point Methods*, SIAM, Philadelphia, USA.

H. Yamashita, H. Yabe and T. Tanabe (2004), 'A globally and superlinearly convergent primal-dual interior point trust region method for large scale constrained optimization', *Math. Program.* Online First DOI 10.1007/s10107-004-0508-9.

Y. Ye (1997), *Interior Point Algorithms, Theory and Analysis*, Wiley-Interscience Series in Discrete Mathematics and Optimization, Wiley, New York, USA.

E. A. Yildirim and S. J. Wright (2002), 'Warm-start strategies in interior-point methods for linear programming', *SIAM J. Optim.* **12**(3), 782–810.

Y. Yuan, ed. (1998), *Advances in Nonlinear Programming*, Kluwer, Dordrecht, Netherlands.

Y. Yuan (2000), 'On the truncated conjugate-gradient method', *Math. Program., Ser. A* **87**(3), 561–573.

Y. Zhang (1994), 'On the convergence of a class of infeasible interior-point methods for the horizontal linear complementarity problem', *SIAM J. Optim.* **4**(1), 208–227.

C. Zhu, R. H. Byrd, P. Lu and J. Nocedal (1997), 'Algorithm 778. L-BFGS-B: Fortran subroutines for large-scale bound constrained optimization', *ACM Trans. Math. Software* **23**(4), 550–560.

Acta Numerica (2005), pp. 363–444
DOI: 10.1017/S096249290400025X

Computational chemistry from the perspective of numerical analysis

Claude Le Bris
CERMICS, École Nationale des Ponts et Chaussées,
6 & 8, Avenue Blaise Pascal, Champs sur Marne,
77455 Marne-La-Vallée Cedex 2, France
and
INRIA Rocquencourt, MICMAC project,
Domaine de Voluceau, B.P. 105, 78153 Le Chesnay Cedex, France
E-mail: lebris@cermics.enpc.fr

We present the field of computational chemistry from the standpoint of numerical analysis. We introduce the most commonly used models and comment on their applicability. We briefly outline the results of mathematical analysis and then mostly concentrate on the main issues raised by numerical simulations. A special emphasis is laid on recent results in numerical analysis, recent developments of new methods and challenging open issues.

CONTENTS

1. Outline

The field of computational chemistry is traditionally less well known by applied mathematicians than other fields of the engineering sciences, such as computational mechanics. Nevertheless, it is undoubtedly a rich source of very difficult problems for numerical simulation, some of which are likely to remain among the most challenging simulation tasks for years to come. Examples are the complete and detailed simulation of protein folding, or the description of the long time radiation damage of materials in nuclear power plants.

Many of the difficult problems have already been tackled, with definite success, by experts in computational chemistry. Thanks to their constant effort and their ever-productive ideas, the field has made great progress since its early days. The birth of quantum chemistry is commonly marked by the publication by Heitler and London (1927) on the electronic structure of the hydrogen molecule. That of *computational* quantum chemistry is fixed around 1950 with the first effective computations of molecular systems consisting of a few (say 5 to 20) electrons on the then newly created computers. Fifty years later, contemporary methods and techniques allow for the simulation of a broad spectrum of systems, ranging from molecules of hundreds of electrons modelled by very precise quantum models, up to samples of billions of particles modelled by molecular dynamics equations with force fields parametrized in advance (on the basis of more precise computations of smaller subsystems). These techniques are implemented in a large variety of softwares, freely distributed or not, that have general purposes or are dedicated to specific applications. Thanks to them, theoretical computational chemistry has gained full recognition in the world of chemistry, a domain traditionally more experimentally oriented. The Nobel prize recently awarded to Walter Kohn and John Pople testifies to this success: see Kohn (1999).

In spite of this great success, some challenging issues remain open, mostly related to the simulation of large systems over long times. New techniques must be invented, otherwise it will not be possible to bridge the size and time gaps up to most of the systems of practical interest (*e.g.*, proteins, nanosystems, crystalline materials). Such new ideas will certainly arise among experts of the field. But they are also likely to come from mathematical contributions. In many respects, computational chemistry is still an art, and relies upon a delicate mix of physical intuition, pragmatic cleverness, and practical know-how. Therefore progress is difficult, and slow. For an applied mathematician, such a situation often indicates that the existing discretization techniques and solution procedures suffer from a lack of numerical analysis, and this is indeed the case for computational chemistry. Some results on the numerical analysis have appeared in the past decade, but a lot remains to be done. We conjecture that a better theoretical knowledge of existing techniques will lead to their enhancement, and will therefore improve their applicability, as demonstrated by the history of scientific computing in the engineering sciences.

The purpose of the present article is to overview the numerical difficulties encountered in computational chemistry, to show how they are solved within the current state of the art, and to indicate the needs for further improvements. Of course, whenever they exist, we will indicate the results of numerical analysis that help to give a sound grounding for these techniques. But most often the article will deal with descriptions of techniques rather than with statements of theorems. We hope that this bias will stimulate further works.

The article is organized in a rather narrative way, without any ordering of scientific priority and/or importance. We begin with a description of the models and the discretization techniques for small systems in a static picture. Then, we progressively and (hopefully) pedagogically proceed to the modelling and simulation of more difficult situations: larger systems, systems *in situ*, time-dependent settings. Of course, as the size of the system increases, the models need to be coarse-grained, otherwise they cannot be tackled in practice. Therefore, the article also proceeds from the finest models to the coarsest ones.

The present state-of-the-art survey focuses on issues in numerical analysis for a readership familiar with such questions in other settings. A more detailed description may be found in the book of Le Bris, ed. (2003), and also, to a smaller extent, in the proceedings volume by Defranceschi and Le Bris, eds (2000). For readers with a background in chemistry, we refer to two survey articles, Defranceschi and Le Bris (1997) and Defranceschi and Le Bris (1999), for an introduction to the mathematical and numerical analysis. On the other hand, questions related to the mathematical analysis of models are overviewed in Le Bris and Lions (2005).

2. A short introduction to modelling for molecular simulation

2.1. A hierarchy of models

The domain called *computational chemistry* is traditionally more focused on the accurate simulation of (rather) small systems in their finest details, the term *molecular simulation* covering the other end of the spectrum. As suggested by its title, the present survey is thus more concentrated on small systems. Nevertheless, the current trend is increasingly to account for precise effects even in large-scale simulations. This can be done in one of the following ways.

- **Sequentially** by precomputing the parameters of a coarse model on the basis of quantum simulations of subsystems: typical examples are the computation of force fields for classical molecular dynamics (see Section 5.2), or the fitting of pseudopotentials for large-scale calculations in the solid phase (see Section 4.1).
- **In parallel** by dividing the system under study into pieces that are modelled at different levels, as for quantum mechanics/molecular mechanics (QM/MM) calculations: *e.g.*, the active site of a protein is simulated at the quantum level to account for the change of electronic structure, while the major part of the protein is modelled classically in order to only simulate the changes in the conformation.
- Also **in parallel** by inserting on-the-fly evaluations of interactions by quantum models in classical models, as is done in the *ab initio* molecular dynamics simulations (see Section 5.1).

In any case, computational chemistry irrigates molecular simulation in such an intimate manner, and the two fields are so strongly entangled, that it would not be giving a fair account to overview the former while ignoring the latter. Consequently, some sections of the present survey are aimed at giving at least a rough idea of simulations of very large systems (see Sections 3.7, 5.2 and 5.3). By no means, these sections, of limited size and scope, are intended to give a complete account. They rather aim only to give a flavour of the field.

In computational chemistry, the most accurate approximations are called *ab initio* approximations, for they involve no parameter except the universal constants of physics. They are only tractable for systems of small size. In order to allow for the simulation of systems of larger size, some further approximations are made, and some quantities are neglected or evaluated on the basis of experimental measures. Then the methods are called *semi-empirical*. Just to give one example, a typical quantity that can be inserted in the calculations is an interatomic distance (a quantity that in principle is an *output* of the computation, namely an optimization of the total energy of the system), or the value of an overlap between two electronic orbitals (*i.e.*, the value of some entries of the 'mass matrix' that again in principle should be calculated; see Section 3.2 for the definition of orbitals). The computational task is then reduced, and larger systems can be addressed. Finally, the models can be so much simplified that all the quantum information is aggregated into force fields for classical mechanics, and one reaches the field of molecular simulation, often subdivided into the domains called *molecular mechanics* and *molecular dynamics*. These domains are overwhelmingly those that have percolated efficiently into other fields of sciences related to chemistry, for instance biology and materials science.

Focusing on the *ab initio* models to start with, we notice that in addition, it is natural to focus on the determination of the ground state (that is, the state of minimal energy) of the system under consideration. Indeed, in the natural environment, systems are usually found in their most stable state. Likewise any chemical system A reacts, spontaneously or with a compound X, to give products B, C, ... according to a chemical reaction if the variation of energy corresponds to a stabilization of the whole system. The above thermodynamic consideration does not suffice, however, to explain all the observations (kinetics comes into play), but the determination of the ground state and of the ground state energy remains a preliminary calculation needed before turning to other questions in computational chemistry: calculation of excited states, linear or nonlinear response theory, *etc.* The latter questions will not be addressed here and we refer to the bibliography.

Before getting to the heart of the matter, let us recall the orders of magnitude for the objects we will be manipulating henceforth, as they are rather unusual. The typical size of the electronic cloud of an isolated atom is the Angström (10^{-10} m). The size of the nucleus embedded therein is 10^{-15} m. The weight of an atom is of the order of 10^{-26} kg. Regarding the time-scale, the typical vibration period of a molecular bond is the femtosecond (10^{-15} s), while the characteristic relaxation time for an electron is 10^{-18} s. Consequently, computational chemistry concerns the behaviour of very small and very light systems over very short time frames.

An introduction to all the quantum models we will deal with can be read in Levine (1991), for example, while the basics of quantum mechanics are recalled in McWeeny (1992). The main mathematical tools for the standard (linear) analysis with an emphasis on physics are contained in Blanchard and Bruning (1982), Cycon, Froese, Kirsch and Simon (1987), Gustafson and Sigal (2003), Reed and Simon (1975), Schechter (1981), Thirring (1983). In addition, the series of Lipkowitz and Boyd, eds (1995–) periodically publishes state-of-the-art surveys by experts in chemistry. We shall give more specific references later.

2.2. Standard ab initio models for molecular systems

In most situations in chemistry, it is legitimate to consider the nuclei as classical objects, and as point-like particles with charges $(z_1, \ldots, z_M)$ at positions $(\bar{x}_1, \ldots, \bar{x}_M)$, while treating the electrons as quantum particles. This is the so-called *Born–Oppenheimer* approximation. In view of this approximation, the determination of the ground state structure of a molecular system consisting of M nuclei and N electrons amounts to solving the following two nested minimization problems:

$$\inf_{(\bar{x}_1, \ldots, \bar{x}_M) \in \mathbb{R}^{3M}} \left\{ W(\bar{x}_1, \ldots, \bar{x}_M) = U(\bar{x}_1, \ldots, \bar{x}_M) + \sum_{1 \leq k < l \leq M} \frac{z_k\, z_l}{|\bar{x}_k - \bar{x}_l|} \right\},$$

$$(2.1)$$

where

$$U(\bar{x}_1, \ldots, \bar{x}_M) = \inf\{\langle \psi_e, H_e^{(\bar{x}_1,\ldots,\bar{x}_M)} \psi_e \rangle, \quad \psi_e \in \mathcal{H}_e, \quad \|\psi_e\|_{L^2} = 1\}. \quad (2.2)$$

The variational problem (2.2) determines the ground state electronic structure when the nuclei are clamped at the positions $(\bar{x}_1, \ldots, \bar{x}_M)$. We concentrate on this problem, and will only briefly address the outer minimization loop (2.1) (that requires techniques of *molecular mechanics*) in Section 3.8 below.

Problem (2.2) consists in finding the lowest eigenvalue of the N-body Hamiltonian $H_e^{(\bar{x}_1,\ldots,\bar{x}_M)}$, parametrized by the positions of the nuclei,

$$H_e^{(\bar{x}_1,\ldots,\bar{x}_M)} = -\sum_{i=1}^{N} \frac{1}{2} \Delta_{x_i} - \sum_{i=1}^{N} \sum_{k=1}^{M} \frac{z_k}{|x_i - \bar{x}_k|} + \sum_{1 \le i < j \le N} \frac{1}{|x_i - x_j|}. \quad (2.3)$$

We easily recognize in (2.3) the kinetic energy of the electrons, the attraction electrostatic energy between the nuclei and the electrons, and the repulsion electrostatic energy between the electrons, respectively.

In order to write $H_e^{(\bar{x}_1,\ldots,\bar{x}_M)}$, we have chosen the *atomic unit system*, commonly used in quantum chemistry:

$$m_e = 1, \quad e = 1, \quad \hbar = 1, \quad \frac{1}{4\pi\epsilon_0} = 1,$$

where m_e, e, $\hbar$, ϵ_0, respectively, denote the electron mass, the elementary charge, the reduced Planck constant, and the dielectric permittivity of vacuum.

On the other hand, the variational space in (2.2) is set to the following subspace of $L^2(\mathbb{R}^{3N})$, the antisymmetrized tensor product

$$\mathcal{H}_e = \bigwedge_{i=1}^{N} H^1(\mathbb{R}^3), \quad (2.4)$$

in order to ensure that the kinetic energy term is finite.[1] The antisymmetry requirement comes from the Pauli exclusion principle that states that the electronic wavefunction indeed needs to be antisymmetric with respect to any permutation of the electrons.

The Euler–Lagrange equation of the minimization problem (2.2) is the celebrated Schrödinger equation,

$$H_e^{(\bar{x}_1,\ldots,\bar{x}_M)} \psi_e = E_e \psi_e, \quad (2.5)$$

[1] Here and below, for clarity of exposition, we omit the spin variable, which has of course a huge practical importance. The introduction of the spin in the models and techniques we shall describe in this review is not conceptually difficult, but may give rise to substantial additional technicalities. Again for simplicity, we also assume that the wavefunctions are real-valued.

with E the lowest possible eigenvalue of the self-adjoint operator $H_e^{(\bar{x}_1,\ldots,\bar{x}_M)}$ on $\mathcal{H}_e$, in fact equal to $U(\bar{x}_1,\ldots,\bar{x}_M)$ given by (2.2).

Unfortunately, for almost all problems of interest, the treatment of the minimization problem (2.2) or alternatively that of equations (2.5) is at the present time essentially impossible, owing to the huge size of the Hilbert space $\mathcal{H}_e$. The only techniques that are indeed tractable at this level are mostly stochastic-like techniques. One class of such methods is called *variational Monte Carlo* and consists in evaluating the multidimensional integrals

$$\langle \psi_e, H_e^{(\bar{x}_1,\ldots,\bar{x}_M)} \psi_e \rangle$$

with adequate sampling techniques.[2] The minimization in (2.2) is then performed by standard tools. We refer, for example, to the chapter by W. Lester in Le Bris, ed. (2003). Another approach, not well developed in the world of chemistry[3] but very promising according to Lions (1996) in view of its success in other fields such as financial mathematics, consists in considering the time-dependent parabolic equation

$$\frac{\partial \psi_e}{\partial t} + H_e \psi_e = 0, \qquad (2.6)$$

and noticing that, as t goes to infinity,

$$\frac{1}{t} \mathrm{Log} |\psi_e(t,x)| = \frac{1}{t} \mathrm{Log} |e^{-tH_e} \psi_e(0,x)| \longrightarrow U \quad \text{(defined by (2.2))}. \qquad (2.7)$$

Then $\psi_e(t,x)$ is computed using the Feynman–Kac representation formula for the solution to (2.6), which requires efficient sampling techniques.

Apart from these stochastic techniques designed to directly attack problem (2.5), there are the seemingly promising techniques of *sparse tensor products* (also known as *sparse grid methods*) that are only emerging in computational chemistry. Equation (2.5) is a partial differential equation set on a vectorial space of high dimension, and most numerical techniques cannot deal with dimensions larger than 6. Nevertheless, the sparse grid approximations precisely aim at dealing with PDEs in high dimension: see Griebel, Oswald and Schlekefer (1999), Bungartz and Griebel (2004) or Schwab and von Petersdorff (2004). These techniques are definitely successful in many contexts. In their present state, however, they require a high regularity of the solution manipulated, in order to use sparse tensor

---

[2] Regarding deterministic methods, see Bokanowski and Lemou (1998, 2001) for a tentative adaptation (still in its early stage but apparently promising) of the fast multipole method to these multidimensional integrals.

[3] What is however used in chemistry is a simplified version of the approach, known as the *Diffusion Monte Carlo* (DMC) method. We refer to Cancès, Jourdain and Lelièvre (2004*d*) for a first mathematical study of this method.

product spaces without losing any information in comparison to the full tensor product. Now, the functions of chemistry may be singular, notably because of the cusp present in the interaction between electrons (the $\frac{1}{|x_i - x_j|}$ terms in $H_e^{(\bar{x}_1, \ldots, \bar{x}_M)}$); we will come back to this point in Section 3.2. Nevertheless, we are allowed to be optimistic, for two reasons. First, in order to deal with cases when there is some lack of regularity, there are works in progress by M. Griebel and colleagues that include the use of *adaptive* sparse tensor products. Second, recent results by Yserentant (2003, 2004a, 2004b) indicate that the wavefunction solution to the (stationary) Schrödinger equation (2.5) is more regular than expected (and in fact has *almost* the regularity needed for state-of-the-art sparse tensor product techniques). The sparse tensor product approximation *should* therefore work nicely (see Garcke and Griebel (2000) for a first step). On the other hand, it is worth mentioning that dealing with the antisymmetry requirement in the context of sparse tensor products is still an issue (see however Hackbusch (2001) for a possible track).

In addition to these emerging techniques that come from other domains of the engineering sciences, an alternative way has appeared a long time ago in the world of chemistry but is still under practical development. It starts from the observation that the Hamiltonian (2.3) only involves one- and two-electron terms, and therefore that the multidimensional integral $\langle \psi_e, H_e^{(\bar{x}_1, \ldots, \bar{x}_M)} \psi_e \rangle$ can be expressed in terms of the second-order reduced matrix

$$\gamma_2(x, y, x', y') = \int_{\mathbb{R}^{3(N-2)}} \psi_e(x, y, x_3, \ldots, x_N) \psi_e(x', y', x_3, \ldots, x_N) \, dx_3 \cdots dx_N.$$

$$(2.8)$$

Next, the minimization problem, possibly under the form of its Euler–Lagrange equation is reformulated, and treated numerically. This is undoubtedly an appealing idea, though still not mature enough (issues both at the theoretical and at the practical levels remain unsolved). Regarding what seems to be a promising track, we refer to Coleman and Yukalov (2000) for the general context (see also many references by the first of these two authors), to Zhao, Braams, Fukuda, Overton and Percus (2004) for an instance of a direct minimization approach, and most of all to the works by D. Mazziotti (see, *e.g.*, Mazziotti (1998a, 1998b, 1999, 2004)), and C. Valdemoro (see, *e.g.*, Valdemoro, Tel and Perez-Romero (2000)).

At this stage, it seems to us important to emphasize that the main challenge for the future of computational chemistry is to directly attack the N-body problem (2.5), or in a less ambitious manner, increasingly account for the N-body interaction itself (and this begins with $N = 2$). The stochastic approach, the use of sparse tensor products, the reduced density matrix approach described above, are instances of methods that go in this direction.

Likewise, the introduction of many-electron wavefunctions in the basis sets (see Section 3.3) and many very recent approaches emerging in computational chemistry are aimed at addressing this timely question. However, as these methods are not yet mature, we will concentrate in the rest of this state-of-the-art survey on methods that are to some extent better established.

The most commonly used approximations for the minimization problem (2.2) can be filed schematically into two main classes.

- **Wavefunction methods**, exemplified by the famous Hartree–Fock model, aim to find an approximation of the ground state electronic wavefunction, *i.e.*, of the minimizer of (2.2). The variational space $\mathcal{H}_e$ is reduced but the 'exact' form of the energy $\langle \psi_e, H_e^{(\bar{x}_1, \ldots, \bar{x}_M)} \psi_e \rangle$ is kept. Wavefunction methods are preferred by chemists who are interested in the precise simulations of systems of small size, when computational time is not the primary concern. We refer to the treatises of Hehre, Radom, Schleyer and Pople (1986) and Szabo and Ostlund (1982) in the chemistry literature. A famous program implementing Hartree–Fock-type methods is the code GAUSSIAN.

- **Density functional methods** originate from density functional theory. They are based on a reformulation of problem (2.2) in such a way that the unknown function is the electronic density

$$\rho(x) = N \int_{\mathbb{R}^{3(N-1)}} |\psi_e(x, x_2, \ldots, x_N)|^2 \, \mathrm{d}x_2 \cdots \mathrm{d}x_N$$

(*i.e.*, a scalar field on $\mathbb{R}^3$) rather than the wavefunction ψ_e (*i.e.*, a scalar field on $\mathbb{R}^{3N}$) as in the original problem (2.2). This is why these methods are widely used by those of the chemists who are interested in large molecular systems (*e.g.*, biological systems) as well as most solid state physicists. The fact that various parameters or even the very form of some terms of the energy functional need to be arbitrarily chosen or tuned for these methods makes the method particularly efficient for some situations but is sometimes seen as a lack of rigour by chemists. Some major references in the chemistry literature are March (1992), Parr and Yang (1989) and Dreizler and Gross (1990).

2.3. Hartree–Fock-type models

The Hartree–Fock (HF) approximation consists in restricting in (2.2) the variational space $\mathcal{H}_e$ to that of functions of variables $(x_1, \ldots, x_N) \in \mathbb{R}^{3N}$ which can be written as a *single* determinant (*i.e.*, an antisymmetrized product) of N functions defined on $\mathbb{R}^3$. Recall that, in the whole generality, an arbitrary element of $\mathcal{H}_e$ is a converging infinite sum of such determinants.

The Hartree–Fock approximation is therefore defined as

$$U^{HF}(\bar{x}_1, \ldots, \bar{x}_M) = \inf\big\{ \langle \psi_e, H_e \psi_e \rangle : \quad \psi_e \in \mathcal{S}_N \big\}, \tag{2.9}$$

with

$$\mathcal{S}_N = \left\{ \psi_e = \frac{1}{\sqrt{N!}} \det(\phi_i(x_j)) : \phi_i \in H^1(\mathbb{R}^3), \int_{\mathbb{R}^3} \phi_i \, \phi_j = \delta_{ij}, \, 1 \le i, j \le N \right\}. \tag{2.10}$$

In quantum chemistry, a function of the form $\frac{1}{\sqrt{N!}} \det(\phi_i(x_j))$ is called a *Slater determinant*, and the ϕ_i are called *molecular orbitals*.

Apart from antisymmetry, the Hartree–Fock approximation heuristically consists in observing that the probability density $|\psi|^2(x_1, \ldots, x_N)$ of finding the N electrons at positions $(x_1, \ldots, x_N)$ can be approximated by the product $|\phi_1|^2(x_1) \cdots |\phi_N|^2(x_N)$. This is equivalent to considering the positions of the electrons as *independent* variables. This simplification causes a certain loss of *correlation* between the positions of the electrons, and is responsible for some error in the result obtained. Indeed, restricting the minimization to *some* specific forms of functions in (2.9) provides us only with an upper bound on the energy (2.2). On the other hand, the fact that it is an upper bound and not only an approximation of the exact energy is of course a substantial practical advantage of the method, in comparison with other, nonvariational, approximations, such that those coming from density functional theory (see Section 2.4 and below).

Explicitly, the computation of $E^{HF}(\phi_1, \ldots, \phi_N) = \langle \psi_e, H_e \psi_e \rangle$ for ψ_e in $\mathcal{S}_N$ leads to

$$I_N^{HF} = \inf \left\{ \sum_{i=1}^{N} \frac{1}{2} \int_{\mathbb{R}^3} |\nabla \phi_i|^2 + \int_{\mathbb{R}^3} \rho\, V + \frac{1}{2} \int_{\mathbb{R}^3} \int_{\mathbb{R}^3} \frac{\rho(x)\, \rho(x')}{|x - x'|}\, \mathrm{d}x\, \mathrm{d}x' \right.$$
$$\left. - \frac{1}{2} \int_{\mathbb{R}^3} \int_{\mathbb{R}^3} \frac{|\tau(x, x')|^2}{|x - x'|}\, \mathrm{d}x\, \mathrm{d}x' : \right.$$
$$\left. \phi_i \in H^1(\mathbb{R}^3), \quad \int_{\mathbb{R}^3} \phi_i \, \phi_j = \delta_{ij}, \quad 1 \le i, j \le N \right\}, \tag{2.11}$$

where

$$V(x) = -\sum_{k=1}^{M} \frac{z_k}{|x - \bar{x}_k|}, \tag{2.12}$$

$$\tau(x, x') = \sum_{i=1}^{N} \phi_i(x)\, \phi_i(x'), \tag{2.13}$$

$$\rho(x) = \sum_{i=1}^{N} |\phi_i(x)|^2. \tag{2.14}$$

The functions τ and ρ are respectively called the *density matrix* and the *density* associated to the state ψ_e.

The Euler–Lagrange equations of (2.11) are the *Hartree–Fock equations*

$$\begin{cases} F_\Phi \phi_i & = \lambda_i\, \phi_i, \\ \int_{\mathbb{R}^3} \phi_i \phi_j & = \delta_{ij}, \end{cases} \qquad (2.15)$$

where F_Φ is the *Fock operator*

$$F_\Phi = -\frac{1}{2}\Delta - \sum_{k=1}^{M} \frac{z_k}{|x - \bar{x}_k(t)|} + \left(\sum_{j=1}^{N} |\phi_j|^2 \star \frac{1}{|x|} \right) - \sum_{j=1}^{N} \left(\cdot\, \phi_j \star \frac{1}{|x|} \right) \phi_j, \quad (2.16)$$

and the λ_i are the Lagrange multipliers of the orthonormality constraints (owing to an invariance property, obvious on (2.10), of the HF energy functional with respect to orthogonal (unitary) transformations of the ϕ_i, the matrix of Lagrange multipliers may be diagonalized without loss of generality).

The above *Hartree–Fock model* (2.11) has been extensively studied by mathematicians, the two landmark papers being those by Lieb and Simon (1977*a*) and Lions (1987), where the existence of a minimizer is demonstrated under convenient assumptions, and the Euler–Lagrange equations are thoroughly studied.

2.4. Density functional theory models

As announced above, the purpose of density functional theory, abbreviated as DFT, is to replace the minimization problem (2.2) defined in terms of the unknown wavefunction ψ_e by a minimization problem set on the unknown density ρ.

To fulfil this goal, it suffices, for instance, to define

$$E(\rho) = \inf \left\{ \left\langle \psi_e, \left(-\sum_{i=1}^{N} \frac{1}{2}\Delta_{x_i} + \sum_{1 \leq i < j \leq N} \frac{1}{|x_i - x_j|} \right) \psi_e \right\rangle : \right.$$

$$\left. \psi_e \in \mathcal{H}_e, \quad \|\psi_e\|_{L^2} = 1, \quad \psi_e \text{ has density } \rho \right\}, \qquad (2.17)$$

on

$$\mathcal{I}_N = \left\{ \rho \geq 0 : \quad \sqrt{\rho} \in H^1(\mathbb{R}^3), \quad \int_{\mathbb{R}^3} \rho = N \right\}, \qquad (2.18)$$

so that

$$U(\bar{x}_1, \ldots, \bar{x}_M) = \inf \left\{ E(\rho) - \int \left(\sum_{k=1}^{M} \frac{z_k}{|\cdot - \bar{x}_k|} \right) \rho : \quad \rho \in \mathcal{I}_N \right\}. \qquad (2.19)$$

The functional E is the *density functional*. However it is derived (the above manner is one among many, all related to a paper by Hohenberg and Kohn (1964) celebrated in quantum chemistry), finding an explicit expression for E is an open problem. In practice, approximations of the density functional have been developed, that rely on exact or very accurate evaluations of different contributions to the energy for reference systems 'close' to the system under study.

The best option for the approximation of the kinetic energy term is today considered to be the model introduced by Kohn and Sham (1965). Their idea was to take N *non-interacting* electrons as the reference system, and made the DFT approach tractable. Under convenient assumptions, the kinetic energy of such a system reads

$$T_{KS}(\rho) = \tag{2.20}$$

$$\inf\left\{\frac{1}{2}\sum_{i=1}^{N}\int_{\mathbb{R}^3}|\nabla\phi_i|^2 : \quad \phi_i \in H^1(\mathbb{R}^3), \quad \int_{\mathbb{R}^3}\phi_i\phi_j = \delta_{ij}, \quad \sum_{i=1}^{N}|\phi_i|^2 = \rho\right\}.$$

This expression is then chosen as an *approximation* of the kinetic energy term for the system of *interacting* electrons under study and added to other terms of electrostatic nature (attraction by the nuclei and inter-electronic repulsion) to form the famous *Kohn–Sham model*

$$I_N^{KS} = \inf\left\{\frac{1}{2}\sum_{i=1}^{N}\int_{\mathbb{R}^3}|\nabla\phi_i|^2 + \int\rho V + \frac{1}{2}\int_{\mathbb{R}^3}\int_{\mathbb{R}^3}\frac{\rho(x)\rho(y)}{|x-y|}\,\mathrm{d}x\,\mathrm{d}y + E_{xc}(\rho) :$$

$$\phi_i \in H^1(\mathbb{R}^3), \quad \int_{\mathbb{R}^3}\phi_i\phi_j = \delta_{ij}\right\}, \tag{2.21}$$

where ρ is a notation for $\sum_{i=1}^{N}|\phi_i|^2$. The functional $E_{xc}(\rho)$, called the *exchange term*, is a correction term, accounting for the non-independence of the electrons, for which approximations are in turn developed for different situations. One of these approximations consists in using as a reference system a uniform non-interacting electron gas. For such a system, Dirac explicitly computed the exchange energy term

$$E_{xc}(\rho) = -C_D \int_{\mathbb{R}^3}\rho^{4/3}, \tag{2.22}$$

where $C_D = \frac{3}{4}\left(\frac{3}{\pi}\right)^{1/3}$.

This approximation of the exchange term is one occurrence of the *local density approximation* (LDA) for which $E_{xc}(\rho) = \int_{\mathbb{R}^3}F(\rho)$. Other more precise expressions have also been developed.

The Euler–Lagrange equations of the problem (2.21) are the *Kohn–Sham equations*

$$\begin{cases} K(\rho_\Phi)\phi_i = \lambda_i\phi_i, \\ \int_{\mathbb{R}^3} \phi_i\phi_j = \delta_{ij}, \end{cases} \tag{2.23}$$

where $\rho_\Phi = \sum_{i=1}^{N} |\phi_i|^2$,

$$K(\rho_\Phi) = -\frac{1}{2}\Delta - \sum_{k=1}^{M} \frac{z_k}{|\cdot - \bar{x}_k(t)|} + \left(\sum_{i=1}^{N} |\phi_i|^2 \star \frac{1}{|x|} \right) + v_{xc}(\rho_\Phi), \tag{2.24}$$

and $v_{xc} = \frac{\partial E_{xc}(\rho)}{\partial \rho}$.

The comparison of the energy functionals in the Hartree–Fock case (2.11) and in the Kohn–Sham case (2.21) (or that of their respective Euler–Lagrange equations (2.15) and (2.23)) reveals the global similarity between the two approaches from a formal viewpoint. We will therefore concentrate on the Hartree–Fock problem, and also indicate when important the necessary modifications for treating the Kohn–Sham model.

Before we get to this, we would like to mention a variant of the Kohn–Sham model, still in the category of density functional theory: the *orbital-free* models. These models are more or less based upon a rather old idea, indeed an ancestor of the DFT, namely the Thomas–Fermi theory. We mention them here because from the standpoint of the implementation and the algorithmic procedure, they exhibit significant differences to the other approaches.

The idea underlying the TF theory is to use as a reference system the uniform non-interacting electron gas (already mentioned above for the approximation of the exchange term, but this time used *also* for the kinetic energy term). For a uniform non-interacting electron gas, one can indeed compute analytically the kinetic energy

$$T_{TF}(\rho) = C_{TF} \int_{\mathbb{R}^3} \rho(x)^{5/3} \, dx, \tag{2.25}$$

where $C_{TF} = \frac{10}{3}(3\pi^2)^{2/3}$ denotes the Thomas–Fermi constant. In addition, a correction term, obtained by studying perturbations generated by small heterogeneities of the density, and due to von Weizsäcker, can be expressed in terms of $C_W \int_{\mathbb{R}^3} |\nabla\sqrt{\rho}|^2$ and added to the kinetic energy. The minimization problem obtained is thus of the form

$$\inf\left\{ E(\rho), \quad \rho \geq 0, \quad \sqrt{\rho} \in H^1(\mathbb{R}^3), \quad \int_{\mathbb{R}^3} \rho = N \right\}. \tag{2.26}$$

with

$$E(\rho) = C_W \int_{\mathbb{R}^3} |\nabla\sqrt{\rho}|^2 + C_{TF} \int_{\mathbb{R}^3} \rho^{5/3} + \int_{\mathbb{R}^3} \rho V$$
$$+ \frac{1}{2} \int_{\mathbb{R}^3} \int_{\mathbb{R}^3} \frac{\rho(x)\rho(y)}{|x-y|} \, \mathrm{d}x \, \mathrm{d}y - C_D \int_{\mathbb{R}^3} \rho^{4/3}.$$

In contrast to the Kohn–Sham theory, this model only involves the density ρ. It is attacked by discretizing the density ρ on a grid in $\mathbb{R}^3$, which is an approach completely different to that we describe below for models ultimately involving wavefunctions. Of course, more elaborated energy functionals, still functions of ρ, can be derived, but the spirit remains. After years during which the Thomas–Fermi approach was considered out of date, and definitely superseded by the Kohn–Sham approach, it seems that, for very specific purposes (when at least some vague information on the electronic structure must be inserted in simulations of systems of very large size), this approach is seeing a revival in the form of orbital-free methods. We refer to Carter (2000), for example. A mathematical analysis of a simple form of the method is developed in Blanc and Cancès (2004).

Note that, on the other hand, from the academic (and in particular mathematical) viewpoint, the Thomas–Fermi approach has been a constant subject of interest. Major contributions include those by Lieb and Simon (1977b) and review articles on all the aspects of these models include Jones and Gunnarsson (1989), Spruch (1991) and Lieb (1983).

3. Discretization of molecular models

The Galerkin approximation procedure consists in approaching the *infinite-dimensional* HF problem (2.11) by a *finite-dimensional* problem where the HF energy is minimized over the set of molecular orbitals ϕ_i that can be expanded with respect to a given finite basis set $\{\chi_\mu\}_{1\le\mu\le N_b}$:

$$\phi_i = \sum_{\mu=1}^{N_b} C_{\mu i} \chi_\mu.$$

The ith column of the rectangular matrix $C \in \mathcal{M}(N_b, N)$ contains the N_b coefficients in the basis $\{\chi_\mu\}_{1\le\mu\le N_b}$ of each of the N molecular orbitals ϕ_i forming the desired Slater determinant. Letting S be the *overlap* matrix with elements

$$S_{\mu\nu} = \int_{\mathbb{R}^3} \chi_\mu \chi_\nu, \tag{3.1}$$

the constraints $\int_{\mathbb{R}^3} \phi_i \phi_j = \delta_{ij}$ become

$$C^* S C = I_N,$$

where I_N denotes the $N \times N$ identity matrix. We next introduce the notation

$$h_{\mu\nu} = \frac{1}{2} \int_{\mathbb{R}^3} \nabla \chi_\mu \cdot \nabla \chi_\nu + \int_{\mathbb{R}^3} V \chi_\mu \chi_\nu \qquad (3.2)$$

for the matrix of the core Hamiltonian $h = -\frac{1}{2}\Delta + V$ with respect to the basis $\{\chi_k\}$, together with

$$J(X)_{\mu\nu} = \sum_{\kappa,\lambda=1}^{N_b} (\mu\nu|\kappa\lambda) X_{\kappa\lambda}, \qquad K(X)_{\mu\nu} = \sum_{\kappa,\lambda=1}^{N_b} (\mu\lambda|\nu\kappa) X_{\kappa\lambda},$$

$$G(X)_{\mu\nu} = J(X)_{\mu\nu} - K(X)_{\mu\nu},$$

where X can be any $N_b \times N_b$ matrix, and

$$(\mu\nu|\kappa\lambda) = \int_{\mathbb{R}^3} \int_{\mathbb{R}^3} \frac{\chi_\mu(x)\chi_\nu(x)\chi_\kappa(x')\chi_\lambda(x')}{|x - x'|} \, \mathrm{d}x \, \mathrm{d}x' \qquad (3.3)$$

are the so-called *bi-electronic integrals*. The HF problem then becomes

$$\inf\{E^{HF}(CC^*): \quad C \in \mathcal{M}(N_b, N), \quad C^*SC = I_N\}, \qquad (3.4)$$

where

$$E^{HF}(CC^*) = \mathrm{Trace}(hCC^*) + \frac{1}{2}\mathrm{Trace}(G(CC^*)CC^*).$$

Alternatively, the HF energy can be written in terms of the symmetric $N_b \times N_b$ *density matrix* $D = CC^*$,

$$\inf\{E^{HF}(D): \quad D \in \mathcal{P}_N\}, \qquad (3.5)$$

where

$$\mathcal{P}_N = \{D \in \mathcal{M}(N_b, N_b): \quad DSD = D, \quad \mathrm{Trace}(SD) = N\}. \qquad (3.6)$$

The associated Euler–Lagrange equations therefore read

$$\begin{cases} F(D)C = SC\Lambda, \\ C^*SC = I_N, \\ D = CC^*, \end{cases} \qquad (3.7)$$

where

$$F(D) = h + G(D)$$

denotes the Fock operator.

Using, as in the continuous case (2.16), the fact that the HF model (3.4) is invariant with respect to the unitary transform $C \mapsto CU$, the matrix of Lagrange multipliers Λ can be diagonalized. In addition, it must be emphasized at this stage that, for a minimizer of the infinite-dimensional HF problem (2.11), the eigenvalues given by Λ are known to be the lowest N eigenvalues of the Fock operator F_Φ (see Lions (1987) and, for the question of the non-degeneracy of the Nth level, Bach, Lieb, Loss and Solovej (1994)). This property is preserved in the finite-dimensional setting for a minimizer C of (3.4),[4] and it is said that C and $D = CC^*$ satisfy the *Aufbau* principle, which is a principle for placing electrons within shells. This property will be strongly exploited below in the design and the analysis of self-consistent field (SCF) algorithms. Therefore the Euler–Lagrange equations read

$$\begin{cases} F(D)C = SCE, \qquad E = \mathrm{Diag}(\epsilon_1, \epsilon_2, \dots, \epsilon_N) \\ C^*SC = I_N \\ D = CC^*, \end{cases} \tag{3.8}$$

where $\epsilon_1 \leq \epsilon_2 \leq \cdots \leq \epsilon_N$ are the lowest N eigenvalues of the Fock operator $F(D)$. In particular, still denoting by Φ_i the ith column of C,

$$F(D)\Phi_i = \epsilon_i S\Phi_i. \tag{3.9}$$

The examination of equations (3.8) gives us the opportunity to emphasize the following key feature: all the models we deal with in the present article are models at *zero temperature*. This is explicit in the Aufbau principle stated above, as the lowest N eigenstates are occupied, while any higher one is empty. In a picture at *positive temperature*, we introduce occupation numbers $\alpha_i \in [0, 1]$, not necessarily equal to zero or one, along each eigenstate Φ_i, including indices $i \geq N+1$. The coefficients α_i are simultaneously optimized, accounting for some entropy term. In fact, the *free energy* is then minimized instead of the energy.

The Kohn–Sham models are discretized in a similar way to the HF one:

$$I^{KS} = \inf\left\{ E^{KS}(CC^*) : \quad C \in \mathcal{M}(N_b, N), \quad C^*SC = I_N \right\} \tag{3.10}$$

with

$$E^{KS}(D) = \mathrm{Trace}(hD) + \mathrm{Trace}(J(D)D) + E_{xc}(D), \tag{3.11}$$

$E_{xc}(D)$ denoting the exchange-correlation energy. If, for instance, an LDA functional is used,

$$E_{xc}(D) = \int_{\mathbb{R}^3} \rho(x)\, \epsilon_{xc}^{LDA}(\rho(x))\, \mathrm{d}x, \quad \text{with} \quad \rho(x) = 2\sum_{i=1}^{N_p} D_{\mu\nu}\chi_\mu(x)\chi_\nu(x).$$

---

[4] Actually, different variants of the HF problem exist, owing to the treatment of the spins. We do not want to go into technicalities here and will assume that the property under examination is always satisfied. Even in the absence of a mathematical proof for some variants, the numerical practice shows that it is always true. In the same vein, we shall assume below that this property is also satisfied by the KS problem, while in this latter case it is even unclear at the infinite-dimensional level, however, again, confirmed by experiment.

Likewise, the KS equations read

$$\begin{cases} F^{KS}(D)C = SCE, & E = \mathrm{Diag}(\epsilon_1, \epsilon_2, \ldots, \epsilon_N) \\ C^*SC = I_{N_p} \\ D = CC^* \end{cases} \qquad (3.12)$$

with $F^{KS}(D) = h + J(D) + \nabla E_{xc}(D)$ and ϵ_i the ith eigenvalue of $F^{KS}(D)$. As announced above, we observe the formal similarity between problems (3.8) and (3.12), which confirms the opportunity to concentrate mainly on the HF problem in the present expository survey.

At this stage, let us also mention that solutions of

$$\begin{cases} FC = SCE, & E = \mathrm{Diag}(\epsilon_1, \epsilon_2, \ldots, \epsilon_N) \\ C^*SC = I_N \\ D = CC^* \end{cases} \qquad (3.13)$$

(where F is a given matrix) are the same as the solutions in $\mathcal{P}_N$ to the equation

$$[F, D] = 0, \qquad (3.14)$$

where $[\cdot, \cdot]$ denotes the 'commutator' defined by $[A, B] = ABS - SBA$. Likewise, the solutions to (3.13) which satisfy in addition the Aufbau principle are the same as the solutions to the problem

$$D = \arg \inf\{\mathrm{Trace}(FD') : \quad D' \in \mathcal{P}_N\}. \qquad (3.15)$$

If in addition there is a positive gap between the Nth and the $(N+1)$th eigenvalue of F (*i.e.*, if $\epsilon_1 \leq \cdots \leq \epsilon_N < \epsilon_{N+1} \leq \cdots \leq \epsilon_{N_b}$), then the Aufbau solutions D to (3.13) are also the solutions to

$$D = \arg \inf\{\mathrm{Trace}(FD') : \quad D' \in \widetilde{\mathcal{P}}_N\}, \qquad (3.16)$$

where

$$\widetilde{\mathcal{P}}_N = \{D \in \mathcal{M}(N_b, N_b) : \quad DSD \leq D, \quad \mathrm{Trace}(SD) = N\}. \qquad (3.17)$$

This latter property will be briefly justified in Section 3.5 below.

3.1. Anticipating the numerical difficulties

The examination of the discrete form (3.5)–(3.6) of the Hartree–Fock minimization problem suffices to measure the main difficulties experienced in the numerical approach.

The minimization is performed over the manifold of density matrices ($DSD = D$) which is nonconvex (a good way to think of the problem is to think of minimizing a function over a sphere). Whatever the properties of the function to be minimized, the problem is likely to be difficult. It is indeed. At the theoretical level, we of course lose the equivalence between the minimization viewpoint and the Euler–Lagrange viewpoint. A rigorous approach would attack the minimization directly. Unfortunately, on the one hand, the enormous number of critical points that have been observed in practice rule out deterministic gradient algorithms, while on the other hand zero-order methods such as stochastic methods and direct search methods lead to an overwhelming number of function evaluations. Therefore, contrary to rigour, we are obliged to attack the problem by solving the Euler–Lagrange equations.[5] This liberty with rigour will of course plague the numerical analysis of existing approaches, as we will see in Section 3.5. The somewhat surprising fact is that it works in practice, provided a number of precautions are taken, such as a proper choice of an initial guess. Here a point must be made. For a mathematician, trusting a lucky star in order to obtain convergence to the global minimizer is of course both frustrating and crazy, and we by no means pretend otherwise here. But a specificity of computational chemistry comes into the picture. The calculation of a molecular system is rarely done *from scratch*. A computation with a coarser model, or with the same model implemented more coarsely, has often been done before. In some cases as well, the solution procedure is an inner loop of a more global simulation: when the evolution of a molecular system is simulated, the nuclei are moved incrementally through Newton's equation, and for each updated position of the set of nuclei a new calculation of the electronic structure is performed (we shall see such methods in Section 5). Then, the natural initial guess is the output of the previous computation. All this contributes to render the Euler–Lagrange approach successful in practice.

Of course, the chances of success are significantly increased by a design of the solution algorithms for the Euler–Lagrange equations that is, in some way or another, reminiscent of the fact that the solution to be determined is not any critical point, but the global minimizer. And this is where a rigorous numerical analysis reveals itself to be efficient. This is done by at least requiring the Aufbau principle for the solution, but also in a more sophisticated manner, by elaborating algorithmic strategies based upon partial minimizations. We will return to this in Section 3.5.

Having accepted that the Euler–Lagrange approach is the only tractable one, we can adopt either finite difference-type methods or variational methods. The former are far less commonly used, unless for very specific purposes, and we refer to the literature for more details (see, *e.g.*, the review article by Beck (2000) and also the chapters by J. L. Fattebert and J. Chelikowski *et al.* in Le Bris, ed. (2003)). In computational chemistry, finite difference methods are often referred to as *real-space* methods.

---

[5] Obviously, our statement describes the state of the art. New and efficient ideas to attack the problem by minimization are welcome.

Basically, they amount to discretizing the Euler–Lagrange equations on a grid, using high-order schemes (typically order 4 or 6) for the Laplacian operator associated with the kinetic energy term. We will only concentrate on the variational methods. A key issue will be the choice of the variational space. We will again see below some specific features of computational chemistry (mostly related to the approximation of singularities), which often lead to 'problem-dependent' basis sets.

A second point on the Euler–Lagrange approach is to observe the algebraic nature of these equations (3.7): they form a *nonlinear eigenvalue problem*. Heuristically, the price to pay for making the *linear* eigenvalue problem (2.5) tractable numerically is *nonlinearity*. Therefore iterations are required, consisting more or less in freezing the operator and diagonalizing it, before updating it. The computational task is first to assemble the matrix, and next to diagonalize it. Here again, comments are in order.

Let us begin with the assembling step. In the Hartree–Fock setting, the computational effort necessary to build the mean-field Hamiltonian matrix (*i.e.*, the Fock matrix) in a basis containing N_b elements *a priori* scales as N_b^4 because of the calculation of the bielectronic integrals (3.3). For small systems, this quartic scaling constitutes another peculiarity of computational chemistry, because constructing the matrix is there more expensive than diagonalizing it, a fact that must of course be borne in mind for the design of methods. For large systems, however, the scaling is much lower in practice because the overlap of two atomic orbitals attached to two nuclei far away from one another is negligible (we will introduce this particular type of basis functions in Section 3.2, but for the time being it is enough to know that these basis functions are attached to the nuclei, and remark that integrals of type (3.3) are small when χ_μ and χ_ν (resp. χ_κ and χ_λ) have a small overlap). Various algorithms taking benefit of *a priori* estimates of the integrals were developed in the late 1970s and the 1980s (see Gill (1994) and the references therein); it is estimated that the scaling of these algorithms is around $N_b^{2.7}$ in practice. The prefactor mainly depends on the choice of the basis set (in fact of the degree of contraction of the Gaussian atomic orbitals, *i.e.*, of the parameter K in (3.22) below). For very large molecules, a much better scaling ($O(N_b)$) can however be obtained with linear scaling algorithms based on the Fast Multipole Method by Greengard and Rokhlin (1997), adapted to Coulomb and exchange matrix computations by M. Challacombe and coworkers: see Schwegler and Challacombe (1997) and further works.

In the Kohn–Sham model, the third term in (3.11) is evaluated by numerical integration on a grid, a computation that has complexity $O(N_b^3)$. For small molecular systems, the calculation is thus dominated by the computation of the second term in (3.11), and still scales in $O(N_b^4)$ (theoretically). On the other hand, for larger systems when the second term approximately scales in $N_b^{2.7}$, the limiting evaluation is that of the third term, thus in $O(N_b^3)$. Note that linear scaling integration methods have been introduced recently (see Scuseria (1999), Challacombe (2000) and references therein).

Let us turn to the diagonalization step. As mentioned above, it basically scales as N_b^3. It must be emphasized that *stricto sensu* the full diagonalization is not needed as we only search[6] for the *lowest* N eigenvectors and eigenvalues of a matrix of size $N_b \times N_b$ (in view of the Aufbau principle). However, for systems of reasonable size, the most efficient basis sets currently are atomic orbital basis sets for which N_b is typically a few times as large as N, thus the question asked cannot be reduced to finding the first few eigenvectors: the number of these eigenvectors might be half the size of the matrix.[7] The full diagonalization procedure is the most efficient choice, and this is all the more true as it benefits from many rapid implementations developed over the years for various applications of scientific computing. For systems of reasonable size (say a few electrons), the cost of the Hartree–Fock approximation is thus dominated by the N_b^4 cost of the construction of the Fock matrix, while it progressively diminishes to a power of roughly 3 for large systems, where the diagonalization step becomes the limiting process.

The above observation along which no complete diagonalization is *stricto sensu* needed.[8] is at the origin of a class of methods, dedicated to very large systems, and called *alternative to diagonalization methods* They will be overviewed in Section 3.7. Together with a rapid construction of the Fock matrix (by FMM), these methods allow us to bring down the complexity to a linear complexity (or slightly more than that) and make HF calculations of about one thousand atoms feasible on today's available workstations.

Let us mention to end this section that the N_b^4 complexity of the Hartree–Fock method is by no means an upper bound on the complexity of methods in the context of computational chemistry: the so-called post-Hartree–Fock methods, such as configuration interaction methods, MCSCF methods (see Section 3.8) and others have a computational complexity that can reach N_b^8 or more. We now measure the practical limitation of such methods for simulation of large systems, and this is one of the reasons of the success of DFT-based methods for the simulation of such systems, the latter scaling cubically (or less if the linear scaling methods mentioned above are employed). As there is no free lunch, this favourable scaling is obtained at the price of approximations in the model itself. Such approximations are not so rigorously founded, despite huge ongoing efforts, and their impact on the

---

[6] And in fact this statement will be further weakened in the next paragraph.

[7] The statement can be somewhat different for the plane waves basis set, for which the matrix is far larger than the number of eigenvectors needed (say 10 to 100 times larger).

[8] and in fact no diagonalization at all, as will be made clear in Section 3.7.

final result is uneasy to measure. This shows the urgent need for further theoretical contributions in the field.

3.2. Basis sets

We now consider the question of the determination of the $\{\chi_\mu\}$, *i.e.*, that of building an efficient finite-dimensional approximation of the space of wavefunctions to be considered for the determination of the electronic structure of the molecular system under study.

It is natural, in order to figure out the difficulty of the approximation, to look at the qualitative properties that are expected for the wavefunctions or the electronic density. For this purpose, an illuminating step is to consider the hydrogen-like atom, whose Hamiltonian reads

$$H_Z = -\frac{1}{2}\Delta - \frac{Z}{|x|}. \tag{3.18}$$

It is simple to see that the positive ground state of this Hamiltonian is

$$\psi_1^Z(x) = \frac{Z^{3/2}}{\sqrt{\pi}}e^{-Z|x|}, \tag{3.19}$$

and from this, two crucial observations stem. First, the electronic density of the molecular system is expected to have *cusps* at each nucleus of the molecule, *i.e.*, singularities in the first derivatives, as is the case for (3.19). Second, the density is expected to decay exponentially fast at large distance.[9] From these two observations, it can be anticipated that general-purpose basis sets, such as finite elements, will not be very well adapted to the problem. This guess is confirmed by numerics. The singularity around each nucleus typically asks for an extensive refinement of the mesh around these points and the dimension of the discrete variational space correspondingly increases, rendering the approach inefficient.[10] Likewise, large distance effects are difficult to reproduce within such methods. This pleads for dedicated (*i.e.*, *problem-dependent*) basis sets, in the spirit of the *component mode synthesis* or the *reduced basis methods*, advocated by A. Patera, Y. Maday and collaborators in various domains of engineering sciences. We refer to Almroth, Stern and Brogan (1978) and Noor and Peters (1980) for pioneering works some decades ago, and Nguyen, Veroy and Patera (2005)

---

[9] The properties of regularity and decay at infinity of the density of the solution to the original Schrödinger equation (2.5) have been studied in Fournais, Hoffmann-Ostenhof, Hoffmann-Ostenhof and Sorensen (2002a, 2002b, 2004) and Hoffmann-Ostenhof, Hoffmann-Ostenhof and Sorensen (2001).

[10] Many electronic structure calculations for large systems, and in particular periodic ones, are done with the help of pseudopotentials that, in addition to other purposes, aim to smear out the singularity at each nucleus; then the above discussion no longer holds and basis sets such as plane waves are tractable: see Section 4.1.

for an up-to-date survey and more references. We also refer to Section 5 for a work in progress in the present context. Actually, an ancestor of such methods was introduced as early as the 1930s in computational chemistry. As the molecular system consists of an assembly of atoms, the natural idea then arose to choose as finite-dimensional space for the approximation the vectorial space generated by some particular functions related to the problem, here a finite number of atomic orbitals (AO), *i.e.*, basis functions used to solve the *same* problem but in the *atomic* or hydrogen-like case (only one nucleus). We speak of an *LCAO approximation*, the acronym LCAO standing for *linear combination of atomic orbitals.*

In so doing, we expect that the size of the variational space will be kept reasonably small, contrary to general-purpose basis sets. Atomic orbital basis sets are thus built by associating to each atom A of the molecule a collection $\{\xi_\mu^A\}_{1 \leq \mu \leq n_A}$ of linearly independent functions of $H^1(\mathbb{R}^3)$, and then by collecting all the ξ_μ^A for the different atoms of which the system is composed.

At this stage, we are left with defining a good basis set for each atom of the molecule. Again, it is useful to consider in detail the simple case of the *hydrogen-like* ion, that is, a system modelled by the Hamiltonian (3.18) consisting of a single electron bound by a single nucleus with charge Z. This system indeed serves as a paradigm for the computations of more complicated molecular systems.[11]

Standard general results of spectral theory and explicit calculations give us a very detailed description of not only the ground state (3.19) but also all of the eigenstates of the operator (3.18) that are the functions

$$\psi_{nlm}^Z(r, \theta, \phi) = Q_{nl}(Zr)\mathrm{e}^{-Zr/n}Y_l^m(\theta, \phi), \qquad (3.20)$$

for $n \in \mathbb{N}$, $0 \leq l \leq n-1$, $-l \leq m \leq l$, where Q_{nl} denotes a polynomial defined by an induction formula, and where the functions $Y_l^m(\theta, \phi)$ denote the spherical harmonics, in turn defined by the first m derivatives of the Legendre polynomial P_l. Such functions, called *hydrogen-like orbitals* when used to form a basis set for a molecular calculation, provide high accuracy results but through tedious computations. Therefore the idea very early arose to modify them slightly, giving birth to *Slater-type orbitals* (STOs): in (3.20), the polynomial Q_{nl} is replaced by the monomial r^l. These orbitals were introduced in Slater (1930) and widely used in the early days of quantum chemistry. They were in turn superseded by another type of

---

[11] An interesting point to make at this stage is that, since they are primarily based upon a *one-electron* model, the basis sets we will now construct have no reason to be able to adequately represent the *electronic* cusp, *i.e.*, the singularity in the *multielectronic* wavefunction due to the $\frac{1}{|x_i - x_j|}$ singular term in the Hamiltonian. We will return to this in the next section.

basis functions. The reason is the overwhelming computational cost of the bielectronic integrals (3.3)

$$(\mu\nu|\kappa\lambda) = \int_{\mathbb{R}^3} \int_{\mathbb{R}^3} \frac{\chi_\mu(x)\chi_\nu(x)\chi_\kappa(x')\chi_\lambda(x')}{|x - x'|} \, \mathrm{d}x \, \mathrm{d}x'.$$

As there are N_b^4 such integrals, this calculation is a bottleneck for the whole computation. Without any further simplification, it is hopeless to calculate each of these N_b^4 integrals by an integration scheme over $\mathbb{R}^3 \times \mathbb{R}^3$. The computation time required would be prohibitive.

The groundbreaking idea by Boys (1950) that has suddenly changed the whole landscape of quantum chemistry, was to replace, in the role of basis functions, STOs by *Gaussian-type orbitals* (GTOs), which are Gaussian functions or successive derivatives of Gaussian functions:

$$\xi(x, y, z) = C \, x^{n_x} y^{n_y} z^{n_z} \mathrm{e}^{-\alpha r^2}. \tag{3.21}$$

The crucial advantage in considering such functions is that the calculation of the overlap matrix (3.1), of the core Hamiltonian matrix (3.2), and, above all, of the bielectronic integrals (3.3) can then be greatly simplified. Indeed, because of some specific properties of Gaussian functions, the computations of the six-dimensional integrals (3.3) are brought down to the numerical computations of one-dimensional integrals of the form $F(w) = \int_0^1 \mathrm{e}^{-w s^2} \, \mathrm{d}s$.

On the other hand, simply using GTOs would not allow for a correct description of the shape of the molecular orbitals both near the nuclei and at infinity, unless a large number of orbitals are employed, which is not desired. The current state of the art of the LCAO approximation is thus to use basis sets made of *contracted Gaussian functions*, which are linear combinations of primitive Gaussian functions,

$$\xi(x, y, z) = \sum_{k=1}^{K} C_k \, x^{n_x^k} y^{n_y^k} z^{n_z^k} \mathrm{e}^{-\alpha_k r^2}, \tag{3.22}$$

in which the C_k are optimized once and for all in order to accurately represent the cusps and the fall-off at infinity. These functions allow both for easy calculation of the bielectronic integrals and for a correct description of the qualitative properties of the wavefunctions. Somehow they constitute the best compromise between STOs and GTOs.

Let us mention for completeness that an alternative to the use of contracted Gaussian functions is that of *fully numerical atomic orbitals*, *i.e.*, basis functions that are solely defined by their numerical values on a grid. They are in general compactly supported in balls centred at the nuclei and whose radii do not exceed a few atomic units. Both the Hamiltonian and the overlap matrix are thus sparse, which is a clear advantage with a view

to designing algorithmic approaches of low complexity. However, the use of
numerical orbitals makes the computation of integrals of the type

$$\int\int \left(\rho_1 \star \frac{1}{|x|} \right) \rho_2$$

more time-consuming, for they must be carried out by solving a Poisson
equation on a large domain, with *ad hoc* boundary conditions, which are
not easy to define, and possibly singular functions ρ_1 in the right-hand side
(typically think of ρ_1 being a Dirac mass at each nucleus).

However they are derived, atomic orbital basis sets are used in most of the
gas or liquid phase calculations, for the determination of the static electronic
structure of the molecular system under study. With a surprisingly low
number of basis functions, typically a few times the number of electrons
(say 2–10), it is then possible to obtain very accurate results.

In some situations, however, the LCAO approximation is not adequate,
or at least causes significant problems. Indeed, a drawback of the LCAO
approximation lies precisely in the fact that the basis *depends* on the system
and is in some sense bound to it. When studying the evolution of a molecular
system where the nuclei move, or when studying the interaction of two
systems (or also a system embedded in a condensed phase), we are obliged
to modify the basis set, either by translating the basis functions according
to the motion of the nuclei, or respectively by adding new basis functions to
account for the presence of more than one single system (the latter situation
gives rise to *basis set superposition error*). In either of these situations, more
intrinsic basis sets are preferred. The most commonly used example is that
of *plane waves*. Then a much larger number of basis functions is needed,
but, as those are fixed, this can still prove to be more efficient than the
LCAO in the very particular settings mentioned above. We will come back
to them in Section 4.1.

3.3. Evaluation of the quality of the basis set

Let us recall some basics. When a space X_δ of finite dimension is fixed,
with a view to approximating the space X, the error between the exact
solution Φ_0 of the problem set in infinite dimension and the solution Φ_δ
found numerically can be split into two components. One component of
this error comes from the fact that the best approximation we can get is
not Φ_0, but the function $\pi_\delta \Phi_0$ in X_δ that is as close as possible to Φ_0. The
second component is due to the fact that the solution procedure will only
provide an approximation Φ_δ of $\pi_\delta \Phi_0$ itself. Therefore we can (formally)
estimate the global error as follows:

$$\|\Phi_0 - \Phi_\delta\| \leq \|\Phi_0 - \pi_\delta \Phi_0\| + \|\pi_\delta \Phi_0 - \Phi_\delta\|. \qquad (3.23)$$

Clearly the first component of the error only depends on the quality of the approximation space X_δ, that is to say, of the basis set of X_δ, while the second component depends on the quality of the solution procedure itself. The latter is said to be optimal if the ratio of the error between the exact solution and the computed solution (second component) by the error between the exact solution and the closest element in the discrete space (first component) does not depend on the size of the basis set.

In the chemistry literature, the theoretical studies on the quality of the basis set are numerous and have all concentrated on the first component of the error, with a view not only to establishing the asymptotic convergence but also to evaluating the rate of convergence of the best approximation $\pi_\delta \Phi_0$ in the given basis to Φ_0, with respect to the size of the basis. Further, in practice, the ratio between the convergence and the complexity of the computations is of greater interest than the rate of convergence itself. Therefore a large body of the chemistry literature has been developed along these lines, particularly in order to enrich the basis sets and improve the convergence rate.

For evaluating the quality of an atomic orbitals basis set, there is no general approach, in contrast to the situation with finite elements or spectral methods such as plane wave basis sets. The choice of an AO basis for solving a given problem mostly relies upon some practical know-how. The lack of rigorous understanding is a pity, because the output of the calculations (typically some molecular property) might be very sensitive to the choice of the basis set. The only available measures of the quality of the basis set are obtained, in the chemistry literature, by choosing test cases, *i.e.*, reference systems, where the solution of the exact Schrödinger equation may be computed, mostly through numerical computations and, when possible, with the help of an analytic calculation.

In the hydrogen-like atom, the system consists of a single electron, and the problem amounts to finding the first eigenvalue and eigenfunction of the Hamiltonian. Klahn and Bingel (1977) were the first in the chemistry literature to investigate the conditions for the convergence to the exact eigenvalue. They established that the basis sets used conventionally[12] were complete for the H^1 topology, thus yielding the asymptotic correctness of the results when the size of the basis goes to infinity. As mentioned above, the critical aspect, as far as the convergence to the exact wavefunction is concerned, is the *representation of the nuclear cusp*. In the chemistry literature, it was shown heuristically by Hill (1985) that a rapid convergence is only possible if the basis set describes correctly the *singularities* of the function to be expanded, a fact that can of course be understood and described on the basis of rigorous, and simple, mathematical arguments.

---

[12] See, however, the comments at the end of the section.

Regarding the convergence rate, Klahn and Morgan, III (1984) have studied the convergence of expansions of the ground state of the H atom in a simplified basis set consisting of Gaussian orbitals, and found that the error of the energy goes as $\sim d^{-3/2}$ if d is the dimension of the basis, which is a very slow convergence. Other studies then aimed at further studying, theoretically (Klopper and Kutzelnigg 1986, Hill 1995) or experimentally (Schmidt and Ruedenberg 1979, Feller and Ruedenberg 1979), this rate of convergence and then improving it by introducing new basis functions.

One important point is that for practical purposes there are many ways to measure the efficiency of a given basis set, depending on the output chosen: *e.g.*, the distance of the expansion to the exact function, the error of the density at the position of the nucleus, the error of the energy value, *etc.* In this direction we cite the series of works by Kutzelnigg (1989, 1994).

In the case of many-electron systems, the question of evaluating the best approximation of the exact ground state wavefunction requires understanding an effect that has been omitted so far in this survey, namely the effect on the convergence of basis expansions of the *correlation cusp*, created by the $\frac{1}{|x_i - x_j|}$ interaction term. Schwartz (1962) was the first to study the rate of convergence of the expansion of correlated wave functions in a one-electron basis. He considered the helium ground state, treating the electron interaction as a perturbation of the one-electron case. More recently Klahn and Morgan, III (1984) and Hill (1985) studied the rate of convergence of variational calculations in a general setting (for a review of this problem see Morgan, III (1984)) while Kutzelnigg and Morgan, III (1992) presented a detailed study of the solution of the Schrödinger equation near $r_{12} = 0$.

It is interesting to note that, with a view to circumventing the difficulty of representing many-electron wavefunctions in terms of one-electron functions, which inevitably lowers the rate of convergence of the expansion, the idea arose to use variational trial wave functions that depend explicitly on the interelectronic distances $|x_i - x_j|$, and allow us to describe the correlation cusp correctly. Hylleraas (1929) initiated the approach with an accurate calculation of the ground state energy of He-like ions using some function $\Psi(x_1, x_2, |x_1 - x_2|)$ with only a small number of parameters. Further works are those of Kinoshita (1957), Pekeris (1958), James and Coolidge (1933), Kolos and Rychlewski (1993), for instance. More recently some progress has been achieved for 3- and 4-electron atoms by Kleindienst and collaborators: see, *e.g.*, Luchow and Kleindienst (1994). The feasibility of the approach for a higher number of electrons is an open issue. Note that, of course, for systems of more than 3 electrons, the coalescence of three particles has to be studied. This is done by the so-called *Fock expansion*: see Fock (1958) and, *e.g.*, Peterson, Wilson, Woon and Dunning, Jr. (1997). It is important to note that these studies certainly need to be complemented on the rigorous mathematical side.

As a conclusion to this section, we would like to point out that atomic orbital basis sets in their contemporary implementation remain the basis sets of choice for electronic structure calculations of small systems. They give an impressive accuracy for a surprisingly small size of the basis. Therefore we believe that the field would definitely benefit from further mathematical studies. One reason is that there is room for improvement in many of the proofs mentioned above, in particular because they sometimes do not apply to the basis sets actually used in practice but rather to an idealized version of them (for instance, exponents in the Gaussian functions are allowed to vary arbitrarily, or contracted Gaussian functions are not addressed). A second reason for this is precisely the incredible efficiency of the atomic orbital basis sets: convergence is obtained long before the asymptotic regime. To some extent, the asymptotic analysis described above is indeed useful for the sake of rigour and as a first step, but not sufficient for shedding light on the practice.

3.4. Convergence analysis

Let us consider now $\pi_\delta \Phi_0$ as being the best fit of Φ_0 by elements of the discrete space X_δ. The question is now to evaluate $\|\pi_\delta \Phi_0 - \Phi_\delta\|$, i.e., the second error term in (3.23). No study in the chemistry literature deals with this question. In the mathematical literature, the question has been addressed by a series of works by Y. Maday and G. Turinici. The following lines are based upon their work.

Above all, some particular precautions have to be taken before evaluating this norm. Owing to the invariance of the HF energy with respect to orthogonal (or unitary) transforms, the error between any two Slater determinants Ψ_1 and Ψ_2 cannot be evaluated naïvely. We need to introduce distances of the type

$$\|\Psi_1 - \Psi_2\| = \inf\{\|U\Psi_1 - \Psi_2\|_{[L^2(\mathbb{R}^3)]^N} : U \in \mathcal{U}(N)\},$$

where $\mathcal{U}(N)$ represents the set of all unitary $N \times N$ matrices. In the same spirit, Maday and Turinici (2003) have introduced, for any $\Phi \in [H^1(\mathbb{R}^3)]^N$, the decomposition

$$[H^1(\mathbb{R}^3)]^N = \mathcal{A}_\Phi \oplus \mathcal{S}_\Phi \oplus \Phi^{\perp\perp},$$

where

$$\mathcal{A}_\Phi = \{C\Phi : \ C \in \mathbb{R}^{N \times N}, C^\star = -C\},$$

$$\mathcal{S}_\Phi = \{S\Phi : \ S \in \mathbb{R}^{N \times N}, S^\star = S\},$$

$$\Phi^{\perp\perp} = \{\Psi = (\psi_i)_{i=1}^N \in [H^1(\mathbb{R}^3)]^N : \ \langle \psi_i, \phi_j \rangle = 0; i, j = 1, \ldots, N\}.$$

Applying this to $\Psi - \Phi$, with Φ and Ψ in $[H^1(\mathbb{R}^3)]^N$ satisfying $\langle \phi_i, \phi_j \rangle = \delta_{ij}$,

$\langle \psi_i, \psi_j \rangle = \delta_{ij}$, we thus have the decomposition

$$\Psi = \Phi + C\Phi + S\Phi + W, \quad W \in \Phi^{\perp\perp}. \tag{3.24}$$

Generally, it can be established that, up to an adequate orthogonal transform, C may be set to zero, and that the symmetric part $S\Phi$ is not the main part of the decomposition (when $\Psi - \Phi$ is presumably small) as there exist constants C_1, C_2 depending only on N such that

$$\|S\Phi\|_{[L^2(\mathbb{R}^3)]^N} \leq C_1 \|\Psi - \Phi\|^2_{L^2(\mathbb{R}^3))^N}, \tag{3.25}$$

$$\|S\Phi\|_{[H^1(\mathbb{R}^3)]^N} \leq C_2 \|\Psi - \Phi\|^2_{\mathcal{H}} \|\Phi\|_{[H^1(\mathbb{R}^3)]^N}. \tag{3.26}$$

With the above preparatory work, it is possible to show that in a sufficiently small neighbourhood of Φ_0 (or $\pi_\delta \Phi_0$), there exists a discrete solution Φ_δ of the HF equation (unique in some weakened sense) and such that the error between Φ_0 and Φ_δ is of the same order as $\Phi_0 - \pi_\delta \Phi_0$. Again, it should be emphasized that all these precautions originate from the fact that there is no uniqueness known on the solution to the HF equations, and thus the usual error estimates established in other contexts need to be adapted.

To prove this claim, the following energy functional parametrized by any $\Phi \in [H^1(\mathbb{R}^3)]^N \cap \mathcal{K}$, is introduced:

$$\mathcal{E}^\Phi(\psi_1, \ldots, \psi_N) = E^{HF}(\psi_1, \ldots, \psi_N) + \sum_{i,j=1}^{N} \langle F_\Phi \phi_i, \phi_j \rangle (\langle \psi_i, \psi_j \rangle - \delta_{ij}), \tag{3.27}$$

where F_Φ denotes the Fock operator. Then, for an arbitrary Ψ decomposed in $\Psi = \pi_\delta(\Phi_0) + S\pi_\delta(\Phi_0) + W$ in view of (3.24), we compute

$$E^{HF}(\Psi) - E^{HF}(\pi_\delta(\Phi_0)) = \frac{1}{2} D^2 \mathcal{E}^{\Phi_0}(W - \Phi_0 + \pi_\delta(\Phi_0), W - \Phi_0 + \pi_\delta(\Phi_0))$$

$$- \frac{1}{2} D^2 \mathcal{E}^{\Phi_0}(\Phi_0 - \pi_\delta(\Phi_0), \Phi_0 - \pi_\delta(\Phi_0))$$

$$+ O(\|W\|^3 + \|\Phi_0 - \pi_\delta(\Phi_0)\|^3).$$

It suffices then to minimize this quantity with respect to W such that Ψ remains in a small neighbourhood of $\pi_\delta(\Phi_0)$, and this yields the correct Φ_δ. It can be shown that such a minimizer is unique (due to the coercivity of the Hessian $D^2 \mathcal{E}^{\Phi_0}$), and that

$$\|W\| \leq c \|\Phi_0 - \pi_\delta \Phi_0\|,$$

thus proving the above claim on the order of the error.

At this stage, we pass from *a priori* considerations (the discrete solution exists and if the discrete space is large enough we obtain an accurate result (even optimal)) to *a posteriori* considerations: in a final stage where one approximate solution has been computed, the need arises to *validate* the result. *A posteriori* analysis and, more precisely, the definition of explicit lower and upper bounds for outputs was introduced in Maday, Patera and Peraire (1999), and first analysed in Maday and Patera (2000).

Let us consider an approximation Φ_δ such that $\|\Phi_0 - \Phi_\delta\|_{[H^1(\mathbb{R}^3)]^N} \leq \varepsilon$, and perform the decomposition $\Phi_0 - \Phi_\delta = S\Phi_\delta + W$, where W belongs to $\Phi_\delta^{\perp\perp}$, $\|W\|_{[H^1(\mathbb{R}^3)]^N} \leq C\varepsilon$ and $\|S\Phi_\delta\|_{[H^1(\mathbb{R}^3)]^N} \leq C\varepsilon^2$. The evaluation of the quality of the approximation is based upon the introduction of the following problem:[13] finding the *reconstructed error* $\hat{W} \in \Phi_\delta^{\perp\perp}$ such that

$$D^2\mathcal{E}^{\Phi_\delta}(\hat{W}, \Psi) + D\mathcal{E}^{\Phi_\delta}(\Psi) = 0, \quad \text{for all } \Psi \in \Phi_\delta^{\perp\perp}, \qquad (3.28)$$

a problem that has, owing to some coercivity of $D^2\mathcal{E}^{\Phi_\delta}$, a unique solution. Then,

$$E^{HF}(\Phi_0) = E^{HF}(\Phi_\delta) - \frac{1}{2}D^2\mathcal{E}^{\Phi_\delta}(\hat{W}, \hat{W}) + \frac{1}{2}D^2\mathcal{E}^{\Phi_\delta}(W - \hat{W}, W - \hat{W})$$
$$+ O(\varepsilon^3)$$

For sufficiently small ε, this yields

$$E^{HF}(\Phi_\delta) \geq E^{HF}(\Phi_0) \geq \mathcal{E}^{HF}(\Phi_\delta) - D^2\mathcal{E}^{\Phi_\delta}(\hat{W}, \hat{W}), \qquad (3.29)$$

the left-hand side holding true because a variational approximation always provides a discrete minimum that is larger than the global one. This provides an explicit upper and lower bound on the Hartree–Fock energy. This bound is effective in the sense that it has been proved in Maday and Turinici (2003) that there exists a constant such that

$$\|\hat{W}\|_{[H^1(\mathbb{R}^3)]^N} \leq c\|W\|_{[H^1(\mathbb{R}^3)]^N}, \qquad (3.30)$$

so that the width of the bound is small and of the same order as

$$\|E^{HF}(\Phi_\delta) - E^{HF}(\Phi_0)\|_{[H^1(\mathbb{R}^3)]^N}.$$

The estimate (3.29) therefore provides on the solution Φ_δ found numerically an *a posteriori* estimate, both rigorous mathematically and tractable in practice. The estimate can serve as an evaluation of the quality of the basis set employed, and, possibly, indicates the need for an enlargement of this basis set. The procedure was implemented and successfully tested some years ago in an academic code for electronic structure calculations. It seems, however, not to be used today in any widely distributed software in the field.

A corollary of the above technique is the following: $\hat{W}$ can actually be shown to be very close to the actual (main part of) the error W, so that an improvement on the solution Φ_δ can be proposed by setting

$$\tilde{\Phi}_\delta = \Phi_\delta + \hat{S}\Phi_\delta + \hat{W}, \qquad (3.31)$$

---

[13] Note that this computation involves a direct problem and not an eigenvalue problem.

where $\hat{S}\Phi_\delta \in S_{\Phi_\delta}$ and $\|\hat{S}\Phi_\delta\| = O(\varepsilon^2)$. This justifies the name *reconstructed error* for $\hat{W}$. Correspondingly, a new evaluation $E^{HF}(\tilde{\Phi}_\delta)$ of the energy can be proposed. Actually, (3.31) can be re-interpreted as follows. It yields the same improvement $\tilde{\Phi}_\delta$ of Φ_δ as that obtained by performing *one* step of a Newton algorithm on $E^{HF}(e^C \Phi_\delta)$ where C is a matrix subject to some constraints.

3.5. SCF cycles

We now concentrate on the strategy to solve the discretized Hartree–Fock equations (3.8). This solution procedure provides the approximation Φ_δ in (3.23). There lies the third type of numerical analysis involved, after that of Section 3.2 and that of Section 3.3: the speed of convergence of the algorithm toward the solution Φ_δ needs to be evaluated.

The first class of algorithms we report on is that of *self-consistent field* (SCF) algorithms, *i.e.*, iterations of the form

$$\begin{cases} \tilde{F}_k C_{k+1} &= S C_{k+1} E_{k+1}, \qquad E_{k+1} = \text{Diag}(\epsilon_1^{k+1}, \ldots, \epsilon_N^{k+1}) \\ C_{k+1}^* S C_{k+1} &= I_N \\ D_{k+1} &= C_{k+1} C_{k+1}^*. \end{cases} \tag{3.32}$$

Here, $\epsilon_1^{k+1} \leq \epsilon_2^{k+1} \leq \cdots \leq \epsilon_N^{k+1}$ are the smallest N eigenvalues of the linear generalized eigenvalue problem

$$\tilde{F}_k \phi = \epsilon S \phi,$$

and C_{k+1} contains the corresponding N orthonormal eigenvectors. The expression of the current Fock matrix $\tilde{F}_k$ characterizes the algorithm. We have, for instance,

$$\tilde{F}_k = F(D_k)$$

for the simplest algorithm we shall see, but more sophisticated forms will be examined. In spirit, these algorithms are more or less fixed-point iterations. The hope is that C_k, D_k and $F(D_k)$ converge, respectively to C, D and $F(D)$, so that we get from (3.32) a solution to (3.8) in the limit $k \longrightarrow +\infty$.

For years, no mathematical analysis was available for the SCF algorithms, however much used in practice. In the chemistry literature, convergence successes and failures were reported, comparisons of rates of convergence between algorithms were experimentally established, remedies and tricks were given – all without any rigorous understanding. Examples of such contributions are Schlegel and McDouall (1991), Seeger and Pople (1976), Stanton (1981*a*), Starikov (1993), Zerner and Hehenberger (1979), Koutecký and Bonacic (1971), Douady, Ellinger, Subra and Levy (1980), Natiello and

Scuseria (1984), Fischer and Almlöf (1992) and Chaban, Schmidt and Gordon (1997). The first mathematical work appeared some years ago and is due to Auchmuty and Wenyao Jia (1994), who studied the convergence of a prototypical algorithm which is unfortunately not used in practice. The situation recently evolved, both from the standpoint of numerical analysis and from that of the construction of more efficient strategies, with the series of works by Cancès and Le Bris (2000a, 2000b), Cancès (2000, 2001), Kudin, Scuseria and Cancès (2002) and Cancès, Kudin, Scuseria and Turinici (2003a), on which the following lines are based.

Before describing the algorithms and elaborating on their theoretical properties, we need to make a few remarks.

First, the SCF algorithm (3.32) above needs to be well defined. This requires there to be no ambiguity on the choice of C_{k+1} and thus the Nth eigenvalue, counted with multiplicity, should be nondegenerate: $\epsilon_N^{k+1} < \epsilon_{N+1}^{k+1}$ (which is true in the limit $k \to \infty$, in view of a theoretical result mentioned above). Actually, in order to be able to prove convergence, the following slightly stronger property was introduced. An SCF algorithm of the form (3.32) with initial guess D_0 is said to be *uniformly well posed* (UWP) if there exists some positive constant γ such that

$$\text{for all } k \in \mathbb{N}, \qquad \epsilon_{N+1}^{k+1} \geq \epsilon_N^{k+1} + \gamma.$$

This property can be shown to be satisfied automatically at least for one algorithm, namely the level-shifting algorithm we will study below. In practice, it seems to be largely satisfied for the algorithms examined below.

Second, recall that in the present context we look for the minimizer of a nonconvex minimization problem by a solution procedure for the Euler–Lagrange equations. Without uniqueness, the convergence of fixed point-like iterations, and even more the convergence towards a global minimizer, are likely to be impossible to establish. Therefore the notion of convergence has to be weakened, making it more practical, but still interesting, in the present context. In this spirit, an SCF algorithm of the form (3.32) is said to *numerically converge toward a solution to the HF equations* if the sequence (D_k) satisfies

(i) $D_{k+1} - D_k \to 0$,

(ii) $[F(D_k), D_k] \to 0$,

the second condition of course being reminiscent of (3.14). Likewise, we shall say it *numerically converges toward an Aufbau solution to the HF equations* if (i) holds together with a condition stronger than (ii), namely

(iii) $\text{Trace}(F(D_k)D_k) - \inf\{\text{Trace}(F(D_k)D) : D \in \mathcal{P}_N\} \to 0$.

Of course, in both cases, one should note that the convergence of D_k up to an extraction is not an issue, since the set $\mathcal{P}_N$ defined in (3.6) is compact, because of the finite-dimensional setting.[14]

The simplest fixed point algorithm was introduced by Roothaan (1951). It is now obsolete, but it serves as a basis for more sophisticated algorithms, and as an explanatory example for the numerical analysis. It consists in setting $F_k = F(D_k)$ in (3.32).

It was very early realized that the convergence properties of the Roothaan algorithm are not satisfactory: it sometimes converges towards a solution to the HF equations and frequently oscillates between two states, neither of which are solutions to the HF equations. In addition, the behaviour may depend on the basis set chosen. But anyway, surprisingly, no case other than convergence or oscillation of the above type (say binary oscillations) were observed.

This behaviour can be fully explained by introducing the auxiliary function

$$E(D, D') = \text{Trace}(hD) + \text{Trace}(hD') + \text{Trace}(G(D)\, D'),$$

and noting that the sequence of D_k generated by the Roothaan algorithm is exactly that generated by the relaxation algorithm

$$D_{2k+1} = \arg\,\inf\{E(D_{2k}, D), \quad D \in \mathcal{P}_N\},$$
$$D_{2k+2} = \arg\,\inf\{E(D, D_{2k+1}), \quad D \in \mathcal{P}_N\}.$$

The functional E, which decreases at each iteration of the relaxation procedure, can therefore be interpreted as a Lyapunov functional of the Roothaan algorithm. This basic remark is the foundation of the proof of the following result.

Theorem 1. Let $D_0 \in \mathcal{P}_N$ be such that the Roothaan algorithm with initial guess D_0 is UWP. Then the sequence (D_k^{Rth}) generated by the Roothaan algorithm either numerically converges toward an Aufbau solution to the HF equations, or oscillates between two states, none of them being an Aufbau solution to the HF equations. In the latter case, $(D_{2k}^{\text{Rth}}, D_{2k+1}^{\text{Rth}})$ converges to (D, D'), $D \neq D'$, where

$$\begin{cases} F(D')C = SCE \\ F(D)C' = SC'E', \end{cases}$$

together with the other obvious constraints.

---

[14] Actually, we can extend the above definitions, and most of the results and proofs will be given below, to an infinite-dimensional setting. This is useful additional information, particularly when we want to assess the impact of the increase of the basis set on the convergence issues. We refer to the bibliography.

Proof. The proof goes by proving that

$$E(D_{k+1}^{\mathrm{Rth}}, D_{k+2}^{\mathrm{Rth}}) + \frac{\gamma}{2}\|D_{k+2}^{\mathrm{Rth}} - D_k^{\mathrm{Rth}}\|^2 \leq E(D_k^{\mathrm{Rth}}, D_{k+1}^{\mathrm{Rth}}),$$

thus $\sum_{k\in\mathbb{N}} \|D_{k+2}^{\mathrm{Rth}} - D_k^{\mathrm{Rth}}\|^2 < +\infty$, which implies in particular that

$$D_{k+2}^{\mathrm{Rth}} - D_k^{\mathrm{Rth}} \longrightarrow 0.$$

Now, either $D_{k+1}^{\mathrm{Rth}} - D_k^{\mathrm{Rth}}$ converges to zero or it does not. The examination of each case allows us to conclude the alternative stated in the theorem. $\square$

A first attempt to stabilize the Roothaan algorithm is the *level-shifting* algorithm due to Saunders and Hillier (1973). It consists in setting

$$\widetilde{F}_k = F(D_k) - bD_k,$$

in (3.32), where b is some sufficiently large positive constant.

Then oscillations disappear, and the algorithm always converges. However, the level-shift parameters b which guarantee convergence are large, so that convergence is very slow, and often the algorithm converges to critical points which do not satisfy the Aufbau principle and are not even local minimizers.

Again, this algorithm can be re-interpreted in a standard way. In view of the analysis of the Roothaan algorithm, it is natural to introduce a simple penalty functional $b\|D - D'\|^2$, where b is a positive constant and where $\|\cdot\|$ denotes the Hilbert–Schmidt norm, in order to enforce $D = D'$ in the limit and thus to obtain a solution of the Euler–Lagrange equation. The relaxation algorithm associated with the minimization problem

$$\inf\{E^b(D, D'), \quad (D, D') \in \mathcal{P}_N \times \mathcal{P}_N\},$$

where

$$E^b(D, D') = \mathrm{Trace}(hD) + \mathrm{Trace}(hD') + \mathrm{Trace}(G(D)\,D') + b\,\|D - D'\|^2$$

is exactly the level-shifting algorithm with shift parameter b. Then we have the following.

Theorem 2. For sufficiently large b, the level-shifting algorithm is UWP. The energy $E^{HF}(D_k^b)$ of the kth iterate D_k^b decreases toward some stationary value of E^{HF} and the sequence (D_k^b) numerically converges toward a solution to the HF equations.

Proof. The proof follows the same lines as that of Theorem 1, and relies upon the inequality

$$E^{HF}(D_{k+1}^b) + \frac{b}{2}\,\|D_{k+1}^b - D_k^b\|^2 \leq E^{HF}(D_k^b).$$

See Cancès (2000) for the details. $\square$

With a view to both enforcing and accelerating the convergence of the iterations, the *direct inversion in the iterated subspace* (DIIS) algorithm has been introduced by Pulay (1982). It is still commonly used in calculations. The basic idea of the algorithm is to make use of the fact that $[F(D), D] = 0$ is equivalent to the HF equations in order to insert damping into the iterations. This is done by setting

$$\widetilde{F}_k = F(\widetilde{D}_k)$$

in (3.32), where

$$\widetilde{D}_k = \sum_{i=0}^{k} c_i^{\text{opt}} D_i,$$

and

$$\{c_i^{\text{opt}}\} = \arg \inf \left\{ \left\| \sum_{i=0}^{k} c_i [F(D_i), D_i] \right\|^2 : \quad \sum_{i=0}^{k} c_i = 1 \right\}.$$

It turns out that the DIIS algorithm works extremely well: in many cases, it typically converges in a dozen iterations. However, the DIIS algorithm suffers from a qualitative drawback: it is not ensured that the Hartree–Fock energy decreases throughout the iterations. In addition, there exist cases where this algorithm does not converge.

Unfortunately, no numerical analysis on this algorithm is available to date, and thus the convergence failures cannot be satisfactorily explained and remedied. This, at least, has motivated the introduction of other algorithms.

Relaxed constrained algorithms (RCAs) have been introduced in Cancès and Le Bris (2000*b*) and Cancès (2000, 2001). They are based on the following remark: all the local minima of $E^{HF}(D)$ on $\widehat{\mathcal{P}}_N$ defined in (3.17) indeed belong to $\mathcal{P}_N$ defined by (3.6), which amounts to saying that the constraint $DSD = D$ may be relaxed while keeping the same local minima. Loosely speaking, this is simply due to a property of concavity of the HF energy with respect to the norm of each of the ϕ_i (think again of the minimization of a concave functional on the unit ball and on the unit sphere, respectively). Therefore, without loss of generality, we may transform the HF problem into a minimization problem set on the convex set $\widetilde{\mathcal{P}}_N$, for which many more techniques are available (in particular, and this will be the case here, we can damp the algorithm by using any convex combination of previously computed iterates). Convergence is then easier to establish.

We will focus here on the simplest RCA, called the *optimal damping algorithm* (ODA). It consists in setting $\widetilde{F}_k = F(\widetilde{D}_k)$ where

$$\widetilde{D}_k = \arg \inf \left\{ E^{HF}(\widetilde{D}) : \quad \widetilde{D} = (1 - \lambda)\widetilde{D}_{k-1} + \lambda D_k, \quad \lambda \in [0, 1] \right\}$$

As E^{HF} is a second-degree polynomial in the density matrix, the computation of $\widetilde{D}_k$ only consists in minimizing a quadratic function of λ in $[0, 1]$, which can be done analytically.

In fact, one can again understand the ODA on the basis of a very simple numerical analysis. Because of the Taylor expansion

$$E^{HF}((1 - \lambda)\widetilde{D}_{k-1} + \lambda D') = E^{HF}(\widetilde{D}_{k-1}) + \lambda \text{Trace}(F(\widetilde{D}_{k-1}) \cdot (D' - \widetilde{D}_{k-1}))$$

$$+ \frac{\lambda^2}{2} \text{Trace}(G(D' - \widetilde{D}_{k-1}) \cdot (D' - \widetilde{D}_{k-1})),$$

for any $\lambda \in [0, 1]$, the direction D_k selected by the Aufbau principle, namely

$$D_k = \arg \inf\{\text{Trace}(F(\widetilde{D}_{k-1})D') : D' \in \mathcal{P}_N\},$$

can be interpreted as the steepest descent direction, while the choice of the damping parameter λ in the ODA is the optimal step along this direction. Therefore the ODA is a representative of a standard descent algorithm in this context.[15,16] This observation underlies the following theorem.

Theorem 3. Let us consider an initial guess $D_0 \in \mathcal{P}_N$ such that the optimal damping algorithm is UWP. Then,

(1) the sequence $E^{HF}(\widetilde{D}_k)$ of energies of the intermediate matrices $\widetilde{D}_k$ decreases toward a stationary value of the HF energy;

(2) the sequence (D_k) numerically converges toward an Aufbau solution to the HF equations.

Proof. One may show that

$$E^{HF}(\widetilde{D}_{k+1}) \leq E^{HF}(\widetilde{D}_k) - \alpha\|D_{k+1} - \widetilde{D}_k\|^2$$

for some $\alpha > 0$, which implies that

$$D_{k+1} - \widetilde{D}_k \longrightarrow 0. \tag{3.33}$$

As $\widetilde{D}_{k+1} \in [\widetilde{D}_k, D_{k+1}]$, it follows that $\widetilde{D}_{k+1} - \widetilde{D}_k \longrightarrow 0$, and then that $D_{k+1} - D_k \longrightarrow 0$. The proof is then easy to complete. $\square$

---

[15] In fact, pursuing the analogy, the algorithm known in computational chemistry as the *mixing algorithm*, where $\widetilde{F}_k = F(\widetilde{D}_k)$ with $\widetilde{D}_k = (1 - \alpha)\widetilde{D}_{k-1} + \alpha D_k$ and α is a fixed damping parameter, can be recast as a steepest descent procedure with a fixed step, which, naturally, performs rather poorly.

[16] Then the following question arises: Since in other fields of optimization it is well known that the gradient direction is not the best direction to take, why not take a more efficient one? A practical answer to this question is that the big advantage of this direction is that it is 'easy' to calculate, since it is that obtained by diagonalization, and quantum chemistry codes use optimized diagonalization routines.

Numerical tests show that

- the solution obtained by the ODA is always the same whatever the initial guess chosen in the list of commonly used initial guesses (this robustness is a very important property),
- the energy of the solution obtained by the ODA is always lower than or equal to that of the solution obtained by any other method,
- the ODA always converges,
- the ODA is less demanding in terms of memory than, *e.g.*, DIIS,
- however, the ODA, like any RCA available today, does not converge as fast as the DIIS algorithm when the latter does converge, and proves to be rather slow in the latest steps of the convergence.

The latter observation motivated the introduction by Kudin *et al.* (2002) of the *energy direct inversion in the iterative subspace* (EDIIS) algorithm as an improvement of the ODA for the latest steps. For the damping step, the HF energy is, in the spirit of DIIS, minimized on the convex set generated by all (or some of) the density matrices computed at the previous iterations:

$$\widetilde{D}_k = \arg\inf\left\{ E^{HF}(\widetilde{D}) : \quad \widetilde{D} = \sum_{i=0}^{k} c_i D_i, \quad 0 \le c_i \le 1, \quad \sum_{i=0}^{k} c_i = 1 \right\}.$$

Since this is exactly the HF energy which is minimized, and not

$$\left\| \sum_{i=0}^{k} c_i [F(D_i), D_i] \right\|^2,$$

the damping step does force convergence.

Let us mention that, for the KS problem, the same algorithms (Roothaan, level-shifting, DIIS, RCA, ODA, EDIIS) can be applied. The main two differences are, first, that there is no proof of convergence, and, second, that relaxing the constraints $DSD = D$ in the KS model modifies the model itself and leads to the *extended* Kohn–Sham model. We refer the reader to Cancès (2001).

3.6. Second-order methods

The SCF iterations are basically first-order methods, as shown by their interpretation given above. In order to accelerate their convergence in the latest steps, one can insert damping, as in the DIIS, ODA, or EDIIS algorithms, or, and this is the purpose of the present section, one may resort to second-order algorithms.

The first Newton-like algorithm for computing HF ground states is due to Bacskay (1961). The basic idea is to make a change of variable in order to remove the constraints and use a standard Newton algorithm for *unconstrained* optimization. The convenient parametrization of the manifold $\mathcal{P}_N$ used by Bacskay is the following: for any $C \in \mathcal{M}(N_b, N_b)$ such that $C^* S C = I_{N_b}$,

$$\mathcal{P}_N = \left\{ C \exp(A) D_0 \exp(-A) C^* : \quad D_0 = \begin{bmatrix} I_N & 0 \\ 0 & 0 \end{bmatrix}, \right.$$

$$\left. A = \begin{bmatrix} 0 & -A_{vo}^* \\ A_{vo} & 0 \end{bmatrix}, \quad A_{vo} \in \mathcal{M}(N_b - N, N) \right\},$$

where, in the language of chemistry, the subscript vo denotes the 'virtually occupied' off-diagonal block of the matrix A. Let us now write

$$E^C(A_{vo}) = E^{HF}(C \exp(A) D_0 \exp(-A) C^*).$$

The problem now reads as the minimization of $E^C(A_{vo})$. Starting from some reference matrix C, the *Bacskay QC algorithm* (QC standing for *quadratically convergent*) consists in applying to this unconstrained minimization problem, *one* Newton step starting from $A_{vo} = 0$, and next to update C. It thus reads:

$$\begin{cases} \text{compute the solution } A_{vo}^k \\ \text{of the Newton equation } \nabla^2 E^{C_k}(0) \cdot A_{vo} + \nabla E^{C_k}(0) = 0, \\ \text{set } C_{k+1} = C_k \exp(A_k) \text{ with } A_k = \begin{bmatrix} 0 & -A_{vo}^{k*} \\ A_{vo}^k & 0 \end{bmatrix}. \end{cases}$$

A natural alternative to Bacskay QC is to use a Newton-like algorithm for *constrained* optimization. We write down the optimality equations for problem (3.8) and then solve them by Newton iterations. Unfortunately, owing to the unitary invariance of the HF energy, the system of equations obtained is not well posed, and some technical modifications are in order. This gives rise to a variety of Newton-type algorithms. We refer to the literature and in particular to Shepard (1993).

The computational costs of the various Newton-type algorithms are particularly high in the present context. Indeed, the construction of many Fock matrices per step is needed, and we have noticed that this construction is especially costly. In order to lower the computational cost, various attempts have been made to build quasi-Newton versions of the Bacskay QC algorithm (see, for instance, Fischer and Almlöf (1992), Chaban *et al.* (1997)), but we are not aware of any work on quasi-Newton methods for solving the constrained optimization problem (3.4).

To conclude this section, we would like to mention that one of the most recent and efficient combinations of a first-order algorithm in the earliest steps of SCF iterations with a second-order algorithm in the latest step is that proposed by Cancès *et al.* (2003a). The setting is that of a KS-type model: the initial value D_0 of the second-order algorithm is the density matrix output of the EDIIS algorithm. In the spirit of the Bacskay QC algorithm, the manifold of density matrices is parametrized by

$$D = \Omega \begin{pmatrix} I_N & 0 & 0 \\ 0 & \Lambda & 0 \\ 0 & 0 & 0 \end{pmatrix} \Omega^*$$

with some matrices Ω and Λ that are updated at each iteration through $\Omega_{k+1} = \Omega_k \exp A_k$ and $\Lambda_{k+1} = \Lambda_k + M_k$. The matrix A_k, of particular form, and the matrix Ω_k are determined by performing one step of the Newton algorithm for the minimization of

$$E^{KS} \left(\Omega_k \exp A \begin{pmatrix} I_N & 0 & 0 \\ 0 & \Lambda_k + M & 0 \\ 0 & 0 & 0 \end{pmatrix} \exp(-A) \Omega_k^* \right).$$

We refer to Cancès et al. (2003a) for the details. More generally, we also refer to Areshkin, Shenderova, Schall and Brenner (2003) for a recent survey of SCF methods and techniques for their acceleration.

3.7. Diagonalization procedure: small and large size systems

In the previous section, we focused on the SCF cycles, i.e., the iterations on the nonlinearity. At each cycle, the current mean-field Hamiltonian $\widetilde{F}_k$ is used to build a new density matrix on the basis of the Aufbau principle. This in principle amounts to solving the minimization problem (3.15), that is,

$$\inf\{\operatorname{Trace}(FD), \quad DSD = D, \quad \operatorname{Trace}(SD) = N\}, \tag{3.34}$$

where F is frozen at the value $\widetilde{F}_k$. It has already been said that, typically, for atomic orbitals basis sets, N_b is of the order of $2N$ to $10N$ (the matrix F then being sparse owing to the localization of atomic orbitals), while for plane wave basis sets, which will be mentioned in Section 4.1, N_b can be one hundred times as large as N, and the matrix is dense.[17]

The direct approach to solving these problems is to diagonalize F. The algorithms in use are standard algorithms, no specificity of computational chemistry arising at this level. Nevertheless, this procedure has complexity N^3 (see, e.g., Demmel (1997)), and cannot be applied to systems of large size. The limitation is all the more of concern as the diagonalization is the inner loop of the SCF procedure, which may itself be one step of an outer

---

[17] Notice, therefore, that when we speak of linear scaling algorithms below, it might be quite different *at the practical level* to consider an algorithmic complexity w.r.t. N or w.r.t. N_b in the case of plane waves basis sets, even if *in the asymptotic regime* the two complexities are the same.

loop. This is the case for calculations in the solid phase (when there are as
many equations as points in the reciprocal lattice – see Section 4.1), or for
geometry optimization routines (when the problem is parametrized by the
positions of nuclei – see Section 3.8), or for time-dependent simulations by
ab initio molecular dynamics (see Section 5.1).

The paradigm of *linear scaling* calculations has therefore arisen in the
past decade, with a view to designing procedures that would scale linearly
with respect to the size of the system. Linear scaling methods are founded
on the following simple remark: in fact, the solution of (3.34) requires the
projector onto the subspace generated by the eigenstates associated to the
lowest N eigenvalues, and not the eigenstates themselves. Diagonalization
can thus be avoided in principle, hence the name *alternative to diagonalization* for such methods, which can significantly reduce the algorithmic
complexity, basically from N^3 to N, at least in some cases. From the physical standpoint, some assumptions justify linear scaling methods, the most
important being *locality of interactions*: two regions of a large molecular
system that are very far away from one another only slightly interact.

The linear scaling methods can be schematically divided into three categories:

- decomposition methods,
- penalization approaches,
- non-variational approaches.

The decomposition methods rely on the *divide and conquer* paradigm.
Loosely speaking, the idea is to partition the molecular system into subsystems, and solve iteratively the subproblems by fine schemes in parallel, and
the global problem by a coarse solver. Surprisingly, it seems that there is
no mature version of such methods in the context of computational chemistry as there is in other fields of scientific computing. Note that we have
in mind decomposition domain methods at the discrete level, but also to
a smaller extent methods at the pure algebraic level, in the vein of Schur
complement techniques. Even if, in the latter case, it could be possible to
apply generic methods, it seems that they have not percolated very much in
computational chemistry either. The state of the art in chemistry seems to
be at best a one-shot algorithm: computing the 'partial' density matrices,
and merging adequately the submatrices to build the global density matrix. More sophisticated algorithms are currently being tested in Barrault,
Cancès, Hager and Le Bris (2004c).

Therefore we refer to the bibliography for the divide and conquer approach
and prefer now to concentrate on the last two categories. For the description
of the main representative methods in each category, we let the overlap
matrix S be the Identity, for simplicity, understanding that algorithms can
be adapted if this were not the case.

In order to make alternatives to diagonalization practical, the problem (3.34) is reformulated in such a way that the constraint $DSD = D$ disappears,[18] and next an algorithm is constructed, which might scale cubically in the whole generality, but scales linearly when F is sparse and when the density matrix D to be determined is assumed to be sparse.[19] This favourable scaling is obtained because the algorithm is deliberately constructed in such a way that only a limited number of products of sparse matrices are performed. Of course, in order to define the sparsity, some cut-off parameters have to be adequately tuned, on the basis of physical assumptions such as that of locality, alluded to above.

The *penalization methods* consist in eliminating the constraint of idempotency by constructing an *exact penalized functional* (any local minimizer to the constrained problem is a minimizer to the unconstrained one). Then a standard algorithm of unconstrained numerical minimization, such as the nonlinear conjugate gradient algorithm for instance, is performed on the latter problem.

For this purpose, one idea is to penalize the constraint $D^2 = D$ in (3.34) by using functionals of the type

$$\text{Trace}(FD) + \text{Trace}(Fg(D)) \qquad\qquad (3.35)$$

for some convenient function g. The simplest choice is $g(D) = 3D^2 - 2D^3 - D$, hence the *Density Matrix Minimization* method, due to Li, Nunes and Vanderbilt (1993):

$$\inf\big\{\text{Trace}\big((F - \varepsilon_F\,\text{Id})(3D^2 - 2D^3)\big) : \quad D \in \mathcal{M}_S(N_b)\big\}. \qquad (3.36)$$

It can be shown that this problem has a unique local minimizer, although the infimum is $-\infty$. The solution of (3.34) is obtained by solving (3.36) with a nonlinear conjugated gradient algorithm. For this purpose, it is necessary that the initial guess of the conjugated gradient is in the 'attraction basin' of the local minimum. Despite the definitely 'risky' nature of this numerical approach, it performs well in practice.

An alternative to the above techniques is provided by *non-variational approximations*. They consist in approaching the solution D to the problem (3.34) as the implicit function

$$D = \mathcal{H}(\varepsilon_F\,\text{Id} - F),$$

where $\mathcal{H}$ denotes the Heaviside function (all states of energy lower than ε_F are occupied, the other ones above ε_F being empty).

---

[18] The constraint $\text{Trace}(SD) = N$ fixing the number of electrons is easy to deal with since it can be associated with *one scalar* Lagrange multiplier ε_F, called the *Fermi level*, determined iteratively by an outer loop; we therefore treat ε_F as known.

[19] The latter assumption is in some sense an *a posteriori* assumption, and not easy to analyse.

A first option is to resort to the *Fermi operator expansion* (FOE), which consists in approaching the Heaviside function $\mathcal{H}$ by a Chebyshev polynomial approximation. Up to a renormalization, we may assume that the Fermi level ε_F is zero and that the eigenvalues of F all lie in the range $[-1, 1]$, so that the minimizer D of (3.34) satisfies $D = \mathcal{H}(-F)$. Decomposing $\mathcal{H}$ on the range $[-1, 1]$ into

$$\mathcal{H}(-x) = \sum_{j=0}^{+\infty} c_j T_j(x),$$

where T_j is the jth Chebyshev polynomial and $(c_j)_{0 \leq j \leq +\infty}$ are the Chebyshev coefficients, we obtain

$$D = \sum_{j=0}^{+\infty} c_j T_j(F).$$

The FOE method consists in truncating the above expansion to a given order k which depends both on the spectral gap and on the required accuracy (see Goedecker (1999), and Liang *et al.* (2003) for a recent improvement). Note that the truncation is a very delicate issue, for which no analysis is known, and that has a crucial impact on the quality of the result.

Note also that the computation of the truncated expansion

$$D_k = \sum_{j=0}^{k} c_j T_j(F)$$

is done by taking advantage of the recursion formula

$$T_{j+1}(F) = 2F T_j(F) - T_{j-1}(F), \quad T_0(F) = I_{N_b}, \quad T_1(F) = F,$$

which allows us to compute *independently* each column of the matrix and makes the method easily parallelizable.

A second instance of a nonvariational approximation technique is provided by the method of *purification of the density matrix*. This method was introduced by Palser and Manopoulos (1998) following the earlier work by McWeeny (1992). The idea is to remark that for $x_0 \in]-1/2, 3/2[$ and $f(x) = 3x^2 - 2x^3$, the algorithm defined by the induction formula $x_{k+1} = f(x_k)$ converges to $\mathcal{H}(1/2 - x_0)$. Now, again up to a renormalization of F, the minimizer D of (3.34) satisfies

$$D = \mathcal{H}(1/2 - F),$$

and thus the sequence defined by

$$D_0 = F, \qquad D_{k+1} = f(D_k) = 3D_k^2 - 2D_k^3$$

converges toward D. This algorithm can be interpreted as the approximation of the Heaviside function by a polynomial, namely that defined by the nth iteration of the function f (which is called the *McWeeny purification function*).

For more details and other linear scaling methods, we refer to the articles of Ordejon, Drabold and Martin (1995), Clementi and Davis (1966), Jay, Kim, Saad and Chelikowski (1999), Kohn (1996), Scuseria (1999), Shepard (1993), Shao, Saravan, Head-Gordon and White (2003) and Head-Gordon, Shao, Saravan and White (2003), and also to the survey articles of Daniels and Scuseria (1999), Galli (2000), Goedecker (1999), Bowler *et al.* (1997), Bowler and Gillan (1999) and Bowler, Miyazaki and Gillan (2002).

Let us, however, conclude this section with a necessarily schematic state of the art for these linear scaling methods. Essentially, one may say that, when employed with localized basis sets of limited size per atom, these methods are remarkably efficient for the modelling of insulators, a physical situation that corresponds to a large enough gap $\epsilon_{N+1} - \epsilon_N = \gamma > 0$ between the Nth eigenvalue of F and the following one (recall that such an assumption also plays a role in the convergence of SCF iterations, as manifested by the UWP property stated above). On the other hand, they experience the worst difficulties when dealing with metallic systems ($\gamma \simeq 0$). For the latter, they are clearly in a nonsatisfactory state and need to be further developed and adapted. From the numerical standpoint, the difficulty of metallic systems is twofold. First, the problem of finding the eigenvalues is ill-conditioned in the sense that the eigenvalues for the Hamiltonian of such systems are very close to one another. Second, the density matrix D for the ground state is dense. The two difficulties together are an overwhelming task for the current algorithms. Note an attempt by Barrault, Bencteux, Cancès and Duwig (2004d) to develop deflation techniques in this setting in order to artificially enlarge the gap γ and consequently enhance the efficiency of the methods. However, definite conclusions about the efficiency of the approach are yet to be obtained.

It is to be emphasized that the numerical analysis of the linear scaling methods overviewed above that would account for cut-off rules and locality assumptions, is not yet available. The efficiency of the different methods has therefore only been investigated on a few benchmark calculations which are far from reproducing all the situations met in practice. In addition, even at the formal level, the interaction between linear scaling procedures for the linear subproblems and the SCF iterations has not been investigated yet.

3.8. Additional issues

We conclude our survey of the methods for molecular systems by addressing here a few additional topics.

Beyond Hartree–Fock

Many post-Hartree–Fock methods exist in the chemical literature. As the Hartree–Fock approximation is a variational approximation of (2.2), that is, an approximation constructed by restricting the variational space to a smaller one, most of its improvements consist in enlarging the variational space.

The *multiconfiguration self-consistent field method* (MCSCF) aims to recover more generality on the wavefunction ψ_e by minimizing on *sums* of determinants:

$$E_N^K = \inf\left\{ \langle \psi_e, H_e \psi_e \rangle : \quad \psi_e = \sum_{I=\{i_1,\ldots,i_N\}\subset\{1,\ldots,K\}} c_I \frac{1}{\sqrt{N!}} \det(\phi_{i_1},\ldots,\phi_{i_N}), \right.$$

$$\left. \phi_i \in H^1(\mathbb{R}^3), \quad \int_{\mathbb{R}^3} \phi_i \phi_j = \delta_{ij}, \quad \sum_I c_I^2 = 1 \right\}, \tag{3.37}$$

where $K \geq N$ is some fixed integer.

The mathematical knowledge on the MCSCF model is now at the level of that on the HF model, owing to a recent work by Lewin (2004), following prior works by Le Bris (1994) and Friesecke (2003).

The numerical practice consists, as in the Hartree–Fock case for which (3.8) is attacked, in solving the Euler–Lagrange equations for (3.37), the *MCSCF equations*, which take the form of the following system:

$$\begin{cases} \left(\left(-\frac{\Delta}{2} + V\right)\Gamma + 2W_\Phi\right) \cdot \Phi + \Lambda\,\Phi = 0, \\ \qquad\qquad\qquad\qquad H_\Phi \cdot c = \beta \cdot c. \end{cases} \tag{3.38}$$

In (3.38), the first line translates the optimality of the wavefunctions Φ_i and is in fact a system of K nonlinear PDEs involving the Lagrange multipliers matrix Λ to account for the orthonormality constraints. The matrix Γ is easily computed from the coefficients c_I of the expansion appearing in (3.37) while V is defined by (2.12) and W_Φ denotes the interelectronic interaction term, also easily obtained from the c_I. On the other hand, the second line translates the optimality of the coefficients c_I, with the Lagrange multipliers β to account for the normalization. The matrix H_Φ is the $\binom{K}{N} \times \binom{K}{N}$ matrix with general term $\langle H_N \Phi_I, \Phi_J \rangle$ with $\Phi_I = \frac{1}{\sqrt{N!}} \det(\phi_{i_1},\ldots,\phi_{i_N})$ in the notation of (3.37).

The recent work by Cancès, Galicher and Lewin (2004a) aims to solve (3.38), particularly in order to rigorously define and efficiently compute excited states in the MCSCF setting.

Relativistic models

In the case when the molecular system under study involves one or many heavy atoms, the relativistic effects need to be accounted for, otherwise erroneous conclusions, even at the qualitative level,[20] can be drawn from the computations.

---

[20] such as, gold is *not* yellow

The huge difference in the relativistic modelling is that the Laplacian operator appearing in the Hamiltonian (2.3) has to be replaced by the *Dirac Hamiltonian*,

$$ H_c = -i\alpha_1 \frac{\partial}{\partial x_1} - i\alpha_2 \frac{\partial}{\partial x_2} - i\alpha_3 \frac{\partial}{\partial x_3} + c^2\beta, \qquad (3.39) $$

where c is the speed of light, while α_k, $k = 1, 2, 3$, and β are 4×4 matrices depending on the Pauli matrices. The introduction of this Hamiltonian, by Dirac, is motivated by the fact that H_c^2 needs to be equal to the operator $-c^2\Delta + c^4$ which is the quantum analogue of the Hamiltonian of classical relativity $p^2c^2 + c^4$ (where p is the momentum operator). The Dirac Hamiltonian H_c acts on 4-spinors, *i.e.*, wavefunctions valued in $\mathbb{C}^4$. The crucial point is that its spectrum $\sigma(H_c) =]-\infty, -c^2] \cup [c^2, +\infty[$, contrary to that of the Laplacian $[0, +\infty[$, is not bounded from below. When inserted in the modelling of an hydrogen-like atom, it therefore leads to a minimization problem that is not well posed, and a good definition of the ground state has to be introduced. Basically, the minimization has to be replaced by adequate saddle-point methods. Some new minimax characterizations have been established by Esteban and Séré (2002), Dolbeault, Esteban and Séré (2000*a*), Desclaux *et al.* (2003), and have given rise to new algorithmic techniques to compute the eigenfunctions and eigenvalues of the Dirac operator in molecules: see Dolbeault, Esteban, Séré and Vanbreugel (2000*b*), Dolbeault, Esteban and Séré (2003). Likewise, in the many-electron case where models such as the Dirac–Fock model, introduced in Swirles (1935, 1936), play the role of the Hartree–Fock model, adequate definitions of the ground state need to be derived. Again, from the numerical viewpoint, an adequate treatment has to be developed. After years of rather brutal techniques, the situation has recently evolved toward more rigour, and also efficiency, with the series of works by the authors cited above.

Molecular mechanics

As mentioned above, the search for the electronic ground state for a fixed set of positions of nuclei might only be an *inner* calculation. The outer loop consists in solving the minimization problem (2.1)

$$ \inf_{(\bar{x}_1, \ldots, \bar{x}_M) \in \mathbb{R}^{3M}} \left\{ W(\bar{x}_1, \ldots, \bar{x}_M) = U(\bar{x}_1, \ldots, \bar{x}_M) + \sum_{1 \le k < l \le M} \frac{z_k \, z_l}{|\bar{x}_k - \bar{x}_l|} \right\}. $$

that is a purely classical optimization problem in dimension $3N$, up to trivial invariance properties (translation and rigid rotation at least). Of course, the potential $U(\bar{x}_1, \ldots, \bar{x}_M)$ can be parametrized on the basis of precomputations performed with the *ab initio* models we have seen above, and this gives rise to the field called *molecular mechanics*. This field is prominent in biology, say, where, for instance, stable conformations of protein must be determined. From the computational viewpoint, the problem is that of minimizing a parametrized function in a space of very high dimension, for which billions of local minimizers are likely to exist. The relevant theory is numerical optimization, or even combinatorial optimization, since in practice substructures are first optimized and then assembled combinatorially to find (at least a good guess for) the most stable global structure.

Here we would like to concentrate on a somewhat different problem, that of finding the optimal configuration of nuclei when the number of nuclei is not so large but when the potential $U(\bar{x}_1, \ldots, \bar{x}_M)$ is indeed that obtained by some of the computations above, *i.e.*, approximations of (2.2).

In most situations, except those when gradient-free (or direct search) algorithms are utilized, the optimization algorithm, in order to be efficient, needs to account for derivatives of $U(\bar{x}_1, \ldots, \bar{x}_M)$ with respect to the $\bar{x}_k$.

The key point is that determining the gradient $\frac{\partial U}{\partial \bar{x}_k}$ (or further derivatives) only requires a small additional computational time. This is not the case in other settings, where the computation of the derivative is generically considered as many times more costly than the function evaluation itself.

To illustrate the situation, let us write (2.1)–(2.2) in the abstract form

$$\inf\{W(x): \quad x \in \Omega\}, \qquad W(x) = \inf\{E(x, \phi): \quad \phi \in \mathcal{H}, \quad g(x, \phi) = 0\}. \tag{3.40}$$

In this formal setting it is indeed easy to recognize x as the collection of co-ordinates of the nuclei, ϕ as the electronic wavefunction, $\mathcal{H}$ as the variational space, $E(x, \phi)$ as the energy functional depending both on the nuclear coordinates and the electronic wavefunction, and $g(x, \phi)$ as the orthonormality conditions on the molecular orbitals. We make the latter depend explicitly on x as it is the case when problem (3.40) is considered at the discrete level and when AO basis sets are used. When $\phi(x)$ denotes the ground state for x (here assumed to be unique for simplicity), we may write formally

$$\frac{\partial W}{\partial x_i}(x) = \frac{\partial}{\partial x_i} E(x, \phi(x)) = \frac{\partial E}{\partial x_i}(x, \phi(x)) + \left\langle \nabla_\phi E(x, \phi(x)), \frac{\partial \phi}{\partial x_i}(x) \right\rangle, \tag{3.41}$$

by the chain rule. Next, as $\phi(x)$ is a minimizer, it satisfies

$$\nabla_\phi E(x, \phi(x)) = d_\phi g(x, \phi(x))^T \cdot \lambda(x)$$

for some Lagrange multiplier $\lambda(x)$. On the other hand, by differentiation of the constraint $g(x, \phi(x)) = 0$, we have

$$\frac{\partial g}{\partial x_i}(x, \phi(x)) + d_\phi g(x, \phi(x)) \cdot \frac{\partial \phi}{\partial x_i}(x) = 0.$$

Thus (3.41) yields

$$\frac{\partial W}{\partial x_i}(x) = \frac{\partial E}{\partial x_i}(x) - \left\langle \lambda(x), \frac{\partial g}{\partial x_i}(x, \phi(x)) \right\rangle.$$

The crucial point is that $\frac{\partial \phi}{\partial x_i}(x)$ has been eliminated. Therefore, the gradient of W can be directly computed from $(x, \phi(x), \lambda(x))$ (that is to say, the set of positions of nuclei considered, the electronic ground state, and the Lagrange multipliers, *i.e.*, the monoelectronic energies) without any further calculations. This property, referred to as *analytical derivatives*, is also used in the context of *ab initio* molecular dynamics in Section 5.1.

For the sake of completeness, let us mention that the existence of the global minimizer of (2.1) has been theoretically investigated in a series of works: Catto and Lions (1992, 1993*a*, 1993*b*, 1993*c*). While the existence is theoretically proved in most academic cases, it is to be emphasized that in the Hartree–Fock case it is still an open question, even for the simplest diatomic molecular systems.

4. The condensed phase

4.1. The solid phase

In the case of a crystalline solid, the formal Hartree–Fock equations are derived through Bloch's theorem (see Ashcroft and Mermin (1976), Kittel (1996)): all sums over j involved in the definitions (2.13)–(2.14) of τ and ρ are replaced by sums over j and integrals over the Brillouin zone BZ, *i.e.*, the Wigner–Seitz cell of the lattice $\mathcal{R}^\star$ reciprocal to the physical lattice $\mathcal{R}$. The wave functions and energies, that are here in infinite (uncountable) number, are labelled by $j \in \mathbb{N}^*$ as in the molecular case, *and* by $k \in BZ$, a specificity of the crystalline solid phase. More precisely, setting

$$\tau(x, x') = \sum_{j \in \mathbf{N}} \int_{BZ} \phi_j^k(x) \phi_j^{k*}(x')(\varepsilon_F - \varepsilon_j^k)_+ \, \mathrm{d}k, \qquad (4.1)$$

where the term $(\varepsilon_F - \varepsilon_j^k)_+$ selects only the states with energy lower than the Fermi energy ε_F, and defining V_{tot} to be the solution to

$$\begin{cases} -\Delta V_{\text{tot}} = -4\pi \left(\sum_{T \in \mathcal{R}} m(\cdot + T) - \rho \right), \\ V_{\text{tot}} \quad \mathcal{R}\text{-periodic,} \end{cases} \qquad (4.2)$$

where $\rho(x) = \tau(x, x)$ and m is the measure defining the nuclei in the primitive unit cell, we may write down the Fock operator

$$F\phi = -\frac{1}{2}\Delta\phi + V_{\text{tot}}\phi - \int_{\mathbb{R}^3} \frac{\tau(x, x')}{|x - x'|} \phi(x') \, \mathrm{d}x'. \qquad (4.3)$$

The Hartree–Fock wavefunctions ϕ_j^k are then defined to be the solutions of

$$
\begin{cases}
F\phi_j^k = \varepsilon_j^k \, \phi_j^k, \\
\text{for all } j \in \mathbb{N} \text{ and } k \in BZ, \quad e^{-ikx}\phi_j^k(x) \quad \mathcal{R}\text{-periodic}, \\
\int_{\mathcal{Q}} \phi_j^k(x)\, \phi_{j'}^{k'}(x)^* \, dx = \delta(k - k')\delta_{jj'}.
\end{cases}
\tag{4.4}
$$

For a study of the rigorous foundation of this model, we refer to a series of works initiated in Catto, Le Bris and Lions (1998) and more particularly to Catto, Le Bris and Lions (2001).

For the Kohn–Sham model, the equations read

$$
\begin{cases}
-\tfrac{1}{2}\Delta\phi_j^k + V_{\text{eff}}\phi_j^k = \varepsilon_j^k\phi_j^k, \quad \text{for all } j, k, \\
V_{\text{eff}} = V_{\text{tot}}(\rho) + v_{xc}(\rho), \\
\rho = \sum_j \int_{BZ} |\phi_j^k|^2(\varepsilon_F - \varepsilon_j^k)_+ \, dk,
\end{cases}
\tag{4.5}
$$

and are treated analogously to the Hartree–Fock case.

In principle, we need to solve an infinite number of Hartree–Fock equations, for system (4.4) is indeed an infinite collection of molecular-like Hartree–Fock-type systems

$$
F\phi_j^k = \varepsilon_j^k\phi_j^k
$$

indexed by the points k of the Brillouin zone. In practice, it turns out that, fortunately, a limited number of points k is generally enough to obtain accurate results. The general trend is that for metals many points (say hundreds of points) are needed, while for insulators a few (or even one) k-points yield realistic values: see Blanc (2000) for an introduction, and, *e.g.*, Dovesi *et al.* (2000) for specific details. Regarding software, we may cite the code CRYSTAL, based on Hartree–Fock-type models, while the code ABINIT is based upon models from density functional theory.

Regarding the basis set used for developing the wavefunctions ϕ_j^k, a peculiarity of the solid phase setting comes into play. Contrary to the molecular case where, as explained above, GTO basis sets are the method of choice (except for very specific applications), the periodicity of the solid phase makes the choice of plane wave (PW) basis sets natural and simple.[21] Notably, they make the kinetic operator diagonal and FFT algorithms can be systematically adopted. The disadvantage of using PWs is that the fine oscillations of some of the orbitals (in fact the valence orbitals) near the nuclei require a huge number of PWs to be described accurately. The latter difficulty is in turn mostly circumvented by the introduction of a *pseudopotential*.

---

[21] Note that a combination of GTO and PW functions to form a basis set presenting the best compromise is also an option.

The technique consists in

- eliminating the explicit consideration of the core states (*i.e.*, in a classic picture those corresponding to electrons orbiting close to the nucleus) by freezing them and aggregating them with the nuclei, while treating their effect upon the valence electrons almost exactly (those orbiting far away),

- replacing the wavefunctions of the valence electrons by pseudowavefunctions (indeed generated by the diagonalization of an operator with pseudopotential) that are less oscillating and more regular, so that the size of the PW basis set needed for accuracy can be reduced.

Over the years, numerous pseudopotentials of increasingly better quality have appeared and are now widely spread: see, *e.g.*, Troullier and Martins (1990) and Vanderbilt (1990). One point is that there is no rigorous foundation, and no mathematical or formal understanding of the derivation of a pseudopotential. Therefore the method lacks a theoretical foundation, a lack that translates into a problem of crucial practical interest, that of *transferability* (a pseudopotential is definitely useful when it can be used for different atomic arrangements). Current efforts in the mathematics community are directed toward giving a sound base to the concept of pseudopotential. In the chemistry literature, the reference that is the most accessible to a mathematical audience, and that is the best attempt to introduce a rigorous formalism in the development of pseudopotentials, is Blöchl (1994).

4.2. The liquid phase

Most of the physical and chemical phenomena of interest in chemistry and biology take place in the liquid phase (see Allen and Tildesley (1987) for an introduction to the modelling issues) and it is well known from experimental evidences that solvent effects play a crucial role in these processes. Accounting for such effects is thus a main concern.

A natural idea is that of building a 'supermolecule' consisting of the solvated molecule under study plus several neighbouring solvent molecules. It seems that the additional work needed to treat the latter ones makes the approach inefficient in most[22] cases of practical interest. In addition it may be remarked that long-range effects of the solute–solvent interaction are not included in such an approach. Coupling a quantum model for the solvated molecule with a solvation continuum model provides an economical, and

---

[22] Not *all* cases: when the solvated molecule is already very large, and thus coarsely modelled, the solvent molecules that need to be added are not so numerous, in comparison.

actually more accurate, alternative. This consists in locating the solute molecule under study inside a cavity Ω, modelling a solvent excluding volume, surrounded by a continuous medium modelling the solvent.

In the standard model, the continuous medium behaves as a homogeneous isotropic dielectric of relative permittivity ϵ_s ($\epsilon_s > 1$). The electrostatic interactions between the charge distributions which compose the solute (point nuclei and electronic cloud) are affected by the presence of the solvent: the standard Coulomb potential $\frac{1}{|x-y|}$, which is the Green kernel $G(x-y)$ of $-\frac{1}{4\pi}\Delta$ in $\mathbb{R}^3$ must be replaced by that of the operator $-\frac{1}{4\pi}\text{div}\,(\epsilon\nabla\cdot)$, with $\epsilon(x) = 1$ inside the cavity Ω and $\epsilon(x) = \epsilon_s$ outside. Correspondingly, a charge ρ creates a potential V solution to[23]

$$-\text{div}\,(\epsilon(x)\nabla V(x)) = 4\pi\rho(x). \tag{4.6}$$

The various terms of the Hartree–Fock (respectively KS) energy functional are changed correspondingly. Note, however, that in practice (with a view to keeping the efficiency of the computations of bielectronic integrals in vacuum) the exchange term is often left unchanged.

When the solvent is an ionic solution, equation (4.6) is replaced by the *linearized Poisson–Boltzmann equation*

$$-\text{div}(\epsilon(x)\nabla V(x)) + \epsilon(x)\kappa^2(x)V(x) = 4\pi\rho(x) \tag{4.7}$$

(with $\epsilon(x) = 1$ and $\kappa(x) = 0$ inside the cavity Ω and $\epsilon(x) = \epsilon_s > 1$ and $\kappa(x) = \kappa_s > 0$ outside), while for a liquid crystal, it keeps the form (4.6) but the dielectric constant $\epsilon(x)$ is no longer a scalar but a 3×3 anisotropic symmetric tensor $\underline{\underline{\epsilon}}(x)$.

In practice, equation (4.6) is most often solved by an integral equation method. The equation is posed on the surface of Ω, called the *molecular surface*. The approach necessitates efficient meshing techniques for this molecular surface. We refer to Cancès, Le Bris, Mennucci and Tomasi (1999) and Le Bris, ed. (2003) for details and extensions.

5. Time-dependent problems

Ideally, the determination of the evolution of a molecular system requires the solution of the time-dependent Schrödinger equation,

$$i\frac{\partial}{\partial t}\Psi = H\Psi, \tag{5.1}$$

---

[23] Notice the analogy with the case of solids where the Coulomb potential is indeed replaced by the Green kernel of the Laplacian, but with periodic boundary condition on the unit cell, see (4.2). This gives a unified setting to all the models addressed here.

where the wavefunction describes the state of the complete system (electrons plus nuclei) and the Hamiltonian is also the complete one,

$$H = - \sum_{k=1}^{M} \frac{1}{2\,m_k} \Delta_{\bar{x}_k} - \sum_{i=1}^{N} \frac{1}{2} \Delta_{x_i} - \sum_{i=1}^{N} \sum_{k=1}^{M} \frac{z_k}{|x_i - \bar{x}_k|}$$
$$+ \sum_{1 \le i < j \le N} \frac{1}{|x_i - x_j|} + \sum_{1 \le k < l \le M} \frac{z_k\, z_l}{|\bar{x}_k - \bar{x}_l|}. \tag{5.2}$$

Even when inserting in the above description the approximation that the wavefunction is *a product* of the wavefunction of the electronic degrees of freedom times that of the nuclear ones, this equation remains intractable for any system consisting of more than a few particles. Indeed, equation (5.1) is a time-dependent partial differential equation set on a vectorial space of high dimension. Here again, the use of sparse grid techniques, already mentioned in the static setting in Section 2.2, can be envisioned. Nevertheless, there is again an issue about the regularity of the function manipulated. For time-dependent equations, such a regularity is typically obtained by supplying the equation with regular data (initial and/or boundary conditions), or by using regularization properties of the equation itself, as is the case for parabolic equations. Now, as mentioned above, the functions of chemistry may be singular, and the mixed parabolic/hyperbolic nature of the Schrödinger equation makes the regularization properties very peculiar and different from those of parabolic equations. Therefore it seems that further efforts are needed to apply such techniques to the case of the Schrödinger equation, at least as efficiently as in the parabolic case.

Fortunately, the solution of (5.1), which considers the nuclei as quantum objects, is not needed in most applications, apart from very particular ones issued from fundamental physics. An example of the latter is provided by the emerging domain of laser control of molecular evolutions where light-matter interactions are to be modelled in the most precise way. An introduction to the physical modelling, as well as the mathematical and numerical challenges of this field, has appeared in the recent books of Le Bris, ed. (2003) and Bandrauk, Delfour and Le Bris, eds (2004). Other instances of applications where the nuclei need to be modelled by quantum mechanics, along with numerical approaches for this purpose, can be read in Worth and Robb (2002).

For almost all of the applications in chemistry and biology, the nuclei can be, as in the time-independent setting, treated as classical objects, and the Schrödinger equation above thus simplifies into a system coupling the Newton equation of motion for the positions $\bar{x}_k(t) \in \mathbb{R}^3$ of the nuclei and the Schrödinger equation for the electronic structure.

In turn, as is the case for the stationary problem, the Schrödinger equation ruling the evolution of the electrons cannot be treated without further approximations. One of them is the time-dependent Hartree–Fock approximation, which is obtained by forcing the wave function ψ_e to evolve on the manifold

$$\mathcal{A} = \left\{ \psi_e(x_1, \ldots, x_n) = \frac{1}{\sqrt{N!}} \det(\phi_i(x_j)) : \quad \phi_i \in H^1(\mathbb{R}^3), \int_{\mathbb{R}^3} \phi_i \cdot \phi_j = \delta_{ij} \right\}$$

of $\mathcal{H}_e$ and in replacing the time-dependent Schrödinger equation for the evolution of the electronic structure by the stationarity condition for the action

$$\int_0^T \langle \psi_e(t), (i\partial_t \psi_e(t) - H_e(t)\psi_e(t)) \rangle \, dt.$$

In so doing, we obtain the following mixed *quantum/classical* system ruling the evolution of the complete molecular system (nuclei and electrons),

$$\begin{cases} m_k \frac{d^2 \bar{x}_k}{dt^2}(t) &= -\nabla_{\bar{x}_k} W(t; \bar{x}_1(t), \ldots, \bar{x}_M(t)), \\ W(t; \bar{x}_1, \ldots, \bar{x}_M) &= -\sum_{k=1}^{M} \sum_{i=1}^{N} z_k \int \frac{|\phi_i(t,x)|^2}{|x - \bar{x}_k|} \, dx + \sum_{1 \le k < l \le M} \frac{z_k \, z_l}{|\bar{x}_k - \bar{x}_l|}, \\ i \frac{\partial \phi_i}{\partial t} &= -\frac{1}{2}\Delta \phi_i - \sum_{k=1}^{M} \frac{z_k}{|\cdot - \bar{x}_k(t)|} \phi_i + \left(\sum_{j=1}^{N} |\phi_j|^2 \star \frac{1}{|x|} \right) \phi_i \\ &\quad - \sum_{j=1}^{N} \left(\phi_j^* \phi_i \star \frac{1}{|x|} \right) \phi_j, \end{cases}$$

(5.3)

supplied with the initial condition $\bar{x}_k(0) = \bar{x}_k^0$, $\frac{d\bar{x}_k}{dt}(0) = \bar{v}_k^0$, $\phi_i(0) = \phi_i^0$. The above system is a prototypical example of a *non-adiabatic simulation*. It was proved to be well posed in Cancès and Le Bris (1999), a work that very much relies on the previous important study by Chadam and Glassey (1975) in a slightly different (uncoupled) setting. The simulation of this system is still a demanding task, necessary in some situations such as collisions of molecular systems. The practical bottleneck consists in the discrepancy between the time-scale of the electronic motion (typically 10^{-18} s) and that of the nuclear motion (typically 10^{-15} s).

Recent works developing numerical algorithms for the solution of quantum dynamics equations as in the above systems are those of Jahnke and Lubich (2003) and Jahnke (2003, 2004). Regarding the specific simulation of the TDHF equations themselves, we refer to the new ideas related to variational integrators developed in Lubich (2004) that deal with the time-dependent multiconfiguration Hartree (not Hartree–*Fock*) equations, in the vein of Section 3.8.

5.1. Ab initio molecular dynamics

Again, for most applications, the above setting can be further simplified. Indeed, it can be considered, within a good level of approximation, that the electrons stay in a well-defined energy surface. This surface, called the *Born–Oppenheimer energy surface*, is parametrized by the set of positions of the nuclei, and is often the ground-state energy surface (and we shall suppose it is henceforth). The system under consideration then reads

$$
\begin{cases}
m_k \frac{\mathrm{d}^2 \bar{x}_k}{\mathrm{d}t^2}(t) & = -\nabla_{\bar{x}_k} W(\bar{x}_1(t), \ldots, \bar{x}_M(t)), \\
W(\bar{x}_1, \ldots, \bar{x}_M) & = U(\bar{x}_1, \ldots, \bar{x}_M) + \sum_{1 \le k < l \le M} \frac{z_k z_l}{|\bar{x}_k - \bar{x}_l|}, \\
U(\bar{x}_1, \ldots, \bar{x}_M) & \quad \text{energy of the electronic degrees of freedom} \\
& \quad \text{evaluated in a given static model.}
\end{cases} \tag{5.4}
$$

The main advantage of this setting is that the time-step for numerical integration of the dynamics can now be chosen of the same order of magnitude as the characteristic evolution time of the *nuclei* rather than that of the *electrons*. But, as will be seen shortly, each time-step is likely to be more costly.

This approximation mainly relies on physical arguments: the characteristic relaxation time of the electrons is so small with respect to that of the nuclei that it can be considered that the electronic wavefunction reacts *adiabatically* to a change in the position of the nuclei.

As far as applications are concerned, the adiabatic approximation turns out to be valid for the simulation of *physical* properties (phase diagrams, surface reconstruction, diffusion in alloys), as well as for the simulation of most chemical reactions. Both theoretically and practically, however, huge difficulties arise when the energy surfaces happen to cross each other for a given particular set of positions of nuclei (see Hagedorn (1996), Teufel (2003) for related mathematical works).

In practice, the potential U in (5.4) has to be approximated, as in the pure time-independent case, by one of the standard (Hartree–Fock- or DFT-type) methods. If the model is the Hartree–Fock approximation, we need to find at each time-step the Hartree–Fock ground state, which in practice (following the discussion of the previous sections) amounts to solving the SCF equations. Thus the system to be simulated reads

$$
\begin{cases}
m_k \frac{\mathrm{d}^2 \bar{x}_k}{\mathrm{d}t^2}(t) & = -\nabla_{\bar{x}_k} W(\bar{x}_1(t), \ldots, \bar{x}_M(t)), \\
W(\bar{x}_1, \ldots, \bar{x}_M) & = U(\bar{x}_1, \ldots, \bar{x}_M) + \sum_{1 \le k < l \le M} \frac{z_k z_l}{|\bar{x}_k - \bar{x}_l|}, \\
U(\bar{x}_1, \ldots, \bar{x}_M) & = E^{HF}(\phi_1, \ldots, \phi_N) \\
\begin{cases} F_\Phi^{\bar{x}_1, \ldots, \bar{x}_M} \phi_i & = \lambda_i \, \phi_i, \\ \int_{\mathbb{R}^3} \phi_i \phi_j & = \delta_{ij} \end{cases}
\end{cases} \tag{5.5}
$$

where we recall that $F_\Phi^{\bar{x}_1, \ldots, \bar{x}_M}$ is the Fock operator (2.16) that depends parametrically on the positions $\bar{x}_k$ of the nuclei and the ϕ_j are the lowest N eigenfunctions of F_Φ.

Likewise, if the static model is, *e.g.*, a Kohn–Sham model, then the last two lines of (5.5) are replaced by the equations (2.23), and thus

$$
\begin{cases}
m_k \frac{\mathrm{d}^2 \bar{x}_k}{\mathrm{d}t^2}(t) & = -\nabla_{\bar{x}_k} W(\bar{x}_1(t), \ldots, \bar{x}_M(t)), \\
W(\bar{x}_1, \ldots, \bar{x}_M) & = U(\bar{x}_1, \ldots, \bar{x}_M) + \sum_{1 \leq k < l \leq M} \frac{z_k z_l}{|\bar{x}_k - \bar{x}_l|}, \\
U(\bar{x}_1, \ldots, \bar{x}_M) & = E^{KS}(\phi_1, \ldots, \phi_N) \\
\begin{cases}
K^{\bar{x}_1, \ldots, \bar{x}_M}(\rho_\Phi)\phi_i = \lambda_i \phi_i, \\
\int_{\mathbb{R}^3} \phi_i \phi_j = \delta_{ij}
\end{cases}
\end{cases}
\tag{5.6}
$$

Even within the above approximation, the coupled problem (5.4) remains very time-consuming since a minimization problem (*i.e.*, a nonlinear eigenvalue problem) has to be solved on the fly for each time-step. The solution procedure of the static electronic problem is thus the *inner loop* of the dynamics. Therefore, from a numerical viewpoint, the task for simulating an adiabatic-type model is a sequence of 3-step iterations on the time variable:

(i) determine the electronic state by solving the nonlinear eigenvalue problem (SCF problem),

(ii) compute the gradient of the interaction potential W, using the techniques of analytical derivatives described in Section 3.8,

(iii) integrate in time the Newtonian dynamics.

With a view to circumventing the difficulty of solving a nonlinear eigenvalue problem at each time-step, Car and Parrinello (1985) introduced the idea of replacing the last two lines of (5.6) by a *virtual* time evolution, thus:

$$
\begin{cases}
m_k \frac{\mathrm{d}^2 \bar{x}_k}{\mathrm{d}t^2}(t) & = -\nabla_{\bar{x}_k} W(\bar{x}_1(t), \ldots, \bar{x}_M(t), t), \\
W(\bar{x}_1, \ldots, \bar{x}_M, t) & = E^{KS}_{\bar{x}_1, \ldots, \bar{x}_M}(\phi_1(t), \ldots, \phi_N(t)) + \sum_{1 \leq k < l \leq M} \frac{z_k z_l}{|\bar{x}_k - \bar{x}_l|}, \\
\mu \frac{\partial^2 \phi_i}{\partial t^2}(t) & = -K^{\bar{x}_1, \ldots, \bar{x}_M}(\rho_{\Phi(t)})\phi_i(t) + \sum_{j=1}^{N} \Lambda_{ij}(t)\phi_j(t), \\
\Lambda_{ij}(t) & = \langle \phi_j(t), K^{\bar{x}_1, \ldots, \bar{x}_M}(\rho_{\Phi(t)})\phi_j(t) \rangle - \mu \int_{\mathbb{R}^3} \frac{\partial \phi_j}{\partial t}(t) \frac{\partial \phi_i(t)}{\partial t}(t),
\end{cases}
\tag{5.7}
$$

where μ is a fictitious mass (the limit $\mu \longrightarrow 0$ formally yields the adiabatic approximation (5.6)). The time-step used needs to be smaller than that used for the adiabatic simulation, but the Car–Parrinello method is usually more efficient because no minimization is required. The method is thus extremely popular and very successful, and is thus widely used in a broad spectrum of contexts. It has allowed for the treatment of definitely larger systems, which is huge progress. However, for our main focus here, regarding numerical analysis, the practical difficulty of the method lies in the proper tuning of the parameter μ, for which no theoretical grounding is known.

The only mathematical work on the approach is due to Bornemann and
Schütte (1998). Notice that in practice, the forces $\nabla_{\bar{x}_k} W$ are determined,
as in the case of (5.4), by the technique of analytical derivatives, this time
apparently without any rigorous foundation.

For the sake of completeness, let us mention another track that is currently investigated in the applied mathematics community. It consists in
keeping the minimization problem as such (and not modifying it as in the
Car–Parrinello approach), but treating it in a rather approximate way,
through the paradigm of reduced basis techniques. As the minimization
is parametrized by the positions of the nuclei, it is natural to envision a
method where the solution is indeed developed on an adequate problem-
dependent basis made of the solutions of the same problem for reference
positions of the nuclei. Ideas in this direction, still in their infancy, are
described in Cancès, Le Bris, Maday and Turinici (2002), Barrault, Maday,
Nguyen and Patera (2004a) and Barrault *et al.* (2004b). The issues under
investigation in particular embody issues regarding reduced basis techniques
for eigenvalue problems, for vectorial problems, and for nonlinear problems
in general, along with the development of adequate error estimators to cer-
tify the results.

5.2. Classical molecular dynamics

We concentrate in this section on the numerical simulation of Newtonian
dynamics, which is a part of any of the above coupled simulations: the non-
adiabatic one (5.3), the adiabatic one (5.4), and even the Car–Parrinello
simulation (5.7). We again emphasize, as we did in the Introduction, that
the present section does not pretend to be a comprehensive exposition of the
state of the art of molecular dynamics, but rather a rapid guided tour of the
challenging issues in the domain. Nor will we go into a description of all the
applications of molecular dynamics, which is indeed the most popular and
commonly used field of molecular simulation. This would require a whole
encyclopaedia.

The literature is rich. On the one hand, we refer to the classical monographs of Hairer, Nørsett and Wanner (1993), Hairer and Wanner (1996)
and Sanz-Serna and Calvo (1994) for the numerical analysis of methods
for ordinary differential equations and Hamiltonian systems. The reference
Griebel, Knapek, Zumbusch and Caglar (2004) is dedicated to molecular
dynamics, and also more algorithmically oriented. On the other hand, from
the application viewpoint, the treatises by Allen and Tildesley (1987), Frenkel and Smit (2001), Schlick (2002), Haile (1992) and Rapaport (1995) are
useful references in the field. In the chemistry literature, there are regular
surveys by experts in the field, and we would like to mention Neumaier
(1997), Tuckerman and Martyna (2000) and Tuckerman (2002). Finally,

proceedings books such as Deuflhard *et al.*, eds (1999) or Nielaba, Mareschal and Ciccotti, eds (2002) collect various contributions and show how lively the field is.

For simplicity we restrict ourselves to the case of the Newtonian equations contained in (5.4) that are autonomous (no explicit dependence of the Hamiltonian with respect to time). For extensions to the non-autonomous case, we refer to the literature. The focus is therefore on the simulation of

$$m_k \frac{\mathrm{d}^2 \bar{x}_k}{\mathrm{d}t^2}(t) = -\nabla_{\bar{x}_k} W(\bar{x}_1(t), \dots, \bar{x}_M(t)), \tag{5.8}$$

supplied with initial conditions on the positions and the velocities. The system may be recognized as a *Hamiltonian system*:

$$\begin{cases} \frac{\mathrm{d}q_k}{\mathrm{d}t} = \frac{\partial H}{\partial p_k}(q_1, p_1, \dots, q_M, p_M), & k = 1, \dots, M, \\ \frac{\mathrm{d}p_k}{\mathrm{d}t} = -\frac{\partial H}{\partial q_k}(q_1, p_1, \dots, q_M, p_M), & k = 1, \dots, M, \end{cases} \tag{5.9}$$

where we have introduced the *Hamiltonian*

$$H(q_1, p_1, \dots, q_M, p_M) = \frac{1}{2} \sum_{k=1}^{M} \frac{p_k^2}{m_k} + W(q_1, \dots, q_M). \tag{5.10}$$

In classical molecular dynamics, the potential W is typically a parametrized potential which gives rise to force fields ∇W that are representable in terms of simple mathematical forms, say, *e.g.*, as explicit functions of the bond lengths, and the dihedral angles, *etc.*, in the molecular system. The analytic form of the functions and the parameters in such potentials are often least-square fitted with *ab initio* computations performed off-line on smaller systems. Two of the most famous force fields are those of the codes AMBER and CHARMM. Parametrized potentials are of course unable to simulate the changes of electronic structure in the molecule (thus in particular chemical reactions) contrary to the coupled quantum/classical simulations (5.4) and, above all, systems such as (5.3).[24]

For such potentials, the computational price of (5.8) is only due to the number of interacting particles: it is only when millions of atoms are simulated that calculating the interactions is a serious task, which can for instance be done with rapid methods such as FMM, and/or multiple time-step methods. For a smaller number of particles it is an easy task. On the other hand, when the potential is of quantum nature and is calculated on the fly, the computational cost of each evaluation of the potential is itself

---

[24] Note, however, the existence of rough approximations, such as the *variable charge molecular dynamics*, that aim to account for changes in the electronic structures while using parametrized potentials.

costly, even for a small system. The state of the art of the technology[25] is as follows: with parametrized force fields, millions of atoms can be simulated over a time frame of 10^{-8} s, while for quantum forces, only a few hundreds atoms can be simulated on 10^{-11} s.

The major point to bear in mind when addressing the construction and numerical analysis of integration schemes for molecular dynamics is that the question asked is *not* to simulate the particular evolution of a single system, starting from a precise initial configuration. Of course, such a task exists, in particular when simulating chemical reactions: details of the dynamics will then be 'observed' that are mostly not accessible to experiment (unless emerging and still difficult techniques such as those of femtochemistry are employed). But the main purpose of molecular dynamics is to simulate the evolution of *a set* of systems, in order to compute *statistical ensemble averages*, with a view to evaluating thermodynamic properties, which is of primary interest because again some of these properties cannot be provided by experiment. This latter objective has its theoretical roots in the (claimed) ergodicity of the system under study. Typically, the average value $\langle A \rangle$ of some observable A on a system of M particles reads as the following integral over the phase space of the position/impulsion of the M particles:

$$\langle A \rangle = \int_{\mathbb{R}^{6M}} A(q_1, p_1, \ldots, q_M, p_M) \, f(q_1, p_1, \ldots, q_M, p_M) \, \mathrm{d}q_1 \, \mathrm{d}p_1 \cdots \mathrm{d}q_M \, \mathrm{d}p_M,$$
(5.11)

where f is the distribution function in the phase space. It can be evaluated through a Monte Carlo sampling method, but it is often more efficient to obtain it by

$$\langle A \rangle = \lim_{T \longrightarrow +\infty} \frac{1}{T} \int_0^T A(q_1(t), p_1(t), \ldots, q_M(t), p_M(t)) \, \mathrm{d}t \qquad (5.12)$$

along a trajectory of the system obeying to the dynamics

$$\begin{cases} \frac{\mathrm{d}q_i}{\mathrm{d}t} = p_i, & i = 1, \ldots, M, \\ \frac{\mathrm{d}p_i}{\mathrm{d}t} = F(q_j, p_j), & i = 1, \ldots, M, \end{cases} \qquad (5.13)$$

where F is some force field, so that the measure

$$f(q_1, p_1, \ldots, q_M, p_M) \, \mathrm{d}q_1 \, \mathrm{d}p_1 \cdots \mathrm{d}q_M \, \mathrm{d}p_M$$

is precisely the invariant measure of the dynamics. In the simplest case, the statistical ensemble to sample is the microcanonical ensemble (N, V, E)

---

[25] We provide here a hopefully representative statement, on figures that of course are highly sensitive to the computing facilities at hand.

and the dynamics to be considered is the Hamiltonian dynamics (5.9). We mostly concentrate on this case.

The numerical challenge can be gauged on (5.12). There is no free lunch, and transforming the difficult sampling of the phase space in order to evaluate (5.11) into the dynamics (5.13) results in the fact that it is only in the long-time limit that the average is obtained. Some comments are then in order.

First, ergodicity is not easy to establish. For most systems it can be at best conjectured, but is rarely proved (see Walters (1982) or Gutzwiller (1990) for the mathematical background). The case of Hamiltonian systems has been examined in Markus and Meyer (1974): they are *generically not* ergodic. In addition, ergodicity may stem from various phenomena: the ergodic nature that integrable Hamiltonian systems might have is very peculiar, and, *e.g.*, different from that of 'chaotic' systems.[26] In fact, the lack of a satisfactory theoretical understanding of ergodicity is not a limitation for the numerical practice. What is indeed a bottleneck is the evaluation of the time T on which the system can reasonably be considered to have visited its whole space phase. For the rare systems for which this time can be evaluated, T can be as large as many times the age of the universe, which makes the ergodicity property useless in practice. One way or another, it must be understood how to circumvent the largeness of T in (5.12). This will be the purpose of some acceleration techniques for molecular dynamics that we will introduce in Section 5.3.

Second, the difficulty is enhanced by the fact that, by nature, the molecular dynamics is a multiscale phenomenon. Recall that, in its simplest occurrence, the evolution of bond lengths, dihedral angles, *etc.*, are simulated. Typically, there are in this set of variables rapidly changing degrees of freedom, oscillating over time periods of the order of 10^{-15} s, and also slower ones, and we wish to simulate all of them over a time frame of a few fractions of a second, to say the least. Therefore, even for accessible times T, there are still challenging issues due to the highly oscillatory character of the system.

Third, as we have mentioned above, it is not a question of simulating the evolution for a precise initial configuration. This clearly advocates a geometric viewpoint, where flows, rather than individual trajectories, are the objects of interest. This is therefore the domain of geometric integration (see Hairer, Lubich and Wanner (2002), or Leimkuhler and Reich (2005)), and as the system often has a Hamiltonian structure, the field of symplectic integration is concerned. In order to reproduce the qualitative properties of the evolution at the continuous level, namely the symplectic nature and pos-

---

[26] The systems of molecular dynamics seem to be equally far from either of these two categories.

sibly the reversibility, integration schemes that conserve these properties are utilized. However, there is no simple way to go beyond this. Schematically, one could say that, from the standpoint of numerical analysis,

- it is well understood how to depart from the question of accuracy for short times and rather turn one's interest to stability and conservation issues in the long time, which is the purpose of geometric and symplectic integration,

- it is still an open question to go further than that in the spirit of the computation of averages: to a certain extent there is no better way in order to compute averages than to follow individual trajectories as precisely as possible (in the sense of the first item above).

Fourth, even when using up-to-date techniques for treating oscillatory systems together with the dedicated tools of geometric integration, the time T that can be reached for systems of practical interest is still often too small to get correct ensemble properties (and this is true even in the simplest versions of molecular dynamics). Therefore, classical molecular dynamics is not enough, and we resort to more efficient techniques. Such techniques basically all rely on stochastic simulations. As the main object is the phase space (a fact already exploited by the geometric viewpoint above), we definitely focus on the energy landscape, and designs techniques, more general than 'simply' following trajectories, that aim at exploring this landscape. This allows us to reach simulation times that are, eventually, of practical interest. The next section is devoted to such techniques for the acceleration of simulations in order to bridge the time gap.

In the rest of this section, we overview some of the most commonly used techniques for integration of the equations of motion over reasonably long times and for the treatment of oscillatory terms.

We begin with the long-time integration. From the numerical viewpoint, the purpose is, as usual, to build algorithms that reproduce the theoretical mathematical properties of the system to be simulated. The main property is *symplecticity* (and also possibly *reversibility in time* when the Hamiltonian is autonomous as in (5.9)).

We simply recall here that symplecticity always implies that the flow keeps the volume constant in phase space, which conservation is indeed related to the conservation of energy. In fact, it can be shown, by *backward analysis*, that algorithms enjoying symplecticity at the discrete level have the following property: their *numerical flow* Φ_n is close to the exact flow Φ of a Hamiltonian system (in fact approximately of order $e^{-1/\Delta t}$ if Δt denotes the discretization time-step). This latter system is not the original system, but its energy $\tilde{H}$ is close to the energy H of the original system (in fact approximately of order $(\Delta t)^p$ if the numerical scheme is of order p).

The flow $\tilde{\Phi}$, being the exact flow of a Hamiltonian system, preserves its energy $\tilde{H}$. Consequently, $\tilde{\Phi}$ almost preserves H. Finally, Φ_n, being close to $\tilde{\Phi}$, behaves accordingly, and indeed nearly conserves the energy at order $(\Delta t)^p$ over time intervals of length $e^{1/\Delta t}$. The symplecticity is thus the key property for the simulation of Hamiltonian systems on large times.

An interesting application of backward error analysis in this framework is performed in the recent works of Cancès *et al.* (2004b, 2004c) that study the speed of convergence of the discretized version of (5.12) toward the average value (5.11) when a symplectic scheme is used.[27] It is proved there that, in the long time limit, the fact that the numerical trajectory generated is the exact trajectory of a modified Hamiltonian allows us to evaluate averages over an isosurface of the Hamiltonian, with a speed of convergence that is $O(\frac{1}{T})$. More precisely, one can show that

$$\left| \left(\frac{1}{T} \int_0^T A(q(t), p(t)) \, dt \right)_{\substack{\text{numerical} \\ \text{approximation}}} - \langle A \rangle \right| = O\left(\frac{1}{T} \right) + O(\Delta t^r) \quad (5.14)$$

where r is the order of the symplectic scheme, and the prefactor in $O(\frac{1}{T})$ indeed depends on the largest oscillations in the system. This estimate can be rigorously established in the (somewhat academic) case of integrable systems (then the average $\langle A \rangle$ of course denotes the average for given values of the invariants of the system), and then extended by KAM theory to the case of near-integrable ones (loosely speaking, such systems that are perturbations of integrable systems behave like integrable systems over periods of time exponentially long w.r.t. the perturbation size). Despite this extension, the result unfortunately covers only a tiny subset of the set of Hamiltonians (see again Markus and Meyer (1974)). Estimate (5.14) in turn leads to an acceleration technique for the computation for averages by (5.12), the acceleration being based on the use of signal filtering techniques and ending up in a convergence at the rate $O(\frac{1}{T^k}) + O(\Delta t^r)$, k arbitrarily large, in (5.14). The technique yields promising results for test cases, but its adaptation to real cases of interest, where Hamiltonians are not near-integrable, is still unclear. It is of course to be emphasized that the technique aims at accelerating the $O(\frac{1}{T})$ convergence *only* when it holds. On the other hand, when, *e.g.*, the convergence only holds at the rate $O(\frac{1}{\sqrt{T}})$, which is the case for many systems of practical interest, then the method does not succeed in improving the rate of convergence. More details can be found in the above references. Despite the limitation of these works, they are among the rare

---

[27] Actually, the use of a *symmetric* scheme for a *reversible integrable* or a *reversible near-integrable* Hamiltonian would allow for the same conclusions (see Hairer *et al.* (2002, Chapter 11)).

ones that, in some weak sense at least, try to assess the accuracy of molecular dynamics simulations on the basis of the output they are primarily used to provide, namely averages. Some related issues are addressed by Tupper (2005).

The prototypical example of an algorithm that is symplectic and reversible and that is commonly used for molecular simulation is the following *leap-frog algorithm*, known in the chemistry literature as the *Verlet algorithm* for it was introduced there by Verlet (1967). It is an explicit algorithm that, for system (5.9) (for $M = 1$), becomes

$$\begin{cases} q_{n+1} & = q_n + \delta t\, p_{n+1/2}, \\ p_{n+1/2} & = p_{n-1/2} + (\delta t)\, \frac{\partial H}{\partial q}(q_n), \end{cases} \tag{5.15}$$

or equivalently

$$\begin{cases} q_{n+1} = q_n + \delta t\, p_n + \frac{(\delta t)^2}{2}\, \frac{\partial H}{\partial q}(q_n), \\ p_{n+1} = p_n + \frac{\delta t}{2}\left(\frac{\partial H}{\partial q}(q_{n+1}) + \frac{\partial H}{\partial q}(q_n)\right), \end{cases} \tag{5.16}$$

the latter version being called *velocity Verlet*.

This algorithm works remarkably well in the context of molecular dynamics. Higher-order schemes, still symplectic and reversible, are used when accuracy is required. For their expression, as well as for their numerical analysis, we refer to the treatises mentioned before. We also refer to Hairer, Lubich and Wanner (2003) for a pedagogic presentation of the Verlet algorithm.

The situation described above is the case when a Hamiltonian dynamics in the (microcanonical) (N, V, E) ensemble must be generated. But in fact, computing ensemble averages in this ensemble is of little interest[28] in comparison to the (canonical) (N, V, T) ensemble. Of course, in the limit of an infinite number of particles, or in an infinite volume, the two averages coincide (at least for local observables A), but this is not the case in practice.[29] Therefore, there is the need to sample the (N, V, T) ensemble and this can be performed by *ad hoc* trajectories. For this purpose, a method in chemistry is that of thermostats. The idea has been developed in Nosé (1984), Hoover (1985) and Nosé (1986). For the purpose of illustration, we only mention the method in its simplest case, namely for one particle in 1D.

[28] Note that we do not claim that simulating (N, V, E) trajectories is also of little interest, as it can help in sampling ensembles different from (N, V, E).

[29] Works in progress by Olla aim at evaluating the speed of convergence of one average to the other when the volume of the system goes to infinity.

The dynamics then reads

$$\begin{cases} \frac{\mathrm{d}q}{\mathrm{d}t} = \frac{p}{m}, \\ \frac{\mathrm{d}p}{\mathrm{d}t} = F - \frac{p_\xi}{Q}\, p, \\ \frac{\mathrm{d}\xi}{\mathrm{d}t} = \frac{p_\xi}{Q}, \\ \frac{\mathrm{d}p_\xi}{\mathrm{d}t} = \frac{1}{m}\, p^2 - k_B\, T, \end{cases} \tag{5.17}$$

where Q is a coupling constant and T denotes the temperature that needs to be fixed (k_B is the Boltzmann constant). The evolution of the additional pair of variables (ξ, p_ξ) aims to measure to what extent the constraint on the temperature is obeyed. This form is in fact only a trivial case of a general form of the so-called *Nosé–Hoover chain*, where more than one additional pair of variables is used. The method is widely used, but there are still open questions on its validity. First, on the very theoretical level, there is no justification of the method, even in the simplest occurrence described above.[30] Second, in practice, some observations are puzzling: for instance the efficiency of the method is highly sensitive to the number of thermostats used, and there is no convincing explanation of this fact. This twofold statement justifies at least the use of alternatives methods. One of them is based upon the Langevin equation. Simply stated, it consists in replacing the Hamiltonian dynamics,

$$\begin{cases} \frac{\mathrm{d}q}{\mathrm{d}t} = p, \\ \frac{\mathrm{d}p}{\mathrm{d}t} = -\nabla V(q), \end{cases} \tag{5.18}$$

by the stochastic dynamics,

$$\begin{cases} \mathrm{d}q = p\,\mathrm{d}t, \\ \mathrm{d}p = -\big(\gamma\, p + \nabla V(q)\big)\,\mathrm{d}t + \sigma\,\mathrm{d}W_t, \end{cases} \tag{5.19}$$

where $\mathrm{d}W_t$ is a Brownian motion, σ depends on the temperature, and $\gamma \neq 0$ is a constant. Here, there exists a theoretical foundation for the computation of the average of an observable using the evaluation along the trajectory. In other words, ergodicity need not be assumed: it is *proved*. There is convergence in law in the long time to the invariant measure for the (N, V, T) ensemble, and this convergence occurs exponentially fast, which is good news for the computational cost.[31] However, in practice, the need to

---

[30] It is only known that, if ergodicity is assumed, then the Nosé–Hoover dynamics does sample the correct ensemble (see Nosé (1984), and also Bond, Leimkuhler and Laird (1999) for a study devoted to the Nosé–Hoover dynamics).

[31] However, nothing is known mathematically on the variation of the speed of convergence with respect to the number of particles N (here taken to one for simplicity), while it is expected that the convergence is more rapid as N grows to infinity.

compute the empirical mean, which requires averaging over Brownian trajectories, counterbalances this gain. In comparison to other approaches, the method can be costly, but the fact that it has a sound theoretical ground is undoubtedly appealing. We refer to Mattingly, Stuart and Higham (2002) for a related mathematical study.

Let us turn to the treatment of oscillatory terms. In this respect, molecular mechanics is a domain very close to domains such as structural mechanics, robotics, chemical engineering, and other domains where systems of ODEs with different time-scales must be handled. Note that the difference in time-scales for the various variables can come from the characteristic time itself (a rapid oscillation of a bond), or also from a viewpoint mixing time-scales and distance-scales (the long-distance potential created by an atom that is far away needs not be updated so frequently). Heuristically, the treatment of the system can be based upon

- a dedicated treatment of the full system by multiple time-step methods,
- an elimination of the rapidly oscillating degrees of freedom using an algebraic constraint,
- the addition of a stochastic modelling, possibly ending in a friction term, in order to damp the rapid oscillations,

all options with a view to adopting a time-step in the simulation limited by the slow degrees of freedom and far larger than that of the rapid degrees. Whichever option is followed, one key issue is the way the information provided by the rapid degrees of freedom is inserted into the dynamics of the slow degrees of freedom, because at some point, some averaging or homogenization technique is required. This issue is indeed intimately related to renormalization techniques.

The toy model is that of the following system:

$$
\begin{cases}
\frac{dy}{dt} = f(y, z), \\
\varepsilon \frac{dz}{dt} = g(y, z),
\end{cases}
\tag{5.20}
$$

where y stands for the slow degrees of freedom and z for the rapid ones, ε denoting the discrepancy between the timescales. The direct treatment of the above system leads to operator splitting methods, where possibly the effect of the rapid variable z onto the slow one y is obtained by homogenization. We refer to the literature. On the other hand, ε can be considered as so small that the second equation of (5.20) is replaced by $0 = g(y, z)$, thus leading to the algebraic-differential system

$$
\begin{cases}
\frac{dy}{dt} = f(y, z), \\
0 = g(y, z).
\end{cases}
\tag{5.21}
$$

In the present context, a prototypical situation is the dynamics for the following Hamiltonian:

$$H(q,p) = \frac{p^2}{2} + V(q) + \frac{1}{\varepsilon} W(q), \tag{5.22}$$

namely

$$\begin{cases} \frac{\mathrm{d}q}{\mathrm{d}t} = p, \\ \varepsilon \frac{\mathrm{d}p}{\mathrm{d}t} = -\varepsilon \frac{\mathrm{d}V}{\mathrm{d}q}(q) - \frac{\mathrm{d}W}{\mathrm{d}q}(q), \end{cases} \tag{5.23}$$

which is approximated by

$$\begin{cases} \frac{\mathrm{d}q}{\mathrm{d}t} = p, \\ 0 = \frac{\mathrm{d}W}{\mathrm{d}q}(q), \end{cases} \tag{5.24}$$

The numerical simulation of a system like (5.21) resorts to the well-known techniques of constrained dynamics. The situation crucially depends on whether the constraint is holonomic or not, *i.e.*, depends only on the position of the system (which is the case in (5.24)), or also on the velocity. For holonomic constraints (think, for instance, of a bond length that is fixed in the molecular dynamics), algorithms such as SHAKE, introduced by Ryckaert, Giccotti and Berendsen (1977), and its improvement RATTLE due to Andersen (1983), are employed. Basically, they consist in running the first line of (5.21) regardless of the constraint, while next imposing the constraint by projection at each time-step. Higher-order variants of these algorithms have been introduced and developed by Jay (1994, 1996).

Regarding the alternative techniques based on an adequate stochastic damping of the rapid degrees of freedom, which is kind of a compromise between the true simulation of the complete system, and the constrained dynamics, the literature is very rich. We refer, *e.g.*, to the works by Schütte, Walter, Hartmann and Huisinga (2004) and Vanden-Eijnden (2003).

Before we get to the next section, we would like to mention a general-purpose technique that can be applied in particular to the molecular dynamics trajectories, in order to improve the applicability of all the above techniques. The technique we allude to is a domain decomposition technique *in time*, called the *parareal method*. It is well known that in scientific computing there have been many attempts to adapt the domain decomposition method, successfully applied to the space variables, in order to apply it to the time variable. The parareal method is the most recent attempt to date in this direction, and seems to be a breakthrough. It is originally due to Lions, Maday and Turinici (2001) and was further developed in a series of works by Maday, Turinici and collaborators. The method is dedicated to PDEs, while other previous attempts were focused on ODEs: see, *e.g.*, Bellen and Zennaro (1989) or Chartier and Philippe (1993). As in the spatial case, the paradigm is based upon adequate iterations between a local solver (fine time-step Δt) and a global, or coarse, solver. The originality of the method resides in the way the iterations are done. Its efficiency heavily relies upon the definition of the coarse solver. It can of course be the same solver as the local one, but with a larger time-step $\Delta T \gg \Delta t$, a choice that does work for parabolic equations and some hyperbolic equations. However, for many hyperbolic equations it is of crucial importance to design the coarse solver adequately, *e.g.*, by solving with the fine time-step Δt a coarse-grained model, otherwise the method is not interesting. An example of a successful application of the method to the present context is discussed in Baffico *et al.* (2002).

We devote the next section to our last point, the acceleration techniques.

5.3. *Methods for bridging the time-scale gap*

Basically, the dynamics of a molecular system consists of a succession of long periods of time where the system oscillates around local minimizers of the energy inside a given basin, and rapid transitions between two nearby basins, usually following a path that goes through a saddle-point. The direct simulation of a dynamics in this context requires long integration times that are mostly uninteresting, but that cannot be skipped without running the risk of missing the *rare events*, that is, escape from the current basin followed by transition to another.

Unfortunately, in many if not all cases, these rare events occur after a time that is not accessible to a direct simulation. Thus alternative techniques have to be developed. As briefly mentioned above, the techniques that have recently appeared in the literature and that aim to enlarge the applicability of molecular dynamics, focus on the energy landscape rather than on the dynamics itself. This has the twofold interest of making possible long time dynamics and identifying the relevant objects (saddle-points = transition states, and local minimizers = metastable states) for the system under study.

Schematically, one can say that the dynamics starting from an energy basin denoted by A can be decomposed in the following steps:

(a) find the list of basins B that are accessible from A,
(b) determine the pathway from A to each B,
(c) (randomly) choose one B in the list,

then set $A=B$ and continue. We now rapidly examine some issues related to each of the above steps.

The first difficulty to overcome when starting a dynamics from a point in a given energy basin is to find an appropriate escape path more quickly than in reality. More or less, the idea is to modify the probability of an escape. As the depth of the basin is indeed responsible for the long time needed to escape, a possibility is to 'modify' this depth.

First, we may notice that implicitly the depth that is seen by the system depends on the temperature: in a simplified picture, the transition state theory relates the energy barrier ΔE to overcome in order to escape the basin with the rate of escape r, at a given temperature θ, through

$$r = (\text{Const.}) \times \exp\left(-\frac{\Delta E}{k_B \theta}\right). \tag{5.25}$$

Therefore heating the system amounts to raising the probability of escaping from energy minimizers. This is the bottom line of the *temperature accelerated dynamics* introduced by Sorensen and Voter (2000) (see also Montalenti and Voter (2002)). Then we need to correct the output of the method, as of course some escapes that are almost impossible in practice will indeed be observed at virtual high temperatures: should the need arise, the dynamics is reflected back in the basin if the transition is inconvenient. Note that, from the practical viewpoint, it is an issue to know whether the upper boundary of the basin has been attained by the dynamics or not. One way to proceed is to perform a descent method, periodically along the dynamics, in order to check whether we get back to the starting basin A, or to another one $B \neq A$, the latter case indicating the transition. Another option is to monitor the lowest eigenvalue of the Hessian. As the *relative* rates of escape are correct but not the rates themselves (see (5.25)), we extrapolate from the virtual time of escape the real time of escape at the real temperature using formula (5.25). Impressive boosting factors of the dynamics can be observed (up to 10^7 in convenient cases) but some practical issues remain, and the method is not applicable to all situations.

Many alternative methods can be quoted, similar in spirit to the temperature accelerated dynamics. They modify the shape of the energy surface in order also to escape more easily. One is the *hyperdynamics method*, again due to Voter (1997), predating the temperature-accelerated method. It designs a biased potential surface with a potential $\tilde{V}$ modified from the original potential V, that ensures that the saddle-points of V and $\tilde{V}$ are the same, but that the basins of $\tilde{V}$ are less deep. We refer to Sanz-Navarro and Smith (2001) for an application. Another method is that introduced by Laio and Parrinello (2002), where a non-Markovian coarse-grained dynamics is performed. The idea is to modify on the fly the potential surface in order not to visit again zones that have already been explored, which amounts to filling in progressively each basin of the energy surface. A related idea is that introduced by Barth, Laird and Leimkuhler (2003), who also modify the energy landscape by generating a modified ensemble dynamics, which amounts to reducing the depths of the basins (in a slightly different way from above, since the method leaves the bottoms of the basins unchanged). Still another idea is explored in Darve, Wilson and Pohorille (2002).

A somewhat different paradigm can be used to deal with item (a) (together indeed with (b) and (c)): there is the possibility of running many trajectories (and this can be done on the original energy surface, or a modified one as above) in order to make the rare event more frequent in terms of wall-clock time. This is the essence of the *method of replicas* introduced in Voter (1998) (see also Voter and Sorensen (1999) for a review). With this method, the general spirit of most acceleration methods for molecular dynamics becomes obvious: the fundamental idea is to replace a long trajectory (that required by (5.12)) by a set of smaller ones, plus some post-treatment of the result. With the help of such methods, we are then able to reach longer times of virtual simulation, only performing simulation on small times. It must of course be understood that 'small' means small in comparison with the time T appearing in (5.12), but 'as large as possible' in view of the current state of the art for the best trajectory simulation algorithms available. In some sense, the dynamics simulation is seen as an inner calculation, inserted in a second step in a method based upon another paradigm.

For the sake of illustration, let us present a schematic description of the method of replicas. Several trajectories are simulated in parallel, and the transition time is modelled by an exponential law. When the first transition is observed in the list of all dynamics generated in parallel, the corresponding wall-clock time is set to the sum of all the individual times of each of the trajectories. In so doing, we obtain correct transition times, and may detect transitions to other basins more easily. The rates of transitions obtained may be used in a second step in a kinetic Monte Carlo simulation, a technique we will come back to below.

Let us turn specifically to step (b) and suppose that we have at hand two basins, respectively called A and B. We now need to discover the correct transition pathway from A to B, which in chemistry is called the *reaction pathway*, the curve abscissa along the pathway being called the *reaction coordinate*. Think typically of an angle in a molecule that modifies in order to oscillate between two enantiomers (stable configurations).

Techniques used at this stage essentially reduce to generating moves in the space of all trajectories linking A and B along the energy surface, with a view to determining the realistic one. Often it is one that goes through the saddle-point(s) separating the basin of A from that of B. The move in the trajectory space can be deterministic, and then it is based on the properties of the first and second derivatives near a saddle-point (following gradient information along the current trajectory to locate the saddle-point), or more stochastic, and not necessarily relying on a saddle-point information. Techniques in this vein are the *nudge elastic band* method by Henkelman and Jonsson (2000), the *string* method by E, Ren and Vanden-Eijnden (2002) (see also E, Ren and Vanden-Eijnden (2004*b*), the *dimer* method by Henkelman and Jonsson (1999), and also E, Ren and Vanden-Eijnden (2004a), the method by Garrahan and Chandler (2002), that by Zuckerman and Woolf (1999), by Ayala and Schlegel (1997), and so on. We also refer to the survey, with a self-explanatory title, by Bolhuis, Chandler, Dellago and Geissler (2002), and to the introductory text by Chandler (1998).

To perform step (c), a method already used in (a) to escape the basin may of course be used, but a new paradigm can be adopted. When minimizers and transition pathways are located,[32] the rate of escapes from each basin can be evaluated (by transition state theory) and a purely stochastic method can be performed, called *kinetic Monte Carlo*, based on the same paradigm as that mentioned above for the method of replicas: independent transition rates following exponential laws are employed. This time, no dynamical trajectory is explicitly used, and the long time dynamics of the system boils down to a sequence of random numbers, monitoring in fact the transitions and non-escapes in a follow-up of basins. An instance of an alternative approach for item (c) is introduced in Mousseau and Barkema (2004).

What about the mathematical understanding of all the above methods? Actually, depending on the viewpoint, it can be considered as either sufficient or poor. As the output of most methods is a simple Monte Carlo simulation, and since this method has long been the method of choice for studies in probability theory because of the wide range of its applications, there is not much to say. On the other hand, the global strategy definitely needs some understanding and mathematical foundation. A series of works with a solid mathematical ground is that by Huisinga and Schütte (2003), which also gives rise at the algorithmic level to reportedly efficient methods for the above problem: see Deuflhard, Huisinga, Fischer and Schütte (2000), Schütte, Fischer, Huisinga and Deuflhard (1999), and other works by the same authors. The formalization is as follows. The dynamics over a time frame of width τ is seen as a transition operator T_τ that associates the initial point to the final one. If the configuration space is discretized, the examination of the spectrum and spectral projector of this transition operator reveals the metastable zones of the configuration space (in brief, the zones where it is mostly likely that over the time τ any trajectory starting from a point in the zone will end up in another point in the same zone) and the transition rates from one metastable state to the other.

We wish to emphasize that, clearly, the above list of methods is by no means comprehensive. For instance, we have not even approached the techniques based upon the renormalization paradigm, where a coarse-grained

---

[32] In fact there is no hope of locating all of them, but the above means that, at some stage, it is considered that a sufficient knowledge of the topology of the energy surface has been reached, at least at the vicinity of the configuration of the system under study, so that one may proceed to the next step.

model featuring only a limited number of degrees of freedom is derived. Because of this limited size, classical approaches such as those of the previous section are efficient. Typically a set of mesoscale (*i.e.*, intermediate scale) particles is simulated that interact through an adequate potential (the determination of which is the key issue). We refer, *e.g.*, to Forrest and Suter (1995) and to the so-called dissipative particle dynamics technique.

The field of acceleration methods for molecular dynamics is under construction, and we are convinced it will witness huge efforts in view of the importance of the applications.

6. Current and future trends

The domain of computational chemistry is now a well-established domain both from the downstream standpoint, that of applications, and from the upstream standpoint, that of theory, modelling and, increasingly, numerical analysis. To some extent, the publication of the present article in *Acta Numerica* testifies to the latter.

The relevance of computational chemistry does not only concern traditional fields such as chemistry, and its companion field biology. Because the size of technological devices is ever shrinking, once irrelevant phenomena at the microscopic level are now considered relevant. At the other end of the spectrum, for macroscopic compounds or devices, understanding of microscopic behaviour is becoming a key issue, in order to reach maximum efficiency. Let us quote two instances of this trend.

In biology, dynamics of proteins, possibly with chemical reactions at some sites, involves the coupled simulation of quantum degrees of freedom with classical ones: see Monard and Mertz, Jr. (1999) for a review of applications in biochemistry.

In materials science, multiscale methods in computational mechanics are growing in importance. There, computational chemistry is involved in the simulation of the microscopic degrees of freedom, and coupled to more usual techniques of the engineering sciences, such as finite element methods for computational continuum mechanics. The simulation of dislocations, fractures, *etc.*, with a view to further understanding, *e.g.*, fatigue phenomena cannot be studied without a pinch of computational chemistry. For instances of works and challenges in this direction, we refer to the books by Barth, Chan and Haimes, eds (2002), Bulatov *et al.* (1999), Deák, Frauenheim and Pederson, eds (2000), Kirchner, Kubin and Pontikis, eds (1996), Kitagawa *et al.*, eds (1998), Raabe (1998), the special issue by Liu *et al.*, eds (2004), and to the emerging techniques coupling the atomistic level to the continuum description such as Tadmor, Smith, Bernstein and Kaxiras (1999), Shenoy *et al.* (1998), Miller, Tadmor, Phillips and Ortiz (1998), Tadmor, Phillips and Ortiz (1996), Tadmor, Ortiz and Phillips (1996) and Shenoy *et al.* (1999).

All this makes computational chemistry an important field for the future. From the numerical analysis viewpoint, however, the field is not completely explored. It can be considered that some challenging issues, such as the convergence of SCF algorithms, have been treated in a satisfactory way, but many well-established techniques still require mathematical understanding. It is of course even clearer for the new techniques that appear almost every day, and that aim to treat problems of outstanding difficulty: linear scaling methods for the static description of large size systems, evolution PDEs in high dimensions, acceleration methods for dynamics over long times, and so on.

Obviously, the competences required are varied: optimization, linear and nonlinear programming, approximations of PDEs, of ODEs, stochastic processes, to name a few.

Acknowledgements

I have benefited from enlightening interactions with a number of colleagues. Some of them also made a careful reading of a draft version of this article, suggested many improvements, or pointed out references that are included in the present version. I would like to thank all of them: X. Blanc, E. Cancès, F. Castella, Ph. Chartier, M. Esteban, E. Faou, M. Griebel, F. Legoll, T. Lelièvre, M. Lewin, C. Lubich, S. Olla, A. Patera, E. Séré, P. Tupper, G. Turinici and G. Zérah. It is also my pleasure to thank my co-authors for the book Cancès, Defranceschi, Kutzelnigg, Le Bris and Maday (2003b). Some sections of the present paper draw much of their inspiration from this earlier work.

REFERENCES

M. P. Allen and D. J. Tildesley (1987), *Computer Simulation of Liquids*, Oxford Science Publications.

B. O. Almroth, P. Stern and F. A. Brogan (1978), 'Automatic choice of global shape functions in structural analysis', *AIAA Journal* **16**, 525–528.

H. C. Andersen (1983), 'Rattle: A "velocity" version of the Shake algorithm for molecular dynamics calculations', *J. Comput. Phys.* **52**, 24–34.

D. A. Areshkin, O. A. Shenderova, J. D. Schall and D. W. Brenner (2003), 'Convergence acceleration scheme for self consistent orthogonal basis set electronic structure methods', *Molecular Simulation* **29**, 269–286.

N. W. Ashcroft and N. D. Mermin (1976), *Solid-State Physics*, Saunders College Publishing.

G. Auchmuty and Wenyao Jia (1994), 'Convergent iterative methods for the Hartree eigenproblem', *Math. Model. Numer. Anal.* **28**, 575–610.

P. Y. Ayala and P. B. Schlegel (1997), 'A combined method for determining reaction paths, minima and transition state geometries', *J. Chem. Phys.* **107**, 375–384.

V. Bach (1992), 'Error bound for the Hartree–Fock energy of atoms and molecules', *Comm. Math. Phys.* **147**, 527–548.

V. Bach, E. H. Lieb, M. Loss and J. P. Solovej (1994), 'There are no unfilled shells in unrestricted Hartree–Fock theory', *Phys. Rev. Lett.* **72**, 2981–2983.

G. B. Bacskay (1961), 'A quadratically convergent Hartree–Fock (QC-SCF) method: Application to closed shell systems', *Chem. Phys.* **61** 385–404.

L. Baffico, S. Benard, Y. Maday, G. Turinici and G. Zerah (2002), 'Parallel in time molecular dynamics simulations', *Phys. Rev. E* **66**, 057701.

A. Bandrauk, M. Delfour and C. Le Bris, eds (2004), *Quantum Control: Mathematical and Numerical Challenges*, AMS, CRM proceedings series.

M. Barrault, Y. Maday, N. C. Nguyen and A. Patera (2004*a*), 'An "empirical interpolation" method: Application to efficient reduced-basis discretization of PDEs', *CR Acad. Sci., Paris, Série I, Math.* **339**, 667–672.

M. Barrault, E. Cancès, C. Le Bris, Y. Maday, N. C. Nguyen, A. Patera and G. Turinici (2004*b*), in preparation.

M. Barrault, E. Cancès, W. Hager and C. Le Bris (2004*c*), 'Multilevel domain decomposition method for electronic structure calculations', in preparation.

M. Barrault, G. Bencteux, E. Cancès and V. Duwig (2004*d*) Toward a linear scaling for ill-conditioned eigenvalue problems in quantum chemistry. Technical report.

E. J. Barth, B. B. Laird and B. J. Leimkuhler (2003), 'Generating generalized distributions from dynamical simulation', *J. Chem. Phys.* **118** 5759.

T. J. Barth, T. Chan and R. Haimes, eds (2002), *Multiscale and Multiresolution Methods: Theory and Applications*, Vol. 20 of *Lecture Notes in Computational Science and Engineering*, Springer.

T. L. Beck (2000), 'Real-space mesh techniques in density-functional theory', *Rev. Modern Phys.* **72**, 1041–1080.

A. Bellen and M. Zennaro (1989), 'Parallel algorithms for initial-value problems for difference and differential equations', *J. Comput. Appl. Math.* **25**, 341–350.

X. Blanc (2000), A mathematical insight into ab initio simulations of solid phase, in *Mathematical Models and Methods for ab initio Quantum Chemistry* (M. Defranceschi and C. Le Bris, eds), Vol. 74 of *Lecture Notes in Chemistry*, Springer, pp. 133–158.

X. Blanc and E. Cancès (2004), Nonlinear instability of density-independent orbital-free kinetic energy functionals, *J. Chem. Phys.*, submitted.

P. Blanchard and E. Bruning (1982), *Variational Methods in Mathematical Physics*, Springer.

P. E. Blöchl (1994), 'Projector augmented-wave method', *Phys. Rev. B* **50**, 17953–17979.

O. M. Bokanowski and M. Lemou (1998), 'Fast multipole method for multidimensional integrals', *CR Acad. Sci., Paris, Série I, Math.* **326** 105–110.

O. M. Bokanowski and M. Lemou (2001), 'Fast multipole method for multi-variable integrals', Laboratoire MIP, Toulouse, preprint.

P. G. Bolhuis, D. Chandler, C. Dellago and P. Geissler (2002), 'Transition paths sampling: Throwing ropes over rough mountain passes, in the dark', *Annu. Rev. Phys. Chem.* **53**, 291–318.

S. D. Bond, B. J. Leimkuhler and B. B. Laird (1999), 'The Nosé–Poincaré method for constant temperature molecular dynamics', *J. Comput. Phys.* **151** 114–134.

F. A. Bornemann (1998), *Homogenization in Time of Singularly Perturbed Mechanical Systems*, Vol. 1697 of *Lecture Notes in Mathematics*, Springer.

F. A. Bornemann and C. Schütte (1998), 'A mathematical investigation of the Car–Parrinello method', *Numer. Math.* **78**, 359–376.

F. A. Bornemann, P. Nettesheim and C. Schütte (1996), 'Quantum-classical molecular dynamics as an approximation to full quantum dynamics', *J. Chem. Phys.* **105**, 1074–1083.

D. Bowler and M. Gillan (1999), 'Density matrices in $O(N)$ electronic structure calculations', *Comput. Phys. Comm.* **120**, 95–108.

D. Bowler et al. (1997), 'A comparison of linear scaling tight-binding methods', *Model. Simul. Mater. Sci. Eng.* **5**, 199–202.

D. Bowler, T. Miyazaki and M. Gillan (2002), 'Recent progress in linear scaling *ab initio* electronic structure theories', *J. Phys. Condens. Matter* **14**, 2781–2798.

S. F. Boys (1950), 'Electronic wavefunctions I: A general method of calculation for the stationary states of any molecular system', *Proc. Roy. Soc. A* **200**, 542–554.

V. V. Bulatov et coll. (1999), *Multiscale Modelling of Materials*, MRS.

H. J. Bungartz and M. Griebel (2004), 'Sparse grids', in *Acta Numerica*, Vol. 13, Cambridge University Press, pp. 147–269.

E. Cancès (2000), SCF algorithms for Hartree–Fock electronic calculations, in *Mathematical Models and Methods for ab initio Quantum Chemistry* (M. Defranceschi and C. Le Bris, eds), Vol. 74 of *Lecture Notes in Chemistry*, Springer, pp. 17–43.

E. Cancès (2001), 'SCF algorithms for Kohn–Sham models with fractional occupation numbers', *J. Chem. Phys.* **114**, 10616–10623.

E. Cancès and C. Le Bris (1998), 'On the perturbation method for some nonlinear Quantum Chemistry models', *Math. Models Methods Appl. Sci.* **8**, 55–94.

E. Cancès and C. Le Bris (1999), 'On the time-dependent electronic Hartree–Fock equations coupled with a classical nuclear dynamics', *Math. Models Methods Appl. Sci.* **9**, 963–990.

E. Cancès and C. Le Bris (2000a), 'On the convergence of SCF algorithms for the Hartree–Fock equations', *Math. Model. Numer. Anal.* **34**, 749–774.

E. Cancès and C. Le Bris (2000b), 'Can we outperform the DIIS approach for electronic structure calculations', *Int. J. Quantum Chem.* **79**, 82–90.

E. Cancès, C. Le Bris, B. Mennucci and J. Tomasi (1999), 'Integral equation methods for molecular scale calculations in the liquid phase', *Math. Models Methods Appl. Sci.* **9**, 35–44.

E. Cancès, C. Le Bris, Y. Maday and G. Turinici (2002), 'Towards reduced basis approaches in *ab initio* electronic computations', *J. Sci. Comput.*, **17**, 461–469.

E. Cancès, K. Kudin, G. Scuseria and G. Turinici (2003a), 'Quadratically convergent algorithm for fractional occupation numbers in density functional theory', *J. Chem. Phys.* **118**, 5364–5368.

E. Cancès, M. Defranceschi, W. Kutzelnigg, C. Le Bris and Y. Maday (2003b), Computational chemistry: A primer, in *Handbook of Numerical Analysis*,

Vol. X, *Computational Chemistry* (C. Le Bris, ed.), North-Holland, pp. 3–270.

E. Cancès, H. Galicher and M. Lewin, (2004a), 'Computing electronic structures: A new multiconfiguration approach for excited states', *J. Comput. Phys.*, submitted.

E. Cancès, F. Castella, P. Chartier, E. Faou, C. Le Bris, F. Legoll and G. Turinici (2004b), 'High-order averaging schemes with error bounds for thermodynamical properties calculations by MD simulations', *J. Chem. Phys.* **121** 10346–10355.

E. Cancès, F. Castella, P. Chartier, E. Faou, C. Le Bris, F. Legoll and G. Turinici (2004c), 'Long-time averaging for integrable Hamiltonian dynamics', *Numer. Math.*, to appear.

E. Cancès, B. Jourdain and T. Lelièvre (2004d), 'Quantum Monte-Carlo simulations of fermions: A mathematical analysis of the fixed-node approximation', *Math. Models Methods Appl. Sci.*, submitted.

R. Car and M. Parrinello (1985), 'Unified approach for molecular dynamics and density functional theory', *Phys. Rev. Lett.* **55**, 2471–2474.

E. Carter and Y. A. Wang (2000), Orbital-free kinetic energy density functional theory, in *Theoretical Methods in Condensed Phase Chemistry* (S. D. Schwartz, ed.), Progress in Theoretical Chemistry and Physics, Kluwer, pp. 117–84.

I. Catto and P.-L. Lions (1992), 'Binding of atoms and stability of molecules in Hartree and Thomas–Fermi type theories I: A necessary and sufficient condition for the stability of general molecular systems', *Comm. Part. Diff. Equ.* **17**, 1051–1110.

I. Catto and P.-L. Lions (1993a), 'Binding of atoms and stability of molecules in Hartree and Thomas–Fermi type theories 2: Stability is equivalent to the binding of neutral subsystems', *Comm. Part. Diff. Equ.* **18**, 305–354.

I. Catto and P.-L. Lions (1993b), 'Binding of atoms and stability of molecules in Hartree and Thomas–Fermi type theories 3: Binding of neutral subsystems', *Comm. Part. Diff. Equ.* **18**, 381–429.

I. Catto and P.-L. Lions (1993c), 'Binding of atoms and stability of molecules in Hartree and Thomas–Fermi type theories 4: Binding of neutral systems for the Hartree model', *Comm. Part. Diff. Equ.* **18**, 1149–1159.

I. Catto, C. Le Bris and P.-L. Lions (1998), *Mathematical Theory of Thermodynamic Limits: Thomas–Fermi Type Models*, Oxford University Press.

I. Catto, C. Le Bris and P.-L. Lions (2001), 'On the thermodynamic limit for Hartree–Fock type models', *Ann. Inst. Henri Poincaré Anal. non linéaire* **18**, 687–760.

I. Catto, C. Le Bris and P.-L. Lions (2002), 'On some periodic Hartree-type models for crystals', *Ann. Inst. Henri Poincaré Anal. non linéaire* **19**, 143–190.

G. Chaban, M. W. Schmidt and M. S. Gordon (1997), 'Approximate second order method for orbital optimization of SCF and MCSCF wavefunctions', *Theor. Chem. Acc.* **97**, 88–95.

J. M. Chadam and R. T. Glassey (1975), 'Global existence of solutions to the Cauchy problem for time-dependent Hartree equations', *J. Math. Phys.* **16**, 1122–1230.

M. Challacombe (2000), 'Linear scaling computation of the Fock matrix V: Hierarchical cubature for numerical integration of the exchange-correlation matrix', *J. Chem. Phys.* **113**, 10037–10043

D. Chandler (1998), Barrier crossings: Classical theory of rare but important events, in *Classical and Quantum Dynamics in Condensed Phase Simulations*, (Berne *et al.*, eds), World Scientific, p. 3.

P. Chartier and B. Philippe (1993), 'A parallel shooting technique for solving dissipative ODEs', *Computing*, **51** (3–4).

E. Clementi and D. R. Davis (1966), 'Electronic structure of large molecular systems', *J. Comput. Phys.* **1**, 223–244.

A. J. Coleman and V. I. Yukalov (2000), *Reduced Density Matrices*, Vol. 72 of *Lecture Notes in Chemistry*, Springer.

H. L. Cycon, R. G. Froese, W. Kirsch and B. Simon (1987), *Schrödinger Operators with Applications to Quantum Mechanics and Global Geometry*, Springer.

A. Daniels and G. Scuseria (1999), 'What is the best alternative to diagonalization of the Hamiltonian in large scale semiempirical calculations?', *J. Chem. Phys.* **110**, 1321–1328.

E. Darve, M. Wilson and A. Pohorille (2002), 'Calculating free energies using scaled-force molecular dynamics algorithm', *Molecular Simulation* **28**, 113.

P. Deák, T. Frauenheim and M. R. Pederson, eds (2000), *Computer Simulation of Materials at Atomic Level*, Wiley, 2000.

M. Defranceschi and P. Fischer (1998), 'Numerical solution of the Schrödinger equation in a wavelet basis for hydrogen-like atoms', *SIAM J. Numer. Anal.* **35**, 1–12.

M. Defranceschi and C. Le Bris (1997), 'Computing a molecule: A mathematical viewpoint', *J. Math. Chem.* **21**, 1–30.

M. Defranceschi and C. Le Bris (1999), 'Computing a molecule in its environment: A mathematical viewpoint', *Int. J. Quantum Chem.* **71**, 257–250.

M. Defranceschi and C. Le Bris, eds (2000), *Mathematical Models and Methods for ab initio Quantum Chemistry*, Vol. 74 of *Lecture Notes in Chemistry*, Springer.

J. W. Demmel (1997), *Applied Numerical Linear Algebra*, SIAM.

J. P. Desclaux, J. Dolbeault, P. Indelicato, M. J. Esteban and E. Séré, (2003), Computational approaches of relativistic models in quantum chemistry, in *Handbook of Numerical Analysis* (P. G. Ciarlet, series ed.), Vol. X, *Computational Chemistry* (C. Le Bris, ed.), North-Holland.

P. Deuflhard *et al.*, eds (1999), *Computational Molecular Dynamics: Challenges, Methods, Ideas, Lecture Notes in Computational Science and Engineering*, Springer.

P. Deuflhard, W. Huisinga, A. Fischer and C. Schütte (2000), 'Identification of almost invariant aggregates in reversible nearly uncoupled Markov chains', *Linear Algebra Appl.* **315**, 39–59.

J. Dolbeault, M. J. Esteban and E. Séré (2000*a*), Variational methods in relativistic quantum mechanics: New approach to the computation of Dirac eigenvalues, in *Mathematical Models and Methods for ab initio Quantum Chemistry* (C. Le Bris and M. Defranceschi, eds), Vol. 74 of *Lecture Notes in Chemistry*, Springer.

J. Dolbeault, M. J. Esteban, E. Séré and M. Vanbreugel (2000*b*), 'Minimization methods for the one-particle Dirac equation', *Phys. Rev. Lett.* **85**, 4020–4023

J. Dolbeault, M. J. Esteban and E. Séré (2003), 'A variational method for relativistic computations in atomic and molecular physics', *Int. J. Quantum Chem.* **93**, 149–155.

J. Douady, Y. Ellinger, R. Subra and B. Levy (1980), 'Exponential transformation of molecular orbitals: A quadratically convergent SCF procedure I: General formulation and application to closed-shell ground states', *J. Chem. Phys.* **72**, 1452–1462.

R. Dovesi, R. Orlando, C. Roetti, C. Pisani and V. R. Saunders (2000), 'The periodic Hartree–Fock method and its implementation in the Crystal code', *Phys. Stat. Sol.* (b) **217**, 63–88.

R. M. Dreizler and E. K. U. Gross (1990), *Density Functional Theory*, Springer.

W. E, W. Ren and E. Vanden-Eijnden (2002), 'String method for the study of rare events', *Phys. Rev. B* **66**, 052301.

W. E, W. Ren and E. Vanden-Eijnden (2004*a*), 'Minimum action method for the study of rare events', *Comm. Pure Appl. Math.* **LVII**, 1–20.

W. E, W. Ren and E. Vanden-Eijnden (2004*b*), 'Finite temperature string method for the study of rare events', preprint.

M. J. Esteban and E. Séré (2002), 'An overview on linear and nonlinear Dirac equations', *Discrete Contin. Dyn. Syst.* **8** (2), 381–397.

D. F. Feller and K. Ruedenberg (1979), 'Systematic approach to extended even-tempered orbital bases for atomic and molecular calculations', *Theoret. Chim. Acta* **52**, 231–251.

T. H. Fischer and J. Almlöf (1992), 'General methods for geometry and wave function optimization', *J. Phys. Chem.* **96**, 9768–9774.

V. Fock (1958), 'On Schrödinger equation for the helium atom', *K. Norske Vidensk. Selsk. Forhandl.* **31**, 138–152.

B. M. Forrest and U. W. Suter(1995), 'Accelerated equilibration of polymer metals by time coarse graining', *J. Chem. Phys.* **18**, 7256–7266.

S. Fournais, M. Hoffmann-Ostenhof, T. Hoffmann-Ostenhof and T. Sorensen (2002*a*), On the regularity of the density of electronic wavefunctions, in *Mathematical Results in Quantum Mechanics: Taxco 2001*, Vol. 307 of *Contemp. Math.*, AMS, Providence, RI, pp. 143–148.

S. Fournais, M. Hoffmann-Ostenhof, T. Hoffmann-Ostenhof and T. Sorensen (2002*b*), 'The electron density is smooth away from the nuclei', *Comm. Math. Phys.* **228**, 401–415.

S. Fournais, M. Hoffmann-Ostenhof, T. Hoffmann-Ostenhof and T. Sorensen (2004), 'Analyticity of the density of electronic wavefunctions', *Ark. Mat.* **42**, 87–106.

D. Frenkel and B. Smit (2001), *Understanding Molecular Simulation*, 2nd edn, Academic Press.

G. Friesecke (2003), 'The multiconfiguration equations for atoms and molecules: Charge quantization and existence of solutions', *Arch. Rat. Mech. Anal.* **169**, 35–71.

G. Galli (2000), 'Large scale electronic structure calculations using linear scaling methods', *Phys. Stat. Sol.* (b) **217**, 231–249.

J. Garcke and M. Griebel (2000), 'On the computation of the eigenproblems of hydrogen and helium in strong magnetic and electric fields with the sparse grid combination technique', *J. Comput. Phys.* **165**, 694–716.

J. P. Garrahan and D. Chandler (2002), 'Geometrical explanation and scaling of dynamical heterogeneities in glass forming systems', *Phys. Rev. Lett.* **89** (3), 035704–035707.

P. M. W. Gill (1994), Molecular integrals over Gaussian basis functions, in *Advances in Quantum Chemistry* (P.-O. Löwdin, J. R. Sabin and M. C. Zerner, eds), Vol. 25, Academic Press, pp. 141–205.

S. Goedecker (1999), 'Linear scaling electronic structure methods', *Rev. Modern Phys.* **71**, 1085–1123.

R. Greengard and V. Rokhlin (1997), A new version of the fast multipole method for the Laplace equation in three dimensions, in *Acta Numerica*, Vol. 6, Cambridge University Press, pp. 229–269.

M. Griebel, P. Oswald and T. Schlekefer (1999), 'Sparse grids for boundary integral equations', *Numer. Math.* **83** (2), 279–312.

M. Griebel, S. Knapek, G. Zumbusch and A. Caglar (2004), *Numerische Simulation in der Moleküldynamik: Numerik, Algorithmen, Parallelisierung, Anwendungen*, Springer.

S. J. Gustafson and I. M. Sigal (2003), *Mathematical Concepts of Quantum Mechanics*, Universitext, Springer.

M. C. Gutzwiller (1990), *Chaos in Classical and Quantum Mechanics*, Vol. 1 of *Interdisciplinary Applied Mathematics*, Springer.

W. Hackbusch (2001), 'The efficient computation of certain determinants arising in the treatment of Schrödinger's equations', *Computing* **67**, 35–56.

G. A. Hagedorn (1996), 'Crossing the interface between chemistry and mathematics', *Notices Amer. Math. Soc.*, March 1996, 297–299.

J. M. Haile (1992), *Molecular Dynamics Simulations*, Wiley.

E. Hairer and G. Wanner (1996), *Solving Ordinary Differential Equations II*, 2nd edn, Springer.

E. Hairer, S. P. Nørsett and G. Wanner (1993), *Solving Ordinary Differential Equations I*, 2nd edn, Springer.

E. Hairer, C. Lubich and G. Wanner (2002), *Geometric Numerical Integration: Structure-Preserving Algorithms for Ordinary Differential Equations*, Springer.

E. Hairer, C. Lubich and G. Wanner (2003), Geometric numerical integration illustrated by the Störmer–Verlet method, in *Acta Numerica*, Vol. 12, Cambridge University Press, pp. 399–450.

M. Head-Gordon, Y. Shao, C. Saravan and C. White (2003), 'Curvy steps for density matrix based energy minimization: Tensor formulation and toy applications', *Mol. Phys.* **101** (1–2), 37–43.

W. J. Hehre, L. Radom, P. v. R. Schleyer and J. A. Pople (1986), *Ab initio Molecular Orbital Theory*, Wiley.

W. Heitler and F. London (1927), 'Wechselwirkung neutraler Atome und homöopolare Bindung nach der Quantenmechanik', *Z. Phys.* **44**, 455–472.

G. Henkelman and H. Jonsson (1999), 'A dimer method for finding saddle points on high dimensional potential surfaces using only first derivatives', *J. Chem. Phys.* **111** (15), 7010–7022.

G. Henkelman and H. Jonsson (2000), 'Improved tangent estimates in the nudged elastic band method for finding minimum energy paths and saddle points', *J. Chem. Phys.* **113** (22), 9978–9985.

S. C. Hill (1985), 'Rates of convergence and error estimation formulas for the Rayleigh–Ritz variational method', *J. Chem. Phys.* **83**, 1173–1196.

R. N. Hill (1995), 'Dependence of the rate of convergence of the Rayleigh–Ritz method on a nonlinear parameter', *Phys. Rev. B* **51**, 4433–4471.

P. Hohenberg and W. Kohn (1964), 'Inhomogeneous electron gas', *Phys. Rev. B* **136**, 864–871.

M. Hoffmann-Ostenhof, T. Hoffmann-Ostenhof and T. Sorensen (2001), 'Electron wavefunctions and densities for atoms', *Ann. Henri Poincaré* **2**, 77–100.

W. H. Hoover (1986), 'Canonical dynamics: Equilibrium phase-space distributions', *Phys. Rev. A* **31**, 1695–1697.

W. Huisinga and C. Schütte (2003), Biomolecular conformations can be identified as metastable sets of molecular dynamics, in *Handbook of Numerical Analysis*, Vol. X, *Computational Chemistry* (C. Le Bris, ed.), North-Holland.

E. A. Hylleraas (1929), 'Neue Berechnung der Energie des Heliums im Grundzustande, sowie des tiefsten Terms von Ortho-Helium', *Z. Phys.* **54**, 347–366.

T. Jahnke (2003), Numerische Verfahren für fast adiabatische Quantendynamik, PhD Dissertation, Universität Tübingen.

T. Jahnke (2004), 'Long-time-step integrators for almost adiabatic quantum dynamics', *SIAM J. Sci. Comput.* **25**, 2145–2164.

T. Jahnke and C. Lubich (2003), 'Numerical integrators for quantum dynamics close to the adiabatic limit', *Numer. Math.* **94** (2), 289–314.

H. M. James and A. S. Coolidge (1933), 'The ground state of the hydrogen molecule', *J. Chem. Phys.* **1**, 825–835.

L. O. Jay (1994), Runge–Kutta type methods for index three differential-algebraic equations with applications to Hamiltonian systems, PhD thesis, Department of Mathematics, University of Geneva.

L. O. Jay (1996), 'Symplectic partitioned Runge–Kutta methods for constrained Hamiltonian systems', *SIAM J. Numer. Anal.* **33**, 368–387.

L. O. Jay, H. Kim, Y. Saad and J. R. Chelikowski (1999), 'Electronic structure calculations for plane wave codes without diagonalization', *Comput. Phys. Comm.* **118**, 21–30.

R. O. Jones and O. Gunnarsson, 'The density functional formalism, its applications and prospects', *Rev. Modern Phys.* **61**, 689–746.

T. Kinoshita (1957), 'Ground state of the helium atom', *Phys. Rev.* **105**, 1490–1502.

O. Kirchner, L. P. Kubin and V. Pontikis, eds (1996), *Computer Simulation in Materials Science*, Kluwer.

H. Kitagawa *et al.*, eds (1998), *Mesoscopic Dynamics of Fracture, Advances in Materials Research*, Springer.

C. Kittel (1996), *Introduction to Solid State Physics*, 7th edn, Wiley.

B. Klahn and W. A. Bingel (1977), 'The convergence of the Rayleigh–Ritz method in quantum chemistry', *Theoret. Chim. Acta* **44**, 26–43.

B. Klahn and J. D. Morgan, III (1984), 'Rates of convergence of variational calculations of expectation values', *J. Chem. Phys.* **81**, 410–433.

W. Klopper and W. Kutzelnigg (1986), 'Gaussian basis sets and the nuclear cusp problem', *J. Mol. Struct. Theochem.* **135**, 339–356.

W. Kohn (1996), 'Density functional and density matrix method scaling linearly with the number of atoms', *Phys. Rev. Lett.* **76**, 3168–3171.

W. Kohn (1999), 'Nobel Lecture: Electronic structure of matter-wave functions and density functionals', *Rev. Modern Phys.* **71**, 1253–1266.

W. Kohn and L. J. Sham, (1965), 'Self-consistent equations including exchange and correlation effects', *Phys. Rev. A* **140**, 1133–1138.

W. Kolos and J. Rychlewski (1993), 'Improved theoretical dissociation energy and ionization potential for the ground-state of the hydrogen molecule', *J. Chem. Phys.* **98**, 3960–3967.

J. Koutecký and V. Bonacic (1971), 'On convergence difficulties in the iterative Hartree–Fock procedure', *J. Chem. Phys.* **55**, 2408–2413.

K. Kudin and G. E. Scuseria (1998), 'A fast multipole algorithm for the efficient treatment of the Coulomb problem in electronic structure calculations of periodic systems with Gaussian orbitals', *Chem. Phys. Lett.* **289**, 611–616.

K. Kudin, G. E. Scuseria and E. Cancès (2002), 'A black-box self-consistent field convergence algorithm: One step closer', *J. Chem. Phys.* **116**, 8255–8261.

W. Kutzelnigg (1989), Convergence expansions in Gaussian basis' in *Strategies and Applications in Quantum Chemistry: A Tribute to G. Berthier* (M. Defranceschi and Y. Ellinger, eds), Kluwer, Dordrecht, pp. 79–102.

W. Kutzelnigg (1994), 'Theory of the expansion of wave functions in a Gaussian basis', *Int. J. Quantum Chem.* **51**, 447–463.

W. Kutzelnigg and J. D. Morgan III (1992), 'Rate of convergence of the partial-wave expansions of atomic correlation energies', *J. Chem. Phys.* **96**, 4484–4508.

A. Laio and M. Parrinello (2002), 'Escaping free-energy minima', *Proc. Nat. Acad. Sci.* **99**, 20.

C. Le Bris (1994), 'A general approach for multiconfiguration methods in quantum molecular chemistry', *Ann. Inst. Henri Poincaré Anal. non linéaire* **11**, 441–484.

C. Le Bris, ed. (2003), *Handbook of Numerical Analysis*, Vol. X, *Computational Chemistry*, North-Holland.

C. Le Bris and P.-L. Lions (2005), 'From atoms to crystals: A mathematical journey', *Bull. Amer. Math. Soc.*, in press.

B. Leimkuhler and S. Reich (2005), *Simulating Hamiltonian Mechanics*, Cambridge Monographs on Applied and Computational Mathematics, Cambridge University Press, to appear.

I. N. Levine (1991), *Quantum Chemistry*, 4th edn, Prentice-Hall.

M. Lewin (2004), 'Solutions of the multiconfiguration equations in quantum chemistry', *Arch. Rat. Mech. Anal.* **171** (1), 83–114.

X. P. Li, R. W. Nunes and D. Vanderbilt (1993), 'Density matrix electronic structure method with linear system-size scaling', *Phys. Rev B* **47**, 10891–10894.

W. Liang, C. Saravan, Y. Shao, R. Baer, A. T. Bell and M. Head-Gordon (2003), 'Improved Fermi operator expansion methods for fast electronic structure calculations', *J. Chem. Phys.* **119** (8), 4117–4125.

E. H. Lieb (1981), 'Thomas–Fermi and related theories of atoms and molecules', *Rev. Modern Phys.* **53**, 603–641.

E. H. Lieb (1983), 'Density functionals for Coulomb systems', *Int. J. Quantum Chem.* **24**, 243–277.

E. H. Lieb (1985), Density functionals for Coulomb systems, in *Density Functional Methods in Physics* (R. M. Dreizler and J. da Providencia, eds), Plenum, New York, pp. 31–80.

E. H. Lieb and B. Simon (1977*a*), 'The Hartree–Fock theory for Coulomb systems', *Comm. Math. Phys.* **53**, 185–194.

E. H. Lieb and B. Simon (1977*b*), 'The Thomas–Fermi theory of atoms, molecules and solids', *Adv. Math.* **23**, 22–116.

J.-L. Lions, Y. Maday and G. Turinici (2001), 'Résolution d'EDP par un schéma en temps "pararéel"' (A 'parareal' in time discretization of PDE's), *CR Acad. Sci., Paris, Série I, Math.* **332** (7), 661–668.

P.-L. Lions (1985), Hartree–Fock and related equations, in *Nonlinear Partial Differential Equations and their Applications, Lect. Coll. de France Semin.*, Vol. IX, *Pitman Res. Notes Math. Ser.*, Vol. 181, pp. 304–333.

P.-L. Lions (1987), 'Solutions of Hartree–Fock equations for Coulomb systems', *Comm. Math. Phys.* **109**, 33–97.

P.-L. Lions (1996), Remarks on mathematical modelling in quantum chemistry, in *Computational Methods in Applied Sciences*, Wiley, pp. 22–23.

K. B. Lipkowitz and D. B. Boyd, eds (1995–) *Reviews in Computational Chemistry*, 18 volumes to date, Wiley.

W. K. Liu *et al.*, eds (2004), Special issue on multiple scale methods for nanoscale mechanics and materials, *Comput. Methods Appl. Mech. Eng.* **193** (17–20).

C. Lubich (2004), 'A variational splitting integrator for quantum molecular dynamics', *Appl. Numer. Math.* **48**, 355–368.

A. Lüchow and H. Kleindienst (1994), 'Accurate upper and lower bounds to the ^{2}S states of the lithium atom', *Int. J. Quantum Chem.* **51**, 211–224.

R. McWeeny (1992), *Methods of Molecular Quantum Mechanics*, 2nd edn, Academic Press.

Y. Maday and A. T. Patera (2000), 'Numerical analysis of *a posteriori* finite element bounds for linear functional outputs', *Math. Models Methods Appl. Sci.* **10** (5), 785–799.

Y. Maday and G. Turinici (2003), 'Error bars and quadratically convergent methods for the numerical simulation of the Hartree–Fock equations', *Numer. Math.* **94**, 739–770.

Y. Maday, A. T. Patera and J. Peraire (1999), 'A general formulation for *a posteriori* bounds for output functionals of partial differential equations: Application to the eigenvalue problem', *CR Acad. Sci., Paris, Série I, Math.* **328** (9), 823–828.

N. H. March (1992), *Electron Density Theory of Atoms and Molecules*, Academic Press.

L. Markus and K. R. Meyer (1974), 'Generic Hamiltonian dynamical systems are neither integrable nor ergodic', *Mem. Amer. Math Soc.* **144**.

J. C. Mattingly, A. M. Stuart and D. J. Higham (2002), 'Ergodicity for SDEs and approximations: Locally Lipschitz vector fields and degenerate noise', *Stochastic Process. Appl.* **101** (2), 185–232.

D. Mazziotti (1998a), '3, 5-contracted Schrödinger equation: Determining quantum energies and reduced density matrices without wave functions', *Int. J. Quantum Chem.* **70**, 557–570.

D. Mazziotti (1998b), 'Contracted Schrödinger equation: Determining quantum energies and two-particle density matrices without wave functions', *Phys. Rev. A* **57**, 4219–4234.

D. Mazziotti (1999), 'Comparison of contracted Schrödinger and coupled-cluster theories', *Phys. Rev. A* **60**, 4396–4408.

D. Mazziotti (2004), 'Realization of quantum chemistry wave functions through first-order semidefinite programming', *Phys. Rev. Lett.* **93**, 21.

R. Miller, E. B. Tadmor, R. Phillips and M. Ortiz (1998), 'Quasicontinuum simulation of fracture at the atomic scale', *Model. Simul. Mater. Sci. Eng.* **6**.

G. Monard and K. Mertz, Jr. (1999), 'Combined quantum mechanical/molecular mechanical methodologies applied to biomolecular systems', *Acc. Chem. Res.* **32**, 904–911.

F. Montalenti and A. F. Voter (2002), 'Exploiting past visits or minimum-barrier knowledge to gain further boost in the temperature-accelerated dynamics', *J. Chem. Phys.* **116** (12), 4819–4828.

J. D. Morgan, III (1984), in *Numerical Determination of the Electronic Structure of Atoms, Diatomic and Polyatomic Molecules* (M. Defranceschi and J. Delhalle, eds), Kluwer, Dordrecht, pp. 49–84

J. D. Morgan, III (1986), 'Convergence properties of Fock's expansions for s-state eigenfunctions of the helium atom', *Theoret. Chim. Acta* **69**, 181–224.

J. D. Morgan, III and W. Kutzelnigg (1993), 'Hund's rules, the alternating rule, and symmetry holes', *J. Phys. Chem.* **97**, 2425–2434.

N. Mousseau and G. T. Barkema (2004), 'Efficient sampling in complex materials at finite temperature: The thermodynamically-weighted activation-relaxation method', submitted.

M. A. Natiello and G. E. Scuseria (1984), 'Convergence properties of Hartree–Fock SCF molecular calculations', *Int. J. Quantum Chem.* **24**, 1039–1049.

A. Neumaier (1997), 'Molecular modeling of proteins and mathematical prediction of protein structure', *SIAM Rev.* **39**, 407–460.

N. C. Nguyen, K. Veroy and A. T. Patera (2005), Certified real-time solution of parametrized PDEs, in *Handbook of Materials Modeling* (R. Catlow, H. Shercliff and S. Yip, eds), Kluwer/Springer, pp. 1523–1558.

P. Nielaba, M. Mareschal and G. Ciccotti, eds (2002), *Bridging the Time Scales: Molecular Simulations for the Next Decade*, Vol. 605 of *Lecture Notes in Physics*, Springer.

A. K. Noor and J. M. Peters (1980), 'Reduced-basis technique for non-linear analysis of structures', *AIAA Journal* **18**, 455–462.

S. Nosé (1984), 'A molecular dynamics method for simulation in the canonical ensemble', *Mol. Phys.* **52**, 255–268.

S. Nosé (1986), 'An extension of the canonical ensemble molecular dynamics method', *Mol. Phys.* **57**, 187–191.

S. Olla, private communication: for an analogous study on canonical and grand-canonical ensembles, see works by N. Cancrini and co-workers.

P. Ordejon, D. A. Drabold and R. M. Martin (1995), 'Linear system-size scaling methods for electronic structure calculations', *Phys. Rev. B* **51**, 1456–1476.

A. Palser and D. Manopoulos (1998), 'Canonical purification of the density matrix in electronic structure calculations', *Phys. Rev. B* **58**, 12704–12711.

R. G. Parr and W. Yang (1989), *Density Functional Theory of Atoms and Molecules*, Oxford University Press.

C. L. Pekeris (1958), 'Ground state of two-electron atoms', *Phys. Rev.* **112**, 1649–1658.

K. A. Peterson, A. K. Wilson, D. E. Woon and T. H. Dunning, Jr. (1997), 'Benchmark calculations with correlated molecular wave functions 12: Core correlation effects on the homonuclear diatomic molecules B_2, F_2', *Theoret. Chim. Acta* **97**, 251–259.

P. Pulay (1982), 'Improved SCF convergence acceleration', *J. Comput. Chem.* **3**, 556–560.

D. Raabe (1998), *Computational Materials Science*, Wiley.

D. C. Rapaport (1995), *The Art of Molecular Dynamics Simulation*, Cambridge University Press.

M. Reed and B. Simon (1975–1980), *Methods of Modern Mathematical Physics*, in 4 volumes, Academic Press.

C. C. J. Roothaan (1951), 'New developments in molecular orbital theory', *Rev. Modern Phys.* **23**, 69–89.

J.-P. Ryckaert, G. Giccotti and H. J. C. Berendsen (1977), 'Numerical integration of the Cartesian equations of motion of a system with constraints: Molecular dynamics of n-alkanes', *J. Comput Phys.* **23**, 327–341.

C. F. Sanz-Navarro and R. Smith (2001), 'Numerical calculations using the hyper-molecular dynamics simulation method', *Comput. Phys. Comm.* **137**, 206.

J. M. Sanz-Serna and M. P. Calvo (1994), *Numerical Hamiltonian Problems*, Chapman and Hall.

V. R. Saunders and I. H. Hillier (1973), 'A "level-shifting" method for converging closed shell Hartree–Fock wavefunctions', *Int. J. Quantum Chem.* **7**, 699–705.

M. Schechter (1981), *Operator Methods in Quantum Mechanics*, North-Holland.

H. B. Schlegel and J. J. W. McDouall (1991), Do you have SCF stability and convergence problems?, in *Computational Advances in Organic Chemistry*, Kluwer Academic, pp. 167–185.

T. Schlick (2002), *Molecular Modeling and Simulation*, Springer.

M. W. Schmidt and K. Ruedenberg (1979), 'Effective convergence to complete orbital bases and to the atomic Hartree–Fock limit through systematic sequences of Gaussian primitives', *J. Chem. Phys.* **71**, 3951–3962.

C. Schütte, A. Fischer, W. Huisinga and P. Deuflhard (1999), 'A direct approach to conformational dynamics based on hybrid Monte Carlo', *J. Comput. Phys.* **151**, 146–168.

C. Schütte, J. Walter, C. Hartmann and W. Huisinga (2004), 'An averaging principle for fast degrees of freedom exhibiting long-term correlations', *SIAM Multiscale Model. Simul.* **2**, 501–526.

C. Schwab and T. von Petersdorff (2004), 'Numerical solution of parabolic equations in high dimensions', *M2AN: Math. Model. Numer. Anal.* **38**, 93–127.

C. Schwartz (1962), 'Importance of angular correlation between atomic electrons', *Phys. Rev.* **126**, 1015–1019.

E. Schwegler and M. Challacombe (1997), 'Linear scaling computation of the Fock matrix', *J. Chem. Phys.* **106**, 5526–5536.

G. E. Scuseria (1999), 'Linear scaling density functional calculations with Gaussian orbitals', *J. Phys. Chem. A* **103**, 4782–4790.

R. Seeger and J. A. Pople (1976), 'Self-consistent molecular orbital methods XVI: Numerically stable direct energy minimization procedures for solution of Hartree–Fock equations', *J. Chem. Phys.* **65**, 265–271.

Y. Shao, C. Saravan, M. Head-Gordon and C. White (2003), 'Curvy steps for density matrix based energy minimization: Application to large-scale self-consistent-field calculations', *J. Chem. Phys.* **118** (14), 6144–6151.

V. B. Shenoy, R. Miller, E. B. Tadmor, R. Phillips and M. Ortiz, (1998), 'Quasicontinuum models of interfacial structure and deformation', *Phys. Rev. Lett.* **80** (4), 742.

V. B. Shenoy, R. Miller, E. B. Tadmor, D. Rodney, R. Phillips and M. Ortiz, (1999), 'An adaptive finite element approach to atomic-scale mechanics: The QuasiContinuum method', *J. Mech. Phys. Solids* **47**, 611.

R. Shepard (1993), 'Elimination of the diagonalization bottleneck in parallel direct-SCF methods', *Theoret. Chim. Acta* **84**, 343–351.

J. C. Slater (1930), 'Atomic shielding constants', *Phys. Rev.* **36**, 57–64.

M. R. Sorensen and A. F. Voter (2000), 'Temperature-accelerated dynamics for simulation of rare events', *J. Chem. Phys.* **112** (21), 9599–9606.

L. Spruch (1991), 'Pedagogic notes on Thomas–Fermi theory (and on some improvements): Atoms, stars and the stability of bulk matter', *Rev. Modern Phys.* **63**, 151–209.

R. E. Stanton (1981*a*), 'The existence and cure of intrinsic divergence in closed shell SCF calculations', *J. Chem. Phys.* **75**, 3426–3432.

R. E. Stanton (1981*b*), 'Intrinsic convergence in closed-shell SCF calculations: A general criterion', *J. Chem. Phys.* **75**, 5416–5422.

E. B. Starikov (1993), 'On the convergence of the Hartree–Fock selfconsistency procedure', *Mol. Phys.* **78**, 285–305.

B. Swirless (1935), 'The relativistic self-consistent field', *Proc. Roy. Soc. A* **152**, 625–649.

B. Swirless (1936), 'The relativistic interaction of two electrons in the self-consistent field method', *Proc. Roy. Soc. A* **157**, 680–696.

A. Szabo and N. S. Ostlund (1982), *Modern Quantum Chemistry: An Introduction to Advanced Electronic Structure Theory*, MacMillan.

E. B. Tadmor, R. Phillips and M. Ortiz (1996), Mixed atomistic and continuum models of deformation in solids, *Langmuir* **12**, 4529–4534.

E. B. Tadmor, M. Ortiz and R. Phillips (1996), 'Quasicontinuum analysis of defects in solids', *Phil. Mag. A* **73**, 1529–1563.

E. B. Tadmor, G. S. Smith, N. Bernstein and E. Kaxiras, (1999), 'Mixed finite element and atomistic formulation for complex crystals', *Phys. Rev. B* **59** (1), 235.

S. Teufel (2003), *Adiabatic Perturbation Theory in Quantum Dynamics*, Vol. 1821 of *Lecture Notes in Mathematics*, Springer.

W. Thirring (1983), *A Course in Mathematical Physics*, in 4 volumes, Springer.

N. Troullier and J. L. Martins (1990), 'A straightforward method for generating soft transferable pseudopotentials', *Solid State Comm.* **74**, 613–616.

M. E. Tuckerman (2002), '*Ab initio* molecular dynamics: Basic concepts, current trends and novel applications', *J. Phys. Condens. Matter* **14**, R1297–R1355.

M. E. Tuckerman and G. J. Martyna (2000), 'Understanding modern molecular dynamics: Techniques and applications', *J. Phys. Chem.* **104**, 159–178.

P. Tupper (2005), 'Ergodicity and the numerical simulation of Hamiltonian systems', *SIAM J. Appl. Dyn. Syst.*, to appear.

C. Valdemoro (1992), 'Approximating the second-order reduced density matrix in terms of the first-order one', *Phys. Rev. A* **45**, 4462–4467.

C. Valdemoro, L. M. Tel and E. Perez-Romero (2000), '*N*-representability problem within the framework of the contracted Schrödinger equation', *Phys. Rev. A* **61**, 032507–032700.

E. Vanden-Eijnden (2003), 'Numerical techniques for multi-scale dynamical system with stochastic effects', *Comm. Math. Sci.*, **1** (2), 385–391.

D. Vanderbilt (1990), 'Soft self-consistent pseudopotentials in a generalized eigenvalue formalism', *Phys. Rev. B* **41**, 7892–7895.

L. Verlet (1967), 'Computer "experiments" on classical fluids I: Thermodynamical properties of Lennard–Jones molecules', *Phys. Rev.* **159**, 98–103.

A. F. Voter (1997), 'Hyperdynamics: Accelerated molecular dynamics of infrequent events', *Phys. Rev. Lett.* **78**, 3908.

A. F. Voter (1998), 'Parallel replica method for dynamics of infrequent events', *Phys. Rev. B, Rapid Comm.* **57** (22), 57–60.

A. F. Voter and M. R. Sorensen, (1999), 'Accelerating atomistic simulations of defect dynamics: Hyperdynamics, parallel replica dynamics and temperature-accelerated dynamics', *Mat. Res. Soc. Symp. Proc.* **538**, 427–439.

P. Walters (1982), *An Introduction to Ergodic Theory*, Vol. 79 of *Graduate Texts in Mathematics*, Springer.

G. A. Worth and M. A. Robb (2002), 'Applying direct molecular dynamics to non-adiabatic systems', *Adv. Chem. Phys.* **124**, 355–431.

H. Yserentant (2003), On the electronic Schrödinger equation, Lecture Notes, Universität Tübingen.

H. Yserentant (2004*a*), 'On the regularity of the electronic Schrödinger equation in Hilbert spaces of mixed derivatives', *Numer. Math.* **98**, 731–759.

H. Yserentant (2004*b*), 'Sparse grid spaces for the numerical solution of the electronic Schrödinger equation', *Numer. Math.*, submitted.

M. C. Zerner and M. Hehenberger (1979), 'A dynamical damping scheme for converging molecular SCF calculations', *Chem. Phys. Lett.* **62**, 550–554.

Z. Zhao, B. J. Braams, M. Fukuda, M. L. Overton and J. K. Percus (2004), 'The reduced density matrix method for electronic structure calculations and the role of three-index representability conditions', *J. Chem. Phys.* **120** (5), 2095–2104.

D. Zuckerman and T. Woolf (1999), 'Dynamic reaction paths and rates through importance-sampled stochastic dynamics', *J. Chem. Phys.* **111** (21), 9475–9484.

Acta Numerica (2005), pp. 445–508
DOI: 10.1017/S0962492904000261

Steady-state convection-diffusion problems

Martin Stynes
Department of Mathematics,
National University of Ireland,
Cork, Ireland
E-mail: m.stynes@ucc.ie

In convection-diffusion problems, transport processes dominate while diffusion effects are confined to a relatively small part of the domain. This state of affairs means that one cannot rely on the formal ellipticity of the differential operator to ensure the convergence of standard numerical algorithms. Thus new ideas and approaches are required.

The survey begins by examining the asymptotic nature of solutions to stationary convection-diffusion problems. This provides a suitable framework for the understanding of these solutions and the difficulties that numerical techniques will face. Various numerical methods expressly designed for convection-diffusion problems are then presented and extensively discussed. These include finite difference and finite element methods and the use of special meshes.

CONTENTS

1. Introduction

1.1. What are convection-diffusion problems?

Our interest is in elliptic operators whose second-order derivatives are multiplied by some parameter ε that is allowed to be close to zero. These derivatives model diffusion while first-order derivatives (which are assumed to be present) are associated with convective or transport processes. In classical problems where ε is not close to zero, diffusion is the dominant mechanism in the model and the first-order convective derivatives play a relatively minor rôle in the analysis. On the other hand, when ε is near zero and the elliptic differential operator has convective terms, it is called a convection-diffusion operator. Such operators, while still satisfying the definition of ellipticity, live dangerously by flirting with the non-elliptic world. Their convective terms have a significant influence on the theoretical and numerical solution of the problem and cannot be summarily dismissed as 'lower-order terms'.

We shall see that the solutions of convection-diffusion problems have a convective nature on most of the domain of the problem, and the diffusive part of the differential operator is influential only in certain narrow subdomains. In these subdomains the gradient of the solution is large: its magnitude is proportional to some negative power of the parameter ε. We describe such behaviour by saying that the solution has a *layer*.

The fact that the elliptic nature of the differential operator is disguised on most of the domain means that numerical methods designed for elliptic problems will not work satisfactorily. In practice they usually exhibit a certain degree of instability. The challenge then is to modify these methods into a stable form without compromising their accuracy.

A second-order differential operator in n variables whose highest-order derivatives are

$$-\sum_{i,j=1}^{n} a_{ij} \frac{\partial^2(\cdot)}{\partial x_i \partial x_j},$$

where the a_{ij} are constants, is said to be elliptic if

$$\sum_{i,j=1}^{n} a_{ij}\xi_i\xi_j \geq \sigma \sum_{i=1}^{n} \xi_i^2 \quad \text{for all } \xi_i \text{ and } \xi_j, \tag{1.1}$$

where $\sigma > 0$ is called the ellipticity constant. The differential operators in convection-diffusion problems stretch this definition as far as they dare: their ellipticity constant is close to zero.

It is often assumed (certainly in introductory textbooks in both theoretical differential equations and numerical analysis) that σ is not close to zero; for example the Laplacian has $\sigma = 1$. This assumption avoids many difficulties. Consider, say, the proof of convergence of a finite difference method for the problem $-\sigma u''(x) + u'(x) = f(x)$ on $(0,1)$ with $u(0) = u(1) = 0$: if you allow the positive constant σ to take a value near zero, does the argument still work? In fact, on a more fundamental level, what happens to the solution u of this boundary value problem when σ becomes small? Taking into account this alteration in the behaviour of u, how can we modify the numerical method so that it remains stable and accurate? It is questions such as these that will preoccupy us for the duration of this survey.

Our task now is to make concrete these suspicions and assertions. We shall begin in Section 2 by recalling some ideas about maximum principles and asymptotic expansions. In Section 3 we use these tools to begin an examination of the asymptotic nature of solutions to convection-diffusion problems. Furthermore, to carry out any numerical analysis we need *a priori* to have some bounds on the derivatives of the solutions of these problems; such estimates, and useful decompositions of the solutions, are also given in this section. Finite difference methods and the accuracy of their solutions are examined in Section 4. This leads naturally to the question of constructing suitable meshes for convection-diffusion problems, and Section 5 is devoted to an epitome of this class: Shishkin meshes. We present in this section a full analysis of a finite difference method on a Shishkin mesh.

The discussion up to this point has dealt only with ordinary differential equations, where the theory is fairly complete. Now we move into deeper waters: in Section 6 we discuss the nature of solutions to convection-diffusion problems posed in two-dimensional domains. *A priori* estimates for such problems are presented in Section 7, then some preliminary comments on numerical methods are given in Section 8. Finite difference methods for such problems are considered in Section 9, but our main emphasis is on Section 10 which is devoted to finite element methods.

This survey cannot, for reasons of length, give a complete account of the many numerical methods used to solve steady-state convection-diffusion problems. Roos, Stynes and Tobiska (1996) give a comprehensive discussion of numerical methods in this area and a new edition of this book is at present in preparation.

1.2. A little motivation and history

Perhaps the most common source of convection-diffusion problems is as linearizations of Navier–Stokes equations with large Reynolds number. Morton (1996) points out that this is by no means the only place where they arise: in his opening chapter he lists ten examples involving convection-diffusion equations that include the drift-diffusion equations of semiconductor device modelling and the Black–Scholes equation from financial modelling. He also observes that 'Accurate modelling of the interaction between convective and diffusive processes is the most ubiquitous and challenging task in the numerical approximation of partial differential equations.'

The numerical solution of convection-diffusion problems goes back to the 1950s (Allen and Southwell 1955), but only in the 1970s did it acquire a research momentum that has continued to this day. A potted history of the development of numerical methods for convection-diffusion problems is presented in Stynes (2003). The field is still very active and, as we shall see in our later sections, much remains to be done.

1.3. Notation

Throughout this article, ε is a small positive parameter and C will denote a generic constant that is independent of ε and of any mesh used – it can take different values in different places (even sometimes in the same calculation). A subscripted C (e.g., C_1) is also a constant that is independent of ε and of any mesh used, but takes one fixed value.

2. Analytical tools

Consider the second-order differential operator L in n variables defined on some bounded domain (open connected set) D by

$$Lu(x) = -\sum_{i,j=1}^{n} a_{ij} \frac{\partial^2 u(x)}{\partial x_i \partial x_j} + \sum_{i=1}^{n} b_i(x) \frac{\partial u(x)}{\partial x_i} + h(x)u(x),$$

where the a_{ij} are constants. We assume that L is elliptic in the sense of (1.1). Denote the closure of D by $\bar{D}$ and its boundary by ∂D, and let $C^k(S)$ denote the space of functions that are defined on a set S and k-times differentiable on S.

Lemma 2.1. (maximum principle) Let $u \in C^0(\bar{D}) \cap C^2(D)$ satisfy the differential inequality $Lu \geq 0$ on D. Suppose that the functions b_i and h are bounded on D, and $h \geq 0$ on D. Suppose also that $u \geq 0$ on ∂D. Then $u \geq 0$ on $\bar{D}$.

This familiar result is proved in Protter and Weinberger (1984). It is the key to analysing the behaviour of solutions to convection-diffusion problems and proving the convergence to these solutions of the outputs of various numerical methods.

A maximum principle can be used to bound a function in absolute value.

Corollary 2.2. (barrier function) Suppose that the functions b_i and h are bounded on D, and $h(x) \geq 0$ on D. Let $u, v \in C^0(\bar{D}) \cap C^2(D)$. Suppose that $|Lu(x)| \leq Lv(x)$ for all $x \in D$ and $|u(x)| \leq v(x)$ for all $x \in \partial D$. Then $|u(x)| \leq v(x)$ for all $x \in \bar{D}$.

Proof. One cannot immediately apply Lemma 2.1 to the functions $|u|$ and v because $|u|$ may not be differentiable. Instead apply this lemma to the functions $u - v$ and $u + v$ and deduce the desired result. □

A function such as v in Corollary 2.2 is called a *barrier function* for u. This corollary is often applied to a function u that is a solution of a boundary value problem – so $u|_{\partial D}$ and Lu are known, but $u|_D$ is unknown. We then try to choose a suitable function v that satisfies the hypotheses of the corollary in order to deduce some worthwhile information about the behaviour of u inside D.

Putting barrier functions aside for the moment, we turn our attention to a useful descriptive tool: asymptotic expansions.

Let $\varepsilon > 0$ be a small parameter. If $f = f(x, \varepsilon)$ and $g = g(x, \varepsilon)$ with x lying in some domain D, we write $f(x, \varepsilon) = \mathcal{O}(g(x, \varepsilon))$ as $\varepsilon \to 0$ if there exist a positive number A that is independent of ε and an $\varepsilon_0 > 0$ such that $|f(x, \varepsilon)| \leq A|g(x, \varepsilon)|$ for $0 < \varepsilon \leq \varepsilon_0$. If in addition A and ε_0 are independent of x, we say that $f(x, \varepsilon) = \mathcal{O}(g(x, \varepsilon))$ as $\varepsilon \to 0$ uniformly for $x \in D$.

This notation is useful for comparing functions of similar size. For functions of greatly differing relative size, we use a 'small o' notation: we write $f(x, \varepsilon) = o(g(x, \varepsilon))$ as $\varepsilon \to 0$ if, given any $\delta > 0$, there exists an $\varepsilon_0 > 0$ such that $|f(x, \varepsilon)| \leq \delta|g(x, \varepsilon)|$ for $0 < \varepsilon \leq \varepsilon_0$. If in addition ε_0 is independent of x, we say that $f(x, \varepsilon) = o(g(x, \varepsilon))$ as $\varepsilon \to 0$ uniformly for $x \in D$.

An asymptotic sequence $\{\phi_n(\varepsilon)\}$, $n = 1, 2, \ldots$, is a sequence of functions of ε such that

$$\phi_{n+1}(\varepsilon) = o(\phi_n(\varepsilon)) \quad \text{as } \varepsilon \to 0 \quad \text{for each } n.$$

Asymptotic sequences are the building blocks from which one constructs asymptotic expansions.

Let $u(x, \varepsilon)$ be defined for all $x \in D$ and all sufficiently small ε. Let $\{\phi_n(\varepsilon)\}$ be an asymptotic sequence. The series $\sum_{n=1}^{N} u_n(x)\phi_n(\varepsilon)$, where N may be finite or infinite, is said to be the *asymptotic expansion* of u with respect to $\{\phi_n\}$ as $\varepsilon \to 0$, if for each $M \in \{1, \ldots, N\}$ we have

$$u(x, \varepsilon) - \sum_{n=1}^{M} u_n(x)\phi_n(\varepsilon) = o(\phi_M) \quad \text{as } \varepsilon \to 0. \tag{2.1}$$

In this case we write $u(x, \varepsilon) \sim \sum_{n=1}^{N} u_n(x)\phi_n(\varepsilon)$. This asymptotic expansion is *uniform in D* if (2.1) holds true uniformly for $x \in D$.

To introduce our final asymptotic concept, we take a simple example involving functions of ε that have no additional dependence on a variable x.

Example 2.3. One can easily show that one solution u_ε of the algebraic equation $u_\varepsilon^2 + \varepsilon u_\varepsilon - 1 = 0$, where ε is a small positive parameter, satisfies $u_\varepsilon = 1 + \mathcal{O}(\varepsilon)$. Thus, as $\varepsilon \to 0$, this solution approaches the solution $u_0^{(1)} = 1$ of the problem $u_0^2 - 1 = 0$. Similarly, the other solution of $u_\varepsilon^2 + \varepsilon u_\varepsilon - 1 = 0$ approaches the other solution $u_0^{(2)} = -1$ of $u_0^2 - 1 = 0$. Thus, as $\varepsilon \to 0$, the solutions of the original problem approach the solutions of the modified problem with ε set equal to zero.

The situation is different for the solutions $v_\varepsilon^{(1)}$ and $v_\varepsilon^{(2)}$ of the equation $\varepsilon v_\varepsilon^2 + v_\varepsilon - 1 = 0$. An application of the quadratic formula and binomial theorem shows that

$$v_\varepsilon^{(1)} = 1 - \varepsilon + 2\varepsilon^2 - 5\varepsilon^3 + \cdots, \qquad v_\varepsilon^{(2)} = -\varepsilon^{-1} - 1 + \varepsilon - 2\varepsilon^2 + \cdots.$$

Hence, as $\varepsilon \to 0$, one has $v_\varepsilon^{(1)} \to 1$ (the solution of the modified problem $v_0 - 1 = 0$ obtained by setting $\varepsilon = 0$) but $v_\varepsilon^{(2)} \to -\infty$.

The first part of this example is a *regular perturbation* problem: the behaviour of the solution when the perturbation parameter ε reaches its limit value of 0 is quite similar to the behaviour when ε is near but not equal to 0. The second part is a *singular perturbation* problem, where reaching the limit value of the parameter causes some significant change in the solution (here $v_\varepsilon^{(2)}$ is not close to $v_0 = 1$). As we shall see, convection-diffusion problems form a class of singular perturbation problems.

3. Convection-diffusion problems in one dimension

In this section we shall examine the asymptotic nature of solutions to convection-diffusion problems in one dimension, which will provide useful insights. The behaviour of the derivatives of these solutions, which is critical for the numerical analysis that follows later, is also discussed. Finally these two lines of attack are combined in the final subsection on Shishkin decompositions of solutions.

3.1. Asymptotic analysis

To avoid excessive detail, we do not begin with the most general situation but work instead with the two-point boundary-value problem

$$Lu(x) := -\varepsilon u''(x) + u'(x) = f(x) \quad \text{for } 0 < x < 1, \tag{3.1a}$$
$$u(0) = u(1) = 0, \tag{3.1b}$$

where we recall from Section 1.3 that ε is a small positive parameter. Assume that $f \in C^\infty[0,1]$. This is a convection-diffusion problem: the coefficient of the first-order derivative is much larger in magnitude than the coefficient of the second-order derivative.

It would be more precise to write $u(x, \varepsilon)$ for the solution of (3.1), but for convenience we use $u(x)$.

If we set $\varepsilon = 0$ then (3.1a) becomes a first-order differential equation – a significant change – so we expect that this problem is singularly perturbed. A more careful definition of singularly perturbed (with respect to

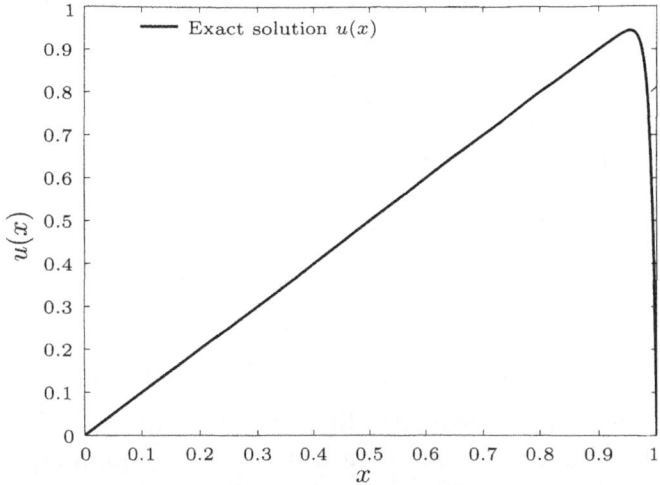

Figure 3.1. Graph of (3.3) with $\varepsilon = 0.01$.

the maximum norm) is that there exists $\hat{x} \in [0, 1]$ (in fact $\hat{x} = 1$ for this problem) such that

$$\lim_{\varepsilon \to 0} \lim_{x \to \hat{x}} u(x) \neq \lim_{x \to \hat{x}} \lim_{\varepsilon \to 0} u(x). \tag{3.2}$$

Example 3.1. To get some immediate insight into (3.1), consider the simple case where $f(x) \equiv 1$. Then

$$u(x) = x - \frac{e^{-(1-x)/\varepsilon} - e^{-1/\varepsilon}}{1 - e^{-1/\varepsilon}} \quad \text{for } 0 \le x \le 1. \tag{3.3}$$

See Figure 3.1.

One can check that (3.2) holds true with $\hat{x} = 1$. We say that $u(x)$ has a *boundary layer* at $x = 1$: this is a narrow region where u is bounded independently of ε but its derivatives blow up as $\varepsilon \to 0$ (differentiate (3.3) to see that $u'(1) \approx -1/\varepsilon$, $u''(1) \approx -1/\varepsilon^2$, *etc.*). All the important features of the general problem (3.1) are also present in (3.3).

For the differential operator L of (3.1), only with certain exceptional combinations of the boundary conditions and f does the problem fail to be singularly perturbed. For example, if $f(x) \equiv 1$ and the boundary conditions were changed to $u(0) = 0$, $u(1) = 1$, then the solution of (3.1) becomes the well-behaved function $u(x) = x$ and (3.2) is no longer satisfied for any $\hat{x} \in [0, 1]$, *i.e.*, (3.1) is now a regular perturbation problem.

The standard way of generating an asymptotic expansion for the solution

$u(x)$ of a boundary-value problem such as (3.1) is to assume that

$$u(x) = \sum_{n=0}^{\infty} u_n(x)\varepsilon^n. \tag{3.4}$$

Substituting this into (3.1a) yields

$$-\varepsilon \sum_{n=0}^{\infty} u_n''(x)\varepsilon^n + \sum_{n=0}^{\infty} u_n'(x)\varepsilon^n = f(x).$$

Comparing coefficients of powers of ε, we get

$$u_0'(x) = f(x), \quad u_1'(x) = u_0''(x), \quad u_2'(x) = u_1''(x), \quad etc.$$

Each of these is a first-order ordinary differential equation and should have associated with it a single boundary condition. But the boundary conditions (3.1b) seem to imply that $u_n(0) = u_n(1) = 0$ for all n: twice as many conditions as we can handle! It turns out that if we require for all n that $u_n(0) = 0$ and place no condition on $u_n(1)$, then we shall be able to build an asymptotic expansion – but no other way of using the boundary conditions (*e.g.*, $u_n(1) = 0$ for all n) works. Recalling (3.3), which is qualitatively similar to the solution of (3.1), in forming the asymptotic expansion (3.4) *one must discard boundary conditions where a layer occurs.*

We can now solve for the $u_n(x)$:

$$u_0(x) = \int_0^x f(t)\,dt, \quad u_1(x) = f(x) - f(0), \quad u_2(x) = f'(x) - f'(0), \quad etc.$$

Thus (3.4) becomes

$$\sum_{n=0}^{\infty} \left(F^{(n)}(x) - F^{(n)}(0)\right)\varepsilon^n, \tag{3.5}$$

where $F(x) := \int_0^x f(t)\,dt$. One can show that

$$u(x) = \sum_{n=0}^{M} \left(F^{(n)}(x) - F^{(n)}(0)\right)\varepsilon^n + o(\varepsilon^M)$$

for each $M \geq 0$, but this expansion is not uniform for $0 \leq x \leq 1$; it is uniform only for $0 \leq x \leq \delta$ where δ is any fixed constant in $(0, 1)$. This situation is unsatisfactory since at $x = 1$ we expect that $u(x)$ has a boundary layer, which is its most interesting feature. Of course the inadequacy of the expansion near $x = 1$ is unsurprising because its construction has ignored the boundary condition $u(1) = 0$ from (3.1b).

What can be done to improve the asymptotic expansion? Consider the special case $f(x) \equiv 1$. Then (3.5) collapses to the function x, but the exact solution is given by (3.3). In this formula the terms $e^{-1/\varepsilon}$ are 'exponentially small' (*i.e.*, negligible compared with any integer power of ε) and can safely be ignored. What is missing from (3.5) is some approximation of $e^{-(1-x)/\varepsilon}$, that is, some function of the variable $(1-x)/\varepsilon$ must be added to (3.5).

A standard systematic way of introducing such a function is as follows: define the *stretched variable* $\rho := (1-x)/\varepsilon$ and rewrite the differential equation as a function of ρ instead of a function of x. (Note: in the formula for ρ, the number 1 appears as the location of the layer, but the division by ε is more subtle – the purpose of the change of variable is to achieve the same dependence on ε in all the relevant terms of the transformed differential operator, but the exact scaling to use in general singular perturbation problems is not always obvious.)

Thus set $\tilde{u}(\rho) \equiv u(x)$ for $0 < \rho < 1/\varepsilon$ (corresponding to $0 < x < 1$). In fact we work with $0 < \rho < \infty$ as it is slightly simpler. Now

$$\frac{\mathrm{d}u}{\mathrm{d}x} = \frac{\mathrm{d}\tilde{u}}{\mathrm{d}\rho} \cdot \frac{\mathrm{d}\rho}{\mathrm{d}x} = -\frac{1}{\varepsilon}\tilde{u}_\rho \quad \text{and} \quad u''(x) = \frac{1}{\varepsilon^2}\tilde{u}_{\rho\rho},$$

so writing the differential operator in terms of ρ we get

$$-\varepsilon u'' + u' = -\frac{1}{\varepsilon}\left(\tilde{u}_{\rho\rho} + \tilde{u}_\rho\right) =: \tilde{L}u.$$

The original asymptotic expansion $\sum_{n=0}^{\infty} u_n(x)\varepsilon^n$ in (3.4) satisfied

$$L\left(\sum_{n=0}^{\infty} u_n(x)\varepsilon^n\right) = f,$$

so the correction $v(\rho)$ that is to be added to this expansion must satisfy $\tilde{L}v = 0$, *i.e.*, $v_{\rho\rho} + v_\rho = 0$. This second-order differential equation needs boundary conditions on $v(\rho)$ at both $\rho = 0$ (which corresponds to $x = 1$) and at $\rho = \infty$. We can now finally enforce the original boundary condition $u(1) = 0$ by requiring that our modified asymptotic expansion satisfies this condition, *i.e.*, that

$$\sum_{n=0}^{\infty} u_n(1)\varepsilon^n + v(0) = 0.$$

We want the function v to act like a boundary layer, which implies that it dies off rapidly as ρ becomes large. Thus it is natural to impose the boundary condition $v(\infty) = 0$.

The two-point boundary value problem that defines v is now completely specified and can be solved explicitly:

$$v(\rho) = e^{-\rho}v(0) = -e^{-(1-x)/\varepsilon}\sum_{n=0}^{\infty} u_n(1)\varepsilon^n$$

$$= -e^{-(1-x)/\varepsilon}\sum_{n=0}^{\infty}\left(F^{(n)}(1) - F^{(n)}(0)\right)\varepsilon^n.$$

Adding this term to (3.5), the new proposed expansion is

$$u_{as}(x) := \sum_{n=0}^{\infty}\big(F^{(n)}(x) - F^{(n)}(0)\big)\varepsilon^n - e^{-(1-x)/\varepsilon}\sum_{n=0}^{\infty}\big(F^{(n)}(1) - F^{(n)}(0)\big)\varepsilon^n.$$

$$(3.6)$$

To show that (3.6) is indeed a valid asymptotic expansion, *i.e.*, that $u(x) \sim u_{as}(x)$, set

$$\theta_M(x) = u(x) - \sum_{n=0}^{M}\big(F^{(n)}(x) - F^{(n)}(0)\big)\varepsilon^n$$

$$+ e^{-(1-x)/\varepsilon}\sum_{n=0}^{M}\big(F^{(n)}(1) - F^{(n)}(0)\big)\varepsilon^n$$

for $M = 0, 1, 2, \ldots$. We shall bound θ_M by means of a suitably chosen barrier function. Now $\theta_M(1) = 0$ and $\theta_M(0) = e^{-1/\varepsilon}\sum_{n=0}^{M}\big(F^{(n)}(1) - F^{(n)}(0)\big)\varepsilon^n = \mathcal{O}\big(\varepsilon^{M+1}\big)$. Also,

$$L\theta_M(x) = f(x) - \sum_{n=0}^{M}\big[-\varepsilon F^{(n+2)}(x) + F^{(n+1)}(x)\big]\varepsilon^n$$

$$= f(x) - F'(x) + \varepsilon^{M+1}F^{(M+2)}(x)$$

$$= \varepsilon^{M+1}F^{(M+2)}(x),$$

where the series telescoped. For each $w \in C[0, 1]$, set

$$\|w\|_\infty = \max_{x\in[0,1]} |w(x)|.$$

Define the barrier function $b(x) = C\varepsilon^{M+1}(1 + x)$, where the constant $C \geq \|F^{(M+2)}\|_\infty$ is chosen such that $b(0) = C\varepsilon^{M+1} \geq |\theta_M(0)|$. Then $Lb(x) = C\varepsilon^{M+1} \geq |L\theta_M(x)|$ for $0 < x < 1$. By Corollary 2.2, $|\theta_M(x)| \leq b(x) \leq 2C\varepsilon^{M+1}$ for $0 \leq x \leq 1$, and this is $o(\varepsilon^M)$ uniformly for $x \in [0, 1]$.

Thus (3.6) is an asymptotic expansion of $u(x)$ that is valid uniformly for $0 \leq x \leq 1$.

Consider now the more general problem

$$-\varepsilon u''(x) + a(x)u'(x) + b(x)u(x) = f(x) \quad \text{for } 0 < x < 1, \qquad (3.7)$$
$$u(0) = A, \quad u(1) = B,$$

where $a(x) > \alpha > 0$ and $b(x) \geq 0$ on [0,1], and A, B are given constants.

Remark 3.2. In fact, given that $a(x) > 0$ on [0,1], one can assume without loss of generality that $b(x) \geq 0$ (so Corollary 2.2 can be invoked), provided that ε is sufficiently small. To see this, set $u(x) = v(x)e^{kx}$ where the constant k is yet to be chosen; then $Lu = f$ is equivalent to

$$-\varepsilon v''(x) + [a(x) - 2\varepsilon k]v'(x) + [b(x) + ka(x) - \varepsilon k^2]v(x) = f(x)e^{-kx},$$

and one can choose k such that the coefficients of v' and v are both positive, so v satisfies a differential equation of the desired type. Any numerical method for v will easily yield a numerical solution for u via the transformation $u(x) = v(x)e^{kx}$.

Lemma 3.3. Let u be the classical solution of (3.7). There exists a constant C such that

$$\|u\|_\infty \leq C \tag{3.8}$$

and

$$|u'(0)| \leq C. \tag{3.9}$$

Proof. Set $z(x) = u(x) - A$ for $0 \leq x \leq 1$. Then $z(0) = 0$, $|z(1)| \leq |B - A|$ and

$$|Lz(x)| = |f(x) - Ab(x)| \leq \|f\|_\infty + |A| \|b\|_\infty.$$

Apply Corollary 2.2 to bound $|z(x)|$ by the barrier function

$$\theta(x) = \frac{x}{\alpha}\left(\alpha|B - A| + \|f\|_\infty + |A| \|b\|_\infty\right).$$

This immediately implies (3.8), and (3.9) follows from

$$|u'(0)| = \lim_{x \to 0^+}\, [|z(x)|/x] \leq \lim_{x \to 0^+}\, [\theta(x)/x]. \qquad \square$$

Inequality (3.9) shows that the solution $u(x)$ of (3.7) has no boundary layer at $x = 0$. It will in general have a boundary layer at $x = 1$, like Example 3.1.

Away from $x = 1$, we have $u(x) \approx u_0(x)$, where $u_0(x)$ is the solution of the *reduced problem*

$$a(x)u_0'(x) + b(x)u_0(x) = f(x) \quad \text{for } 0 < x < 1, \qquad u(0) = A. \tag{3.10}$$

This is the same $u_0(x)$ as the first term in (3.4). An analysis similar to that for (3.1) will construct functions $u_n(x)$ and $v_n(x)$ such that, for $k = 0, 1, 2, \ldots$,

$$u(x) = \sum_{n=0}^{k} u_n(x)\varepsilon^n + \sum_{n=0}^{k} v_n(x)\varepsilon^n + \varepsilon^{k+1}R(x, \varepsilon, k), \tag{3.11}$$

where for all i and n we have

$$|u_n^{(i)}(x)| \leq C = C(i, n), \qquad |v_n^{(i)}(x)| \leq C\varepsilon^{-i}e^{-\alpha(1-x)/\varepsilon}$$

with $C = C(i, n)$, and $|R(x, \varepsilon, k)| \leq C = C(k)$ uniformly for $0 \leq x \leq 1$. Hence

$$\sum_{n=0}^{\infty} u_n(x)\varepsilon^n + \sum_{n=0}^{\infty} v_n(x)\varepsilon^n$$

is an asymptotic expansion of $u(x)$ that is valid uniformly for $0 \leq x \leq 1$.

Remark 3.4. If in (3.7) we have $a(x) < 0$ on [0,1], then the change of variable $x \mapsto 1 - x$ reduces the problem to the case $a(x) > 0$ already considered. Thus the essential nature of $u(x)$ remains unaltered except that the boundary layer is now at $x = 0$.

If $a(x)$ changes sign on [0,1] then the solution $u(x)$ may have interior layers and/or boundary layers; see Roos *et al.* (1996, §I.1.2).

Further examples of asymptotic expansions of solutions of singularly perturbed problems can be found in Kevorkian and Cole (1996). For a comprehensive discussion of the construction of asymptotic expansions for a large variety of convection-diffusion problems in n dimensions, see Il'in (1992).

3.2. Bounds on derivatives

Asymptotic expansions of the solution u of a convection-diffusion problem such as (3.1) give us a good idea of how u behaves. To analyse numerical methods, information on the derivatives of u is also needed, and this is now presented.

Consider the general convection-diffusion problem

$$Lu(x) := -\varepsilon u''(x) + a(x)u'(x) + b(x)u(x) = f(x) \quad \text{for } 0 < x < 1, \quad (3.12)$$
$$u(0) = u(1) = 0,$$

where $a(x) > \alpha > 0$ and $b(x) \geq 0$ on [0,1]. Assume that a and b lie in $C^\infty[0, 1]$.

We already know from Lemma 3.3 that $|u'(0)| \leq C$; the next result, which is due to Kellogg and Tsan (1978), tells us what happens on all of [0,1].

Theorem 3.5. For $i = 0, 1, 2, \ldots$ and ε sufficiently small, there exists a constant $C = C(i)$ such that

$$|u^{(i)}(x)| \leq C\left(1 + \varepsilon^{-i}e^{-\alpha(1-x)/\varepsilon}\right) \quad \text{for } 0 \leq x \leq 1. \quad (3.13)$$

Proof. The case $i = 0$ is covered by Lemma 3.3. The case $i = 1$ is proved by a clever but elementary argument using integrating factors. Then the result can be deduced for $i = 2, 3, \ldots$ by an inductive argument. See Kellogg and Tsan (1978) or Roos *et al.* (1996, p. 9) for the details. □

Remark 3.6. If in (3.12) we replace the Dirichlet boundary condition $u(1) = 0$ at the layer by a *Neumann boundary condition* $u'(1) = k$ (for some constant k), then (3.13) becomes

$$|u^{(i)}(x)| \leq C\left(1 + \varepsilon^{1-i}e^{-\alpha(1-x)/\varepsilon}\right) \quad \text{for } i = 0, 1, 2, \ldots \text{ and } 0 \leq x \leq 1.$$

That is, the first-order derivative of u is bounded at $x = 1$ as $\varepsilon \to 0$, but higher-order derivatives still blow up. On a plot of $u(x)$ there is no obvious layer at $x = 1$, but the function is nevertheless not entirely tame.

One might ask: Can we not obtain bounds on derivatives of u simply by differentiating uniform asymptotic expansions, such as (3.11)? This is tempting, but we have developed no theory that controls the difference between a derivative of u and the same derivative of its asymptotic expansion. In general the differentiation of asymptotic expansions of functions is not rigorously justified, but for solutions of elliptic differential equations a theory can be established. This approach is outlined in Theorem 3.7 below and leads not only to bounds on the derivatives of u but also to a convenient decomposition of u.

3.3. Decompositions of the solution

In Theorems 3.7 and 3.9 we show that $u(x)$ can be written as the sum of a well-behaved term and a layer term. Such decompositions of u aid our insight when constructing accurate numerical methods and are often needed in the rigorous analysis of such methods.

Theorem 3.7. (standard decomposition of u) Let u be the solution of (3.12). Let q be a positive integer. Then there is a splitting $u = S + E$ such that, for $0 \le j \le q$, the inequalities

$$\|S^{(j)}\|_\infty \le C \quad \text{and} \quad |E^{(j)}(x)| \le C\varepsilon^{-j}e^{-\alpha(1-x)/\varepsilon} \quad \text{for } 0 \le x \le 1$$

hold true for some constant $C = C(q)$.

Proof. Recall the standard asymptotic expansion of $u(x)$ given in (3.11), and for convenience write $R(x)$ for the remainder $R(x, \varepsilon, k)$. Observe that we have a bound only on $\|R\|_\infty$: no information is available on the derivatives of $R(x)$. As the u_n and v_n are computed explicitly and $Lu = f$, one can determine $LR(x)$ from (3.11). Now the deep *a priori* estimates of Schauder for elliptic differential equations (Ladyzhenskaya and Ural'tseva 1968, p. 110) will yield the bound $\|R^{(j)}\|_\infty \le C\varepsilon^{-j}$ for $0 \le j \le q$.
   Choosing $k = q - 1$ in (3.11), set

$$S = \sum_{n=0}^{q-1} u_n(x)\varepsilon^n + \varepsilon^q R(x) \quad \text{and} \quad E(x) = \sum_{n=0}^{q-1} v_n(x)\varepsilon^n.$$

The result now follows immediately from what is known about the terms in S and E.   □

   In this theorem and other similar results, S is called the *smooth part* of u and E the *layer part*. In the literature dealing with singularly perturbed differential equations, 'smooth' is generally used in this non-standard way to mean that a function has certain low-order derivatives bounded independently of the perturbation parameter.
   Theorem 3.5 is adequate when proving convergence of some numerical methods for (3.12), but for others it is convenient to invoke Theorem 3.7 in

order to analyse separately the smooth and layer parts of u. At first sight
Theorem 3.7 seems the stronger of the two results, but this is not the case,
as Linß (2001) showed.

Theorem 3.8. Theorems 3.5 and 3.7 are equivalent.

Proof. Clearly Theorem 3.7 implies Theorem 3.5.

For the converse implication, assume that (3.13) holds true and let q be
an arbitrary but fixed positive integer. Set $x^* = 1 - (q\varepsilon/\alpha)\ln(1/\varepsilon)$ and
define $S(x) = u(x)$ for $0 \le x \le x^*$. Then (3.13) and the choice of x^* ensure
that $|S^{(j)}(x)| \le C$ for $0 \le j \le q$ and $0 \le x \le x^*$. Consequently one can
(using a Taylor expansion of $S(x)$ about $x = x^*$) extend S to $[0, 1]$ with
$|S^{(j)}(x)| \le C$ for $0 \le j \le q$ and $0 \le x \le 1$.

Now set $E = u - S$. Then $E(x) \equiv 0$ for $0 \le x \le x^*$, and for $x^* < x \le 1$
we have

$$|E^{(q)}(x)| \le |u^{(q)}(x)| + |S^{(q)}(x)| \le C\left(1 + \varepsilon^{-q}\mathrm{e}^{-\alpha(1-x)/\varepsilon}\right) \le C\varepsilon^{-q}\mathrm{e}^{-\alpha(1-x)/\varepsilon}$$

from the definition of x^*. Using induction, we integrate $E^{(k)}(x)$ for $k = q, q-1, \ldots, 1$ to get

$$|E^{(k-1)}(x)| \le \left|\int_{x^*}^{x} E^{(k)}(s)\,\mathrm{d}s\right|$$
$$\le C\int_{x^*}^{x} \varepsilon^{-k}\mathrm{e}^{-\alpha(1-s)/\varepsilon}\,\mathrm{d}s$$
$$\le C\varepsilon^{-(k-1)}\mathrm{e}^{-\alpha(1-x)/\varepsilon}$$

for $x^* < x \le 1$.   □

For the analysis of certain finite difference methods on Shishkin meshes
(which we shall meet in Section 6), we need a decomposition of u with a further property that is originally due to Shishkin: see the references in Farrell,
Hegarty, Miller, O'Riordan and Shishkin (2000) and Miller, O'Riordan and
Shishkin (1996). By a modification of the construction of the asymptotic
expansion (3.11) as described in Dobrowolski and Roos (1997) and Miller
et al. (1996), we can prove the following strengthening of Theorem 3.7.

Theorem 3.9. (Shishkin decomposition of u) Let u be the solution
of (3.12). Let q be a nonnegative integer. Then there is a splitting $u = S + E$
such that, for $0 \le j \le q$, the inequalities

$$\|S^{(j)}\|_{\infty} \le C \quad \text{and} \quad |E^{(j)}(x)| \le C\varepsilon^{-j}\mathrm{e}^{-\alpha(1-x)/\varepsilon} \quad \text{for } 0 \le x \le 1 \qquad (3.14)$$

hold true for some constant $C = C(q)$, and in addition

$$LS(x) = f(x) \quad \text{and} \quad LE(x) = 0 \quad \text{for } 0 \le x \le 1.$$

4. Finite difference methods in one dimension

Consider the convection-diffusion problem

$$Lu(x) := -\varepsilon u''(x) + a(x)u'(x) + b(x)u(x) = f(x) \quad \text{for } 0 < x < 1, \quad (4.1)$$
$$u(0) = u(1) = 0,$$

where $0 < \varepsilon \ll 1$, $a(x) > \alpha > 0$ and $b(x) \geq 0$ on $[0,1]$. Assume that a and b lie in $C^\infty[0,1]$.

Let N be a positive integer. Partition $[0,1]$ by the equidistant mesh $x_i = ih$ for $i = 0, \ldots, N$, where $h := 1/N$. We aim to compute an approximation $\{u_i^N\}_{i=0}^N$ of $\{u_i\}$; here and subsequently we write u_i for $u(x_i)$, a_i for $a(x_i)$, etc.

Standard discretizations of differential equations use a *central difference approximation* of the convective term. That is, one approximates $u'(x_i)$ by $(u_{i+1}^N - u_{i-1}^N)/(2h)$. Using this discretization and the standard approximation $(u_{i-1}^N - 2u_i^N + u_{i+1}^N)/h^2$ of $u''(x_i)$ produces a difference scheme whose matrix B is tridiagonal with ith row

$$\left(0 \ldots 0 \quad -\frac{\varepsilon}{h^2} - \frac{a_i}{2h} \quad \frac{2\varepsilon}{h^2} + b_i \quad -\frac{\varepsilon}{h^2} + \frac{a_i}{2h} \quad 0 \ldots 0\right) \quad (4.2)$$

for $i = 1, \ldots, N-1$. The 0th and Nth rows of B, which incorporate the boundary conditions, are $(1\ 0 \ldots 0)$ and $(0 \ldots 0\ 1)$. The right-hand side of the scheme is $(0\ f_1\ f_2 \ldots f_{N-1}\ 0)^T$.

In the particular case where $a(x) \equiv f(x) \equiv 1$ and $b(x) \equiv 0$, the solution of this difference scheme is

$$u_i^N = x_i - \frac{r^{N-i} - r^N}{1 - r^N}, \quad \text{where } r = \frac{2\varepsilon - h}{2\varepsilon + h}.$$

In practice one usually has $N \ll 1/\varepsilon$, so $\varepsilon \ll h$ and $r \approx -1$. Consequently the computed solution will oscillate as i varies, quite unlike the true solution (3.3). See Figure 4.1.

Remark 4.1. To see that the computed solution is inaccurate near $x = 1$ in the general case (4.1), consider (4.2) with $j = N - 1$. Taking $\varepsilon \ll h^2$, this equation is essentially

$$f_{N-1} = \frac{a_{N-1}(u_N^N - u_{N-2}^N)}{2h} + b_{N-1}u_{N-1}^N = -\frac{a_{N-1}u_{N-2}^N}{2h} + b_{N-1}u_{N-1}^N,$$

on applying the boundary condition. That is, $u_{N-2}^N = \mathcal{O}(h)$; but because of the boundary layer in $u(x)$ at $x = 1$ we expect that u_{N-2} is not close to zero. Thus u_{N-2}^N is far from u_{N-2}, and this is due to oscillations in the computed solution.

What has gone wrong?

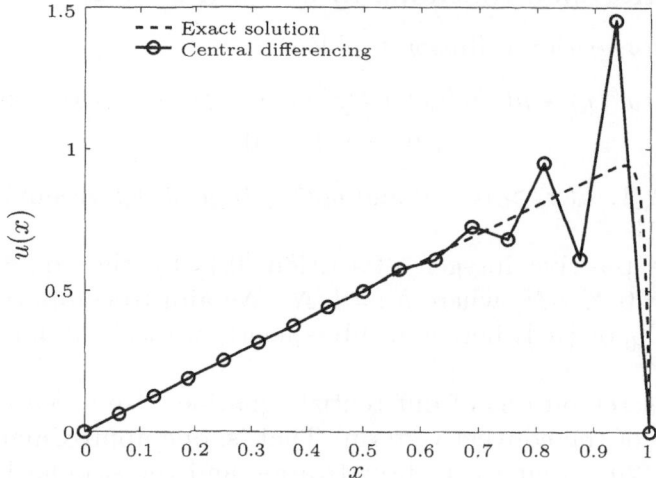

Figure 4.1. Solution to (4.1) with $\varepsilon = 0.01$, $a \equiv 1$, $b \equiv 0$, $f \equiv 1$ computed by central differencing with $N = 16$.

A square matrix $A = (A_{ij})$ is said to be an *M-matrix* if $A_{ij} \leq 0$ for all $i \neq j$ and A^{-1} exists with $(A^{-1})_{ij} \geq 0$ for all i, j. Difference schemes that employ M-matrices are common because they are desirable: they are generally stable, and more amenable to analysis. Our central difference scheme above fails to satisfy the M-matrix sign condition on the off-diagonal entries since $B_{i,i+1} > 0$ when ε is small relative to h. If $h\|a\|_\infty \leq 2\varepsilon$, then the sign condition is satisfied and it turns out that the difference method gives an acceptable computed solution, but to enforce this inequality when ε is small is impractical in many problems (especially in partial differential equations) since it can lead to an intolerably large number of mesh points.

The second M-matrix requirement – that A^{-1} exists with $(A^{-1})_{ij} \geq 0$ for all i and j – does not seem easy to verify in practice. Fortunately there are more tractable alternatives, as stated in the next result.

Lemma 4.2. Suppose that the $n \times n$ matrix $A = (A_{ij})$ satisfies $A_{ij} \leq 0$ for all $i \neq j$. Then A^{-1} exists and $(A^{-1})_{ij} \geq 0$ for all i, j if either of the following two conditions is satisfied:

(i) there exists a vector $\mathbf{v} > 0$ such that $A\mathbf{v} > 0$ (here and below, inequalities like this are understood to hold true component-wise)

(ii) A is strictly diagonally dominant with $a_{ii} > 0$ for all i.

Proof. For a proof that the first condition is sufficient, see Bohl (1981); for the second, see, *e.g.*, Quarteroni and Valli (1994, Lemma 2.1.1). □

One can often construct a vector that satisfies condition (i) of Lemma 4.2 by finding a function $w(x)$ such that $w > 0$ and $Lw > 0$, then forming $\mathbf{v}$ by restricting w to the mesh.

For M-matrices we have discrete analogues of Lemma 2.1 and Corollary 2.2.

Lemma 4.3. (discrete maximum principle) Let A be an M-matrix. If $\mathbf{v}$ is a vector with $A\mathbf{v} \geq 0$, then $\mathbf{v} \geq 0$.

Proof. $\mathbf{v} = (A^{-1})(A\mathbf{v}) \geq 0$, because $A^{-1} \geq 0$ and $A\mathbf{v} \geq 0$. $\qquad\square$

Lemma 4.4. (discrete barrier function) Let A be an M-matrix. If $\mathbf{v}_1, \mathbf{v}_2$ are vectors such that $|A\mathbf{v}_1| \leq A\mathbf{v}_2$, then $|\mathbf{v}_1| \leq \mathbf{v}_2$.

Proof. Now $A(\mathbf{v}_2 - \mathbf{v}_1) \geq 0$, so $\mathbf{v}_2 - \mathbf{v}_1 \geq 0$ by Lemma 4.3. Similarly $\mathbf{v}_2 + \mathbf{v}_1 \geq 0$, and the result follows. $\qquad\square$

The boundary data requirement of Corollary 2.2 seems to be absent from Lemma 4.4, but this is deceptive: the first and last rows of A include this information (see the construction of the matrix B above).

Returning to our difference scheme, we see that the 'incorrect' sign of $B_{i,i+1}$ comes from the central difference approximation of $u'(x_i)$. This approximation is generally recommended in basic courses in numerical methods because it gives $O(h^2)$ consistency error when $\varepsilon = 1$, but this is irrelevant when the method is (as we found here) unstable. To cure the instability, for convection-diffusion problems one can approximate $u'(x_i)$ by the *simple upwinding* formula $(u_i^N - u_{i-1}^N)/h$. Although the consistency error is now only $O(h)$ when $\varepsilon = 1$, the ith row of the scheme is

$$\left(0 \ldots 0 \quad -\frac{\varepsilon}{h^2} - \frac{a_i}{h} \quad \frac{2\varepsilon}{h^2} + \frac{a_i}{h} + b_i \quad -\frac{\varepsilon}{h^2} \quad 0 \ldots 0 \right).$$

Hence, writing A for the associated $(N+1) \times (N+1)$ matrix that incorporates the boundary conditions, we have $A_{ij} \leq 0$ for $i \neq j$, as desired.

Lemma 4.5. The matrix A associated with the simple upwind scheme is an M-matrix.

Proof. Clearly $A_{ij} \leq 0$ for $i \neq j$. Define the vector $\mathbf{v}$ by $v_i = 1 + x_i$ for $i = 0, \ldots, N$. Then a simple calculation shows that $(Av)_i \geq \min\{1, \alpha\} > 0$ for all i. The result now follows from Lemma 4.2. $\qquad\square$

Note that upwinding for (3.12) uses the one-sided difference $(u_i^N - u_{i-1}^N)/h$ to approximate $u'(x_i)$, but the alternative choice of $(u_{i+1}^N - u_i^N)/h$ would not give the correct sign pattern in the matrix. Upwinding means taking a one-sided difference *on the side away from the layer*, so for ε small relative to h^2 the scheme essentially decouples the boundary condition at $x = 1$

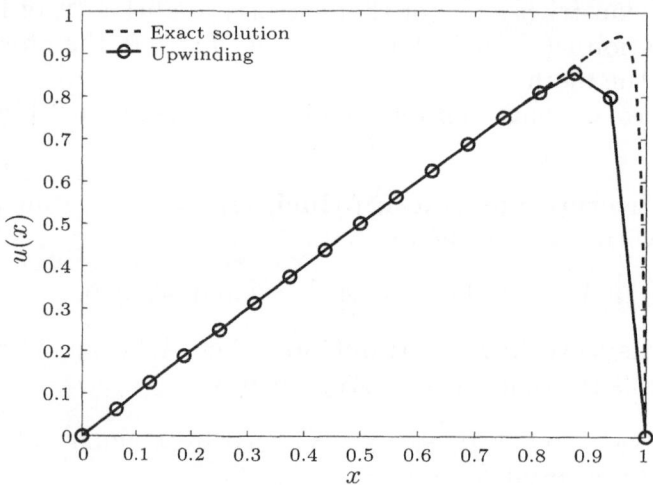

Figure 4.2. Solution to (4.1) with $\varepsilon = 0.01$, $a \equiv 1$, $b \equiv 0$,
$f \equiv 1$ computed by simple upwinding with $N = 16$.

from the values at the interior nodes; this is exactly what we need to avoid
computational infelicities like that of Remark 4.1.

The first satisfactory investigation into the accuracy of simple upwinding
is due to Kellogg and Tsan (1978). Their delicate analysis derived a tight
bound on the consistency error of the method, then converted this to the
following convergence result by means of discrete barrier functions.

**Theorem 4.6. (error bound for simple upwinding on an equidistant
mesh)** Let $\{u_i^N\}_{i=0}^N$ be the solution to (4.1) computed using simple upwinding on an equidistant mesh with N subintervals. Suppose that $h \geq \varepsilon$.
Then there exists a constant C such that

$$|u_i - u_i^N| \leq C\left[h + \exp\left(\frac{-\alpha(1 - x_i)}{\alpha h + 2\varepsilon}\right)\right] \quad \text{for } i = 0, \ldots, N.$$

No proof of this result is given here since it can be found in Kellogg and
Tsan (1978) or Roos *et al.* (1996, §I.2.1.2), and in any case we shall present
a related analysis in Section 5.

If x_i is bounded away from 1, then Theorem 4.6 implies that

$$|u_i - u_i^N| \leq C\left[h + \exp\left(\frac{-\alpha(1 - x_i)}{\alpha h + 2h}\right)\right] \leq Ch. \qquad (4.3)$$

That is, the upwind scheme yields an $\mathcal{O}(h)$-accurate solution away from
$x = 1$. But at interior mesh points that lie close to or inside the layer, the
scheme is only $\mathcal{O}(1)$-accurate. See Figure 4.2.

Remark 4.7. The error bound (4.3) is sharp and can lead to disconcerting and puzzling results in numerical experiments. Suppose that for a given convection-diffusion problem, initially we have an equidistant mesh with $h \gg \varepsilon$, so all mesh points in (0,1) lie well outside the layer. Now consider what happens if we repeatedly bisect each interval and compute a fresh solution. At first the interior mesh points remain outside the layer, so by (4.3) the numerical results show first-order convergence of the maximum nodal error. But as we continue to bisect the mesh, eventually mesh points begin to move into the layer – where the accuracy of the computed solution is only $O(1)$ – so at this stage mesh bisection causes the maximum nodal error to *increase*.

While upwinding does remove unnatural oscillations from the computed solution, we pay a price for this: the layers in the computed solution are excessively smeared, *i.e.*, are not as steep as they should be. See Figure 4.2. To put this another way, upwinding seems to produce an accurate solution for a different problem where the diffusion coefficient is much greater than ε. We now make this visual observation more precise.

The simple upwinding discretization is

$$(-\varepsilon u'' + au' + bu)(x_i)$$

$$\mapsto \frac{-\varepsilon}{h^2}\left(u_{i+1}^N - 2u_i^N + u_{i-1}^N\right) + \frac{a_i}{h}\left(u_i^N - u_{i-1}^N\right) + b_i u_i^N$$

$$= -\left(\varepsilon + \frac{ha_i}{2}\right)\frac{1}{h^2}\left(u_{i+1}^N - 2u_i^N + u_{i-1}^N\right) + \frac{a_i}{2h}\left(u_{i+1}^N - u_{i-1}^N\right) + b_i u_i^N.$$

That is, upwinding applied to $Lu = f$ is the same method as standard central differencing applied to the modified differential equation $\tilde{L}u := -(\varepsilon + ha/2)u'' + au' + bu = f$. The diffusion coefficient in this modified differential equation is so large (relative to ε) that central differencing produces an M-matrix and yields an accurate approximation of the true solution of $\tilde{L}u = f$, but of course near $x = 1$ this solution is not close to the solution of $Lu = f$.

The amount $ha(x)/2$ by which the diffusion coefficient was apparently increased is called the *artificial diffusion* introduced by upwinding.

This relationship between simple upwinding, $Lu = f$ and $\tilde{L}u = f$ opens the door to a flood of possibilities: one can choose a certain amount of artificial diffusion to add to the problem $Lu = f$, then apply a standard (non-upwinded) numerical method, with the aim of retaining stability (*i.e.*, excluding oscillations) while minimizing the smearing of layers in the computed solution. Pursuing this approach turns out to be quite fruitful; in fact, stable numerical methods on uniform meshes for convection-diffusion ODEs are usually equivalent to modifying the diffusion in the original differential equation, then applying a standard method such as central differencing – but for PDEs, the connection may be less straightforward.

To summarize: when a standard numerical method is applied to a convection-diffusion problem, if there is too little diffusion then the computed solution is often oscillatory, while if there is too much diffusion, then the computed layers are smeared.

We now consider difference schemes that are accurate both outside and inside the boundary layer. A difference scheme on a family of meshes is said to be *robust* or *uniformly convergent (with respect to ε) of order $\beta > 0$ in the discrete L^∞ norm* if its solution $\{u_i^N\}$ satisfies $|u_i - u_i^N| \leq CN^{-\beta}$ for $i = 0, \ldots, N$ and all sufficiently small H, independently of ε. Here N is the number of mesh intervals, H is the mesh diameter and β is some positive constant that is independent of the mesh and of ε.

A uniformly convergent scheme must address explicitly the exponential nature of the layer part of the solution u, as the next result shows.

Theorem 4.8. (necessary conditions for uniform convergence on an equidistant mesh) Assume that we have an equidistant mesh of width h. Suppose that a difference scheme for the problem $-\varepsilon u'' + au' = f$, $u(0) = u(1) = 0$, with a and f positive constants, can be written in the form

$$\theta_- u_{i-1}^N + \theta_0 u_i^N + \theta_+ u_{i+1}^N = hf_i \quad \text{for } i = 1, \ldots, N-1, \qquad u_0^N = u_N^N = 0,$$
$$(4.4)$$

where each $\theta = \theta(h, \varepsilon)$ depends only on the ratio h/ε. If the scheme is uniformly convergent for some $\beta > 0$, then

$$\theta_- + \theta_0 + \theta_+ = 0 \quad \text{and} \quad e^{-ah/\varepsilon}\theta_- + \theta_0 + e^{ah/\varepsilon}\theta_+ = 0. \qquad (4.5)$$

Proof. The idea is to use uniform convergence to replace the u_j^N in (4.4) by u_j, then investigate what happens as $h \to 0$ in the special case where h/ε is held constant, so each θ remains constant. See Roos *et al.* (1996, p. 40) for the details. □

The hypothesis of Theorem 4.8 that each θ depend only on the ratio h/ε is not restrictive. The first condition in (4.5) is satisfied by all plausible difference schemes; it is the second condition that distinguishes uniformly convergent schemes. Simple upwinding fails to satisfy that second condition.

Example 4.9. On equidistant meshes, the best-known uniformly convergent scheme for (4.1) is the *Il'in–Allen–Southwell difference scheme*. Allen and Southwell (1955) proposed it without any analysis of its behaviour, then it was independently rediscovered by Il'in (1969), who gave a complicated analysis of its convergence. The scheme is

$$-\frac{a_i e^{\rho_i}}{h(e^{\rho_i} - 1)} u_{i-1}^N + \left[\frac{a_i(e^{\rho_i} + 1)}{h(e^{\rho_i} - 1)} + b_i\right] u_i^N - \frac{a_i}{h(e^{\rho_i} - 1)} u_{i+1}^N = f_i$$
$$\text{for } i = 1, \ldots, N-1,$$

where $\rho_i = ha_i/\varepsilon$, with $u_0^N = u_N^N = 0$. It computes $\{u_i\}$ exactly in the special case where a, b and f are constants. Recalling our discussion above of adding artificial diffusion, this scheme is obtained if central differencing is applied to the modified differential equation

$$-\varepsilon\left(\frac{ha(x)}{2\varepsilon}\coth\frac{ha(x)}{2\varepsilon}\right)u''(x) + a(x)u'(x) + b(x)u(x) = f(x).$$

Il'in's scheme can be generated in a wide variety of ways (Roos 1994). In Kellogg and Tsan (1978) discrete barrier functions were used for the first time in the convection-diffusion literature to show that the solution $\{u_i^N\}$ computed by this scheme is uniformly convergent: $|u_i - u_i^N| \le CN^{-1}$ for all i.

The more complicated El Mistikawy–Werle 3-point scheme has the form

$$r_i^- u_{i-1}^N + r_i^0 u_i^N + r_i^+ u_{i+1}^N = q_{i-1}f_{i-1} + q_i^0 f_i + q_{i+1}^+ f_{i+1} \quad \text{for } i = 1, \ldots, N-1.$$

It achieves second-order uniform convergence on equidistant meshes, *i.e.*, $\max_i |u_i - u_i^N| \le CN^{-2}$.

See Roos *et al.* (1996, §I.2.1.3) for more information on both of these schemes.

Numerical methods like these, whose coefficients involve exponential functions of h/ε, are known collectively as *exponentially fitted* schemes. While they have become less popular in recent years, nevertheless exponential fitting is the mainstay of the FEM package PLTMG and is still widely used in semiconductor device modelling (where the Il'in scheme is known as the Scharfetter–Gummel scheme).

Remark 4.10. In the case of a Neumann boundary condition the layer is weaker (Remark 3.6). Simple upwinding on an equidistant mesh then yields (Linß 2005)

$$|u_i - u_i^N| \le Ch \quad \text{for } i = 0, \ldots, N.$$

5. Shishkin meshes

When numerically solving a convection-diffusion problem, it seems reasonable to cluster mesh points in the layer – where the solution $u(x)$ is most troublesome – instead of spreading them equidistantly over [0,1]. Graded meshes, where the mesh width gets finer and finer as one moves closer and closer to $x = 1$, have been advocated by several authors; see Roos *et al.* (1996, §I.2.4.2) for references. Since the early 1990s a simpler piecewise-equidistant mesh has been enthusiastically propagated by Shishkin and other authors (Farrell *et al.* 2000, Miller *et al.* 1996).

Consider the convection-diffusion problem (4.1). Set

$$\sigma = \min\{1/2, (2/\alpha)\varepsilon \ln N\}.$$

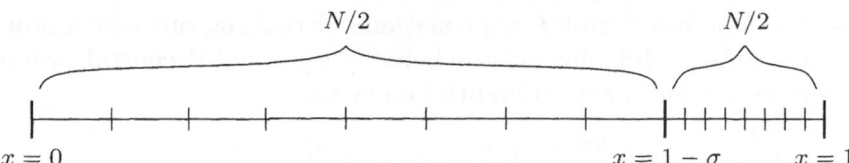

Figure 5.1. Shishkin mesh with $N = 16$.

We shall assume that $\sigma = (2/\alpha)\varepsilon \ln N$, as the other value of σ occurs only when N is exponentially large relative to ε, which is rare in practice. Then the *mesh transition point* is defined to be $1 - \sigma$. Let N be an even integer. Divide each of $[0, 1 - \sigma]$ and $[1 - \sigma, 1]$ by an equidistant mesh with $N/2$ subintervals; see Figure 5.1.

The coarse part of this Shishkin mesh has spacing $H = 2(1 - \sigma)/N$, so $N^{-1} \le H \le 2N^{-1}$. The fine part has spacing $h = 2\sigma/N = (4/\alpha)\varepsilon N^{-1} \ln N$, so $h \ll \varepsilon$. On the mesh, $x_i = iH$ for $i = 0, \ldots, N/2$ and $x_i = 1 - (N - i)h$ for $i = N/2 + 1, \ldots, N$. Set $h_i = x_i - x_{i-1}$ for each i. Note that the mesh width h_i changes abruptly at $i = N/2$, and $H/h = \alpha(1 - \sigma)/(2\varepsilon \ln N)$ can be very large.

Remark 5.1. Nonequidistant meshes for convection-diffusion problems are sometimes described as 'layer-resolving' meshes. One might presume that this terminology means that wherever the derivatives of $u(x)$ are large, the mesh is sufficiently fine to control the truncation error of the difference scheme. But the Shishkin mesh does not fully resolve the layer: $|u'(x)| \approx C\varepsilon^{-1}\exp(-\alpha(1 - x)/\varepsilon)$, so $|u'(1 - \sigma)| \approx C\varepsilon^{-1}\exp(-2\ln N) = C\varepsilon^{-1}N^{-2}$, which in general is large since typically $\varepsilon \ll N^{-1}$. Thus $|u'(x)|$ is still large on part of the first coarse-mesh interval $[x_{N/2-1}, x_{N/2}]$.

This is not a drawback: it is in fact the genius of the Shishkin mesh. For if one set out to construct a two-stage piecewise-equidistant mesh as we have done, but with the additional requirement that the mesh be fine enough to control the local truncation error wherever $|u'(x)|$ is very large, then the number of mesh points required would have to grow like $\ln(1/\varepsilon)$ as ε got smaller. Shishkin's insight was that one could achieve satisfactory theoretical and numerical results without resolving all of the layer. His construction enables us to work with a fixed number $(N + 1)$ of mesh points that is independent of the value of ε.

We apply simple upwinding. For each mesh function $\{v_i\}_{i=0}^N$, set $D_- v_i = (v_i - v_{i-1})/h_i$ and

$$\delta^2 v_i = \frac{2}{h_i + h_{i+1}}\left(\frac{v_{i+1} - v_i}{h_{i+1}} - \frac{v_i - v_{i-1}}{h_i}\right).$$

Our difference scheme is

$$-\varepsilon\delta^2 u_i^N + a_i D_- u_i^N + b_i u_i^N = f_i \quad \text{for } i = 1, \ldots, N-1, \qquad u_0^N = u_N^N = 0.$$
$$(5.1)$$

It is straightforward to check (*cf.* Lemma 4.5) that the matrix L^N associated with (5.1) is an M-matrix. To analyse the convergence of the method, recall the Shishkin decomposition $u = S + E$ of Theorem 3.9 and split the discrete solution $\{u_i^N\}$ in an analogous manner: define $\{S_i^N\}$ and $\{E_i^N\}$ by

$$L^N S_i^N = (LS)_i \quad \text{for } i = 1, \ldots, N-1, \qquad S_0^N = S(0), \quad S_N^N = S(1),$$
$$L^N E_i^N = (LE)_i = 0 \quad \text{for } i = 1, \ldots, N-1, \qquad E_0^N = E(0), \quad E_N^N = E(1).$$

Then $u_i^N = S_i^N + E_i^N$ for all i, and

$$|u_i - u_i^N| = |(S+E)_i - (S_i^N + E_i^N)| \le |S_i - S_i^N| + |E_i - E_i^N|. \qquad (5.2)$$

We shall bound each difference separately.

Lemma 5.2. There exists a constant C_0 such that

$$|S_i - S_i^N| \le C_0 N^{-1} \quad \text{for } i = 0, \ldots, N.$$

Proof. As the derivatives of S are bounded, a standard consistency error analysis shows that

$$
\begin{aligned}
|L^N(S_i - S_i^N)| &= |L^N S_i - (LS)_i| \\
&\le 2\varepsilon \int_{x_{i-1}}^{x_{i+1}} |S'''(x)| \, dx + a_i \int_{x_{i-1}}^{x_i} |S''(x)| \, dx \\
&\le C(x_{i+1} - x_{i-1}) \\
&\le C N^{-1}
\end{aligned}
$$
$$(5.3)$$

for $i = 1, \ldots, N-1$. Set $w_i = C_0 N^{-1} x_i$ for all i, where the positive constant C_0 will be chosen so that $\{w_i^N\}$ is a discrete barrier function for $\{S_i - S_i^N\}$. Now

$$L^N w_i = a_i C_0 N^{-1} + b_i w_i > \alpha C_0 N^{-1} \ge |L^N S_i - (LS)_i|$$

by (5.3), provided that C_0 is a sufficiently large constant. Clearly $w_0 = 0 = |S_0 - S_0^0|$ and $w_N = C_0 N^{-1} \ge 0 = |S_N - S_N^N|$. Thus Lemma 4.4 can be applied and we get $|S_i - S_i^N| \le w_i \le C_0 N^{-1}$ for all i, as desired. $\qquad\square$

To bound $|E_i - E_i^N|$ one again invokes Lemma 4.4, but the approach is less direct because $E(x)$ has large derivatives on part of the coarse mesh (see Remark 5.1). We show first that $|E_i|$ and $|E_i^N|$ are small on $[0, 1-\sigma]$ because they decay rapidly away from $x = 1$, then on $[1-\sigma, 1]$ the mesh is so fine that $|E_i - E_i^N|$ can be bounded by a consistency error analysis like that of Lemma 5.2.

From (3.14),

$$|E_i| \leq Ce^{-\alpha(1-(1-\sigma))/\varepsilon} = CN^{-2} \leq CN^{-1} \quad \text{for } i = 0, \ldots, N/2. \quad (5.4)$$

In the next lemma a discrete barrier function is used to show that $|E_i^N|$ also is small when $i \leq N/2$. Set

$$Z_i = \prod_{j=1}^{i} \left(1 + \frac{\alpha h_j}{2\varepsilon}\right) \quad \text{for } i = 0, \ldots, N.$$

Lemma 5.3. There exists a constant C such that

$$|E_i^N| \leq CN^{-1} \quad \text{for } i = 0, \ldots, N/2.$$

Proof. For $i = 1, \ldots, N$, a calculation shows that there exists a constant $C_1 > 0$ such that

$$L^N Z_i \geq \frac{C_1}{\max\{\varepsilon, h_i\}} Z_i. \quad (5.5)$$

Now $e^t \geq 1 + t$ for all $t \geq 0$, so

$$\frac{Z_i}{Z_N} = \prod_{j=i+1}^{N} \left(1 + \frac{\alpha h_j}{2\varepsilon}\right)^{-1} \geq \prod_{j=i+1}^{N} e^{-\alpha h_j/(2\varepsilon)} = e^{-\alpha(1-x_i)/(2\varepsilon)}. \quad (5.6)$$

Set $Y_i = C_2 Z_i / Z_N$ for $i = 0, \ldots, N$. Then $L^N Y_i = (C_2/Z_N) L^N Z_i \geq 0 = |L^N E_i^N|$ for $i = 1, \ldots, N-1$, by (5.5) and the definition of $\{E_i^N\}$. Also $Y_N = C_2 \geq |E(1)| = |E_N^N|$ if the constant C_2 is chosen sufficiently large, by the bound on $|E(x)|$ given by inequality (3.14). Finally, (5.6) implies that

$$Y_0 = \frac{C_2 Z_0}{Z_N} \geq C_2 e^{-\alpha/(2\varepsilon)} \geq C_2 e^{-\alpha/\varepsilon} \geq |E(0)| = |E_0^N|$$

provided that the constant C_2 is chosen sufficiently large, where we appealed again to (3.14). Thus we can choose C_2 so that the conditions of Lemma 4.4 are satisfied, *i.e.*, $\{Y_i\}$ is a discrete barrier function for $\{E_i^N\}$, and it follows that

$$|E_i^N| \leq Y_i = \frac{C_2 Z_i}{Z_N} \quad \text{for all } i. \quad (5.7)$$

But for $i = 0, \ldots, N/2$,

$$\frac{Z_i}{Z_N} \leq \frac{Z_{N/2}}{Z_N} = \prod_{j=1+N/2}^{N} \left(1 + \frac{\alpha h}{2\varepsilon}\right)^{-1}$$

$$= \left(1 + 2N^{-1} \ln N\right)^{-N/2}$$

$$\leq N^{-1} e^{(\ln^2 N)/N} \leq CN^{-1}$$

for some constant C (to prove the penultimate inequality, take a logarithm of the left-hand side and notice that $\ln(1+t) \geq t - t^2/2$ for $t \geq 0$). Combining this inequality with (5.7), the proof is complete. $\qquad\square$

Corollary 5.4. There exists a constant C such that

$$|E_i - E_i^N| \leq CN^{-1} \quad \text{for } i = 0, \ldots, N/2.$$

Proof. This is immediate from (5.4) and Lemma 5.3. $\qquad\square$

It remains only to bound $|E_i - E_i^N|$ for $i > N/2$.

Lemma 5.5. There exists a constant C such that

$$|E_i - E_i^N| \leq CN^{-1} \ln N \quad \text{for } i = N/2 + 1, \ldots, N.$$

Proof. We shall apply a discrete barrier function argument at the nodes $\{x_i\}_{i=N/2}^N$ by considering the discretization of a two-point boundary value problem on the interval $[1 - \sigma, 1]$. Observe that when L^N is restricted to the interior nodes of this interval it still yields an M-matrix.

Recalling the bounds on $|E^{(j)}(x)|$ in (3.14), a standard consistency error analysis shows that for $i = N/2 + 1, \ldots, N - 1$,

$$|L^N(E_i - E_i^N)| = |L^N E_i - (LE)_i|$$

$$\leq 2\varepsilon \int_{x_{i-1}}^{x_{i+1}} |E'''(x)| \, dx + a_i \int_{x_{i-1}}^{x_i} |E''(x)| \, dx$$

$$\leq C \int_{x_{i-1}}^{x_{i+1}} \varepsilon^{-2} e^{-\alpha(1-x)/\varepsilon} \, dx$$

$$= C\varepsilon^{-1} e^{-\alpha(1-x_i)/\varepsilon} \sinh(\alpha h/\varepsilon)$$

$$\leq C\varepsilon^{-1} N^{-1} (\ln N) e^{-\alpha(1-x_i)/\varepsilon},$$

since $\sinh(\alpha h/\varepsilon) = \sinh(4N^{-1} \ln N) \leq CN^{-1} \ln N$ for all $N \geq 2$.

Set $\phi_i = C_3 N^{-1} (\ln N)(1 + Z_i/Z_N)$ for $i = N/2, \ldots, N$, where the constant C_3 will be chosen later. By (5.5) and (5.6),

$$L^N \phi_i \geq C_3 N^{-1} (\ln N)(L^N Z_i)/Z_N$$

$$\geq C_3 C_1 \varepsilon^{-1} N^{-1} (\ln N) Z_i / Z_N$$

$$\geq C_3 C_1 \varepsilon^{-1} N^{-1} (\ln N) e^{-\alpha(1-x_i)/(2\varepsilon)}$$

for $i = N/2 + 1, \ldots, N$. Consequently $L^N \phi_i \geq |L^N(E_i - E_i^N)|$ if the constant C_3 is sufficiently large. Furthermore, we can choose C_3 such that

$$\phi_{N/2} = C_3 N^{-1} (\ln N)(1 + Z_{N/2}/Z_N) \geq C_3 N^{-1} (\ln N) \geq |E_{N/2} - E_{N/2}^N|$$

by Corollary 5.4, and $\phi_N = 2C_3 N^{-1} (\ln N) > 0 = |E_N - E_N^N|$.

Thus $\{\phi_i\}$ is a discrete barrier function for $\{E_i - E_i^N\}$, and Lemma 4.4 now implies that for $i = N/2, \ldots, N$ we have $|E_i - E_i^N| \leq \phi_i \leq 2C_3 N^{-1} \ln N$. $\square$

The final convergence result can now be stated.

Theorem 5.6. (uniform convergence of simple upwinding on a Shishkin mesh) There exists a constant C such that the solution $\{u_i^N\}$ of (5.1) satisfies

$$|u_i - u_i^N| \leq CN^{-1} \ln N \quad \text{for } i = 0, \ldots, N.$$

Proof. Combine (5.2), Lemma 5.2, Corollary 5.4 and Lemma 5.5. $\qquad\square$

Observe that uniform convergence is attained even though the consistency error in the maximum norm is not bounded uniformly in ε.

Roos (1996) shows that the condition number of the discrete linear system associated with (5.1) is $O(\varepsilon^{-2} N^2 \ln^{-2} N)$, which is uncomfortably large when ε is small, but that an easy preconditioning by diagonal scaling (approximate equilibration) reduces this condition number to $O(N^2 \ln^{-1} N)$.

Remark 5.7. The precise choice of mesh transition point $1 - \sigma$ in the Shishkin mesh is of both theoretical and computational interest. A careful examination of the proof of Theorem 5.6 reveals that σ should have the form $(k/\alpha)\varepsilon\phi(N)$, where $\phi(N) \to \infty$ but $N^{-1}\phi(N) \to 0$ as $N \to \infty$, and k is some constant. The simplest choice for $\phi(N)$ is $\ln N$. The choice $k = 2$ used in our definition of σ subtly enters the proof of Lemma 5.3 during the final chain of inequalities that bound Z_i/Z_N. How to choose k in an optimal way is discussed in Stynes and Tobiska (1998). It is shown there, using an argument close to our proof of Theorem 5.6, that for a variant of simple upwinding one has

$$|u_i - u_i^N| \leq C \max\{N^{-k}, kN^{-1} \ln N\} \quad \text{for } i = 0, \ldots, N.$$

The sharpness of this bound is confirmed by numerical experiments. Consequently choosing k larger than 1 only slightly diminishes the numerical accuracy of the method, but choosing k smaller than 1 causes a noticeable deterioration in the numerical rate of convergence.

Andreev and Kopteva (1996) show that for central differencing on a Shishkin mesh, the computed solution $\{u_i^N\}$ satisfies $|u_i - u_i^N| \leq CN^{-2} \ln^2 N$ for all i. The proof is difficult as the scheme does not satisfy a discrete maximum principle. Numerical experience (Linß and Stynes 2001*b*) with analogues of this approach for two-dimensional problems reveals that it is quite expensive to solve the discrete linear system efficiently, so we shall not pursue it further.

Remark 5.8. Error estimates in various norms for numerical methods on Shishkin meshes usually include a multiplicative factor $\ln^\beta N$ for some $\beta > 0$. This factor is asymptotically unimportant relative to the main convergence factor N^{-k}, where $k > 0$, but its effect is evident in numerical experiments.

If we work with certain graded meshes (*e.g.*, Bakhvalov meshes) then the $\ln N$ factor disappears so these meshes yield a higher rate of convergence but they are more complicated to construct.

The result of Theorem 5.6 can be extended to more general forms of upwinding and to other non-equidistant layer-adapted meshes that are designed for convection-diffusion problems. For an excellent survey of such generalizations for problems in one and two dimensions, see Linß (2003).

6. Convection-diffusion problems in two dimensions

In two dimensions, the convection-diffusion equation takes the form

$$Lu(x,y) := -\varepsilon \Delta u(x,y) + \mathbf{a}(x,y).\nabla u(x,y) + b(x,y)u(x,y) = f(x,y) \quad (6.1a)$$

on $\Omega \subset \mathbb{R}^2$, with

$$u(x,y) = g(x,y) \quad \text{on } \partial\Omega, \quad (6.1b)$$

where $0 < \varepsilon \ll 1$, and the functions $\mathbf{a}, b$ and f are assumed to be Hölder continuous on $\bar{\Omega}$, the closure of Ω. We also assume that $b \geq 0$ on $\bar{\Omega}$. Here Ω is any bounded domain in $\mathbb{R}^2$ with a piecewise Lipschitz-continuous boundary $\partial\Omega$ (*e.g.*, a rectangle or a domain with differentiable boundary). Assume that g is continuous except perhaps for a jump discontinuity at a single point. Il'in (1992) gives asymptotic expansions of the solutions to several specific cases of (6.1).

The differential operator L is elliptic, so (6.1) has a solution in $C^2(\Omega)$; see for example Gilbarg and Trudinger (2001). Recall that L satisfies the maximum principle of Lemma 2.1.

Assume that $|\mathbf{a}| \approx 1$, so that convection dominates diffusion. In the problems that we consider, the solution $u(x,y)$ of (6.1) has an asymptotic structure similar to that for one-dimensional problems. That is, analogously to the case $k = 0$ in (3.11), one can write u as the sum of the solution to a first-order PDE, plus layer(s), plus an $\mathcal{O}(\varepsilon)$ term.

To make this more precise, divide the boundary $\partial\Omega$ into 3 parts:

$$\text{inflow boundary } \partial^-\Omega = \{x \in \partial\Omega : \mathbf{a}.\mathbf{n} < 0\}, \quad (6.2a)$$

$$\text{outflow boundary } \partial^+\Omega = \{x \in \partial\Omega : \mathbf{a}.\mathbf{n} > 0\}, \quad (6.2b)$$

$$\text{characteristic (tangential) flow boundary } \partial^0\Omega = \{x \in \partial\Omega : \mathbf{a}.\mathbf{n} = 0\}, \quad (6.2c)$$

where $\mathbf{n}$ is the outward-pointing unit normal to $\partial\Omega$. See Figure 6.1.

A typical solution u will have *boundary layers* – narrow regions close to $\partial\Omega$ where $|\nabla u|$ is large – along $\partial^+\Omega$ and $\partial^0\Omega$. As in one-dimensional problems, exceptional Dirichlet boundary conditions g can eliminate these layers; recall the comments following Example 3.1. Also, Neumann boundary conditions on some or all of $\partial^+\Omega$ and $\partial^0\Omega$ mean that layers are no longer

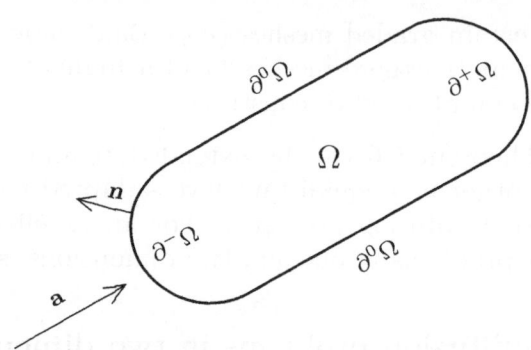

Figure 6.1. Partition of $\partial\Omega$.

visible there (*cf.* Remark 3.6). We shall exploit this property in some numerical examples where Neumann boundary conditions are introduced so that boundary layers will not distract the reader from other visual phenomena.

On most of Ω, u is approximately equal to $u_0(x,y)$, the solution of the *reduced problem*

$$\mathbf{a}(x,y).\nabla u_0(x,y) + b(x,y)u_0(x,y) = f(x,y) \quad \text{on } \Omega, \qquad u_0 = g \quad \text{on } \partial^-\Omega. \tag{6.3}$$

This first-order problem is the two-dimensional analogue of (3.10). Following the standard theory of such PDEs, the *characteristic traces* or *characteristic curves* or *characteristics* of (6.3) are the parametrized curves $(x(t), y(t))$ in Ω defined by

$$x'(t) = a_1(x,y), \quad y'(t) = a_2(x,y), \tag{6.4}$$

with initial data $(x(0), y(0)) = (\hat{x}, \hat{y})$, where $(\hat{x}, \hat{y})$ is any point in $\partial^-\Omega$. Thus one such curve emanates into Ω from each point in $\partial^-\Omega$. The function $u_0(x,y)$ propagates itself along these curves: on each characteristic, (6.3) simplifies to the ordinary differential equation

$$\frac{du_0(t)}{dt} + bu_0 = f \tag{6.5}$$

with initial data $u_0(0) = g(\hat{x}, \hat{y})$, where we have abused the notation by writing u_0 as a function of t along each characteristic. As in fluid dynamics, the direction of propagation $\mathbf{a}$ is often called the *flow*; this explains the terminology of (6.2).

We shall refer to the characteristics of (6.3) as the *subcharacteristics* of (6.1).

Just like in one dimension, boundary layers occur where there is a mismatch between the reduced solution u_0 and the boundary data. This can happen only along $\partial^+\Omega$ and $\partial^0\Omega$. While all layers look much the same when plotted, nevertheless there can be significant analytical differences between them.

Layers along $\partial^+\Omega$ are called *regular* or *exponential boundary layers*. Writing $\vec{n} = (n_1, n_2)$ for the unit outward-pointing normal to $\partial\Omega$, then near $\partial^+\Omega$, exponential layers are essentially multiples of the function

$$\exp[-(\mathbf{a}.\mathbf{n})\, d((x,y), \partial^+\Omega)/\varepsilon],$$

where $d((x,y), \partial^+\Omega)$ denotes the distance from the point (x, y) to the outflow boundary. Thus in cross-section perpendicular to $\partial^+\Omega$ these layers are very similar to the boundary layers that we met in one dimension. Their first-order derivatives in the direction perpendicular to the boundary have magnitude $\mathcal{O}(1/\varepsilon)$, and the width of the layer (*i.e.*, the distance one must travel from the boundary before all first-order derivatives are bounded by some constant C) is $\mathcal{O}(\varepsilon \ln(1/\varepsilon))$.

Layers along $\partial^0\Omega$ are called *parabolic* or *characteristic boundary layers*. In asymptotic expansions of u, these layers can be written as the solution of a parabolic PDE but not as the solution to an ODE; they have a much more complicated structure than exponential boundary layers. Their first-order derivatives in the direction perpendicular to the boundary are $\mathcal{O}(1/\sqrt{\varepsilon})$ – not as large as for exponential layers – but the width of the layer is $\mathcal{O}(\sqrt{\varepsilon} \ln(1/\varepsilon))$, so they are wider than exponential layers.

Example 6.1.   In Figure 6.2 we plot the solution $u(x, y)$ to the problem

$$-\varepsilon\Delta u(x,y) + u_x(x,y) = 1 \quad \text{on } \Omega := (0,1) \times (0,1), \qquad u(x,y) \equiv 0 \quad \text{on } \partial\Omega,$$

where $\varepsilon = 0.01$.

The inflow boundary $\partial^-\Omega$ is the side $x = 0$ of $\bar{\Omega}$; the tangential flow boundary comprises the sides $y = 0$ and $y = 1$; the outflow boundary is the remaining side $x = 1$.

From (6.4) each subcharacteristic is parametrized by $x'(t) = 1$, $y'(t) = 0$, so we can take $x = t$ and the subcharacteristics are the lines $y = k$ for arbitrary constant k. Then by (6.5) the reduced problem u_0, written as a function of the parameter t, satisfies $u_0'(t) = 1$, with initial data $u_0(0) = 0$. Hence $u_0(t) = t$, *i.e.*, $u_0(x, y) = x$ for all $(x, y) \in \Omega$.

On most of Ω one therefore has $u(x, y) \approx x$. The side $x = 1$ of $\bar{\Omega}$ is the outflow boundary $\partial^+\Omega$ and an exponential layer appears there. The tangential flow boundaries $y = 0$ and $y = 1$ have characteristic boundary layers that grow in strength as x moves from 0 to 1 because of the increasing discrepancy between u_0 and the boundary condition.

In an asymptotic expansion of u, the leading term describing the layer along $y = 0$ (the layer along $y = 1$ is of course analogous) is

$$v_0\left(x, \frac{y}{\sqrt{\varepsilon}}\right) = -\sqrt{\frac{2}{\pi}} \int_{s=y/\sqrt{2\varepsilon x}}^{\infty} e^{-s^2/2}\, u_0\left(x - \frac{y^2}{2\varepsilon s^2}, 0\right) ds.$$

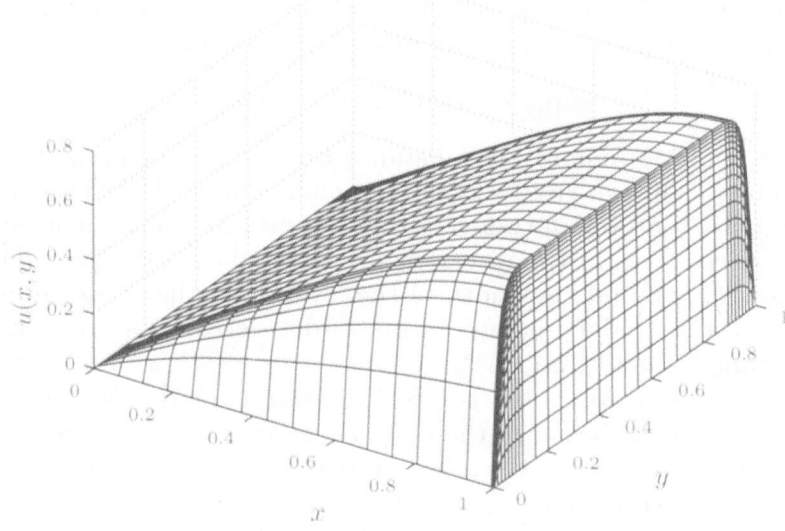

Figure 6.2. Exponential boundary layer with two
characteristic boundary layers.

This is much more complicated than for an exponential layer, but at least
we can see that when deriving this term the correct choice for the local
stretched variable is $(x,\, y/\sqrt{\varepsilon})$.

As well as boundary layers, solutions of convection-diffusion problems in
two-dimensional domains can have *interior layers* if there is a discontinuity
in the boundary data on $\partial^{-}\Omega$. This phenomenon has no analogue in one-
dimensional problems. From the theory of first-order PDEs, if g has a jump
discontinuity at a point $(\hat{x}, \hat{y}) \in \partial^{-}\Omega$, then u_0 will be discontinuous across
the subcharacteristic $\Gamma(\hat{x}, \hat{y})$ that passes through $(\hat{x}, \hat{y})$. Now first-order
PDEs preserve Dirichlet boundary data discontinuities but second-order
elliptic PDEs smooth out such discontinuities, so the solution $u(x, y)$ of
(6.1) will be continuous across $\Gamma(\hat{x}, \hat{y})$. At the same time, u must be close
to u_0 once we are a small distance away from $\Gamma(\hat{x}, \hat{y})$. Combining these
facts, we deduce that u has an interior layer along the subcharacteristic
$\Gamma(\hat{x}, \hat{y})$. Such layers have an asymptotic structure similar to characteristic
boundary layers; they are often referred to as *parabolic* or *characteristic*
interior layers.

CONVECTION-DIFFUSION PROBLEMS

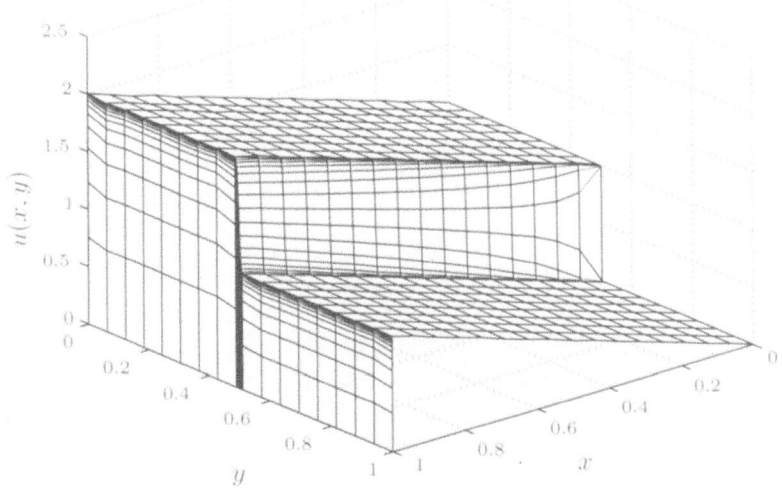

Figure 6.3. Straight interior layer.

Example 6.2. In Figure 6.3 we use the same differential operator as in Example 6.1, with $\varepsilon = 10^{-6}$. A jump discontinuity has been introduced in the inflow boundary data:

$$g(0, y) = \begin{cases} 1 & \text{for } 0 \leq y < 0.5, \\ 0 & \text{for } 0.5 < y \leq 1. \end{cases}$$

Consequently the reduced solution is

$$u_0(x, y) = \begin{cases} 1 + x & \text{for } 0 \leq y < 0.5, \\ x & \text{for } 0.5 < y \leq 1. \end{cases}$$

This yields an interior layer along the subcharacteristic passing through the discontinuity at $(0.5, 0)$, that is, along the line $y = 0.5$. Neumann boundary conditions have been applied on the sides $y = 0$ and $y = 1$ so no layers are visible there, unlike Figure 6.2. A homogeneous Dirichlet boundary condition is still assumed at $x = 1$, and again produces an exponential outflow layer, but this layer is sharper than in Figure 6.2 because ε is much smaller in the present example.

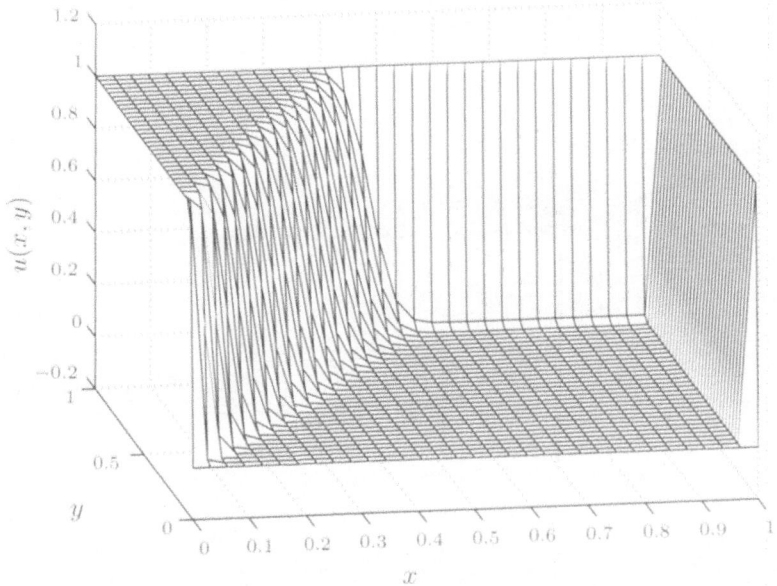

Figure 6.4. Solution of Example 6.3.

Example 6.3. Consider the problem

$$Lu(x, y) := -\varepsilon \Delta u(x, y) + u_x(x, y) + 2u_y(x, y) = 0 \quad \text{on } \Omega := (0, 1) \times (0, 1),$$

where the boundary condition is $u(x, y) = g(x, y)$ with

$$g(x, y) = \begin{cases} 0 & \text{when } y = 0, \\ 1 & \text{otherwise.} \end{cases}$$

There is no tangential flow boundary. The inflow boundary $\partial^- \Omega$ comprises the sides $x = 0$ and $y = 0$ of $\bar{\Omega}$. In (6.5) the functions b and f are both zero, so the reduced solution $u_0(x, y)$ is just the initial data on $\partial^- \Omega$ propagated along the subcharacteristics of L without change. These subcharacteristics are the lines $y = 2x + k$ for arbitrary constant k.

The solution $u(x, y)$ is as usual very close to u_0 away from layers. The outflow boundary $\partial^+ \Omega$ comprises the sides $x = 1$ and $y = 1$ of $\bar{\Omega}$. Along the portion $0 \leq x \leq 1/2$ of the side $y = 1$ there is no layer because $u_0 = g$ there. There are exponential boundary layers along the rest of $\partial^+ \Omega$. An interior layer emanates across Ω from the discontinuity in g at the point $(0, 0)$, *i.e.*, along the line $y = 2x$. See Figure 6.4, where $\varepsilon = 0.001$. The slightly diffuse nature of the interior layer in this figure is an artifact of the method used to compute u; see Remark 10.8.

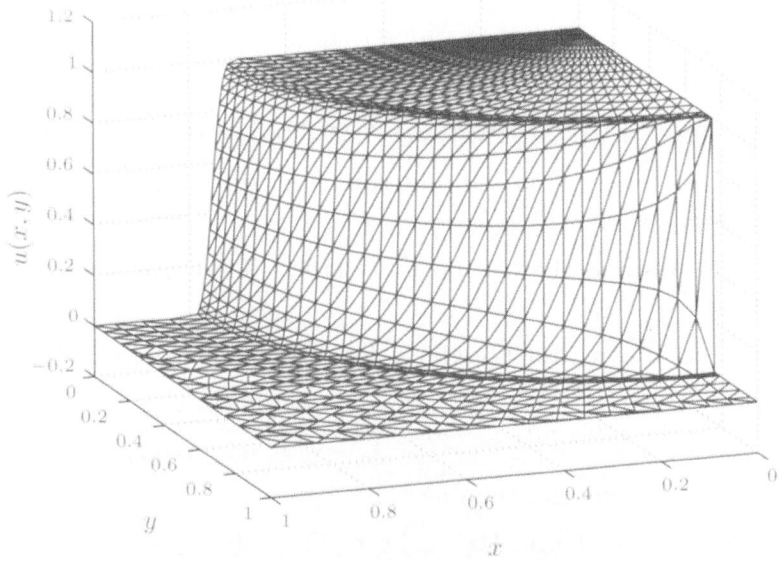

Figure 6.5. Curved interior layer.

Example 6.4. Finally we consider a problem with a curved interior layer:

$$-\varepsilon\Delta u + \mathbf{a}.\nabla u = 0 \quad \text{on } \Omega := (0,1) \times (0,1),$$
$$\nabla u.\mathbf{n} = 0 \quad \text{on } \{(x,0) : 0 \le x \le 1\} =: \partial\Omega_1,$$
$$u = g \quad \text{on } \partial\Omega \setminus \partial\Omega_1,$$

where $\mathbf{n}$ is the outward-pointing unit normal to $\partial\Omega$ and

$$\mathbf{a}(x,y) = (\sin\theta, -\cos\theta)$$

with θ the argument of the point (x,y) in polar coordinates. The function g is defined by

$$g = \begin{cases} 1 & \text{for } x = 0, \ 0 \le y \le 0.75, \\ 0 & \text{for } x = 0, \ 0.75 < y \le 1, \\ 0 & \text{for } 0 \le x \le 1, \ y = 1, \\ 0 & \text{for } x = 1, \ 0 \le y \le 1. \end{cases}$$

The Neumann condition on $\partial\Omega_1$ ensures that no outflow boundary layer appears there.

The subcharacteristics are quarter-circles centred at the origin. Since $b = f = 0$, the reduced solution merely propagates the inflow boundary values along these quarter-circles without changing their values. A computed solution to this problem is shown in Figure 6.5 with $\varepsilon = 0.0001$.

7. *A priori* estimates

In this section various *a priori* results for the solution of (6.1) are presented.
Many *a priori* analyses in the literature assume the condition

$$\mathbf{a}(x,y) = \big(a_1(x,y), a_2(x,y)\big) > (\alpha_1, \alpha_2) > (0,0) \quad \text{on } \Omega, \qquad (7.1)$$

which in the case where Ω is the unit square ensures that no characteristic
boundary layers are present.

Lemma 7.1. Assume that (7.1) holds true. Then the following results
hold.

(i) There exists a constant C, which depends on the domain Ω, such that

$$\|u\|_{L^\infty(\Omega)} \le \|g\|_{L^\infty(\partial\Omega)} + \frac{C\|f\|_{L^\infty(\Omega)}}{\max\{\alpha_1, \alpha_2\}}. \qquad (7.2)$$

  If Ω is the unit square, then $C = 1$.

(ii) For each $\delta > 0$, define $\Omega_\delta = \{x \in \Omega : \text{dist}(x, \partial^+\Omega \cup \partial^0\Omega) > \delta\}$.
  Let $g \in C(\partial\Omega)$. Then there exists a constant $C = C(\delta)$ such that
  $|u(x,y) - u_0(x,y)| \le C\varepsilon$ for all $(x,y) \in \Omega_\delta$.

Proof. The proof of (i) is similar to the proof of Lemma 3.3.

The hypothesis of (ii) ensures that there are no interior layers. The proof
can be found in Goering, Felgenhauer, Lube, Roos and Tobiska (1983). □

Let $\|\cdot\|_k$ and $|\cdot|_k$ denote the usual norm and seminorm on the Sobolev
space $H^k(\Omega)$ for all nonnegative integers k. In particular $\|\cdot\|_0 = \|\cdot\|_{L^2(\Omega)}$.

The presence of layers in u means that one does not have $\|u\|_k \le C$ for
any $k \ge 1$. Even in one dimension, the H^k norm of the function $e^{-(1-x)/\varepsilon}$
is easily seen to be $\mathcal{O}(\varepsilon^{-k+1/2})$, and exponential layers in two-dimensional
problems have a similar magnitude. This observation motivates the follow-
ing definition of a weighted energy norm that is commonly used in finite
element analyses of convection-diffusion problems: for all $w \in H^1(\Omega)$, set

$$\|w\|_{1,\varepsilon} = \sqrt{\varepsilon|w|_1^2 + \|w\|_0^2}.$$

Then typically $\|u\|_{1,\varepsilon} \le C$, uniformly in ε.

Lemma 7.2. Let u be the solution of (6.1). Assume that $b - (\text{div } \mathbf{a})/2 \ge C_5 > 0$ on $\bar\Omega$ for some constant C_5. Assume also that Ω is convex or has
smooth boundary. Then there exists a constant C such that

$$\varepsilon^{3/2}|u|_2 + \varepsilon^{1/2}|u|_1 + \|u\|_0 \le \varepsilon^{3/2}|u|_2 + \sqrt{2}\,\|u\|_{1,\varepsilon} \le C.$$

Proof. Let G be the solution of the problem $\Delta G = 0$ on Ω, $G = g$ on
$\partial\Omega$. Then the hypotheses on the domain Ω ensure that $\|G\|_2 \le C$ by a
classical inequality (see, *e.g.*, Gilbarg and Trudinger (2001)). Subtract G

from u to reduce the problem to the case of homogeneous Dirichlet boundary conditions. Now use a standard energy norm argument: multiply $Lu = f$ by u then integrate by parts, obtaining

$$\varepsilon|u|_1^2 + \int_\Omega (b - \tfrac{1}{2}\operatorname{div}\mathbf{a})u^2 = \int_\Omega fu \leq \|f\|_0\|u\|_0 \leq \frac{1}{2C_5}\|f\|_0^2 + \frac{C_5}{2}\|u\|_0^2$$

and $\|u\|_{1,\varepsilon} \leq C$ follows.

The PDE (6.1) and this inequality now yield

$$\varepsilon\|\Delta u\|_0 \leq C(|u|_1 + \|u\|_0 + \|f\|_0) \leq C(\varepsilon^{-1/2} + 1) \leq C\varepsilon^{-1/2},$$

so $\varepsilon^{3/2}\|\Delta u\|_0 \leq C$. But the classical inequality $|u|_2 \leq C(\|\Delta u\|_0 + \|u\|_0)$ holds true (Gilbarg and Trudinger 2001), and we get $\varepsilon^{3/2}|u|_2 \leq C$. $\qquad\square$

Remark 7.3. Analogously to Remark 3.2, if (7.1) holds true then one can assume without loss of generality that $b - (\operatorname{div}\mathbf{a})/2 \geq C_5 > 0$ on $\bar{\Omega}$ also holds true.

We now give some idea of the behaviour of derivatives of the solution u of (6.1) near exponential boundary layers and corners. Suppose that Ω is the unit square and the differential operator is as in Example 6.3, so that (7.1) holds true. Then the sides $x = 1$ and $y = 1$ form the outflow boundary $\partial^+\Omega$. Assuming that no extra complications such as interior layers are present, near $x = 1$ one expects the solution u to satisfy the bound

$$\left|\frac{\partial^{i+j}u(x,y)}{\partial^i x \partial^j y}\right| \leq C\big(1 + \varepsilon^{-i}e^{-(1-x)/\varepsilon}\big), \tag{7.3}$$

while near $y = 1$ one expects

$$\left|\frac{\partial^{i+j}u(x,y)}{\partial^i x \partial^j y}\right| \leq C\big(1 + \varepsilon^{-j}e^{-2(1-y)/\varepsilon}\big). \tag{7.4}$$

Close to the corner $(1,1)$ there will be an *outflow corner layer*, which is like a product of exponential boundary layers, and satisfies the bound

$$\left|\frac{\partial^{i+j}u(x,y)}{\partial^i x \partial^j y}\right| \leq C\big(1 + \varepsilon^{-(i+j)}e^{-(1-x)/\varepsilon}e^{-2(1-y)/\varepsilon}\big). \tag{7.5}$$

Despite the extra negative powers of ε in (7.5), corner layers of this type rarely cause difficulty for numerical methods because they decay so rapidly as one moves away from the corner.

A rigorous proof of bounds such as (7.3)–(7.5) is a delicate and lengthy matter. Such a proof is given by Linß and Stynes (2001a) for problems like the one under discussion, but with the extra assumptions that the Dirichlet boundary condition $g(x, y)$ is a continuous function and that a sufficient number of *compatibility conditions* hold true at the corners of $\bar{\Omega}$.

Compatibility conditions are relationships between the data of the problem and the differential operator that ensure that derivatives of u up to a desired order are continuous on $\bar{\Omega}$. They arise only at corners and are not caused by the singularly perturbed nature of the problem. Grisvard (1985) provides a general exposition of compatibility conditions for elliptic operators on polygonal domains and Han and Kellogg (1990) write down the precise form that they take when applied to convection-diffusion problems posed on the unit square.

If compatibility conditions beyond a certain order are not satisfied at a corner of a domain, then certain derivatives of that order and higher orders must blow up as one approaches this corner. Kellogg and Stynes (2005) derive bounds on the derivatives of the solution of a generalization of Example 6.1 in terms of the number of compatibility conditions that are satisfied at each corner. Near $x = 1$, but away from corners, we have (7.3). Near the characteristic boundary $y = 1$, we find that

$$\left| \frac{\partial^{i+j} u(x,y)}{\partial^i x \partial^j y} \right| \leq C \left[1 + (\sqrt{\varepsilon})^{-j} e^{-2(1-y)/\sqrt{\varepsilon}} \right]$$

provided we stay away from corners. Near the corners, singularities in the derivatives begin to appear; we do not give the details here.

The data of Example 6.3 are not fully compatible at the corner (1,1) with the differential operator L. This incompatibility will cause singularities in the derivatives of u at (1,1). The interaction between these singularities and the exponential and corner layers is not yet fully understood. That is, we are currently unable to write down reliable sharp pointwise bounds on the derivatives of u near the point (1,1), but one expects that sharp bounds are at least as bad as (7.5) and will blow up as (x, y) approaches $(1, 1)$.

It is in general difficult to derive bounds on derivatives of solutions of convection-diffusion problems inside characteristic boundary and interior layers. Although such bounds are of great interest to numerical analysts, few rigorous results appear in the literature. Kellogg and Stynes (2005) provide pointwise bounds for characteristic boundary layers. In a subsequent paper (Kellogg and Stynes 2004) they consider a convection-diffusion problem in a half-plane with a discontinuity in an arbitrary specified derivative of the boundary data and derive pointwise bounds on derivatives of the solution, including the behaviour along the interior layer emanating from the point of discontinuity.

Dörfler (1999) gives bounds on u and its derivatives in various norms (both isotropic and anisotropic) and for a variety of convection-diffusion problems on bounded domains. Shishkin (1990) contains pointwise bounds on derivatives of u for many variants of (6.1) but the arguments are presented in a very concise style and it is difficult to ascertain the precise assumptions made.

8. General comments on numerical methods

Numerical methods (such as central differencing on equidistant meshes) that contain no mechanism for stabilizing solutions in exponential layers will usually have wild oscillations in their computed solutions on much of Ω, like in Section 4. As we shall see, this problem can be handled by modifying the approximation of the convective terms (*e.g.*, using some form of finite difference upwinding or special choices of finite element trial and test spaces) or by modifying the mesh (*e.g.*, a two-dimensional Shishkin mesh). When this is done correctly, one can compute accurate solutions inside these layers.

Characteristic layers, on the other hand, differ in both respects:

- if the method has no stabilizing mechanism specifically designed to address characteristic layers, then the layer will induce small oscillations in the computed solution, but these oscillations usually appear only inside and near the characteristic layer, so the solution can still be computed accurately on the rest of Ω;

- it is often difficult – at least in the case of interior layers – to compute accurate solutions inside characteristic layers.

Thus one could use some form of upwinding (*i.e.*, some discrete approximation of $\mathbf{a}.\nabla u$ that is skewed away from the outflow boundary) to stabilize the method for exponential layers, combined with some heuristic mesh refinement near characteristic layers. Whether or not the mesh refinement yields an accurate solution inside the characteristic layers, nevertheless the solution elsewhere will be accurate.

The following pair of examples are related to our observation that one can to a certain extent neglect characteristic layers but not exponential layers.

Consider again Example 6.3 but with $g(x,y) \equiv 1$. Then the solution $u(x,y)$ has exponential boundary layers along $x = 1$ and $y = 1$. The reduced solution $u_0(x,y)$ will of course ignore these layers, and we find that $\|u - u_0\|_{1,\varepsilon} = \mathcal{O}(1)$.

On the other hand the solution u of Example 6.1 has two characteristic layers and one exponential layer. Schieweck (1986) proves that if one sets $v(x,y) = u_0(x,y) - u_0(1,y)\mathrm{e}^{-(1-x)/\varepsilon}$ (this is the reduced solution plus an appropriate exponential layer term, so it ignores only the parabolic layers), then $\|u - v\|_{1,\varepsilon} \leq C\varepsilon^{1/4}$.

Nevertheless, in some applications characteristic layers cannot be neglected.

9. Finite difference methods in two dimensions

Assume that Ω is the unit square and the mesh $\{(x_i, y_j)\}$ is rectangular and equidistant in each coordinate direction: $x_i = ih$ and $y_j = jk$ for

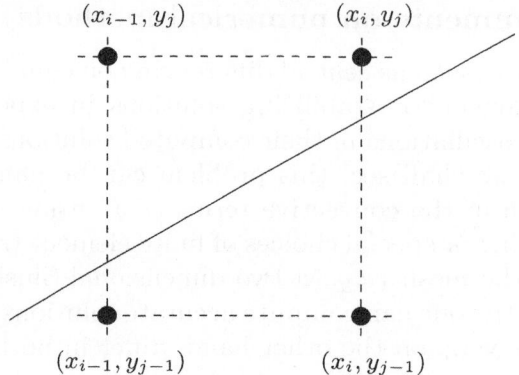

(x_{i-1}, y_j) (x_i, y_j)

(x_{i-1}, y_{j-1}) (x_i, y_{j-1})

Figure 9.1. Mesh points and line
indicating nearby interior layer.

$i = 0, \ldots, N$ and $j = 0, \ldots, M$ with $h := 1/N$ and $k := 1/M$. We use a standard approximation of the second-order derivatives:

$$u_{xx}(x_i, y_j) \approx \frac{u_{i+1,j}^N - 2u_{ij}^N + u_{i-1,j}^N}{h^2}, \tag{9.1}$$

$$u_{yy}(x_i, y_j) \approx \frac{u_{i,j+1}^N - 2u_{ij}^N + u_{i,j-1}^N}{k^2},$$

where u_{ij}^N is the computed solution at each mesh point (x_i, y_j).

As for one-dimensional problems, approximating the first-order derivatives in (6.1) by central differences

$$u_x(x_i, y_j) \approx \frac{u_{i+1,j}^N - u_{i-1,j}^N}{2h} \quad \text{and} \quad u_y(x_i, y_j) \approx \frac{u_{i,j+1}^N - u_{i,j-1}^N}{2k}$$

leads to an unstable method. Instead one can use simple upwinding,

$$u_x(x_i, y_j) \approx \frac{u_{i,j}^N - u_{i-1,j}^N}{h} \quad \text{and} \quad u_y(x_i, y_j) \approx \frac{u_{i,j}^N - u_{i,j-1}^N}{k},$$

and this yields an M-matrix. Combining this with (9.1) and the approximation $u(x_i, y_j) \approx u_{ij}^N$ for the zero-order term in (6.1), the resulting method is stable but we expect from our experience with ODEs that it will smear exponential boundary layers.

In fact, one can foresee heuristically that this method will also smear interior layers. In Figure 9.1, the value of $u(x_i, y_j)$ depends strongly on the u values along the upstream portion of the subcharacteristic that passes through (x_i, y_j) – this is a line through (x_i, y_j) parallel to the line drawn – but simple upwinding makes $u(x_i, y_j)$ depend on $u(x_i, y_{j-1})$, which introduces inaccuracies because the value of $u(x_i, y_j)$ has little to do with the values of u on the other side of the interior layer indicated by the line in Figure 9.1.

A difference scheme on a family of arbitrary rectangular meshes of $(N+1)^2$ points (we take the same number of mesh points in each coordinate direction for simplicity) is said to be *robust* or *uniformly convergent (with respect to ε) of order $\beta > 0$ in the discrete L^∞ norm* if its solution $\{u_{ij}^N\}$ satisfies

$$|u_{ij} - u_{ij}^N| \leq CN^{-\beta} \quad \text{for } i, j = 0, \ldots, N$$

and all sufficiently small H, independently of ε. Here we take $N + 1$ mesh points in each coordinate direction for simplicity, H is the mesh diameter, β is some positive constant that is independent of the mesh and of ε, and we write u_{ij} instead of $u(x_i, y_j)$ (we shall do likewise for all other functions in $C(\bar{\Omega})$).

For uniform convergence on an equidistant mesh, an analogue of Theorem 4.8 shows that once again the coefficients in the scheme must have a certain exponential character (Roos *et al.* 1996, p. 194). One can define a five-point scheme that is a two-dimensional analogue of the Il'in scheme of Example 4.9. When the data of (6.1) are smooth and some compatibility conditions are satisfied at the corners of Ω, this scheme can be proved to achieve uniform convergence of order β, where β is almost $1/2$, in the discrete L^∞ norm (Roos *et al.* 1996, p. 195). Nevertheless this scheme, which is a form of upwinding, smears interior layers quite badly and is rarely used.

Continuing in the footsteps of our earlier sections, we now consider a two-dimensional Shishkin mesh for a problem on the unit square that satisfies (7.1) and consequently has exponential boundary layers along $x = 1$ and $y = 1$. Let N, an even integer, be the number of mesh intervals in each coordinate direction. Define the transition points on the x- and y-axes to be $1 - \lambda_x$ and $1 - \lambda_y$ respectively, where $\lambda_x = (2\varepsilon/\alpha_1) \ln N$ and $\lambda_y = (2\varepsilon/\alpha_2) \ln N$. The fine and coarse mesh regions on the coordinate axes each contain $N/2$ mesh intervals. See Figure 9.2 for the mesh with $N = 8$.

One can define simple upwinding on non-equidistant meshes similarly to the formulas of Section 5. Writing u_{ij}^N for the solution computed using this method on the Shishkin mesh, then under compatibility assumptions from Linß and Stynes (2001a) guaranteeing that the solution u can be decomposed as a sum of the reduced solution, an exponential layer at $x = 1$, an exponential layer at $y = 1$ and a corner layer at $(1,1)$, where these layers satisfy bounds similar to (7.3)–(7.5), an analysis similar to that of Section 5 shows that

$$|u_{ij} - u_{ij}^N| \leq CN^{-1} \ln N \quad \text{for all } i, j.$$

That is, we get almost first-order uniform convergence in the discrete L^∞ norm.

If we modify this scheme by using central differencing instead of upwinding wherever the Shishkin mesh is fine in the relevant coordinate direction, then the M-matrix property is retained and a variant of the upwind analysis

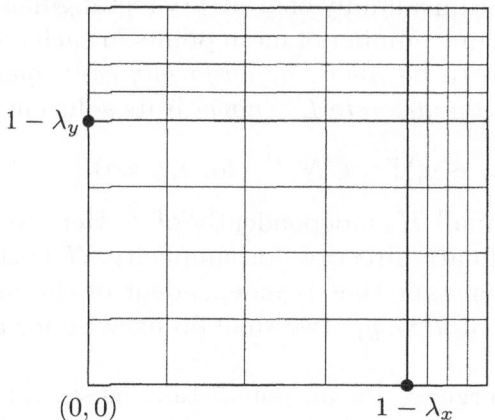

Figure 9.2. Shishkin mesh with $N = 8$.

yields (Linß and Stynes 1999) the improved bound

$$|u_{ij} - u_{ij}^{N,\text{hybrid}}| \leq CN^{-1} \quad \text{for all } i, j,$$

where $u_{ij}^{N,\text{hybrid}}$ is the solution computed by this hybrid scheme.

Kopteva (2003) shows, under some extra compatibility assumptions at the corners, that one iteration of Richardson extrapolation applied to the simple upwind solution u_{ij}^N on the Shishkin mesh yields a solution v_{ij}^N for which

$$|u_{ij} - v_{ij}^N| \leq CN^{-2} \ln^2 N \quad \text{for all } i, j.$$

Approximation of the first-order derivatives of u is also discussed in this paper.

Remark 9.1. (Shishkin's obstacle theorem) The above convergence results are all proved under hypotheses that exclude characteristic layers. The difficulty of accurately approximating characteristic boundary layers is underlined by a remarkable result of Shishkin (1989): suppose that one has a problem whose solution has a characteristic boundary layer. Suppose also that one applies any difference scheme on an equidistant mesh whose coefficients are drawn from a fixed class of functions (*e.g.*, the Il'in scheme, whose coefficients are all exponentials and polynomials; the point is that one is forbidden to vary the difference scheme by choosing the type of coefficients to correspond exactly to the precise nature of each new set of boundary data). Then *this scheme cannot yield uniform convergence of any positive order in the discrete L^∞ norm inside the characteristic boundary layer for all smooth and compatible boundary data g.* The essential reason for this negative result is that at each point (x, y) near $\partial^0 \Omega$ a characteristic boundary layer depends on all the data along that connected component of $\partial^0 \Omega$; this is

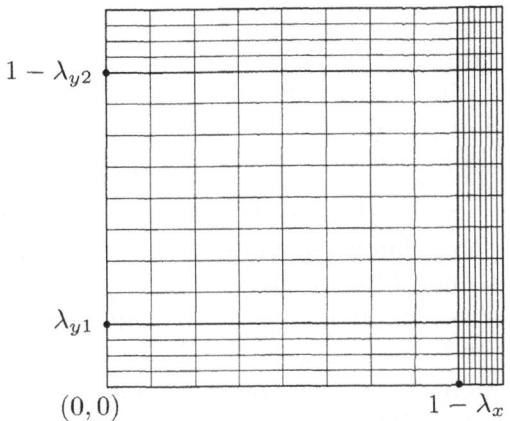

Figure 9.3. Shishkin mesh for Example 6.1
with $N = 16$.

quite unlike an exponential boundary layer, whose behaviour at (x, y) near $\partial^+\Omega$ depends only on the difference between the reduced solution u_0 and the boundary data at the nearest boundary point – a much simpler situation. A consequence of Shishkin's result is that special schemes on equidistant meshes are unsatisfactory inside characteristic boundary layers; instead, we must use meshes that are adapted either *a priori* or *a posteriori*.

For a problem on the unit square (such as Example 6.1) that has an exponential boundary layer at $x = 1$ and characteristic boundary layers at $y = 0$ and $y = 1$, a suitable Shishkin mesh is constructed as follows: use an x-axis transition point exactly as in Figure 9.2. Place y-axis transition points at λ_{y1} and $1 - \lambda_{y2}$ where each λ_{yk} is $\mathcal{O}(\varepsilon^{1/2} \ln N)$, then use $N/4$ equidistant mesh intervals in each of $[0, \lambda_{y1}]$ and $[1 - \lambda_{y2}, 1]$ and $N/2$ equidistant mesh intervals in $[\lambda_{y1}, 1 - \lambda_{y2}]$. See Figure 9.3. Then for simple upwinding, Shishkin (1990) shows that under certain fairly strong hypotheses on the smoothness and compatibility of the data of the problem, simple upwinding yields

$$|u_{ij} - u_{ij}^N| \leq C N^{-1} \ln N \quad \text{for all } i, j,$$

where u_{ij}^N is the computed solution.

A large collection of numerical computations on Shishkin meshes for various problems can be found in Farrell *et al.* (2000). While the assembly of Shishkin meshes for boundary layers along straight portions of $\partial\Omega$ is straightforward once the asymptotic nature of the layer has been ascertained, for general (curved) interior layers there are practical difficulties in the construction of these meshes and the only examples in the literature are for problems like Example 6.4, where the layer has a certain symmetry. Nevertheless one can achieve satisfactory visual results with heuristic approximations of Shishkin meshes in such situations; see Madden and Stynes (1997). Shishkin's second doctoral thesis (Shishkin 1990) contains a wealth of theoretical results for finite differences applied on piecewise uniform meshes for many convection-diffusion problems. It is at present being translated into English, but is written in an extremely condensed style.

Remark 9.2. (Defect correction method) This technique seeks to generate a useful higher-order scheme by combining a stable low-order scheme with a higher-order but unstable scheme.

Consider an arbitrary rectangular mesh. Compute an initial approximation $\hat{u}^N$ using simple upwinding: $L_{up}^N \hat{u}^N = f^N$. Obtain the 'defect' σ^N by means of the formally higher-order central difference scheme L_c^N: set $\sigma^N = f^N - L_c^N \hat{u}^N$. Compute the defect correction δ^N by solving $L_{up}^N \delta^N = \sigma^N$. Form the final solution $u^N := \hat{u}^N + \delta^N$.

This method avoids instability by solving only discrete systems that involve the upwind operator L_{up}^N, yet aims to attain the higher-order convergence associated with the operator L_c^N. The idea can be placed in a more general setting and has been applied to many problems unrelated to convection-diffusion (Bohmer and Rannacher 1984). For convection-diffusion the only satisfactory analysis of the method, which shows that it does indeed achieve second-order convergence on a Shishkin mesh, is in Fröhner, Linß and Roos (2001) where a one-dimensional problem is treated. Defect correction is related to Richardson extrapolation, and to obtain a rigorous proof of its validity in two dimensions on a Shishkin mesh like that of Figure 9.2 would require, *e.g.*, some extension of the delicate analysis in Kopteva (2003). Nevertheless numerical results for the method are encouraging (see Remark 10.2).

Finally, we point out that when one no longer assumes hypotheses such as $\mathbf{a}(\cdot,\cdot) > (0,0)$, then although simple upwinding remains stable (*i.e.*, the computed solution is free of non-physical oscillations), it can give dangerously misleading results. Brandt and Yavneh (1991) give an example of linearized recirculating flow in an annulus where the subcharacteristics are circles and, except near the boundary of the domain, the solution computed by a version of simple upwinding is $\mathcal{O}(1)$ distant from the true solution!

10. Finite element methods

If one attempts to solve a convection-diffusion problem by means of a standard Galerkin finite element method with linear or bilinear elements on an equidistant mesh, then a typical computed solution will display large oscillations. This is analogous to our experience in Section 9 with central differencing. Thus some mechanism is needed to stabilize a FEM: a special choice of trial or test functions, or a special mesh, or a modification of the standard bilinear form, or a combination of these devices. In the subsections that follow we discuss each in turn.

Throughout this section we shall assume (*cf.* Remark 7.3) that

$$b(x, y) - \frac{\operatorname{div} \mathbf{a}(x, y)}{2} \geq C_5 > 0 \quad \text{on } \bar{\Omega} \text{ for some constant } C_5. \qquad (10.1)$$

For convenience also assume that $u \equiv 0$ on $\partial \Omega$.

10.1. L^*-splines

We did not discuss finite element methods for one-dimensional convection-diffusion problems such as (3.12) since often they are merely an alternative way of generating finite difference schemes.

For example, one can generate the Il'in scheme of Example 4.9 by a finite element method on the same equidistant mesh. It is a Petrov–Galerkin FEM, that is, the trial space S^N and test space T^N are not identical, unlike standard (Bubnov–)Galerkin methods. One takes S^N to be the standard space of piecewise linear functions on the mesh $x_i = i/N$, for $i = 0, 1, \ldots, N$, that vanish at $x = 0, 1$ to satisfy the boundary conditions in (3.12). Recall that the differential equation in (3.12) is $Lu(x) := -\varepsilon u''(x) + a(x)u'(x) + b(x)u(x) = f(x)$, with $a(\cdot) > 0$. Define the test space T^N to be the space of approximate L^*-splines spanned by $\{\psi_i\}_{i=1}^{N-1}$, where

$$\bar{L}^*(\psi_i)(x) := -\varepsilon \psi_i''(x) - \bar{a}(x)\psi_i'(x) + \bar{b}(x)\psi_i(x) = 0 \qquad (10.2)$$
$$\text{on each subinterval } (x_{j-1}, x_j)$$

and $\psi_i(x_j) = \delta_{ij}$, the discrete Kronecker delta. Here $\bar{a}$ is some approximation of $a(x)$ that is constant on each mesh subinterval, and b and f are approximated by $\bar{b}$ and $\bar{f}$ in a similar way. As usual in FEMs, the computed solution $u^N(x) \in S^N$ is generated by a weak form of the differential equation:

$$\int_0^1 \left[\varepsilon(u^N)'(x)\psi_i'(x) + \bar{a}(x)(u^N)'(x)\psi_i(x) + \bar{b}u^N(x)\psi_i(x) \right] \, \mathrm{d}x$$

$$= \int_0^1 \bar{f}(x)\psi_i(x) \, \mathrm{d}x \quad \text{for } i = 1, \ldots, N-1.$$

If one defines $\bar{a}$ by the quadrature rule

$$\int_0^1 \bar{a}(x)(u^N)'(x)\psi_i(x) \, \mathrm{d}x = a_i \int_0^1 (u^N)'(x)\psi_i(x) \, \mathrm{d}x,$$

with similar definitions for $\bar{b}$ and $\bar{f}$, then one obtains the Il'in scheme. The alternative choice

$$\bar{a}\Big|_{(x_{j-i}, x_j)} = \frac{a_{j-1} + a_j}{2} \quad \text{for each } j$$

(with similar definitions for $\bar{b}$ and $\bar{f}$) yields the El Mistikawy–Werle scheme of Section 4.

Both of these are successful schemes, and the only special construction we made when generating them in a FEM context was to use L^*-splines. Why do L^*-splines make such good test functions?

The explanation is to be found by considering Green's functions for the differential operator L. For each mesh point $x_i \in (0,1)$ let $G(\cdot, x_i)$ denote the Green's function associated with that point, that is,

$$L^* G(\xi, x_i) = \delta(\xi - x_i) \quad \text{for } 0 < \xi < 1, \qquad G(0, x_i) = G(1, x_i) = 0,$$

where we define

$$L^* G(\xi, x_i) := -\varepsilon G_{\xi\xi}(\xi, x_i) - \big(a(\xi) G(\xi, x_i)\big)_\xi + b(\xi) G(\xi, x_i).$$

Then

$$
\begin{aligned}
u_i &= \int_0^1 f(\xi) G(\xi, x_i) \, \mathrm{d}\xi \\
&= \int_0^1 (Lu)(\xi) G(\xi, x_i) \, \mathrm{d}\xi \\
&= \int_0^1 \big[\varepsilon u'(\xi)) G_\xi(\xi, x_i) + a(x) u'(\xi) G(\xi, x_i) + bu(\xi) G(\xi, x_i) \big] \, \mathrm{d}\xi.
\end{aligned}
$$

Note both the resemblance between this identity and the weak form of the differential equation that was used above to generate the FEM and the similarity between the definitions of G and ψ_i. The key idea of this FEM was to choose the ψ_i in such a way that the test space T^N was capable of producing a decent approximation of the Green's function, and this property can be exploited in the analysis of the method.

The Green's function exhibits layers at $\xi = 0$ and at $\xi = x_i$; on each subinterval $[0, x_i]$ and $[x_i, 1]$ these layers occur at the left-hand end, unlike the layer in $u(x)$ at $x = 1$, because of the negative coefficient $-a(\xi)$ in the convective term appearing in the definition of L^*. See Figure 10.1.

Remark 10.1. When piecewise linears or bilinears are used as the trial space for convection-diffusion problems in one or two dimensions, useful numerical methods on general meshes are based on some test space that is constructed to approximate the Green's function of the continuous operator. This Green's function is skewed away from the outflow boundary; see Morton (1996) for a discussion of its properties in two dimensions.

Alternatively, one can shift the work from the test space to the trial space by using trial functions ϕ that are approximate L-splines (*i.e.*, satisfy some approximate version of $L\phi = 0$), together with some standard space of test functions such as piecewise linears. The relationship between this dual

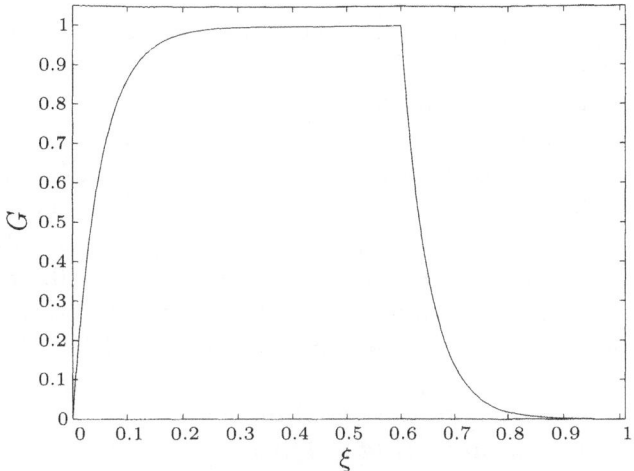

Figure 10.1. Green's function $G(\xi, x_i)$ with $a \equiv 1$, $b \equiv 0$, $x_i = 0.6$ and $\varepsilon = 0.05$.

approach and the use of L^*-spline test functions is discussed at length in Roos *et al.* (1996, §I.2.2.3).

Some authors have generalized the L^*-splines of (10.2) to two dimensions by taking their tensor product on rectangular grids, but this method is applicable only on domains whose boundary comprises straight-line segments each of which is parallel to one of the coordinate axes, and so negates one of the main advantages of finite element methods over finite differences. Consequently we do not discuss this approach here but refer the reader to Roos *et al.* (1996, §II.3.4).

A more useful generalization that is genuinely two-dimensional is found in Sacco, Gatti and Gotusso (1999): to solve (6.1) on an arbitrary triangular mesh one uses a trial space with local basis

$$1, \quad \mathrm{e}^{(\bar{a}_1 x + \bar{a}_2 y)/\varepsilon}, \quad \bar{a}_1 y - \bar{a}_2 x,$$

where $(\bar{a}_1, \bar{a}_2)$ is a piecewise-constant approximation of $\mathbf{a} = (a_1, a_2)$. Here the functions 1 and $\mathrm{e}^{\bar{a}_1 x + \bar{a}_2 y}$ come from the functions that appear in approximate L-splines for the corresponding one-dimensional problem (3.12), but the third function $\bar{a}_1 y - \bar{a}_2 x$ is new. Observe that all three functions lie in the null space of the operator $-\varepsilon \Delta(\cdot) + \bar{a}_1 (\cdot)_x + \bar{a}_2 (\cdot)_y$, *i.e.*, they are approximate L-splines. Piecewise linears are used in the test space. It is shown in Sacco and Stynes (1998) that this method is essentially equivalent to the unusual exponentially upwinded scheme used in PLTMG.

10.2. Shishkin meshes

FEMs can of course be implemented on Shishkin meshes like those of Figures 9.2 and 9.3 (the mesh rectangles can be bisected into triangles to permit the use of, *e.g.*, a piecewise linear FEM). Note that some mesh rectangles have a high aspect ratio, *i.e.*, their length greatly exceeds their width. To analyse such methods, the highly anisotropic nature of the mesh necessitates the use of sharp anisotropic interpolation estimates like those of Apel and Dobrowolski (1992) and Apel (1999), which we now describe.

Suppose that each element τ (triangle or rectangle) of the mesh is contained in a rectangle with side lengths (h_x, h_y) and contains a rectangle with side lengths (Ch_x, Ch_y) for some fixed constant $C > 0$. In the case of triangles, assume also a maximum angle condition: the interior angles are bounded away from π. (A triangular Shishkin mesh satisfies this maximum angle condition.)

Let $v \in H^2(\tau)$. Let v^I denote the nodal interpolant (linear or bilinear) of v. Write $\| \cdot \|_{0,\tau}$ for the norm in $L^2(\tau)$. Then

$$\|v - v^I\|_{0,\tau}^2 \le C \sum_{|\alpha|=2} h^{2\alpha} \|D^\alpha v\|_{0,\tau}^2,$$

$$\|\partial_x(v - v^I)\|_{0,\tau}^2 \le C \sum_{|\alpha|=1} h^{2\alpha} \|D^\alpha \partial_x v\|_{0,\tau}^2,$$

$$\|\partial_y(v - v^I)\|_{0,\tau}^2 \le C \sum_{|\alpha|=1} h^{2\alpha} \|D^\alpha \partial_y v\|_{0,\tau}^2.$$

Here α is the multi-index (α_1, α_2), $|\alpha| = \alpha_1 + \alpha_2$, $h^\alpha = h_x^{\alpha_1} h_y^{\alpha_2}$, and

$$D^\alpha = \frac{\partial^{\alpha_1}}{\partial x^{\alpha_1}} \frac{\partial^{\alpha_2}}{\partial y^{\alpha_2}}.$$

Bounds of this type are useful on Shishkin meshes because of the very small mesh width in precisely the coordinate direction whose derivative is large, and because no term v_{yy} appears in the bound on $\|\partial_x(v - v^I)\|_{0,\tau}$. Standard isotropic interpolation error estimates use only the diameter of the element and thereby lose the benefit of the Shishkin mesh in the analysis of interpolation error.

With these estimates, Dobrowolski and Roos (1997) show that if Ω is the unit square and the solution u of (6.1) can be written as the sum of a reduced solution and exponential boundary and corner layers, then for piecewise linear or bilinear interpolation on a Shishkin mesh,

$$\|u - u^I\|_{L^\infty(\Omega)} \le CN^{-2} \ln^2 N \quad \text{and} \quad \|u - u^I\|_0 \le CN^{-2} + C\sqrt{\varepsilon}N^{-2} \ln^2 N$$

so

$$\|u - u^I\|_0 \le CN^{-2} \text{ when } \sqrt{\varepsilon} \le C\ln^{-2}N, \quad \text{and} \quad \|u - u^I\|_{1,\varepsilon} \le CN^{-1} \ln N.$$

These bounds give us some idea of what convergence rates one can hope for when devising FEMs for convection-diffusion problems on Shishkin meshes.

Define the bilinear form

$$B(v, w) = (\varepsilon \nabla v, \nabla w) + (\mathbf{a}.\nabla v, w) + (bv, w) \quad \text{for all } v, w \in H^1(\Omega), \quad (10.3)$$

where $(\cdot, \cdot)$ denotes the $L^2(\Omega)$ inner product. Then (10.1) implies that

$$B(v, v) \geq \min\{1, C_5\} \|v\|_{1,\varepsilon}^2 \quad \text{for all } v \in H_0^1(\Omega). \quad (10.4)$$

Remark 10.2. Linß and Stynes (2001b) perform numerical experiments that compare several methods on the same Shishkin mesh for a test problem on the unit square that has exponential outflow layers along $x = 1$ and $y = 1$. The methods considered are central differencing, simple upwinding, the hybrid difference scheme of Linß and Stynes (1999), defect correction (see Remark 9.2), linear and bilinear Galerkin FEMs, and linear and bilinear streamline-diffusion FEMs (which we will discuss in Section 10.3). Graphs of the computed solutions, errors and convergence rates in the discrete $L^\infty(\Omega)$ norm are given, and known theoretical convergence results for the various methods are listed. It is concluded that, taking into account any difficulties that arise in solving the discrete linear systems, the methods that performed best for this problem are the defect correction method and the two streamline-diffusion FEMs, and that inside the layers bilinears yield more accuracy than linears.

10.3. The streamline-diffusion FEM

With linear or bilinear Galerkin methods, one has coercivity only with respect to the norm $\| \cdot \|_{1,\varepsilon}$ as shown in (10.4). This alone is insufficient to guarantee the stability of the method: numerical experiments on equidistant meshes will produce large oscillations like those seen for central differencing. Thus several finite element practitioners have devised FEMs that are coercive with respect to a stronger norm. Of these, the most commonly used is the *streamline-diffusion FEM* (SDFEM), which dates from 1979 (Hughes and Brooks 1979); it is also called the *streamline upwind Petrov–Galerkin* (SUPG) method.

Given a partition Ω^N of Ω, let S^N be a conforming space of piecewise polynomials of degree $k \geq 1$ defined on Ω^N. Define the SDFEM solution $u_{SD} \in S^N$ by

$$B_{SD}(u_{SD}, w^N)$$

$$:= B(u_{SD}, w^N) + \sum_{\tau \in \Omega^N} \delta_\tau (-\varepsilon \Delta u_{SD} + \mathbf{a}.\nabla u_{SD} + b u_{SD}, \mathbf{a}.\nabla w^N)_\tau$$

$$= (f, w^N) + \sum_{\tau \in \Omega^N} \delta_\tau (f, \mathbf{a}.\nabla w^N)_\tau \quad \text{for all } w^N \in S^n. \quad (10.5)$$

Here $B(\cdot, \cdot)$ is the standard bilinear form defined in (10.3), $(\cdot, \cdot)_\tau$ is the $L^2(\tau)$ inner product, and δ_τ is a nonnegative user-chosen piecewise constant that will be used to stabilize the method (if $\delta_\tau = 0$ for all $\tau \in \Omega^N$ then we return to the standard Galerkin method). The term $\sum_{\tau \in \Omega^N} \delta_\tau(f, \mathbf{a}.\nabla w^N)$ is included in the right-hand side of (10.5) to give the standard FEM property of *Galerkin orthogonality*:

$$B_{SD}(u - u_{SD}, w^N) = 0 \quad \text{for all } w^N \in S^N. \tag{10.6}$$

In the particular case when S^N comprises piecewise linears and $b \equiv 0$, the bilinear form simplifies to

$$B_{SD}(u_{SD}, w^N) = (\varepsilon \nabla u_{SD}, \nabla w^N) + (\mathbf{a}.\nabla u_{SD}, w^N)$$
$$+ \sum_{\tau \in \Omega^N} \delta_\tau(\mathbf{a}.\nabla u_{SD}, \mathbf{a}.\nabla w^N)_\tau,$$

which is the same as the standard Galerkin bilinear form $B(\cdot, \cdot)$ associated with the differential operator $-\varepsilon \Delta u - \delta|\mathbf{a}|^2 u_{\mathbf{aa}} + \mathbf{a}.\nabla u$, where δ is a piecewise-constant function and $u_{\mathbf{a}}$ denotes the directional derivative in the subcharacteristic direction. That is, we have added artificial diffusion to the PDE, but only in the direction of the subcharacteristics, which for stationary problems are the same as the so-called *streamlines* of the differential operator. This is the explanation of the name SDFEM.

The SDFEM can be regarded as a Petrov–Galerkin method with trial space S^N and test space $\{w^N + \sum_{\tau \in \Omega^N} \delta_\tau \mathbf{a}.\nabla w^N : w^N \in S^N\}$, *i.e.*, the test functions are obtained by 'upwinding' the trial functions along the subcharacteristics. For this reason it is also known as the SUPG method.

Assume that the mesh is quasi-uniform, so that (Brenner and Scott 2002, §4.4) on each element $\tau \in \Omega^N$ one has the standard interpolation property

$$|u - u^I|_{m,\tau} \le C h_\tau^{k+1-m} |u|_{k+1,\tau} \quad \text{for } m = 0, 1, 2 \tag{10.7}$$

and the local inverse inequality

$$\|\Delta w^N\|_{0,\tau} \le C_{\text{inv}} h_\tau^{-1} |w^N|_{1,\tau} \quad \text{for all } w^N \in S^N, \tag{10.8}$$

where the $|\cdot|_{\ell,\tau}$ are local Sobolev seminorms on the element τ, the norm on $L^2(\tau)$ is $\|\cdot\|_{0,\tau}$, and h_τ denotes the diameter of τ.

Define a norm that is stronger than $\|\cdot\|_{1,\varepsilon}$ and natural for the analysis of the SDFEM: for each $v \in H^1(\Omega)$, set

$$\|v\|_{SD} = \left(\varepsilon |v|_1^2 + \sum_{\tau \in \Omega^N} \delta_\tau \|\mathbf{a}.\nabla v\|_{0,\tau}^2 + C_5 \|v\|_0^2 \right)^{1/2}.$$

Lemma 10.3. Suppose that the SDFEM parameter δ_τ satisfies

$$0 \le \delta_\tau \le \frac{1}{2} \min \left\{ \frac{C_5}{\|b\|_{L^\infty(\tau)}^2}, \frac{h_\tau^2}{\varepsilon C_{\text{inv}}^2} \right\} \quad \text{for each } \tau \in \Omega^N. \tag{10.9}$$

Then the bilinear form $B_{SD}(\cdot, \cdot)$ is coercive with respect to $\|\cdot\|_{SD}$ over

$S^N \times S^N$, that is,

$$B_{SD}(w^N, w^N) \geq \frac{1}{2} \|w^N\|_{SD}^2 \quad \text{for all } w^N \in \Omega^N.$$

Proof. For each $w^N \in \Omega^N$, we get easily

$$B_{SD}(w^N, w^N) \geq \varepsilon |w^N|_1^2 + C_5 \|w^N\|_0^2 + \sum_{\tau \in \Omega^N} \delta_\tau \|\mathbf{a}.\nabla w^N\|_{0,\tau}^2$$

$$+ \sum_{\tau \in \Omega^N} \delta_\tau (-\varepsilon \Delta w^N + b w^N, \mathbf{a}.\nabla w^N)_\tau. \qquad (10.10)$$

Now the inequality $st \leq s^2 + t^2/4$ for s and $t \geq 0$, inequality (10.8) and the hypothesis on δ_τ yield

$$\left| \sum_{\tau \in \Omega^N} \delta_\tau (-\varepsilon \Delta w^N + b w^N, \mathbf{a}.\nabla w^N)_\tau \right|$$

$$\leq \sum_{\tau \in \Omega^N} \left[\varepsilon^2 \delta_\tau \|\Delta w^N\|_{0,\tau}^2 + \delta_\tau \|b\|_{L^\infty(\tau)}^2 \|w^N\|_{0,\tau}^2 + \frac{1}{2} \delta_\tau \|\mathbf{a}.\nabla w^N\|_{0,\tau}^2 \right]$$

$$\leq \frac{1}{2} \left[\varepsilon |w^N|_1^2 + C_5 \|w^N\|_0^2 + \sum_{\tau \in \Omega^N} \delta_\tau \|\mathbf{a}.\nabla w^N\|_{0,\tau}^2 \right].$$

Applying this bound in (10.10), the lemma is proved. $\qquad \square$

One can exploit this result to derive an error estimate in a fairly standard way. Let $u^I \in S^N$ denote the nodal interpolant of u. Then, under the hypothesis of Lemma 10.3,

$$\|u^I - u_{SD}\|_{SD}^2 \leq 2 B_{SD}(u^I - u_{SD}, u^I - u_{SD}) = 2 B_{SD}(u^I - u, u^I - u_{SD}),$$

by the Galerkin orthogonality property (10.6). Applying Cauchy–Schwarz-type inequalities to the right-hand side here and invoking (10.7) and $\varepsilon \delta_\tau \leq C h_\tau^2$ from (10.9), we arrive at (Roos *et al.* 1996, p. 232)

$$\|u^I - u_{SD}\|_{SD} \leq C h^k \left[\sum_\tau (\varepsilon + \delta_\tau + \delta_\tau^{-1} h_\tau^2 + h_\tau^2) |u|_{k+1,\tau}^2 \right]^{1/2},$$

where $h := \max_\tau h_\tau$ is the mesh diameter. In order to extract the best possible rate of convergence from this inequality while honouring the constraint on δ_τ in (10.9), set

$$\delta_\tau = \begin{cases} \delta_0 h_\tau & \text{for } \mathrm{Pe}_\tau > 1, \\ \delta_1 h_\tau^2/\varepsilon & \text{for } \mathrm{Pe}_\tau \leq 1, \end{cases} \qquad (10.11)$$

where we define the *mesh Péclet number* $\mathrm{Pe}_\tau := \|\mathbf{a}\|_{L^\infty(\tau)} h_\tau/\varepsilon$. Here δ_0 and

δ_1 are user-chosen positive constants. The more important case $\text{Pe}_\tau > 1$ is usually referred to as the convection-dominated case.

Remark 10.4. No precise general formula for an 'optimal' (in some sense) value of the SDFEM parameter δ_τ is known; the choice (10.11) seems to be the best statement that one can make. There has been much research into this question. For discussions of how to choose δ_τ see, *e.g.*, Akin and Tezduyar (2004), Brezzi and Russo (1994), Fischer, Ramage, Silvester and Wathen (1999), Houston and Süli (2001), Madden and Stynes (1996) and Roos *et al.* (1996).

The above analysis leads to the following bound (Roos *et al.* 1996, p. 233).

Theorem 10.5. Let each δ_τ be chosen according to (10.11) while satisfying the hypotheses of Lemma 10.3. Then there exists a constant C such that

$$\|u - u_{SD}\|_{SD} \le \|u - u^I\|_{SD} + \|u^I - u_{SD}\|_{SD} \le C(\varepsilon^{1/2} + h^{1/2})h^k|u|_{k+1},$$
$$(10.12)$$

where u^N is the solution of the SDFEM method (10.5).

In a very technical paper Sangalli (2003) shows that in the one-dimensional case (3.12), on an equidistant grid the SDFEM yields a solution that is quasi-optimal with respect to a certain interpolated norm that is roughly similar to our norm $\|\cdot\|_{SD}$.

When the mesh is coarse everywhere, so we are in the convection-dominated case on all elements, then $\varepsilon \le Ch_\tau$ for all $\tau \in \Omega^N$ and the bound (10.12) becomes

$$\|u - u_{SD}\|_{SD} \le Ch^{k+1/2}|u|_{k+1}.$$

This implies that

$$\|u - u_{SD}\|_0 + \left(\sum_{\tau \in \Omega^N} \delta_\tau \|\mathbf{a}.\nabla(u - u_{SD})\|_{0,\tau}^2 \right)^{1/2} \le Ch^{k+1/2}|u|_{k+1}, \quad (10.13)$$

Here the term $|u|_{k+1}$ is typically $\mathcal{O}(\epsilon^{-k-1/2})$. In general this will dominate the $h^{k+1/2}$ term and consequently (10.13) does not imply that the error $u - u_{SD}$ is small in some norm. Thus this estimate is of limited value. Nevertheless one can choose some maximal subset $\hat{\Omega}$ of Ω that excludes all layers, restrict the norms in (10.13) to $\hat{\Omega}$, then prove essentially the same bounds again (in terms of the new norms) by means of cut-off functions (Roos *et al.* 1996, §II.3.2.1).

Recalling that $\delta_\tau = \mathcal{O}(h_\tau)$ in the convection-dominated case, we see that in (10.13) the error bound for the *streamline derivative* $\mathbf{a}.\nabla u$ is of optimal order, but the estimate of $\|u - u^N\|_0$ is order $1/2$ less than optimal. This apparent loss of accuracy in the L^2 norm has attracted much attention.

Whether or not the bound on $\|u - u^N\|_0$ was sharp remained unresolved for many years until Zhou (1997) constructed a simple example for piecewise linears on a special mesh where the SDFEM converged with order only 1.5. For bilinears the situation is different. (Recall the comments on the numerical results for linears versus bilinears in Linß and Stynes (2001b).) For the SDFEM on the unit square Ω, under the usual hypotheses that u has only exponential boundary and corner layers, Stynes and Tobiska (2003) prove convergence results on a Shishkin mesh that imply, *inter alia*, $\|u - u^N\|_0 \leq CN^{-2} \ln^2 N$. The fundamental difference between bilinears and linears in the analysis is that for bilinears one has sharp interpolation error identities (Lin 1991) that enable the analysis to be carried out separately on each rectangle, while for triangles the corresponding identities require one to combine neighbouring elements to obtain an optimal error bound and this is not feasible on, *e.g.*, a Shishkin mesh.

Remark 10.6. Lemma 10.3 implies an *a priori* estimate for the SDFEM solution u_{SD}:

$$\|u_{SD}\|_{SD} \leq C \left(\|f\|_0^2 + \sum_\tau \delta_\tau \|f\|_{0,T}^2 \right)^{1/2}. \tag{10.14}$$

Thus the method retains some control over the streamline derivative $\mathbf{a}.\nabla u_{SD}$ of the computed solution. In the more interesting convection-dominated case, with $\delta_\tau = \delta_0 h_\tau$, inequality (10.14) says essentially that, on a quasi-uniform mesh, $\|\mathbf{a}.\nabla u_{SD}\|_{0,\tau}$ can be at most $\mathcal{O}(h_\tau^{1/2})$. It is this property that distinguishes the SDFEM from a standard Galerkin method, for whose oscillatory solution u^N one can have $\|\mathbf{a}.\nabla u^N\|_{0,\tau} = \mathcal{O}(1)$, since the slope of u^N is locally $\mathcal{O}(h_\tau^{-1})$.

This enhanced stability in the subcharacteristic direction means that the SDFEM can compute fairly satisfactory exponential layers in solutions of convection-diffusion problems, provided that δ_τ is chosen carefully. Note however that the method contains no mechanism for stabilization perpendicular to the subcharacteristics, so along characteristic layers the computed solution typically displays oscillations; as usual with such layers, these oscillations are confined to a fairly small neighbourhood of the layer.

Kopteva (2004) gives a detailed analysis of the accuracy of the SDFEM inside characteristic layers.

Figure 10.2 shows a solution computed by the SDFEM for a problem with an interior layer and two outflow exponential layers. The computed solution has oscillations along the interior layer and at one of the outflow boundary layers. In this example δ_τ is for simplicity set equal to the same value on all triangles of the equidistant mesh, but this is not in general the best approach. The same problem is solved again in Figure 10.3 but the common value of δ_τ has been increased judiciously to the value recommended

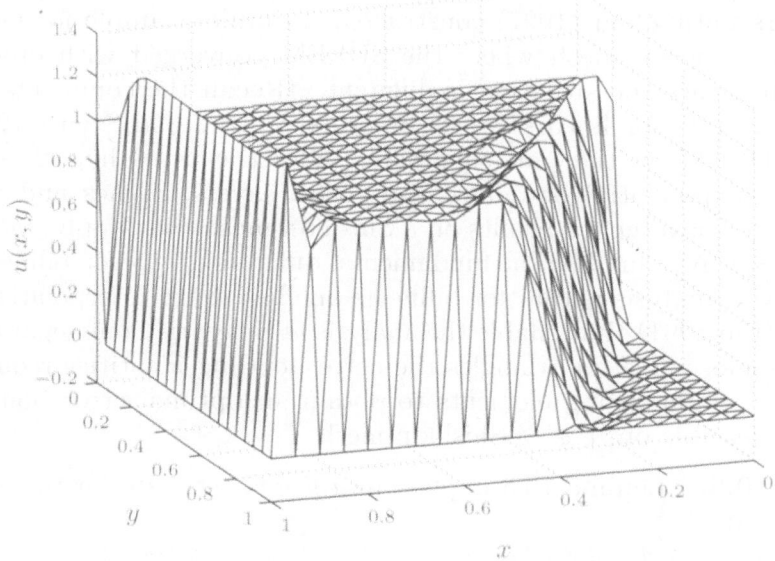

Figure 10.2. SDFEM I: δ_τ is the same for all τ.

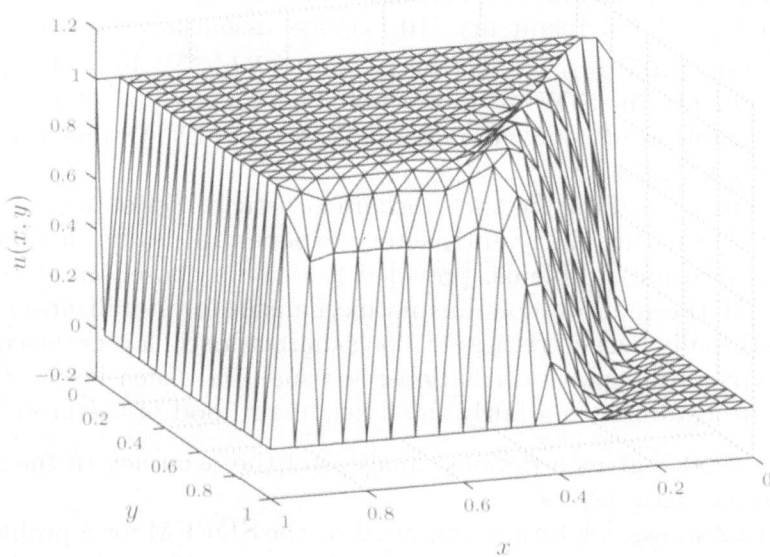

Figure 10.3. SDFEM II: increased value for δ_τ.

in Madden and Stynes (1996) to compute a sharp layer without oscillations along $x = 1$; unfortunately the outflow layer at $y = 1$ is smeared since the value of δ_τ is now larger than optimal for that layer. In line with the discussion in Remark 10.6, the increase in δ_τ has had little effect on the computed interior layer compared with the dramatic changes in the computed outflow layers.

Remark 10.7. In order to reduce or remove any oscillations that appear along characteristic layers, several authors have modified the SDFEM by adding artificial crosswind diffusion to the PDE or even by introducing nonlinear 'shock-capturing' terms into the SDFEM formulation. Expert opinion is divided on the value of this approach; for contrasting views see Shih and Elman (2000) and Knopp, Lube and Rapin (2002).

Remark 10.8. The SDFEM can of course be combined with a Shishkin mesh, and this technique was used to compute $u(x, y)$ in Figures 6.2, 6.3 and 6.5. In Figure 6.4 the SDFEM was also used but on an equidistant mesh; one can see that the interior layer in this figure is less sharp.

10.4. Discontinuous Galerkin finite element method

Recently, the *discontinuous Galerkin FEM* (DGFEM) has attracted a great deal of attention from many distinguished researchers. Like the SDFEM it achieves stability by a judicious choice of bilinear form, but the details of the construction are very different from Section 10.3.

Its name comes from its use of a standard piecewise polynomial trial space that is not required to be continuous across element boundaries. This local nature means the method is more readily parallelizable than (say) the SDFEM, and clearly permits the use of polynomials of different degrees on different elements, which can be exploited to gain increased accuracy when the problem is quite smooth on only part of the domain – as is usually the case with convection-diffusion problems. A drawback is the much larger number of degrees of freedom compared with finite element spaces that lie in $C(\Omega)$.

Methods of this type were first introduced in the 1970s and today there are several prominent variants. Arnold, Brezzi, Cockburn and Marini (2001/02) consider the problem $-\Delta u = f$ on Ω with $u = 0$ on $\partial\Omega$ and show that nine distinct versions of the DGFEM can be placed in the framework of a mixed-method weak formulation. They go on to analyse the stability of these methods, but this is of limited value in the context of convection-diffusion problems where the Laplacian is multiplied by a small parameter. This paper also gives an account of the historical development of DGFEMs that includes methods specifically designed for convection-diffusion problems.

Given the diversity of methods described as DGFEMs, we shall not attempt to give a thorough survey of this area. Instead we concentrate on one variant and the references appearing in this subsection will assist the reader who wishes to broaden his or her knowledge of the DGFEM.

Consider the nonsymmetric interior penalty DGFEM (NIPD) from Houston, Schwab and Süli (2002); related methods appear in, *e.g.*, Oden, Babuška and Baumann (1998) and Rivière, Wheeler and Girault (2001).

Assume that Ω is polygonal. Let $\mathcal{T}$ be a partition of Ω into elements κ (*e.g.*, triangles or rectangles). Houston *et al.* (2002) permit up to one hanging node for each κ, but for simplicity we shall assume that our partition has no hanging nodes. Assume also that each $\kappa \in \mathcal{T}$ is an affine image of a fixed master element $\hat{\kappa}$, *i.e.*, that $\kappa = F_\kappa(\hat{\kappa})$ where $\hat{\kappa}$ is either the open unit simplex or the open unit square in $\mathbb{R}^2$. For each nonnegative integer k, let $\mathcal{P}_k(\hat{\kappa})$ denote the set of polynomials of total degree k on $\hat{\kappa}$. (If $\hat{\kappa}$ is the unit square, one can also consider $\mathcal{Q}_k(\hat{\kappa})$, the set of all tensor-product polynomials on $\hat{\kappa}$ of degree k in each coordinate direction.) For each $\kappa \in \mathcal{T}$ write p_κ for the local polynomial degree. Set $\mathbf{p} = \{p_\kappa : \kappa \in \mathcal{T}\}$ and $\mathbf{F} = \{F_\kappa : \kappa \in \mathcal{T}\}$ and define the finite element space

$$S^{\mathbf{p}}(\Omega, \mathcal{T}, \mathbf{F}) = \{v \in L^2(\Omega) : v|_\kappa \circ F_\kappa \in \mathcal{R}_{p_\kappa}(\hat{\kappa})\},$$

where $\mathcal{R}$ is either $\mathcal{P}$ or $\mathcal{Q}$.

For $s = 0, 1$ define the broken Sobolev spaces

$$H^s(\Omega, \mathcal{T}) = \{v \in L^2(\Omega) : v|_\kappa \in H^s(\kappa) \text{ for all } \kappa \in \mathcal{T}\}.$$

Let $\partial\kappa$ denote the boundary of κ for each $\kappa \in \mathcal{T}$. Define the inflow and outflow parts of $\partial\kappa$ by

$$\partial^-\kappa = \{(x, y) \in \partial\kappa : \mathbf{a}(x, y).\mu_\kappa(x, y) < 0\},$$
$$\partial^+\kappa = \{(x, y) \in \partial\kappa : \mathbf{a}(x, y).\mu_\kappa(x, y) \geq 0\}$$

respectively, where $\mu_\kappa(x, y)$ denotes the outward-pointing unit normal to $\partial\kappa$ at $(x, y) \in \partial\kappa$.

Let $v \in H^1(\Omega, \mathcal{T})$. For each $\kappa \in \mathcal{T}$, denote by v_κ^+ the inner trace of $v|_\kappa$ on $\partial\kappa$. If $\partial^-\kappa \setminus \partial\Omega$ is nonempty, then for almost every point $(x, y) \in \partial^-\kappa \setminus \partial\Omega$ there exists a unique $\kappa' \in \mathcal{T}$ (which depends on (x, y)) such that $x \in \partial^+\kappa'$ and $\kappa' \cap (\partial^-\kappa \setminus \partial\Omega)$ has nonzero one-dimensional measure, and we define the outer trace v_κ^- of v on $\partial^-\kappa \setminus \partial\Omega$ relative to κ to be the inner trace $v_{\kappa'}^+$ relative to κ'. Then define the jump of v across $\partial^-\kappa \setminus \partial\Omega$ by $\lfloor v \rfloor_\kappa = v_\kappa^+ - v_\kappa^-$.

We shall drop the subscript κ from the above notation when it is clear from the context what is intended.

Let $\mathcal{E}_{\text{int}}$ be the set of all open one-dimensional edges of the partition $\mathcal{T}$ that lie in Ω. Set $\Gamma_{\text{int}} = \{x \in \Omega : x \in e \text{ for some } e \in \mathcal{E}_{\text{int}}\}$. Numbering the elements κ consecutively, for each $e \in \mathcal{E}_{\text{int}}$ there exist indices i and j such that $i > j$ and the elements κ_i and κ_j share the interface e. Define the (element-numbering-dependent) jump of $v \in H^1(\Omega, \mathcal{T})$ across e and the mean value of v on e by

$$[v]_e = v|_{\partial \kappa_i \cap e} - v|_{\partial \kappa_j \cap e} \quad \text{and} \quad \langle v \rangle_e = \tfrac{1}{2}\left(v|_{\partial \kappa_i \cap e} + v|_{\partial \kappa_j \cap e}\right)$$

respectively. Furthermore, for each $e \in \mathcal{E}_{\text{int}}$ let ν denote the unit normal vector pointing from κ_i to κ_j; if $e \subset \partial \Omega$, take $\nu = \mu$.

The bilinear form associated with the NIPD for $(6.1a)$ with $u \equiv 0$ on $\partial \Omega$ is

$$B_{DG}(v, w) = \sum_{\kappa \in \mathcal{T}} \left(\varepsilon \int_{\kappa} \nabla v . \nabla w \, \mathrm{d}x + \int_{\kappa} (\mathbf{a}.\nabla v + bv)w \, \mathrm{d}x \right.$$

$$\left. - \int_{\partial^- \kappa \cap \partial^- \Omega} (\mathbf{a}.\mu)v^+ w^+ \, \mathrm{d}s - \int_{\partial^- \kappa \backslash \partial \Omega} (\mathbf{a}.\mu_\kappa) \lfloor v \rfloor w^+ \, \mathrm{d}s \right)$$

$$+ \varepsilon \int_{\partial \Omega} \left(v(\nabla w.\mu) - (\nabla v.\mu)w \right) \mathrm{d}s + \int_{\partial \Omega} \sigma v w \, \mathrm{d}s$$

$$+ \varepsilon \int_{\Gamma_{\text{int}}} \left([v]\langle \nabla w.\nu \rangle - \langle \nabla v.\nu \rangle [w] \right) \mathrm{d}s + \int_{\Gamma_{\text{int}}} \sigma [v][w] \, \mathrm{d}s,$$

for all $v, w \in H^1(\Omega, \mathcal{T})$. Here σ, the user-chosen nonnegative *discontinuity-penalization parameter*, is defined by

$$\sigma|_e = \sigma_e \quad \text{for each } e \in \mathcal{E}_{\text{int}} \cup \partial \Omega.$$

Houston *et al.* (2002) choose $\sigma_e = \mathcal{O}(\varepsilon/h_e)$ where h_e is the length of edge e.

The NIPD method is then: find $u_{DG} \in S^{\mathbf{P}}(\Omega, \mathcal{T}, \mathbf{F})$ such that

$$B(u_{DG}, w^N) = \sum_{\kappa \in \mathcal{T}} \int_{\kappa} f w^N \, \mathrm{d}x \, \mathrm{d}y \quad \text{for all } w^N \in S^{\mathbf{P}}(\Omega, \mathcal{T}, \mathbf{F}). \tag{10.15}$$

Existence and uniqueness of a solution to (10.15) are shown in Houston *et al.* (2002) by combining results from earlier papers of these authors.

Assuming that $u \in H^2(\Omega, \mathcal{T})$ and ∇u is continuous across each edge $e \in \mathcal{E}_{\text{int}}$, one can deduce the Galerkin orthogonality property

$$B_{DG}(u - u_{DG}, w^N) = 0 \quad \text{for all } w^N \in S^{\mathbf{P}}(\Omega, \mathcal{T}, \mathbf{F}).$$

For all $v \in H^2(\Omega, \mathcal{T})$ define the norm $\| \cdot \|_{DG}$ by $\|v\|^2_{DG} = B_{DG}(v, v)$. Setting $(v, w)_e = \int_e |\mathbf{a}.\mu_\kappa| v w \, \mathrm{d}s$ for each $e \subset \partial \kappa$ and $\|v\|^2_e = (v, v)_e$, after some manipulation we get

$$\|v\|^2_{DG} = \sum_{\kappa \in \mathcal{T}} \left(\varepsilon \|\nabla v\|^2_{0,\kappa} + \|c_0 v\|^2_{0,\kappa} \right) + \int_{\partial \Omega} \sigma v^2 \, \mathrm{d}s + \int_{\Gamma_{\text{int}}} \sigma [v]^2 \, \mathrm{d}s$$

$$+ \tfrac{1}{2} \sum_{\kappa \in \mathcal{T}} \left(\|v^+\|^2_{\partial^- \kappa \cap \partial \Omega} + \|v^+ - v^-\|^2_{\partial^- \kappa \backslash \partial \Omega} + \|v^+\|^2_{\partial^+ \kappa \cap \partial \Omega} \right),$$

where $\| \cdot \|_{0,\kappa}$ is the $L^2(\kappa)$ norm and we set

$$c_0(x, y) = \sqrt{b(x, y) - \operatorname{div} \mathbf{a}(x, y)/2};$$

by (10.1) the function c_0 is well defined. Clearly $\| \cdot \|_{DG}$ is stronger than $\| \cdot \|_{1,\varepsilon}$.

Now Houston *et al.* (2002) write $u - u_{DG} = (u - \Pi u) + (\Pi u - u_{DG})$ where Π is the orthogonal projector in L^2 into $S^{\mathbf{P}}(\Omega, \mathcal{T}, \mathbf{F})$. From Galerkin orthogonality we have

$$\|\Pi u - u_{DG}\|_{DG}^2 = B_{DG}(\Pi u - u_{DG}, \Pi u - u_{DG}) = B_{DG}(\Pi u - u, \Pi u - u_{DG}),$$
$$(10.16)$$

and, under the assumption that $\mathbf{a}.\nabla w^N|_\kappa$ lies in $S^{\mathbf{P}}(\Omega, \mathcal{T}, \mathbf{F})$ for all $w^N \in S^{\mathbf{P}}(\Omega, \mathcal{T}, \mathbf{F})$, some analysis of the right-hand side of (10.16) enables $\|\Pi u - u_{DG}\|_{DG}$ to be estimated in terms of various norms of $u - \Pi u$. Invoking the triangle inequality $\|u - u_{DG}\|_{DG} \leq \|u - \Pi u\|_{DG} + \|\Pi u - u_{DG}\|_{DG}$ then leads to a bound on $\|u - u_{DG}\|_{DG}$. In the particular case where the mesh elements are rectangles, piecewise polynomials of degree p are used, h is the mesh diameter and the solution u lies in $H^{p+1}(\Omega)$, the bound becomes

$$\|u - u_{DG}\|_{DG} \leq C(\varepsilon^{1/2} h^p + h^{p+1/2}) \|u\|_{H^{p+1}(\Omega)}, \quad \text{where } C = C(p).$$

Note that the right-hand side here depends on a Sobolev norm of u that is typically $\mathcal{O}(\varepsilon^{-p-1/2})$. It may be possible to use cut-off functions to localize this result away from layers, removing this undesirable feature.

The above analysis from Houston *et al.* (2002) assumes that the mesh is nondegenerate (Brenner and Scott 2002, §4.4), which excludes the long thin elements one expects in any mesh that is specifically designed to improve the behaviour of the method inside layers. Roos and Zarin (2003) apply this DGFEM to a problem on the unit square that has exponential layers along $x = 1$ and $y = 1$ and no other layers. Working with piecewise bilinears on a rectangular Shishkin mesh like that of Figure 9.2 with N mesh intervals in each coordinate direction, they adapt the analysis of Houston *et al.* (2002) to this situation (which entails a different choice for σ_e on part of the mesh) and prove that

$$\|u - u_{DG}\|_{DG} \leq C N^{-1} \ln^{3/2} N. \qquad (10.17)$$

A related paper (Zarin and Roos 2005) considers a problem similar to Example 6.1 and, using a Shishkin mesh similar to the one in Figure 9.3 with N mesh intervals in each coordinate direction, again obtains the bound (10.17).

We remind the reader that there is no universal agreement on a 'best' form of the DGFEM. For example, Gopalakrishnan and Kanschat (2003) consider a symmetric version of our bilinear form $B_{DG}(v, w)$ that is obtained by changing the signs of the terms $\varepsilon \int_{\partial\Omega} v(\nabla w.\mu)\,\mathrm{d}s$ and $\varepsilon \int_{\Gamma_{\text{int}}} [v]\langle \nabla w.\nu \rangle\,\mathrm{d}s$. A good sense of the breadth of interest in the DGFEM and the variety of its manifestations can be inferred from the collection of papers in Cockburn, Karniadakis and Shu (2000).

10.5. *Adaptive methods*

Adaptive FEMs compute a solution to a boundary-value problem on some conventional (*e.g.*, equidistant) mesh using some stable method such as SDFEM, then use this solution to compute *a posteriori* some local error estimator that gives guidance on where one should refine or coarsen the mesh to obtain a mesh better suited to the boundary-value problem. On this new mesh one then computes a fresh solution to the problem, then the mesh is again modified based on the local error estimator. The process is continued iteratively until some stopping criterion is reached. See Ainsworth and Oden (2000) or Brenner and Scott (2002, Chapter 9) for a more precise description.

There is perhaps a general consensus that in the long run adaptive methods will provide the most satisfactory approach to solving convection-diffusion problems, but today their behaviour when applied to such problems is still poorly understood, despite many published numerical experiments. John (2000) gives numerical examples of how apparently reasonable error estimators can yield inaccurate solutions to convection-diffusion problems.

A difficulty with the theory of *a posteriori* error estimators for convection-diffusion problems is that published inequalities relating the estimator to the true error frequently contain multiplicative factors that depend badly on the small diffusion parameter ε. This seriously undermines the validity of the estimator. Below we shall confine our discussion to a few ε-independent results that have been obtained.

For the one-dimensional problem (3.12), an adaptive-mesh algorithm that is based on arc-length equidistribution (where mesh points are moved but no points are created or deleted) is analysed by Kopteva and Stynes (2001), using earlier *a posteriori* bounds from Kopteva (2001). It is shown that, starting from an equidistant mesh with N subintervals, after $\mathcal{O}(\ln(1/\varepsilon)/(\ln N))$ iterations one obtains a computed solution u^N that resolves the layer with moreover $|u(x_i) - u_i^N| \leq CN^{-1}$ for all i. The underlying numerical method is simple upwinding so this is a finite difference approach, but we include it here since it is a clear convergence result for an adaptive method and few such results exist for convection-diffusion problems. It seems difficult to extend this type of result to two-dimensional problems.

In Sangalli (2001) the *residual-free bubble* FEM is considered; this method is related to the SDFEM (Brezzi, Marini and Süli 2000). An error estimator based on element residuals and jumps in the normal derivative of the solution across edges is shown to be robust for (6.1), *i.e.*, the global value of the estimator is equivalent to the true error up to a constant factor that is independent of ε, but the norm in which the true error is measured is

$$w \mapsto \varepsilon |w|_{H^1(\Omega)} + \|\mathbf{a}.\nabla w\|_{H^{-1}(\Omega)},$$

which is weak: the factor multiplying $|\cdot|_{H^1(\Omega)}$ is ε, not the more natural $\varepsilon^{1/2}$ that appears in the weighted energy norm $\|\cdot\|_{1,\varepsilon}$ of Section 7.

The *dual-weighted-residual* method for goal-oriented error estimation has been successfully applied to convection-diffusion problems by various authors; see Eriksson, Estep, Hansbo and Johnson (1996) and Bangerth and Rannacher (2003). Here the aim is to adapt the mesh in order to compute accurately some functional of the solution but not the solution itself. The theoretical basis for this method has recently been surveyed in *Acta Numerica* (Giles and Süli 2002) so we shall not discuss it further here.

Finally, Verfürth (2004) shows that for the SDFEM the error in the computed solution is equivalent (up to a constant factor that is independent of ε) to the global value of each of three different estimators (one based on element and edge residuals; one based on the solution of local Dirichlet problems; one based on the solution of local Neumann problems). The true error is measured in a norm

$$w \mapsto \|w\|_{1,\varepsilon} + \|w\|_*,$$

where $\|\cdot\|_*$ is the dual norm on $H^{-1}(\Omega)$ defined by

$$\|w\|_* = \sup_{v \in H_0^1(\Omega) \setminus \{0\}} \frac{(w, v)}{\|v\|_{1,\varepsilon}},$$

with $(\cdot, \cdot)$ the corresponding duality pairing. (This special norm is used to bound the convective term.) But the paper assumes that the mesh is quasi-uniform, which excludes the long thin elements that one expects an adaptive code to construct when solving a convection-diffusion problem.

In summary, we do not have today a satisfactory adaptive method for two-dimensional convection-diffusion problems that, starting from an ordinary coarse mesh, is guaranteed to produce a layer-adapted mesh with a bound on the error in the computed solution in some reasonably strong norm.

11. Concluding remarks

Our survey has not been exhaustive. For example, the hp finite element method appeared only in an incidental way in the title of Houston *et al.* (2002) in Section 10.4. For general surveys of methods for convection-diffusion problems see Morton (1996) and Roos, Stynes and Tobiska (2005). (For the hp finite element method see Schwab (1998), and also Melenk (2002), where singularly perturbed linear reaction-diffusion problems are examined in great detail.)

Time-dependent convection-diffusion problems are of great practical importance but space constraints did not allow their discussion here. As well as the general references cited above, see Ewing and Wang (2001) and Hundsdorfer and Verwer (2003).

The numerical analysis and solution of convection-diffusion problems on polygonal regions, where the solution is assumed to exhibit boundary but not interior layers and one has sufficient compatibility of the data at the corners of the domain, is by now fairly well understood in the framework of Shishkin meshes combined with finite difference or finite element methods. When we consider interior layers (and the effects of data incompatibilities at corners) our grasp is much less sure and there are several competing methods. In the long run the view of this author is that adaptive methods will triumph over all types of convection-diffusion problem, but much work remains to be done.

Acknowledgements

A first version of this survey was written while the author was on a sabbatical visit at the Department of Mathematics and the Industrial Mathematics Institute at the University of South Carolina, USA, for whose support I am deeply grateful. I am greatly indebted to Niall Madden, who cheerfully and efficiently provided all the figures. Hans-Görg Roos and Lutz Tobiska kindly read the entire article and offered helpful criticisms, for which I wish to express my gratitude.

REFERENCES

M. Ainsworth and J. T. Oden (2000), *A Posteriori Error Estimation in Finite Element Analysis*, Pure and Applied Mathematics (New York), Wiley-Interscience, New York.

J. E. Akin and T. E. Tezduyar (2004), 'Calculation of the advective limit of the SUPG stabilization parameter for linear and higher-order elements', *Comput. Methods Appl. Mech. Engrg.* **193**, 1909–1922.

D. N. d. G. Allen and R. V. Southwell (1955), 'Relaxation methods applied to determine the motion, in two dimensions, of a viscous fluid past a fixed cylinder', *Quart. J. Mech. Appl. Math.* **8**, 129–145.

V. B. Andreev and N. V. Kopteva (1996), 'Investigation of difference schemes with an approximation of the first derivative by a central difference relation', *Zh. Vychisl. Mat. i Mat. Fiz.* **36**(8), 101–117.

T. Apel (1999), *Anisotropic Finite Elements: Local Estimates and Applications*, Advances in Numerical Mathematics, B. G. Teubner, Stuttgart.

T. Apel and M. Dobrowolski (1992), 'Anisotropic interpolation with applications to the finite element method', *Computing* **47**, 277–293.

D. N. Arnold, F. Brezzi, B. Cockburn and L. D. Marini (2001/02), 'Unified analysis of discontinuous Galerkin methods for elliptic problems', *SIAM J. Numer. Anal.* **39**, 1749–1779 (electronic).

W. Bangerth and R. Rannacher (2003), *Adaptive Finite Element Methods for Differential Equations*, Lectures in Mathematics ETH Zürich, Birkhäuser, Basel.

E. Bohl (1981), *Finite Modelle gewöhnlicher Randwertaufgaben*, Teubner, Stuttgart.

K. Bohmer and R. Rannacher (1984), *Defect Correction Methods: Theory and Applications*, Springer, Berlin.

A. Brandt and I. Yavneh (1991), 'Inadequacy of first-order upwind difference schemes for some recirculating flows', *J. Comput. Phys.* **93**, 128–143.

S. C. Brenner and L. R. Scott (2002), *The Mathematical Theory of Finite Element Methods*, Vol. 15 of *Texts in Applied Mathematics*, 2nd edn, Springer, New York.

F. Brezzi and A. Russo (1994), 'Choosing bubbles for advection-diffusion problems', *Math. Models Methods Appl. Sci.* **4**, 571–587.

F. Brezzi, D. Marini and E. Süli (2000), 'Residual-free bubbles for advection-diffusion problems: the general error analysis', *Numer. Math.* **85**, 31–47.

B. Cockburn, G. E. Karniadakis and C.-W. Shu, eds (2000), *Discontinuous Galerkin Methods: Theory, Computation and Applications*, Vol. 11 of *Lecture Notes in Computational Science and Engineering*, Springer, Berlin. Papers from the 1st International Symposium held in Newport, RI, May 24–26, 1999.

M. Dobrowolski and H.-G. Roos (1997), '*A priori* estimates for the solution of convection-diffusion problems and interpolation on Shishkin meshes', *Z. Anal. Anwendungen* **16**, 1001–1012.

W. Dörfler (1999), 'Uniform *a priori* estimates for singularly perturbed elliptic equations in multidimensions', *SIAM J. Numer. Anal.* **36**, 1878–1900 (electronic).

K. Eriksson, D. Estep, P. Hansbo and C. Johnson (1996), *Computational Differential Equations*, Cambridge University Press, Cambridge.

R. E. Ewing and H. Wang (2001), 'A summary of numerical methods for time-dependent advection-dominated partial differential equations', *J. Comput. Appl. Math.* **128**, 423–445. Numerical analysis 2000, Vol. VII, Partial differential equations.

P. Farrell, A. Hegarty, J. Miller, E. O'Riordan and G. Shishkin (2000), *Robust Computational Techniques for Boundary Layers*, Chapman & Hall/CRC, Boca Raton.

B. Fischer, A. Ramage, D. J. Silvester and A. J. Wathen (1999), 'On parameter choice and iterative convergence for stabilised discretisations of advection-diffusion problems', *Comput. Methods Appl. Mech. Engrg.* **179**, 179–195.

A. Fröhner, T. Linß and H.-G. Roos (2001), 'Defect correction on Shishkin-type meshes', *Numer. Algorithms* **26**, 281–299.

D. Gilbarg and N. S. Trudinger (2001), *Elliptic Partial Differential Equations of Second Order*, Classics in Mathematics, Springer, Berlin. Reprint of the 1998 edition.

M. B. Giles and E. Süli (2002), Adjoint methods for PDEs: *a posteriori* error analysis and postprocessing by duality, in *Acta Numerica*, Vol. 11, Cambridge University Press, pp. 145–236.

H. Goering, A. Felgenhauer, G. Lube, H.-G. Roos and L. Tobiska (1983), *Singularly Perturbed Differential Equations*, Vol. 13 of *Mathematical Research*, Akademie, Berlin.

J. Gopalakrishnan and G. Kanschat (2003), 'A multilevel discontinuous Galerkin method', *Numer. Math.* **95**, 527–550.

P. Grisvard (1985), *Elliptic Problems in Nonsmooth Domains*, Vol. 24 of *Monographs and Studies in Mathematics*, Pitman (Advanced Publishing Program), Boston, MA.

H. Han and R. B. Kellogg (1990), 'Differentiability properties of solutions of the equation $-\epsilon^2 \Delta u + ru = f(x, y)$ in a square', *SIAM J. Math. Anal.* **21**, 394–408.

P. Houston and E. Süli (2001), 'Stabilised hp-finite element approximation of partial differential equations with nonnegative characteristic form', *Computing* **66**, 99–119.

P. Houston, C. Schwab and E. Süli (2002), 'Discontinuous hp-finite element methods for advection-diffusion-reaction problems', *SIAM J. Numer. Anal.* **39**, 2133–2163 (electronic).

T. J. R. Hughes and A. Brooks (1979), A multidimensional upwind scheme with no crosswind diffusion, in *Finite Element Methods for Convection Dominated Flows (Papers, Winter Ann. Meeting Amer. Soc. Mech. Engrs., New York, 1979)*, Vol. 34 of *AMD*, Amer. Soc. Mech. Engrs. (ASME), New York, pp. 19–35.

W. Hundsdorfer and J. Verwer (2003), *Numerical Solution of Time-Dependent Advection-Diffusion-Reaction Equations*, Vol. 33 of *Springer Series in Computational Mathematics*, Springer, Berlin.

A. M. Il'in (1969), 'A difference scheme for a differential equation with a small parameter multiplying the highest derivative', *Mat. Zametki* **6**, 237–248.

A. M. Il'in (1992), *Matching of Asymptotic Expansions of Solutions of Boundary Value Problems*, Vol. 102 of *Translations of Mathematical Monographs*, AMS, Providence, RI. Translated from the Russian by V. Minachin [V. V. Minakhin].

V. John (2000), 'A numerical study of *a posteriori* error estimators for convection-diffusion equations', *Comput. Methods Appl. Mech. Engrg.* **190**, 757–781.

R. B. Kellogg and M. Stynes (2005), 'Corner singularities and boundary layers in a simple convection-diffusion problem', *J. Differential Equations*, to appear.

R. B. Kellogg and M. Stynes (2004), A singularly perturbed convection-diffusion problem in a half-plane, Technical Report 13, Industrial Mathematics Institute, University of South Carolina.

R. B. Kellogg and A. Tsan (1978), 'Analysis of some difference approximations for a singular perturbation problem without turning points', *Math. Comp.* **32**, 1025–1039.

J. Kevorkian and J. D. Cole (1996), *Multiple Scale and Singular Perturbation Methods*, Vol. 114 of *Applied Mathematical Sciences*, Springer, New York.

T. Knopp, G. Lube and G. Rapin (2002), 'Stabilized finite element methods with shock capturing for advection-diffusion problems', *Comput. Methods Appl. Mech. Engrg.* **191**, 2997–3013.

N. Kopteva (2001), 'Maximum norm *a posteriori* error estimates for a one-dimensional convection-diffusion problem', *SIAM J. Numer. Anal.* **39**, 423–441 (electronic).

N. Kopteva (2003), 'Error expansion for an upwind scheme applied to a two-dimensional convection-diffusion problem', *SIAM J. Numer. Anal.* **41**, 1851–1869 (electronic).

N. Kopteva (2004), 'How accurate is the streamline-diffusion FEM inside characteristic (boundary and interior) layers?', *Comput. Methods Appl. Mech. Engrg.* **193**, 4875–4889.

N. Kopteva and M. Stynes (2001), 'A robust adaptive method for a quasi-linear one-dimensional convection-diffusion problem', *SIAM J. Numer. Anal.* **39**, 1446–1467 (electronic).

O. A. Ladyzhenskaya and N. N. Ural'tseva (1968), *Linear and Quasilinear Elliptic Equations*, Translated from the Russian by Scripta Technica, Inc. (translation editor, Leon Ehrenpreis), Academic Press, New York.

Q. Lin (1991), A rectangle test for finite element analysis, in *Proc. Syst. Sci. Eng.*, Great Wall (H.K.) Culture Publish Co., pp. 213–216.

T. Linß (2001), 'The necessity of Shishkin decompositions', *Appl. Math. Lett.* **14**, 891–896.

T. Linß (2003), 'Layer-adapted meshes for convection-diffusion problems', *Comput. Methods Appl. Mech. Engrg.* **192**, 1061–1105.

T. Linß (2005), 'On a convection-diffusion problem with a weak layer', *Appl. Math. Comput.* **160**, 791–795.

T. Linß and M. Stynes (1999), 'A hybrid difference scheme on a Shishkin mesh for linear convection-diffusion problems', *Appl. Numer. Math.* **31**, 255–270.

T. Linß and M. Stynes (2001a), 'Asymptotic analysis and Shishkin-type decomposition for an elliptic convection-diffusion problem', *J. Math. Anal. Appl.* **261**, 604–632.

T. Linß and M. Stynes (2001b), 'Numerical methods on Shishkin meshes for linear convection-diffusion problems', *Comput. Methods Appl. Mech. Engrg.* **190**, 3527–3542.

N. Madden and M. Stynes (1996), 'Linear enhancements of the streamline diffusion method for convection-diffusion problems', *Comput. Math. Appl.* **32**, 29–42.

N. Madden and M. Stynes (1997), 'Efficient generation of oriented meshes for solving convection-diffusion problems', *Int. J. Numer. Methods Engrg.* **40**, 565–576.

J. M. Melenk (2002), *hp-Finite Element Methods for Singular Perturbations*, Vol. 1796 of *Lecture Notes in Mathematics*, Springer, Berlin.

J. Miller, E. O'Riordan and G. Shishkin (1996), *Fitted Numerical Methods for Singular Perturbation Problems*, World Scientific, Singapore.

K. W. Morton (1996), *Numerical Solution of Convection-Diffusion Problems*, Vol. 12 of *Applied Mathematics and Mathematical Computation*, Chapman & Hall, London.

J. T. Oden, I. Babuška and C. E. Baumann (1998), 'A discontinuous *hp* finite element method for diffusion problems', *J. Comput. Phys* **146**, 491–519.

M. H. Protter and H. F. Weinberger (1984), *Maximum Principles in Differential Equations*, Springer, New York. Corrected reprint of the 1967 original.

A. Quarteroni and A. Valli (1994), *Numerical Approximation of Partial Differential Equations*, Springer, Berlin.

B. Rivière, M. F. Wheeler and V. Girault (2001), 'A priori error estimates for finite element methods based on discontinuous approximation spaces for elliptic problems', *SIAM J. Numer. Anal.* **39**, 902–931 (electronic).

H.-G. Roos (1994), 'Ten ways to generate the Il'in and related schemes', *J. Comput. Appl. Math.* **53**, 43–59.

H.-G. Roos (1996), 'A note on the conditioning of upwind schemes on Shishkin meshes', *IMA J. Numer. Anal.* **16**, 529–538.

H.-G. Roos and H. Zarin (2003), The discontinuous Galerkin finite element method for singularly perturbed problems, in *Challenges in Scientific Computing: CISC 2002* (E. Bänsch, ed.), Vol. 35 of *Lecture Notes in Computational Science and Engineering*, Springer, Berlin, pp. 246–267.

H.-G. Roos, M. Stynes and L. Tobiska (1996), *Numerical Methods for Singularly Perturbed Differential Equations*, Springer, Berlin/Heidelberg/New York.

H.-G. Roos, M. Stynes and L. Tobiska (2005), *Numerical Methods for Singularly Perturbed Differential Equations*, 2nd edn, Springer, Berlin/Heidelberg/New York, in preparation.

R. Sacco and M. Stynes (1998), 'Finite element methods for convection-diffusion problems using exponential splines on triangles', *Comput. Math. Appl.* **35**, 35–45.

R. Sacco, E. Gatti and L. Gotusso (1999), 'A nonconforming exponentially fitted finite element method for two-dimensional drift-diffusion models in semiconductors', *Numer. Methods Partial Diff. Equations* **15**, 133–150.

G. Sangalli (2001), 'A robust a posteriori estimator for the residual-free bubbles method applied to advection-diffusion problems', *Numer. Math.* **89**, 379–399.

G. Sangalli (2003), 'Quasi optimality of the SUPG method for the one-dimensional advection-diffusion problem', *SIAM J. Numer. Anal.* **41**, 1528–1542 (electronic).

F. Schieweck (1986), Eine asymptotische angepaßte Finite-Element-Methode für singulär gestörte elliptische Randwertaufgaben, PhD thesis, Technische Hochschule Magdeburg, GDR.

C. Schwab (1998), *p- and hp-Finite Element Methods: Theory and Applications in Solid and Fluid Mechanics*, Numerical Mathematics and Scientific Computation, The Clarendon Press (Oxford University Press), New York.

Y.-T. Shih and H. C. Elman (2000), 'Iterative methods for stabilized discrete convection-diffusion problems', *IMA J. Numer. Anal.* **20**, 333–358.

G. I. Shishkin (1989), 'Approximation of solutions of singularly perturbed boundary value problems with a parabolic boundary layer', *Zh. Vychisl. Mat. i Mat. Fiz.* **29**, 963–977, 1102.

G. I. Shishkin (1990), Grid approximation of singularly perturbed elliptic and parabolic equations, second doctoral thesis, Keldysh Institute, Moscow.

M. Stynes (2003), Numerical methods for convection-diffusion problems or the 30 years war, in *20th Biennial Conf. on Numerical Analysis* (D. F. Griffiths and G. A. Watson, eds), Numerical Analysis Report NA/217, University of Dundee, UK, pp. 95–103.

M. Stynes and L. Tobiska (1998), 'A finite difference analysis of a streamline diffusion method on a Shishkin mesh', *Numer. Algorithms* **18**, 337–360.

M. Stynes and L. Tobiska (2003), 'The SDFEM for a convection-diffusion problem with a boundary layer: optimal error analysis and enhancement of accuracy', *SIAM J. Numer. Anal.* **41**, 1620–1642.

R. Verfürth (2004), Robust *a posteriori* error estimates for stationary convection-diffusion equations, Technical report, University of Bochum.

H. Zarin and H.-G. Roos (2005), 'Interior penalty discontinuous approximations of convection-diffusion problems with parabolic layers', *Numer. Math.*, to appear.

G. Zhou (1997), 'How accurate is the streamline diffusion finite element method?', *Math. Comp.* **66**, 31–44.

Acta Numerica (2005), pp. 509–573
DOI: 10.1017/S0962492904000273

Total variation and level set methods in image science

Yen-Hsi Richard Tsai

Department of Mathematics,
University of Texas at Austin, TX 78712, USA
E-mail: ytsai@math.utexas.edu

Stanley Osher*

Department of Mathematics,
University of California, Los Angeles,
Los Angeles, CA 90095-1555, USA
E-mail: sjo@math.ucla.edu

We review level set methods and the related techniques that are common in many PDE-based image models. Many of these techniques involve minimizing the total variation of the solution and admit regularizations on the curvature of its level sets. We examine the scope of these techniques in image science, in particular in image segmentation, interpolation, and decomposition, and introduce some relevant level set techniques that are useful for this class of applications. Many of the standard problems are formulated as variational models. We observe increasing synergistic progression of new tools and ideas between the inverse problem community and the 'imagers'. We show that image science demands multi-disciplinary knowledge and flexible, but still robust methods. That is why the level set method and total variation methods have become thriving techniques in this field.

Our goal is to survey recently developed techniques in various fields of research that are relevant to diverse objectives in image science. We begin by reviewing some typical PDE-based applications in image processing. In typical PDE methods, images are assumed to be continuous functions sampled on a grid. We will show that these methods all share a common feature, which is the emphasis on processing the level lines of the underlying image. The importance of level lines has been known for some time. See, *e.g.*, Alvarez, Guichard, Morel and Lions (1993). This feature places our slightly general definition of the level set method for image science in context. In Section 2 we describe the building blocks of a typical level set method in the continuum

---

* Research supported by ONR N00014-03-1-0071, NIH U54Rl021813, NSF DMS-0312223, NSF ACI-032197, NSF NYU Subcontract and ONR MURI Stanford Subcontract.

setting. Each important task that we need to do is formulated as the solution to certain PDEs. Then, in Section 3, we briefly describe the finite difference methods developed to construct approximate solutions to these PDEs. Some approaches to interpolation into small subdomains of an image are reviewed in Section 4. In Section 5 we describe the Chan–Vese segmentation algorithm and two new fast implementation methods. Finally, in Section 6, we describe some new techniques developed in the level set community.

CONTENTS

1. Level set methods and image science

The level set method for capturing moving fronts was introduced in Osher and Sethian (1988). (Two earlier conference papers containing some of the key ideas have recently come to light (Dervieux and Thomasset 1979, 1981).) Over the years, the method has proved to be a robust numerical device for this purpose in a diverse collection of problems. One set of problems lies in the field of image science. In this article, we will emphasize not only what has been done in image science using level set techniques, but also in other areas of science in which level set methods are applied successfully – the idea is to point out the related formulations and solution methods to the image science communities. These communities include image/video processing, computer vision, and graphics. These are diverse, with specialities such as medical imaging and Hollywood-type special effects.

We begin with a quick examination of what constitutes a classical level set method: an implicit data representation of a hypersurface (codimension 1 object), a set of PDEs that govern how the surface moves, and the corresponding numerical methods for implementing this on computers. In fact, a typical application in image science will need all these features. We will illustrate this point with some classical applications.

The term 'image science' is used here to denote a wide range of problems related to digital images. It is generally referred to problems related to acquiring images (imaging), image processing, computer graphics, and computer vision. The type of mathematical techniques involved include discrete math, linear algebra, statistics, approximation theory, partial differential equations, quasi-convexity analysis related to solving inverse problems, and even algebraic geometry. The role of a level set method often relates to PDE techniques involving one or more of the following features: (1) regarding an image as a function sampled on a given grid with the grid values corresponding to the pixel intensity in suitable colour space, (2) regularization of the solutions, (3) representing boundaries, and (4) numerical methods. It is not hard to seek an application of the level set method for image segmentation or to model obstacles in inverse problems, since boundaries and level contours are fundamental objects in image science.

In a later section, we will examine some essential fundamentals of the level set methodology. We refer the reader to the original paper, Osher and Sethian (1988), and a new book, Osher and Fedkiw (2002), for detailed exposition of the level set method. A set of presentation slides is also available from the first author's home page.[1]

An image is considered as a function $u : \Omega \mapsto X$, where Ω is typically a rectangular domain in $\mathbb{R}^2$ and X is some compact space that is determined by the imaging device; e.g., $X = [0,1]$ if u is a grey value image, and $X = S^1 \times [0,1]$ if the chromaticity and intensity is used for a colour image. Unless otherwise noted, we will discuss grey level images here.

We write a typical PDE method as

$$\lambda L u = R u, \tag{1.1}$$

or

$$u_t + \lambda L u = R u, \tag{1.2}$$

where L is some operator applied to the given image, $\lambda \geq 0$ is a predetermined parameter, and R denotes the regularization operator. For example, in the TV deblurring of Rudin and Osher (1994),

$$L u = K * (K u - f),$$

where K is a compact integral operator, f is the given image, and the restored image is the limit $u(t)$ as $t \longrightarrow \infty$. When L is not invertible, as in the above deblurring model, or when a certain regularity is needed in the image u, a regularization term will be added. In the usual version of total variation methods, regularization usually appears in a form similar to

$$R u = \left(\nabla \cdot \frac{\nabla u}{|\nabla u|} \right). \tag{1.3}$$

---

[1] http://www.math.princeton.edu/~ytsai

Typically, equations (1.1) and (1.2) are derived either by directly writing down some PDEs whose solutions possess the desired properties, or by devising an energy functional $\mathcal{E}(u)$ and solving for a minimizer. For example, the shock filter of Rudin and Osher (1990) and the inpainting algorithm of Bertalmio, Sapiro, Caselles and Ballester (2000) fall into the first category. The variational approaches seem to be the mainstream for many important problems nowadays, partly owing to the existing mathematical tools, involving calculus of variations and Γ-convergence, available to study such kind of models. The Mumford–Shah multiscale segmentation model (Mumford and Shah 1989) and the total variation (TV) denoising model of Rudin–Osher–Fatemi (ROF) (Rudin, Osher and Fatemi 1992) are successful variational models. Both models have inspired much research activity in this field and will be discussed frequently in this article. The ROF model can be written as

$$\min_u \mathcal{E}_{TV}(u) = \frac{\lambda}{2} \int (f - u)^2 \, \mathrm{d}x + \int |\nabla u| \, \mathrm{d}x, \qquad (1.4)$$

where f is the given noisy image. In this set up, the Euler–Lagrange equation for (1.4) defines Lu as $(u - f)$, and $R = \nabla \cdot \frac{\nabla u}{|\nabla u|}$, which is the curvature of the level curve at each point of the image u. We remark that in many other image applications, the unregularized energy functional is nonconvex, and its global minimizer corresponds to the trivial solution. Only a local minimizer is needed. However, in (1.4), we obtain a useful global minimizer.

In the development of this type of method, one often qualitatively studies the solutions of the governing PDEs by investigating what action occurs on each of the level sets of a given image. In the TV regularization of Marquina and Osher (2000), for example, $Ru(x)$ actually denotes the mean curvature of the level set of u passing through x. The effects of (1.3) in noise removal can be explained as follows: the level curves in the neighbourhoods of noise on the image have high curvatures. The level curves of the viscosity solution to

$$u_t = \left(\nabla \cdot \frac{\nabla u}{|\nabla u|} \right) |\nabla u|$$

shrink with the speed of the mean curvature and eventually disappear. Consequently, the level curves with very high curvature (noise) disappear much more rapidly than those with relatively lower curvatures (this helped motivate the approach taken in Marquina and Osher (2000)). If the $|\nabla u|$ term is dropped (as it usually is) the velocity is inversely proportional to the gradient. This means relatively flat edges do not disappear. The analysis of motion by curvature and other geometric motions are all important consequences of viscosity solution theory, originally devised for Hamilton–Jacobi equations and a wide class of second-order nonlinear equations. The

Figure 1.1. Image obtained from
`http://mountains.ece.umn.edu/~guille/inpainting.htm`.

viscosity solution theory describes how evolution extends beyond singularities, including the pinching-off of level curves. Chambolle and Lions (1997) provide some analysis of the total variation denoising model. See Chen, Giga and Goto (1991), Crandall, Ishii and Lions (1992) and Evans and Spruck (1991, 1992a, 1992b, 1995) for more general viscosity theory applied to a wide class of second order equations.

Another interesting category of applications is data interpolation. In the problem of inpainting (see, *e.g.*, Bertalmio *et al.* (2000) and Figure 1.1), the challenge is to repair images which have regions of missing information. The algorithms are motivated in part by connecting the level curves over the 'inpainting domain' in an 'appropriate way'. In a rather orthogonal way, the AMLE (absolutely minimizing Lipschitz extension) algorithm (see, *e.g.*, Caselles, Morel and Sbert (1998)) assumes a given set of level curves of an image, and fills in the regions in between the given level curves while trying to minimize the variation of the new data generated.

In many applications such as image segmentation or rendering, level set methods are used to define the objects of interest. For example, a level set function is used to single out desired objects such as the land mass of Europe (Chan and Vese 2001a). The land mass is defined to be the connected region where the level set function is of one sign (see Figure 1.2). There are many successful algorithms of this type. Examples also include Chan and Vese (2001b) and Paragios and Deriche (1997). In a different, but related, context, Zhao *et al.* use a level set function to interpolate unorganized data sets (Zhao, Osher, Merriman and Kang 2000, Zhao, Osher and Fedkiw 2001).

Many of the above methods rely on the variational level set calculus similar to that of Zhao, Chan, Merriman and Osher (1996) to formulate the energies whose minimizers are interpreted as the solution to the problems,

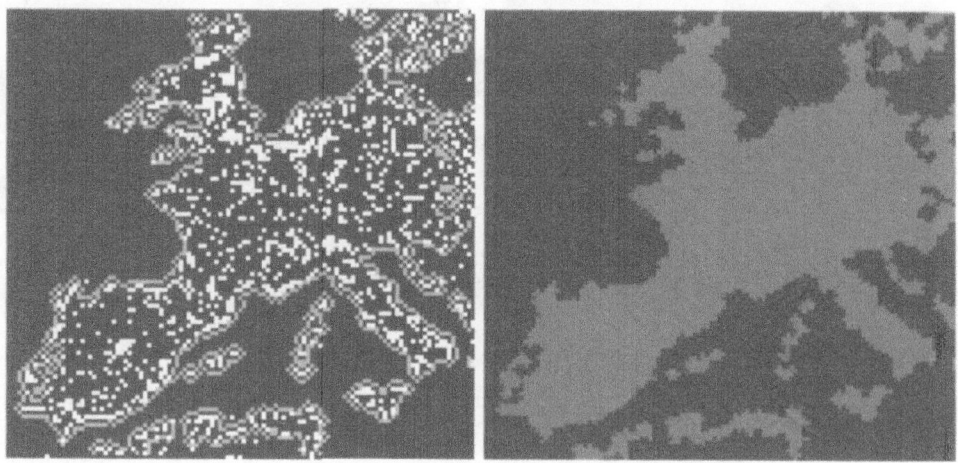

Figure 1.2. Land mass of Europe found using active contours.

and the solutions are level set functions. In general, the energies are variants
of the surface integral

$$\int_\Omega F(\phi, u)\delta(\phi)|\nabla\phi|\,\mathrm{d}x,$$

and the volume integral

$$\int_\Omega G(\phi, u)H(-\phi)\,\mathrm{d}x.$$

See Zhao *et al.* (1996) for details and definitions. We shall return to this in
a later section on image segmentation.

   We notice that in some of the above applications, level set functions are
used to separate the domain into different regions. The interfaces separating
those regions are defined as the zeros of the level set function. The PDEs
that govern the motion of the interface can be derived from a variational
principle. In many other cases, the interface motion is governed by classical
laws of physics. In fact, in the original level set paper (Osher and Sethian
1988), a level set function was used to distinguish burnt and unburnt regions
in flame propagation problems. Fedkiw and collaborators used level set
methods to simulate diverse physical phenomena such as splashing water,
flame propagation, and detonation waves. When the results are rendered
on the screen, they become very effective and realistic rendering of natural
phenomena suitable for special effects in movie productions. The reader
can find a detailed description and references in Osher and Fedkiw (2001).
Figure 1.3 provides two such simulations.

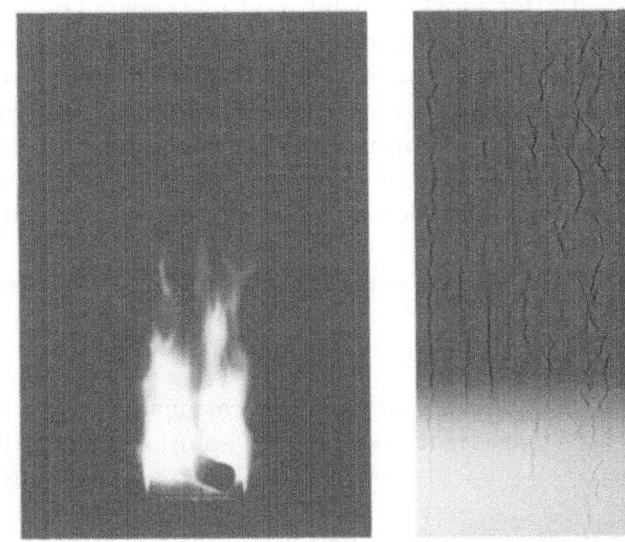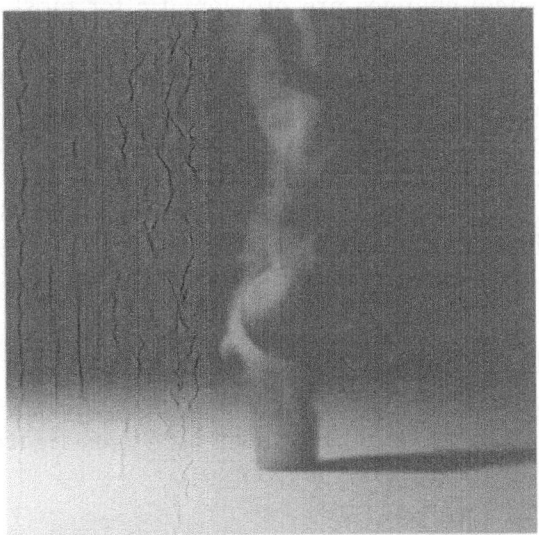

Figure 1.3. Image obtained from `www.cs.stanford.edu/~fedkiw`.

Finally, there is a collection of level set numerics, consisting mostly of approximations to general Hamilton–Jacobi equations and compressible and incompressible fluid dynamics. These methods are not limited only to pure level set formulations. They can also be used to solve other PDE-based image models. The basic numerics started in Osher and Sethian (1988) and Osher and Shu (1991), and generalizations have been carefully documented in Osher and Fedkiw (2002). Some new ones can be found in Enright, Fedkiw, Ferziger and Mitchell (2002), Kao, Osher and Tsai (2002), Sethian and Vladimirsky (2001), Tsai (2002), Tsai, Cheng, Osher and Zhao (2003a), Tsai, Giga and Osher (2003b) and Tsitsiklis (1995). Additionally, we mention Tornberg and Engquist (2003), which addresses the issue of regularization.

Ideas originating in this type of numerics, for instance, ENO interpolation (Harten, Engquist, Osher and Chakravarthy 1987), have been used to develop wavelet-based methods which minimize ringing, or Gibbs' phenomena at edges (Chan and Zhou 2002).

2. Brief review of the level set method

A significant number of problems in science reduce to the study of the evolution of curves, which are usually the boundaries between different media. These curves (or interfaces) move according to their own geometries or according to the laws of physics associated with the problem. They break up, merge, or disappear during the course of time evolution. These topological changes are problematic for most conventional methods. The level set method, however, handles these topological changes *'with no emotional involvement'* (Osher and Sethian 1988). Since its introduction, there has developed a powerful level set calculus used to solve a great variety of problems in fluid dynamics, materials sciences, computer vision, computer graphics, to name a few topics. We refer to Osher and Fedkiw (2002) for an extensive exposition of the level set calculus. See also Giga (2002) for a related theoretical exposition.

Typically, one can write a general level set algorithm in three steps enumerated below.

(1) Initialize/re-initialize ϕ at $t = t^n$.

(2) Construct/approximate $H(t, x, \phi, D\phi, D^2\phi)$. (Occasionally higher derivatives also appear for which rigorous viscosity solution theory definitely does not apply).

(3) Evolve

$$\phi_t + H(t, x, \phi, D\phi, D^2\phi) = 0,$$

for $t = t_n + \Delta t$.

For image applications, ϕ above can either be the image itself (*e.g.*, deblurring applications) or an extra function that is used to process the given image (*e.g.*, segmentation applications).

We will discuss the key components of the three steps in the following sections. More precisely, we will follow convention and start our exposition for step (2). Steps (1) and (3) are implemented by suitable numerical methods that will be reviewed in the next section.

2.1. Basic formulation

For simplicity, we discuss the conventional level set formulation in two dimensions. The interfaces represented by a level set function are thus also referred to as curves. However, the methodology presented in this section can be naturally extended to any number of space dimensions. There, the interface that is represented is generally called a hypersurface (in three dimensions, it is simply called a surface). We will use the words interface and curves interchangeably.

In the level set method, the curves are implicitly defined as the zeros of a Lipschitz-continuous function ϕ. This is to say that $\{(x, y) \in \mathbb{R}^2 : \phi(x, y) = 0\}$ define the embedded curve Γ. In many situations, we will also regard Γ as the boundary of the sublevel sets $\Sigma = \{\phi \leq 0\}$. See Figures 2.1 and 2.2 for some examples. If we associate a continuous velocity field v whose restriction onto the curve represents the velocity of the curve, then, at least locally in time, the evolution can be described by solving the Cauchy problem

$$\phi_t + v \cdot \nabla\phi = 0, \qquad \phi(x, 0) = \phi_0(x),$$

where ϕ_0 embeds the initial position of the curve. To derive this, let us look at a parametrized curve $\gamma(s,t)$ and assume that $\partial\gamma/\partial t$ is the known dynamics of this curve. If we require that $\gamma(s,t)$ be the zero of the function ϕ for all time, $i.e.$, $\phi(\gamma(s,t),t) = 0$ for all $t \geq 0$, then, at least formally, the equation

$$\phi_t + \frac{\partial\gamma}{\partial t} \cdot \nabla\phi(\gamma,t) = 0$$

is satisfied along γ. Extending $\partial\gamma/\partial t$ continuously to the whole domain will create the velocity field v.

In general, the velocity v can be a function of position x, t, and some other geometrical properties of the curve, or of other physical quantities

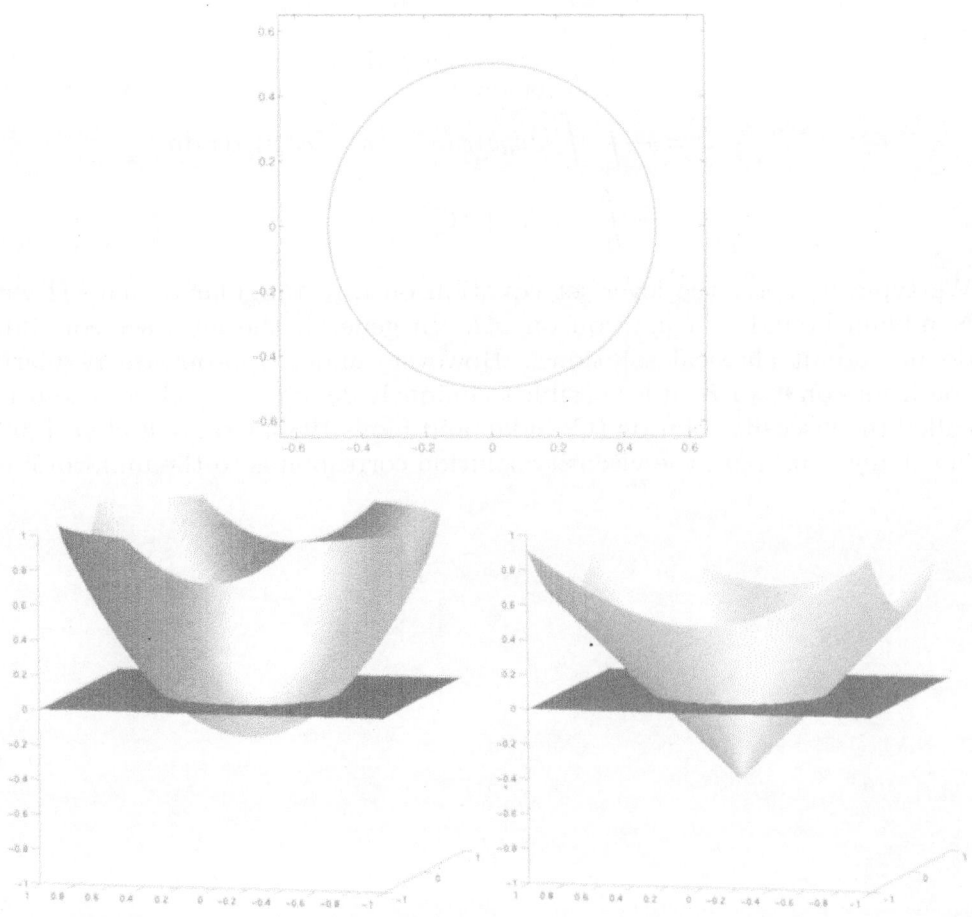

Figure 2.1. A circle embedded by different continuous functions.

that come with the problem. The equation can be written using the normal velocity:

$$v_n = v \cdot \frac{\nabla\phi}{|\nabla\phi|}, \phi_t + v_n|\nabla\phi| = 0. \tag{2.1}$$

We note that these equations are usually fully nonlinear first-order Hamilton–Jacobi or second-order degenerate parabolic equations, and in many cases, the theory of viscosity solutions (Crandall and Lions 1984) can be applied to guarantee well-posedness of the Cauchy problem.

It is instructive to derive the level set equation via a weak formulation using the area and co-area formula (Evans and Gariepy 1992). Let w be a test function, and let v_n be the normal velocity of $\Gamma = \partial\Sigma = \{\phi \le 0\}$:

$$\int_{\mathbb{R}^2} \frac{\partial\phi}{\partial t} w \,\mathrm{d}x = \frac{\mathrm{d}}{\mathrm{d}t} \int_{\mathbb{R}^2} \phi w \,\mathrm{d}x = -\frac{\mathrm{d}}{\mathrm{d}t} \int_{\mathbb{R}} \int_{\{\phi(\cdot,t)<\eta\}} w \,\mathrm{d}x \,\mathrm{d}\eta$$

$$= -\int_{\mathbb{R}} \int_{\partial\{\phi(\cdot,t)<\eta\}} wv_n \,\mathrm{d}s \,\mathrm{d}\eta$$

$$= -\int_{\mathbb{R}} \int wv_n\delta(\phi(x) - \eta)|\nabla\phi(x)| \,\mathrm{d}x \,\mathrm{d}\eta$$

$$= -\int_{\mathbb{R}^2} v_n|\nabla\phi|w \,\mathrm{d}x.$$

We typically solve the level set equation on a rectangular domain Ω with Neumann boundary condition on $\partial\Omega$. In general, the level set equations do not admit classical solutions. However, under appropriate regularity conditions on v_n or H, it is possible to uniquely define a special weak solution called the viscosity solution (Crandall and Lions 1983, Crandall *et al.* 1992). For many equations, the viscosity solution corresponds to the uniform limit

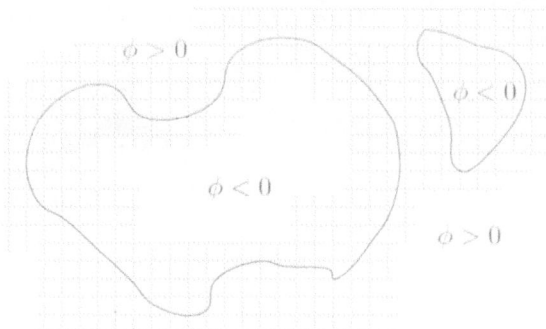

Figure 2.2. Two closed curves that are implicitly embedded by a single level set function defined on the grid.

of the vanishing viscosity solution. It can be shown that the motion of the zero level set of the viscosity solution is a generalization of a smooth motion in the normal direction, and the motion is uniquely defined if no *fattening* occurs; *i.e.*, if $\{\phi = 0\}$ remains a set of measure zero for all time. We refer the interested readers to Evans (1998) and Giga (2002) for more information on this aspect of the level set method. Corresponding to viscosity solution theory, there is a set of simple finite difference methods to construct approximation solutions (Barles and Souganidis 1991, Crandall and Lions 1984).

Finally, in the level set formulation, the surface integral of function f along the zero level set is defined via the surface integral

$$\int_{\mathbb{R}^d} f(x)\delta(\phi)|\nabla\phi|\,\mathrm{d}x.$$

If $f \equiv 1$, this integral yields the arc length for curves in two dimensions, and surface area in three dimensions. Volume integrals are defined as

$$\int_{\mathbb{R}^d} f(x)H(\phi)\,\mathrm{d}x,$$

where $H(x) = 1$ for $x \geq 0$ and $H(x) = 0$ for $x < 0$. In Sections 3.8 and 3.9, we will review the related numerics proposed by Engquist, Tornberg and Tsai (2004) related to approximating the delta and Heaviside functions.

2.2. Reshaping the level set function

In many situations, the level set function will develop steep or flat gradients leading to problems in numerical approximations. It is then needed to reshape the level set function to a more useful form, *while keeping the zero location unchanged.* One way to do this is to perform what is called distance re-initialization (Sussman, Smereka and Osher 1994) by evolving the following PDE to steady state:

$$\phi_\tau + \mathrm{sgn}(\phi_0)(|\nabla\phi| - 1) = 0, \quad \phi(x, \tau = 0) = \phi_0(x). \tag{2.2}$$

Here ϕ_0 denotes the level set function before re-intialization. If we evolve the solution to steady state over the computational domain, the solution ϕ becomes the signed distance function to the interface $\{\phi_0 = 0\}$. One can understand the mechanism of this approach from the following scenario: in the region in which ϕ_0 is positive, $\phi_\tau < 0$ whenever $|\nabla\phi| > 1$; therefore, the value of ϕ will decrease, and consequently, $|\nabla\phi|$ will become closer to 1. Notice that $\phi_\tau \equiv 0$ wherever $\phi_0 \equiv 0$, since $\mathrm{sgn}(0) = 0$. See Figure 2.3. We will come back to issues related to how proper discretizations of the discontinuous signum function should be carried out in order to achieve efficiency and accuracy.

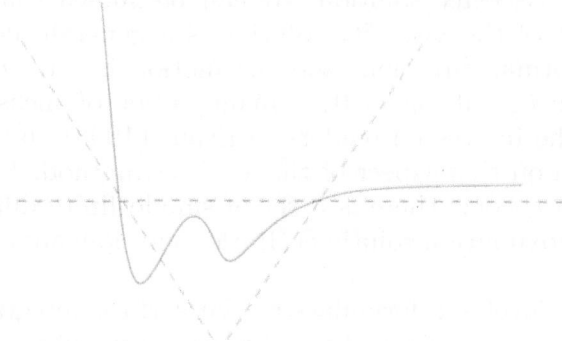

Figure 2.3. Re-initialization.

Another equivalent approach is to solve the eikonal equation

$$|\nabla \phi| = 1$$

with the boundary condition $\phi = 0$ on $\{\phi_0 = 0\}$. A common numerical approach, *e.g.*, Peng, Merriman, Osher, Zhao and Kang (1999*a*), is to run distance re-initialization (2.3) with a high-order accurate method for a short amount of time, so that in a thin tube around $\{\phi_0 = 0\}$, ϕ is now the distance function. Then fix the values of ϕ in this tube as boundary conditions, and use fast sweeping or fast marching methods to solve the eikonal equations. We shall discuss the sweeping method in Section 3.7.

We remark that for most applications, the re-initialization is only needed in a neighbourhood around the zero level set, and the diameter of this neighbourhood depends on the discretization of the partial derivatives in the PDE. This implies that only a few time-steps in τ are needed. We also note that it is important to solve (2.2) using a high-order discretization method. Otherwise, the location of the original interface will be perturbed noticeably by numerical error. Finally, re-initialization globally in the computational domain will prevent new zero contours from appearing. Thus, one needs to be careful if emergence of new level contours is of interest. In many image segmentation tasks, this is important, and we shall comment on this in a later section.

2.3. *Extending quantities off the normals of the interface*

In many models, one can only derive the interface velocity v_n in equation (2.1) along Γ. It is necessary to create a continuous velocity field defined on the whole domain Ω, or at least in a tubular neighbourhood of Γ whose restriction on Γ agrees with the known interface velocity. One common way to obtain such a velocity field is to solve the following boundary value problem:

$$\text{sgn}(\phi)\nabla w \cdot \nabla \phi = 0, \quad \text{with} \quad w|_\Gamma = v_n, \tag{2.3}$$

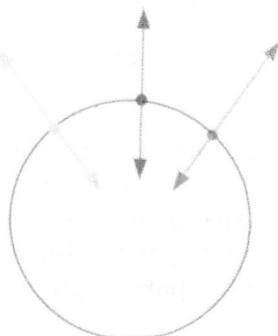

Figure 2.4. Quantities are extended off the
zero level set in the normal directions.

or equivalently, to solve for the steady state of the time-dependent equation:

$$w_t + \mathrm{sgn}(\phi)\nabla w \cdot \nabla \phi = 0, \qquad (2.4)$$

with any initial data w_0 whose restriction on Γ matches v_n.

The interpretation of this approach is that v_n will be propagated as a constant along the characteristics of the PDE (2.3), emanating from Γ, parallel to the surface normals. See Figure 2.4. Fast sweeping (Kao *et al.* 2002, Kao, Osher and Qian 2004, Tsai 2002, Tsai *et al.* 2003*a*, Zhao 2005) or fast marching (Tsitsiklis 1995, Sethian 1996) can be used to solve the first equation while a higher-order accurate Hamilton–Jacobi solver can be used for the second (Osher and Shu 1991). In the next section, we will briefly describe some popular discretizations.

2.4. Tracking quantities defined on the fronts using level set method

So far we have described the basic level set method that enables us to move curves and surfaces normal to themselves by the prescribed velocities. We have concentrated on describing how the physical location of the curves and surfaces change. In many applications, including image processing and computer vision, we need to track quantities that are defined on the surfaces. In this section, we review some techniques for doing this.

Let $\tilde{f} : \Gamma \mapsto X$ denote the quantity defined on Γ, the zero level set of ϕ, and $\tilde{f}$ satisfies

$$\tilde{f}_t + Q_\Gamma \tilde{f} = 0, \qquad \tilde{f}(x, t = 0) = \tilde{f}_0(x), \qquad (2.5)$$

where Q_Γ denotes the differential operator on Γ. This equation determines how $\tilde{f}$ is changing on Γ. Let $f : U \subset \mathbb{R}^d \mapsto X$ be a function defined in a neighbourhood U of Γ, and $f|_\Gamma \equiv \tilde{f}$. Here $\mathbb{R}^d$ is the ambient space of Γ; *i.e.*, $\phi : \mathbb{R}^d \mapsto \mathbb{R}$, and $\Gamma = \{x : \phi(x) = 0\}$. In a typical level set method, instead of solving (2.5) directly on Γ, one solves the corresponding PDE in $\mathbb{R}^d$,

$$f_t + Qf = 0,$$

so that the restriction of $f(t)$ to Γ matches with $\tilde{f}(t)$ for $t \geq 0$. At this point, it is natural to ask what Q is, given the Q_Γ? In many applications, the form of Q *is* the centre of the study, and it might be more convenient to track an alternative quantity g in order to obtain an equation that is easier to solve. See the recent paper by Jin, Liu, Osher and Tsai (2005*b*) for such an example. In the next paragraph, we discuss another example of this situation.

Assume that we are interested in quantities defined and parametrized on the surfaces, and we need to know how these quantities redistribute during the evolution of the surfaces. Harabetian and Osher (1998) introduced a method for doing this. Let ϕ denote the level set function that embeds the surface of interest. The idea is to introduce an auxiliary function ψ such that (ϕ, ψ) forms a coordinate system near the zero level set of ϕ.

Let the family of closed curves $\Gamma(s,t) = (x(s,t), y(s,t))$ be parametrized by s and t. We want to evolve, for example, $\Gamma(s,0)$ to time t, by the level set functions

$$\phi(x(s,t), y(s,t), t) \equiv 0, \qquad \psi(x(s,t), y(s,t), t) \equiv s.$$

However, ψ is not a single-valued function over a closed curve if it is defined this way. The authors then proposed to evolve the Jacobian

$$J = \det \begin{bmatrix} \varphi_x & \varphi_y \\ \psi_x & \psi_y \end{bmatrix}$$

instead of ψ to circumvent this problem. J has to be nonzero and finite so that we can express (x_s, y_s) by $(-\phi_y, \phi_x)/J$. Thus, in order to track the tangential motion we evolve

$$J_t + \nabla \cdot (Jv) = 0$$

in addition to

$$\phi_t + v \cdot \nabla \phi = 0.$$

Finally, we briefly describe the systematic approach that began in Cheng (2000), and was developed in Bertalmio, Cheng, Osher and Sapiro (2001*b*) for solving PDEs on surfaces for image processing and more general applications. A similar approach was later adopted by Xu and Zhao (2003) to study surfactants on interfaces that move in time. For simplicity, we assume the zero level set to be fixed in time.

Consider the surface gradient $Q_\Gamma = \nabla_\Gamma$ that maps scalar functions defined on Γ to the tangent bundle of Γ. The key notion is to replace ∇_Γ by a suitable projection of the gradient operator ∇ in $\mathbb{R}^d$. The corresponding projection operator is a linear operator defined by

$$\mathcal{P}_v = \mathcal{I} - \frac{v \otimes v}{|v|^2},$$

or equivalently, as a matrix, $\mathcal{P}_v$ can be written as

$$(\mathcal{P}_v)_{ij} = \delta_{ij} - \frac{v_i v_j}{|v|^2},$$

where v is a vector in $\mathbb{R}^d$, and δ_{ij} is the Kronecker delta function. For $x \in \Gamma$, and v the normal of Γ at x, $\mathcal{P}_v$ projects vectors onto the tangent plane of Γ at x.

Recall that $\Gamma = \{\phi = 0\}$, and $\nabla\phi$ is parallel to the normal of Γ. It can be proved that ∇_Γ and $\mathcal{P}_{\nabla\phi}\nabla$ are equivalent on Γ. Thus, for scalar functions f,

$$\nabla_\Gamma f = P_{\nabla\phi}\nabla f,$$

and for surface divergence of vector fields F,

$$\nabla_\Gamma \cdot F = P_{\nabla\phi}\nabla \cdot F.$$

Let us illustrate this approach with a few examples. Consider a continuous function $\tilde{f}$ defined on Γ, a surface in $\mathbb{R}^3$, and a given vector field v defined on the tangent bundle of Γ. If the zeros of $\tilde{f}$ embed the curve of interest (call it C) on Γ, then by solving

$$\tilde{f}_t + v \cdot \nabla_\Gamma \tilde{f} = 0,$$

one obtains the evolution of the curve constrained to the surface. Correspondingly, the extension f of $\tilde{f}$ in $\mathbb{R}^3$ is another level set function, whose zero level set intersects with that of ϕ on C, and the corresponding PDE in $\mathbb{R}^3$ is

$$f_t + v \cdot P_{\nabla\phi}\nabla f = 0,$$

or

$$f_t + P_{\nabla\phi}v \cdot \nabla f = 0.$$

To perform distance re-initialization on $\tilde{f}$, one can evolve

$$f_\tau + \operatorname{sgn}(f_0)(|P_{\nabla\phi}\nabla f| - 1) = 0.$$

As an example of solving PDEs on surfaces, we consider total variation diminishing flow of an image u, defined on a surface Γ, takes the form

$$\mathcal{E}(u) = \int_{\mathbb{R}^3} |P_{\nabla\phi}\nabla u|\delta(\phi)|\nabla\phi|\,\mathrm{d}x,$$

and the corresponding gradient descent equation becomes

$$u_t = P_{\nabla\phi}\nabla \cdot \left(\frac{P_{\nabla\phi}\nabla u}{|P_{\nabla\phi}\nabla u|}\right),$$

where the right-hand side corresponds to the geodesic curvature, and can also be written as

$$\nabla \cdot \left(\frac{\mathcal{P}_{\nabla\phi} \nabla u}{|\mathcal{P}_{\nabla\phi} \nabla u|} |\nabla\phi| \right) \frac{1}{|\nabla\phi|}.$$

The function u is extended outside Γ as described in Section 2.3. For time-dependent problems, this extension is redone every few time iterations. The PDE needs to be solved only in a small neighbourhood of Γ, as described in Peng *et al.* (1999*a*).

2.5. Level set methods involving variational approaches

Assume that the energy functional $\mathcal{E}$ is an integral operator on u over $\Sigma \subset \Omega \subset \mathbb{R}^d$,

$$\mathcal{E}(u, \Sigma) = \int_{\Sigma} F(u(x)) \, \mathrm{d}x,$$

and the non-positive region of ϕ defines Σ; *i.e.*, $\{\phi \le 0\} = \Sigma$. The key idea of the variational level set method formulated in Zhao *et al.* (1996) is that the above integral can be written as

$$\int_{\Sigma} F(u(x)) \, \mathrm{d}x = \int_{\mathbb{R}^2} \chi_{\Sigma}(x) F(u(x)) \, \mathrm{d}x = \int_{\mathbb{R}^2} H(-\phi) F(u) \, \mathrm{d}x,$$

where H is the Heaviside function: $H(x) = 1$ if $x \le 0$ and $H(x) = 0$ elsewhere. One can then try to find the minimizer ϕ for this energy. Variational calculus reveals that that the change in ϕ on this functional can be quantified through the boundary integral over $\partial\Sigma = \{\phi = 0\}$.

We follow the review of Burger and Osher (2005) and describe how sensitivity of this type of energies can be studied in the context of level set methods.

Level set method and shape calculus

Shape sensitivity analysis is a classical topic in shape optimization, and defines a natural calculus on shapes. For sufficiently regular shapes (*i.e.*, with C^1 boundary), there are two equivalent ways of introducing shape sensitivities, namely the *deformation method* and the *speed method* (Sokołowski and Zolésio 1992). Owing to its relation to the level set method, we shall use the latter as the basis of the following presentation.

Given a set $\Sigma(t)$ evolving in a velocity field V. Consider an energy $\mathcal{E}(\Sigma)$ that depends on the shape of Σ. The shape sensitivity of $\mathcal{E}$ in the direction of a perturbation V is then given by

$$\mathrm{d}\mathcal{E}(\Sigma; V) = \frac{\mathrm{d}}{\mathrm{d}t} \mathcal{E}(\Sigma(t))|_{t=0}.$$

$\mathrm{d}\mathcal{E}(\Sigma; \cdot)$ is called the shape differential. In the level set framework, $\Sigma(t)$

may be embedded as $\{\phi(\cdot, t) \leq 0\}$. Thus the shape sensitivity is

$$\mathrm{d}\mathcal{E}(\Sigma; V) = \frac{\mathrm{d}}{\mathrm{d}t}\mathcal{E}(\{\phi(\cdot, t) \leq 0\})|_{t=0}.$$

Typically, the energies that appear in the image processing applications are either volume integrals,

$$\mathcal{E}(\Sigma) = \int_{\Sigma} g \, \mathrm{d}x,$$

or boundary integrals,

$$\mathcal{E}(\Sigma) = \int_{\partial\Sigma} g \, \mathrm{d}S.$$

For example, in the former case, direct calculation shows that

$$\frac{\mathrm{d}}{\mathrm{d}t}\mathcal{E}(\Sigma(t))|_{t=0} = \int_{\mathbb{R}^d} gV \cdot \nabla\phi\delta(\phi) \, \mathrm{d}x = \int_{\partial\Sigma(0)} gV_n \, \mathrm{d}S.$$

The shape derivative is hence related back to the variational level set method (Zhao *et al.* 1996). Hence, in a variational level set model, one can choose many different V_n to decrease $\mathcal{E}$. Different choices of V_n may result in minimizing $\mathcal{E}$ in different normed spaces. If $\mathcal{E}$ is nonconvex, the choice of V_n and consequently the descent path might lead to different local minimizers.

In many applications involving shape optimization, *e.g.*, image segmentation, it is desirable to test the sensitivity of the energy function with respect to topological changes in a given shape. The topological derivative of a shape Σ with respect to a spherical perturbation at $x \in \Omega$ (Ω is the computational domain) is given by

$$D_\tau\mathcal{E}(\Sigma; x) = \lim_{R \to 0} \frac{\mathcal{E}(\Sigma \setminus B_R(x)) - \mathcal{E}(\Sigma)}{|B_R(x) \bigcap \Omega|},$$

if the limit on the right-hand side exists. Here, $B_R(x)$ denotes the closed ball of radius R centred at x, while $d_\tau\mathcal{E}(\Sigma; x)$ measures the variation with respect to the nucleation of an infinitesimal hole at x. Thus, if $d_\tau\mathcal{E}(\Sigma; x) < 0$, then the nucleation of a hole at x will decrease the objective energy functional. One can respectively define the topological derivative of the complement of Σ by

$$D_\tau\mathcal{E}(\Sigma; x) = \lim_{R \to 0} \frac{\mathcal{E}(\Sigma \bigcup B_R(x)) - \mathcal{E}(\Sigma)}{|B_R(x) \bigcap \Sigma|}.$$

In this case, we are interested in the sensitivity of the energy function with respect to the introduction of a new connected component to the given shape Σ. One can see the link between the shape derivative and the topological derivative by evaluating $D_\tau\mathcal{E}(\Sigma; x)$ at $\partial\Sigma$.

Burger, Hackl and Ring (2004) successfully incorporate this idea above to solve a class of shape optimization problems. Their idea is to add $d_\tau\mathcal{E}(\Sigma; x)$ as a forcing term in the gradient descent.

Preserving topology

In some applications, one may be interested in preserving the topology of an given initial zero level set of ϕ, *e.g.*, in mapping of brain images, or the optimization of microstructured optical fibres. In particular, one usually needs to prevent disconnected components of $\{\phi < 0\}$ from merging with each other when they get close to each other. This is common behaviour for many level set simulations that compute the viscosity solution (Crandall *et al.* 1992).

An automatic way to incorporate this additional property was recently devised by Alexandrov and Santosa (2005). They proposed adding a penalty term to the original energy,

$$H(\Sigma) = -\int_{\partial\Sigma} \left(\log(d_\Sigma(x + \sigma\nabla d_\Sigma(x)) + \log(-d_\Sigma(x - \sigma\nabla d_\Sigma(x)))\right) \mathrm{d}s$$

for some small constant $\sigma > 0$. Here, d_Σ denotes the signed distance function to $\partial\Sigma$ with $d_\Sigma(x) \leq 0$ for $x \in \Sigma$. Note that for $x \in \Sigma$, $x + \sigma\nabla d_\Sigma(x)$ is a point projected a distance σ outside Σ, while $x - \sigma\nabla d_\Sigma(x)$ is a point that is projected a distance σ inside Σ. Hence this penalty term forces a minimum distance of σ between connected components of $\{d_\Sigma < 0\}$, and respectively $\{d_\Sigma > 0\}$. Therefore, no topological change can arise. However, this penalty also indirectly regulates the curvatures of $\partial\Sigma$. One can conjure up a scenario in which the desired shape has many slender fingering components such that the thickness of each 'finger' is less than σ. This added penalty will unfortunately prevent structures of the type from being computed.

Another more general method developed to prevent merging can be found in Han, Xu and Prince (2003). This appears to be quite useful in brain mapping.

2.6. Limitations of the level set methods

The original idea in the level set method is to use the sign of a given function to separate the given domain into two disjoint regions, and use the continuity of the level set function near its zero to define the boundary of these disjoint regions. One realizes that it can be more complicated to extend this idea to handle non-simple curves, and multiple phases. An equally important issue is to solve the problem at hand in obtaining reasonable quality without excessive complexity. We refer the readers to Smith, Solis and Chopp (2002), Vese and Chan (2002) and Zhao *et al.* (1996) for level set methods for multiple phases, Burchard, Cheng, Merriman and Osher (2001) and Osher, Cheng, Kang, Shim and Tsai (2002*a*) for higher codimensions, Smereka (2000) for open curves, and Peng *et al.* (1999*a*) and Strain (1999*a*, 1999*b*) for localization. We also refer to Enright *et al.* (2002) for a hybrid particle level set method that is designed to lessen the numerical diffusion effect for some class of problems, particularly two-phase incompressible flows.

3. Numerics

The numerical solution of conservation laws has been an active field of research for quite some time. The finite difference methods commonly used in the level set methods (in particular, those related to Hamilton–Jacobi equations) are developed under the general philosophy of the Godunov procedure and the nonlinear ENO reconstruction techniques for avoiding oscillations in calculations. As a result, upwinding and ENO interpolation become the indispensable parts of the algorithms documented here.

In what follows, we will first describe the Godunov procedure in the context of solving conservation laws and Hamilton–Jacobi equations. We will also describe the ENO interpolation and compare the differences between its usage in conservation laws schemes and in Hamilton–Jacobi solvers. We refer the details to the book of Osher and Fedkiw (2002) and the extensive references therein. (For simplicity of exposition, we again restrict our discussion to two space dimensions.)

Let us introduce some notations that we shall use in this section. Let $\phi_{i,j}^n$ denote the value of $x_{i,j} = (x_0 + i\Delta x, y_0 + j\Delta y) \in \Omega$ at time $t_n = t_0 + \Delta t$. We shall assume that $\Delta x = \Delta y$.

Definition 3.1. (Finite difference operators) Given the values of u on the grid we first define the forward and backward difference operators,

$$D_x^{\pm} u_{i,j} := \pm \frac{u_{i\pm 1,j} - u_{i,j}}{\Delta x}$$

and

$$D_y^{\pm} u_{i,j} := \pm \frac{u_{i,j\pm 1} - u_{i,j}}{\Delta y},$$

and the central difference operators,

$$D_x^0 u_{i,j} := \frac{u_{i+1,j} - u_{i-1,j}}{2\Delta x}$$

and

$$D_y^0 u_{i,j} := \frac{u_{i,j+1} - u_{i,j-1}}{2\Delta y}.$$

3.1. The Godunov procedure

The Godunov procedure (Godunov 1959) developed for conservation laws begins by regarding grid values as cell averages of the solution at time t_n. We then 'build' a piecewise constant function whose value in each cell is the cell average. We solve the Riemann problem at cell boundaries 'exactly' for an appropriate time-step Δt. This involves following the characteristics and making sure that the Rankine–Hugoniot and entropy conditions are satisfied. Finally, we average the function at $t = t_n + \Delta t$ in each cell, and repeat the above steps.

In the context of certain conventional Hamilton–Jacobi equations, piece-wise constant cell averages are replaced by a piecewise linear function that is continuous at the cell boundaries, and point values are updated. This is described in Bardi and Osher (1991) and Osher and Sethian (1988).

In high-order schemes, cell averages are replaced by more accurate non-oscillatory reconstruction on the functions or the fluxes. We perform this reconstruction by ENO/WENO methods.

3.2. ENO/WENO interpolation

We want to approximate the value of the function f in the interval $I_i :=$ $[x_{i-\frac{1}{2}}, x_{i+\frac{1}{2}}]$, using the given values (or averaged values) of f on the grid nodes x_i and its neighbours. Two commonly used methods to get a kth-order approximation of f in I_i are spectral interpolation, *e.g.*, based on Fourier expansions, and fixed-order polynomial interpolation. Both approaches produce oscillations near the jumps in the function values or their derivatives. We will not comment on the Fourier-based methods since they are not particularly useful in this connection. Conventional polynomial interpolations usually use the function values on *all the grid points* within a certain fixed distance from x_i, *regardless of the smoothness of the interpolated function*. ENO interpolation, on the other hand, is a nonlinear procedure that is built on a 'progression' of Newton's divided differences. By 'progression', we mean that the procedure starts by building a linear reconstruction of f in I_i using *either* $f(x_i)$ and $f(x_{i-1})$ or $f(x_i)$ and $f(x_{i+1})$, depending on which pair of values will give a smoother reconstruction. Suppose the reconstruction from $f(x_i)$ and $f(x_{i-1})$ is selected; we then carry out the reconstruction using the values of f on either x_{i-2}, x_{i-1}, x_i or x_{i-1}, x_i, x_{i+1}. This procedure is iterated until the desired order of approximation is achieved. Newton's interpolation is natural in this framework, since one can incrementally compute the divided differences for interpolation. In addition, we can use the values of the divided differences as an indicator of the smoothness of the functions in the intervals formed by the grid points that are considered as possible points in the stencil.

For conservative schemes approximating conservation laws, this ENO reconstruction is performed on the flux function f or the cell averages $\bar{u}$ by first reconstructing the integral of the solution u. For Hamilton–Jacobi equations, we perform the ENO reconstruction on the solution u.

In the ENO reconstruction procedure, only one of the k candidate stencils (grid points used for the construction of the scheme) covering $2k - 1$ cells is actually used. If the function is smooth in a neighbourhood of these $2k - 1$ cells, we can actually get a $(2k - 1)$th-order approximation if we use all these grid values. This is the idea behind the WENO reconstruction. In short, WENO reconstruction uses a convex linear combination of all the potential stencils. The weights in the combination are determined so that the WENO reconstruction procedure behaves like ENO near discontinuities. As a result, WENO method use smaller stencils to achieve the same order of accuracy as ENO in smooth regions. Currently, our choice of scheme is fifth-order WENO. For details, we refer to the original papers by Engquist, Harten and Osher (1987), Harten *et al.* (1987), Jiang and Peng (2000) and Liu, Osher and Chan (1994), and the review article by Shu (1997). Recently, Balsara and Shu (2000) developed even higher-order WENO reconstructions.

There are successful adaptations of this ENO idea/philosophy to other frameworks. See Chan and Zhou (1999, 2002) for ENO wavelet decompositions for image processing, and Cockburn and Shu (1989) for an application of the ENO philosophy in discontinuous Galerkin methods.

3.3. Numerics for equations with Hamiltonians $H(x, u, p)$ nondecreasing in u

We repeat here that any discussion of the numerical schemes cannot be detached from the solution theory of the equations in questions. This is especially important for nonlinear hyperbolic equations, since, in general, discontinuities in the function values or in the derivatives develop in finite time. We are usually seeking a particular type of weak solution.

Crandall and Lions (1983) introduced viscosity solution theory for a class of Hamilton–Jacobi equations requiring Lipschitz-continuous initial data and for which the Hamiltonian $H(x, u, p)$ is Lipschitz-continuous and *non-decreasing in* u. Later, in Crandall and Lions (1984), they proved the convergence to the viscosity solution of monotone, consistent schemes for Hamilton–Jacobi equations with H independent of x and u. Souganidis (1985) extended the results to include variable coefficients. Osher and Sethian (1988) contributed to the numerics of Hamilton–Jacobi equations in their level set paper. This was later generalized and completed in the paper by Osher and Shu (1991), in which they provided a family of numerical Hamiltonians related to the ENO schemes for conservation laws. WENO schemes using the numerical Hamiltonians described in Osher and Shu (1991) were introduced in Jiang and Peng (2000). The method of lines using TVD Runge–Kutta time discretization is used (Shu and Osher 1988). We first discretize the spatial derivatives and compute the appropriate approximation to the Hamiltonians,

$$\hat{H}(p_-, p_+; q_-, q_+),$$

with $p_\pm$, $q_\pm$ representing the left/right approximations of the derivatives, obtained from ENO/WENO reconstruction of the solution. They are higher-order versions of the forward and backward divided differences of the grid functions:

$$p_\pm \sim D_x^\pm u_{i,j} := \pm \frac{u_{i\pm1,j} - u_{i,j}}{\Delta x},$$

and

$$q_\pm \sim D_y^\pm u_{i,j} := \pm \frac{u_{i,j\pm1} - u_{i,j}}{\Delta y}.$$

3.4. The Lax–Friedrichs schemes for the level set equation

Following the methods originally conceived for HJ equations $\phi_t + H(D\phi) = 0$ in Osher and Shu (1991) (see also Osher and Sethian (1988)), and suppressing the dependence of H on x and y, we recommend using the Local Lax–Friedrichs (LLF) numerical Hamiltonian:

$$\hat{H}^{LLF}(p^+, p^-, q^+, q^-) = H\left(\frac{p^+ + p^-}{2}, \frac{q^+ + q^-}{2}\right) \tag{3.1}$$

$$-\frac{1}{2}\alpha^x(p^+, p^-)(p^+ - p^-) - \frac{1}{2}\alpha^y(q^+, q^-)(q^+ - q^-),$$

for the approximation of H. In the above scheme,

$$\alpha^x(p^+, p^-) = \max_{p \in I((p^+, p^-)), C \le q \le D} |H_{\phi_x}(p, q)|,$$

$$\alpha^y(q^+, q^-) = \max_{q \in I((q^+, q^-)), A \le p \le B} |H_{\phi_y}(p, q)|,$$

$$I(a, b) = [\min(a, b), \max(a, b)],$$

and $p^\pm$, $q^\pm$ are the forward and backward approximations of ϕ_x and ϕ_y respectively, and the intervals $[A, B]$ and $[C, D]$ are *a priori* bounds of ϕ_x and ϕ_y. This Hamiltonian is used together with ENO or WENO interpolation to obtain higher-order methods.

3.5. Curvature

In many applications, the mean curvature term

$$\nabla \cdot \frac{\nabla \phi}{|\nabla \phi|} \quad \text{or} \quad \nabla \cdot \frac{\nabla u}{|\nabla u|}$$

for the level set function ϕ or the image function u appears as a regularization. We will use u in our following discussion. This term is usually approximated by finite differencing centred at each grid point. For convenience, let $(n_{i,j}^x, n_{i,j}^y)$ denote the values of $\nabla u / |\nabla u|_\epsilon$ at the grid point $x_{i,j}$, and $\nabla u / |\nabla u|_\epsilon$ is a smooth approximation of $\nabla u / |\nabla u|$. (This avoids the issue of singularity at $|\nabla u|$ and is useful for numerical computations.) A popular choice would be $|\nabla u|_\epsilon = (|\nabla u|^2 + \epsilon^2)^{1/2}$, $0 < \epsilon \ll 1$. Under these settings, the curvature $\kappa_{i,j}$ is approximated by

$$\kappa_{i,j}^{\epsilon} := \frac{n_{i+1/2,j}^{x} - n_{i-1/2,j}^{x}}{\Delta x} + \frac{n_{i,j+1/2}^{y} - n_{i,j-1/2}^{y}}{\Delta y},$$

and

$$n_{i\pm1/2,j}^{x} := \frac{D_{x}^{\pm} u_{i,j}}{\sqrt{(D_{x}^{\pm} u_{i,j})^2 + D_{y}^{0}(S_{x}^{\pm} u_{i,j})^2 + \epsilon^2}},$$

$$n_{i,j\pm1/2}^{\epsilon} := \frac{D_{y}^{\pm} u_{i,j}}{\sqrt{D_{x}^{0}(S_{y}^{\pm} u_{i,j})^2 + (D_{y}^{\pm} u_{i,j})^2 + \epsilon^2}},$$

where

$$S_{x}^{\pm} u_{i,j} = \frac{u_{i\pm1,j} + u_{i,j}}{2} \quad \text{and} \quad S_{y}^{\pm} u_{i,j} = \frac{u_{i,j\pm1} + u_{i,j}}{2}$$

are the averaging operators in the x and y direction. In practice, we choose ϵ to be the same scale as the mesh size.

It is important to point out that one can not prove convergence to the viscosity solution from this discretization using $\epsilon = a\Delta x + b\Delta y$, for two fixed nonnegative constants a and b. However, in practice, this approximation seems to work well. To be safe, we recommend taking $\epsilon = \mathcal{O}(\Delta x^p, \Delta y^p)$ for $0 < p < 1$. A general approximation theory for this type of degenerate elliptic or parabolic equations is outlined by Barles and Souganidis (1991). Following this theory, a numerical discretization needs to be monotone, consistent and stable in order to achieve convergence. Recently, Oberman proposed a convergent numerical discretization for the mean curvature term on two-dimensional Cartesian grids (Oberman 2004). In his work, an extra degree of freedom is introduced: the curvature term is not only discretized with Δx and Δy, but also with $\Delta\theta$, which is the angle between two adjacent vectors formed by the grid points in the stencil. The last term discretizes the angle of the normal of the level sets of u. Hence, the resulting scheme enlarges the stencil as one refines the grid, which makes it a bit impractical.

3.6. Time discretization

From the previous subsections, we know how to discretize the terms involving spatial derivatives. What remains is to discretize in time in order to evolve the system; *i.e.*, we need to solve the following ODE system:

$$\frac{\partial}{\partial t}\phi_{i,j} = -\tilde{H}(\phi_{i-1,j}, \phi_{i+1,j}, \phi_{i,j}, \phi_{i,j-1}, \phi_{i,j+1}),$$

where $\tilde{H}$ is the numerical approximation of $H(x, \phi, D\phi, D^2\phi)$. For example, if we use local Lax–Friedrichs for $H(\phi_x, \phi_y)$, and forward Euler for time, we end up having

$$\phi_{i,j}^{n+1} = \phi_{i,j}^n - \Delta t \, H^{LLF^n}(x_i, y_j, D_+^x \phi_{i,j}^n, D_-^x \phi_{i,j}^n, D_+^y \phi_{i,j}^n, D_-^y \phi_{i,j}^n). \qquad (3.2)$$

Typically, we use the third-order TVD Runge–Kutta scheme of Shu and Osher (1988), or the fourth-order schemes of Spiteri and Ruuth (2005) to evolve the system, since higher-order accuracy can be achieved while using larger time-steps. To keep this description self-contained, we describe the third-order TVD RK scheme below. We wish to advance $u_t = L(u)$ from t_n to t_{n+1}:

(1) $u_1 = u^n + \Delta t \cdot L(u^n)$;
(2) $u_2 = \frac{3}{4}u^n + \frac{1}{4}u_1 + \frac{1}{4}\Delta t \cdot L(u_1)$;
(3) $u_{n+1} = \frac{1}{3}u^n + \frac{2}{3}u_2 + \frac{2}{3}\Delta t \cdot L(u_2)$.

3.7. Algorithms for constructing the distance function

In the following subsections, we review some of the solution methods for the eikonal equation:

$$|\nabla u| = r(x, y), \quad u|_\Gamma = 0.$$

We present a fast Gauss–Seidel-type iteration method which utilizes a monotone upwind Godunov flux for the Hamiltonian. We show numerically that this algorithm can be applied directly to equations of the above type with variable coefficients.

Solving eikonal equations
In geometrical optics (Keller 1962), the eikonal equation

$$\sqrt{\phi_x^2 + \phi_y^2} = r(x, y) \qquad (3.3)$$

is derived from the leading term in an asymptotic expansion

$$e^{i\omega(\phi(x,y)-t)} \sum_{j=0}^{\infty} A_j(x, y, t)(i\omega)^{-j}$$

of the wave equation

$$w_{tt} - c^2(x, y)(w_{xx} + w_{yy}) = 0,$$

where $r(x, y) = 1/|c(x, y)|$, is the function of slowness. The level sets of the solution ϕ can be thus be interpreted as the first arrival time of the wave front that is initially Γ. It can also be interpreted as the 'distance' function to Γ.

We first restrict our attention to the case in which $r = 1$. Let Γ be a closed subset of $\mathbb{R}^2$. It can be shown easily that the distance function defined by

$$d(x) = \text{dist}(x, \Gamma) := \min_{p \in \Gamma} |x - p|, \quad x = (x, y) \in \mathbb{R}^2$$

is the viscosity solution to equation (3.3) with the boundary condition

$$\phi(x, y) = 0 \quad \text{for } (x, y) \in \Gamma.$$

Rouy and Tourin (1992) proved the convergence to the viscosity solution of an iterative method solving equation (3.3) with the Godunov numerical Hamiltonian approximating $|\nabla \phi|$. They also noticed that the Godunov numerical Hamiltonian can be written in the following simple form for this eikonal equation:

$$H_G(p_-, p_+, q_-, q_+) = \sqrt{\max\{p_-^+, p_+^-\}^2 + \max\{q_-^+, q_+^-\}^2}, \qquad (3.4)$$

where $p_\pm = D_\pm^x \phi_{i,j}$, $q_\pm = D_\pm^y \phi_{i,j}$, and $x^+ = \max(x, 0)$, $x^- = -\min(x, 0)$. The task is then to solve

$$H_G = 1$$

on the grid.

Osher (1993) provided a link to the time-dependent eikonal equation by proving that the t-level set of $\phi(x, y)$ is the zero level set of the viscosity solution of the evolution equation at time t

$$\psi_t + |\nabla \psi| = 0$$

with appropriate initial conditions. In fact, the same is true for a very general class of Hamilton–Jacobi equations (see Osher (1993)). As a consequence, one can try to solve the time-dependent equation by the level set formulation (Osher and Sethian 1988) with high-order approximations to the partial derivatives (Osher and Shu 1991, Jiang and Peng 2000). Crandall and Lions (1984) proved that the discrete solution obtained with a consistent, monotone Hamiltonian converges to the desired viscosity solution.

Tsitsiklis (1995) combined heap sort with a variant of the classical Dijkstra algorithm to solve the steady state equation of the more general problem

$$|\nabla \phi| = r(x).$$

This was later rederived in Sethian (1996) and Helmsen, Puckett, Colella and Dorr (1996). It has become known as the fast marching method, and has complexity $\mathcal{O}(N \log N)$, where N is the number of grid points. Osher and Helmsen (2005) have extended the fast marching type method to somewhat more general Hamilton–Jacobi equations. Since the fast marching method is by now well known, we will not give details here on its implementation in this paper.

The sweeping idea

Danielsson (1980) proposed an algorithm to compute Euclidean distance to a subset of grid points on a two-dimensional grid by visiting each grid node in some predefined order. Boué and Dupuis (1999) suggested a similar 'sweeping' approach to solve the steady state equation which, experimentally, results in a $\mathcal{O}(N)$ algorithm for the problem at hand. This 'sweeping' approach has recently been used in Tsai (2002) and Zhao *et al.* (2000) to compute the distance function to an arbitrary data set in computer vision. Zhao (2005) proved that the fast sweeping algorithm achieves reasonable accuracy in a (small) finite number of iterations independent of grid size. Using this 'sweeping' approach, the complexity of the algorithms drops from $\mathcal{O}(N \log N)$ in the fast marching to $\mathcal{O}(N)$, and the implementation of the algorithms becomes a bit easier than the fast marching method in that no heap sort is needed.

This sweeping idea is best illustrated by solving the eikonal equation in $[0, 1]$:

$$|u_x| = 1, \qquad u(0) = u(1) = 0.$$

Let $u_i = u(x_i)$ be the grid values and $x_0 = 0$, $x_n = 1$. We then solve the discretized nonlinear system

$$\sqrt{\max(\max(D_- u_i, 0)^2, \min(D_+ u_i, 0)^2)} = 1, \quad u_0 = u_n = 0 \qquad (3.5)$$

by our sweeping approach. Let us begin by sweeping from -1 to 1, *i.e.*, we update u_i from $i = 0$ increasing to $i = n$. This is 'equivalent' to following the characteristics emanating from x_0. Let $u_i^{(1)}$ denote the grid values after this sweep. We then have

$$u_i^{(1)} = \begin{cases} i/n, & \text{if } i < n, \\ 0, & \text{if } i = n. \end{cases}$$

In the second sweep, we update u_i from $i = n$ decreasing to 0, using $u_i^{(1)}$. During this sweep, we follow the characteristics emanating from x_n. The use of (3.5) is essential, since it determines what happens when two characteristics cross each other. It is then not hard to see that after the second sweep,

$$u_i = \begin{cases} i/n, & \text{if } i \leq n/2, \\ (n - i)/n, & \text{otherwise.} \end{cases}$$

Thus, to update u_o, one only uses the immediate neighbouring grid values and does not need the heap sort data structure. More importantly, the algorithm follows the characteristics with certain directions simultaneously, in a parallel way, instead of a sequential way as in the fast marching method. The Godunov numerical Hamiltonian is essential in the algorithm as described here, since it determines what neighbouring grid values should be used to update u on a given grid node o. At least in the examples presented, we only need to solve a simple quadratic equation and run some simple tests to determine the value to be updated. This simple procedure is performed in each sweep, and solution is obtained after a few sweeps. For sweeping applied to very general class of Hamilton–Jacobi equations, we recommend the simple and versatile Lax–Friedrichs method, which we mention in Section 3.7 below. See Kao *et al.* (2004) for details.

Generalized closest point algorithms

In this subsection, we describe an algorithm that can be applied for constructing a level set implicit representation for a surface which is defined explicitly. It can also be used to extend the interface velocity to the whole computational domain.

In the spirit of the Dynamic Surface Extension of Steinhoff, Fang and Wang (2000), we can define functions that map each point in $\mathbb{R}^3$ to the space of (local) representations of surfaces (previously referred to as surface elements). We can further define the distance between a point P and a surface element S by

$$\text{dist}(P, S) := \min_{y \in S}(P, y).$$

The 'surface element' can be, for example, the tangent plane, the curvature, or a NURB description of the surface.

Instead of propagating distance values away from the interface, *we propagate the surface element information along the characteristics* and impose conditions that *enforce the first arrival property* of the viscosity solution to the eikonal equation. The challenge is to compute the exact distance from a given surface element and to derive the 'upwinding' criteria for propagating the surface information throughout the grids.

Given a smooth parametrized surface $\Upsilon : I_s \times I_t \mapsto \mathbb{R}^3$, our algorithm provides a good initial guess for Newton's iterations on the orthogonality identity:

$$F(s_*, t_*; x) = \begin{pmatrix} (x - \Upsilon(s_*, t_*)) \cdot \Upsilon_s(s_*, t_*) \\ (x - \Upsilon(s_*, t_*)) \cdot \Upsilon_t(s_*, t_*) \end{pmatrix} = 0,$$

where $\Upsilon(s_*, t_*)$ is the closest point on the surface to x. The initial guess in this case is simply the closest point of the neighbours of x.

Let W denote the function that maps each point in space to its closest surface element on S. We can then write the algorithm as follows.

Algorithm. Let u be the distance function on the grids, and W be the corresponding generalized closest point function.

(1) Initialize: give the exact distance to u, and the exact surface elements to W at grids near Γ. Mark them so they will not be updated. Mark all other grid values as ∞.

(2) Iterate through each grid point E with index (i, j, k) in each sweeping direction or according to the fast marching heap sort.

(3) For each neighbour P_l of E, compute $u_l^{\mathrm{tmp}} = \mathrm{dist}(E, W(P_l))$.

(4) If $\mathrm{dist}(E, W(P_l)) < \min_k u(P_k)$, set $u_l^{\mathrm{tmp}} = \infty$. This is to enforce the monotonicity of the solution.

(5) Set $u(E) = \min_l u_l^{\mathrm{tmp}} = u_\lambda^{\mathrm{tmp}}$ and $W(E) = W(P_\lambda)$.

This procedure can be used, *e.g.*, to convert triangulated surfaces to implicit surfaces.

In general, if only the level set function is available, one can construct a suitable interpolant of the level set function and try to compute the closest points. This was proposed in Chopp (2001), where a bicubic interpolation of the level set function is constructed and Newton's method is used to find the closest points on the zero level set of the interpolant.

Further generalizations

For further generalizations of the sweeping method to solve more complicated Hamilton–Jacobi equations, such as those which arise in computing distance on a manifold,

$$H(u_x, u_y) = \sqrt{au_x^2 + bu_y^2 + 2cu_xu_y} = r(x, y), \quad \text{for} \quad a, b > 0, \quad ab > c^2,$$

and the equations using Bellman's formulae for convex Hamiltonians, we refer readers to the recent papers by Tsai *et al.* (2003*a*) and Kao *et al.* (2002). Recently, a simple sweeping algorithm, based on the Lax–Friedrichs scheme (3.1), has been shown to work in great generality (Kao *et al.* 2004). Special conditions at the grid boundaries must be enforced in order for this central scheme to compute the correct solution. Accurate estimates of the bounds on the partial derivatives of the Hamiltonian increase the resolution and the efficiency of this algorithm. The main advantage of this algorithm is in the ease of implementation, especially for equations involving complicated and nonconvex Hamiltonians.

Higher-resolution sweeping methods have also been devised (Zhang, Zhao and Qian 2004). Essentially, the idea is to reconstruct the derivatives of the solution using the grid values that have been updated as a correction in the new approximation. The higher-order approximations of the derivatives require a larger stencil, leading to a larger numerical domain of dependence; together with the non-monotonicity in the reconstruction, more iterations are needed for convergence to the discretized nonlinear system. The complexity of these algorithms is still an open question. Nevertheless, it seem to be lower than that of a straightforward time marching to steady state.

3.8. Discretization of delta functions supported along the zero level set

In the level set formulation, the evaluation of a surface integral along the zero level set of ϕ requires singular integrals involving Dirac delta functions. Careless quadratures for this type of integrals might lead to error that prevents convergence (Tornberg and Engquist 2003). Here we review the approaches proposed by Engquist *et al.* (2004).

Let s be a parametrization of Γ and let $\mathrm{d}s$ be the corresponding surface area measure. Define $\delta(\Gamma, x)$, $x \in \mathbb{R}^d$ as a delta function supported on Γ such that

$$\int_{\mathbb{R}^d} \delta(\Gamma, x) \, f(x) \, \mathrm{d}x = \int_{\mathbb{R}^d} f(x)\delta(\phi(x))|\nabla\phi(x)| \, \mathrm{d}x = \int_{\Gamma} f(X(s)) \, \mathrm{d}s, \quad (3.6)$$

where $X(s) \in \Gamma$. The following techniques are based on replacing the distribution function δ by a class of continuous functions δ_ε in the approximation of integral defined in (3.6), and replacing the integral over the domain by a Riemann sum. δ_ε is chosen to be the linear hat function that has two discrete moments:

$$\delta_\varepsilon^L(x) = \begin{cases} \frac{1}{\epsilon}(1 - |\frac{x}{\epsilon}|), & 0 \le |x| \le \varepsilon, \\ 0, & |x| > \varepsilon. \end{cases} \quad (3.7)$$

Here, discrete moments of a function are defined in analogy to the usual notion of moments at continuous level; δ_ε is said to have q discrete moments if

$$h \sum_{j=-\infty}^{\infty} \delta_\varepsilon(x_j - \bar{x})(x_j - \bar{x})^r = \begin{cases} 1, & r = 0, \\ 0, & 1 \le r < q, \end{cases} \quad (3.8)$$

for any $\bar{x} \in \mathbb{R}$, and grid points $\{x_j\}$. It is shown in Tornberg (2002) and Tornberg and Engquist (2003) that the overall approximation is of first order in h if $\varepsilon = \sqrt{h}$. For a very narrow support, such as $\epsilon = C_0 h$, the δ_ε function is not sufficiently resolved and the error must instead be analysed directly by taking into account discrete effects of the computational grid.

Engquist *et al.* (2004) proposed two regularized delta functions built from the linear hat function (3.7). One is the product formula, following Peskin (2002), that requires explicit parametrization of Γ:

$$\delta_\varepsilon(\Gamma, x) = \int_{\Gamma} \prod_{k=1}^{d} \delta_{\varepsilon_k}(x^{(k)} - X^{(k)}(s)) \, \mathrm{d}s. \quad (3.9)$$

Here δ_{ε_k} corresponds to the one-dimensional regularized δ function, and $X(s) = (X^{(1)}(s), \ldots, X^{(d)}(s))$ is a point on Γ. The other method is the level set formulation

$$\delta_\varepsilon(\Gamma, x) = \delta_\varepsilon(\phi(x))|\nabla\phi(x)|.$$

Both approaches use a pointwise variable regularization parameter dependent on the gradient of the level set function; *i.e.*, $\epsilon = \epsilon(x, \phi_x, \phi_x)$. The authors showed that with these approaches and with δ_ε^L as the building block, it is possible to approximate the singular integrals (3.6) on a uniform

Cartesian grid with at least first-order accuracy in h, while keeping minimum support (with $|\epsilon(x, \nabla\phi)| \leq Ch$). The first approach seems to yield approximations that are second-order accurate if ϕ is the distance function to Γ. We refer readers to Engquist *et al.* (2004) for the explicit formula derived from their first approach. We describe their second approach here owing to its simplicity. In short, $\delta(\phi(x))$ is approximated pointwise by

$$\delta^L_{\epsilon(x,\nabla\phi)}(\phi(x)),$$

where

$$\epsilon(x, \nabla\phi) = h|\nabla\phi(x)|_{\ell_1}, \tag{3.10}$$

and $|\nabla\phi(x)|_{\ell_1} = \sum_{j=1}^{d} |\phi_{x_j}|$.

3.9. Regularization of characteristic functions

In the level set method, the average of a function g over the set $\{\phi \geq 0\}$ translates to an integral involving the Heaviside function:

$$\int_\Omega g(x)H(\phi(x))\,dx.$$

Following the discussion in the previous subsection, one can regularize the Heaviside function by

$$H_\epsilon(x) = \begin{cases} 1, & x \geq \epsilon, \\ \frac{1}{2}(1 + \frac{x}{\epsilon}), & |x| < \epsilon, \\ 0, & x \leq -\epsilon, \end{cases} \tag{3.11}$$

with the same type of pointwise scaling:

$$\epsilon(x, \nabla\phi) = \frac{h}{2}|\nabla\phi(x)|_{\ell_1}.$$

It can be shown that the resulting approximation to the volume integral is second-order accurate in the mesh size h.

The signum function used in equation (2.2) is discontinuous:

$$\text{sgn}(z) = \begin{cases} 1, & x > 0, \\ 0, & x = 0, \\ -1, & x < 0, \end{cases}$$

and may introduce grid effects when discretized improperly on the grid. Ideally, a smooth monotone function that passes through zero should replace the signum function, since we only care about the direction of the characteristics and the steady state of the solution in a neighbourhood of its zero level set. With a bounded smooth function such as

$$\tanh(\gamma_0 x), \quad \gamma_0 > 0$$

the accuracy of the solution to (2.2), for smooth zero level set, is then determined by the order of the discretization. However, the characteristics emanate from the zero level set at a speed that is 0 at the interface and smoothly increases as the bicharacteristics are getting farther away. On a grid with N grid points, this usually implies that the number of time-steps needed for steady state on this grid is proportional to N. In many applications, one is only interested in a thin band of width C/N around the zero level set. Therefore, if other operations involve $\mathcal{O}(N)$ operations, this regularization might be an attractive option. (See Section 2.2.)

4. Image interpolation

Consider an old photo with scratches. One can try to restore the original photo by filling in the scratched regions with certain values so that the overall image looks 'right'. This is a complicated interpolation problem. The main difficulties are as follows.

(1) The interpolation domain may be non-simply connected and have irregular boundaries.
(2) The interpolation procedure must allow discontinuities along some meaningful geometrical structures.
(3) Ultimately, the interpolation result is subject to human psycho-visual inspection.

Classically, (2) and (3) relate to the discussion of the function space to which images belong and which norm should be used. A severe problem would be that the interpolation domain is too large so that one essentially has to 'generate' new information.

In this article, we will call the problem of interpolating over 'narrow' domains the inpainting problem, and the other the 'disocclusion' problem. Essentially, interpolation algorithms rely on the regularity of certain *suitable quantities*. Considering grey-scale images, a natural quantity of consideration would be the level lines of the given image functions. One would think of properly connecting the level lines from a neighbourhood of the inpainting domain into it. Pioneering works of importance to this area are Caselles *et al.* (1998) and Masnou and Morel (1998).

Bertalmio *et al.* (2000) proposed an algorithm designed to project the gradient of the smoothness of the image intensity in the direction of the image level lines. The resulting model is a third-order PDE,

$$u_t = \nabla^{\perp} u \cdot \nabla(\Delta u),$$

where u is the image intensity, and $\nabla^{\perp}$ denotes the differential operator $(-\partial_y, \partial_x)$ and Δ is the standard Laplacian operator. At steady state,

$$\nabla^{\perp} u \cdot \nabla(\Delta u) = 0$$

inside the inpainting domain, implying that the gradient of Δu has to be perpendicular to the level line of u. In other words, the image value u is convected along the level curves of the quantity Δu. Later, Bertalmio, Bertozzi and Sapiro (2001 a) established the connection of the image intensity u in this model to the stream function in a 2D incompressible fluid, where Δu can be interpreted as the vorticity of the fluid.

Chan and Shen proposed a variational model for inpainting:

$$ J_\lambda[u] = \int_{E \bigcup D} |\nabla u| \, \mathrm{d}x + \lambda \int_E |f - u|^2 \, \mathrm{d}x. \qquad (4.1) $$

Here, E is the region which is not to be interpolated, and D is the region with missing data. Imposing Neumann boundary condition at the boundary of $E \bigcup D$, the gradient equation is

$$ u_t = \nabla \cdot \left(\frac{\nabla u}{|\nabla u|} \right) - \chi_E(x)\lambda(u - f), \quad \text{for} \ x \in E \bigcup D, $$

where χ_E is the characteristic function of E. One immediately sees the clear connection to the TV denoising model 1.4. This algorithm interpolates a given image so that the total variation in the inpainting domain is minimized. A mental application of the co-area formula reveals that the level lines stemming from E are connected in D with minimal arc lengths. This algorithm performs denoising and inpainting simultaneously.

The models that we have just described use the regularity of some local geometrical quantities for interpolation over the inpainting regions. In reality, the human vision may use more global quantities of a given image for judging whether any particular inpainting algorithm generates suitable solutions. One good test is to see how an inpainting algorithm connects the missing boundaries of a given set of shapes; whether a straight horizontal bar will be reconnected from the image with its middle part removed; or where a curved boundary can be restored. Therefore, many current efforts in devising new inpainting algorithms or in comparing different algorithms concentrate on this aspect. Of course, it is also possible to propose an inpainting model that is based on the regularity of statistical properties of a given image or images, especially when inpainting textures.

There have been efforts to incorporate more global quantities for inpainting. For example, Chan, Kang and Shen (2002) and Esedoglu and Shen (2002) replaced the total variation term in (4.1) by Euler's elastica:

$$ e(\Gamma) = \int_\Gamma (\alpha + \beta\kappa^2) \, \mathrm{d}s = \alpha \, \mathrm{length}(\Gamma) + \beta \int_\Gamma \kappa^2 \, \mathrm{d}s. $$

Figure 4.1 shows an inpainting result from Esedoglu and Shen (2002).

In Bertalmio, Vese, Sapiro and Osher (2003), texture is first separated from a given image (described in Section 6.1), leaving a 'cartoon'-like

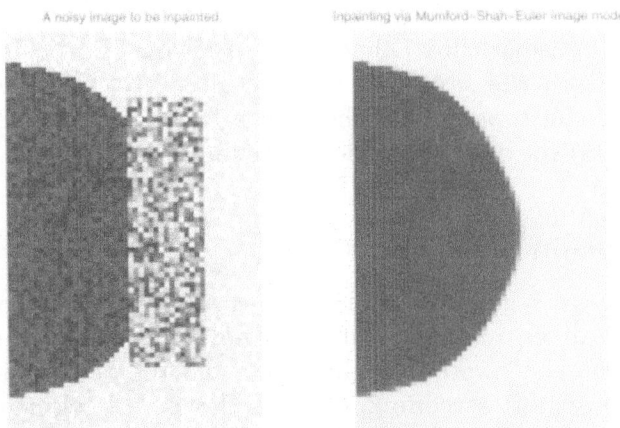

Figure 4.1. This is an inpainting result
from Esedoglu and Shen (2002).

component of the original image. A statistical approach is adopted to 'synthesize' texture for the inpainting domain so that some statistical regularity across the whole image is maintained. After the separation of texture, an inpainting procedure based on local geometric regularity can be performed on the remaining cartoon-like image. Finally, the inpainting is done by putting the synthesized texture together with the cartoon inpainting result. See also Ballester, Bertalmio, Caselles, Sapiro and Verdera (2001) and Ballester, Caselles and Verdera (2003) for related work in inpainting and disocclusion. It is also possible to 'inpaint' in the time-frequency domain with a regularity constraint on the spatial domain. See, *e.g.*, Chan, Shen and Zhou (2004).

Finally, the level set method has also been used for interpolation of unorganized points, curves and/or surface patches by Zhao *et al.* (2000, 2001). Briefly, one finds a level set function whose zero level set passes through a given unorganized set S. The unsigned distance function d_S to the data set is used for fast visualization and analysis. This distance function can be efficiently constructed using the generalized closest point algorithm described in Section 3.7. Then a minimal surface/convection-type model, resembling geodesic snakes, is used for shape reconstruction from the data set. More precisely, Zhao *et al.* (2000, 2001) construct a local minimizer of the following energy:

$$\mathcal{E}(\Gamma) = \left(\int_\Gamma d_S^p(x) \, \mathrm{d}s \right)^{1/p},$$

using gradient descent with an initial guess constructed from $\{d_S = \epsilon_0\}$ by a fast tagging algorithm (Zhao *et al.* 2001). The positive constant ϵ_0 is determined from the sampling density of S. No *a priori* knowledge is assumed about the topology of the shape to be reconstructed. See Zhao and Osher (2003) for a recent review article.

5. Segmentation algorithms

The task of image segmentation is to find a collection of non-overlapping subregions of a given image. In medical imaging, for example, one might want to segment the tumour or the white matter of a brain from a given MRI image. In airport screening, one might wish to segment certain 'sensitive' shapes, such as weapons. There are many other obvious applications. Mathematically, given an image $u : \Omega \subset \mathbb{R}^2(\text{or } \mathbb{R}^3) \mapsto \mathbb{R}^+$, we want to find closed sets Ω_i satisfying

$$\Omega = \bigcup_{i=1}^{N} \Omega_i, \quad \text{and} \quad \bigcap_{i=1}^{N} \Omega_i^{(0)} = \emptyset,$$

such that $\mathcal{F}(u, \Omega_i) = 0$, where $\mathcal{F}$ is some functional that defines the segmentation goals and $\Omega_i^{(0)}$ denotes the interior of Ω_i. As in the example of finding tumours, typically, N is taken to be 2 (sometimes $N = 3$ when volumetric data is given), and Ω_1 is taken to be the region corresponding to the tumour, while Ω_2 contains everything else. It is then natural to devise a level set method to perform this task, by representing, for example, Ω_1 as the region in which ϕ is nonnegative. A slightly more general statement would be to perform segmentation from a given set of images u_j that come from different sources. For example, one might be interested in segmenting stealth fighter jets from both conventional radar signals and also infrared images.

Very often, the definition of what belongs to the 'desired' regions depends on the grey-scale intensity of the given image, and the problem of finding such regions is formulated as a variational problem; *i.e.*, the solution minimizes some 'energy'. In a standard level set method, ϕ is used to represent Ω_i and $\partial\Omega_i$. This is the setting of our discussion. In this section, we describe some level set segmentation methods based on this type of definition.

5.1. The Chan–Vese algorithm

This is closely related to the classical Mumford–Shah algorithm (Mumford and Shah 1989), but uses a simple level set framework for its implementation. We present the original Chan–Vese segmentation algorithm (Chan and Vese 2001*a*), and discuss various aspects of this algorithm.

Basic formulation

The minimization problem is

$$\min_{\phi \in BV(\Omega), c_1, c_2 \in \mathbb{R}^+} \mathcal{E}(\phi, c_1, c_2; u_0),$$

where the energy is defined as

$$\mathcal{E}(\phi, c_1, c_2; f) = \mu \int_{\Omega} \delta(\phi) |\nabla \phi| \, \mathrm{d}x \tag{5.1}$$

$$+ \lambda_1 \int_{\Omega} |f - c_1|^2 H(\phi) \, \mathrm{d}x + \lambda_2 \int_{\Omega} |f - c_2|^2 (1 - H(\phi)) \, \mathrm{d}x.$$

Intuitively, one can interpret from this energy that each segment is defined as the subregions of the images over which the average of the given image is 'closest' to the image value itself in the L^2-norm. The first term in the energy measures the arc length of the segment boundaries. Thus, minimizing this quantity provides some stability to the algorithm as well as preventing fractal-like boundaries from appearing.

If one regularizes the δ function and the Heaviside function by two suitable smooth functions δ_ϵ and H_ϵ, then formally the Euler–Lagrange equations can be written as

$$\partial_\phi \mathcal{E} = -\delta_\epsilon(\phi) \left[\mu \nabla \cdot \frac{\nabla \phi}{|\nabla \phi|} - \lambda_1 (f - c_1)^2 + \lambda_2 (f - c_2)^2 \right] = 0, \tag{5.2}$$

with natural boundary condition

$$\frac{\delta_\epsilon(\phi)}{|\nabla \phi|} \frac{\partial \phi}{\partial \bar{n}} = 0 \quad \text{on} \quad \partial \Omega.$$

$$c_1(\phi) = \frac{\int_\Omega f(x) H_\epsilon(\phi(x)) \, \mathrm{d}x}{\int_\Omega H_\epsilon(\phi(x)) \, \mathrm{d}x}, \tag{5.3}$$

and

$$c_2(\phi) = \frac{\int_\Omega f(x)(1 - H_\epsilon(\phi(x))) \, \mathrm{d}x}{\int_\Omega (1 - H_\epsilon(\phi(x))) \, \mathrm{d}x}. \tag{5.4}$$

Discretization

A common approach to solving the minimization problem is to perform gradient descent on the regularized Euler–Lagrange equation (5.2); that is, solving the following time-dependent equation to steady state:

$$\frac{\partial \phi}{\partial t} = -\partial_\phi \mathcal{E}$$

$$= \delta_\epsilon(\phi) \left[\mu \nabla \cdot \frac{\nabla \phi}{|\nabla \phi|} - \lambda_1 (f - c_1)^2 + \lambda_2 (f - c_2)^2 \right]. \tag{5.5}$$

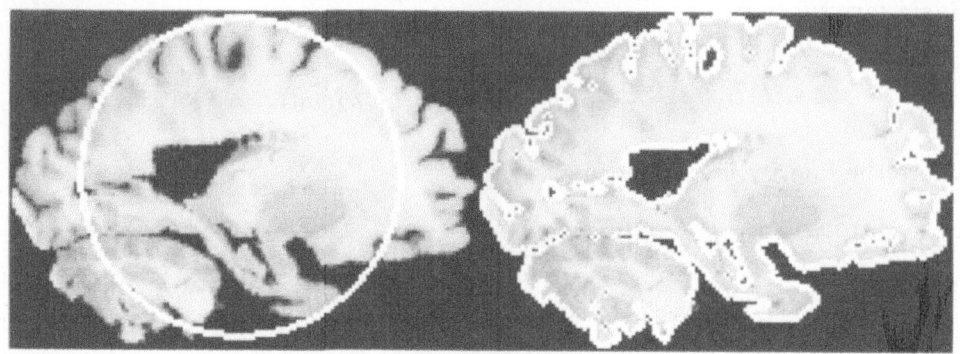

Figure 5.1. Brain segmentation in 2D.

Here, we remind the readers that $c_1(\phi)$ and $c_2(\phi)$ are defined in (5.3) and (5.4).

In the Chan–Vese algorithm, the authors regularized the Heaviside function used in (5.3) and (5.4),

$$H_{2,\epsilon}(z) = \frac{1}{2}\left(1 + \frac{2}{\pi}\arctan\left(\frac{z}{\epsilon}\right)\right),$$

and defined the delta function as its derivative:

$$\delta_{2,\epsilon}(z) = H'_{2,\epsilon}(z).$$

Equation (5.5) is then discretized by a semi-implicit scheme; *i.e.*, to advance from $\phi^n_{i,j}$ to $\phi^{n+1}_{i,j}$, the curvature term right-hand side of (5.5) is discretized as described in the previous section using the value of $\phi^n_{i\pm1,j\pm1}$, except for the diagonal term $\phi_{i,j}$, which uses the implicitly defined $\phi^{n+1}_{i,j}$. The integrals defining $c_1(\phi)$ and $c_2(\phi)$ are approximated by a simple Riemann sum with the regularized Heaviside function defined above. ϕ_t is discretized by the forward Euler method, $(\phi^{n+1}_{i,j} - \phi^n_{i,j})/\Delta t$. Therefore, the final update formula can be written as

$$\phi^{n+1}_{i,j} = \frac{1}{1+\alpha_\kappa}\left(\phi^n_{i,j} + G(\phi^n_{i-1,j},\phi^n_{i+1,j},\phi^n_{i,j-1},\phi^n_{i,j+1})\right),$$

where $\alpha_\kappa \geq 0$ comes from the discretization of the curvature term. If the scheme were fully explicit, $\alpha_\kappa = 0$ and G would depend on $\phi^n_{i,j}$. In the original paper, the authors used $\Delta x = \Delta y = 1$, $\epsilon = 1$, and $\Delta t = 0.1$. This implies that the delta function is really a regular bump function that puts more weight on the evolution of the zero level set of ϕ. See Figures 5.1 and 5.2 for some results of this algorithm applied to brain segmentation.

Finally, it is also possible but usually not advisable in this (unusual) case because new zero level sets are likely to develop spontaneously, to replace the δ function in front of the curvature term by $|\nabla\phi|$ (Marquina and

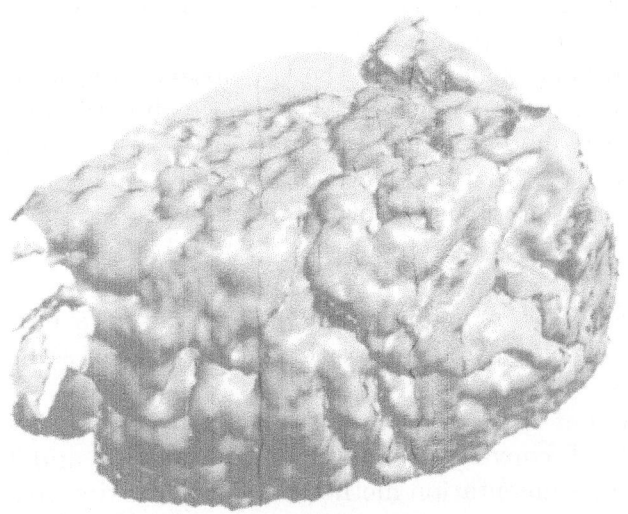

Figure 5.2. Brain segmentation in 3D.

Osher 2000). The equation then becomes independent of the choice of the level set function used, *i.e.*, the problem becomes morphological (Alvarez *et al.* 1993).

5.2. Fast segmentation algorithms

Recently, Gibou and Fedkiw (2005), and later Song and Chan (2002), proposed some fast methods that are based on the Chan–Vese level set segmentation formulation. These algorithms are built upon flipping the values of ϕ at each grid point/pixel from positive to negative or *vice versa* according to a rule $\mathcal{R}$, and contain 4 main steps.

(1) Initialize $\phi^0 : \Omega \mapsto \{-1, 1\}$.

(2) Advance: for each node, set $\phi^{n+1}(x) = -\phi^n(x)$ if $\mathcal{R}(\phi^{n+1}, \phi^n, x) = 1$.

(3) (Perform regularization if needed.)

(4) Repeat until $\phi^{n+1} \equiv \phi^n$.

For example, in Gibou and Fedkiw's algorithm, $\mathcal{R}(\phi^{n+1}, \phi^n) = 1$ if

$$V(\phi^n) \cdot \text{sign}(\phi^n) < 0;$$

here V corresponds to the fitting term in the Euler–Lagrange equation

$$V(\phi^n, x) := -\lambda_1 (f - c_1(\phi^n))^2 + \lambda_2 (f - c_2(\phi^n))^2.$$

(Note that the case $V = 0$ is implicitly defined.) In this algorithm, step (3) is essential for regularizing the segment boundaries. Without it, fractal-like boundaries may develop.

In Song and Chan's algorithm, the key observation is that only the signs of the level set function matter in the energy functional. This can easily be seen from the model defined in equation (4.1), in which the energy is a function of $H(-\phi)$. In this algorithm, $\mathcal{R}(\phi^{n+1}, \phi^n)$ can be interpreted as the logical evaluation of the following inequality:

$$\mathcal{E}(\phi^{n+1}, c_1, c_2; f) \le \mathcal{E}(\phi^n, c_1, c_2; f).$$

Hence, the sign of $\phi^n(x)$ is flipped *only if the energy* (5.1) *is non-increasing.* This provides stability of the algorithm at the cost of some speed of implementation.

We remark that there is a close connection between these two 'level set' methods to the 'Γ-convergence' methods of Ambrosio and Tortorelli (1990). The Chan–Vese segmentation method can be approximated by the following variational problem:

$$E_\epsilon(u, c_1, c_2; f) := \mu \int \epsilon |\nabla u|^2 + \frac{1}{\epsilon} W(u) \, \mathrm{d}x \qquad (5.6)$$

$$+ \lambda_1 \int u^2 (f - c_1)^2 + \lambda_2 \int (1 - u)^2 (f - c_2)^2 \, \mathrm{d}x,$$

where $w(u) = u^2(1 - u)^2$, and ϵ is a small positive number. Due to the strong potential $\epsilon^{-1} W(u)$, u will quickly be attracted to either 1 or 0, and consequently, the term u^2 and $(1 - u)^2$ correspond respectively to $H(\phi)$ and $1 - H(\phi)$ in (5.1), and $\epsilon |\nabla u|^2$ corresponds to the regularization of of the length of $\partial\Omega_i$. Intuitively, one can interpret the Gibou–Fedkiw or Song–Chan algorithm as performing a one-step projection to the steady state that results from the stiff potential W.

5.3. Segmentation of multiple 'phases'

There are successful efforts to generalize the level set methods for multiphase computation. For example, in Zhao *et al.* (1996), each partition Ω_i is represented by a level set function ϕ_i. It is then important to enforce the constraints that (1) the regions represented do not overlap ($\bigcap_{i=1}^N \{\phi_i < 0\} = \emptyset$), and (2) there are no unclaimed regions; *i.e.*, every point in Ω belongs to certain Ω_i ($\Omega = \bigcup_{i=1}^N \{\phi_i \le 0\}$). Interesting formulae are derived in the variational setting to enforce these two conditions. However, this approach is expensive when the number of phases is large.

Vese and Chan (2002) use the sign of the level set functions ϕ_j as a binary coding for the phases, each assigned a nonnegative integer value. Suppose there are four phases, Ω_i, $i =, \cdots, 3$, and two level set functions ϕ_0 and ϕ_1

are used for their representation. One can then write, for instance,

$$\Omega_0 = \{\phi_0 \geq 0\} \bigcap \{\phi_1 \geq 0\}, \qquad \Omega_1 = \{\phi_0 \leq 0\} \bigcap \{\phi_1 \geq 0\},$$

$$\Omega_2 = \{\phi_0 \geq 0\} \bigcap \{\phi_1 \leq 0\}, \qquad \Omega_3 = \{\phi_0 \leq 0\} \bigcap \{\phi_1 \leq 0\}.$$

In full generality, write the phase number i in binary format $i = \sum_{k=0}^{n-1} c_k \cdot 2^k$, where c_k takes on either 0 or 1. Then one way of using $\{\phi_k\}_{k=0}^{n-1}$ level set functions to represent Ω_i is to identify

$$\Omega_i = \bigcap_{k=0}^{n-1} \{x \in \Omega : (1 - c_k) \cdot \phi_k(x) \geq 0\}.$$

It appears that the gradient descent algorithm using this formulation is quite sensitive to the initial configurations and tends to get stuck in some undesirable local minima. There is also the potential misidentification of what is supposed to be categorized as one single phase to two or more 'different' phases, since the formulation really comes with 2^n phases with n level set functions. In the Chan–Vese algorithm, for example, it is possible that the image u has the same average in two different segments. Another drawback is the possible miscalculation of the arc length/surface area of each phase, when two phase boundaries are forced to collapse into one and may be given more weight than others. An important but so far untouched (to the best of our knowledge) problem in the level set world is to determine the optimal number of phases in certain segmentation problems.

5.4. Discussion

One of the successful features reported in Chan and Vese (2001a) is the emergence of new interior contours. As we mentioned earlier, if one enforces the level set function to be the distance function to the existing interfaces or replace the delta function by $|\nabla\phi|$ and computes locally, then the existing interfaces are only allowed to merge or disappear. The authors attributed the possibility of new interior contour emergence to their particular choice of delta function that has non-compact support. One common approach in getting around this problem is to initially seed many small circles that are densely distributed throughout the given image and let them gradually merge and evolve to a number of larger contours. See Figure 5.3.

This approach seems to capture the interior contour pretty well. While the statements about the nonlocal effect of the particular delta function used in Chan–Vese are valid, more careful study is called for to compare the degree of regularization, and diameter of the interior of any segmentation, with the possibility of the emergence of a new interior contour. We would also like to comment that the iterative approach adopted by Chan and Vese can be regarded as a version of Gauss–Jacobi iterations for the nonlinear

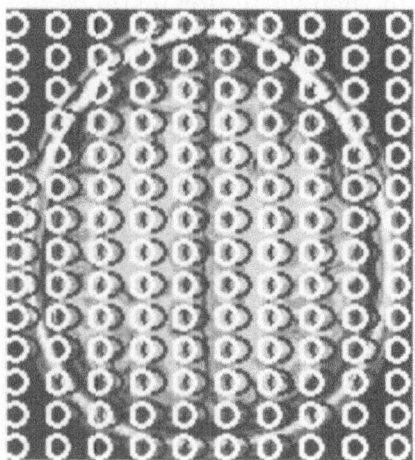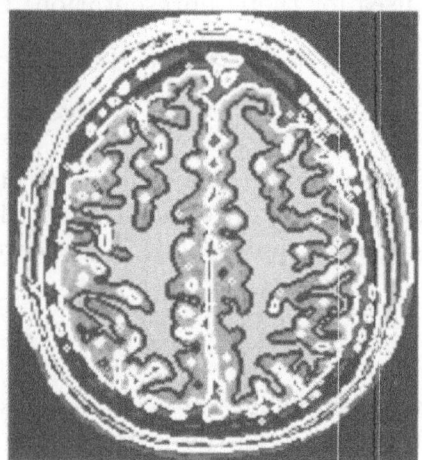

Figure 5.3. Initialization.

Euler–Lagrange equation (5.2). This statement can be supplemented by looking at the same approach applied to the linear equation:

$$u_t = \Delta u.$$

The complexity of both approaches is proportional to N^2, the total number of pixels. We remark that it is possible to speed up the gradient flow in the Chan–Vese algorithm by a splitting method described in Eyre (1998).

There are many new (and old) 'level set' segmentation algorithms that discard the continuity of the level set function and propose, instead, to model the segmentation problem as a completely discrete, pixel-by-pixel, algorithm. As in Gibou and Fedkiw (2005) and Song and Chan (2002), these types of methods typically appear to be faster, and in some cases more flexible in handling multiple phases. These trends seem to be going against the original spirit and *raison d'être* of PDE-based level set methods for image processing – the geometry of the interface is approximated at higher-order accuracy through the assumed continuity of the level set function over the grid. This fact resonates with the criticism of phase field models for segmentation, that there is no accurate representation of the interface, unless one refines the grid and resolves the stiff parameter ϵ^{-1} (something that is typically impossible to do for many image applications). See Merriman, Bence and Osher (1994) for a precise analysis of this.

One should ask whether accurate representation of the phase boundaries is really needed for the problem at hand. Of course, there are applications in which geometrical quantities of the phase boundaries play important roles in the model; *e.g.*, in the disocclusion application of Nitzberg, Mumford and Shiota (1993) and also in the applications related to Euler's elastica. In these types of applications, the 'conventional' level set approach certainly has an advantage. In the cases where the geometrical quantities are not of importance, the piecewise constant model may be quite useful.

Our last comment is on the regularization term of the level set segmentation methods. So far, popular choices have been those variants arising from minimizing the length of the interface. In denoising, as we have seen, this corresponds to L^1 regularization of the image gradient. It is possible that the features to be segmented, owing to their origin, retain special orientations and are anisotropic. This application appears, for example, in material sciences. In this case, one should look into the possible alternatives. We point out that Wulff energy is one such possible candidate. There, the regularization operator $\mathcal{R}$ is a function of the normal of the interface, *i.e.*, $\mathcal{R}(n) = \mathrm{div}(\gamma(n))$. (In the common TV regularization, $\gamma(n) = n$.) We refer to Soravia (1994), Osher and Merriman (1997), Peng, Osher, Merriman and Zhao (1999*b*) and Esedoglu and Osher (2005) for more details.

6. Pushing the limit

In this section, we describe recent work corresponding to the classical applications we listed above.

6.1. Image decomposition

Many important tasks in image science involve the decomposition of given images into different components. Again, we start with total variation denoising model (1.4)

$$\min_u \mathcal{E}(u) = \lambda \int (u - f)^2 \, \mathrm{d}x + \int |\nabla u| \, \mathrm{d}x.$$

One can re-interpret this model as finding a decomposition of the given image f into a sum of two functions: $f = u + v$, with u corresponding to the 'clean' image that one wishes to reconstruct from f, and v contains the unwanted noise that is separated from f. The segmentation model of Mumford and Shah essentially proposes a similar decomposition, with the additional constraint on a lower-dimensional set that is interpreted as the edge of the resulting segmentation. If one considers the special setting in which images take on only two values c_1 and c_2, and the boundaries between the two constant regions are rectifiable, then the total variation of u corresponds to the length of the boundaries weighted by the jump $|c_1 - c_2|$. In this context, the link between the two models is especially clear. This connection was pointed out by Vese and Osher (2002) and was described in Osher (2003).

In his inspiring book of 2001, Meyer examined the total variation model of Rudin *et al.* (1992) more closely and proposed a decomposition in which the noise and texture part, v, is written as the divergence of a vector field; *i.e.*, $v = \mathrm{div}\, g$ with the norm $\|v\|_*$ defined as the infimum of L^∞ norms of such vectors g. The proposed decomposition finds u as the solution to the following minimization problem:

$$\min_u \lambda \|f - u\|_* + \int |\nabla u| \, \mathrm{d}x.$$

The motivation of Meyer is that the L_2 norm used in the first integral in (1.4) to measure the noise and texture part of f can be improved by using the dual norm of $\| \cdot \|_{BV}$ (with proper completion of the space BV). This book triggered a sequence of interesting studies and useful algorithms.

Vese and Osher (2003) approximated Meyer's $\| \cdot \|_*$ norm by an L^p norm, and proposed a modified variational model:

$$\min_{u,g_1,g_2} \mathcal{E}(u, g_1, g_2) = \lambda \int \left(f - \left(u + \frac{\partial g_1}{\partial x} + \frac{\partial g_2}{\partial y} \right) \right)^2 \mathrm{d}x + \mu \left(\int (g_1^2 + g_2^2)^{\frac{p}{2}} \right)^{\frac{1}{p}}$$
$$+ \int |\nabla u| \, \mathrm{d}x.$$

Osher, Sole and Vese (2003) assumed the Hodge decomposition of the vector field g: $g = \nabla P + Q$, where Q is divergence-free. With this assumption, and the H^{-1} norm square in place of $\| \cdot \|_*$, they proposed the model

$$\min_{u,g_1,g_2} \mathcal{E}(u, g_1, g_2) = \lambda \int |\nabla \Delta^{-1}(u_0 - u)|^2 \, \mathrm{d}x + \int |\nabla u| \, \mathrm{d}x.$$

Later, the first decomposition was combined with other texture synthesis technique to inpaint textured images (Bertalmio *et al.* 2003).

See Haddad and Meyer (2004) for a recent review of the related variational models.

6.2. Inverse scale space and PDE-based multi-resolution image analysis

It is possible to construct a hierarchical decomposition of a given image using the 'length scale' parameter λ in the TV denoising model. Tadmor, Nezzar and Vese (2004) study the convergence properties of this type of decomposition using $\lambda = \lambda_0 2^j$. More precisely, the decomposition starts with $f = u_0 + v_0$, where u_0 is the minimizer of the standard TV denoising model

$$\min_{u \in L^2(\Omega)} \mathcal{E}_{TV}(f, u, \lambda_0) = \int_\Omega |\nabla u| \, \mathrm{d}x + \lambda_0 \int_\Omega |f - u|^2 \, \mathrm{d}x$$

and then iteratively performs the same decomposition for the residual v_j:

$$u_{j+1} = \mathrm{arginf}_{u \in L^2(\Omega)} \mathcal{E}_{TV}(v_j, u, \lambda_0 2^j), \quad \text{and} \quad v_{j+1} = v_j - u_{j+1}.$$

This procedure thus leads to a nonlinear hierarchical decomposition

$$f = \sum_{j=0}^{k} u_j + v_k.$$

The same strategy was first proposed in Scherzer and Groetsch (2001).

Instead of using the L^2 norm in the fitting term, Esedoglu and Chan (Esedoglu and Chan 2004) reported interesting results for the model

$$\mathcal{E}_{EC} = \min_{u \in L^1(\Omega)} \int_\Omega |\nabla u|\, \mathrm{d}x + \lambda \|f - u\|_{L^1},$$

which was studied in Alliney (1996) and Nikolova (2002).

Figure 6.1 shows the graph of the fidelity term in $\mathcal{E}_{EC}$ and $\mathcal{E}_{TV}$ as functions of λ, computed from a given image containing features of different sizes (length scales). The fidelity term in $\mathcal{E}_{EC}$ appears to be a piecewise smooth function of λ, and, strikingly, the discontinuities seem to correspond to some visually drastic change in scale space, *e.g.*, to the disappearances of objects of certain fixed length scales. The intensity of the remaining parts seems to remain constant within each connected smooth part of the graph of $\mathcal{E}_{EC}$. Figure 6.1 shows an example of such decomposition. It is worth noting that in the case of the L^2 decomposition, the intensity of every part of u fades gradually with the increment in λ. See Figure 6.2. It is especially intriguing to realize the inferred connection of this decomposition with human perception of the size of objects in images. As it is pointed out by Esedoglu and Chan (2004), the L^1 scale space suggests a way to select the scale parameter λ_j using the discontinuities in the fidelity term.

Bregman distance and inverse scale space

Recently, Osher, Burger, Goldfarb, Xu and Yin (2005) proposed a different approach to multi-resolution image analysis in scale space. Their proposed method, surprisingly, can be interpreted as a rather unique application of a powerful method, known as Bregman iteration, for constructing the minimizer of convex problems.

Rather than varying λ in the total variation denoising model (1.4), the authors proposed to iteratively 'fortify' selected parts of a given image and subsequently perform the standard TV decomposition using the modified image. Their algorithm can be described as follows.

- Let $u_1 = \arg\min \mathcal{E}(u) + \frac{\lambda}{2}\|f - u\|_{L^2}^2.$
- Define $f = u_1 + v_1.$
- Then inductively, let

$$u_k = \arg\min \mathcal{E}(u) + \frac{\lambda}{2}\|f + v_{k-1} - u\|_{L^2}^2$$

and $f + v_{k-1} = u_k + v_k.$

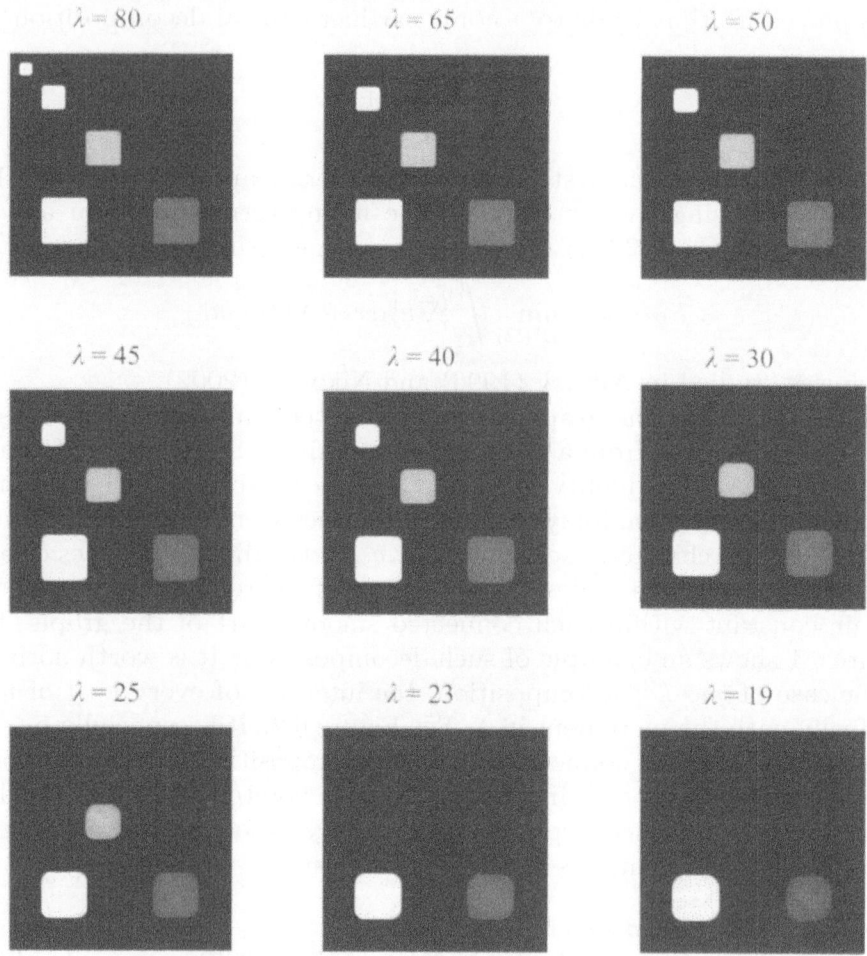

Figure 6.1. Inverse scale space using L^1-BV decomposition.

In other words, the 'noise' v_{k-1} is added back to f and ROF minimization is performed with f replaced by $f + v_{k-1}$ to decompose this function into 'signal' (u_k)+ 'noise' (v_k). Intuitively, if v_k contains both some structural information (*e.g.*, edges) of the optimized cleaned image u as well as the noise, then in the subsequent ROF decomposition, the fitting term will effectively have an inhomogeneous weighting on the locations of the support of v_{k-1}. Noise should be cleaned out from u more rapidly than the structural parts. The authors proved that as $k \to \infty$, $u_k \to f$ monotonically in L_2. In other words, k can be regarded as a parameter for scale; the larger k, the finer the scale information incorporated in u_k. For denoising purpose, the authors observed that one can find a k_0 such that u_{k_0} resembles the ideal

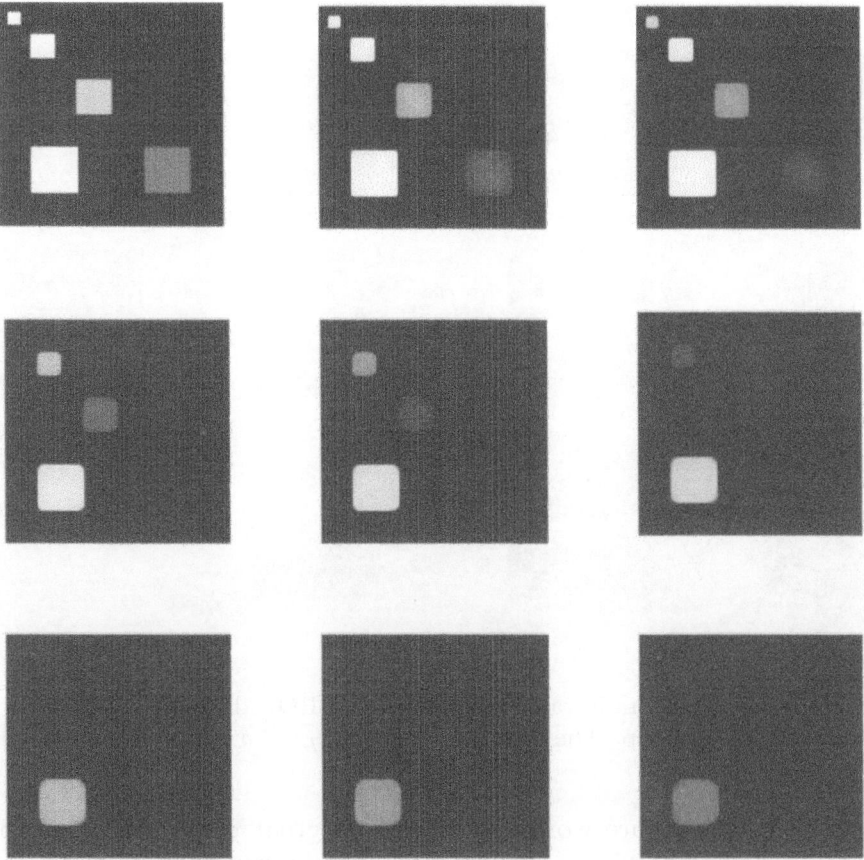

Figure 6.2. Inverse scale space using L^2-BV decomposition.

cleaned image much better than the standard ROF solution. Figure 6.3 shows a decomposition of this sort.

As mentioned above, the procedure outlined above is identical to an iterative procedure using the so-called Bregman distance. Briefly, define a sequence $\{u_k\}$ defined by: Let $u_0 = 0$, $p_0 = 0$, for $k = 1, 2, \ldots$

Compute $u_k = \arg\min Q_k(u)$

$$Q_k : u \to \mathcal{E}(u) - \mathcal{E}(u_{k-1}) - \langle p_{k-1}, u - u_{k-1} \rangle$$
$$+ \frac{\lambda}{2} \|f - u\|_{L^2}^2$$

where $\langle \cdot, \cdot \rangle$ denotes the usual duality product and p_k is the subgradient of $\mathcal{E}(u_k)$. Then compute using the update equation

$$p_k = p_{k-1} + \lambda(f - u_k). \tag{6.1}$$

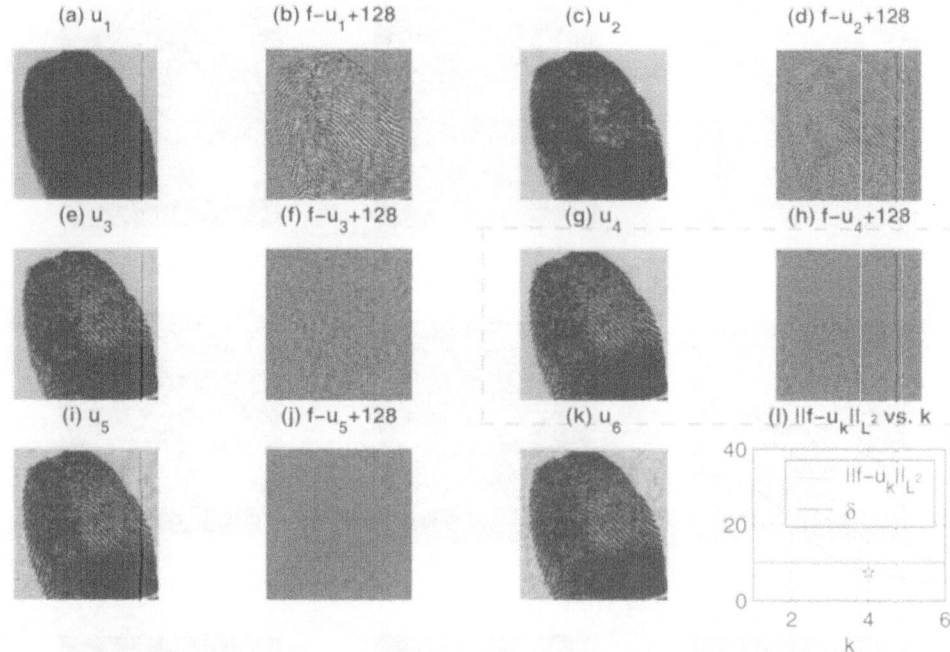

Figure 6.3. Bregman iterative refinement on ROF denoising, 2D finger example. Multi-step. The first k for $\|f - u_k\|_{L^2} < \delta$ is the optimal result.

Actually this procedure works effectively in great generality, for example, deblurring/denoising of images, recovering unknown coefficients for elliptic equations. Of course $\|f - u\|_{L^2}^2$ is replaced by another appropriate fitting term in those examples.

Here we recognize that we are using the Bregman distance between u, u_{k-1}, defined as follows,

$$D(u, v) = \mathcal{E}(u) - \mathcal{E}(v) - \langle u - v, p \rangle, \quad p \in \partial \mathcal{E}(v),$$

where $\partial J(v)$ is the subgradient of the (perhaps) nonstrictly convex function $J(u)$. We have

$$Q_k(u) = D(u, u_{k-1}) + \frac{\lambda}{2} \|f - u\|_{L^2}^2.$$

It was shown in Osher *et al.* (2005) that we obtain a unique sequence of minimizers u_k and subgradients p_k satisfying (6.1) above. The Bregman distance and the associated iteration was not typically used in this fashion in the past. Rather it was used to minimize functions $H(u, f)$ where H is a (usually complicated) convex function of u having a unique minimum: see, *e.g.*, Cetin (1989).

Osher *et al.* (2005) showed that $\{u_k\}$ defined in the sequence satisfies

$$\|u_k - f\|_{L^2}^2 \leq \|u_{k-1} - f\|_{L^2}^2$$

and if $f \in BV(\Omega)$, then

$$\|u_k - f\|_{L^2}^2 \leq \frac{\mathcal{E}(f)}{k},$$

i.e., u_k converges monotonically to f in L^2 with L^2 rate $O\big(\frac{1}{\sqrt{k}}\big)$.

Of course this convergence is not particularly useful to us as a denoising algorithm. The function f is typically noisy. The key denoising result obtained in Osher *et al.* (2005) is as follows.

Let $g \in BV(\Omega)$. Then

$$D(g, u_k) < D(g, u_{k-1})$$

as long as

$$\|f - u_k\|_{L^2}^2 \geq \tau \|g - f\|_{L^2}^2$$

for any $\tau > 1$.

This gives a stopping rule for our iterative procedure. If we have an estimate of the variance of the noise, *i.e.*,

$$f = g + n,$$

where $g \in BV(\Omega)$ is the denoised image and n is the noise, with

$$\|n\|_{L^2} = \sigma,$$

then we stop at the first k for which

$$\|f - u_{k+1}\|_{L^2} \leq \sigma.$$

Burger, Osher and Xu (2005) elegantly reformulated their new algorithm into a continuous flow in scale space, involving the solution of an integro-differential equation. For $\lambda = \Delta t$, $k\Delta t = t$, consider the Bregman iterations written in the form

$$\frac{p(t) - p(t - \Delta t)}{\Delta t} = f - u(t),$$

where

$$p = -\nabla \cdot \frac{\nabla u}{|\nabla u|} = \partial \mathcal{E},$$

and $\mathcal{E}(u) = \int |\nabla u|\, \mathrm{d}x$; *i.e.*, p is the subgradient of $\mathcal{E}$. Letting $\Delta t \downarrow 0$, we arrive at the differential equation

$$\frac{\mathrm{d}p}{\mathrm{d}t} = f - u(t)$$

$$u(t) = u(p(t)), \quad u(0) = p(0) = 0.$$

So if the flow exists and is well behaved, we have an inverse scale space. This means that we start at $u(0) = 0$ and converge as $t \to \infty$, i.e., $\lim_{t \to \infty} u(t) = f$. We go from the smoothest possible image to the noisy image f. The goal is to use the flow to denoise the image, i.e., to get closer initially to the denoised image g, until t crosses a threshold.

As an example, consider the analytically easier case $\mathcal{E}(u) = \frac{1}{2} \int |\nabla u|^2$ that is detailed in Scherzer and Groetsch (2001). Then $p = -\Delta u$ with $\frac{\partial u}{\partial n} = 0$ on $\partial \Omega$. There is a unique solution for u, given $\int_\Omega p = 0$, $u = -\Delta^{-1} p$. (We also normalize so $\int_\Omega u = \int_\Omega f = 0$.) Simple manipulation leads us to the equation

$$\frac{\mathrm{d}}{\mathrm{d}t}(u - f) = \Delta^{-1}(u - f)$$

or

$$u = f - e^{\Delta^{-1}t} f \to f.$$

For example, if Ω is the unit square, then we may expand

$$f = \sum_{i,j=1}^{\infty} \tilde{f}_{ij} \cos(\pi i x) \cos(\pi j y),$$

$$u = \sum_{i,j=1}^{\infty} \tilde{u}_{ij} \cos(\pi i x) \cos(\pi j y),$$

where $\tilde{u}_{ij} = \tilde{f}_{ij} \left(1 - e^{-\frac{t}{(i^2 + j^2)\pi}} \right)$.

We refer the reader to Burger et al. (2005) for an extension to the important case $\mathcal{E}(u) = \sqrt{|\nabla u|^2 + \epsilon^2}$.

6.3. Diffusion-generated motion and the Esedoglu–Tsai algorithm

Recently, Esedoglu and Tsai, partially motivated by the algorithms presented above, proposed a type of fast segmentation algorithm (Esedoglu and Tsai 2004). Their main algorithm can be regarded as a splitting scheme for the Modica–Mortola functional (5.6) using a thresholding approach similar to the MBO scheme (Merriman et al. 1994). The segmentation will be represented by a function v such that $\{v = 0\}$ and $\{v = 1\}$ represent the disjoint partitions in the segmentation. Their algorithm consists of three steps.

(1) Evolve

$$w_t = \Delta w - \frac{\lambda}{\sqrt{\pi \delta t}} \left(w(c_1 - f)^2 + (w - 1)(c_2 - f)^2 \right)$$

for $t \in (t_n, t_n + \delta t]$ using $w(t_n) = v_n$ and the periodic boundary condition.

(2) Set

$$v_{n+1} = \begin{cases} 0 & \text{if } w(x, t_n + \delta t) \in (-\infty, \tfrac{1}{2}], \\ 1 & \text{if } w(x, t_n + \delta t) \in (\tfrac{1}{2}, \infty). \end{cases}$$

(3) Update c_1 and c_2 by

$$c_1 = \frac{\int_D vf \, dx}{\int_D v \, dx}, \quad \text{and} \quad c_2 = \frac{\int_D (1-v)f \, dx}{\int_D (1-v) \, dx}.$$

The stiffness of the phase field model (5.6) is resolved by the splitting and the projection to equilibrium (step (2)). Step (1) involves solving linear PDEs with standard Laplacian and can be solved using any mature numerical scheme such as a Fourier method or a multigrid method. The authors studied the consistency of this algorithm by using an asymptotic expansion near the boundary ($\{v = 1/2\}$) and proposed a modified scaling $\tilde{\lambda} = \lambda/\sqrt{\pi \delta t}$ so that the length parameter in the final algorithm scales independently of any other parameters. See Figure 6.4.

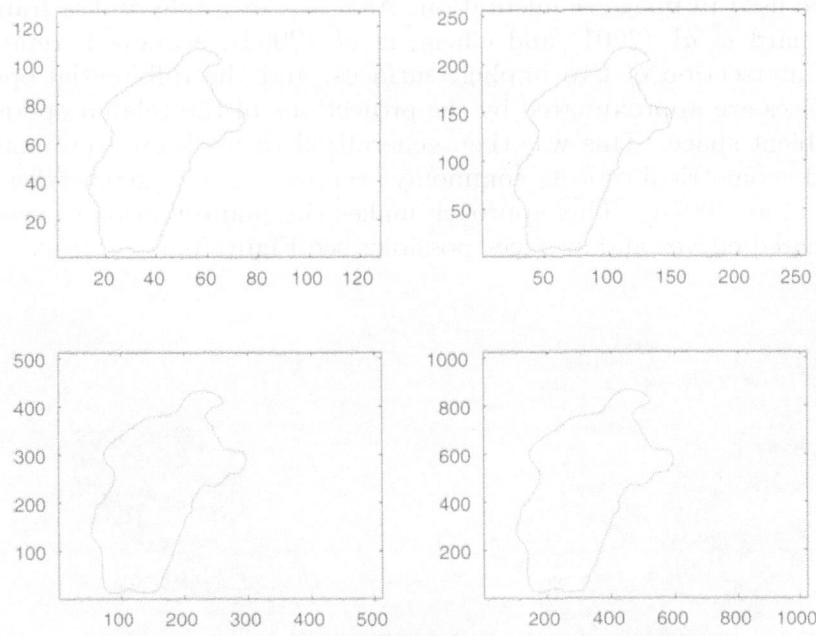

Figure 6.4. Segmentation results in a succession of grid refinements. With the same parameters δt and λ, the algorithms converge in three iterations and produce virtually identical segmentations.

6.4. Computer graphics and beyond

We will see that these efforts combine many different ideas to manipulate more complicated geometrical objects. However, the basic principle and spirit remains unchanged. Burchard *et al.* (2001) and Cheng, Burchard, Merriman and Osher (2002) provided a level set framework to represent and move curves on implicit surfaces or in three-dimensional space. This framework was then generalized to process images and even more general quantities such as vector fields that are defined on nonflat surfaces (Bertalmio *et al.* 2001*b*). Figure 6.5 shows inpainting over a sphere. This is one of the pioneering works on more complicated geometries in the level set framework. Generally speaking, the key is to raise the space dimension and/or the number of level set functions. For example, Zhao *et al.* (1996) used multiple level sets to solve a multiphase minimal surface problem. Vese and Chan (2002) further generalized the idea and applied it to image segmentations. This was discussed in the previous section. Smereka (2000) used multiple level sets to define spirals and study the formation of screw dislocations in crystal growth. Liao, Bergsneider, Vese, Huang and Osher (2002) used this approach in brain morphing. Additionally, Smith *et al.* (2002) also had an interesting level set approach to the multiphase computation that could be used in image segmentation. As a last example, in the framework of Burchard *et al.* (2001) and Cheng *et al.* (2002), a curve is represented as the intersection of two implicit surfaces, and the differential operators on surfaces are approximated by the projections of the related operators in the ambient space. This was then generalized to work on even more complicated geometrical objects commonly seen in dynamic geometrical optics (Osher *et al.* 2002*a*). This approach makes the manipulation of even more complicated curves and surfaces possible: see Figure 6.6.

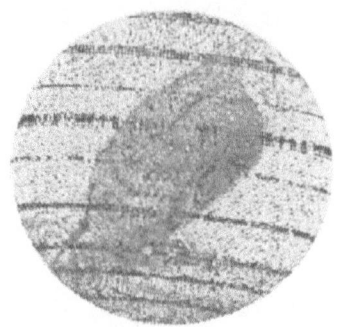 

Figure 6.5. The image on the right is the denoised and inpainted result from the left.

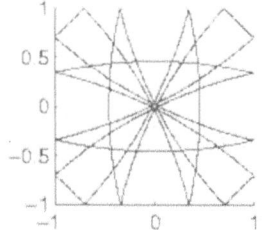 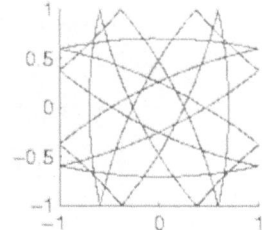

Figure 6.6. These figures shows some complicated curves with self-intersections using the approach in Osher *et al.* (2002 *a*).

Visibility

The problem of visibility involves the determination of regions in space visible to a given observer when obstacles to that sight are present. When the observer is replaced by a light source in the simplified geometrical optics setting with perfectly absorbing boundary condition at the obstacles, the problem translates to that of finding illuminated regions.

One of the most straightforward applications is in surface rendering. Typically, explicit ray tracing techniques have been used to render a 'realistic' projection of the visible part of the given surfaces on the image plane. Not surprisingly, some areas related to the accumulation on surfaces of quantities that propagate like light also need visibility information. Examples include etching (Adalsteinsson and Sethian 1997), the formation of huge ice spikes on the Peruvian Andes mountains (Betterton 2001), and shape from shading models (Jin, Yezzi, Tsai, Cheng and Soatto 2005 *a*): see Figure 6.7.

We point out here that in many of the applications listed above, the data (*i.e.*, surfaces) are given implicitly. It is therefore natural to work directly with the implicit data without converting to a different explicit representation. A very versatile level set method for the visibility problem has recently been developed by the authors and collaborators (Tsai, Cheng, Burchard, Osher and Sapiro 2004). The underlying basic algorithm can be regarded as a multi-level implicit ray tracer that works with volumetric data. Given a level set function ψ describing the obstacles D that obstruct the lines-of-sight, the visibility function $\phi(y; x_0)$ constructed by the algorithm in Tsai *et al.* (2004) takes the form

$$\phi(y; x_0) = \min_{z \in \mathcal{L}(y, x_0)} \psi(z), \qquad (6.2)$$

where $\mathcal{L}(y, x_0)$ is the integral curve of the vector field r, connecting y and x_0. The simplicity of this formulation and the associated algorithm facilitates many further extensions and applications.

Figure 6.7. The picture on the right shows the reconstructed surface from multiple images on the right (Jin *et al.* 2005*a*).

These algorithms have been applied successfully in reconstructing surfaces from multiple images of different views (Jin *et al.* 2005*a*). They can also be applied directly to some surface renderers, *e.g.*, the 'non-photo-realistic' renderer of Hertzmann and Zorin (2000). In the algorithm defined in Tsai *et al.* (2004), the boundaries of visible and invisible regions, both silhouette and swath[2] (Duguet and Drettakis 2002), are implicitly represented in the framework of Burchard *et al.* (2001) and Cheng *et al.* (2002), mentioned above. Figure 6.8 shows an accumulated visibility result of a path above the Grand Canyon. Figure 6.9 shows a result and the silhouette.

This implicit framework for visibility offers many other advantages. For example, the visibility information can be interpreted as the solution of a simple Hamilton–Jacobi equation and Tsai *et al.* (2004) offers a near-optimal solution method on the grid. The dynamics of the visibility with respect to moving vantage points or dynamic surfaces can be derived and tracked implicitly within the same framework. Furthermore, using the same framework and the well-developed level set calculus and numerics, one can start solving variational problems involving the visibility numerically and efficiently (Cheng and Tsai 2004).

Let D be the non-reflecting occluders in a domain Ω. Cheng and Tsai (2004) considered the following three central questions that are important in a variety of applications.

- *What is the optimal location x_0 for an observer such that a maximum volume of Ω is visible?*

---

[2] A swath is a set consisting of the points of intersection of rays which are tangent to an occluder with yet another occluder.

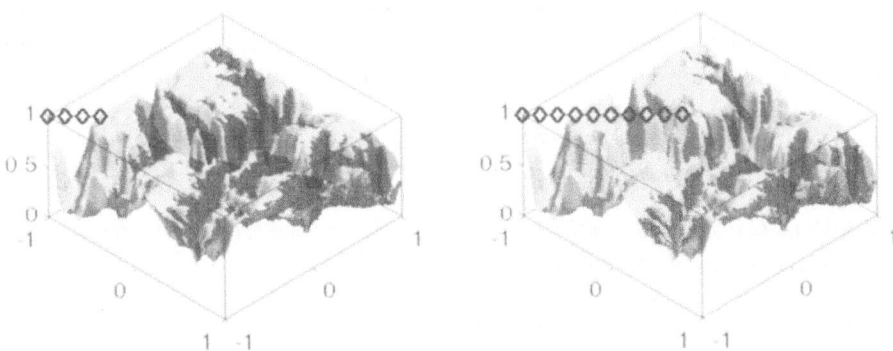

Figure 6.8. The black regions are invisible to the path indicated by the diamonds.

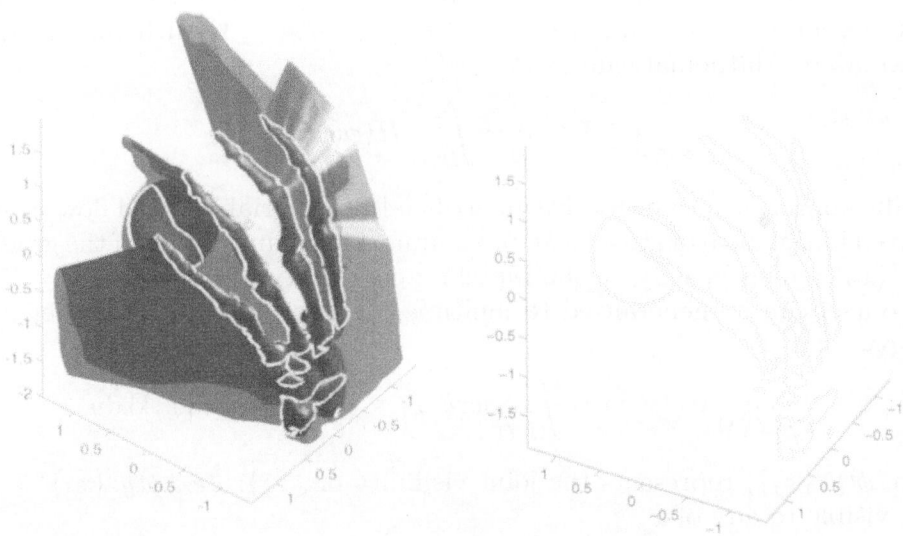

Figure 6.9. The surface borders the visible and invisible regions. The curves indicate the silhouettes and swaths.

A larger class of problems emerges when variations and extensions involving the observer and the space – multiple observers, moving observers, optimality under different measures – are taken into account. For example:

- *What are the optimal locations $\{x_i\}$ for a collection of observers, so that jointly a maximum volume of Ω is visible?*

- *What is the optimal path $\gamma(t)$ of an observer, travelling from A to B, so that a maximum volume of Ω is visible?*

In most situations, it is useful to think of an observer as a light source. Consequently, Cheng and Tsai (2004) approach solving the three central questions by *maximizing the volume of illuminated regions in Ω, or maximizing the averaged illumination (exposure) in Ω.* Two ideas are formulated as the two main variational problems below.

Problem 6.1. (Volume-based visibility optimization)   Define $V(x_o)$ as the volume of Ω visible from x_o. Find $x_o \in \mathcal{A} \subseteq \Omega$ such that $V(x_o)$ is maximized. Mathematically,

$$\max_{x_o \in \mathcal{A} \subseteq \Omega} V(x_0) = \int_{\Omega \setminus D} H(\phi(y; x_0)) \, dy.$$

The approach is to introduce an artificial time variable τ and flow x_o from a given initial location to a local maximum. The computation of the gradient of $V(x_o)$ replies heavily on the Lipschitz-continuity of ϕ for accuracy. This, of course, can be generalized to multiple observers and different weighting in space:

$$\max_{x_J \in \mathcal{A} \subseteq \Omega} V(\{x_j\}) = \int_{\Omega \setminus D} w(y, x_1, \cdots x_j) H(\phi(y; \{x_j\})) \, dy.$$

Here $\phi(y; \{x_j\})$ represents the joint visibility of $\{x_j\}$; *i.e.*, $\phi(y; \{x_j\}) \geq 0$ is y is visible to any of x_j.

Define a function $\mathcal{X}$ that counts how many times a given point y can be seen from a collection of observers. This concept can be extended to construct an optimal path for surveillance. Consider the amount of time a point y is exposed to an observer travelling at unit speed along a path $\gamma : [0, 1] \to \mathbb{R}^d$, parametrized by τ,

$$\mathcal{X}(y; \gamma) = \int_0^1 H \circ \phi(y; \gamma(\tau)) |\gamma'(\tau)| \, d\tau,$$

which we will refer to as the exposure due to γ on x. Points outside obstacles can be said to be viewed in a more uniform manner by an observer moving along γ if the deviation of the exposure from being constant is small for some constant C, as we see in the following problem.

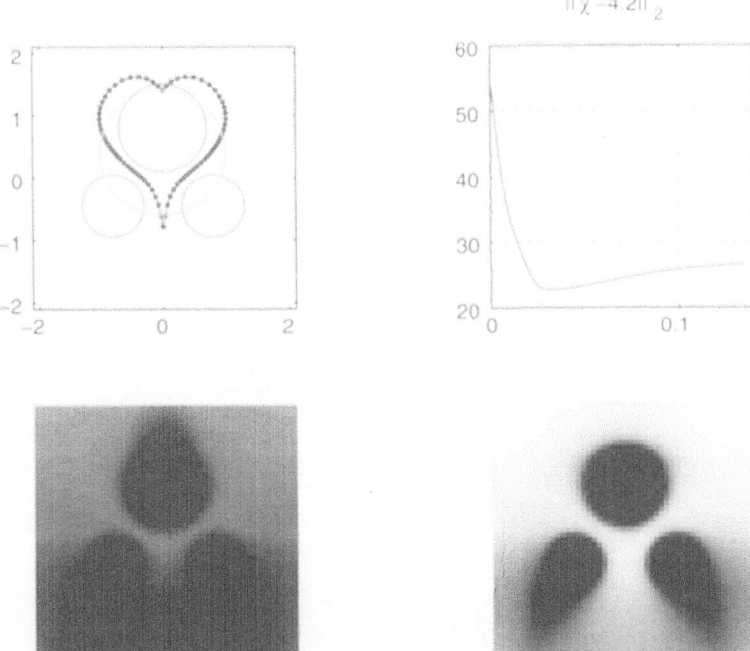

Figure 6.10. The upper left figure shows the occluders
(three disjoint circles), the initial curve (the curve
intersecting the three occluders), and the optimized curve
(dotted curve). The constant C is chosen to be 4.2, the
curvature regularization term is 0.05. The images in the
second row show the exposure of the initial and the
optimized paths.

Problem 6.2. (Exposure-based visibility optimization) Given p_0, p_1
$\in \mathbb{R}^d$, and a constant C, find $\gamma : [0, 1] \mapsto \mathbb{R}^d$ with $\gamma(0) = p_0$ and $\gamma(1) = p_1$
minimizing the energy

$$E(\gamma, C) = \frac{1}{2} \|\mathcal{X}(\cdot; \gamma) - C\|_{L^2}^2 + \lambda \int_0^1 |\gamma'(\tau)| \, d\tau. \qquad (6.3)$$

Finally, Cheng and Tsai (2004) considered a time-dependent problem
driven by the presence of an evader $y(t)$. The objective is to keep the evader
from vanishing into the occlusion. The 'inescapability' of the evader from
the pursuer is quantified as the distance between the evader and the observer, and the distance the evader is from the occlusion. Again, taking
the advantage of the continuity of the visibility representation, Cheng and

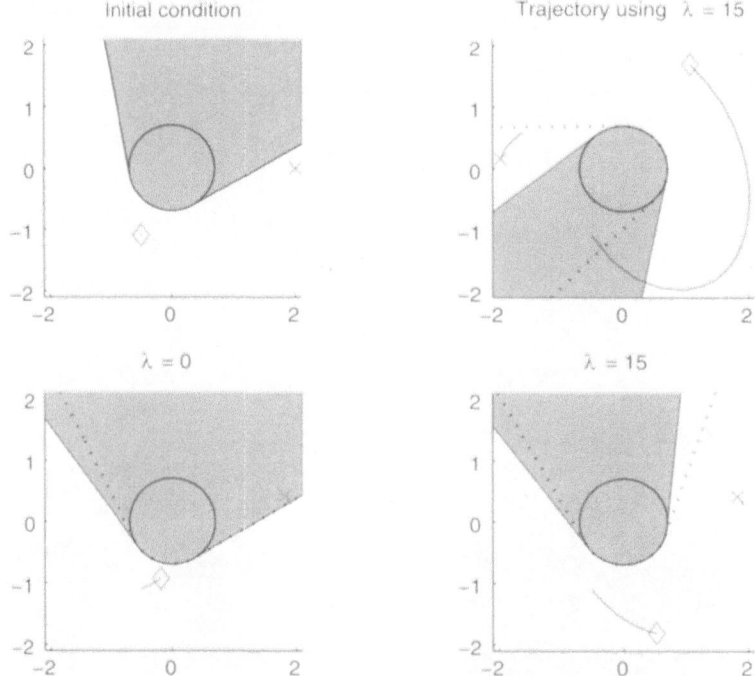

Figure 6.11. Past trajectories due to the absence and presence of the visibility gradient term. The diamonds and crosses indicate the current locations of the observer and the evader, respectively. The lower left plot is with the absence of the gradient term and should be compared to the lower right plot. The upper right plot is a longer time simulation when the gradient term is present.

Tsai defined

$$\mathcal{I}(x_o, y) = \frac{1}{2}|x_0 - y|^2 - \lambda\phi(y; x_0),$$

and formulated the corresponding problem.

Problem 6.3. (Inescapability) Find $x(t)$ so that $I(x_o(t), y(t))$ is strictly decreasing with a prescribed rate.

Various aspects of Problems 6.1 and 6.2 are studied in Cheng and Tsai (2004). Figure 6.10 shows a circular initial path being deformed to a locally optimized path for uniform visibility. Figure 6.11 shows a comparison of the past trajectories of x_o with and without the consideration of maximizing inescapability.

7. Current trends

Currently, higher-order nonlinear PDEs are increasingly appearing in image science. For example, in image inpainting of Chan *et al.* (2002), Esedoglu and Shen (2002) and Lysaker, Osher and Tai (2004), a fourth-order PDE is derived by regularizing the level set curvature a given image. In computer graphics, Tasdizen, Whitaker, Burchard and Osher (2003) proposed performing anisotropic diffusion on the normals of a given level set surface model. In general, fourth-order equations are much harder to analyse, since they rarely have a maximum principle.

An interesting paper of Burchard (2002) discusses the diffusion operators constrained in colour space, which involves a new vector-valued extension of TV minimization.

There are many imaging applications formulated as inverse obstacle problems using level set formulations. Medical imaging contains many such applications. Recently, level set optimization methods have been used for the morphological registration of medical images by Droske and Rumpf (2003/04) and by Vemuri, Ye, Chen and Leonard (2003), where objective functionals similar to elastic energies have been minimized using level set gradient methods. We expect to see more advances in this area.

In many computer graphics simulations using level set formulations, we see an emergence of semi-Lagrangian methods owing to the ease of incorporating them into some adaptive gridding (see Enright, Losasso and Fedkiw (2005), Falcone and Ferretti (2002) and also Losasso, Fedkiw and Osher (2004)). There are efforts to develop Newton method optimization techniques for finding the minima of variational image models. We refer the interested readers to the recent review paper of Burger and Osher (2005) for these problems and applications involving optimal design.

We anticipate increasing efforts in the analysis of the mathematical image models as well as numerical analysis of various aspects of the corresponding algorithms.

Acknowledgements

The authors thank Li-Tien Cheng, Selim Esedoglu, Frédéric Gibou, Jackie Shen, and Luminita Vese for providing their results for this paper.

The first author would like to thank the National Center for Theoretic Study, Taiwan for hosting his stay while parts of this research was being performed.

REFERENCES

D. Adalsteinsson and J. Sethian (1997), 'An overview of level set methods for etching, deposition, and lithography development', *IEEE Trans. Semiconductor Manufacturing* **10**(1), 167–184.

O. Alexandrov and F. Santosa (2005), 'A topological preserving level set method', *J. Comput. Phys.*, to appear.

S. Alliney (1996), 'Recursive median filters of increasing order: A variational approach', *IEEE Trans. Signal Process.* **44**(6), 1346–1354.

F. Alvarez, F. Guichard, J.-M. Morel and P.-L. Lions (1993), 'Axioms and fundamental equations of image processing', *Arch. Rat. Mech. Anal.* **123**, 199–257.

L. Ambrosio and V. M. Tortorelli (1990), 'Approximation of functionals depending on jumps by elliptic functionals via Γ-convergence', *Comm. Pure Appl. Math.* **43**(8), 999–1036.

C. Ballester, M. Bertalmio, V. Caselles, G. Sapiro and J. Verdera (2001), 'Filling-in by joint interpolation of vector fields and gray levels', *IEEE Trans. Image Process.* **10**(8), 1200–1211.

C. Ballester, V. Caselles and J. Verdera (2003), 'Disocclusion by joint interpolation of vector fields and gray levels', *Multiscale Model. Simul.* **2**(1), 80–123 (electronic).

D. Balsara and C.-W. Shu (2000), 'Monotonicity preserving weighted essentially non-oscillatory schemes with increasingly high order of accuracy', *J. Comput. Phys.* **160**(2), 405–452.

M. Bardi and S. Osher (1991), 'The nonconvex multi-dimensional Riemann problem for Hamilton–Jacobi equations', *SIAM J. Math. Anal.* **22**(2), 344–351.

G. Barles and P. E. Souganidis (1991), 'Convergence of approximation schemes for fully nonlinear second order equations', *Asymptotic Anal.* **4**(3), 271–283.

M. Bertalmio, A. L. Bertozzi and G. Sapiro (2001*a*), 'Navier–Stokes, fluid dynamics, and image and video inpainting', *Proc. of ICCV, IEEE* **1**, 1355–1362.

M. Bertalmio, L.-T. Cheng, S. Osher and G. Sapiro (2001*b*), 'Variational problems and partial differential equations on implicit surfaces', *J. Comput. Phys.* **174**(2), 759–780.

M. Bertalmio, G. Sapiro, V. Caselles and C. Ballester (2000), Image inpainting, in *ACM SIGGRAPH*, ACM, pp. 417–424.

M. Bertalmio, L. Vese, G. Sapiro and S. Osher (2003), 'Simultaneous structure and texture image inpainting', *IEEE Trans. Image Process.* **12**(8), 882–889.

M. D. Betterton (2001), 'Formation of structure in snowfields: Penitentes, suncups, and dirt cones', *Phys. Rev. E* **63**, 05629-1-12.

M. Boué and P. Dupuis (1999), 'Markov chain approximations for deterministic control problems with affine dynamics and quadratic cost in the control', *SIAM J. Numer. Anal.* **36**(3), 667–695.

P. Burchard (2002), 'Total variation geometry I: Concepts and motivation', UCLA CAM Report # 02-01.

P. Burchard, L.-T. Cheng, B. Merriman and S. Osher (2001), 'Motion of curves in three spatial dimensions using a level set approach', *J. Comput. Phys.* **170**, 720–741.

M. Burger and S. Osher (2005), 'A survey on level set methods for inverse problems and optimal design', *Inverse Problems*, to appear.

M. Burger, B. Hackl and W. Ring (2004), 'Incorporating topological derivatives into level set methods', *J. Comput. Phys.* **194**(1), 344–362.

M. Burger, S. Osher and J.-J. Xu (2005), 'Iterative refinement and inverse scale space methods for image restoration', in preparation.

V. Caselles, J. Morel and C. Sbert (1998), 'An axiomatic approach to image interpolation', *IEEE Trans. Image Process.* **7**(3), 376–386.

A. Cetin (1989), 'Reconstruction of signals from Fourier transform samples', *Signal Processing* **16**, 129–148.

A. Chambolle and P.-L. Lions (1997), 'Image recovery via total variation minimization and related problems', *Numer. Math.* **76**(2), 167–188.

T. Chan and H. M. Zhou (1999), 'Adaptive ENO-wavelet transforms for discontinuous functions', UCLA CAM Report # 99-21.

T. F. Chan and L. A. Vese (2001*a*), 'Active contours without edges', *IEEE Trans. Image Process.* **10**(2), 266–277.

T. F. Chan and L. A. Vese (2001*b*), 'A level set algorithm for minimizing the Mumford and Shah model functional in image processing', *Proc. 1st IEEE Workshop on 'Variational and Level Set Methods in Computer Vision'*, pp. 161–168.

T. F. Chan and H.-M. Zhou (2002), 'ENO-wavelet transforms for piecewise smooth functions', *SIAM J. Numer. Anal.* **40**(4), 1369–1404.

T. Chan, S. H. Kang and J. Shen (2002), 'Euler's elastica and curvature based inpainting', *SIAM J. Appl. Math.* **63**(2), 564–592.

T. Chan, J. Shen and H.-M. Zhou (2004), 'Total variation wavelet inpainting', UCLA CAM Report # 04-47.

Y. G. Chen, Y. Giga and S. Goto (1991), 'Uniqueness and existence of viscosity solutions of generalized mean curvature flow equations', *J. Differential Geom.* **33**(3), 749–786.

L.-T. Cheng (2000), 'The level set method applied to geometrically based motion, materials science, and image processing', UCLA CAM Report # 00-20.

L.-T. Cheng and Y.-H. R. Tsai (2004), 'Visibility optimization using variational approaches', UCLA CAM Report # 04-03.

L.-T. Cheng, P. Burchard, B. Merriman and S. Osher (2002), 'Motion of curves constrained on surfaces using a level set approach', *J. Comput. Phys.* **175**, 604–644.

D. L. Chopp (2001), 'Some improvements of the fast marching method', *SIAM J. Sci. Comput.* **23**(1), 230–244 (electronic).

B. Cockburn and C.-W. Shu (1989), 'TVB Runge–Kutta local projection discontinuous Galerkin finite element method for conservation laws II: General framework', *Math. Comp.* **52**(186), 411–435.

M. G. Crandall and P.-L. Lions (1983), 'Viscosity solutions of Hamilton–Jacobi equations', *Trans. Amer. Math. Soc.* **277**(1), 1–42.

M. G. Crandall and P.-L. Lions (1984), 'Two approximations of solutions of Hamilton–Jacobi equations', *Math. Comp.* **43**, 1–19.

M. G. Crandall, H. Ishii and P.-L. Lions (1992), 'User's guide to viscosity solutions of second order partial differential equations', *Bull. Amer. Math. Soc.* (N.S.) **27**(1), 1–67.

P.-E. Danielsson (1980), 'Euclidean distance mapping', *Computer Graphics and Image Processing* **14**, 227–248.

A. Dervieux and F. Thomasset (1979), A finite element method for the simulation of Rayleigh–Taylor instability, in *Approximation Methods for Navier–Stokes Problems*, Vol. 771 of *Lecture Notes in Mathematics*, Springer, pp. 145–158.

A. Dervieux and F. Thomasset (1981), Multifluid incompressible flows by a finite element method, in Vol. 11 of *Lecture Notes in Physics*, pp. 158–163. Lectures Notes in Physics, Vol.141, 158-163 (1981)

M. Droske and M. Rumpf (2003/04), 'A variational approach to nonrigid morphological image registration', *SIAM J. Appl. Math.* **64**(2), 668–687 (electronic).

F. Duguet and G. Drettakis (2002), Robust epsilon visibility, in *Proc. ACM SIGGRAPH 2002* (J. Hughes, ed.), Annual Conference Series, ACM Press/ACM SIGGRAPH, pp. 567–575.

B. Engquist, A. Harten and S. Osher (1987), A high order essentially nonoscillatory shock capturing method, in *Large Scale Scientific Computing: Oberwolfach, 1985*, Birkhäuser, Boston, MA, pp. 197–208.

B. Engquist, A.-K. Tornberg and Y.-H. Tsai (2004), 'Discretization of Dirac-δ functions in level set methods', UCLA CAM Report # 04-16. Under review.

D. Enright, R. Fedkiw, J. Ferziger and I. Mitchell (2002), 'A hybrid particle level set method for improved interface capturing', *J. Comput. Phys.* **183**, 83–116.

D. Enright, F. Losasso and R. Fedkiw (2005), 'A fast and accurate semi-Lagrangian particle level set method', UCLA CAM Report # 03-58. *Computers and Structures* **83**, 479-490.

S. Esedoglu and T. Chan (2004), 'Aspects of total variation regularized L^1 function approximation', UCLA CAM Report # 04-07.

S. Esedoglu and S. Osher (2005), 'Decomposition of images by the anisotropic Rudin–Osher–Fatemi model', *Comm. Pure Appl. Math.*, to appear.

S. Esedoglu and J. Shen (2002), 'Digital inpainting based on the Mumford-Shah-Euler image model', *European J. Appl. Math.* **13**, 353–370.

S. Esedoglu and Y.-H. Tsai (2004), 'Threshold dynamics for the piecewise Mumford-Shah functional', UCLA CAM Report # 04-63.

L. C. Evans (1998), *Partial Differential Equations*, AMS, Providence, RI.

L. C. Evans and R. F. Gariepy (1992), *Measure Theory and Fine Properties of Functions*, Studies in Advanced Mathematics, CRC Press, Boca Raton, FL.

L. C. Evans and J. Spruck (1991), 'Motion of level sets by mean curvature I', *J. Differential Geom.* **33**(3), 635–681.

L. C. Evans and J. Spruck (1992*a*), 'Motion of level sets by mean curvature II', *Trans. Amer. Math. Soc.* **330**(1), 321–332.

L. C. Evans and J. Spruck (1992*b*), 'Motion of level sets by mean curvature III', *J. Geom. Anal.* **2**(2), 121–150.

L. C. Evans and J. Spruck (1995), 'Motion of level sets by mean curvature IV', *J. Geom. Anal.* **5**(1), 77–114.

D. J. Eyre (1998), Unconditionally gradient stable time marching: The Cahn–Hilliard equation, in *Computational and Mathematical Models of Microstructural Evolution: San Francisco, CA, 1998*, Vol. 529 of *Mater. Res. Soc. Sympos. Proc.*, MRS, Warrendale, PA, pp. 39–46.

M. Falcone and R. Ferretti (2002), 'Semi-Lagrangian schemes for Hamilton–Jacobi equations, discrete representation formulae and Godunov methods', *J. Comput. Phys.* **175**(2), 559–575.

F. Gibou and R. Fedkiw (2005), A fast level set based algorithm for segmentation, in *Proc. 4th Hawaii International Conference on Statistics, Mathematics and Related Fields*, pp. 281–291. Stanford Technical Report (2002).

Y. Giga (2002), Surface evolution equations: A level set method, Vol. 44 of *Lipschitz Lecture Notes*, University of Bonn.

S. K. Godunov (1959), 'A difference method for numerical calculation of discontinuous solutions of the equations of hydrodynamics', *Mat. Sb.* (N.S.) **47 (89)**, 271–306.

A. Haddad and Y. Meyer (2004), 'Variational methods in image processing', UCLA CAM Report # 04-52.

X. Han, C. Xu and J. Prince (2003), Topology preserving geometric deformable models for brain reconstruction, in *Geometric Level Set Methods in Imaging, Vision, and Graphics* (S. Osher and N. Paragios, eds), Springer, New York, pp. 421–438.

E. Harabetian and S. Osher (1998), 'Regularization of ill-posed problems via the level set approach', *SIAM J. Appl. Math.* **58**(6), 1689–1706 (electronic).

A. Harten, B. Engquist, S. Osher and S. R. Chakravarthy (1987), 'Uniformly high-order accurate essentially nonoscillatory schemes III', *J. Comput. Phys.* **71**(2), 231–303.

J. Helmsen, E. Puckett, P. Colella and M. Dorr (1996), Two new methods for simulating photolithography development in 3D, in *SPIE 2726*, pp. 253–261.

A. Hertzmann and D. Zorin (2000), Illustrating smooth surfaces, in *ACM SIGGRAPH*, ACM, pp. 517–526.

G.-S. Jiang and D. Peng (2000), 'Weighted ENO schemes for Hamilton–Jacobi equations', *SIAM J. Sci. Comput.* **21**(6), 2126–2143 (electronic).

H. Jin, A. Yezzi, Y.-H. Tsai, L. T. Cheng and S. Soatto (2005a), 'Estimation of 3D surface shape and smooth radiance from 2D images: A level set approach', *J. Sci. Comput.*, to appear.

S. Jin, H. Liu, S. Osher and R. Tsai (2005b), 'Computing multivalued physical observables for the semiclassical limit of the Schrödinger equation', www.levelset.com/download/density.pdf, *J. Comput. Phys.*, to appear.

C. Y. Kao, S. Osher and J. Qian (2004), 'Lax–Friedrichs sweeping scheme for static Hamilton–Jacobi equations', *J. Comput. Phys.* **196**, 367–391.

C. Y. Kao, S. Osher and Y.-H. Tsai (2002), 'Fast sweeping methods for Hamilton–Jacobi equations', UCLA CAM Report # 02-66.

J. B. Keller (1962), 'Geometrical theory of diffraction', *J. Opt. Soc. Amer.* **52**, 116–130.

W.-H. Liao, M. Bergsneider, L. Vese, S.-C. Huang and S. Osher (2002), 'From landmark matching to shape and open curve matching: A level set approach', UCLA CAM Report # 02-59.

X.-D. Liu, S. Osher and T. Chan (1994), 'Weighted essentially non-oscillatory schemes', *J. Comput. Phys.* **115**(1), 200–212.

F. Losasso, R. Fedkiw and S. Osher (2004), 'Spatially adaptive techniques for level set methods and incompressible flow', UCLA CAM Report # 04-67.

M. Lysaker, S. Osher and X.-C. Tai (2004), 'Noise removal using smoothed normals and surface fitting', *IEEE Trans. Image Process.* **13**(10), 1345–1357.

A. Marquina and S. Osher (2000), 'Explicit algorithms for a new time dependent model based on level set motion for nonlinear deblurring and noise removal', *SIAM J. Sci. Comput.* **22**(2), 387–405 (electronic).

S. Masnou and J.-M. Morel (1998), Level-lines based disocclusion, in *Proc. 5th IEEE International Conference on Image Processing: Chicago, IL*, pp. 259–263.

B. Merriman, J. K. Bence and S. J. Osher (1994), 'Motion of multiple functions: A level set approach', *J. Comput. Phys.* **112**(2), 334–363.

Y. Meyer (2001), *Oscillating Patterns in Image Processing and Nonlinear Evolution Equations*, AMS, Providence, RI. The fifteenth Dean Jacqueline B. Lewis memorial lectures, Rutgers University.

D. Mumford and J. Shah (1989), 'Optimal approximations by piecewise smooth functions and associated variational problems', *Comm. Pure Appl. Math.* **42**(5), 577–685.

M. Nikolova (2002), 'Minimizers of cost functionals involving nonsmooth data fidelity terms', *SIAM J. Numer. Anal.* **40**(3), 965–994.

M. Nitzberg, D. Mumford and T. Shiota (1993), *Filtering, Segmentation and Depth*, Vol. 662 of *Lecture Notes in Computer Science*, Springer, Berlin.

A. Oberman (2004), 'A convergent upwind difference scheme for motion by mean curvature', *Numer. Math.* **99**, 365–379.

S. Osher (1993), 'A level set formulation for the solution of the Dirichlet problem for Hamilton–Jacobi equations', *SIAM J. Math. Anal.* **24**(5), 1145–1152.

S. Osher (2003), Level set methods, in *Geometric Level Set Methods in Imaging, Vision, and Graphics* (S. Osher and N. Paragios, eds), Springer, New York, pp. 3–20.

S. Osher and R. Fedkiw (2002), *Level Set Methods and Dynamic Implicit Surfaces*, Springer, New York.

S. Osher and R. P. Fedkiw (2001), 'Level set methods: an overview and some recent results', *J. Comput. Phys.* **169**(2), 463–502.

S. Osher and J. Helmsen (2005), 'A generalized fast algorithm with applications to ion etching', in preparation.

S. Osher and B. Merriman (1997), 'The Wulff shape as the asymptotic limit of a growing crystalline interface', *Asian J. Math.* **1**(3), 560–571.

S. Osher and J. A. Sethian (1988), 'Fronts propagating with curvature-dependent speed: algorithms based on Hamilton–Jacobi formulations', *J. Comput. Phys.* **79**(1), 12–49.

S. Osher and C.-W. Shu (1991), 'High-order essentially nonoscillatory schemes for Hamilton–Jacobi equations', *SIAM J. Numer. Anal.* **28**(4), 907–922.

S. Osher, M. Burger, D. Goldfarb, J.-J. Xu and W. Yin (2005), 'An iterative regularization method for total variation based image restoration', *Multiscale Model. Simul.*, to appear.

S. Osher, L.-T. Cheng, M. Kang, H. Shim and Y.-H. Tsai (2002*a*), 'Geometric optics in a phase-space-based level set and Eulerian framework', *J. Comput. Phys.* **179**(2), 622–648.

S. Osher, A. Sole and L. Vese (2003), 'Image decomposition and restoration using total variation minimization and the H^{-1} norm', *Multiscale Model. Simul.* **1**, 344–370.

N. Paragios and R. Deriche (1997), A PDE-based level set approach for detection and tracking of moving objects, Technical Report 3173, INRIA, France.

D. Peng, B. Merriman, S. Osher, H. Zhao and M. Kang (1999a), 'A PDE-based fast local level set method', *J. Comput. Phys.* **155**(2), 410–438.

D. Peng, S. Osher, B. Merriman and H.-K. Zhao (1999b), The geometry of Wulff crystal shapes and its relations with Riemann problems, in *Nonlinear Partial Differential Equations: Evanston, IL, 1998*, AMS, Providence, RI, pp. 251–303.

C. S. Peskin (2002), The immersed boundary method, in *Acta Numerica*, Vol. 11, Cambridge University Press, pp. 479–517.

E. Rouy and A. Tourin (1992), 'A viscosity solutions approach to shape-from-shading', *SIAM. J. Numer. Anal.* **29**(3), 867–884.

L. Rudin and S. Osher (1994), 'Total variation based restoration with free local constraints', *Proceedings ICIP, IEEE, Austin, TX* pp. 31–35.

L. I. Rudin and S. Osher (1990), 'Feature-oriented image enhancement using shock filters', *SIAM J. Numer. Anal.* **27**(4), 919–940.

L. Rudin, S. Osher and E. Fatemi (1992), 'Nonlinear total variation based noise removal algorithms', *Phys. D* **60**(1–4), 259–68.

O. Scherzer and C. Groetsch (2001), Inverse scale space theory for inverse problems, in *Scale-space and Morphology in Computer Vision: Proc. 3rd International Conference Scale Space*, Springer, pp. 317–325.

J. Sethian (1996), Fast marching level set methods for three dimensional photo-lithography development, in *SPIE 2726*, pp. 261–272.

J. A. Sethian and A. Vladimirsky (2001), 'Ordered upwind methods for static Hamilton–Jacobi equations', *Proc. Natl. Acad. Sci. USA* **98**(20), 11069–11074 (electronic).

C.-W. Shu (1997), Essentially non-oscillatory and weighted essentially non-oscillatory schemes for hyperbolic conservation laws, ICASE Report 97-65, NASA.

C.-W. Shu and S. Osher (1988), 'Efficient implementation of essentially non-oscillatory shock-capturing schemes', *J. Comput. Phys.* **77**(2), 439–471.

P. Smereka (2000), 'Spiral crystal growth', *Phys. D* **138**(3-4), 282–301.

K. A. Smith, F. J. Solis and D. L. Chopp (2002), 'A projection method for motion of triple junctions by levels sets', *Interfaces Free Bound.* **4**(3), 263–276.

J. Sokołowski and J.-P. Zolésio (1992), *Introduction to Shape Optimization*, Vol. 16 of *Springer Series in Computational Mathematics*, Springer, Berlin.

B. Song and T. Chan (2002), 'Fast algorithm for level set based optimization', UCLA CAM Report # 02-68.

P. Soravia (1994), 'Generalized motion of a front propagating along its normal direction: a differential games approach', *Nonlinear Anal.* **22**(10), 1247–1262.

P. E. Souganidis (1985), 'Approximation schemes for viscosity solutions of Hamilton–Jacobi equations', *J. Differential Equations* **59**(1), 1–43.

R. Spiteri and S. Ruuth (2005), 'A new class of optimal high-order strong-stability-preserving time discretization methods', preprint.

J. Steinhoff, M. Fang and L. Wang (2000), 'A new Eulerian method for the computation of propagating short acoustic and electromagnetic pulses', *J. Comput. Phys.* **157**, 683–706.

J. Strain (1999a), 'Fast tree-based redistancing for level set computations', *J. Comput. Phys.* **152**(2), 664–686.

J. Strain (1999b), 'Semi-Lagrangian methods for level set equations', *J. Comput. Phys.* **151**(2), 498–533.

M. Sussman, P. Smereka and S. Osher (1994), 'A level set method for computing solutions to incompressible two-phase flow', *J. Comput. Phys.* **114**, 146–159.

E. Tadmor, S. Nezzar and L. Vese (2004), 'A multiscale image representation using hierarchical (BV, L^2) decompositions', *Multiscale Modeling and Simulation: A SIAM Interdisciplinary Journal* **02**(4), 554–579.

T. Tasdizen, R. Whitaker, P. Burchard and S. Osher (2003), 'Geometric surface processing via normal maps', *ACM Trans. Graphics* **22**(4), 1012–1033.

A. Tornberg (2002), 'Multi-dimensional quadrature of singular and discontinuous functions', *BIT* **42**, 644–669.

A. Tornberg and B. Engquist (2003), 'Regularization techniques for numerical approximation of PDEs with singularities', *J. Sci. Comput.* **19**, 527–552.

Y.-H. R. Tsai (2002), 'Rapid and accurate computation of the distance function using grids', *J. Comput. Phys.* **178**(1), 175–195.

Y.-H. R. Tsai, L.-T. Cheng, P. Burchard, S. Osher and G. Sapiro (2004), 'Visibility and its dynamics in a PDE based implicit framework', *J. Comput. Phys.* **199**(06), 260–290.

Y.-H. R. Tsai, L.-T. Cheng, S. Osher and H.-K. Zhao (2003a), 'Fast sweeping methods for a class of Hamilton–Jacobi equations', *SIAM J. Numer. Anal.* **41**(2), 673–699.

Y.-H. R. Tsai, Y. Giga and S. Osher (2003b), 'A level set approach for computing discontinuous solutions of Hamilton–Jacobi equations', *Math. Comp.* **72**(241), 159–181 (electronic).

J. Tsitsiklis (1995), 'Efficient algorithms for globally optimal trajectories', *IEEE Trans. Automatic Control* **40**(9), 1528–1538.

B. Vemuri, J. Ye, Y. Chen and C. Leonard (2003), 'Image registration via level set motion: applications to atlas-based segmentation', *Medical Image Analysis* **7**, 1–20.

L. A. Vese and T. F. Chan (2002), 'A multiphase level set framework for image segmentation using the Mumford and Shah model', *Internat. J. Computer Vision*.

L. A. Vese and S. J. Osher (2003), 'Modeling textures with total variation minimization and oscillating patterns in image processing', *J. Sci. Comput.* **19**(1–3), 553–572.

L. Vese and S. Osher (2002), 'The level set method links active contours, Mumford-Shah segmentation, and Total Variation restoration', UCLA CAM Report # 02-05.

J.-J. Xu and H.-K. Zhao (2003), 'An Eulerian formulation for solving partial differential equations along a moving interface', *J. Sci. Comput.* **19**(1–3), 573–594.

Y.-T. Zhang, H.-K. Zhao and J. Qian (2004), 'High order fast sweeping methods for static Hamilton–Jacobi equations', UCLA CAM Report # 04-37.

H.-K. Zhao (2005), 'Fast sweeping method for eikonal equations', *Math. Comp.*, to appear.

H.-K. Zhao and S. Osher (2003), Visualization, analysis and shape reconstruction of sparse data, in *Geometric Level Set Methods in Imaging, Vision, and Graphics* (S. Osher and N. Paragios, eds), Springer, New York, pp. 361–380.

H.-K. Zhao, T. Chan, B. Merriman and S. Osher (1996), 'A variational level set approach to multiphase motion', *J. Comput. Phys.* **127**, 179–195.

H.-K. Zhao, S. Osher and R. Fedkiw (2001), Fast surface reconstruction using the level set method, in *First IEEE Workshop in Variation and Level Set Methods in Computer Vision*, ICCV, Vancouver, Canada, pp. 194–202.

H.-K. Zhao, S. Osher, B. Merriman and M. Kang (2000), 'Implicit and non-parametric shape reconstruction from unorganized points using a variational level set method', *Computer Vision and Image Understanding* **80**, 295–319.